VOLUME FIVE

DEVELOPMENTS IN MARINE GEOLOGY

QUATERNARY CORAL REEF SYSTEMS

VOLUME FIVE

DEVELOPMENTS IN MARINE GEOLOGY

QUATERNARY CORAL REEF SYSTEMS: HISTORY, DEVELOPMENT PROCESSES AND CONTROLLING FACTORS

By

L. F. MONTAGGIONI

University of Provence, Marseille, France

and

C. J. R. BRAITHWAITE

University of Glasgow, Scotland, UK

ELSEVIER

AMSTERDAM • BOSTON • HEIDELBERG • LONDON • NEW YORK • OXFORD
PARIS • SAN DIEGO • SAN FRANCISCO • SINGAPORE • SYDNEY • TOKYO

Elsevier
30 Corporate Drive, Suite 400, Burlington, MA 01803, USA
Radarweg 29, PO Box 211, 1000 AE Amsterdam, The Netherlands
Linacre House, Jordan Hill, Oxford OX2 8DP, UK

First edition 2009

British Library Cataloguing in Publication Data
A catalogue record for this book is available from the British Library

Library of Congress Cataloging-in-Publication Data
A catalog record for this book is available from the Library of Congress

ISBN: 978-0-444-53247-3
ISSN: 1572-5480

For information on all Elsevier publications
visit our website at books.elsevier.com

Printed and bound in Great Britain

09 10 11 12 13   10 9 8 7 6 5 4 3 2 1

Contents

Preface

One of the central objectives of geologists working on carbonate sediments and rocks is to understand the structure, composition and evolution of organic buildups ('reefs'), to determine the nature of the major biotic and physico-chemical forcing functions that have driven the construction of these features through time, and to understand how the various driving forces have interacted. Recent coral reefs have been used as analogues with which to interpret ancient organic buildups since the 18th century, although for many years the origins, growth patterns and controlling biotic and environmental factors of living reefs were poorly documented and in some cases are still questioned. Since the pioneering works of Darwin (1842), Dana (1875), Daly (1915) and Gardiner (1936) among others, a number of reviews defining the main attributes of modern coral reefs and/or ancient organic buildups have been published, including work by Stoddart (1969a), Stoddart and Yonge (1971, 1978), Jones and Endean (1973a, 1973b, 1976, 1977), Wilson (1975), Chevalier (1977), Fisher (1977), Frost, Weiss, and Saunders (1977), Toomey (1981), Hopley (1982), Barnes (1983), Glynn and Wellington (1983), Fagerström (1987), Guilcher (1988), Dubinski (1990), Sorokin (1993), Birkeland (1997), Wood (1999), Stanley (2001), Kiessling and Flügel (2002), Cortés (2003), Aronson (2007) and Hopley et al. (2007). This list is by no means exhaustive and does not include proceedings from conferences, symposia and workshops that have dealt with both modern and fossil carbonate buildups under the auspices of the International Society for Reef Studies, the International Association for the Study of Fossil Cnidaria and Porifera, the Society of Economic Paleontologists and Mineralogists and the International Association of Sedimentologists.

There has been an increasing interest in recent coral reefs worldwide, in part related to their importance as sensitive indicators of climatic change, and as a consequence research devoted to reef-related topics has increased exponentially over the last two decades. This book is designed to present and combine a multitude of old and new field observations and laboratory analyses from both the Indo-Pacific and Caribbean provinces. It aims at reflecting the state of knowledge of the geology of coral reef ecosystems that have developed in the recent geological past. Our intention is to provide descriptions of representative studies within the diverse fields covered by the coral reef record.

The title of this book, *Quaternary Coral Reef Systems*, accurately describes the contents. As the chapter headings show, the text is concerned with the origins and palaeobiogeography of coral reefs, the distribution of the coral-dominated communities involved in reef building, their ecology and palaeoecology. The factors controlling both coral and reef growth, together with depositional patterns, reef geometry, anatomy, stratigraphy and diagenesis, are examined, together with the evidence provided by coral reefs as indicators of climate change. An additional objective is to present an appropriate database with which to assess the preservation status of modern reefs, currently subject to climatic and human-induced perturbations. In contrast to the vast majority of more ancient organic buildups, most Quaternary reef systems, and particularly those of the late Pleistocene and early Holocene, can be regarded as the counterparts of modern reef ecologies and types. As suggested by Greenstein (2007) and Pandolfi and Jackson (2007), careful analysis of Quaternary reefs may therefore provide data on the likely responses of reef communities to environmental disturbances over geological time scales. This approach may contribute to a better understanding of the problems with which an increasing number of modern coral reefs are confronted, and help to identify and differentiate natural long-term cycles and diverse ecological shifts from possible anthropogenically driven deterioration.

The text is divided into 10 chapters.

In *Chapter 1*, which is the introduction, the reef concept is briefly addressed through the chronicle of two centuries of exploration and research. The geographical distribution of scleractinian corals and coral reefs, together with the major climatic patterns in the tropics, are presented. These serve as a baseline to understand the role played by biotic and environmental factors in reef growth history, and to analyse the environmental parameters encapsulated in coral colonies or reef bodies. The Quaternary epoch is defined, in part on the basis of the major climatic attributes. The chapter ends with a description of the principle dating and sampling methods used to reconstruct Quaternary reef growth histories.

Chapter 2 explains how coral-dominated communities and buildups have evolved throughout the late Tertiary to result in modern coral reefs. It examines the extent to which environmental changes (tectonics, sea-level oscillations, climate changes and nutrient input) have influenced scleractinian coral and reef diversification in space and time, reviewing phases of extinction and recovery with special emphasis on corals, coralline algae and the green alga *Halimeda*.

Chapter 3 compares the community structure and biological zonation of modern and Quaternary coral reefs, using selected examples in the Western Atlantic and Indo-Pacific regions. The question of reef-community dynamics, which account for the time over which the community remains stable or evolved, is investigated at a variety of time scales.

Chapter 4 explores the potential controls on the distribution, development and preservation of coral-dominated communities and reefs through the Quaternary. An array of biotic and environmental factors has been implicated in distribution and development but what is the fate of these communities after death? The answer to this question requires estimations of the roles played by taphonomic processes that have controlled preservation of the composition of fossil communities. The discussion also addresses the issue of the degree of similarity between fossil and modern assemblages.

Chapter 5 examines patterns of carbonate production and deposition. Living coral reefs are significant participants in the global carbon cycle, and are one of the most efficient marine carbonate producers. It is therefore of prime importance to determine growth and carbonate production rates of reef builders and associated calcifying organisms, noting their respective contributions to sediment production during episodes of ecological shifts and disturbances. An additional challenge is to identify the depositional environments of fossil reefs. One way to achieve this is to estimate the degree to which different sediment types are a reflection of adjacent benthic communities. Finally, the major controls on the distribution and deposition rates of major reef components (framework vs. detrital) are defined.

Chapter 6 accounts for the anatomy and stratigraphy of reef systems. First, the composition of Holocene sedimentary piles beneath the different reef zones are presented with conceptual models of reef deposition. Second, the structure and stratigraphy of barrier reefs and atolls, together with those of emerged and submerged reef units are described from selected case studies. Finally, the value of numerical modelling to reproduce reef growth and architecture is discussed on the basis of comparisons between field and computer data.

Chapter 7 examines oceanic and coastal hydrodynamics, together with the physicochemical characteristics of seawater. Both are dominant controls on the latitudinal distribution, zonation and structure of reef communities and on reef geometry and anatomy. In particular, storms, cyclones and tsunamis play a significant in role in restructuring reef tracts. Water circulation through carbonate piles, particularly salty density-driven flows through modern reefs and freshwater flows through emerged fossil reefs, exert important controls on the diagenetic evolution of reef components, changing not only the porosity and permeability but also the mechanical properties of the accumulation. Finally, *Chapter 7* examines the effect of hydrodynamic forcing on the fate of modern to Pleistocene reef systems.

Chapter 8 is devoted to the fundamental diagenetic processes affecting carbonate deposits, with special reference to coral reefs. The mineralogy of the major sediment components is presented. The contrasting attributes of marine and freshwater cements are compared, together with diagenetic features characteristic of replacement, dissolution and compaction. Reference is also made to later reef-associated dolomitization and phosphate deposition, and to the timing and rates of diagenesis, especially in relation to sea-level oscillations.

Chapter 9 presents a survey of the methods used in climate reconstruction. One of the most fundamental contributions of corals and coral reefs to Earth Sciences is the reconstruction of climate history over at least the last tens of thousands of years. A selection of significant records in contrasting tropical regions over intervals from the late Holocene to middle Pleistocene is discussed. Sections are devoted to (1) the climatic significance of coral-based climatic proxies by reference to specific climate reconstructions, and (2) the value and significance of reef-related depositional and erosional features for reconstructing the timing and amplitude of sea-level variations. A brief history of sea-level change based on the reef record over the last 400 ka is presented.

The concluding chapter is a brief attempt to link the past history of coral reef systems to their future in a context of global warming.

A wide range of disciplines, including marine geology, sedimentology, palaeontology, palaeoecology, marine biology and ecology, oceanography, geochemistry and geophysics, is concerned with the study of Pleistocene and Holocene coral reefs. This book should be of interest to post-graduate students and new researchers. It may also be of value to teachers in geosciences, marine biology, oceanography and palaeoclimatology who wish to extend their knowledge and update their views on the Reef Phenomenon, and indeed anyone interested in aspects of environmental change.

The authors are particularly grateful to the Series Editor, Hervé Chamley, for assistance in revising earlier drafts of the book, Sara Pratt and Hannah Russel for managing successive writing and editing steps. The many reef workers with whom we have kept close contact or collaborated during our respective careers have been essential to the existence of this book.

Lucien Montaggioni offers his thanks to the following colleagues who have contributed to increase his experience and knowledge or participated to joint research programmes: to Gérard Faure (formerly of the University of Réunion Island) and Michel Pichon (formerly of the University of Perpignan) for their expertise in the taxonomy and ecology of reef-building corals; to Odile Naïm (formerly also of the University of Réunion Island) for many discussions on reef ecology; to Bernard Salvat (of the Ecole Pratique des Hautes Etudes, Perpignan) for introducing him to French Polynesian Reefs; to Paolo Pirazzoli (of the CNRS, Paris) for collaboration on studies of sea-level history in French Polynesia and the Mediterranean; to Claude Payri (of the French University of the Pacific, Tahiti) for her help with the taxonomy of coralline algae; to Peter Davies (University of Sydney) and David Hopley (University of Townsville) for introducing him to the Australian Great Barrier Reef; to Guy Cabioch (Institut de

Recherche pour le Développement, Nouméa) for collaboration with the study of French Polynesian and New Caledonian Reefs; to Gilbert Camoin (CNRS, University of Aix-en-Provence), for collaboration with the study of Western Indian and Tahitian Reefs; to Colin Braithwaite (University of Glasgow) for collaboration with the study of the Great Barrier Reef and Western Indian Ocean Reefs; to Wolf-Christian Dullo (University of Kiel) for collaboration in the study of the Red Sea and Western Indian Ocean Reefs; to Thierry Corrège (University of Bordeaux) and Florence Le Cornec (Institut de Recherche pour le Développement, Paris) for their help with coral geochemistry.

Colin Braithwaite acknowledges the help given early in his research career by Robin Bathurst and the encouragement from him and from Wally Pitcher. James Taylor provided the first opportunity to study present-day carbonate environments in the Seychelles, collaborating with John Taylor and Brian Rosen (both now of the Natural History Museum in London). More recent support has been from Lucien Montaggioni (University of Provence, Marseilles), Gilbert Camoin (CNRS, Aix-en-Provence), Chris Dullo (GEOMAR, Kiel) and Dick Kroon (University of Edinburgh). Fieldwork in Aldabra, Australia, the Bahamas, Florida, Kenya, Mauritius, the Seychelles, Saudi Arabia, the Sudan and Tobago, has been aided by the support of many others. Work with students, on limestones in Iraq, Ireland, Libya, Norway, Saudi Arabia, Turkey and the United Kingdom, on projects related to the oil, minerals and engineering industries, has provided stimulation and valued experience. Braithwaite is pleased to be able to acknowledge the financial support from the Royal Society of London, the UK Natural Environment Research Council, the Carnegie Trust for the Universities of Scotland, the European Community, The Royal Academy of Engineering, the Leverhulme Trust and others, without which none of this would have been possible.

Unless otherwise indicated, all the photographs have been provided by the authors. Ludovic Laugero is thanked for drafting the figures for most of the chapters.

CHAPTER ONE

Introduction: Quaternary Reefs in Time and Space

1.1. The Reef Phenomenon: Definitions and History of Discovery and Research

The nature of 'reefs' and 'reef communities' has been so diverse throughout geological history that there is no general agreement on what exactly is or is not 'a reef'. The reasons for this are complex but lie to a large extent in the diversity of the scientific disciplines and contrasting perspectives brought to bear on aspects of both ancient and modern structures. There have been fewer problems for biologists where the focus has been on the content rather than on the nature of the 'reef' entity as a whole, but for geologists different approaches have led to a multiplicity of misinterpretations and continuing arguments. A number of attempts have been made to address the problem by proposing definitions of the 'reef' (for instance, see Longman, 1981; Fagerström, 1987; Hallock, 1997; Wood, 1999; Stanley, 2001; Riding, 2002), but no consensus has been reached so far.

The existence of 'coral reefs' was well established by the time European exploration of tropical seas began in the 17th century. Although there was European speculation on the nature of corals as early as the 16th century, it was not until the 19th century that there was any serious scientific evaluation of the characteristics of reefs. One of the key outcomes of the early oceanographic exploration was the description of coral reefs as geological entities. Lyell (1832) described coral reefs in early editions of his *Principles of Geology* from previous observations in the Indo-Pacific regions. It was against this background that Darwin (1842) published his observations on the morphology of Polynesian Islands in *The Structure and Distribution of Coral Reefs*, in which he defined the genetic model of reef development, relating reef growth to subsidence (subsidence-controlled theory). The Darwinian model for the evolution of coral reefs, from fringing to barrier and atoll types, has been widely accepted, following the clear evidence provided by deep drilling through Funafuti Atoll (Cullis, 1904; Finkh, 1904; Ohde et al., 2002). The borehole encountered a substantial thickness of shallow-water limestones (339.5 m), thus implying considerable subsidence. Similar results were obtained from scattered boreholes elsewhere, including the Bahamas (Field & Hess, 1933),

Kita-Daito-Jima (Sugiyama, 1934, in Japanese with some results in Ladd & Schlanger, 1960; Ladd, Tracey, & Gross, 1970; Suzuki et al., 2006) and the Great Barrier Reef (Richards & Hill, 1942). Notwithstanding the dominance of the Darwinian model, there have been other explanations of reef origins and other theories. This was the starting point over one century and half of scientific controversy. In particular, Daly (1915, 1948) was prominent among the detractors and the key initiator and proponent of the so-called 'Glacial Control Theory'. He argued that far from reflecting simple subsidence, reef morphology was a reflection of changing sea levels. Part of the explanation for the formation of reef platforms was based on the pattern of reef erosion during low sea-level stands. Murray (1880) regarded dissolution of the backreef lagoon as an important part of the outward growth of atolls advancing over their own talus. Stearns (1946), Kuenen (1947), Schlanger (1963) and Ladd et al. (1970) suggested that modern barrier reefs and/or atolls formed postglacially on reef surfaces that had been subaerally eroded by dissolution during Pleistocene low sea-level stands. Glacio-eustatic sea-level changes have since been generally recognized as one of the key factors controlling reef development. Similar views were reached by Purdy (1974), Purdy and Bertram (1993) and Purdy and Winterer (2001, 2006). In spite of the indisputable evidence of subsidence supported by deep reef drilling, a subaerial solution-induced relief is suggested to have been accentuated by reef building to produce the typical modern barrier and atoll morphologies.

Perhaps because reefs were regarded as geological features, geologists were quick to seize on them as an explanation for the complex relationships of many ancient limestones. The only constant to emerge subsequently has been a gradual geological restriction of the term to carbonate rocks. But structures described as 'reefs' have regrettably included a number of organic to inorganic carbonate deposits (Norris, 1953; Lees, 1964; Terry & Williams, 1969; Conaghan, Mountjoy, Edgecombe, Talent, & Owen, 1976) with problems of scale where accumulations on a metre (Kirtley & Tanner, 1968) or centimetre (Ager, 1963) scale have been misguidedly referred to as 'reefs'.

Many of the subsequent investigations from the end of the 19th to the middle of the 20th century were based on a zoological approach. Later reviews were published by Stoddart (1969a), Lewis (1977), Dubinski (1990) and Birkeland (1997). By contrast, geology had little use for reefs until the early 1950s when a burgeoning oil industry recognized them as forming important reservoir rocks. Many of the giant fields in the Middle East are within Mesozoic so-called 'reefs', although it is fair to say that the reservoir properties of many of these examples owe as much or more to their diagenetic history than to their depositional characteristics. Cumings and Schrock (1928) had tried to clarify the reef concept by defining two new terms: 'bioherm' and 'biostrome'. The main distinction here is essentially answered by the question: Does the structure have significant relief?

However, the term 'bioherm' in particular seems to have suffered much the same fate as 'reef' and has also been misused in such ways as to raise doubts wherever it appears.

Some of the confusion was generated in the oil industry, because descriptions are commonly based on geophysical or borehole evidence. Also, even when visible in outcrop it may be difficult to differentiate between mound-like forms that had significant relief at the time they were deposited and circumscribed structures that lacked relief and were the result of the local persistence of a distinctive laterally restricted facies over a long period. From a sedimentological perspective this is an important distinction, but the problem was not formally addressed until, in 1970, Dunham proposed two new definitions. 'Thick laterally restricted masses of pure or largely pure carbonate rock long have been called 'reefs'. Such masses … are here termed 'stratigraphic reefs' in contrast to organically bound 'ecologic reefs'. Heckel (1974) proposed a new definition in which a 'reef' is a carbonate buildup; that is a structure that has relief above the surrounding seafloor but which displays evidence of potential wave resistance or growth in turbulent water and evidence of control over the surrounding environment. Various subgroups were recognized including structures in which the principle binding agents were inorganic but these were subject to the same taxonomic inertia as others, and 'reef' continues to be used in an ill-defined way.

The comparison with recent reefs that is implied by the name overlooks important questions regarding their architecture and growth history that need to be addressed. This may seem straightforward but here also there is sometimes disparity between concepts of 'reefs' adopted by geologists and those by biologists, and it is fair to say that there remain differences in opinion as to what constitutes 'the reef', in part because our understanding of processes and reef history is incomplete.

Leaving aside the issue of deep water coral mounds, the structures that we see are typically close to the surface. This places them in steep environmental gradients in which rapidly varying factors such as depth, light penetration and hydrodynamic energy (see Chapter 8) have an important influence. The net result is that reefs are characterized by a distinctive biological zonation (see Chapter 3) and, because the organisms concerned are responsible for the generation of much if not most sediment in the area, there is a parallel sedimentological zonation (see Chapter 5).

1.2. Types of Coral Reefs

Since Darwin's adoption of the tripartite division of fringing reefs, barrier reefs and atolls, there has been a long history of treatment of reefs from a purely morphological point of view. A leading figure clarifying this area has

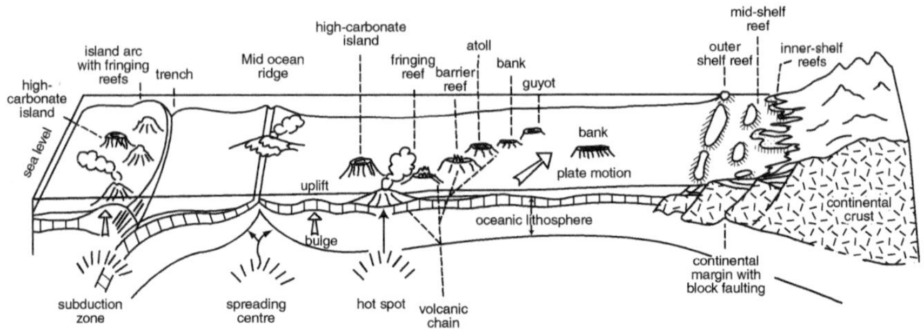

Figure 1.1 Distribution of reef morphotypes in relation to plate tectonics. Modified and redrawn from Scoffin and Dixon (1983).

been Guilcher (1988) who has summarized a considerable number of earlier observations. Apart from fringing reefs, few have precisely the kind of development that Darwin envisaged and there are many examples that do not fit the defining criteria. For example, large reef complexes such as the Australian Great Barrier Reef, the Belize platform system and the New-Caledonian barrier reef system are spectacular examples of disparity between the definitions and the application of the terms. With these exceptions, there has been a proliferation of descriptive terms (for instance, see Andrefouët et al., 2006). Some have not always been well constrained and have commonly led to confusion. Revisiting Darwin's reef morphology concepts, Scoffin and Dixon (1983) provided a convenient classification of reef types based on their relation with plate tectonics (Figure 1.1).

1.2.1. Fringing Reefs

Fringing reefs represent the most basic reef form, although, as Kennedy and Woodroffe (2002) pointed out, they may develop in a variety of ways. Suffice it to say here that the growth framework resulting in the present-day reef flat is regarded as having been established at a relatively short distance from the shore, depending on slope and wave energy, and is usually separated from it by an inboard lagoon or incipient channel. Asymmetrical forms are probably in part reflections of antecedent foundations, but owe their irregularities to a differential response to incident waves and sediment transport. Kennedy and Woodroffe (2002) outline six general models of Holocene fringing reef development based on the use of isochrons to reconstruct the successive stages of accretion.

1.2.2. Barrier Reefs

Barrier reefs (sometimes referred to as mid- to outer-shelf reefs) are essentially linear features separated from the coast of an island or a continent by a relatively deep channel and reflecting differential growth. Only barrier reefs that are associated with volcanic islands are in accordance with Darwin's model. For barrier reefs and platforms associated with continental masses, the origin appears to be more complex since the reefs commonly overlie antecedent tectonic structures (Hopley, 1982; Cabioch, Corrège, Turpin, Castellaro, & Récy, 1999; Purdy, Gischler, & Lomando, 2003). The Australian Great Barrier Reef, the Belize, Maldives and New-Caledonian barrier reef systems only partially meet Darwin's assumption. Purdy (1974), and Purdy and Winterer (2001, 2006) demonstrated that, in most cases, one of the major controls on the physiography and structure of barrier reefs as well as platforms and atolls is dissolution by meteoric freshwater during Pleistocene low sea-level stands. In addition, Chappell (1983), partly following Daly's assumptions, claimed that barriers may be derived from fringing reefs in response to changes in the rate of sea-level rise. When the rate of sea level equates with that of reef accretion, differential growth occurs between the outer reef edge and the backreef areas. Although the reef margin may follow rising sea level, the inner reef parts experiencing more intensive environmental disturbances tend to drown. A depression develops behind the edge finally resulting in the formation of a barrier reef.

1.2.3. Atolls

Atolls are ring-like coral reefs. They may be almost enclosed as on Taiaro in the Tuamotu, or relatively open to the ocean like Mopelia in the Society Islands and Ouvea in New Caledonia. The mid-shelf low-carbonate islands scattered over the northeastern Australian Great Barrier Reef margin comply with this definition. However, genetically speaking, they do not correspond to that of Darwin, since their development history was not subsidence-controlled, but has depended upon changing sea level. A number of Polynesian terms have been used to describe the islands and channels characteristic of atolls and some barrier reef margins and have been reviewed by Stoddart and Fosberg (1994). The islands are referred to as 'motu' and the channels between, essentially shallow overwash channels, as 'hoa'.

Aprons of sand may be present facing these passages and prograding into the lagoon. There is great variability in the distribution of motus. Chevalier (1972) described them without reference to the effects of high-energy events such as hurricanes, but Bourrouilh Le Jan and Talandier (1985) referred to the importance of these events in the formation of motus and hoas. Storm embankments or ramparts are a common feature of Pacific atolls and, where dated, deposits cluster around 2–5.5 ka (Scoffin, 1993; Montaggioni & Pirazzoli, 1984). The lagoons of larger atolls are commonly occupied by coral pinnacles or patch reefs. These are locally numerous. More than 2000 are recorded in the Enewetok lagoon (Stoddart, 1969a). In some areas coral patches amalgamate to form a distinctive mesh-like or reticulate pattern figured by Delesalle (1985) on Mataiva Atoll in the Tuamotus.

1.2.4. Bank Reefs

This reef type refers to isolated submerged reefs in shallow to deep waters. The Sahul Shelf off northwest Australia is an example of a relatively shallow bank reef (Teichert & Fairbridge, 1948). Such reefs are generally only a few kilometres in diameter and typically of circular form, rising from depths of 50–115 m. This area appears to have been subsiding in the late Cenozoic, influenced by subduction along the Indonesian arc (Van Andel & Veevers, 1967). The Saya de Malha Bank in the Indian Ocean is one of the largest submerged banks, forming a ring of approximately 40 km diameter. The rim on the outer margin is only about 8 m deep. The central lagoon varies from 70 to 140 m deep. The rim carries living coralgal communities (Fedorov, Rubinshteyn, Danilov, & Lanin, 1980). A speculative explanation is that the basic morphology of the bank is a result of karst erosion and subsidence, and that the coral growth is a recent addition that may yet reach the surface. Similar features have been described in the Caribbean (Macintyre, 1972). Examples of deep bank reefs include the Darwin Guyot, forming part of the mid-Pacific mountains north of the Marshall Islands. This lies in water 1266 m deep and retains a rim with a central lagoon-like basin about 18 m deep.

In both the tropical Pacific and Atlantic regions, the calcareous alga *Halimeda* forms submerged banks and biohermal structures (Roberts & Macintyre, 1988).

1.3. Geographical Distribution of Corals and Coral Reefs

The regions in which shallow-water, reef-building scleractinian corals are living today are restricted to the intertropical Indo-Pacific and Atlantic provinces (Figures 1.2 and 1.3). As noted by Veron (1995), more than 700 coral species have been described in the Indo-Pacific. There is a well-defined centre of higher coral species diversity, the boundaries of which are represented by Sumatra and Java (southern Indonesia) in the southwest; Sabah (north Indonesia) and the Philippines in the northwest; and the Philippines and Papua New Guinea in the northeast. In the Indo-Pacific centre of diversity, coral species exceed 450 in number, decreasing significantly eastwards. Central Pacific areas (Samoa, French Polynesia and the Cook Islands) range from 50 to less than 150 species. Hawaiian Islands

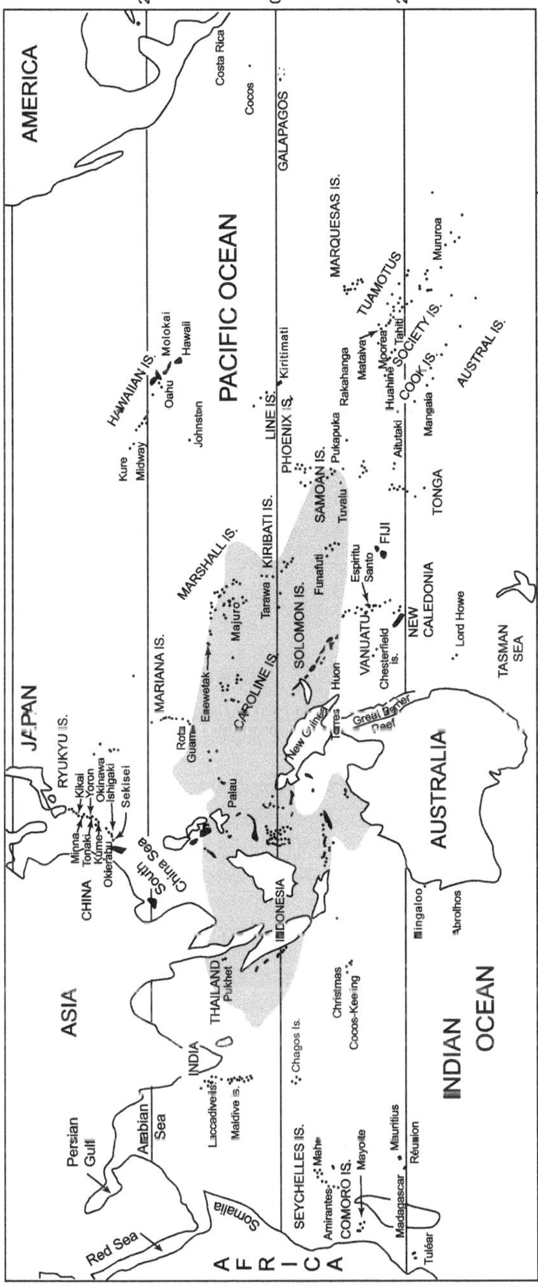

Figure 1.2 General map of the Indo-Pacific region showing the geographical extension (dark area) of the tropical Indo-Pacific Warm Pool (modified from Gagan et al., 2004) and the location of the major reef sites mentioned in the text.

Figure 1.3 General map of the western tropical Atlantic Ocean and the Caribbean with location of the major reef sites mentioned in the text.

and the easternmost part of the Pacific are the most depauperate Indo-Pacific areas with about 10 species. In the tropical Indian Ocean, species diversity tends to be uniform. Species numbers range from 200 to 250 in Western Australia and from 50 to 200 in the Western Indian Ocean. A slightly higher species diversity is found in the central Red Sea (up to 200). In the Caribbean, the number of coral species is very close to that observed from the central to the far eastern Pacific (from less than 10 up to 50).

Species diversity has been slightly modified in both the Indo-Pacific and the Caribbean during the Quaternary reflecting dramatic environmental changes (see Chapter 2). However, the latitudinal distribution of coral reefs has probably remained constant although the number of reef systems may have diminished dramatically during low sea-level stands in response to reduced substrate availability (see Chapter 4, Section 4.4.2).

1.4. Modern Tropical Climate Modes

Climate is regarded as a paramount determinant of reef species distributions. Ecological and palaeoecological studies of coral reefs have established that dynamics at the community level are directly determined by decade- to century-scale climatic changes (see Chapters 3 and 4).

Most of the major climate modes are generated in the equator and tropics; therefore, the intertropical zone constitutes a key region in which to understand the functioning of the earth's climate system. The central tropical Pacific is a controlling forcing source in decadal variability throughout the tropical belt and in some subtropical and temperate areas (Corrège, 2006; Grottoli & Eakin, 2007). The largest reservoir of heat (water temperature $>28°C$) on the planet by far is the Indo-Pacific Warm Pool (IPWP) extending from 90°E to about 175°E and from 10°N to about 18°S along the equatorial belt (Figure 1.2). The IPWP is the engine of the global climate system and profoundly influences heat and moisture exchange in the tropics and higher latitudes. It is associated with deep atmospheric convection and precipitation in the tropics (Gagan, Hendy, Haberle, & Hantoro, 2004). The most important phenomenon linked to IPWP activity is the El Niño/Southern Oscillation (ENSO), a coupled instability of the ocean–atmosphere system centred in the tropical Pacific. ENSO events are known to play a major role in governing the climate outside the tropics over large parts of the globe, through teleconnections, and occur today at about 3–7 year frequency. The warm phase of ENSO is referred to as El Niño, whereas the cold phase is referred to as La Niña (Philander, 1990; Cane, 2005).

The equator is girdled by a zone of low pressure and high moisture content, the Intertropical Convergence Zone (ITCZ) — a region of heavy precipitation. In the southern part of the IPWP, there is a branch of the ITCZ, the South Pacific Convergence Zone (SPCZ) that is a major feature of subtropical Southern Hemisphere climate. This zone interacts with circulation patterns across the Pacific and varies in location with ENSO-related expansion and contraction of the IPWP, migrating eastwards or westwards depending on the ENSO phases (Trenberth, 1984).

In addition to ENSO, there are other tropical to extratropical climate modes that influence climates regionally around the world. Monsoons are one of the major climate components in the tropics. They are governed by differential heating of continents and oceans and are accompanied by a seasonal reversal of surface winds and precipitation regimes. Of prime importance are the North-East (NE) and South-West (SW) monsoons in the control of climate variability in the southern Asian and Australasian regions. Recent studies have also highlighted the importance of the Pacific Decadal Oscillation (PDO) to tropical climates (Mantua & Hare, 2002). The PDO is expressed as a low-frequency co-variability of Pacific sea surface temperature (SST) and pressure patterns that resemble those of ENSO. In contrast to ENSO, PDO has periodicities ranging from about 15–25 and 50–70 yr, and has its largest amplitude in the mid-latitude North Pacific. The Interdecadal Pacific Oscillation (IPO) exhibits SST patterns similar to those of the PDO, but operates on a Pacific basin-wide scale. In the Indian Ocean, one of the most efficient ENSO-related climate modes is the Indian Ocean Dipole. This involves a reversal of the SST gradient and winds across the entire equatorial Indian Ocean (Saji, Goswami, Vinayachandran, & Yamagata, 1999). In the North Atlantic, the Artic Oscillation (AO) is the leading mode of variability in extratropical regions. The North Atlantic Oscillation (NAO), the most important climate driver during the boreal winter in the extratropical Atlantic sectors, may be regarded as a regional manifestation of the AO. The NAO is generated by differences in sea-level atmospheric pressure between the Icelandic low and the Azores high (Delworth, 1996). During positive NAO events, pressure anomalies between low and high centres increase strongly, but become weaker during negative NAO phases. An important mode of low-frequency climatic variability operating in the North Atlantic is the Atlantic Multidecadal Oscillation (AMO), which has its principal expression in the SST field and affects the frequency of severe Atlantic hurricanes forming in the northern tropical Atlantic. In addition, the tropical Atlantic climate experiences seasonal and inter-annual variations regulated by the migration of the ITCZ. This phenomenon is referred to as the Atlantic Dipole and is expressed as a difference between the mean monthly water surface temperatures in the northern and southern tropical Atlantic.

1.5. Quaternary Time Scales

The base of the Quaternary was originally regarded as marked by the onset of widespread glaciations but these have since been shown to have also characterized much of the later Tertiary. The status of the Quaternary (to be or not to be an era distinct from the Cenozoic) is therefore is still awaiting settlement. Quaternary scientists define the Quaternary as the time span covering the past 2.6–2.7 million years (Ma) of geological time (Gibbard, Cohen, & Ogg, 2008). However, the majority of stratigraphers agree to conventionally place the beginning of the Quaternary at 1.806 Ma, considering the term 'Quaternary' informal for time and including it as a subperiod in the Neogene (Figure 1.4). Notwithstanding these differences, it has long been recognized that the Quaternary interval is typified by an alternation of glacial and interglacial episodes reflecting prominent climatic cycles. From the perspective of coral reefs, these are reflected in turn in sometimes dramatic changes in sea level and in ocean circulation. Cycles have been equated with changes in solar insolation allied to changes in the earth's orbital behaviour (Hays, Imbrie, & Shackleton, 1976). This astronomical theory of climatic change was developed by Milankovitch (1941). Three separate components are involved: (1) Variations in the eccentricity of the earth's orbit occur with a period of approximately 100 ka. (2) The angle of tilt of the earth's axis of rotation varies between 21°39′ and 24°36′ over a period of 41 ka. As tilt increases the seasons become more marked and alternate poles spend longer facing away from the sun. (3) The third variable reflects the combined pull of the sun and moon, causing the spin axis to wobble or precess. This has the effect that the seasons also cycle, switching between alternate hemispheres

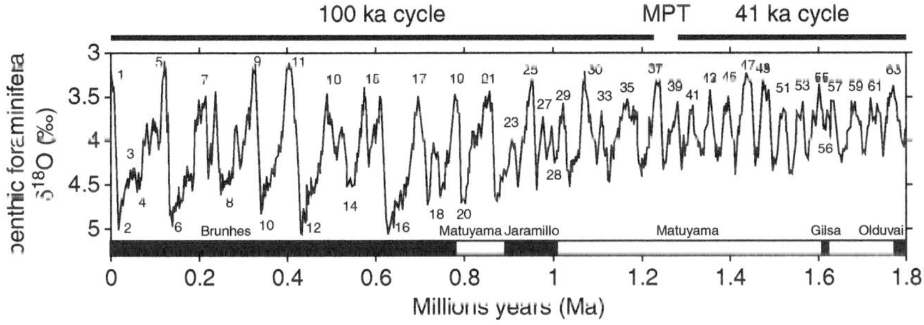

Figure 1.4 Oxygen isotope record as a proxy of global climate and sea-level change for the past 1.8 Ma. MIS stages are indicated. From 1.8 to about 1 Ma, changes in $\delta^{18}O$ are primarily controlled by the 41-ka period typical of obliquity and integrated insolation. From about 1 Ma, climate varies with a roughly 100-ka eccentricity period (based on Raymo & Huybers, 2008). The palaeomagnetic time scale with the relevant normal and reverse polarity events is given.

with a complete cycle in about 21 ka. It has since been shown that this may represent two superimposed cycles at 23 and 19 ka.

The terminology of subdivisions within the Pleistocene, that is the 1.8–0.01 Ma interval, applied to the tropics is largely unsatisfactory because the terms used have been largely defined in northern Europe and North America (Lowe & Walker, 1997). There are broad correlations with marine isotope stages (MIS) back to approximately MIS 20 (0.79 Ma) but equivalents are uncertain beyond this. Although comparison of absolute dates or proxy timescales between the tropics and Northern Europe and America can be made, the use of Northern Hemisphere names in tropical sequences is best avoided.

Finally, by international agreement, the Quaternary covers the time interval of glacial–interglacial events classified as the Pleistocene and the last postglacial to present interglacial period, the Holocene. The early Pleistocene is defined as starting at 1.806 Ma and ending at the Matuyama–Brunhes magnetic boundary (0.781 ± 0.005 Ma). The middle Pleistocene and late Pleistocene cover the 781–130 ka and 130–10 ka time spans respectively. The Holocene is generally accepted to have started approximately 10,000 years (10 ka) before present.

1.6. Trends in the Quaternary Climate Dynamics

The 'saw-tooth' asymmetrical shape of glacial cycles appeared about 3–2.5 Ma ago, at the time when the major Northern Hemisphere glaciation had begun to be established. From about 3.0 to about 1 Ma, the timing of climate variations, and particularly of global ice volume in the Northern Hemisphere, corresponds to the low-amplitude, high-frequency (41-ka) period of orbital obliquity (Figure 1.4). However, according to Lisiecki and Raymo (2007), the 41-ka cycles began to respond sensitively to obliquity forcing before 1.4 Ma.

A 23-ka signal, controlled by the precession of the equinoxes, is superimposed on the obliquity modulation. Long-term climate variations, especially of summer insolation in the Northern Hemisphere, are believed to respond linearly to obliquity and precession forcings. Glaciations and ice-volume fluctuations are likely to have been driven by insolation integrated over the duration of the summer during the early Pleistocene (Raymo & Nisancioglu, 2003; Huybers, 2006).

During the middle Pleistocene, there has been an important internal change in climate system dynamics. The dominant glacial oscillations have changed from 41 ka to a lower-frequency, higher-amplitude variability of about 100 ka, but were not accompanied by a significant change in orbital forcing. This climate change, is referred to as the Mid-Pleistocene Transition (MPT). Whereas the 41-ka oscillations tended to have been slightly asymmetrical prior to the transition, the time series became strongly asymmetrical with long glaciation and short deglaciation phases for each cycle since the MPT. The precise time when the MPT started remains controversial. The expected age of the MPT onset ranges from 1.5–1.25 Ma (Rutherford & D'Hondt, 2000; Clark et al., 2006) to 0.9–0.8 Ma (Lisiecki & Raymo, 2007) but it was well established by 0.70 Ma.

The onset of the MPT was marked by a sudden decline in SSTs, particularly in tropical upwelling areas, and by an increase in monsoon intensity. At the onset of the, tropical semi-precessional periods (with cyclicity of about 11.5 ka) began to shift to higher latitudes, coincident with an increasing amplitude of the 100-ka periods (Rutherford & D'Hondt, 2000). Then a gradual increase in long-term average ice volume occurred during the MPT, reaching 50 m sea-level equivalent. From about 0.90 Ma, there has been a strengthening of glaciation, an 80-ka event of extreme SST cooling followed by recovery and stabilization of long-term tropical and North Atlantic SST, the rise of the global deep-sea circulation and low-frequency variability in the Pacific SST (Clark et al., 2006). Changes in the climate system over the past 0.90 Ma appear to have been responses to non-linear orbital and ice-sheet constraints. The 100 ka cycles actually lasted 124 ka for the last two glaciations and 83 ka for the two earlier glacial stages (Servant, 2001). Controversial hypotheses have been invoked to explain the MPT. For instance, Rutherford and D'Hondt (2000) suggested that heat flow across the equator or from low latitudes was enhanced at about 1.5 Ma and thus promoted the propagation of the semi-precessional period in the Northern Hemisphere. This event may have caused the transition to the 100-ka glacial cycles. By contrast, the MPT was itself believed to have been triggered by a significant, long-term cooling resulting from a decrease in atmospheric CO_2 levels related to an increase in the rates of continental silicate weathering (Clark et al., 2006). Since 1.8 Ma, the Pleistocene period has included about 35 major glaciations and deglaciations (Figure 1.4).

A number of abrupt climate changes have occurred in the last 10 ka as clearly recorded in the GISP2 (Greenland Ice Sheet Project) ice core. The most significant of these include the Younger Dryas, a period of rapid cooling occurring at the transition from the late Pleistocene to the Holocene (around 11–10 ka) (Dansgaard, White, & Johnsen, 1989); the 8.2 ka event, a sudden decline in global temperature (Alley & Ágústsdóttir, 2005); the Holocene Climatic Optimum, centred at around 6 ka (Kaufman et al., 2004); the Medieval Climatic Optimum (or Medieval Climatic Anomaly), lasting from about the 9th to the 14th centuries and initially identified in the North Atlantic (Bradley, Hughes, & Diaz, 2003); and the Little Ice Age, regarded as a series of three colder episodes from approximately the 16th to the mid-19th centuries, each interrupted by slight warming intervals (Broecker, 2000). Some of these events have also been identified from coral reef records (see Chapter 9). Different explanations have been suggested to account for such rapid climate changes. The cooling events (the Younger Dryas, 8.2-ka event and the Little Ice Age) may have resulted from a significant reduction or shutdown of the North Atlantic thermohaline circulation due to sudden release of large amounts of freshwater into the North Atlantic (Broecker, 2000, 2006; Alley & Ágústsdóttir, 2005). Alternative causes identified for the Little Ice Age are lower solar activity and higher volcanic activity (Crowley, 2000a). The Holocene Climatic Optimum is usually regarded as the continuation of changes responsible for the end of the last glaciation and caused by the maximum Northern Hemisphere warming at 9 ka in response to predictable variations in the earth's orbit (Masson et al., 2000). The origin of the Medieval Climatic Warming remains unclear. However, this event may be defined as the upper boundary of the recent natural climatic variability and reflects changes in climate controls such as sunspot variability and internal variability, that is, random variations in the circulation of the atmosphere and oceans (Solomon, Qin, & Mannin, 2007).

1.7. Establishing the Chronology of Quaternary Coral Reefs

A key issue in trying to unravel the Quaternary reef history is the problem of determining the age of deposition from particular units. A number of methods are in use but all suffer from the diagenetic alteration that occurs in the rocks when exposed to meteoric waters (see Chapter 8). All methods, within the limits of experimental error, potentially provide a precise basis for correlation. But, because of the lack of consistency of application, the analysis of Quaternary deposits has commonly been on the basis of intervals of deposition or erosion. The dating methods applied to Quaternary deposits are presented and discussed by Walker (2005).

1.7.1. Oxygen Stable Isotopes

The definitive scale that has emerged over the past decades has been that of a stable isotope chronology based on $^{18}O/^{16}O$ ratios. Typically, analyses are now by accelerator mass spectrometry (AMS) (Linick, Damon, Donahue, & Jull, 1989) and results are given relative to deviations from a laboratory standard ($\delta^{18}O$‰). The standard originally used for carbonates was the PeeDee Belemnite (referred to as PDB), whereas that used for ocean waters and ice was of Standard Mean Ocean Water (SMOW).

Following Emiliani (1955), Shackleton (1967, 1977, 1987) and Broecker (1994) established a pattern of SST extending to 600 ka marked by repeated asymmetric cycles. The overall pattern has been found to match, in general terms, the 100, 41 and 23 ka cycles predicted from astronomical theory by Milankovitch (Hays et al., 1976) with the principal divisions now referred to as marine isotope stages (MIS), numbered from MIS 1, the Holocene, back to at least MIS 63 with an absolute age approaching 1.8 Ma (Figure 1.4). Odd numbers reflect warm periods and even numbers glacial intervals. Because the basic frequencies of the cycles are known they can be used to calculate the age of each isotopic stage (Berger, 1978).

The link between the oxygen isotopic signal and ice volume supports a more general correlation with sea level and thus the record can be read as indicating high and low sea-level stands. As indicated above, the principal correlation in Quaternary reef deposits is between lowstands and erosion. The minor isotopic fluctuations indicated by data such as those of Waelbroeck et al. (2002) may be expected to have had effects on deposition but so far these have not been generally recognized. There is evidence of some variation, and in isotopic stage 7, for example, a double peak is reflected in an additional erosion surface in sequences on Eleuthera (Hearty, 1998), the Great Barrier Reef (Braithwaite et al., 2004) and Mururoa (Camoin, Ebren, Eisenhauer, Bard, & Faure, 2001).

1.7.2. Uranium-Series Dating

Small amounts of uranium are incorporated into crystals of both calcite and aragonite. Unlike the stable oxygen isotopes described above, the radiogenic isotopes of uranium form a decay series from ^{238}U to ^{235}U and ^{232}Th (thorium) to lead. Decay of ^{238}U to ^{234}U and of ^{234}U to ^{230}Th have half lives of 4.47 Ga and 245.5 ka respectively. Analyses have traditionally been made using alpha spectrometry, originally with an accuracy of $\pm$8%, but around $\pm$1.5% is now obtained routinely. However, thermal ionization mass spectrometry (TIMS) has been found to provide more reliable results (better than 0.5% of the age) on far smaller samples (Edwards, Chen, Ku, & Wasserberg, 1987; Li et al., 1989).

Corals typically contain 2–3 ppm uranium and are thus suitable for $^{230}Th/^{234}U$ dating. There must be no measurable ^{232}Th and the $^{234}U/^{238}U$ ratio must be similar to that in present-day corals. These are all indicators that the sample has escaped diagenetic alteration and remained a closed system since its formation. In a closed system, ^{238}U with a half-life of 4.47 Ga decays to ^{234}U (half-life of 245.5 ka) and this in turn becomes ^{230}Th with a half life of 75.4 ka. ^{234}U is present in seawater as it is readily soluble but ^{230}Th is relatively insoluble and is virtually absent. When ^{234}U is incorporated into the carbonate of animal skeletons, the ^{230}Th that is generated accumulates and provides a measure of the time since the skeleton formed. Material from raised terraces in Papua New Guinea (Veeh & Chappell, 1970; Chappell, 1974), Barbados (James, Mountjoy, & Omura, 1971; Broecker et al., 1968) and the Ryukyu Islands

(Konishi, Omura, & Nakamichi, 1974) was among the first used to demonstrate the efficacy of these methods to date corals and also the close correspondence between the cyclic behaviours observed and Milankovitch cycles of sea-level change. Analyses have now become so sensitive that late Quaternary corals can be dated to within a few years (Bard, Hamelin, Fairbanks, & Zindler, 1990).

Although the reliability of these methods has been amply demonstrated for the majority of areas, there are examples where uncertainties have emerged. Cobb, Charles, Cheng, Kastner, & Edwards (2003) described the results of U/Th dating of living and young fossil corals from Palmyra Island in the central Pacific, ranging in age from 50 to 700 yr. Importantly, Palmyra is an atoll and there is thus no obvious rock source for the thorium. Evidence points to a range of $^{230}Th/^{232}Th$ values for fossil corals that overlaps that of living corals, suggesting that the thorium is either primary or is added in some way while the coral is still alive. These results are important because uncertainties in the correction that should be applied for non-radiogenic ^{230}Th may lead to significant errors in U/Th dates.

Results can be adjusted using the $^{232}Th/^{230}Th$ ratio (Schwarcz & Latham, 1989) but remain unreliable. Models have been proposed that allow uranium-series ages to be calculated in what were apparently open systems, and have been applied with some success in Barbados (Thompson & Goldstein, 2005) and New Caledonia (Frank et al., 2006). Attempts by Thompson, Spiegelman, Goldstein, and Speed (2003) to model open-system behaviour based on the calculation of model ages using the decay series ^{238}U–^{234}U–^{230}Th (Villemant & Feuillet, 2003) and work by Scholz, Mangini, and Felis (2004) have been reviewed by Scholz and Mangini (2007). The results presented show that the errors of conventional Th/U dating and the uranium-series method of Thompson et al. (2003) do not account for the true age variability that lies within the range of errors indicated by the models of Villemant and Feuillet (2003). The criteria that are widely used to demonstrate reliability are insufficient to identify all diagenetic alteration, and the authors suggest that the analysis of subsamples of a single specimen provides a better estimate of age variability and diagenetic alteration.

1.7.3. Radiocarbon Dating

This was one of the earliest dating methods to be developed and applied to carbonates. Like other methods, radiocarbon dating has seen a dramatic increase in precision. AMS now only requires a milligram or less of sample. However, the limit of practical counting is approximately 45 ka. In rocks older than this, the amount of ^{14}C present is $<1\%$ of its original value. Greater ages have been measured using a technique to enhance the amount of ^{14}C present, and by this means ages of 60 ka have been recorded. Errors are estimated to be about 1%, equivalent to ±80 yr around 5.5 ka. Calibration is possible up to 10 ka using dendrochronology and there are also direct comparisons with uranium-series (Fairbanks et al., 2005) and other results (Van der Plicht, 2002).

There are several other factors that require adjustment. The first is the so-called reservoir effect. Because of fractionation effects, concentrations of ^{14}C vary between reservoirs such as the oceans, the atmosphere and the biosphere; even in the oceans, there is variation between surface and deep waters (Southon, Kashgarian, Fontugne, Metivier, & Yim, 2002). The present levels of ^{14}C in the atmosphere have been significantly altered by the testing of thermonuclear bombs and thus the concentration before AD 1950 is referred to as the modern standard. The flux of cosmic rays reaching the earth has varied with time and so also has the distribution of carbon in the various reservoirs, but calculations assume that concentrations were initially those of the modern standard. The circulation of carbon within the marine system is a particular problem. As noted above, the transfer of ^{14}C from the atmosphere to surface waters and between surface and deep waters is very slow. Thus, different water bodies have different apparent ages that are transferred to the minerals precipitating within them. In surface waters of the North Atlantic (Bard, Arnold, & Duplessy, 1991), the present apparent age is 400 years and a similar correction factor of 400 years must be applied to ^{14}C dates from corals from Barbados (Fairbanks, 1989), but in the deep oceans the apparent age may be >2000 years (Ostlund & Stuiver, 1980).

In parallel with the contamination of samples by detrital thorium, samples may be compromised by the addition of detrital carbon or by percolating humic acids, giving rise to spurious (older or younger) ages.

1.7.4. Aminostratigraphy

Whole rock aminostratigraphy is based on the progressive racemization of amino acids preserved in biominerals. All biominerals contain varying proportions of organic molecules, typically in the form of nanoscale filaments extending through the crystal structures. In time these begin to break down and L-amino acids racemize (isolucine epimerizes) to their D-isomer form. Analyses are of the ratio of D/L (or isoleucine to alloisoleucine, A/I) that measures the extent of racemization. The A/I ratio is initially 0 but increases to an equilibrium value of about 1.3 with time; it is temperature dependent. Analytical methods are described in Miller and Brigham-Grette (1989). The technique has been applied with some success by Hearty (1998) to estimate ages on Eleuthera in the Bahamas. Samples attributed to MIS 13 generally have amino acid ratios that are too low to be accurately measured but stages 9/11, 7, 5e, 5a and 1 are clearly differentiated. Age estimates of A/I ratios based on an assumed apparent parabolic kinetic pathway (Mitter & Kriausakul, 1989; Hearty & Dai Pra, 1992) compare well with mean ages of MIS and those derived by U-series methods. It must be said, however, that there has been some criticism of the method (Carew & Mylroie, 1994) and in some areas where application has been attempted (Braithwaite et al., 2004) the breakdown of amino acids has proceeded to the point where no results can be obtained, reflecting both the age and degree of alteration of the material.

1.7.5. Electron Spin Resonance

In an effort to avoid problems related to diagenetic changes one of the methods applied has been electron spin resonance (ESR). Electrons orbiting molecules have an intrinsic momentum, referred to as spin. If the sample is placed in a magnetic field, the intrinsic magnetic dipoles of the electrons align in one of two ways, either parallel to or in the opposite direction to the field, with the latter state of lower energy. Background radiation dislodges electrons from their normal positions in atoms and these become trapped in the crystalline structure of the material. When odd numbers of electrons are separated, there is a measurable change in the magnetic field (or spin) of the atoms. The magnetic field changes progressively with time as a result of this process. When radiation of a particular frequency is applied, it raises these electrons to the higher-energy state in which the magnetic dipoles are parallel to the magnetic field. As they fall back to the lower-energy state, they emit photons. Under continued radiation, the electrons resonate between the two energy states with the cycle referred to as electron spin resonance. The method was first applied to corals in the 1980s but recent improvements in the technique have provided results comparable with those from ^{14}C and TIMS U-series dating (Radtke, Grun, & Schwarcz, 1988; Schellmann, Radtke, Potter, Esat, & McCulloch, 2002) and may be applicable to materials as old as 2 Ma.

1.7.6. Magnetostratigraphy

Although carbonate rocks have only weak magnetic intensities, these are easily measured on modern cryogenic magnetometers and magnetization is stable. Measurements therefore provide an alternative time scale that can be used as a control on other age determinations or indeed as a reference sequence where other dates are unobtainable. The past 1.8 Ma can be divided into two general polarity episodes (Figure 1.4), the Matuyama (starting at 2.6–2.47 to 0.78 Ma), mostly of reversed polarity compared with the present, and the Brunhes (from 0.78 Ma to present), mostly normal in polarity. However, each includes intervals of longer or shorter duration in which the dominant polarity is reversed. Within the portion of the Matuyama extending into the Quaternary, there are three such excursions

(Olduvai, Gilsa and Jaramillo). Two polarity excursions are present in the Brunhes: the Emperor dated at 420 ka, the Laschamp at 40 ka, and the Blake at 12 ka that might be used to tie stratigraphic determinations. This method has contributed successfully to date several Pleistocene reef sequences. For instance, in Ribbon Reef 5 core extracted from the Australian Great Barrier Reef, the lower boundary of the Brunhes was used as a control point in attempts by Braithwaite et al. (2004).

1.7.7. Strontium Ratios

Strontium is incorporated into aragonite up to about 2000 ppm and also into calcite at levels of a few hundred ppm. Derived from weathering of the continents, it has a relatively long residence time in the oceans. Hodell, Mead, and Mueller (1990) calculated it as 2.5 Ma and thus, as the relative mixing time of ocean waters is considerably less (1 ka, Broecker, 1963), values of $^{87}Sr/^{86}Sr$ can be considered to be homogeneous and independent of latitude or depth. Therefore, in principle, minerals precipitated from ocean waters should accurately record the $^{87}Sr/^{86}Sr$ ratio at the time of their formation. Hodell et al. (1990) used this principle to construct correlation plots of data for the past 8 Ma incorporating analyses from planktonic foraminifera. The data show a progressive increase of about 25% that is explained by an increase in uplift, principally of the Himalayan–Tibetan region, and weathering. In addition, there is an upturn at 2.5 Ma that is attributed to increased glacial activity in the Northern Hemisphere. Zachos, Obdyke, Quinn, Jones, and Halliday (1999) were subsequently able to tie more detailed variations to climatic changes. The calculated regression during the last 2.5 Ma has relatively high (95%) confidence limits and can potentially be used to provide stratigraphic resolution. However, owing to uncertainties in variability the use of the method has shown no great expansion in Quaternary deposits in the last decade. Ohde et al. (2002), then the International Consortium (2001) and Braithwaite et al. (2004) were able to use this method to infer apparent ages for the upper sections of Funafuti Atoll borehole and to the base of the Great Barrier Reef borehole respectively.

1.7.8. Other Dating Methods

The chronology of Pleistocene reef units can be revealed using additional, but less reliable methods including thermoluminescence (Ninagawa et al., 2001), cosmogenic beryllium (Maejima, Matsuzaki, & Higashi, 2005) and nannofossil-based stratigraphy (Yamamoto et al., 2006; Cabioch, Montaggioni, Thouveny et al., 2008). These methods are used occasionally to complement or replace the more classical procedures.

1.8. Methods of Obtaining Data

The methodologies applied to reef investigation fall into two groups: those related to the surface observation of recent reefs and those related to internal structures or Pleistocene and Holocene deposits.

1.8.1. Surface Observations

Ecologists use a variety of techniques in the study of modern reef communities. These techniques include quadrat and line intercept transect sampling. A quadrat is a frame of known size randomly placed on the reef surface; the numbers of and areas occupied by individual species within this are counted and recorded. Line transects apply the same principle to counting (sampling) along a linear transect of known length and width. Both of these provide reproducible objective measures of the organisms present, the areas they occupy and their relative distribution. The line transect technique was pioneered by Loya (1972, 1978), Done (1977), Pichon (1978a, 1978b) and Scheer (1978), and quadrat technique by Maragos (1974) and Laxton and Stablum (1974). Methods of sampling soft-bottom communities are discussed by Thomassin (1978). Palaeoecologists and sedimentologists have used similar techniques to quantify fossil assemblages and assess growth frameworks and sediment characteristics. Divers have been able to deploy instruments to monitor environmental conditions, including temperature, salinity and current activity, and set up experiments aimed at determining the physiological behaviour of organisms *in vivo*.

Small free-diving submersibles have been developed that allow direct observation of deep slopes (to several hundred metres) and limited sampling. These have made important contributions to our understanding of lowstand deposits and erosional notches. In addition, the development of remotely operated vehicles (ROVs) has allowed similar investigations with high-definition cameras. A key limitation of all underwater investigations has lain in the description of gross morphology. The soundings of the early days of exploration provided crude profiles that became increasingly detailed with the development of various echosounders. Only in the last decade, has the appearance of multibeam soundings allowed detailed surveys of submerged reef morphology. These have yet to be widely used.

1.8.2. Pleistocene and Recent Reef Structures

The examination of raised Pleistocene sequences at outcrop does not differ markedly from the geological investigation of any rock outcrop. Preservation of corals and other organisms may be an issue, but in many areas large clean limestone faces lend themselves to the same census methods applied to living reefs.

There has been some effort to extend the record of deposition by drilling. The basic design of a land-based or barge-mounted drilling systems has been described by Thom (1978) but a large number of manufacturers are able to provide equipment with similar capabilities. The nature of the site provides important constraints on the equipment deployed and the strategy adopted. Access to submerged reefs requires the deployment of a drilling vessel. This strategy is currently adopted by Integrated Ocean Drilling Program (IODP).

Individual massive corals and reef surfaces can both be cored to shallow depths underwater. Light handheld drills using either hydraulic pumps or a pneumatic (compressed air) power source have successfully produced cores of several metres length. A hydraulic system that could also be used in surface investigations was described by Macintyre (1978). As in all coring, it is important that the diameter of the core barrel is sufficient to provide stability. In Macintyre's system, the core was 54 mm but for corals at least 25.4 mm has been found to be satisfactory.

CHAPTER TWO

Palaeobiogeography: Evaluation of the Inheritance from the Tertiary

2.1. Introduction

The present distribution of reef-building corals and reefs can only be properly understood in the light of both the succession of palaeobiogeographic events and the evolution of reef biotas from the Cretaceous–Cenozoic transition to the early Pleistocene. However, the reconstruction of the geographical distribution and evolutionary patterns of ancient reef-building scleractinian corals depends largely on the spatial and stratigraphical distribution of exposures, together with the style, preservation and intensity of sampling at any given reef site (Perrin, 2002; Lopez-Pérez, 2005). For example, the apparent paucity of Tertiary reefs in the eastern and north-eastern Pacific may be explained by a patchy record of data. By comparison, Miocene reefs in Southeast Asia are well documented as a result of intensive hydrocarbon exploration (Wilson, 2002; Fournier, Montaggioni, & Borgomano, 2004). In addition, the poor state of scleractinian and coralline algal taxonomy brings an additional bias to estimates of diversity patterns at generic and species levels and may preclude the emergence of any comprehensive picture of their overall evolution in reef ecosystems (Braga, Bosence, & Steneck, 1993; Veron, 1995). The taxonomy and evolutionary patterns of scleractinian corals are currently in question, following recent phylogenetic analyses of mitochondrial and nuclear genes (Romano & Palumbi, 1996; Romano & Cairns, 2000; Chen, Wallace, & Wolstenholme, 2002; Medina, Collins, Takaoka, Kuehl, & Boore, 2006). Contrary to traditional concepts, there is robust molecular evidence that families do not necessarily belong to single monophyletic groups and do not relate to morphologically based suborders. They may have been derived from distinct clades that differentiated as early as 300 million years (Ma) ago rather than the usually assumed 240 Ma. Many reef coral species are hybridizing forms, belonging to complexes (so-called 'syngameons') that are composed of a number of genetically distinct species or lineages (Romano & Cairns, 2000; Chen et al., 2002; Stanley, 2003).

Problems may also arise from the inaccuracy of dating most Tertiary reefal sequences. Imprecise age assignments of reef coral occurrences are related to their preferential growth in shallow-water settings, which are generally lacking reliable stratigraphic markers (Perrin, 2002; Kiessling & Baron Szabo, 2004).

The Tertiary period (an interval of about 63 Ma) was a time of major expansion in the size and scale of taxonomic diversity gradients (Crame & Rosen, 2002). More specifically, it registered the progressive emergence of scleractinian coral reefs as the dominant marine, shallow-water ecosystems in the tropics and subtropics. The evolution of Quaternary and present-day reef systems took place through the gradual development of modern reef patterns, such as community structure and reef anatomy, punctuated by rapid turnovers of benthic biotas. In many ways, the Tertiary history of reef building is anomalous, and one of the most striking aspects of scleractinian coral diversification events is that they occurred against a backdrop of global climatic cooling and falls in sea level (Crame & Rosen, 2002). Corals survived most of these inimical events and finally formed large framework-dominated reefs. Their amazing resilience to global climatic deterioration is regarded as having been promoted by the remarkable success of the scleractinian–algal symbiosis and by the acquisition of a range of specializations and competitive and defensive adaptations (Wood, 1993).

2.2. Development Patterns of Tertiary Coral Reefs

From the beginning of the Tertiary to the late Pliocene, about 2 Ma ago, reef systems were generally located within a latitudinal belt broadly centred on the equatorial to subtropical zones and slightly shifted northwards. This reef belt varied in width through time and became wider in the middle Miocene (16–11 Ma) (Figure 2.1). Buildups appear first to have been distributed longitudinally, occupying the margins of the ancient Tethys Ocean, a vast, circum-equatorial marine seaway, extending westwards from southern Asia through the Middle East and southern Europe, through the proto-Atlantic Ocean and the American land masses, to the proto-Pacific Ocean. Subsequently, reef growth has gradually migrated towards the present-day boundaries. A wide range of hypotheses has been offered for explaining the distribution of modern coral reefs (see Rosen, 1988; Veron, 1995; Perrin, 2002, for detailed reviews).

2.2.1. From the End-Cretaceous Extinction to the Cenozoic Recovery

Late Cretaceous shallow-water, tropical environments were usually dominated by rudists, forming biostromal structures rather than true framework reefs (Gili, Skelton, Vicens, & Obrador, 1995). Although they

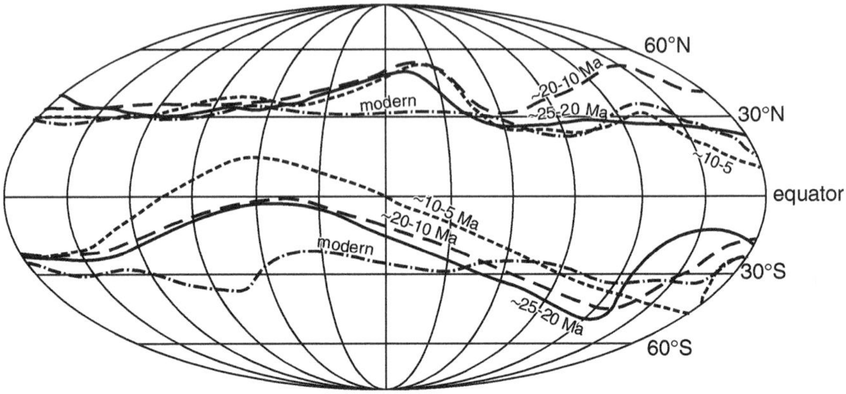

Figure 2.1 Changes in the latitudinal extension of the reef belt throughout the Cenozoic, from about 25 Ma ago to the present day. Redrawn and modified from Perrin (2002).

were as common as rudists, scleractinian corals rarely developed dense, reef-building communities, but the relative paucity of coral buildups was not, apparently, directly due to rudist competition (Gili et al., 1995) and may have been driven by inimical oceanographic conditions (Scott, 1995).

2.2.1.1. Extinction patterns

The late Cretaceous interval was a time of major environmental disturbances including intense tectonic activity, marked climatic changes, reduction of shallow-water habitats and, finally, the Cretaceous–Tertiary (K/T) boundary catastrophe (asteroid impact?). In this scenario, shallow-water reef systems were affected by a significant global scale reduction with the demise of reef constructors. The end-Cretaceous extinction preferentially affected highly specialized warm-shallow-water organisms, particularly bioconstructors. Best estimates of end-Cretaceous scleractinian extinction are slightly higher than 45% at the species level (Kiessling & Baron-Szabo, 2004). Corals inferred to have hosted zooxanthellae in their tissues (zooxanthellate forms) suffered more severely than species inferred to have been devoid of these symbionts (azooxanthellate forms) (Figures 2.2 and 2.3). The morphological complexity of corals, their coloniality and modular colony organization, appears to have acted as a selective criterion for survivorship. The extinction risk for corals was higher also for colonies with high corallite integration. Feeding strategy also played a major selective role in the extinction of corals and probably also other reef-inhabiting taxa. The combination of photo-autotrophy and heterotrophy (i.e. predation on zooplankton) was likely to be less advantageous to survival than simple heterotrophy.

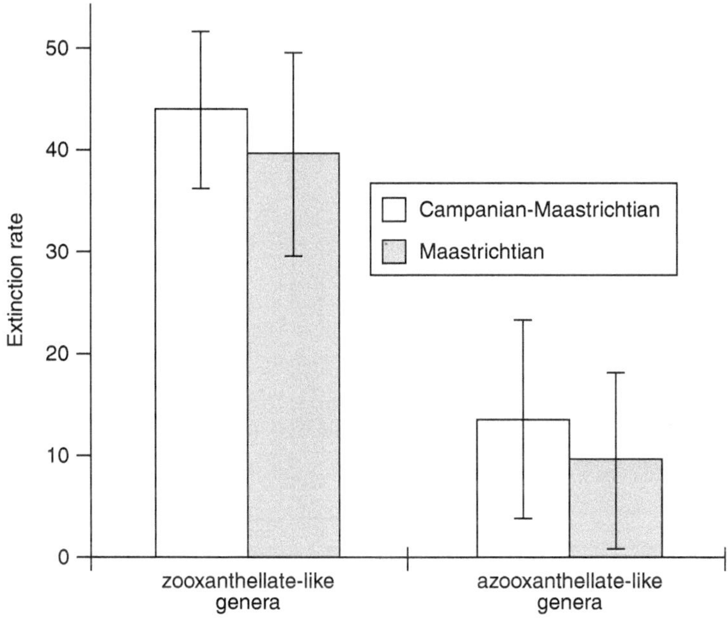

Figure 2.2 Extinction rates of scleractinian coral genera as a function of physiological constraints at time intervals close to the Cretaceous/Tertiary boundary (Campanian–Maastrichtian: 83–65 Ma; Maastrichtian: 72–65 Ma). Susceptibility of zooxanthellate forms to extinction was higher than that of azooxanthellate genera. Vertical lines represent binomial error bars. Modified from Kiessling and Baron-Szabo (2004).

The extinction of zooxanthellate corals is randomly distributed geographically. There was apparently no direct latitudinal control on extinction rates and no hot spots of extinction. However, there was a marked relationship between geographical patterns and extinction risk. Widespread distribution at the end of the Cretaceous was an insurance against disappearance at the K/T boundary (Kiessling & Baron-Szabo, 2004). On a regional scale, extinction rates were quite similar to the global mean: $33 \pm 11\%$ in North America, $35 \pm 10\%$ in Europe and $30 \pm 11\%$ in Africa and India.

At the K/T boundary, the restriction of most scleractinian communities to the tropics, and the higher susceptibility of zooxanthellate corals to extinction, resulted in the near complete disappearance of shallow-water buildups from the tropical belt. Given that the late Cretaceous was a period of attenuation of reef-building capacity, differences in reef patterns immediately before and immediately after the K/T boundary were smoothed. Apart from biotic extinctions, there was no sharp disruption in overall diversity, in rates of carbonate production or total number of buildups. As far as reefs were concerned, the transition to the Tertiary

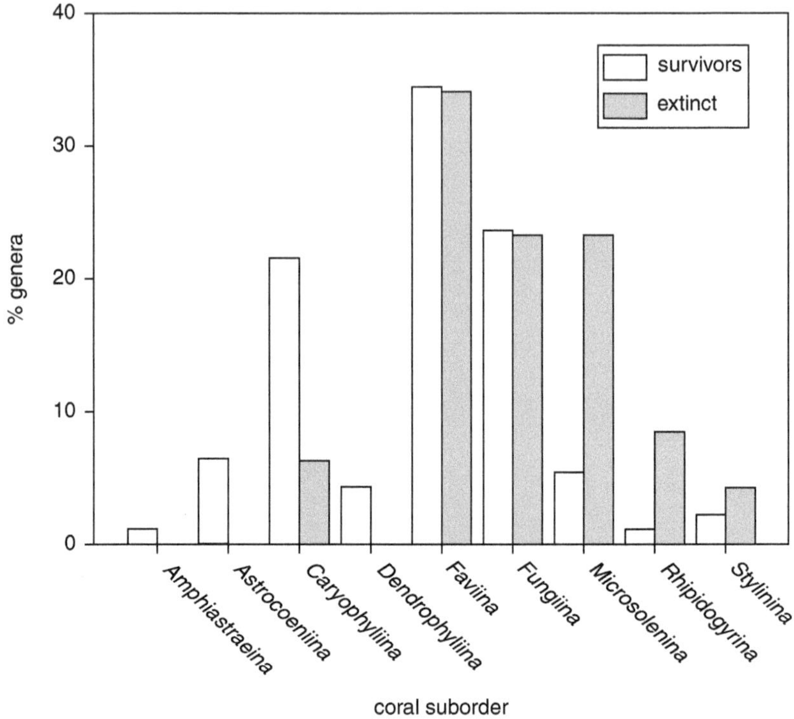

Figure 2.3 Proportion of scleractinian coral genera affected by extinction after the Cretaceous/Tertiary boundary in the main scleractinian orders. Orders Caryophyliina and Dendrophyliina, exclusively composed of azooxanthellate corals, suffered markedly less reduction than Faviina, Fungiina and Microsolenina, dominantly including zooxanthellate forms. Modified from Kiessling and Baron-Szabo (2004).

simply recorded the final stage of a gradual decline of reef systems that started in the Maastrichtian at around 67 Ma (Cooper, 1994).

2.2.1.2. Recovery patterns

The recovery of reefs after biotic crises is usually regarded as delayed in comparison to that of other marine systems (Cooper, 1994), mainly because of the difficulty in re-acquiring an efficient building capacity. The recovery of scleractinian reef communities was apparently gradational, operating through successive phases of coral appearance and the rejuvenation of reef-building capacity.

The Cenozoic history of coral reefs began with the few corals that survived extinction at the K/T boundary. The only reef-dominating Mesozoic survivors that were still abundant in Cenozoic reefs were Faviidae. About 6 of 16 faviid genera had escaped extinction. However, there is no robust fossil record for assessing the fate of the survivors, and when reef coral communities were restored in the Eocene, no species level continuity with the Mesozoic can be identified. Twenty one percent of the new genera that emerged appeared as early as the Danian (65–59 Ma). At that time, genera were nearly uniformly distributed at the global scale. When averaged over the whole of the Paleocene, the number of Paleocene genera relative to Cretaceous survivors seems to be higher in low latitudes. The tropics are therefore commonly considered to have been the source of evolutionary novelty in the post-extinction episode (Jablonski, 1993).

The entire Cenozoic was marked by a progressive global cooling. That this long-term state did not result in any severe disruption of reef development is shown by the occurrence of flourishing reef-building communities and by reef expansion. However, pronounced steps in taxonomic gradients appear to have existed throughout the Cenozoic and especially during the Neogene (23.5–1.8 Ma). Since their recovery following the K/T event, scleractinians developed an increasing capacity for space competition, particularly in nutrient-poor environments, resulting in their overall dominance as reef builders throughout the Cenozoic. The competitive strategies of zooxanthellate corals include biochemical defences against potential competitors, resistance to predators, high plasticity of colonies, high survival ability owing to modular clonal organization and high degree of colony integration (Wood, 1995). Corals became better competitors concurrent with the rise of many new consumer taxa. The appearance of herbivorous groups from the early Cenozoic onwards (e.g. teleost fish during the mid-Eocene) has facilitated coral colonization of suitable substrates to the detriment of fast-growing fleshy macrophytes

2.2.2. Coral and Reef Diversification in Time and Space

2.2.2.1. Mechanisms of diversification

The combination of plate tectonics, climate change, sea level fluctuations and oceanic circulation has been identified as the first-order control on the evolution of reef biotas, the delimitation of biogeographical provinces and in producing disjunct tropical distributions (Frost, 1977a; Rosen, 1984; McManus, 1985; Potts, 1985; Rosen & Smith, 1988; Pandolfi, 1992a, 1992b; Veron, 1995; Paulay, 1997; Wilson & Rosen, 1998; Roy & Pandolfi, 2005). Changes in atmospheric and ocean chemistry (e.g. atmospheric CO_2 and oceanic Ca^{2+} concentrations, oceanic Mg/Ca ratios) are also regarded as promoters of successive changes of the dominant carbonate producers through the Cenozoic (Pomar & Hallock, 2008, and references herein).

Tectonics and climate. Tertiary tectonic events directly affected oceanic circulation and climatic patterns through changes in palaeogeography. At the beginning of the Cenozoic, the eastern Pacific and western Atlantic regions

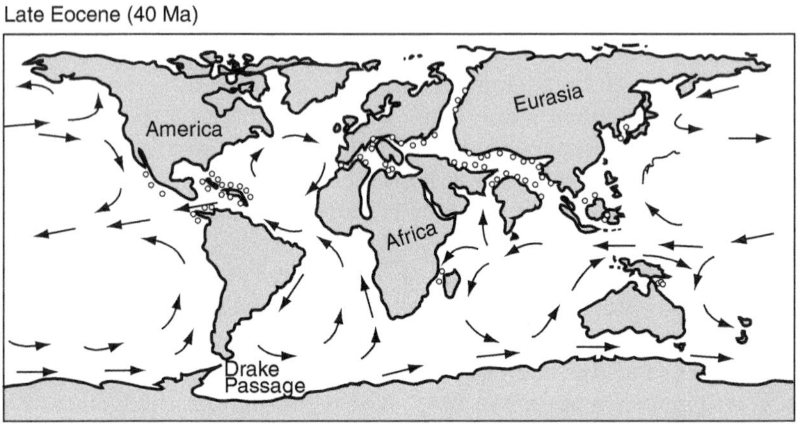

Figure 2.4 Global palaeogeography in the late Eocene (40 Ma). The main features were the maintenance of the tropical circumglobal circulation through the Tethyan seaway and the central America seaway, the intensification of Pacific and Atlantic gyres and the development of a circum-Antarctic circulation, with the opening of the Drake passage. Open circles indicate shallow-water reef sites, dominated by zooxanthellate corals. Arrows indicate major current directions. Redrawn and modified from Veron (1995, 2000) and Crame and Rosen (2002).

were connected via a narrow tropical seaway. The circum-tropical belt of shallow waters was interrupted by the wide proto-Pacific Ocean and this vast oceanic realm was probably subdivided into wide biogeographical regions, some of which were unfavourable to reef growth, thus limiting coral dispersal. During the Paleocene and Eocene, partial isolation of the western and eastern areas of the Tethyan realm may have induced differentiation of some benthic biota (Figure 2.4). Throughout the Cenozoic, the displacement of lithospheric plates progressively widened the Atlantic Ocean. The Mediterranean Tethyan region gradually separated from the Indian part between the late Oligocene and middle Miocene (Figure 2.5). From the early Eocene (55 Ma), known to have been a time of climatic optimum, global climate suffered from a series of pronounced deterioration events. At the Eocene–Oligocene boundary (33–34 Ma), sea surface temperatures dropped by 5 °C due to the onset of Antarctic glaciation (Figure 2.6). Paradoxically, Cenozoic coral diversification occurred as climate was deteriorating, and as tropical areas significantly diminished in size and temperate regions expanded (Crame & Rosen, 2002; Rosen, 2002).

The constriction of the Tethyan realm and its partial closure in the early Miocene (around 20–16 Ma), as a result of the gradual northward shift of the African and Arabian plates and their final collision with Eurasia, were of particular importance. The western Atlantic–Caribbean tropical reef biotas are thought to have been separated from those of the Mediterranean by major variations in the central Atlantic circulation at that time

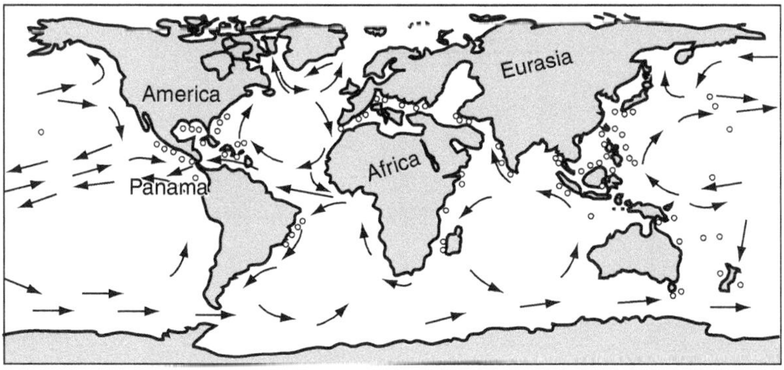

Figure 2.5 Global palaeogeography in the middle Miocene (15 Ma). Tethys is limited to a narrow band connecting the Indian Ocean with the Mediterranean region. The main features were the closure of the Tethyan seaway in the Mediterranean region and of the Indonesian seaway, the onset of the rise of the Isthmus of Panama, formation of the circum-Atlantic current and the initiation of high-diversity centres in the Indo-West Pacific, eastern Atlantic, Caribbean and eastern Pacific. Open circles indicate shallow-water reef sites, dominated by zooxanthellate corals. Arrows indicate major current directions. Redrawn and modified from Veron (1995, 2000) and Crame and Rosen (2002).

(Chevalier, 1977). During the late Miocene, the continuous displacement of the African plate, combined with climatic cooling, transported Mediterranean areas out of the tropics and brought about the demise of the Mediterranean as a coral reef subprovince. Similarly, the collision of the Australian and Indonesian plates in the early-middle Miocene (15 Ma) blocked the open sea seaway between the Indian Ocean and the western Pacific (Grigg, 1988) (Figure 2.5). The progressive rise of the Isthmus of Panama during the middle Miocene–latest Pliocene interval (13–3.5 Ma) separated the eastern Pacific from the western Atlantic province, and thus exerted a major control on oceanic current patterns from around 4 Ma (Haug & Tiedemann, 1998). This event may have limited faunal dispersal as early as the middle-late Miocene (15–11 Ma), until shallow-water circulation between the eastern Pacific and western Atlantic through the central American seaway was interrupted around 3.5–3 Ma. In Southeast Asia, the major diversification of hermatypic corals from the Neogene onwards was probably triggered by the lateral motion of the Australian plate northwards and its collision with Southeast Asia. This brought about an increase in shallow-water shelf areas and in the lengths of coasts. In addition, it led to the movement of Australia within the tropical zone from the early to middle Miocene.

From a palaeobiogeographical perspective, the impact of Cenozoic plate tectonics was to extend land masses, particularly in the tropical Pacific and, combined with global cooling, to promote the isolation of the four

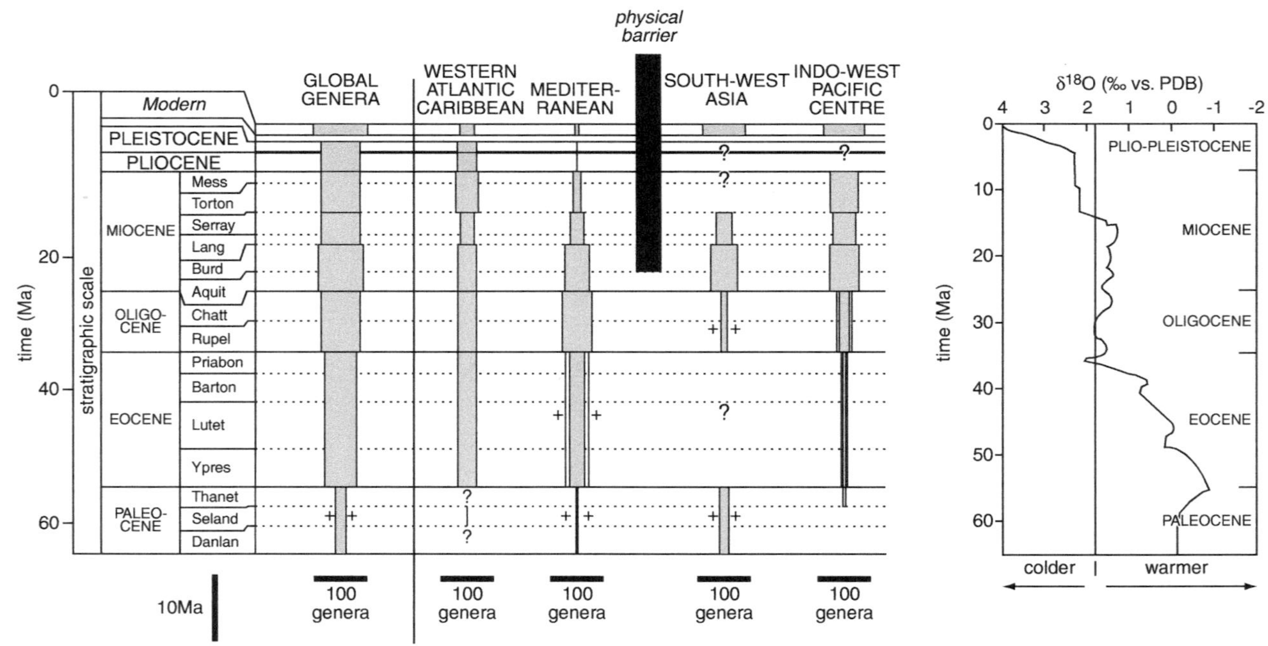

Figure 2.6 Variations in genus diversity of zooxanthellate scleractinian corals at global and regional scales through the Cenozoic. Four key regional pools are identified. Coral diversity is compared with a $\delta^{18}O$ curve used as a proxy for relative sea surface temperature. Note that Indo-West Pacific coral diversity increased, while the climate became colder from the Miocene onwards. Redrawn and simplified from Rosen (2002).

present-day reef provinces: the Indo-West Pacific, western Atlantic–Caribbean, eastern Atlantic and eastern Pacific provinces.

In addition, climate-related events may have controlled substrate availability for reef settlement and promoted reef faunal diversification. Attempts have been made to link phases of rapid coral diversification to periods of rapid climate change (Figure 2.6). Rosen (1981, 1984, 1988) suggested how the intensification of glacio-eustatic cyclicity led to enhanced tropical speciation by creating a 'species diversity pump'. This mechanism may have allowed the transport of new taxonomic groups that appeared in the more remote islands of the western Pacific and eastern Indian Ocean regions during sea-level lowstands into the central Indonesian zone during sea-level highstands. As a consequence this came to include the refuge sites for a number of competitive (sympatric) species. On longer time scales (over at least several million years), repeated cycles of temperature and sea-level variations may have produced gradual geographical (allopatric) speciation among ecological isolates. Processes of this kind are related to the isolate formation model proposed by Dynesius and Jansson (2000). They are supposed to have acted in the western Atlantic–Caribbean–eastern Pacific region as well as in the Indo-Pacific, but on a somewhat smaller scale (Crame & Rosen, 2002).

Nutrification and ocean chemistry. Changes in nutrient supply may also have contributed to reef coral evolution and turnover (Kauffman & Fagerström, 1993; Wood, 1993). Upwelling of nutrients from recycling of deep water masses and/or nutrient-rich land runoff has been considered to be a major control in limiting reef settings and growth (see Hallock & Schlager, 1986; Montaggioni, 2005, for reviews). The observation that modern tropical reef communities preferentially flourish in low-nutrient waters has led to the conclusion that their Cenozoic counterparts required similar conditions. In the eastern Pacific, the turnovers in coral faunas since the closure of the central American Isthmus are likely to have been linked to changes in temperature, salinity and nutrient levels (Budd, Johnson, & Stemann, 1996; Budd & Johnson, 1999). In the western Atlantic, according to Allmon (2001), even though there were variations in temperature, these were only of secondary influence in the evolution of regional reef diversity patterns. Extinction and speciation events were already occurring by around 2.4 Ma, well before glaciations in the Northern Hemisphere. It is suggested that the major reorganization of oceanic circulation caused by the closure of the Panama Isthmus led to a reduction in upwelling activity and thus in primary productivity. A decrease in productivity may have favoured the development of isolates and thus local speciation. In time, this would have resulted in an important decline in the rate of isolate survival, decreased speciation and increased extinction. The almost complete demise of reef corals in the eastern Pacific is consistent with an increase in nutrient levels in this area.

However, the nutrient supply concept may not explain the structure and distribution of most Tertiary buildups (Perrin, 2002). Episodic upwelling within restricted stratified basins during the late Miocene in southern Spain may have favoured the development of benthic communities adapted to high nutrient levels. The development of large *Halimeda* mounds is regarded as having been catalysed by the input of nutrient-rich upwellings (Mankiewicz, 1988).

Estimates of atmospheric CO_2 and oceanic Ca^{2+} concentrations indicate a decreasing trend from the Paleogene onwards, with CO_2 approaching pre-industrial levels by the latest Oligocene. pCO_2 remained significantly low throughout the Miocene and Pliocene, despite warming during the middle Miocene. By contrast, Mg/Ca ratios in sea water increased consistently. Calcifying organisms have adapted to chemical changes, using strategies that efficiently linked photosynthesis and calcification (Pomar & Hallock, 2008, and references herein).

Biotic controls. In addition to environmental and biogeographical factors, coral diversification can also be linked to biotic factors. Intense ecological interactions between corals and various other reef-inhabiting taxa may have controlled coral radiation. In particular, predation has been invoked as a significant selective driving force for the evolution of reef communities and reefs throughout the Cenozoic. Macroborers, microborers and grazers increased in species numbers, and, concurrently, bioerosion pressure on reef communities also increased (Wood, 1993).

2.2.2.2. History of coral reef evolution

Veron (1995) suggested that the overall history of modern zooxanthellate corals can be divided into three intervals: the Paleogene (65.5–23.03 Ma) with the proliferation of Mesozoic survivors into a cosmopolitan fauna; the Miocene (23.03–5.33 Ma), with the distribution of this fauna within the present biogeographic provinces, and the evolution of most of the extant coral species; and the Plio-Pleistocene to present, with the extension of the polar ice world and the emergence of the modern distribution patterns (Figures 2.7 and 2.8).

Early to late Paleocene (about 65–55 Ma). As a result of the end-Cretaceous mass extinction, the early Paleocene (from about 65 to 59 Ma) has sometimes been considered to represent a hiatus in carbonate deposition and especially in tropical shallow-water coral reef life. This observation is related to the fact that most coral-dominated buildups were composed of azooxanthellate scleractinians, red algae and bryozoans (Wilson & Rosen,

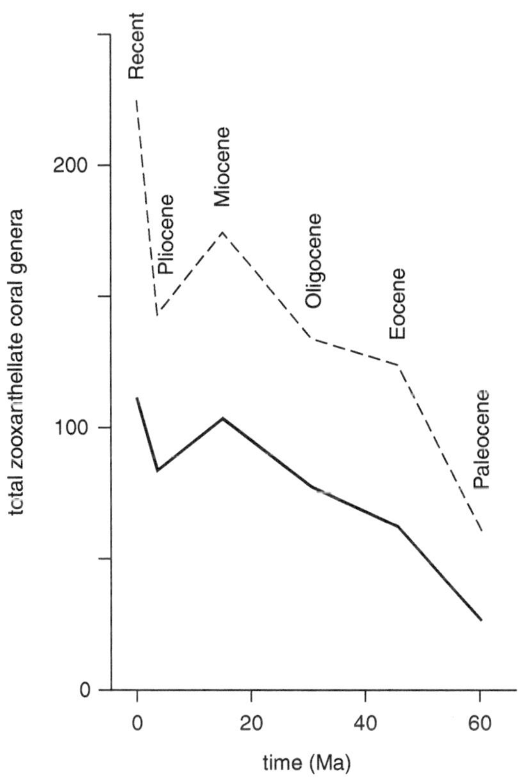

Figure 2.7 Variations in the numbers of zooxanthellate scleractinian coral genera throughout the Cenozoic. Solid and dotted lines refer to mean and maximum values respectively. Simplified from Veron (1995).

1998; Kiessling, 2002). The richest Paleogene reef coral faunas developed in Europe and the Caribbean (Wilson & Rosen, 1998), but buildups did not match the distribution and complexity of those of the Cretaceous until the Oligocene–Miocene (Hallock, 1997; Perrin, 2002). A clear recovery trend is reported for the central Tethyan region, culminating in the development of thick and laterally extensive reef systems during the late Danian, approximately 3–4 million years after the K/T boundary. Buildups are found in southern and northern Europe, the western Atlantic and northern Pacific Oceans. By contrast, in Southeast Asia, corals and reefs were scarce during the Paleogene, an interval referred to as 'the Paleogene Gap' (Wilson & Rosen, 1998). During the late Paleocene, bioconstruction occurred in low-latitude, shallow and deep waters, locally forming atoll-like structures. Most consisted chiefly of low-diversity (less than 5 species) scleractinian populations, but some consisted mainly of coralline algae associated with corals and bryozoans.

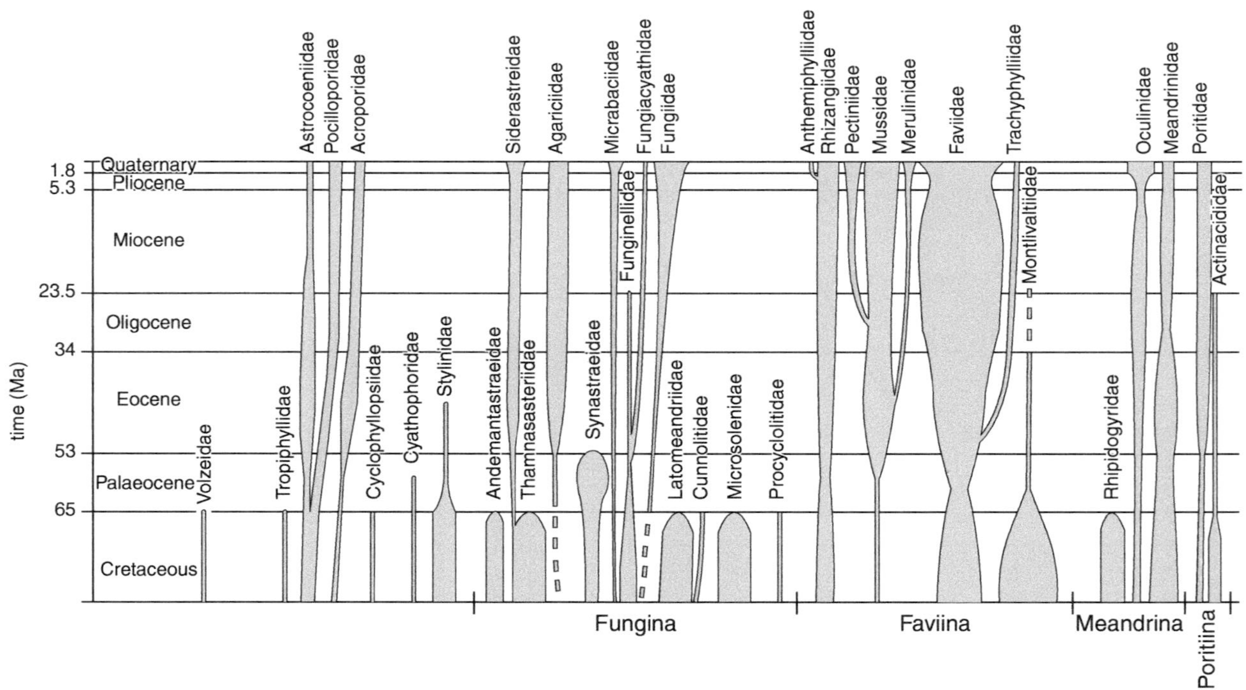

Figure 2.8 Family tree of scleractinian corals based on current coral taxonomy and the fossil record. Branch widths vary as a function of the numbers of genera per family within each stratigraphical interval. Redrawn and simplified from Veron (1995).

Late Paleocene to early Eocene (about 55–46 Ma). During this interval, reef structures remained relatively scattered, mainly developing in the western-central Atlantic, in central Tethys, and in the Pacific (northeastern Australia, and on western to mid-oceanic seamounts). The oldest Tertiary reef community in the Caribbean is of early Eocene age and consists of several genera that subsequently became reef builders (*Favia*, *Goniopora*, *Astrocoenia*, *Montastraea*, *Siderastrea* and *Stylophora*) (Budd, 2000). The distribution of scleractinian faunas during the Paleocene and Eocene is usually thought of as cosmopolitan and controlled by circum-tropical oceanic circulation, despite the fact that the western Atlantic and central to eastern Tethyan realms were already differentiated at that time.

With the exception of the Tethyan region, the spatial distribution of early Eocene reefs is known mainly from subsurface investigations. These structures are typically represented by low-diversity, framework reefs and reef mounds, developed within a tropical to subtropical belt, on shallow shelves and platforms and along upper foreslopes and ramps or in epeiric seas. Clear latitudinal gradients in coral diversity appear as a result of the emergence of the scleractinians as the dominant reef builders in the tropics (Perrin, 2002). However, although framework reefs were present mainly composed of scleractinian corals, reef mounds dominated, chiefly made up of ahermatypic corals and red algae.

Early to late Eocene (about 46–36 Ma). From the early Eocene onwards (about 46–37 Ma), reef structures tended to disperse, with centres of reef growth spreading westwards from central Tethys to the Caribbean. The diversification of coral communities increased significantly from the middle to late Eocene, but scleractinian framework reefs were relatively weakly developed and formed only small structures. These included shallow-marine biostromal banks, fringing and barrier reefs. Most coral families with modern representatives evolved during the late Eocene (Budd, Stemann, & Stewart, 1992; Budd, 2000). However, many scleractinian species, particularly older Mesozoic-like forms, had disappeared by the late Eocene.

No buildups are recorded eastwards in Southeast Asia, although the tropical belt was probably larger than today (Adams, Lee, & Rosen, 1990). Apparently most zooxanthellate coral genera and reef structures are absent from Southeast Asia until the latest Oligocene to earliest Miocene (about 26–22 Ma) (Wilson & Rosen, 1998). In these regions, the scarcity of reef corals is thought to be due to their geographical isolation from areas with rich coral biotas. However, some reefs have been described from deep drilling in the western and central Pacific and around the Indian Ocean. Thus, the apparent lack of reefs in the eastern Pacific may be explained by the relative lack of subsurface data.

In parallel with their dispersal, Eocene framework reefs increased in size and number to become the most common type of buildup. They exhibited a well-differentiated coral zonation from fore-reef slopes to back-reef areas. During the mid-late Eocene, reef biotas were dominated by high-diversity scleractinian communities, locally associated with red algae. Although no clear latitudinal trend in diversity is recognized on the basis of available data (Perrin, 2002), latitudinal gradients in reef characteristics became better expressed from the mid-Eocene onwards as scleractinians became the main tropical reef builders. The zooxanthellae (*Symbiodinium* forms) associated with corals are assumed to have emerged in the early Eocene, and diversified during Eocene cooling, after the late Paleocene thermal maximum (Pochon, Montoya-Burgos, Stadelmann, & Pawlowski, 2006).

Late Eocene to early Oligocene (about 36–28 Ma). The Eocene–Oligocene boundary was a time of marked change in climate and marine life.

During the late Eocene-early Oligocene, the latitudinal extension of the reef belt tended to be reduced, probably as a result of climatic deterioration. High-latitude regions experienced cooling, thus promoting the restriction of warm-water biota to lower latitudes. However, whereas other groups decreased in diversity and abundance, scleractinian corals maintained high diversity as observed in the Eocene, and greatly expanded in their potentialities as reef builders (Perrin, 2002). A rise in Mg/Ca ratios is speculated to have played a significant role in the rapid expansion of reef-building capacity during the Oligocene (Stanley, 2006). Coevally, coralline red algae also increased in diversity (Aguirre, Riding, & Braga, 2000).

From the beginning of the Oligocene, reef distribution extended to eastern Tethys, Southeast Asia and the eastern Pacific, although the centre of the Tethyan realm still exhibited a higher concentration of buildups. The zooxantellate coral fauna of the Oligocene is ultimately regarded as cosmopolitan. From western Tethys to the western Atlantic–Caribbean region, there is a high degree of similarity at the species level of faviids (*Favia*, *Diploria*, *Platygyra*, *Colpophyllia*, *Antiguastrea*, and *Agathiphyllia*) and other forms (*Astreopora*, *Stylophora*, *Stylocoenia* and *Astrocoenia*).

In the early Oligocene, following a dramatic drop in seawater temperature at the Eocene–Oligocene boundary, highly tolerant *Actinacis* dominated in low-diversity buildups in western Tethys, probably as a result of a hypothermic effect (Bosellini & Russo, 1988). At the same time, reefs in the Caribbean were only weakly developed. From the late Oligocene (Chattian), reef coral faunas again began to diversify and taxonomic richness gradually increased.

Late Oligocene to earliest Miocene (about 28–20 Ma). While the latitudinal distribution of late Oligocene–early Miocene reefs remained broadly similar to that in the late Eocene–early Oligocene, there was a major surge in coral reef development representing one of the largest in the Tertiary.

The divergence of Mediterranean and Indo-Pacific coral reef faunas is believed to have occurred during the early Miocene (Chevalier, 1977; Schuster & Wielandt, 1999), but may have taken place prior to the final closure of the relevant seaway (Rosen & Smith, 1988; Perrin, Plaziat, & Rosen, 1998; Rögl, 1998). At that time, the Tethyan Ocean was subdivided into three relatively isolated biogeographical regions: the Mediterranean, the Middle East, connected to the western Pacific, and the western Atlantic. Two tropical high biodiversity foci were differentiated: the Indo-West Pacific region, and the Atlantic, Caribbean and eastern Pacific region (Crame & Rosen, 2002). In the western Pacific, new shallow-marine areas emerged as a result of the collision between the Australian plate and the Southeast Asian craton, thus promoting shallow-water reef initiation (Wilson & Rosen, 1998; Wallace & Rosen, 2005). In the central Pacific, intense reef development appears to have been triggered by the northward shift of the Australian plate into the tropical zone, accompanied by gradual warming from the late Oligocene to early Miocene (Mackenzie & Davies, 1993).

At the generic level, there was a still marked similarity between Mediterranean (i.e. western Tethys) and western Atlantic–Caribbean corals (Frost, 1981; Veron, 1995). Two dominating zooxanthellate coral assemblages emerged in the eastern Mediterranean Tethys in the late Oligocene–early Miocene: a deeper-water *Leptoseris–Stylophora* assemblage and a shallower water *Porites*–faviid association, representing a mixture of typical Mediterranean and Indo-West Pacific elements (Chevalier, 1977). From the Oligocene–Miocene transition, assemblages of reef bioeroders and bioerosional patterns appear to have become similar to those observed in modern reef environments (Wood, 1999).

Most buildups were located along the margins of shelves and platforms or developed in the form of atolls. Reef sequences range from a few metres to several hundreds of metres in thickness. The lateral extension of reefs increased greatly; 20% of recorded reef systems exceed 100 km in length. More than 75% of bioconstructions were true framework coral and coralgal reefs. High-diversity reefs are frameworks dominated by zooxanthellate scleractinians (Perrin, 2002).

Early to late Miocene (about 20–6.5 Ma). In contrast to the Eocene and early Oligocene, the early Miocene records a fourfold increase in the number of coral genera in the Indo-West Pacific centre relative to the Caribbean region. The high-diversity Indo-Pacific centre of Southeast Asia seems to have emerged during the Miocene as a consequence of local speciation and migration of taxa into the region (Wilson & Rosen, 1998). The coral faunas of both the Indo-Pacific and the central Tethyan realms are represented by 40–50 genera and more than 100 species. The increase in coral species richness compared to the late Oligocene may be related to the fact that during the early Miocene, the latitudinal belt within which sea surface temperatures were similar to those of the tropics today was much wider (Adams et al., 1990).

The early-middle Miocene period (20 to about 11 Ma) reflects the acme of reef development in the Tertiary with the highest abundance of buildups. In particular, this period is typified by the widespread deposition of coral reef-derived carbonates in a variety of Pacific areas (Southeast Asia, northern Australia, New Guinea and the western and central Pacific). In the tropical belt, reef sequences usually exceed 100 m in thickness in the Pacific, particularly in French Polynesia (Montaggioni, 1988a), northeastern Australia (Davies, Powell, & Stanton, 1989) and Southwest Asia (Wu, 1994), but are thinner in the Mediterranean and the Red Sea. During the mid-Miocene, coral reefs flourished in the Red Sea, favoured by an arid climate that prevented production of fine-grained particles through weathering and dramatically reduced the flux of terrigenous sediment from the adjacent mountains (Perrin et al., 1998).

However, even though the early-mid Miocene was an acme of reef expansion, reef environments with high taxonomic richness appear not to have been widespread. Only about 20% of individual reef sites display high coral species diversity (>100), and more than 50% of sites have coral communities with moderate (<25) to low (<5) species diversity. At the global scale, high-diversity communities that formed framework reefs are usually dominated by zooxanthellate scleractinians or by calcareous algae, and locally by hydrozoan-scleractinian assemblages.

From the early to middle Miocene, there were marked latitudinal gradients in the distribution of the dominant reef biota at scales varying from global to regional. Reef-building scleractinians tended to occur within a median tropical to subtropical belt, whereas red algae and bryozoans became the main reef builders along the northern and southern borders of this belt. The latitudinal extent of the reef coral belt varied within each province, depending on local environmental constraints (e.g. the location of continents, oceanic circulation regimes, ecological conditions). The tropical Southeast Asian–western Pacific province had the most extensive reef coral areas, probably driven by the greater number of island and continental shallow-water areas. By contrast, the Mediterranean province possessed only a narrow reef coral belt.

During the late Miocene, all three reef regions (the western Atlantic–Caribbean, Mediterranean and Indo-Pacific) experienced a latitudinal contraction of reef belts and a marked decrease in the number of reef sites. The western Atlantic–Caribbean coral faunas became substantially different from those of the Indo-West Pacific and the Mediterranean regions. The regional shallow-water biota responded to environmental changes coinciding with the rise of the Panama Isthmus, through both extinction and a reorganization of local benthic food webs. Differences between the tropical American and Indo-Pacific regions were a result of the evolution of endemic forms (*Agaricia*) and the gradual regional extinction of genera such as *Stylophora*, *Goniopora* and *Goniastrea* (Frost, 1977b,c) that are still present in the Indo-Pacific realm. Paradoxically, in the Caribbean, coral diversification increased from approximately 16 to 4 Ma (Budd, 2000).

In the Mediterranean basin, marine areas were reduced as a consequence of closure of the seaway connecting to the Red Sea. Analysing the distributional patterns of Miocene coral reefs in the Mediterranean region, Pomar and Hallock (2007, 2008) pointed out that coral habitat experienced a bathymetric upward migration through the Tortonian. The ability of hermatypic corals to build shallow-water reefs in high-hydrodynamic-energy and well-illuminated settings was not acquired before the late Tortonian. In pre-late Tortonian times, small coralgal patches and mounds developed on shelf tops and toes of slopes without reaching the sea surface.

At the global scale, framework reefs mostly consisted of scleractinian corals, occasionally dominated by coralline algae and hydrozoans and, less commonly, by bivalves and bryozoans. In tropical American areas, shallow-water (less than 50 m) bivalve assemblages exhibit low diversity, in relation to an increased dominance of a few superabundant groups within each assemblage (Johnson, Todd, & Jackson, 2007).

In addition to coralline algae, other algal forms, notably the green alga *Halimeda*, have locally contributed to the development of buildups, particularly along upper and middle shelf slopes. *Halimeda* mounds have been described from several Mediterranean platforms (Esteban, 1996) where deposits have promoted the generation of relief by producing a substrate suitable for the settlement and growth of microbial crusts. These are regarded as *Halimeda* microbial mounds rather than true reefs.

Latest Miocene to early Pliocene (about 6.5–4.5 Ma). Although carbonate platforms older than the late Tortonian have been assumed to have little similarity with modern reef systems, in the late Tortonian–early Messinian, barrier reefs with typical reef-crest structures built to sea level became relatively common (Pomar & Hallock, 2007, 2008). Paradoxically, the generic diversity of zooxanthellate corals decreased at this time. These drastic changes are suggested to have resulted from the coevolution of corals and *Symbiodinium* zooxanthellae, coeval with global cooling and, at least regionally, changes in seawater chemistry promoting an increase in coral calcification.

During the late Neogene, the reef coral belt continued to narrow in response to the reduction of tropical–subtropical areas and the correlated southward shift of the northern limit of reef-building corals to its present-day position (Rosen, 1988). With the exception of generic extinctions in the Caribbean, there were few changes in the compositions of scleractinian faunas at that time (Frost, 1977b,c). High-diversity reef communities appear to have occurred preferentially in the South Asian–central Pacific regions and even in the Caribbean. There, the most diverse reef coral populations are of late Miocene to early Pliocene age (Budd, 2000). Little is known regarding coral richness in the Indian Ocean and eastern Pacific at that time.

During the latest Miocene, low-diversity communities were dominant in Mediterranean coral reefs (see review by Esteban, 1996). Those of Messinian age (6.5–5.3 Ma) were chiefly built by only two scleractinian genera (*Porites* and *Tarbellastraea*) in association with encrusting coralline algae (dominated by *Spongites* and *Lithophyllum*), foraminifera (*Miniacina*, *Nubecularia* and haddonids), bryozoans and microbial crusts. This low diversity was most likely to have resulted from biogeographical causes, that is, from a gradual impoverishment of coral species richness since the late Oligocene, reflecting the relative isolation of the Mediterranean pool, rather than from inimical environmental factors. However, the common occurrence of stromatolitic/microbialitic mounds and/or *Halimeda* bioherms in Messinian reef systems strongly suggests that nutrient supplies had increased and thus the structure of coral communities might have been disturbed by mesotrophic conditions, resulting in the survival of only the most tolerant coral taxa (Martin, Braga, & Riding, 1997; Bosellini, 2006).

In parallel with their coral hosts, the endosymbiotic zooxanthellae also experienced radiation since the Miocene–Pliocene transition (LaJeunesse, 2005). The symbiont partner of modern reef corals (*Symbiodinium*) includes several lineages or subgeneric clades. Clade C dominates in both the western Atlantic–Caribbean and Indo-Pacific host faunas today but each oceanic province possesses a diverse clade C assemblage that has evolved independently through host specialization and allopatric diversification. The selective expansion of clade C may have taken place before the separation of the two major oceanic realms, in response to major climatic changes and low CO_2 concentrations.

Early to late Pliocene (about 4.5–1.8 Ma). During this period, the total number of reef sites appears to have drastically diminished compared to those of the late Miocene. Such a decrease is most likely to have resulted mainly from the extinction of those of the late Miocene reef biota and the corresponding disappearance of coral reefs and associated organisms in the Mediterranean province. However, the apparent poor development of coral reefs worldwide, especially in the central Pacific and Indian Oceans, may be an artefact. It may in part reflect the lack of directly accessible outcrops presently overlain by younger reef systems, or the difficulty in dating and distinguishing Pliocene and early Pleistocene reefs. More than 25% of the total known Pliocene reefs have been identified subsurface in Southeast Asia, the central to eastern Pacific and the Bahamas. Framework reefs are by far the most common type. High-diversity reefs, mainly composed of scleractinian-dominated assemblages, and commonly associated with coralline algae, have been reported from both the Indo-Pacific and the western Atlantic Caribbean provinces.

By the mid-Pliocene, the physical separation of the four biogeographical provinces recognized in the tropics today (the Indo-West Pacific, eastern Pacific, western Atlantic–Caribbean and eastern Atlantic provinces) had become reality and was accompanied by the evolutionary divergence of regional reef biotas (Rosen, 1988). The isolation of these regions has made inter-regional expansion of their biotas difficult between the Caribbean and eastern Pacific, and virtually impossible, between the western Pacific and Caribbean or between the eastern Atlantic and Indo-West Pacific provinces. The extinction of previously widespread groups within regions, in association with regional post-isolation diversification, has resulted in the emergence of the endemic biotas recognized today.

During the late Pliocene and/or early Pleistocene, corals and clade B of the *Symbiodinium zooxanthellae* appear to have experienced a coevolution in the Caribbean, probably promoted by regional environmental changes such as the closure of the Panama Isthmus, associated with a substantial drop in sea surface temperature (LaJeunesse, 2005).

2.3. Temporal and Spatial Variations in Coral and Calcareous Algal Diversity

The Tertiary–early Quaternary record of some major reef biotas and especially coral taxa (Figures 2.7 and 2.8) provides valuable information on the rates of speciation and extinction and the origin of the compositional and distributional patterns of Recent reef communities.

2.3.1. Reef-Building Corals

2.3.1.1. The western Atlantic–Caribbean province

Three main peaks of coral speciation have been recognized (Figures 2.9 and 2.10) closely related to periods of maximum reef development: in the middle to late Eocene (40–36 Ma), in the late Oligocene to earliest Miocene (28 22 Ma), and in the late Miocene to late Pliocene (5–2 Ma) (Budd, 2000). A decrease in rates of generic diversification occurred throughout the Tertiary, beginning as larval recruitment from the Mediterranean region ceased.

A total of 36 genera and 77 species are reported from the entire Eocene of the region. These were probably part of a cosmopolitan fauna that escaped the late Cretaceous extinction and diversified across both the western and central Mediterranean Tethyan regions during the Paleocene and Eocene. More than half of the total genera identified in

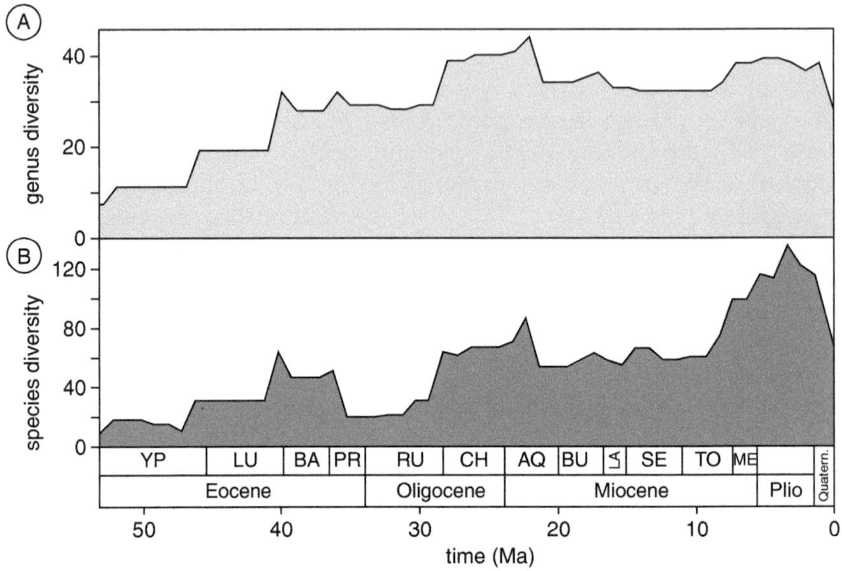

Figure 2.9 Reef coral diversity at the genus and species levels in the western Atlantic–Caribbean regions, from the early Eocene onwards. Modified from Budd (2000).

the entire Caribbean Cenozoic were already living there during the Eocene (Budd, 2000).

The Oligocene–early Miocene represents the acme of Caribbean reef coral generic diversity (Hallock, 1997). The region was typified by the emergence of new coral genera that are now extinct in the area. The main reef-crest and fore-reef builders included massive *Porites*, *Diploastrea*, *Goniopora* and *Astreopora*, whereas intermediate to deeper reef slopes were dominated, by branching *Stylophora*, *Acropora* and *Porites* and by encrusting *Hydnophora*, *Leptoseris* and plate-like *Porites*, respectively. Back-reef zones were colonized chiefly by *Favites* and *Colpophyllia*. In sheltered settings, the dominant forms were branching *Porites*, *Montastraea* and *Agathiphyllia*.

The Neogene history of Caribbean reef corals appears to have been typified by a repeated diversification and restructuring of communities via episodic reduction in response to environmental changes. Faunal turnover may have taken 5 Ma or more, whereas speciation and extinction may have operated over relatively short time ranges, typically of less than a million years. The early Miocene coral record suggests that Caribbean faunas were transitional in composition between a cosmopolitan late Oligocene assembly and a later Miocene assembly including numerous endemic forms (Veron, 1995; Crame & Rosen, 2002). From the early to middle Miocene, 33 genera and 80 species are known in the Caribbean. During the late Miocene, many new species of Agariciidae, Faviidae and Meandrinidae

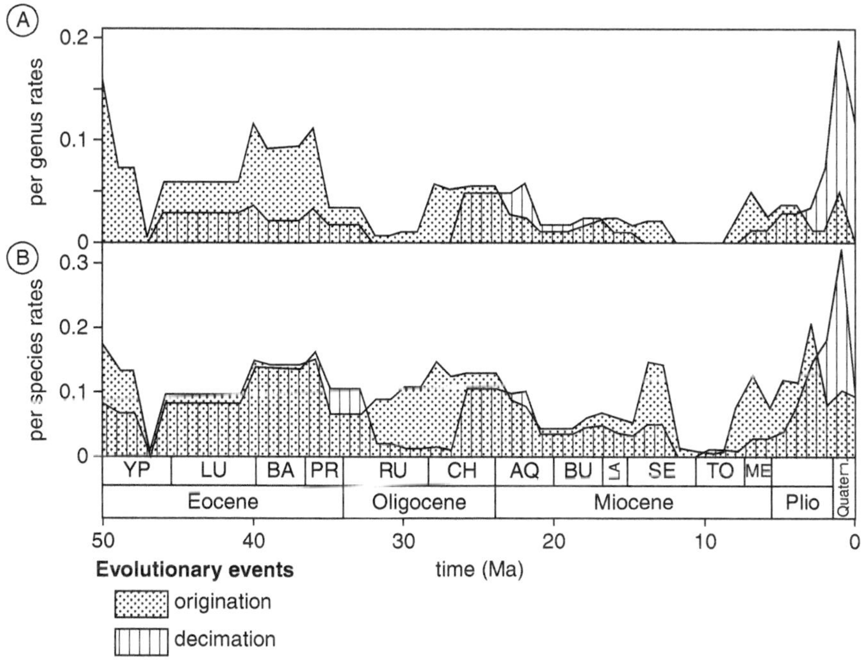

Figure 2.10 Normalized rates of evolutionary events (origination, extinction) affecting the western Atlantic–Caribbean reef coral taxa from the early Eocene onwards. Rate estimates were made using 1 Ma time slices and occurrences within each slice were weighted relative to the duration of reef development in the site in which corals occurred. (A) Genus level; (B) species level. Modified from Budd (2000).

emerged. About 41 genera and 115 species are known from the late Miocene to early Pliocene. Eight genera appeared during the Mio-Pliocene but only two new genera appeared in the late Pliocene. As emphasized by Budd (2000), the last two events represent the youngest highpoints of generic diversification of Cenozoic corals.

From the Pliocene to early Pleistocene (4–1 Ma), diversification was highest in the Acroporidae, Poritidae, Faviidae and Mussidae, generating a total of 38 genera and 133 species. Most of the main reef builders in present-day Caribbean reefs emerged at this time, including *Acropora palmata, Diploria strigosa, Porites astreoides* and the *Montastraea annularis* complex (Budd, Stemann, & Johnson, 1994). Taxa with a higher resistance (growing in the form of large, long-lived colonies and reproducing by fragmentation) became dominant in coral communities (*A. palmata, Acropora cervicornis* and *M. annularis* complex).

There were no major extinction events in the western Atlantic province until after the late Oligocene. The first extinction event affecting scleractinian species occurred in the latest Oligocene and continued through to the early to middle Miocene (about 24–14 Ma). This period was of prime importance in partitioning a formerly cosmopolitan coral fauna into several provinces (Edinger & Risk, 1994). After the decimation of the early Miocene and further turnover in the mid to late Miocene, the coral fauna remained relatively unchanged from around 8 Ma until the Plio-Pleistocene. About 50% of reef coral genera disappeared. Extinction affected genera including *Astrocoenia, Astreopora, Pironastrea, Goniastrea, Antiguastrea* and *Agathiphyllia.* Of these, only *Astreopora* and *Goniastrea* still survive in the Indo-Pacific; the others are globally extinct (Budd et al., 1994). During the turnover event, cold-tolerant, eurytopic species that brood larvae survived preferentially (Edinger & Risk, 1994). Coevally, four new genera emerged, but these were confined to the Caribbean, indicating that the regional coral fauna had lost its cosmopolitan trans-Atlantic composition and had become more provincial (Budd, 2000). It is noteworthy that most of the corals that disappeared in the Caribbean are still extant in the Indo-West Pacific, indicating that they were affected by geographic restriction rather than by extinction *per se* (Edinger & Risk, 1994). Of the 41 genera present in the late Miocene and Pliocene, less than 70% are still living in the region today. For example, *Galaxea* and *Psammocora*, both widespread in the Indo-Pacific region today, became extinct. The two earlier extinction peaks were somewhat selective relative to both the physiological and anatomical properties of the corals and the intensity of environmental disruptions. Highly tolerant forms survived the late Oligocene–early Miocene reduction. The Plio-Pleistocene extinction was again selective and resulted in a significant shift from small gracile forms, characteristic of soft-bottom and/or sheltered substrates, to the dominance of large robust reef-building species, particularly in reef-front settings. Massive and tabular colony forms seem to be better adapted to survive than gracile branching or free-living colonies. During faunal turnover, as reef-coral niches experienced severe stress triggered by glaciations in the Northern Hemisphere, only ecological generalists, i.e., corals capable of colonizing different reef environments, were likely to escape extinction. This resulted in more than a simple species replacement; coral assemblages were totally restructured (Klaus & Budd, 2003). Higher-hydrodynamic-energy reef zones seem to have experienced more rapid faunal turnover than less-agitated or more-sheltered environments. Plio-Pleistocene assemblages evolved gradually, as new species were added to existing populations. No similar effects are seen in Indo-Pacific corals at this time, suggesting that driving factors may have been regional (Budd et al., 1994).

Extinction patterns appear to have affected nearly all coral families. The reduction event was not taxonomically selective, and is also seen to have affected both molluscs and bryozoans. The average extinction rate of coral species at that time, as reported from Central American sites, was approximately 10% per million years throughout the Pliocene and culminated at 33% per million years in the Pleistocene (Getty, Asmerom,

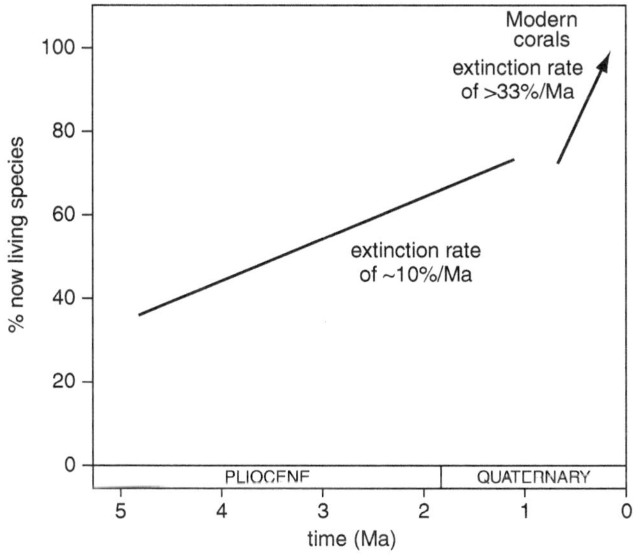

Figure 2.11 Average extinction rates of corals in Costa Rica reef sequences during the Plio-Pleistocene. The rate was approximately 10% per million years between 5 and 1 Ma. By comparison with species diversity of the regional coral fauna, the rate is estimated at around 33% per million years over the past 0.9 Ma. Simplified from Getty et al. (2001).

Quinn, & Budd, 2001) (Figure 2.11). Reef-building corals in southern Florida also suffered a marked decline. Well-developed reefs of Pliocene age are found further north than the present ones (Allmon, Emslie, Jones, Morgan, 1996; Budd et al., 1996). Various lines of evidence suggest that in most hermatypic coral families extinctions were approximately synchronous (Veron, 1995; Budd et al., 1996). About 40% of species and more than 50% of the genera living during the Pliocene are now extinct. Generic diversity declined regionally and only 25 of the original genera are living now. *Pocillopora* survived in the southern Caribbean until the late Pleistocene (Frost, 1977c). Regionally, reef coral faunas did not assume a distinctly modern composition until the early to middle Pleistocene (Budd et al., 1994).

Accelerated speciation and extinction occurred almost simultaneously in the northern and southern Caribbean. High species richness was maintained throughout the turnover episode, suggesting that reef coral communities did not collapse during faunal replacement (Budd & Johnson, 1997). For instance, in pre-turnover late Miocene assemblages, forms belonging to the *M. annularis* species complex, one of the most prolific reef builders in modern Caribbean reefs, were less diverse than they were in post-turnover sequences (Budd & Klaus, 2001; Klaus & Budd, 2003). Thus, the three present-day species, represented by closely related forms dominating some reef zones for at least the past 22 Ma, seem to have emerged before the high extinction peak at the Pliocene–Pleistocene transition (2–1.5 Ma), which accompanied the reef faunal turnover in the region. Other species probably coexisted for at least 5 Ma and these explain the high diversity of *M. annularis*-like corals during the turnover and their survival.

The three significant species extinction events were directly related to severe oceanographic disturbances. As underlined by Budd (2000), the Eocene extinction may be attributed to a large-scale temperature decline. Patterns of selective extinction and the distribution of coral-associated bioeroders across the Oligocene–Miocene boundary indicate that the concomitant decline in reef diversity and reef growth in the early Miocene was probably tied to eutrophication (Edinger & Risk, 1995) in response to an increasing extension of a strong thermocline following the intensification of glaciation in the Southern Hemisphere (Johnson, 2001). The potential controls implicated in the Plio-Pleistocene extinction include oceanographic changes linked to the final closure of the Panama Isthmus, the intensification of Northern Hemisphere glaciation, and changes in the primary production pattern from high planktonic to primarily benthonic regimes (Hallock & Schlager, 1986; Allmon, 2001). The influence of climate deterioration was probably enhanced in the Caribbean due to the areal restriction of the dispersal pool following the rise of the central American Isthmus. As a result, although zooxanthellate corals in the Caribbean are almost as diverse as those from the Indo-Pacific at the family level, they are significantly depauperate at the genus and species levels.

2.3.1.2. The eastern Pacific

There is still a relative lack of studies of zooxanthellate coral-bearing sequences of Cenozoic age in the eastern Pacific and this has limited our understanding of the biogeography and evolutionary history of reefs and coral faunas in the region (Glynn & Wellington, 1983; Lopez-Pérez, 2005). Current knowledge regarding the regional history of reef coral taxa can be summarized as follows.

During the Cretaceous and early Tertiary, about 89% of eastern Pacific corals were also present in the tropical Atlantic region, whereas only 40% of this fauna occurred in the Indo-West Pacific. The eastern Pacific experienced some isolation from the rest of the Pacific as early as the end of the Mesozoic, but maintained a connection with the western Atlantic–Caribbean region until the rise of the central American Isthmus (3.5–3 Ma). Thirty-six reef coral genera were present from the end of the Cretaceous to the Oligocene, eighteen during the Miocene and ten now (Glynn & Wellington, 1983; Cortés, 1997; Glynn & Ault, 2000). During the early to late Pliocene, reef coral faunas suffered a rapid, large-scale turnover over a 2–3 Ma time span (Rosen & Smith, 1988). Although the compositions of the coral faunas remained very similar throughout the central American region in this period, during much of the Tertiary there were marked changes as the eastern Pacific and western Atlantic regions separated (Budd, 1989). The number of coral species declined dramatically in the eastern Pacific, but coral communities were depleted as they changed, maintaining a higher species diversity (55 species on average). In shallow- to intermediate-depth settings, the pre-turnover assemblages are mostly composed of *Stylophora* and *Acropora*, whereas *Acropora* species have dominated since the turnover.

Several hypotheses have been suggested to explain the origin of the eastern Pacific reef coral fauna through vicariance and dispersion. The vicariance hypothesis, the replacement of one taxon by another through successive speciations and extinctions, suggests that eastern Pacific reef corals were derived from a single, widely distributed, early Neogene Caribbean pool (Heck & McCoy, 1978; Budd, 1989). Some eastern Pacific species are present in the fossil record, back to the Plio-Pleistocene, suggesting that they have lived there continuously since before the rise of the Isthmus. From this point of view, Recent eastern Pacific corals represent a relict fauna that became isolated from the Caribbean basin as a result of closure of the Panama Isthmus. Indirect evidence is supported by the fact that 8 of the 10 coral genera still living in the eastern Pacific (among them, *Psammocora*, *Pavona*, *Pocillopora* and *Gardineroseris*) are also recorded in the Caribbean during the Plio-Pleistocene, whereas several other genera (*Leptoseris, Porites* and *Siderastrea*) occur in the modern reefs in both the eastern Pacific and the western Atlantic provinces. Present differences between the eastern Pacific and western Atlantic–Caribbean coral faunas may have resulted partly from the reduction of contrasting taxa in the two regions during the late Pliocene and Pleistocene, following intra-regional environmental disturbances (Budd, 1989).

The dispersion hypothesis assumes that the only available source for coral recolonization was that of the central Pacific, although this was separated westwards by a vast open-oceanic barrier (Grigg & Hey, 1992; Cortés, 1997). This may explain the fact that more than 90% of the species living today in the eastern Pacific, belonging mostly to the genera *Pocillopora, Acropora, Porites, Psammocora, Siderastrea, Leptoseris* and *Pavona*, are also present in modern reefs in the Indo-West Pacific. By contrast, only a third of the genera and none of the species are shared with the Caribbean (Veron, 1995; Paulay, 1997). The dispersal connection between the western and eastern Pacific areas may have existed for a long time, since coral distribution patterns are long standing (Veron, 1995). Probably it even operated prior to the movement of the volcanic Line Islands that are considered to have been the central Pacific source for reef coral biota. These islands were carried by seafloor spreading towards the northwest, passing into the eastward flowing equatorial countercurrent by the late Pliocene, and thus favouring the eastward displacement of coral larvae. Evidence for this comes from the similarity of pre-Neogene and Miocene coral faunas in both the American and Indo-West Pacific regions (Budd et al., 1992; Paulay, 1997; Glynn & Ault, 2000). Glynn and Ault (2000) suggest that the dispersal and vicariance hypotheses are not mutually exclusive because each explains different biogeographical events that actually occurred.

In the eastern Pacific, the restriction of reef development during the Pliocene might have been caused by high-intensity El Niño-Southern Oscillation (ENSO) disruptions, linked to the rise of the Panama Isthmus (Colgan, 1990 in Glynn, 1997). In the Pleistocene, the alternating exposure of coral communities to stressful cooling and falls in sea level during glacial episodes, and to warming and rising sea level during interglacials, may have affected reef growth. All of the warmer, interglacial episodes would have been punctuated by sporadic ENSO events (Glynn, 1997).

2.3.1.3. The eastern Atlantic

The eastern Atlantic reef coral province appears to have suffered vicissitudes similar to those of the eastern Pacific (Paulay, 1997). Most of the original Tethyan corals disappeared after closure of the seaway between the Mediterranean and the Atlantic Ocean during the Oligocene–Miocene (Boekschoten & Best, 1988). This probably resulted from a deterioration of oceanographic conditions to the east and dispersion of corals from the western Atlantic. From a pool of about 50 genera in the Mediterranean Tethys and the eastern Atlantic, there were only 7 survivors (*Madracis*, *Stylaraea*, *Siderastrea*, *Schizoculina*, *Cladocora*, *Favia*, and *Montastraea*).

Although the biotic affinities between the eastern Atlantic and the Indo-West Pacific regions are much less marked than those with the Caribbean, as a result of the earlier separation (Paulay, 1997), 27 of the original genera survive in the Indo-Pacific, whereas only 13 are still present in the Caribbean. Approximately 70% of extant eastern Atlantic species are also living in the Caribbean. This strongly suggests that the presence of modern corals in the eastern Atlantic is due to long-distance dispersion. The remaining 30% are considered to be endemic.

2.3.1.4. The Indo-West Pacific province

Because of its large areal extent, great variety of reef habitats and high numbers of genera and species, the evolutionary history of the scleractinian corals of the Indo-Pacific province, has proven difficult to document at various taxonomic levels and therefore remains partly speculative (Veron, 1995; Paulay, 1997).

Rosen and Smith (1988) and Pandolfi (1992a, 1992b) suggested that species diversification of the corals in Indo-Pacific reefs was a response to geological events that resulted in the progressive isolation of communities during the Tertiary. The diversification was derived from the Paleocene or earlier (Cretaceous) ancestral taxa with a widespread distribution in the Tethys realm, and spread from the Mediterranean region to the central Pacific. Regionally, three types of geological and biogeographical processes are believed to have been involved in modifying reef coral distribution: (1) passive longitudinal displacement of biotas by seafloor spreading, (2) rise or collapse of land and oceanic barriers and (3) high-amplitude sea-level and related climatic changes. The displacement of coral faunas may have occurred in relation to the accretion of islands and terranes onto continental land masses. The rise or collapse of barriers may have prevented or enhanced the dispersal of larvae, and thus promoted speciation through the isolation of taxa and subsequent fragmentation of species ranges, or have impeded the process. Finally, climatically driven sea-level changes during the Cenozoic certainly resulted in changes in the configurations of continental margins and fragmented island regions. All three processes, the isolation of relicts, migration of taxa and diversification within the region, are likely to have been important in the enhancement of coral richness. Most lineages of zooxanthellate coral faunas currently living in the region probably originated in the western Indian Ocean, Australia or the southwest Pacific, through vicariant events triggered by continental breakup and the displacement of island arcs, in association with the effects of changing sea level. In this view, Indonesian diversity is mainly the result of an amalgamation of different faunas (Santini & Winterbottom, 2002).

On the basis of a phylogenetic analysis of modern coral species, Pandolfi (1992a, 1992b) demonstrated that the present coral biogeography of the Indo-Pacific experienced a predominantly stepwise progression from west to east with adjacent areas more closely linked to each other than to areas further apart. This progression would have been in relation to a number of factors, including the submergence of the Ninetyeast Ridge, the separation between the Indian and Pacific Oceans as a result of the collision between the Australian plate and Southeast Asian land masses, the opening of western Indian Ocean–Red Sea seaway, and Plio-Quaternary temperature and sea-level changes. The Ninetyeast Ridge in the central Indian Ocean has probably controlled long-distance larval dispersal. It emerged in the Eocene and Oligocene, was flooded in the early Miocene, and may later have served as an oceanic barrier between the eastern and western parts of the Indian Ocean. The coral faunas that earlier used the Ridge as a refuge to maintain genetic continuity across the Indian Ocean became isolated. The Southeast Asian microcontinents, and the associated rotation and northward drift of both Australia and New Guinea, formed land barriers acting as an *in situ* diversity pump and as a filter between the Indian and Pacific Oceans throughout the Tertiary. In the middle Miocene, the collision between northern Australia and Southeast Asia may have separated faunas of the eastern Indian Ocean from those of the Pacific and influenced coral biogeography in New Guinea. The Red Sea was populated by dispersal from the western and central Indian Ocean coral pools when the Gulf of Aden seaway opened in the early Pliocene. The mid-Oligocene onset of Antarctic glaciation and the Plio-Pleistocene commencement of Northern Hemisphere glaciations caused marked drops in global sea levels and changes in regional current regimes. All of these events produced thermal vicariant barriers, resulting in the formation of widely separated endemic populations in the Indo-Pacific. Coral distribution was fragmented into several subprovinces (e.g. Red Sea; western–central and eastern India; western and central Pacific; southwest and southeast Australia). For example, in Western Australia, both *Coscinaraea* and *Symphyllia* species exhibit a high degree of endemicity (Figure 2.12). They would have

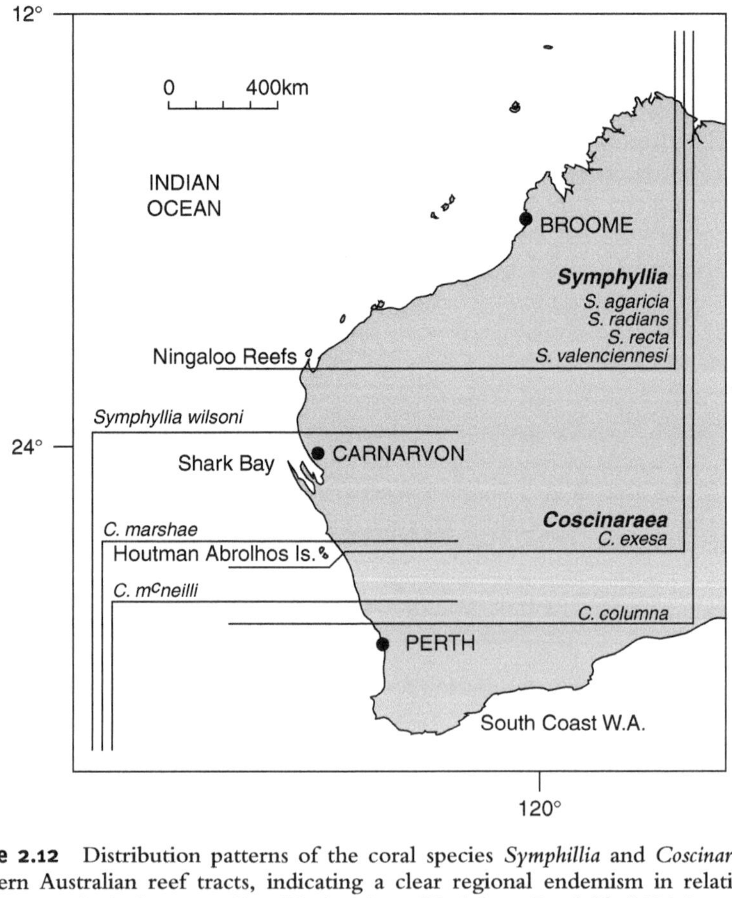

Figure 2.12 Distribution patterns of the coral species *Symphillia* and *Coscinaraea* in Western Australian reef tracts, indicating a clear regional endemism in relation to Cenozoic geological events. Simplified and modified from Pandolfi (1992a).

emerged at or close to the Western Australian continental shelf, as a result of a thermal vicariant event in the early Pleistocene, outside the so-called Indo-Pacific centre-of-origin. Peripheral endemicity accords with the hypothesis of successive isolation in the evolutionary history of Indo-West Pacific corals rather than the existence of evolutionary centres. As in the Acroporidae (Wallace, Pandolfi, Young, & Wolstenholme, 1991), speciation may have been followed by dispersion to enlarge the range of the taxa.

The historical biogeography of some other coral families is relatively well documented. Fungiids and Pectiniids are largely confined to the Indo-Pacific region, having originated in the Paleogene (*Cycloseris*), Miocene (*Fungia*, *Echinophyllia* and *Mycedium*), Pliocene (*Herpolitha* and *Oxypora*) and Pleistocene (*Physophyllia*) (Paulay, 1997). The blue coral *Heliopora* (Octocorallia), which originally had a circum-tropical distribution with occurrences in both the western and central Tethys since the Cretaceous, was restricted to the Indo-West Pacific by the end of the Tertiary. In contrast to the high rates of species turnover reported from the eastern Pacific, western Atlantic–Caribbean and eastern Atlantic provinces, rates of extinction in reef corals seem to have been similar to background levels in the Indo-West Pacific, at least since the Pliocene (Veron & Kelley, 1988; Paulay, 1991, 1997). No extant coral genus is regarded as having suffered regional extinction in the Indo-West Pacific region. Some 13 genera (the Mussid *Mussa*, *Mussismilia*, *Isophyllia*, *Isophyllastrea*, *Mycetophyllia*; the meandrinid *Meandrina*, *Dichocoenia*, *Dendrogyra*; the faviid *Cladocora*, *Colpophyllia*, *Manicina*, *Solenastrea*; and the astrocoeniid *Stephanocoenia*) are today restricted within the eastern Pacific and/or western Atlantic, but also occurred in the Mediterranean Tethys, and may earlier have lived in the Indo-West Pacific. These forms may reflect regional Indo-West Pacific extinctions. However, this picture is confusing as genetic analyses of Indo-West Pacific and Caribbean corals indicate that most Caribbean Faviids and Mussids are not regional representatives of Indo-Pacific lineages of these families, but belong to distinct clades. Only the polyphyletic genus *Montastraea* is closely tied to the group containing Pacific faviids (Budd, 2006). In contrast with western Tethys, Indo-West Pacific coral reefs were taxonomically poor throughout the Paleogene. The isolation of Mediterranean Tethys and the proto-Indian Ocean, possible seaways for faunal exchanges between the western Tethyan areas and the western Pacific, occurred from the late Oligocene to the middle Miocene (Rosen, 1988), and was coeval with or pre dated the emergence of these genera (Paulay, 1997). From the early Miocene, the Indo-West Pacific region remained the richest centre for coral faunas throughout the remainder of the Cenozoic.

The coral faunas of the Indo-West Pacific province show a surprising relative homogeneity, only disturbed by subregional endemism. Diversity is highest between Southeast Asia (Indonesia) and Australia and remains relatively high westwards across the Indian Ocean. It decreases dramatically to the east across the central Pacific. According to Wilson and Rosen (1998) and Wallace and Rosen (2006), the central Indo-Pacific, rather than being a global locus of origin of almost all Indo-Pacific zooxanthellate coral genera, as previously believed, was the focus for only a regional subset of speciation events. Patterns of coral distribution and diversity probably result from a variety of modes of speciation and extinction (Paulay, 1997).

2.3.1.5. Inter-regional comparison

Globally, reef coral biotas appear to have been characterized by marked endemism, reflecting the regional elimination or survival of formerly pantropical Tethyan forms, and the appearance of new species within restricted areas. Thus, the present Indo-Pacific and western Atlantic–Caribbean provinces share only eight living genera, which include most modern prolific reef builders (*Acropora*, *Porites*, *Siderastrea*, *Favia*, *Montastraea*) that diversified worldwide since the Cretaceous to Eocene. By contrast, there are no common corals at the species level. This results largely from the extinction in the western Atlantic of 21 genera that remained extant in the Indo-Pacific province, and the diversification of new endemic genera in both regions from the Eocene to the Miocene. Thus, among others, the branching pocilloporid *Seriatopora*, the encrusting siderastreid *Pseudosiderastrea*, the encrusting or domal faviids *Echinopora*, *Leptastrea* and *Oulophyllia* appeared and remained strictly confined to the Indo-Pacific, whereas the massive faviids *Solenastrea* and *Diploria* were restricted to the western Atlantic (Paulay, 1997). The number of coral genera that were shared between the two provinces fell from about half to one-third of the Caribbean fauna during the Plio-Pleistocene extinction event that was responsible for the loss of a third of Caribbean corals (Budd et al., 1994).

2.3.2. Case Study: The Historical Biogeography of the Genus *Acropora*

A powerful technique in understanding the evolutionary history of Recent coral biotas is to focus on a taxonomic subset of representative forms for which comprehensive information is available. Thus, as the most diverse and widespread genus of present-day reef-building corals, *Acropora* serves to illustrate the long-term evolutionary history and origins of modern biodiversity patterns of tropical reef corals (Figure 2.13). It now includes more than 120 valid species worldwide, dominating shallow reef zones (Wallace, 1999).

The earliest acroporid fossils are found in eastern Tethys, west of the proto-Indian Ocean, in Somalia. Several species of *Acropora* have been reported from Paleocene reef assemblages and occur throughout the Eocene in Mediterranean Tethys and the western Atlantic. *Acropora* emerged in the late Paleocene, in western Tethys (northeast Africa),

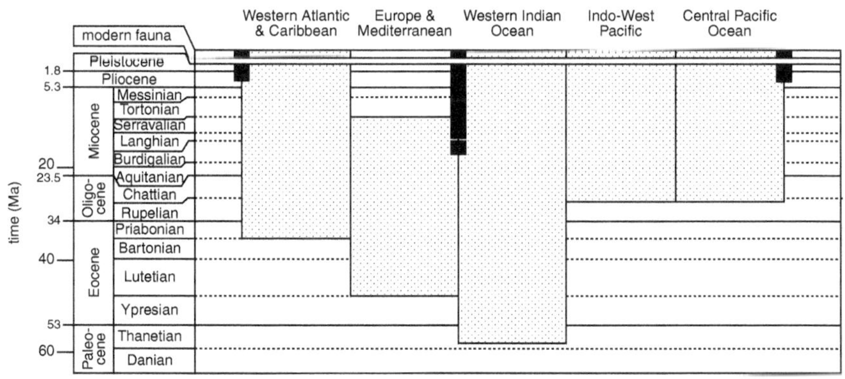

Figure 2.13 Distribution of *Acropora* species groups throughout the Cenozoic in major world areas. Dark bars indicate the location of land barriers. Simplified from Wallace and Rosen (2006).

possibly extending eastwards to the western Indian Ocean. At this time, the Tethyan seaway between Southwest Asia and Arabia is believed to have opened (Wallace & Rosen, 2006). During the middle Eocene, the seaway offered a passage north of the Indian subcontinent and through the Mediterranean, with no separation of the Indian and Atlantic Oceans. Although there is no record of acroporids from India at that time, Wallace and Muir (2005) speculated that such corals may have developed on reefs to the north of India and could have contributed to seed the western Tethyan reefs. An early acroporid fauna is likely to have been shared between the Mediterranean and the proto-Indian Ocean and several species groups remained after the closure and desiccation of the Mediterranean and the formation of the Indian Ocean at the end of the Miocene.

Acropora material of Eocene age (49–34 Ma), collected in western Europe, has been assigned by Wallace and Rosen (2006) to nine of the currently living species groups. Palaeolatitudinal reconstruction indicates that these developed at far higher latitudes than today (51° north).

The first record of *Acropora* as the main framework builder in reef structures comes from late Oligocene (28–23 Ma) rocks in western Tethys (Greece) (Schuster, 2000). But by this time it was already present in the Indo-Australian arc (Wilson & Rosen, 1998). *Acropora* seems to have developed preferentially in low-hydrodynamic-energy, shallow-water lagoonal or lagoon-like environments where it formed widespread, dense thickets. No example has been recorded from the Eocene to early Oligocene carbonates of the Pacific and central Indo-Pacific regions. This implies that the diverse and widespread *Acropora* assemblages of the Indo-Pacific region in the late Oligocene to Miocene originated from an eastern Mediterranean–western Indian source. The 'Paleogene gap' hypothesis of Wilson and Rosen (1998) supports the idea that, within Southeast Asia and the so-called 'Indonesian centre of diversity', new *Acropora* species could only have emerged in the late Oligocene. Several species group lineages appeared earlier during the Eocene outside the central Indo-Pacific regions, probably as a result of both radiation from existing species groups and diversification of new groups (Wallace & Rosen, 2006).

From the middle Miocene (16–12 Ma) onwards, the genus is no longer present in Mediterranean Tethys; it disappeared during the gradual Miocene extinction that affected all coral reef biotas throughout this region (Rosen, 1999). However, by the mid-Miocene, it was already living throughout Indo-Pacific areas. Although some authors (Veron, 1995; Fukami, Omori, & Hatta, 2000) assumed that all living *Acropora* species are derived from a single Pliocene ancestor, following the extinction of other lineages by the mid-Miocene, Wallace and Rosen (2006) suggested that species group survivors, already present in the Eocene in the eastern Mediterranean–western Indian regions, were involved in the Indo-Pacific radiation. In the central Indo-Pacific the diversification of *Acropora* species groups would have occurred within the past 2 Ma.

In the Caribbean, the earliest fossil of the gracile branching, low-hydrodynamic-energy *A. cervicornis* is of early-late Miocene (6.6 Ma) age (Budd & Johnson, 1999). The first appearance of robust branching *A. palmata*, the major reef edge builder in the Caribbean province today, is in the early-late Pliocene (*ca.* 3.6–2.6 Ma) (McNeill et al., 1997). The latter appearance correlates well with the transition phase of a Pliocene–early Pleistocene (4–1 Ma) faunal turnover that was typified by widespread reduction and diversification of coral species in the Caribbean. *A. palmata* appeared early during the turnover event and was directly associated with coral communities that were dominating reef edges and fronts composed mostly of *Pocillopora* and *Stylophora* and forms now extinct in the region (*Caulastrea*, *Pavona*, and *Goniopora*). *A. palmata* did not start to become the dominant form in reef edges and upper fore-reef zones until after the extinction pulse at the end of the turnover event at 2–1 Ma (Jackson, Budd, & Pandolfi, 1996 in McNeill et al., 1997).

The emergence of *A. palmata* was coeval with the rise of the Panama Isthmus and with climatic and sea-level fluctuations related to the onset of Northern Hemisphere glaciations. The origin of this species at a time of climatic reorganization raises the question of its adaptation to severe environmental disruption. *A. palmata* appeared during the early stages of climate deterioration and developed during successive glacial and interglacial episodes. The growth pattern of the species, particularly its high extension rate (50–100 mm yr^{-1}), enables reef tops to keep pace with rapid sea-level rise and thus is considered to be adapted to rapidly changing environmental conditions (MacNeill et al., 1997).

The persistence of several species groups is consistent with the continuous occurrence of *Acropora* in Plio-Pleistocene reefs and the present-day, and widespread distribution in Indo-Pacific reefs as a dominant builder (Wallace & Rosen, 2006). Since the Eocene, typical growth forms have mostly been similar to those found in modern reefs, including arborescent, tabular and branching forms. By reference to modern analogues, these colony shapes are diagnostic of habitat type; they have grown mainly in shallower reef zones, including reef flats, upper reef slopes and inter-tidal back-reef patches, usually at depths of less than 10–15 m (Done, 1982; Wallace, 1999; Montaggioni, 2005).

Van Oppen, McDonald, Willis, and Miller (2001) suggested that the impact of major vicariant events on the evolutionary history of the genus *Acropora* (with the distinction between *Acropora Acropora* and *A. Isopora* subgenera) may be detected using molecular relationships between species (Figure 2.14). Thus, assuming the Caribbean *A. Acropora* represents the extant ancestral species, on the basis of mitochondrial DNA analyses, the internode of the phylogenetic tree from which other Caribbean species (e.g. *A. palmata, A. cervicornis*) emerged is thought to coincide with the effective separation between the Caribbean and Indo-Pacific *A. Acropora* species. This node expresses an event that occurred before the appearance of the earliest Caribbean acroporid species (e.g. *A. cervicornis*, 6.6 Ma), in other words, well before the closure of the Panama Isthmus (3.5–3.0 Ma). The latest possible

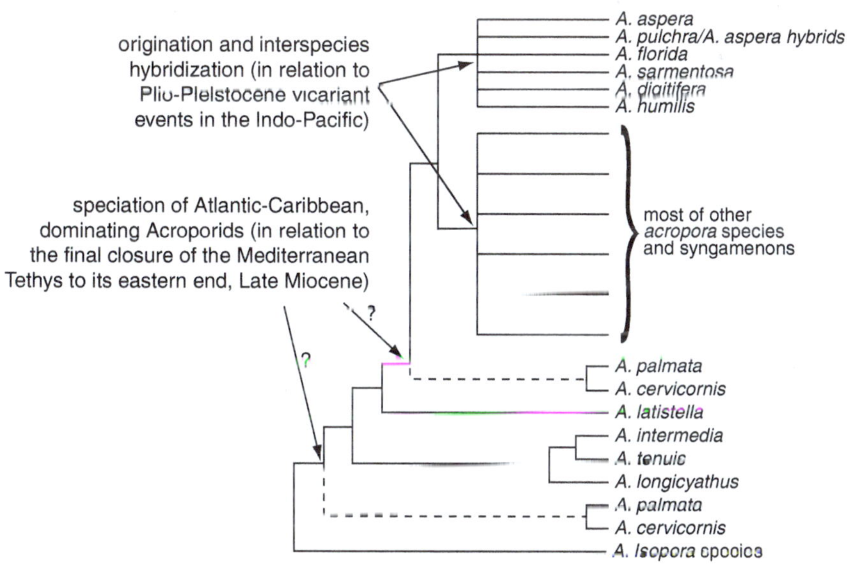

Figure 2.14 Expected phylogenetic relationships between a range of *Acropora* species groups based on analysis of mitochondrial and nuclear markers. Radiation of some species groups is interpreted as correlating with major tectonic and climatic events, both in the Caribbean and in the Indo-West Pacific. Modified from van Oppen et al. (2001).

connection between the Caribbean and Indo-Pacific coral faunas through the Mediterranean and the Middle East was open until approximately 12 Ma, that is, the time of final closure of the Tethys Ocean. If correct, this may indicate that most Indo-Pacific acroporids evolved over the past 10 Ma. Major sea-level changes that took place during the Neogene and onwards probably resulted in repeated isolation and reconnection of coral populations (Veron, 1995; Van Oppen et al., 2001). These events could have triggered diversification of *A. Acropora* species through fragmentation of population, hybridization and recombination processes. In the Caribbean, the third extant *Acropora* species (*A. prolifera*) has no fossil record and probably emerged recently (in the Holocene) (Budd et al., 1994). This species is regarded as resulting from hybridization between *A. palmata* and *A. cervicornis* and as surviving under favourable conditions, in marginal, shallow-water, reef-crest and lagoonal niches (Wallace, 1999; Willis, van Oppen, Miller, Vollmer, & Ayre, 2006). The genus *Acropora* is therefore regarded as including a large number of morphospecies, presenting varying levels of reproductive compatibilities with each other. Many of these species are therefore hybridized forms rather than genetically distinct evolutionary units (Van Oppen et al., 2001). This is consistent with the process of reticulate evolution proposed by Veron (1995).

2.3.3. Coralline Red Algae

The palaeobiogeographical history of reef-building corallinaceans, although still partly obscure and confusing due to the poor knowledge of their taxonomy and distributional pattern in the geological record, can be summarized as follows (Aguirre et al., 2000) (Figure 2.15). Coralline algae originated in the early Cretaceous (Barremian, 116–114 Ma). They are thought to have experienced the extinction of two-thirds of their species during the Maastrichtian, but became primary carbonate producers of shallow-marine communities throughout the Cenozoic.

The most important adaptative radiation of coralline algae, accompanied by a marked increase in diversity, occurred from the late Cretaceous to early Cenozoic. Diversification may have been directly favoured by the coeval radiation of herbivorous organisms that greatly enhanced herbivory pressure and removed soft algal overgrowth, and by the decline of calcifying solenoporacean algae that disappeared in the late Paleocene (Wood, 1995). The evolutionary history of the group also coincides with significant environmental changes, particularly the decrease in temperature and fluctuations in sea level.

The success of the coralline algae over the solenopores probably rests on their particular adaptative resistance to intense excavating herbivory (Perrin, 2002). Coralline algae play major roles as reef encrusters (Macintyre, 1997) and as substrates for larval settlement of reef-dwelling organisms

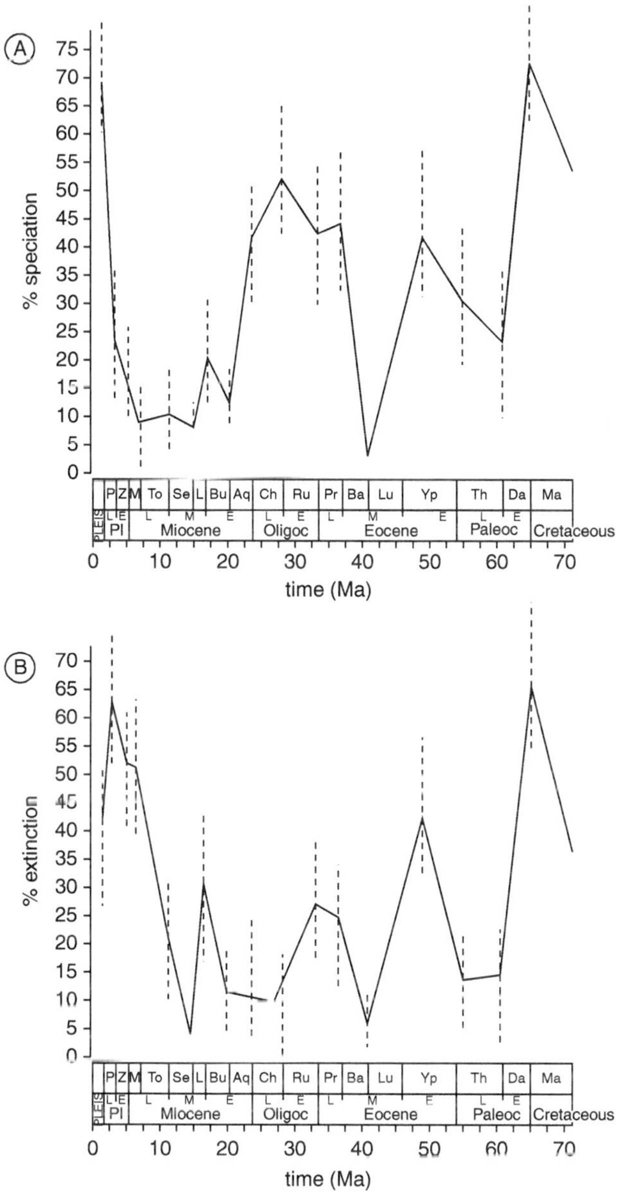

Figure 2.15 Rates of evolutionary events affecting coralline algal species from the end of the Cretaceous to the Pliocene–Pleistocene transition. Vertical lines represent binomial error bars. (A) Origination rates (plotted at the beginning of each geological stage); (B) extinction rates (plotted at the end of each stage). Simplified from Aguirre et al. (2000).

(Fabricius & De'ath, 2001). The impoverishment of hermatypic coral groups in the earliest Cenozoic is thought to have been related to the evolutionary diversification of encrusting algae from the Paleocene onwards. Most well-documented Paleocene reef tracts exhibit rich coralline algal assemblages (Aguirre et al., 2000).

The ancestors of the corallinaceans are believed to lie within the family Sporolithaceae, and especially the extant genus *Sporolithon*. Coralline-like forms developed and evolved in shallow tropical settings, prior to rapidly colonizing the deeper parts of carbonate platforms by the middle Cretaceous (Albian, 108–96 Ma). Typical corallines emerged within the tropical belt of the western Tethyan realm, successively including lithophylloid, mastophoroid and melobesioid types. The opening of a south Atlantic marine seaway during the late-early Cretaceous promoted the expansion of the group towards areas of western and central Tethys. Sporolithaceans reached their acme in the late Cretaceous (Coniacean, 88–87 Ma) and their decline started in the Danian. However, during the Langhian (16–14.5 Ma), coinciding with the Miocene climatic optimum, the Cenozoic richness of *Sporolithon* species was at a maximum (Figure 2.16). Subsequently, and following the global cooling event that began at approximately 14 Ma (Braga & Bassi, 2007), the number of species decreased markedly. Modern sporolithaceans mainly occur in tropical seas.

A contrasting pattern typifies the history of corallinaceans. These diversified rapidly in the Paleocene, became more abundant and expanded in the early Miocene and at present occupy both low and high latitudes. The most significant speciation event, reflected in a 68% increase in new species, occurred in the Danian (65–59 Ma). Additional significant speciation stages (>35%) are known in the earliest Eocene (53–46 Ma), early Oligocene (34–28 Ma), earliest Miocene (23.5–20 Ma) and Pleistocene, but they are typically followed by significant extinctions.

During the Paleocene to Eocene, as sea temperatures declined globally, the cool-water melobesioid subgroup flourished and became dominant, especially in the Pacific Ocean. Free-living coralline algal (rhodolith) deposits have been described from a number of Paleogene sites (Halfar & Mutti, 2005).

From the Oligocene to early Miocene, lithophylloid and mastophoroid subgroups increased in species numbers in shallow, warm-water habitats, along with zooxanthellate corals. This expansion is thought to have been linked to the partitioning of shallow-water environments as a result of the latitudinal climatic demarcation following the onset of Southern Hemisphere glaciations near the Eocene–Oligocene boundary. The high-diversity coral communities in reefs may have offered lithophylloids and mastophoroids new ecological niches. Although sporolithaceans decreased dramatically in the early Oligocene as climate became cooler, melobesioids continued to increase in species number.

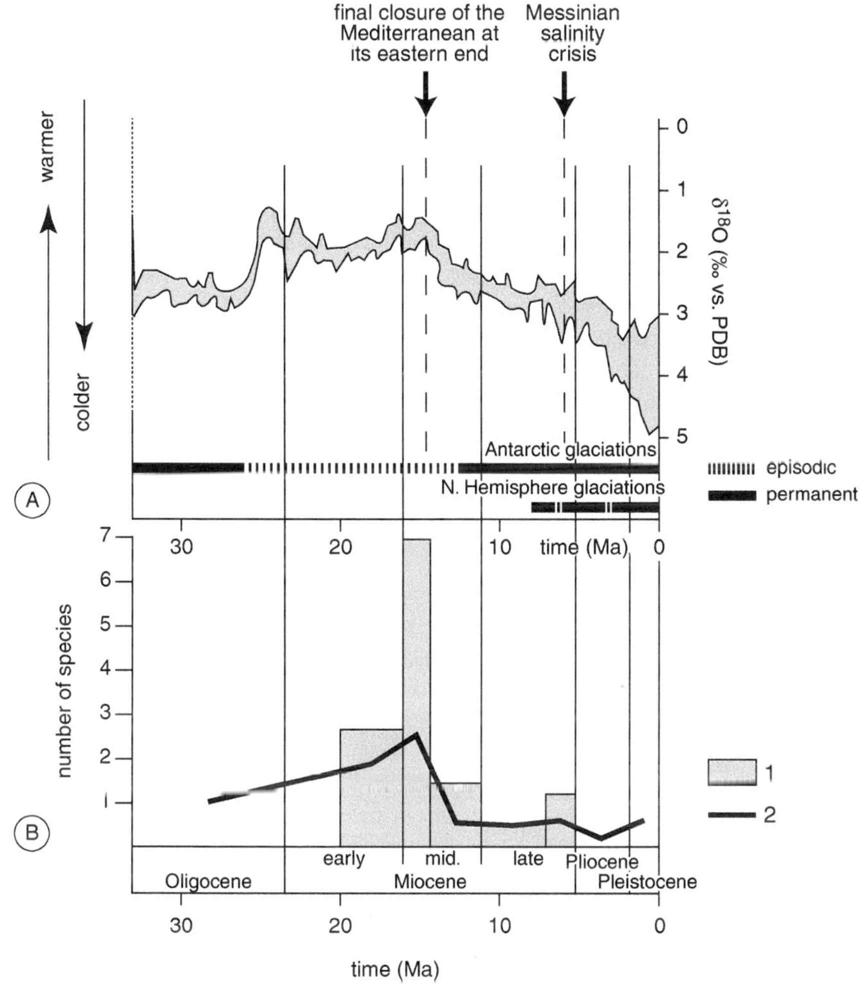

Figure 2.16 Variations in species diversity of the coralline algal genus *Sporolithon* in the Mediterranean from the Oligocene onwards (A), compared to a global $\delta^{18}O$ curve used as a proxy for relative seawater temperature (B). 1 = stage-level data, that is number of species recorded in each geological stage; 2 = intra basin level data, that is number of species recorded in a given sedimentary basin. Estimates of species numbers from both data sets are normalized according to the duration of the geological stages. The mid-Miocene (Langhian) peak appears to correlate with the Miocene climatic optimum. The timing of different major environmental events is indicated in (B). Simplified from Braga and Bassi (2007).

The diversity of corallinaceans as a whole peaked in the earliest to middle Miocene (more than 240 identified species). During the Miocene, thick deposits of rhodolith facies formed in a variety of localities in the Tethyan and Paratethyan realms and in numerous areas in the Caribbean and the Pacific. In the Mediterranean Tethys, rhodalgal accumulations are volumetrically more important than other coral reef sediments. From the Burdigalian to early Tortonian (about 20–10 Ma), rhodoliths became major components of Tertiary shallow-water carbonate environments, particularly in the tropics (Bourrouilh-Le Jan & Hottinger, 1988). In parallel with the diversification of corallinaceans and the expansion of rhodalgal facies, reef scleractinians and other photosymbiont-bearing animals adapted to oligotrophic conditions (e.g. larger foraminifera) suffered severe declines. Thus, in the tropical belt, coralline-dominated facies commonly replaced coral reefs during the late-early to early-late Miocene. The global dominance of coralline algal facies at this time is regarded as having been triggered by an oceanographic event (Halfar & Mutti, 2005). This probably resulted from a global enhancement of trophic resources in relation to a marked increase in marine productivity during the Burdigalian. In the mid-Miocene, following the early-middle Miocene climatic optimum, increases in upwelling- and weathering-derived nutrient supply into shallow-water ecosystems, together with a drop in temperatures, promoted further development of rhodalgal facies and prevented the recovery of coral reefs.

From the middle-late Miocene, there was a slight decrease in coralline species numbers. Melobesioids suffered a similar gradual decrease in diversity in the late Pliocene to Pleistocene. By contrast, lithophylloids and mastophoroids experienced a marked increase in diversity, reaching a maximum in the Pleistocene. This may reflect differing latitudinal responses to climatic deterioration. Although the onset of glaciation in the Northern Hemisphere probably resulted in high-latitude habitat disruption, in tropical and subtropical areas, shallow-water environments may have escaped marked disturbances, promoting new speciation of lithophylloids and mastophoroids.

From the Cretaceous to the Pleistocene, the extinction rates of sporolithaceans and corallinaceans varied widely from 20% to 67% of identified species. The highest mortalities are in the late Cretaceous (67%) and late Miocene (58%). Additional extinctions occurred in the early Eocene (53–46 Ma), late Eocene (40–34 Ma), late Miocene (from 14. 5 Ma) and Pliocene.

2.3.4. Green alga *Halimeda*

The genus *Halimeda* (Chlorophyta, Order Bryopsidales) is an important contributor to the calcareous sediments of Recent reefs (Roberts & Macintyre, 1988). The earliest recorded *Halimeda* remains date back to at least the Permian, but the earliest high richness and wide occurrence of the genus seems to have been in the late Cretaceous. These algae survived extinction at the K/T boundary, presumably with most of them intact. They displayed high species diversity in the Paleocene (Flügel, 1988) and surpassed the former dominance of Dasycladales to become the main calcifying

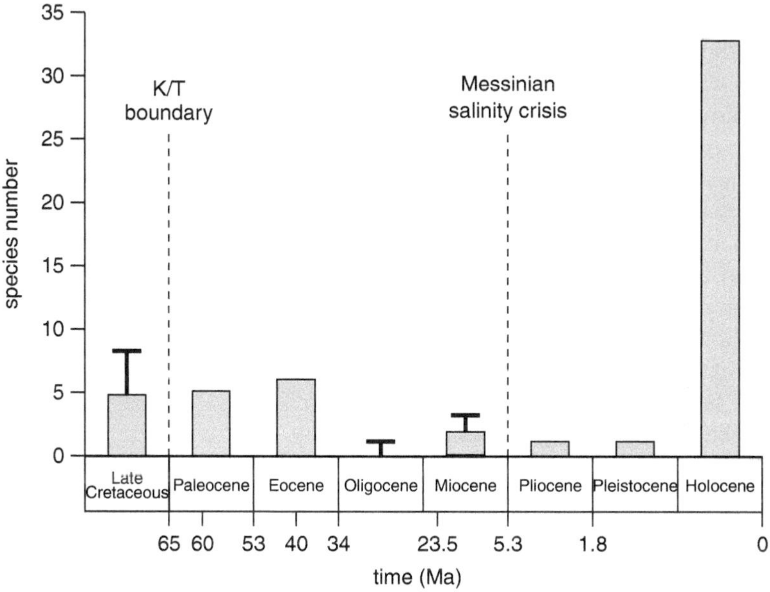

Figure 2.17 Global variations in species diversity of the calcareous green alga *Halimeda*, from the end of the Cretaceous onwards. Vertical lines represent uncertainty. Modified from Hillis (2001).

chlorophyte in tropical reef systems (Hillis, 2001). Throughout the Eocene to the Pleistocene, the genus is only known from sparse, taxonomically poorly constrained descriptions, providing little information on species diversity. Considering the present-day high species richness (more than 30), the apparent scarcity of *Halimeda* in Pliocene to early Pleistocene reefs probably reflects poor documentation rather than a real paucity (Figure 2.17). Species diversity peaked in the late Cretaceous, Paleocene and Eocene, from the last 30 Ma of the Mesozoic through about the first 30 Ma of the Cenozoic. During the first half of the Cenozoic, the major morphological and functional attributes of the genus became differentiated. The occurrence of the oldest form, still living in modern reefs and known since the Miocene in the Tethyan seaway (*Halimeda opuntia*), indicates that the major diversification into clades colonizing distinct functional habitats in reefs was completed by this epoch. The second half of the Cenozoic was apparently a time of very low diversity before a new intensive radiation in the Holocene. Extinction events occurred in the Paleocene and in transitions between the Eocene and Oligocene, and the Miocene and Pliocene, seemingly in step with the extinctions of other reef organisms.

As emphasized previously, this picture surely results from the differential quality of data collected and from a differential collecting effort. Records of unidentified *Halimeda* detritus in coral reef environments, like those described from the late Miocene of southeast Spain (Mankiewicz, 1988), underline the conviction that *Halimeda* remained a significant taxon in coral reef biotas and a major carbonate producer throughout the Cenozoic.

Following Hillis (2001), the phylogenetic analysis of *Halimeda* species shows two major evolutionary events. The first relates to the early differentiation of the genus into three main lineages, each adapted to particular habitats (unconsolidated, moving sands to high hydrodynamic energy environments and, hard substrates). These functional adaptations resulted in the ability of *Halimeda* to successfully occupy a wider range of ecological niches. The radiations of the genus into new clades is regarded as important to the overall development and economy of coral reefs, leading to the division of the reef system into functional clades of *Halimeda* and related radiations of accompanying reef-dwellers.

Once the functional clades were in place, differentiation within clades was associated with major vicariance events. These relate to the second major radiative event, the geographic diversification of the sand-growing lineage into Atlantic and Indo-Pacific species groups. Hillis (2001) noted a strong morphological and anatomical resemblance between the Atlantic *Halimeda monile* and the Indo-Pacific *Halimeda cylindracea* that probably reflects parallel or convergent evolution. The Atlantic and Pacific varieties of *Halimeda discoidea* may not be a pantropical form, but separate clades. If so, and unusually, the event responsible for their separation is not linked to the closure of the Panama Isthmus. Divergence from their ancestor probably occurred much earlier, at approximately 15–12 Ma, promoted by the interruption of the circumglobal Tethyan oceanic circulation, isolating Atlantic and Indo-Pacific groups.

2.4. Conclusions

For the past 65 Ma, there have been significant variations in the nature and composition of hermatypic scleractinaian corals and their associated biota inhabiting and forming shallow-water reefs. These variations have operated relatively rapidly (through intervals of less than 1 million years) following relatively long periods of stability of community structure (1–5 million years). During the periods of turnover, high diversity was maintained, especially during the Neogene, suggesting that reef coral communities did not collapse. The coral diversity patterns observed today are mainly functions of biogeographical provinces/regions sizes and climate. These result from dramatic plate tectonic displacements that gave the tropical areas their present-day physiography.

Throughout the Tertiary, different reef-building biotic assemblies produced in turn different growth fabrics, frameworks and reef types in shallow-marine, tropical reef systems. Basically, Paleogene reefs are characterized by a relatively homogeneous reef-building biota that was largely dominated by a low species richness, but abundant zooxanthellate coral fauna. Although barrier reefs are known from the central Tethys Ocean, buildups developed mostly in the form of patches and banks along shelf margins. From the early Neogene, growth fabrics and reef types appear to have increased in complexity with the significant contribution of secondary builders and sediment-producers (*Halimeda*).

The effects of historical factors linked to changes in global climate (sea surface temperatures, sea levels) and geography (land distribution patterns, oceanic circulation regimes) are likely to have induced changes in the diversity of reef biotas in the tropics. Throughout the Tertiary, the distributional patterns of zooxanthellate coral species at a global level were mainly controlled by climatic cooling in response to periodic intensification of glaciation. However, regionally, only the Mediterranean appears to have been directly affected by climate deterioration as a major control of biodiversity in response to a northward displacement. By contrast, the Indo-West Pacific and western Atlantic–Caribbean regions, mostly situated in the central part of the tropics, escaped reef demise, although there were important faunal turnovers, especially in the Caribbean. Most of the ecological characteristics observed in Recent Caribbean reefs were acquired from intervals of rapid turnover, which affected coral faunas from the early Neogene to the Plio-Pleistocene. Colony size appears to have been the most important trait controlling extinction rates. Species with large colonies resisted impoverishment, in comparison to those with smaller ones. This explains why modern Caribbean reef-coral communities are dominated by large, long-lived colonies.

In the Indo-West Pacific, one of the most striking features in the evolution of reef biotas at a global scale is the apparent out of phase relationship between diversity and climate. Coral richness was highest when the climate was coolest in the Neogene to Recent, but the reverse in the Paleogene. This apparent paradox is likely explained through regional geodynamic history, where tectonics-related palaeogeographical constraints outweighed the influence of an adverse climatic trend.

The Oligo-Miocene closure of Tethys (the isolation of the Mediterranean region at both ends), followed by the Pliocene rise of the Panama Isthmus, created four distinct reef provinces. Faunal interchange between the western Atlantic and eastern Pacific was interrupted by the Pliocene. Extinction of most coral faunas occurred at this time. The depauperate coral communities in Recent eastern Pacific reefs arrived from the Indo-West Pacific across the oceanic eastern Pacific barrier. Similarly, the recent characteristics of the eastern Atlantic reef biotas has resulted from regional extinction following the isolation created by the closure of Tethys, and its recolonization from the Caribbean across the Atlantic Ocean. While the Caribbean, eastern Atlantic and eastern Pacific have suffered severe turnovers of coral faunas during the late Cenozoic, there is no evidence of similar large-scale reduction in the Indo-West Pacific.

Through the Quaternary, luxuriant reefs not only developed in the high-diversity core tropics, but also in remote, low-diversity areas so long as environmental constraints were appropriate. Reefs dominated and continue to dominate wide areas of the Indo-West Pacific and western Atlantic–Caribbean, but are restricted in area in the eastern Pacific and eastern Atlantic. In both the Indo-Pacific and Caribbean provinces, the rise to dominance of branching *Acropora*, together with the decline of massive forms in the Caribbean, has resulted in coral communities structured as they are today by about the mid-Pleistocene.

CHAPTER THREE

Structure, Zonation and Dynamic Patterns of Coral Reef Communities

3.1. Introduction

To understand the palaeoecology of Quaternary coral reefs one primarily needs to describe how the diversity and taxonomic composition of coral reef communities have varied over different temporal and spatial scales. With the increasing disruptions that have affected many coral reefs since the Industrial Revolution, from the middle of the 19th century, access to the Quaternary record has become an important issue in interpreting the recent past history of reef populations (Aronson, 2007) and their future (Greenstein & Pandolfi, 2008). Indeed detailed examination of both Pleistocene and Holocene reefs from a number of reef sites worldwide has shown the usefulness of fossil communities as long-term ecological analogues for understanding the distributional patterns and dynamics of modern assemblages. The main reasons for using Quaternary data set are threefold: (1) most Quaternary reefs have escaped human impact and thus preserved truly pristine biological populations; (2) given reef corals usually develop and deposit in growth position or are reworked within a short range, the abundance of a given taxa in a fossil reef deposit probably reflects the abundance of this taxa during the time of reef growth (Pandolfi & Jackson, 2001, 2007); and (3) the impact of time-averaging on community structure remains insignificant due to the rapid deposition rates associated with coral reefs and the limited effects of compaction in most reefal deposits (Stemann & Johnson, 1992). However, palaeoecological interpretations based on comparison with modern reef biota can be limited by spatial heterogeneity of community structure (particularly in sites with high species diversity), degree of variability in relation to reef growth stages (incipient to mature reef stages) and taphonomic alterations controlled by intensity of diagenesis and differential susceptibility of skeletons to diagenesis (Greenstein, Harris, & Curran, 1998; Humblet & Iryu, 2006).

The present study provides qualitative as well as quantitative data on the community structure and zonation of Quaternary coral reefs from the western Atlantic–Caribbean and Indo-Pacific provinces over a wide range of spatial scales, and by comparison with modern counterparts. However, important biases might result from recovery and preservation rates. The interpretation of reef-core data in terms of depth zonation can be difficult because of approximations imposed by both the limited number and narrow diameter of cores extracted from a given reef site (Blanchon & Perry, 2004). Methodological differences can affect measured percent abundance of coral species; especially, data from cores cannot be accurately equivalent to those obtained from quadrats (Hubbard, Zankl, Van Heerden, & Gill, 2005) or transects. Different quantitative approaches used for reconstruction of reef palaeozonation and palaeocommunity structure were reviewed and their value was discussed by Perrin, Bosence, and Rosen (1995) and Pandolfi (2001). Although some differences were observed between diversity and abundance patterns at the species and genus levels, they have apparently little effect on palaeoecological interpretation. Anyway, all these methods allow three main issues to be explored: (1) What is the species richness and abundance of corals and some associated calcifying organisms in Pleistocene and Holocene reefs?; (2) What is the degree of similarity or variability in the composition and diversity of corals or other reef dwellers within and between ancient habitats?; and (3) What is the degree of similarity and variability between ancient and modern assemblages?

3.2. Structure and Zonation of Modern Reef Communities

Coral reefs are known to be partitioned into a variety of habitats (or ecological zones) in which community structure (total cover, spatial organization, diversity and dominance) is controlled by an array of physical factors and gradients (principally, water-energy regime and light) and biotic interactions (see Chapters 4 and 7). Reef zones generally develop as narrow belts roughly parallel to the reef front line and/or the coastline. Patterns in modern reef-coral zonation at local to regional scales were provided by many reef ecologists (see Stoddart, 1969a; Done, 1983, for review).

3.2.1. The Western Atlantic–Caribbean Province

As early as the 1950s, the modern reefs of the western Atlantic–Caribbean areas were typified by a distributional pattern with three dominating, reef-building scleractinian species (Goreau, 1959; Goreau & Goreau, 1973; Glynn, 1973; Adey, 1975; Bak, 1975; Adey & Burke, 1977; Geister, 1977; Zlatarski & Estalella, 1982; Hubbard, 1988; Graus & Macintyre, 1989) (Figure 3.1). These species include the robust-branching (elkhorn) *Acropora palmata*, arborescent (staghorn) *Acropora cervicornis* and massive *Montastraea annularis* species complex. In high-to-moderate wave-energy settings, *Acropora palmata* is the primary frame builder in the reef-crest and the upper

Figure 3.1 Major coral forms of Caribbean reefs, St. Croix, Virgin Islands (photographs by L. Montaggioni): (A) Robust-branching *Acropora palmata*. (B) Arborescent *Acropora cervicornis*. (C) Massive form of *Montastraea annularis* species complex. (D) Domal *Diploria strigosa*. (E) Branching *Porites porites*. (F) Foliaceous *Millepora* sp.

fore-reef zones, at depths usually not exceeding 5 m. In more protected settings, *A. cervicornis* is the most common form, in both back-reef and fore-reef areas, at depths from about 5 to up to around 25 m. *Montastraea annularis* species complex is prevalent on most Caribbean reefs from near the sea surface to depths greater than 30 m in different reef zones, and possesses a remarkable phenotypic plasticity in colony growth shape. This complex appears to include three different species, each having preferential habitats and niche partitioning (Knowlton & Jackson, 1994) and distributed differentially in accordance with water depth gradients. The columnar form (*M. annularis sensu stricto*) occurs between 3 and 15 m, with its greatest abundance at around 6 m; the domal form (*M. faveolata*) extends to a maximum depth of 25 m, with its greatest abundance at around 9–10 m; and the flattened *M. franksi* attains its greatest abundance at about 20 m.

Besides, Geister (1977) proposed a classification of the modern Caribbean reefs based on the degree of wave strength to which the reef front is exposed and the relevant biotic associations observed in the breaker (reef-crest) zone (Figure 3.1). Although this zonal scheme can be occasionally difficult to be applied, six basic reef types can be found in the whole Caribbean, which are defined as follows:

(1) *Melobesieae* reef type (i.e. coralline algal-dominated community), typified by 1–3 m thick crusts of coralline algae and scattered corals that overcap the largest parts of reef margins;
(2) *Palythoa–Millepora* reef type (i.e. zoanthid-arborescent hydrocoral-dominated community), displaying dense covers of the soft coral *Palythoa* and the hydrocoral *Millepora* (*M. alcicornis*) along with domal *Diploria* spp., *Porites astreoides* and scattered *Acropora palmata*;
(3) *Strigosa–palmata* reef type (i.e. domal-robust branching mixed community), composed of the space dominant by *A. palmata* and *Diploria strigosa* accompanied with *P. astreoides*, *Diploria clivosa*, *Siderastrea siderea* and *Favia fragum*;
(4) *Cervicornis* reef type (i.e. arborescent coral-dominated community), dominated by dense thickets of *Acropora cervicornis* with isolated colonies of domal *Montastraea annularis* and *Diploria* spp. The transition to the seaward and landward (*M. annularis*) zones is usually gradual.
(5) *Porites* reef type (i.e. branching poritid-dominated community), exhibiting dense growths of *Porites porites*. Subordinate builders include branching coralline algae, *Porites astreoides*, *Siderastrea radians* and *Favia fragum*. This community is indicative of low-energy settings.
(6) *Annularis* reef type (domal coral-dominated community), consisting mostly of *Montastraea annularis* together with *Mussa angulosa*, *Isophyllia* spp., *Colpophyllia* spp., *Dendrogyra cylindrus* and *Eusmilia fastigiata*. This kind of community is indicative of very sheltered settings or of habitats below fair-weather wave base.

The Geister's classification fits well the computer-simulated zonation established by Graus and Macintyre (1989) (Figure 3.2). Additional information on the coral composition and zonation of the modern reefs in the Caribbean and western Atlantic was provided by Cortés (2003). It is noteworthy that in high-latitude areas, above 25° north, the structure of coral communities may be highly variable and appears not consistent with the classical Caribbean reef classification and zonation patterns (Goreau, 1959; Moyer, Riegl, Banks, & Dodge, 2003). The domal *Montastraea cavernosa* tends to prevailed over other coral forms.

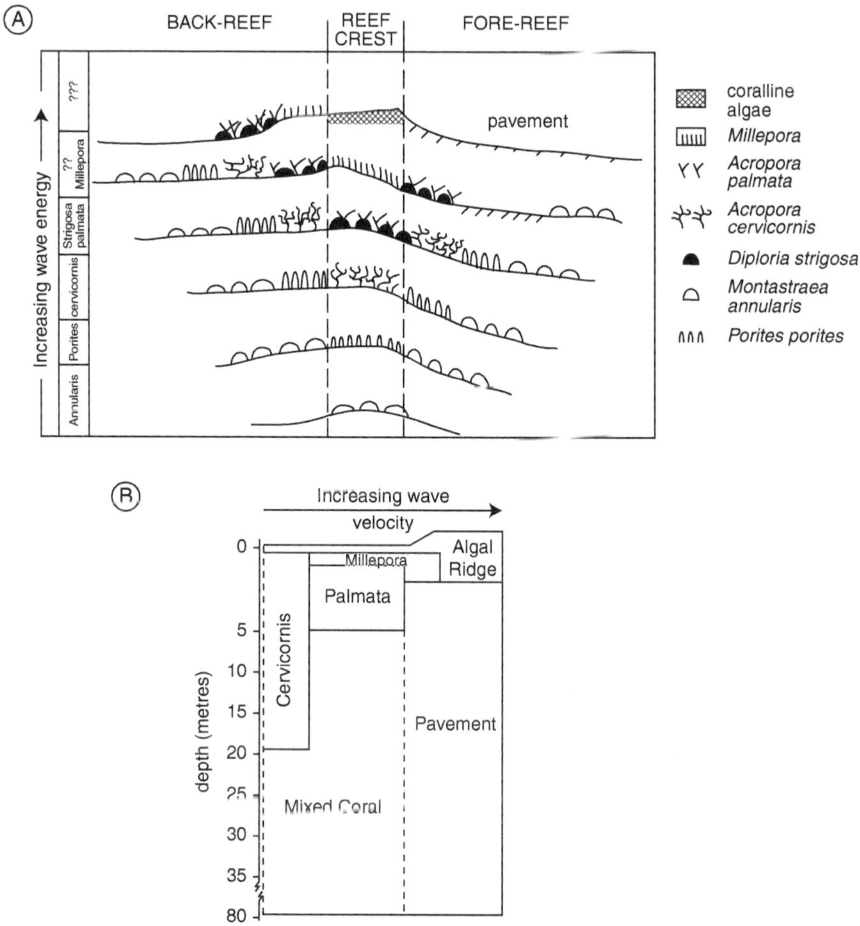

Figure 3.2 Zonation patterns of Caribbean coral reefs. (A) Idealized zonation of the basic reef communities according to increasing exposure to water energy (modified and redrawn from Geister, 1977). (B) Simulated zonation of the major reef-builders according to water energy and depth (modified and redrawn from Graus and Macintyre, 1989). There is a close similarity between the empirically defined model and that based on computer modelling. The *Melobesieae*, *Palythoa–Millepora* and *Cervicornis* reef communities that successively dominate the reef crest are homologous to the (coralline) algal ridge, *Millepora* and *Cervicornis* zones. The *strigosa–palmata* reef community refers to both the *palmata zone* and the transitional, more exposed portion of the *mixed coral zone*. The *Porites* and the *Annularis* reef communities are both included in the *mixed coral zone*, in fore-reef and back-reef environments as well. The *pavement zone* relates to an abrasional surface, that is the breaking point of winter storm waves.

3.2.2. The Indo-Pacific Province

Detailed descriptions on the major zonal features of living corals were provided regionally throughout the province, in the western Indian Ocean (Barnes, Bellamy, Jones, & Whitton, 1971; Braithwaite, 1971; Rosen, 1971; Pichon, 1978a; Montaggioni & Faure, 1980; Faure, 1982; Hamilton & Brakel, 1984), the Red Sea (Loya & Slobodkin, 1971; Scheer, 1971; Loya, 1972; Mergner, 1971; Bouchon, 1980; Sheppard & Sheppard, 1991; Riegl & Velimirov, 1994; Riegl & Piller, 1997; Dullo & Montaggioni, 1998), central and eastern Indian Ocean (Scheer, 1971; Pillai, 1971; Pillai & Scheer, 1976; Veron, 1994), western Pacific (Chevalier, 1971, 1975; Sy, Herrera, & McManus, 1981; Done, 1982; Takahashi, Koba, & Nakamori, 1985; Nakamori, 1986; Tribble & Randall, 1986; Done & Navin, 1990; Titlyanov & Latypov, 1991; Nakamori, Campbell, & Wallensky, 1995; Veron, 1986, 1992a; Iryu, Nakamori, Matsuda, & Abe, 1995; Van Woesik & Done, 1997; Wallace, 1999; Edinger, Kolasa, & Risk, 2000; Ikeda, Iryu, Sugihara, Ohba, & Yamada, 2006), central Pacific (Wells, 1954; Maragos, 1974; Chevalier, 1974, 1979, 1980; Grigg, 1983; Faure & Laboute, 1984; Bouchon, 1985; Maragos & Jokiel, 1986; Veron, 1993; Wallace, 1999) and eastern Pacific (Glynn & Wellington, 1983; Macintyre, Glynn, & Cortés, 1992; Cortés, Macintyre, & Glynn, 1994; Grigg, 1998; Glynn & Ault, 2000; Cortés, 2003). Due to the high degree of species overlap between habitats, it is difficult to delineate distinct reef zones on the basis of species composition. By contrast, the distribution of coral growth forms is more diagnostic in terms of zonation because coral species tend to develop growth forms in accordance with ambient physical conditions (Done, 1983). Based on the predominance of a single genus to groups of species, with characteristic growth forms, reliable zonal schemes across reef profiles were established (Braithwaite, 1971; Rosen, 1971, 1975; Pichon, 1978a; Riegl & Piller, 2000). At the scale of the Indo-Pacific, Montaggioni (2005) conveniently identified six types of coral assemblages, in relation to both wave exposure and habitat-depth: robust-branching, domal, tabular-branching, arborescent, foliaceous and encrusting coral-dominated respectively (Figures 3.3 and 3.4).

(1) The robust-branching (elkhorn, stout-branching) coral assemblages are composed of thick-branched, wave-resistant growth forms, dominated by the genera *Acropora*, *Pocillopora* and *Stylophora*. They are found distinctly on exposed reef settings, that is windward reef-crests, upper fore-reef zones at depths less than 6 m and, less commonly, on reef flat environments. The dominant coral species include *Acropora robusta* group, *A. humilis* group, *A. palifera*, *A. cuneata*, *Pocillopora damicornis*, *P. verrucosa*, *P. eydouxi*, *P. meandrina* and *Stylophora pistillata*. Subordinate corals are domal (*Porites lutea*, *P. lobata*, *Leptoria phrygia*, *Platygyra daedala*, *Goniastrea retiformis*, *G. favulus*, *Favia* spp.), tabular (*Acropora hyacinthus*) and encrusting (*Montipora tuberculosa*, *Echinopora gemmacea*)

Figure 3.3 Major types of coral assemblages in the Indo-Pacific reefs (photographs by L. Montaggioni). (A) Robust-branching *Acropora*-dominated assemblage, outer reef flat: *Acropora robusta* group (top left), *Acropora humilis* group (middle) and a variety of domal faviids (Rodrigues Island, western Indian Ocean). (B) Domal *Porites*-dominated assemblage, inner reef flat (Sanganeb atoll, Red Sea). (C) Mixed, tabular and arborescent *Acropora*-dominated assemblage (tabular *Acropora hyacinthus* group, arborescent *A. muricata* group), upper fore-reef zone, mid-shelf Wheeler Reef (Australian Great Barrier Reef). (D) foliaceous *Pachyseris*-dominated assemblage, lower fore-reef zone, mid-shelf reef (Heron Island, Australian Great Barrier Reef). (E) Encrusting *Millepora*-dominated pavement, upper fore-reef zone, Moorea Island (French Polynesia).

along with the hydrocorals *Millepora platyphylla* and *M. dichotoma*. There are some geographic variations in the composition of this assemblage throughout the Indo-Pacific. In the western Indian Ocean, the dominants are *A. robusta* and *A. humilis* groups, whereas *Pocillopora verrucosa* and *P. meandrina* are the most common species in the eastern side of this ocean. In the Red Sea, *Stylophora* is the most efficient builder along reef margins, associated with *Acropora hyacinthus*,

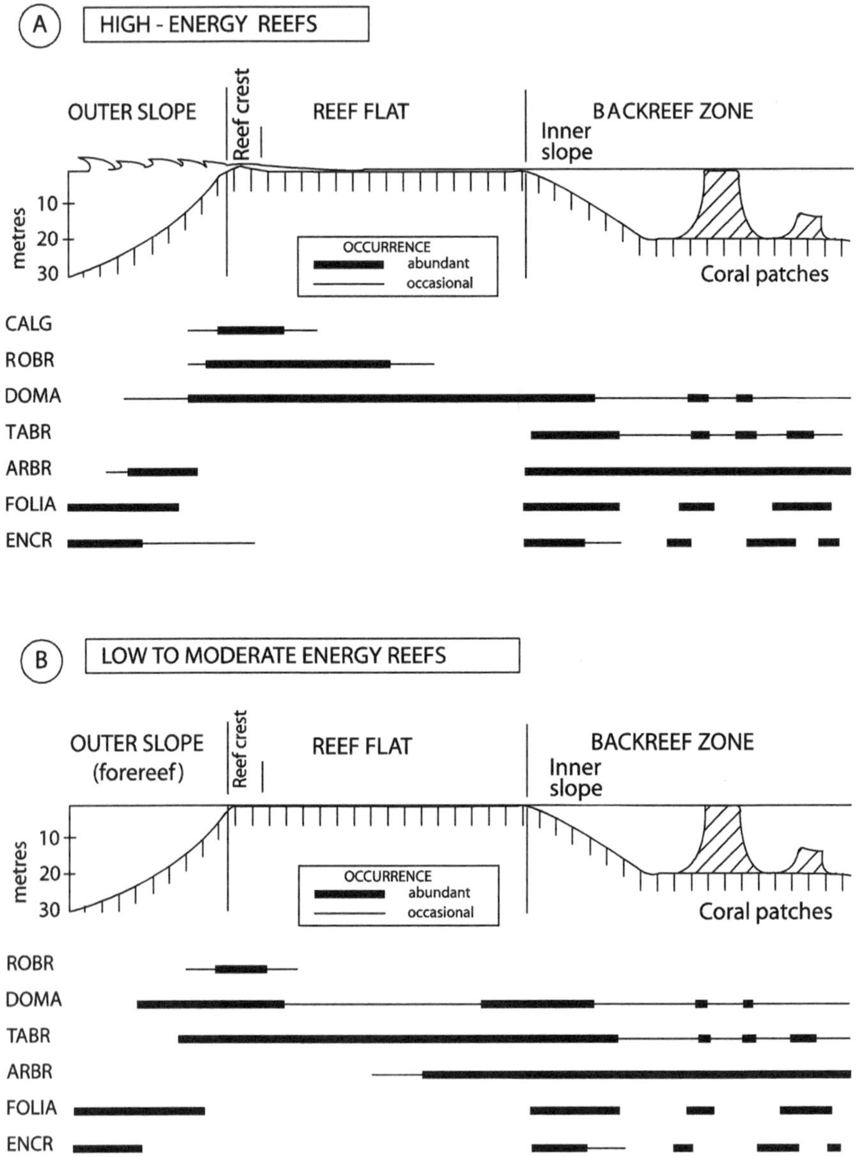

Figure 3.4 Schematic distributional patterns of coral assemblages (and coralline algal) on the Indo-Pacific reefs, in relation to water energy and depth: (A) in ocean-facing, higher-energy settings; (B) in protected, open and mid- to inner-shelf settings. Assemblages: CALG = coralline algal; ROBR = robust-branching coral; DOMA = domal coral; TABR = tabular-branching; ARBR = arborescent coral; FOLIA = foliaceous coral; ENCR = encrusting coral. From Montaggioni (2005).

A. horrida and *A. humilis*. In the western Pacific (Australia, New Caledonia, Papua New Guinea), *A. palifera* locally is one of the major contributors, whereas *Acropora gemmifera* and *Pocillopora verrucosa* are among the most abundant species on Japanese reefs. In western Pacific oceanic islands, the dominant builders belong to *Acropora humilis* group. In French Polynesia, reefs crests and upper reef slopes are mainly colonized by *A. robusta* group and *Pocillopora damicornis*. By contrast, in Hawaii islands, the dominant robust-branching corals include pocilloporids (*Pocillopora meandrina*). A similar scheme was described from the eastern Pacific with the prevalence of *Pocillopora damicornis* and *P. elegans* locally accompanied with *Psammocora stellata*.

(2) The domal (massive, head) coral assemblages contain mainly poritids and faviids and are widespread throughout the province. They occur on semi-exposed to sheltered, windward to leeward fore-reef zones, on reef flats in both outer- and inner-shelf settings and in back-reef slopes and bottoms within the 0–25 m depth range. The dominant species include *Porites lutea, P. lobata, P. cylindrica, Favia favus, F. stelligera, F. speciosa, Favites abdita, Cyphastrea* spp., *Goniastrea* spp., *Diploastrea heliopora, Montastrea curta, Hydnophora microconos* and *Symphillia recta*. Locally, these assemblages incorporate a variety of other coral growth forms, reflecting differing water agitation. In shallower and higher wave energy areas, the communities are composed mostly of *Porites lobata* or enriched in robust-branching forms (*Acropora robusta* and *A. humilis* groups, *A. palifera, Stylophora pistillata*). In less agitated or deeper waters, domal forms are accompanied with tabular (*A. hyacinthus* group) and delicate branching (*Acropora divaricata, A. muricata, A. pharaonis, A. splendida, Seriatopora hystrix*) and/or foliaceous (*Montipora capitata, M. aequitubercu-lata*), laminar (*M. verrucosa*) and columnar (*Porites nigrescens*). These communities exhibit a low regional variability throughout the province except in settings subjected to extreme conditions (lower temperature, higher turbidity) or in relatively remote areas. In this case, the fauna is severely depauperate. Thus, in the Marquesas archipelago, *Porites lobata* predominate. In the far eastern Pacific, the domal coral community comprise *P. lobata, Pavona gigantea* and *Pavona clavus*.

(3) The tabular-branching (tabulate, plate-shaped, corymbose) coral assemblages are dominated by a number of acroporidae (*Acropora hyacinthus* group along with *A. splendida, A. intermedia, A. humilis, A. digitifera, A. nobilis, A squarrosa* and *Montipora digitata*). The assemblages contain other growth forms including pocilloporids (*Pocillopora verrucosa, P. damicornis, P. eydouxi*), poritids (*Porites lutea, P. nigrescens*) together with domal *Lepastrea* and *Platygyra*, columnar *Alveopora* and laminar *Echinophyllia* and *Echinopora*. Tabular-branching corals occur preferentially in semi-exposed to sheltered areas from upper and mid-fore-reef zones, reef crests and flats to adjacent back-reef slopes and patches,

usually in mid-shelf situations. They grow intertidally to subtidally at depths not exceeding 20 m. These assemblages are widely distributed all the way from the western Indian Ocean to the central Pacific. There is no significant change in their composition from locality to locality. The *A. hyacinthus* species group remains the major framework contributor. On the eastern Pacific reefs, tabular-branching forms are missing.

(4) The arborescent (ramose, staghorn) coral assemblages are composed of relatively gracile branching colonies that house lower to middle parts of fore-reef zones, inner reef flats and nearby back-reef areas in semi-exposed to protected environments at depths ranging between the surface and around 20 m. Along fore-reef slopes, the assemblages usually comprise a variety of acroporid species (*Acropora divaricata* group, *A. aculeus*, *A. valenciennesi*, and *A. tenuis*). On reef-flat and back-reef settings, the arborescent assemblages are dominated by large thickets of *Acropora muricata* (formerly *formosa*) group, *A. aspera* group, *A. cerealis*, *A. valida*, *A. tortuosa*, *A. austera*, *A. intermedia*, *A. microphtalma*, *A. lovelli* group with the pocilloporid *Seriatopora hystrix* and the faviid *Echinopora horrida*. A number of other growth forms participate locally in the community: robust-branching *Stylophora*, *Pocillopora damicornis* and *Acropora squarrosa*. In both types of habitat, the subordinate forms consist of tabulate *A. hyacinthus* group, domal *Goniastrea pectinata*, *Galaxea fascicularis* and *Porites lobata*. These assemblages show little variations throughout the province. In Hawaiian Islands, the arborescent assemblage is dominated by *Porites compressa* forming dense stands in protected, back-reef areas.

(5) The foliaceous (lamellar platy to frondose) coral assemblages are dominated by agariciids, dendrophyliids and some acroporids. They occupy protected zones usually suffering suspended sediment loading, or deep fore-reef zones, both zones experiencing low light levels. Along reef slopes facing open shelves (20 to greater than 30 m deep) or in mid- to inner-shelf settings (less than 20 m deep), the assemblages are composed chiefly of *Pachyseris speciosa*, *P. rugosa*, *Turbinaria mesenterina*, *T. reniformis*, *T. frondens*, *Merulina ampliata*, *Montipora aequituberculata*, *M. foliosa* and *Montipora* spp. The foliaceous assemblages often inhabit the upper parts of the niches colonized by the arborescent coral assemblage at depths from 0 to 15 m. On inner reef flats and in shallow back-reef settings, they are typified by the abundance of *Montipora* (*M. tuberculosa*, *M. verrucosa*, *M. danae*) and *Pavona* (*P. cactus*, *P. decussata*, *P. varians*). The nature of associated corals varies from site to site, including domal (*Porites lobata*, *P. solida*, *Favia pallida*, *F. speciosa*, *Favites abdita*, *Plesiastrea versipora*, *Lentastrea purpurea*, *L. transversa*, *Cyphastrea ocellina*, *C. seraila*, *Astreopora myriophthalma*) and branching forms (*Pocillopora verrucosa*, *Psammocora contigua*, *Stylophora pistillata*, *Acropora muricata*, *A. valida*). This type of coral assemblage is absent from the eastern Pacific.

(6) The encrusting (lamellar, platy-encrusting) coral assemblages are observed in a variety of environments ranging from strongly water-agitated or highly turbid and low-lighted. At depths from surface to around 10 m, on high-energy reef crests, fore-reef and inner slopes of ocean-facing fringing reefs and of mid- to inner-shelf reefs, the crustose coral assemblages exhibit a varying composition. According to the region considered, these may contain the acroporids *Montipora monasteriata*, *M. capitata* , *M. undata*, *M. patula*, *M. danae*, the agariciid *Leptoseris mycetoseroides*, the hydrocoral *Millepora platyphylla*, the pectiniid *Echinophyllia aspera*, the faviids *Leptastrea purpurea*, *Echinopora lamellosa*, *E. gemmacea* and the poritid *Alveopora daedala*. These encrusting forms are locally mixed with colonies regarded as initially foliaceous and forming crusts under extreme conditions: *Cyphastrea seraila*, *C. microphthalma*, *C. ocellina*, *Pachyseris speciosa* and *Merulina ampliata*. In deeper or more sheltered habitats from about 20 m downwards, shelf-reef slopes are occupied usually by assemblages typified by the predominance of *Montipora*, *Leptoseris* spp., *Cycloseris* spp., *Diaseris*, *Pachyseris* and/or *Echinophyllia*. Along steep walls, these assemblages can extend upwards to around 8 m deep in response to marked decrease in light supply. The dominants are *Montipora aequituberculata*, *M. verrucosa*, *Leptoseris incrustans*, *L. hawaiiensis*, *L. scabra*, *L. mycetoseroides*, *Pachyseris speciosa*, *Echinophyllia aspera*, *E. echinata* and *Oulophyllia crispa*.

In addition to the coral assemblages mentioned above, the Indo-Pacific province, like the Caribbean, displays calcareous alga dominated communities mainly living at windward reef-crest settings exposed to strong oceanic swells or along fore-reef slopes. Coralline algal crusts occur preferentially on barrier reef and atoll margins in the western and central Pacific where they form the so-called 'algal (*Melobesieae*) ridges or pavements' (Littler & Doty, 1975; Adey, 1986; Steneck, 1986; Macintyre, 1997). Irrespective of their thickness (less than 0.10 to up to 3 m), they are mostly composed of *Hydrolithon* (formerly *Porolithon*) *onkodes*, *Neogoniolithon* spp., *Mesophyllum* spp., *Sporolithon* sp. and *Lithophyllum* spp. in association with encrusting foraminifera (mostly *Homotrema*, *Miniacina*, *Carpenteria* and/or *Acervulina*), vermetid gastropods (*Serpulorbis*, *Dendropoma*) and bryozoans. Coralline algal-dominated crusts also occur at depths of 50–150 m along fore-reef slopes where they develop in association with other encrusters (mainly, foraminifera *Acervulina*, bryozoan and scleractinian corals). The dominating algal forms belong to lithophylloids (*Lithophyllum*), melobesioids (*Mesophyllum*) and *Sporolithon* (Dullo, Moussavian, & Brachert, 1990; Davies, Braga, Lund, & Webster, 2004; Flamand, Cabioch, Payri, & Pelletier, 2008).

The coral zonation schemes for high-latitude regions of the Indo-Pacific are not conform to typical reef classifications. Coral communities are dominated usually by domal forms, including faviids mainly (Stoddart, 1969a; Done, 1982).

3.3. Structure and Zonation of Quaternary Reef Communities

Concern over the faunal composition of Quaternary coral reefs has greatly intensified in the last two decades, especially in order to address the questions about the structure and dynamics of reef community at different temporal and spatial scales. Exposures and coring through modern reefs have revealed distributional patterns of the fossil reefs.

3.3.1. The Western Atlantic–Caribbean Province

Due to the limited number of coral species present in the Caribbean reef communities, the modern dominating corals and their fossil analogues can be identified relatively easily at the species level with comparable reliability. This allows robust comparison between the structure and zonation of modern and Pleistocene assemblages (Pandolfi & Jackson, 2001).

3.3.1.1. The Pleistocene

Comprehensive studies on the composition of reef-coral and/or associated organisms were conducted in many localities of the western Atlantic. These localities possess series of fossil fringing or barrier reef tracts as raised terraces or overlain by recent deposits. The best-documented reefs are those that developed during the last interglacial episode, approximately 125 ka ago (Figure 3.5). Geister (1980) refound the six fundamental coral-dominated communities he defined previously from modern reefs, in response to gradual decrease in water energy.

Jackson's coral-community model. Using the emerged Pleistocene reefs of Barbados as examples, Jackson (1992) revisited coral reef zonation defined by Mesolella (1967), taking into account the cover rate and habitat-depth ranges of three dominant coral species (Figure 3.6). Thus, five coral assemblages were delineated: (1) an 'upper elkhorn' (robust-branching) coral assemblage composed of up to 90% of *Acropora palmata* colonies and restricted between 0 and 3 m depth; (2) a 'lower elkhorn' (robust-branching) assemblage made up of about 50% of *Acropora palmata*, ranging from 3 to 6 m deep; (3) a 'mixed' (arborescent-domal) coral assemblage containing 25–50% *Acropora cervicornis* with occasional *Montastraea annularis* and occurring between 5 and 10 m deep; (4) a 'staghorn' (arborescent) coral assemblage composed of more than 50% of *Acropora cervicornis* and scattered large

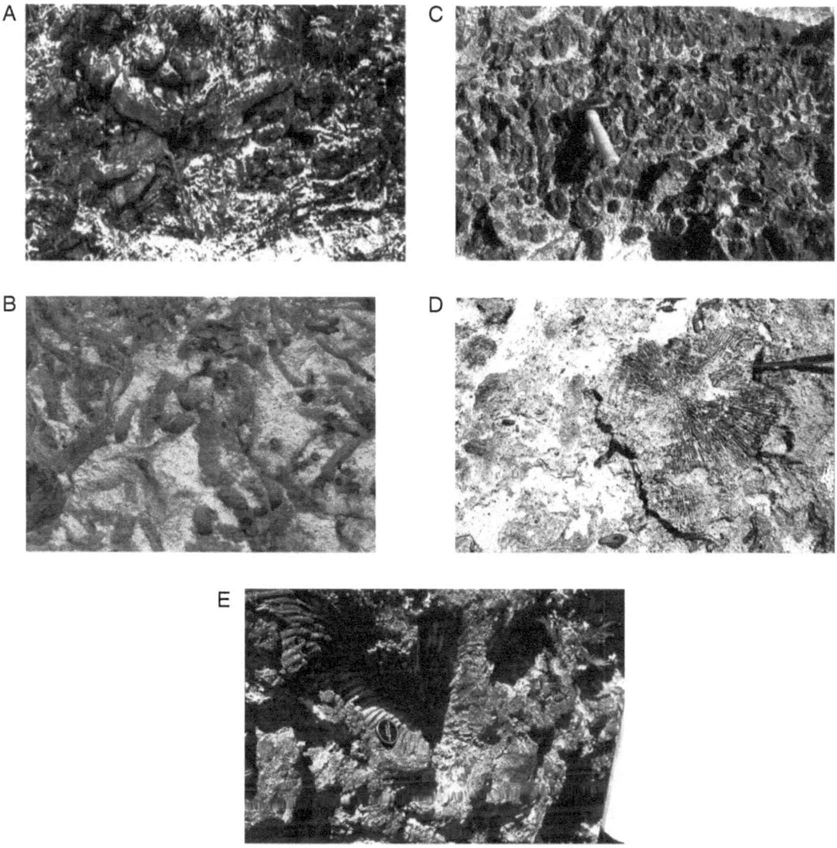

Figure 3.5 Typical corals found in the Caribbean Pleistocene reefs. (A) Branches of *Acropora palmata*, 125-ka terrace, Barbados (photograph by L. Montaggioni). (B) Branches of *Acropora cervicornis*, 125-ka terrace, Guadeloupe Island, French Antilles (photograph courtesy by G. Conesa). (C) Sections of organ-pipe *Montastraea*, 125-ka terrace, Pointe des Châteaux, Guadeloupe Island (photograph by L. Montaggioni). (D) Section of *Diploria cf. strigosa*, 125-ka terrace, Barbados (photograph by L. Montaggioni). (E) Section of *Diploria cf. labyrinthiformis*, 125-ka terrace, Barbados (photograph by L. Montaggioni).

colonies of *Montastraea annularis*, extending from 7 to 25 m deep; and (5) a 'head coral' (domal) coral assemblage typified by 50% of *Montastraea annularis* and a variety of other massive corals found from 15 to 25 m deep. Compared to modern reefs, the only significant difference in the Pleistocene community structure appears to have been the greater amount of *Acropora cervicornis* in both lower robust-branching and arborescent coral assemblages, This difference is thought to have been caused by the faster growth rate of arborescent colonies and its subsequent higher skeletal production compared to other coral forms, especially during cyclonic events.

Regional case studies. In Belize, Tebbutt (1975) provided valuable information about the composition of coral populations encountered in exposures. But, little is known regarding the coral community structure of the Pleistocene barrier reefs and atolls since most of them are overlain by Holocene deposits (Macintyre & Toscano, 2004; Gischler, 2007). Apart

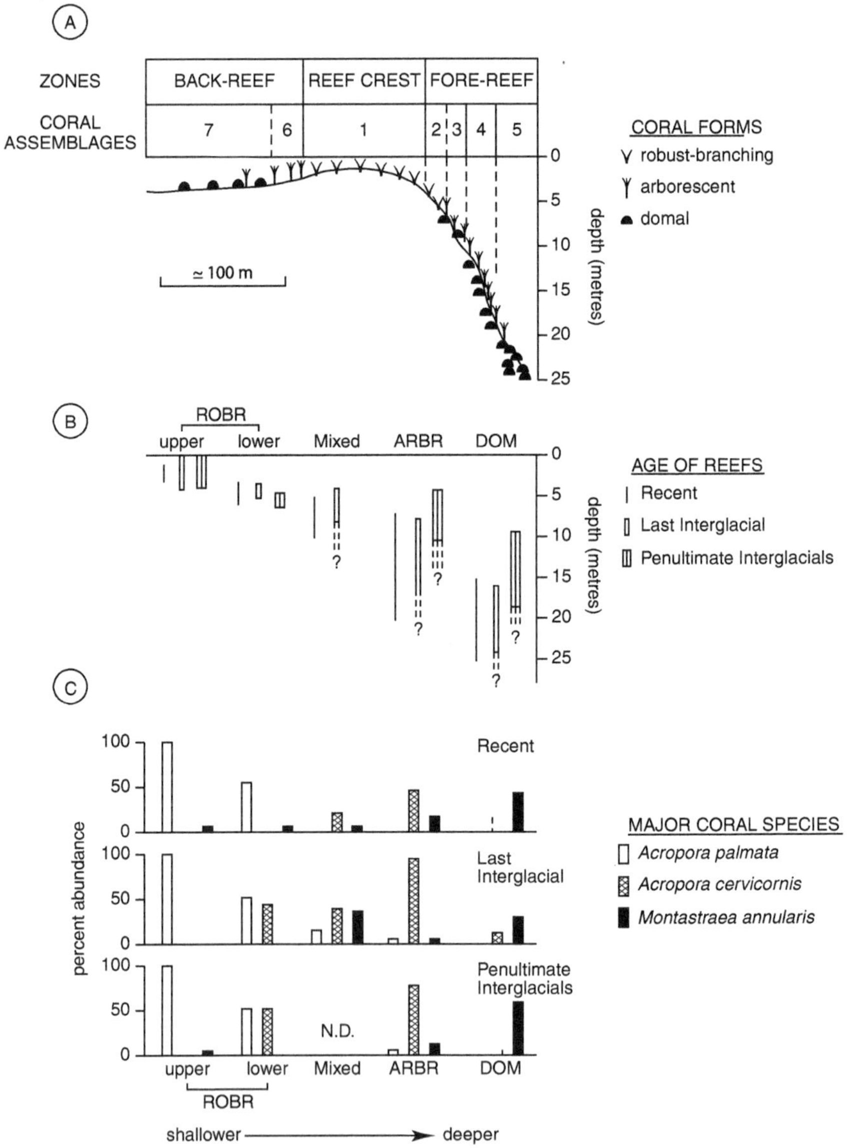

from the typical *A. palmata*-rich reef-crest and *A. cervicornis*-dominated inner reef-flat environments, a palaeo-shelf lagoon environment was recognized (Figure 3.7). This lagoon was subdivided into three subzones: (1) an outer-shelf subzone, including patches reefs, is typified by the dominance of *M. annularis* together with *A. cervicornis*, *Diploria*, *Porites furcata* and *Millepora*. The macrofauna also includes molluscs (*Strombus gigas*, *Bulla cf. striata*). (2) The middle-shelf subzone incorporates sparse molluscs (*Bulla*, *Cerithium*, *Conus*, *Chione*) and scattered thickets of *Porites*. (3) The inner-shelf subzone contains a typical macrofauna with bivalves (*Chione cancellata*, *Tellina interrupta*, *Laevicardium* spp., *Anadara* spp.), gastropods (*Xancus angulatus*, *Cerithidea costata*, *Cerithium*, *Polinices*) and few corals (*Porites*, *A. cervicornis*, *Siderastrea radians*, *Agaricia*). The fourth environment recognized is a mudbank apparently devoid of corals, but containing isolated moulds of *Chione* and *Bulla*. The first two environments and the outer-shelf subzone compare well with nearby modern deposits on the Belize shelf. The reef-crest, inner reef-flat and outer-shelf faunas are virtually identical to their homologues in the Holocene. By contrast, the middle-shelf subzone, widely occupied by oolitic and pelletal sediments, has no recent counterpart. Additional information regarding the composition of coral and macrofaunal assemblages in Belize is derived from drilling investigations by Macintyre and Toscano (2004) and Gischler (2007) through the central and southern barrier reefs and offshore and through nearby atolls respectively. The uppermost portions of the Pleistocene foundations beneath the central barrier reef platform appear to relate to a lagoonal environment. The most striking feature is the abundance of *A. cervicornis* deposits in atoll interior cores. Although this coral is locally a major patch reef builder in the studied areas, *A. cervicornis*-dominated patch reefs are typically rare in the Caribbean Pleistocene. One can speculate that the apparent absence of this coral species on lagoonal patches in most

Figure 3.6 Zonation and composition of coral assemblages from Pleistocene reef tracts on Barbados, Lesser Antilles. (A) Schematic reconstructed zonation with estimated water depth (adapted from Mesolella, Sealy, & Matthews, 1970). Numbers 1–5 refer to the different coral assemblages defined by Jackson (1992): 1 = upper (robust-branching) elkhorn; 2 = lower (robust-branching) elkhorn; 3 = mixed (arborescent–domal); 4 = (arborescent) staghorn; 5 = domal (head coral). Numbers 6 and 7 refer to the transitional *Acropora cervicornis*-dominated and coral head, *Montastraea annularis*-dominated assemblages respectively, found behind the reef-crest zone (according to Mesolella et al., 1970). (B) Depth zonation of recent and Pleistocene coral assemblages. Uncertainty remains about the maximum depth range of some fossil assemblages. Coral assemblages: ROBR = robust branching; ARBR = arborescent; DOM = domal (Modified and redrawn from Jackson, 1992). (C) Relative abundance of the three dominating coral species in recent and Pleistocene reef assemblages from the Caribbean (modified and redrawn from Jackson, 1992).

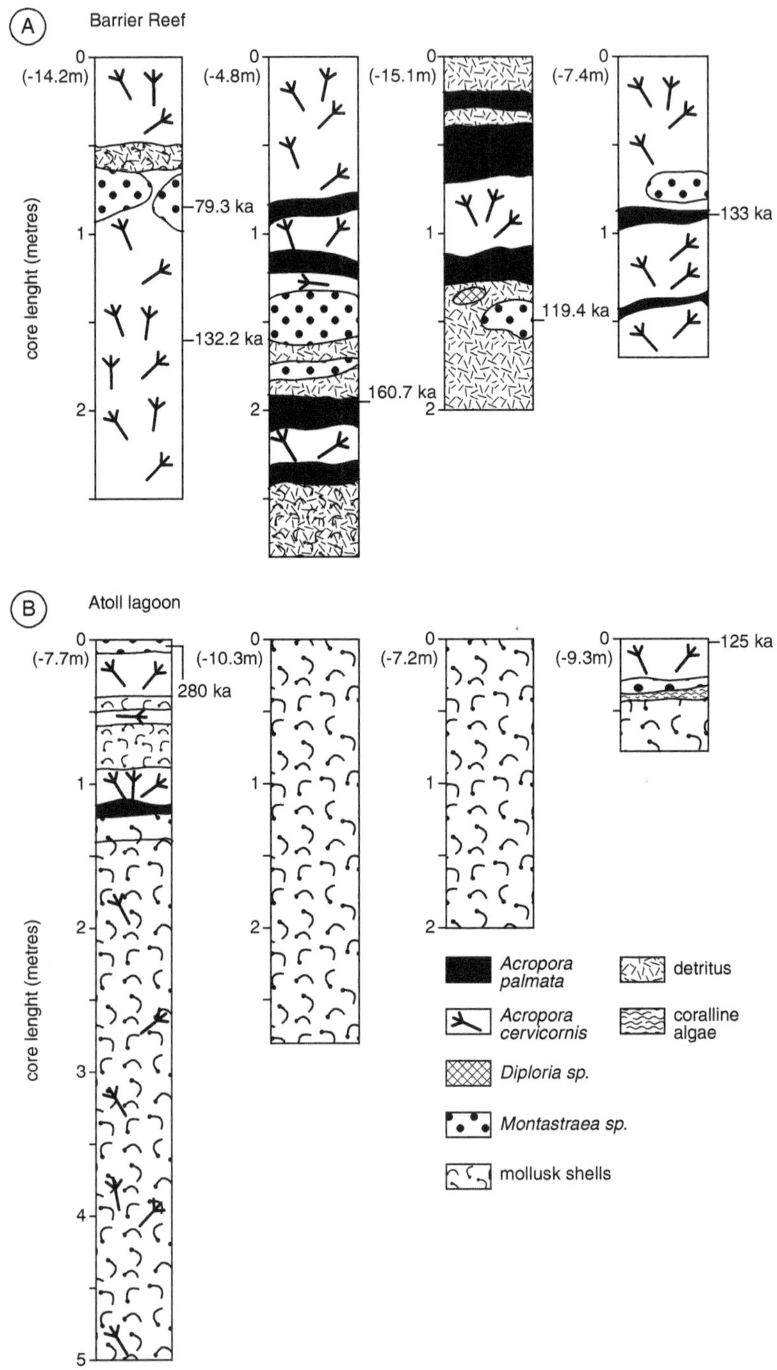

Figure 3.7 Pleistocene coral assemblages in core sections extracted from the barrier reef at Belize. Depth of the Pleistocene reef surface below present sea level is given in brackets. Modified and redrawn from Gischler (2007).

Caribbean reefs may be an artefact of faunal replacement since some studies have demonstrated that the replacement of *A. cervicornis* by other taxa may operate frequently at time scales of decades.

In Florida, the palaeoecologically best described sequence is restricted to the uppermost parts of the Pleistocene 'Key Largo Limestone' that was deposited at about 125 ka (Hoffmeister & Multer, 1968; Multer, Gischler, Lundberg, Simmons, & Shinn, 2002). Precht and Miller (2007) provided a comprehensive overview of the palaeoecological changes that affected the composition of the Florida coral communities throughout the late Pleistocene. Thus, the 125-ka reef appears to have been a major exception to the general pattern of Caribbean reef zonation. Unlike Holocene to modern analogues, this reef lacks *Acropora palmata* and is poor in *Acropora cervicornis*. It is dominated by a community mostly composed of *Montastraea annularis*, *Diploria strigosa* and *Porites astreoides*. According to Harrison and Coniglio (1985), the Key Largo Limestone is most probably the remnant of a bank-barrier complex that was composed of concentrically distributed shallow-water reef units dominated by *Montastraea annularis*. Given the fact that the Florida Peninsula is located at the latitudinal extreme of reef growth in the western Atlantic, the lack of *Acropora palmata* and paucity of *A. cervicornis* are likely to have been caused by a contraction of the species ranges in response to changes in environmental constraints. A modern counterpart, totally lacking acroporids and dominantly made up of *Montastraea*, *Diploria* and *Porites*, is found in Bermuda. Suffering low sea surface temperature in the winter, these coral populations are paucispecific when compared to most provincial sites, but have high cover (Precht & Miller, 2007). Younger reef tracts dated respectively at approximately 112–106 ka and 86–78 ka were discovered beneath the modern reefs in southeast Florida (Toscano & Lundberg, 1999; Lidz, 2004). These developed in the form of shelf-margin units, overlapping the 125-ka reef surface. Unlike the Key Largo Limestone, these outlier reefs contain dense populations of *Acropora palmata*. The reappearance of acroporids strongly supports a recovery of favourable environmental conditions at the scale of the Florida Peninsula.

In Jamaica, the coral assemblages found in the raised last interglacial Pleistocene fringing reefs show striking similarities to those observed on the adjacent modern reefs (Liddell, Ohlhorst, & Coates, 1984; Boss & Liddell, 1987b). Along the southeastern coast, palaeo-reef crest, adjacent back-reef zone and palaeolagoonal areas were identified (James, 2006). The reef crest was classified as a *strigosa–palmata* community according to the Geister's (1977) model.

Inter-regional coral-community comparison. In order to test the validity of Geister's qualitative model (1977, 1980) and predictability in coral species richness, Pandolfi and Jackson (2007) compared the distributional patterns of reef-coral communities of the last interglacial episode from three southern Caribbean islands (Curaçao, San Andrés and Barbados) that extend latitudinally over 2,500 km and rise far from continental influences. These coral assemblages were interpreted as having grown in upper fore-reef and reef-crest zones. Quantitative surveys of species diversity indicate that the community composition varied significantly among these islands (Figure 3.8). At San Andrés, the extinct, organ-pipe *Montastraea nancyi* (Pandolfi, 2007) and the extant, arborescent *Acropora cervicornis* dominate. In association with domal *Diploria strigosa*, *Diploria labyrinthiformis*, *Montastraea faveolata* and *Montastraea annularis*, they represent up to 85% of the total coral assemblage. The communities found on Curaçao are dominated by robust-branching *Acropora palmata* (about 25%) along with *M. nancyi* colonies (about 30%). These species, associated with subordinate forms (*M. annularis*, *D. strigosa*, *A. cervicornis*) represent up to 98% of the coral fauna. In Barbados, about 80% of the assemblage consists of *A. palmata*. Like that of *Acropora palmata*, the abundance of *Montastraea nancyi* in a given assemblage bears witness to local high hydrodynamic-energy conditions (Pandolfi, Jackson, & Geister, 2001). By reference to the present-day wave-energy regime, there was probably a decreasing wave-energy gradient from west to east over the 2,500 km during the 125-ka high sea-stand episode. This scheme satisfactorily explains the late Pleistocene patterns of community composition. Coral assemblages are easily predictable and vary in species richness according to wave exposure.

3.3.1.2. The latest Pleistocene to Holocene

A large body of information on the composition of coral communities over the past 18 ka has been gained in the western Atlantic, mainly by coring modern shallow-water reefs and relict submerged reefs (see Macintyre, 1988, 2007, for review). Below are presented some of the most representative case studies.

The first detailed description of Holocene coral-dominated communities was given by Macintyre and Glynn (1976) from drilled sequences at Galeta Point Reef (Panama). Settlement commenced at around 7.5 ka, the reef is composed of three distinct *in situ* coral assemblages, the composition and distribution patterns of which fit well with the reef zonation model of Geister (1977). These patterns contrast strongly with that of the nearby living reefs. In particular, the present-day reef-flat zone is devoid of corals and supports the fleshy red alga *Acanthophora* and the sea grass *Thalassia*. The outer edge is colonized by zoanthids, as well as fleshy and crustose red algae. On nearby modern reefs, the elimination of the zonation pattern regarded as typical of the Caribbean is likely to be due primarily to the mortality of the dominant reef-building acroporids. As seen in many

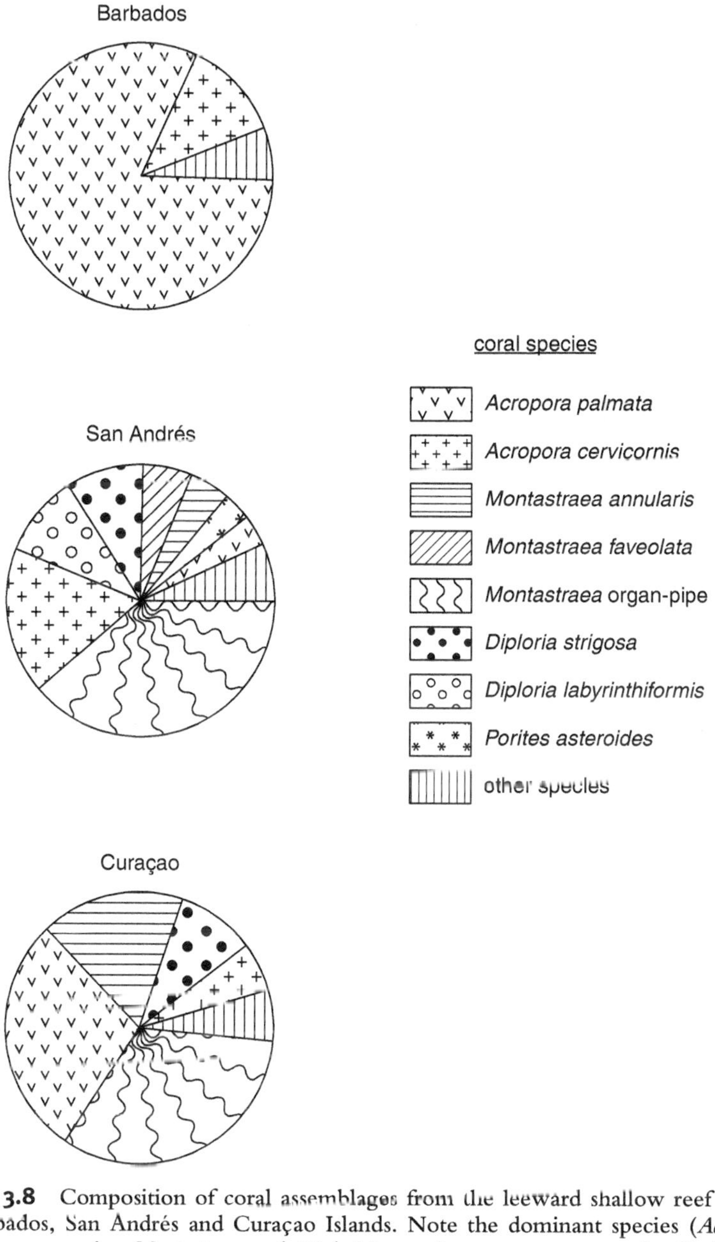

Figure 3.8 Composition of coral assemblages from the leeward shallow reef zones of Barbados, San Andrés and Curaçao Islands. Note the dominant species (*Acropora palmata*, organ-pipe *Montastraea* and *Diploria*) remain constant among the Caribbean islands. Redrawn from Pandolfi et al. (1999) and Pandolfi and Jackson (2007).

regional sites (Aronson & Ellner, 2007), the macroalgal-dominated pattern observed today at Galeta Point may have resulted from the effects of a variety of natural or human-induced disturbances.

In southeastern Florida, the compositional traits of Holocene reef-coral assemblages have been described by many workers (reviews by Macintyre, 1988, 2007; Precht & Miller, 2007). One of the best known in terms of species diversity and composition is that forming a relict ridge along the shelf edge off the southeastern Florida coast (Lighty, 1977; Lighty, Macintyre, & Stuckenrath, 1978; Toscano & Lunberg, 1998). An exposed section of the ridge revealed the occurrence of a typical Caribbean shallow-water *Acropora palmata* reef.

On Barbados, using offshore drilling, Fairbanks (1989) extracted a set of cores from submerged reefs lying along the southern foreslopes (Figure 3.9). A series of three *Acropora palmata*-dominated reefs was found to have developed successively at around 17–12, 11.8–10 and 9.4–7 ka. The reef-crest *A. palmata* assemblage was locally replaced up or downcore by deeper or more sheltered communities mostly composed of *Acropora cervicornis* or a variety of domal forms (*Montastraea annularis*, *Porites astreoides*). These changes in community structure were primarily interpreted as reflecting upward-deepening sequences, triggered by variations in water depth in relation to the postglacial rise in sea level (see Chapter 9, Section 9.4). On Barbados today, reefs are restricted to the leeward western coast, forming well-developed, but discontinuous fringing reefs (Stearn, Scoffin, & Martindale, 1977). Contrary to its forebears, the modern reef-crest zone appears to be devoid of *A. palmata*. It is noteworthy that monitoring of the modern reef was conducted before the onset of the white-band disease that devastated branching acroporids throughout the Caribbean in the late 1970s and early 1980s (Gladefelter, 1982). This indicates that the lower contribution of *A. palmata* to modern reef building is a natural event and does not reflect a human-induced, ecological shift in coral community composition.

In Belize, knowledge on the composition of coral assemblages during the Holocene development of the barrier reef and nearby platforms comes chiefly from drilling investigations by Macintyre, Burke, and Stuckenrath (1981) and Gischler et al. (Gischler, 2003; Gischler & Hudson, 1998, 2004; Gischler & Lomando, 2000). In the outer-rim Holocene sections, dated at 8.5–6.7 ka at the base, the coral assemblages are dominated by *Acropora palmata* and the *Montastraea annularis* species complex (Figure 3.10). The Holocene sections extracted from the interior lagoonal areas are typified by scattered coral populations mainly composed of *Diploria strigosa*, *Manicina areolata* and *Porites astreoides*. Macintyre, Precht, and Aronson (2000) demonstrated that beneath the lagoonal reefs in Belize, the Holocene deposits, formed during the past 9–8 ka, were mostly composed of

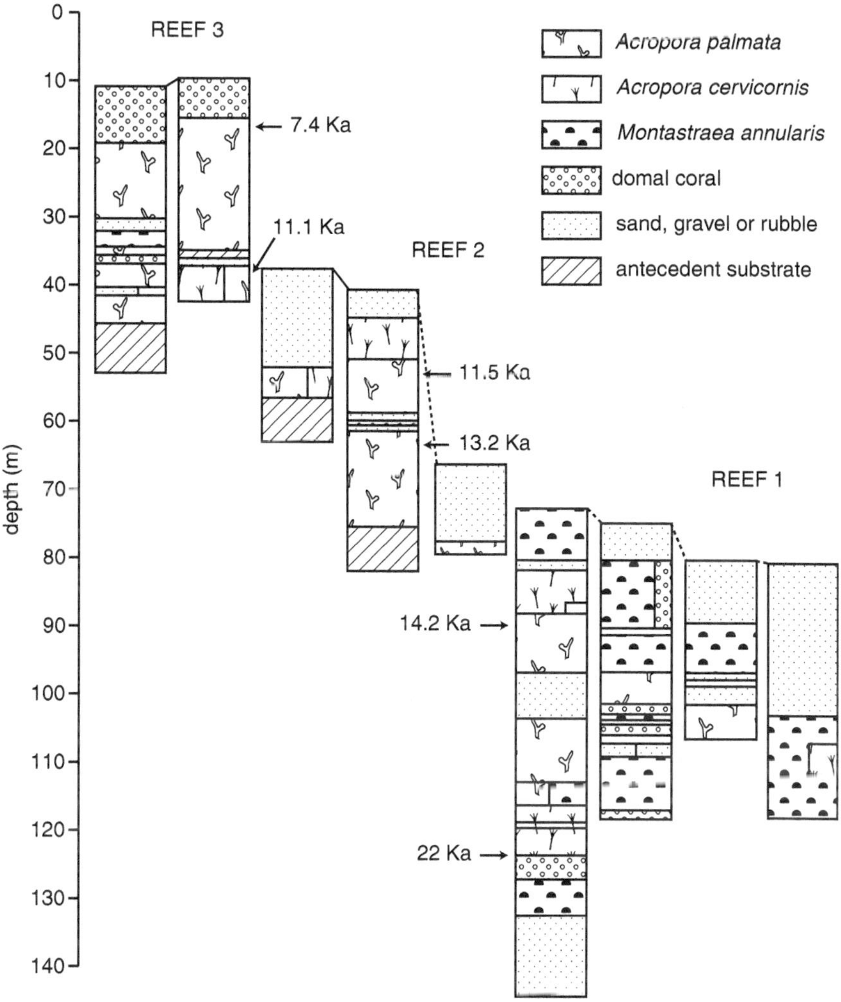

Figure 3.9 Coral assemblages in core sections extracted from Late Pleistocene to Holocene submerged reefs, off the south coast of Barbados. Simplified and redrawn from Fairbanks (1989).

A. cervicornis for at least the past 3 ka. *Agaricia tenuifolia* occurred as a minor component and occasionally replaced *A. cervicornis* during small-scale environmental shift events. At about 0. 5 ka, as the lagoonal reefs grew to within 2 m of present sea level, the *A. cervicornis*-rich community changed into a *Porites*-rich one. The acroporid-to-poritid transition is considered to be a natural event (i.e. a shallowing-upward, ecological succession), in response to changes in physical conditions.

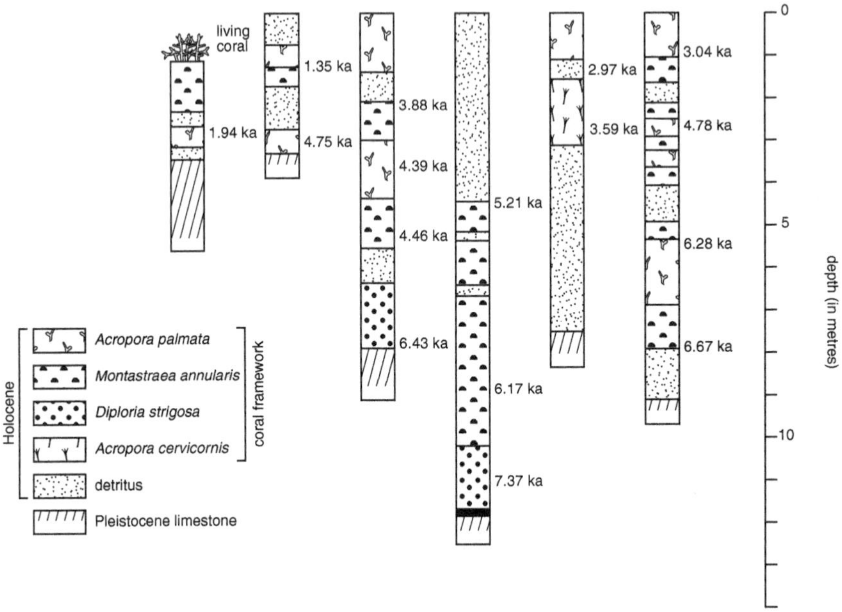

Figure 3.10 Holocene coral assemblages in core sections extracted from isolated carbonate platforms in offshore Belize. Modified from Gischler and Hudson (1998).

3.3.1.3. The recent past

From the late 1970s, the face of Caribbean reefs has changed. In particular, acroporids have suffered high mortality caused by white-band disease (Gladefelter, 1982), resulting in significant alteration of the original coral zonation patterns on many reefs (Aronson & Precht, 2001; Precht & Miller, 2007; Aronson & Ellner, 2007). Coral communities have been destroyed and replaced by fleshy and filamentous macroalgae. The major question posed by all of these disruptions is whether the ecological shifts over the past 25 years are indicative of a new equilibrium in coral community structure or the starting point of long-term, repeated events. The analysis of mass-mortality events in the recent past may help in addressing this crucial question.

Aronson, Precht, and Macintyre (1998), Aronson, Macintyre, Precht, Murdoch, and Wapnick (2002), Aronson, Macintyre, Lewis, and Hilburn (2005) and Aronson and Ellner (2007) identified biotic turnover events over the past 3.5 ka in both the Belize and Panama lagoonal systems and demonstrated that variations in the structure of coral assemblages have operated over the last millennia at two levels, between different depth zones (habitat level) and between geographic localities. Since the 1980s in Belize and Panama, lagoonal bottoms at different depths have displayed a monotypic dominance by *Agaracia tenuifolia*. At Belize, all the cored sections show an uppermost bed about 0.25-m thick of *Agaricia tenuifolia* plates

(Figure 3.11). This bed is believed to represent current, postmortem accumulation. Just below this *Agaricia* layer, is a thin deposit of severely altered and encrusted *A. cervicornis* in growth position. This acroporid layer is interpreted as recording a coral mass-mortality event. Beneath the degraded acroporid layer, the subsurface deposits are homogeneous and dominated by well-preserved, upward-oriented *A. cervicornis* branches. This acroporid was the major framework-builder in the central lagoon of Belizian barrier reef system during the Holocene. In Panama (Bahía Almirante), *Porites* dominance was maintained from 3–2 ka until the shallowest lagoonal areas developed within 0.25 m of present sea level and *Agaricia tenuifolia* replaced *Porites* over the last decades. The recent community dynamics in both geographic locations may have been controlled by intense perturbations (white-band disease in Belize, and declining water quality in the Bahía).

From Buck Island (US Virgin islands), Hubbard et al. (2005) provided a detailed description of the coral assemblages that grew from around 7.7 to *ca.* 1.2 ka. The abundance and species diversity of corals in the cores over the past 7.7 ka (total coral = 20–30% of the core volume, dominated by *A. palmata*) compare well with data on coral cover from the late 1970s, but are markedly richer than those measured in the 1980s and early 1990s (total coral cover = 7–14%; *A. palmata* ≤2%). This recent drastic change in composition is thought to relate to the devastation event that has affected acroporids throughout the Caribbean. By contrast, the overall prevalence of *A. palmata* in most parts of the Holocene section apparently expresses a continuity of benign conditions over periods of hundreds to thousands years. However, the abundance of *A. palmata* in the cores should not obscure the presence of significant hiatuses in its record from Buck island and many other Caribbean localities. There were apparently Caribbean-wide gaps in *A. palmata* growth from 5.9 to 5.2 ka and 3.0 to 2.2 ka respectively. These gaps are thought to have been caused either by disease or by bleaching. Similarly, in a lagoonal reef in Discovery Bay, in North Jamaica, Wapnick, Precht, and Aronson (2004) demonstrated that healthy *Acropora cervicornis* communities developed over the past 1.26 ka and there is no evidence of a near-surface, acroporid bed in the area. This suggests a loss of this coral for about the past three decades in response to both natural and anthropogenic impacts (hurricanes, white-band disease).

3.3.2. The Indo-Pacific Province

3.3.2.1. The Pleistocene

Although Pleistocene reefal remains are widespread in the Indo-Pacific province, they have received limited attention. There are few detailed studies of palaeoecology and distributional patterns of reef-building communities and little comparison of the coral fauna to that of the

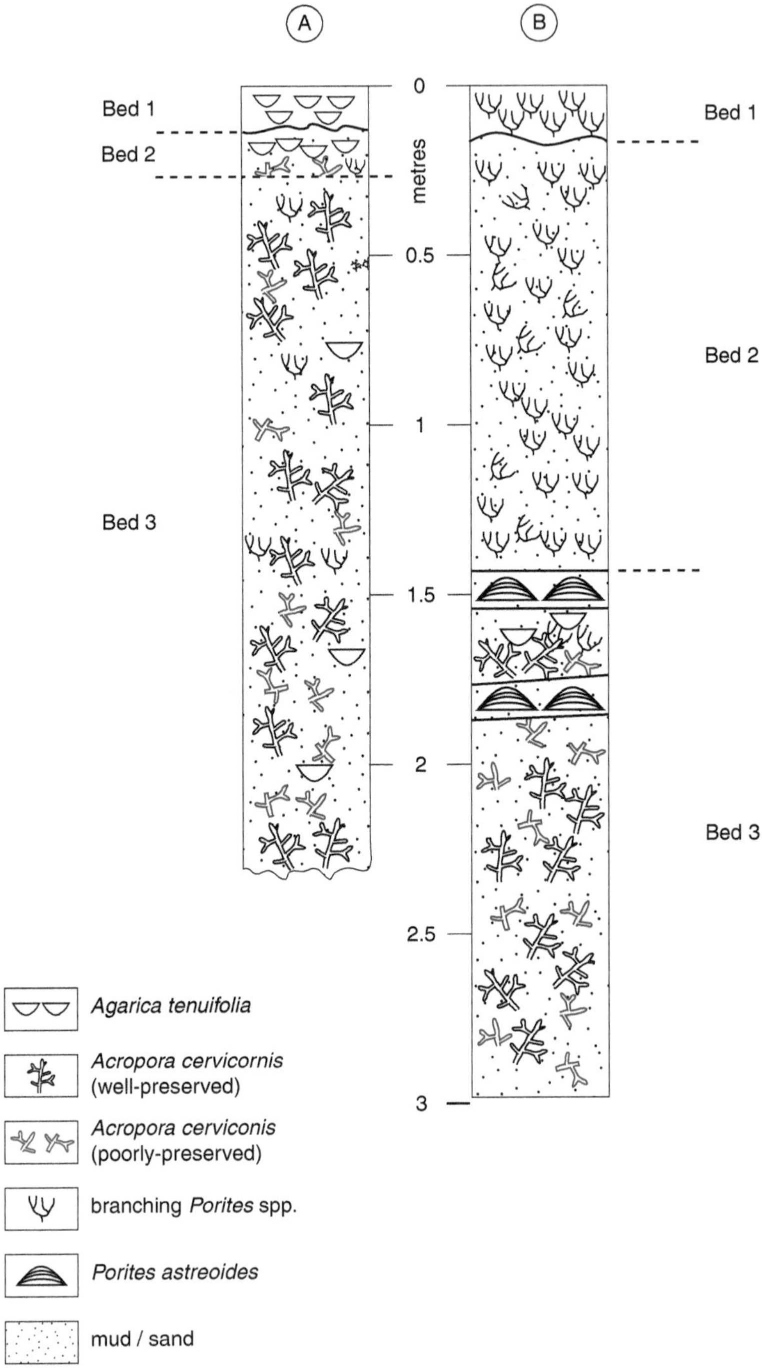
A
B
Bed 1
Bed 2
Bed 3
0
metres
0.5
1
1.5
2
2.5
3
Bed 1
Bed 2
Bed 3
Agarica tenuifolia
Acropora cervicornis (well-preserved)
Acropora cerviconis (poorly-preserved)
branching Porites spp.
Porites astreoides
mud / sand

western Atlantic. This may be due, at least in part, to the higher coral diversity in reef communities, thus making identification of coral colonies at the specific level difficult for most the taxa. In addition, most Pleistocene reef systems are accessible only in the subsurface. A selection of the most significant case studies is presented below.

Kenyan coast. Along the Kenyan coast, exposed reef terraces have been considered to reflect a single reef-building event which took place during the last interglacial highstand (*ca.* 125 ka) (Crame, 1980, 1981; Braithwaite, 1984). These terraces probably represent the remnants of the inner half (shallow-water, back-reef areas) of the original reef systems. The areas include small coral patches or isolated *Acropora*-rich banks. They resemble the quiet-water environments observed today in the inner zones of many western Indian ocean fringing reefs. The local occurrence of large individual colonies and knolls (up to 4 m high) suggests water depths exceeding 10 m in some places. Crame (1980, 1981) identified two distinct water depth-dependent coral successions (i.e. shallow-water *Acropora*-dominated and deeper-water *Porites*-dominated) from the Pleistocene sections (Figure 3.12). In a more general way, the shallow-water succession shows an early phase in which domal corals were dominating, followed by a later phase of arborescent and tabular-branching species. These successions were interpreted as expressing competitive interactions between individuals. Ecological successions also took place in the deeper-water settings (depths in excess of 10 m). Two distinct types of pioneering assemblages were recognized. One consists of domal corals, locally encrusted by thick veneers of coralline algae, while the other comprises robust-branching *Acropora* species. These early communities were overgrown prominently by foliaceous and encrusting *Pachyseris* and *Montipora.* By reference to modern analogues, both deeper instances can be related to former reef slopes. Locally, free-living ahermatypic scleractinians (*Heteropsammia, Heterocyathus*) were also encountered.

Figure 3.11 A comparative analysis of changes in the compositions of coral assemblages from core sections, in the central part of the shelf lagoon, of the barrier reef complex of Belize. (A) Idealized section extracted from 6 to 11 m water depth. Bed 1 = imbricated *Agaricia tenuifolia* plates; Bed 2 = *Agaricia tenuifolia* plates and fragments of poorly preserved *Acropora cervicornis* in mud; Bed 3 = well-preserved *Acropora cervicornis* in growth position along with pieces of *Agaricia tenuifolia* and branches of *Porites* spp. floating in a mud matrix. The replacement *of Acropora cervicornis* by *Agaricia tenuifolia* as the dominant species is interpreted as resulting from ecological disturbance. (B) Section extracted from 0.5 m water depth. Bed 1 = stands of *Porites divaricata*; Bed 2 = stands of *Porites divaricata* in sand; Bed 3 = stands of *Acropora cervicornis* together with some *Agaricia tenuifolia* plates and branches of *Porites* spp. in mud. This shift is regarded as a natural, shallowing-upward succession. Modified and redrawn from Aronson et al. (1998).

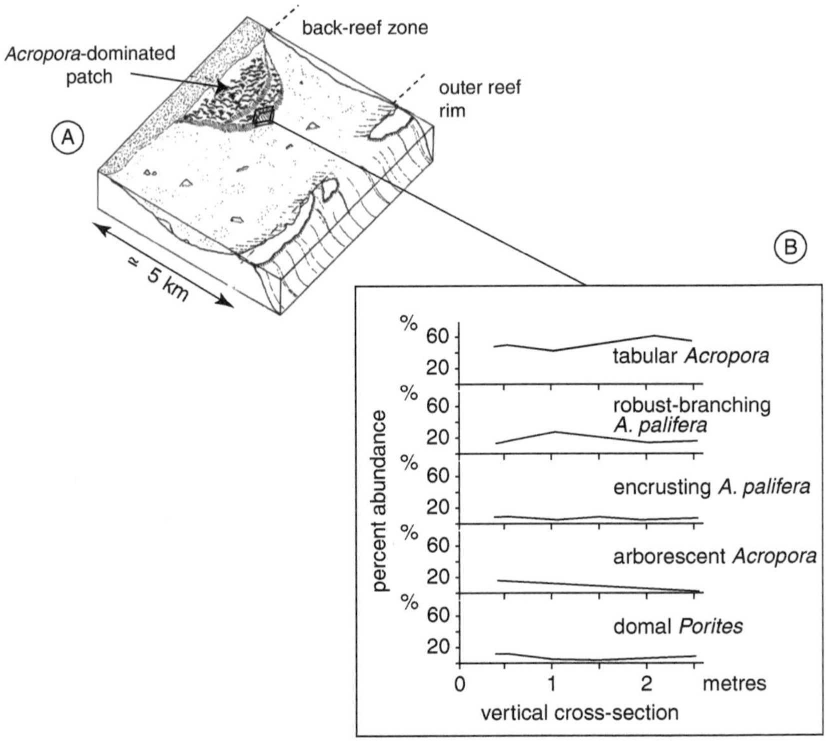

Figure 3.12 Composition and vertical distribution of coral assemblages from a section of the exposed last Interglacial reef terraces, Kenyan coast, east Africa. (A) A three-dimensional reconstruction of a portion of the Pleistocene barrier reef system with location of the studied *Acropora*-rich patch reef. (B) Abundance of the dominant coral forms expressed as percentages of the total coral fauna occupying successive half-metre intervals, along the lower part (about 10 m deep) of the flank of an inner patch reef. Modified and redrawn from Crame (1980).

In addition, Crame (1986) described the composition of the molluscan faunas which inhabited the late Pleistocene reef environments of the Kenya coast. Four molluscan assemblages were recognized and interpreted as deposited in sheltered leeward environments at depths ranging from 10 to 30 m.

(1) Participating in the shallow-water *Acropora*-dominated communities are two prominent groups, hard substrate-associated Trochacean gastropods and epifaunal bivalves, including Arcidae. Gastropods *Turbo argyrostomus* and *Trochus maculatus* are particularly abundant. The bivalves include mostly *Barbatia fusca, Cardita variegata, Semipallium radula, Decatopecten flabelloides, Cryptopecten pallium, Chlamys* spp. and occasional *Tridacna.* Assemblages rich in Trochacea and Arcidae are typical of shallow reef-flat settings. Locally in the arborescent acroporid

assemblage, conid gastropods (*Conus sponsalis* with *C. musicus*, *C. lividus*, *C. rattus* and *C. flavidus*) are very common. Conidae-dominated faunas are characteristic of near-surface *Acropora* habitats on modern Indo-Pacific reefs.

(2) Some portions of the inner palaeo-reef areas are composed mostly of deposits of the green algal *Halimeda* in which Turbinidae and Trochidae dominate with *Leptothrya filifera*, *Turbo argyrostomus*, *Trochus maculatus*, *T. flammulatus* and *Tectus mauritianus*. Cerithiidae (*Cerithium salebrosum*, *Rhinoclavis articulata* and *R. sinensis*) are common. Bivalves are scarce, mainly represented by *Barbatia fusca*. The relevant depositional environment was analogous to the lagoonal reef flats of some Pacific atolls.

(3) Associated with the deeper-water low-energy *Porites*-faviid-dominated community, a variety of molluscan assemblages reflect a wide range of environments. The Arcidae represent the dominant epifaunal bivalve family with *Arca ventricosa*, *Arca plicata* and *Barbatia helblingii*, whereas the infaunal bivalves are mostly represented by Veneridae (*Timoclea marica*, *Pitar* and *Comus platyaulax*). In sandy to gravelly bottoms between coral buildups, the fauna is predominantly composed of epifaunal bivalves (*Barbota helblingii*, *B. caelata*, *B. fusca*, *Arca navicularis*, *A. ventricosa* and rare *Tridacna gigas*), sessile bivalves (*Spondylus*, *Hyotissa hyotis* and *Chama*), infaunal bivalves (*Trachycardium*, *Pitar*, *Clementia papyracea* and *Periglypta puerpera*) and gastropods Cypraeidae (*Cypraea erosa*). All these assemblages show similarities with the bivalve-dominated assemblages characteristic of modern subtidal flats and shallow lagoons.

(4) In *Heteropsammia*- and *Heterocyathus*-rich sandy patches, the main feature is the prominence of gastropods: Strombidae (*Strombus gibberelus*) together with *Turbo*, *Terebra* and *Conus*. Truly epifaunal bivalves are restricted to a few *Barbatia*, *Gloripallium*, *Chlamys* and *Lima*. Semi-infaunal types are represented by *Modiolus*, *Pinna* and *Anadara*. The true infauna consists of limopsids, lucinids, cardiids, tellinids and venerids (*Trachycardium*, *Codakia*, *Tellinella*, *Circe*, *Fragum* and *Timoclea*).

Comparison between the fossil assemblages and modern molluscs living in the nearby fringing reefs indicates a drop in species diversity over the past 125 ka. In fact, soft substrate-associated species of both bivalves (venerids) and gastropods (strombids, mainly) have experienced a significant decline. This process was interpreted as resulting from a reduction in the number of habitat types through time.

Mauritius Island. Rising in the western Indian Ocean, the volcanic island of Mauritius shows two distinct generations of Pleistocene reefs (Montaggioni, 1982). Poorly preserved exposures of an older generation have been tentatively assigned to the penultimate interglacial period

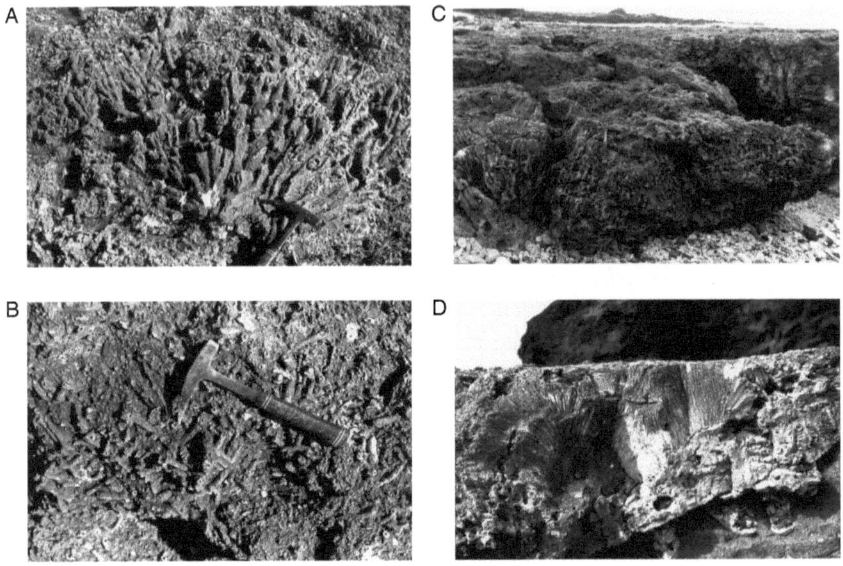

Figure 3.13 Coral forms from the late Pleistocene reefs of Mauritius Island (western Indian Ocean). (A) Robust-branching *Acropora cf. robusta*. (B) Arborescent *Acropora cf. muricata*. (C) Domal *Porites* sp.; domal *Goniastrea cf. retiformis* (photographs by L. Montaggioni).

(*ca.* 200–250 ka). The sections consist of domal faviid-dominated assemblages representing a sheltered reef-flat environment. The younger reef units, dated at 125 ka, have been referred to a single depositional event. The exposures are counterparts of modern fringing reef-flat zones with shingle spreads and boulder ridges. The community pattern is typified by the dominance of a low-diversity, robust-branching coral assemblage representing 60% of the total framework. Coral forms include *Acropora robusta* group, *A. cf. humilis* (both acroporid species representing 28% of the total framework), *Leptoria phrygia* (20%), *Goniastrea retiformis* and *Favites abdita* (both faviid species representing 9%), *Pocillopora damicornis* and *Porites* sp. (3%) (Figure 3.13). The remaining 40% consist of coralline algal crusts (*Hydrolithon*, *Lithophyllum*, *Lithothamnium* and subordinate *Lithoporella*, and *Sporolithon*). Associated encrusters can locally be volumetrically important (4–11% of the framework); they include foraminifera (*Miniacina*, *Carpenteria*), gastropods vermetids, bivalves (*Modiolus*) and bryozoans (*Cheilostomata*). Compositional evidence indicates this exposure can be attributed to a high-energy, outer reef-flat environment.

Western Australia. Last interglacial reef sequences have been drilled along the western coast of Australia, but the compositions of the relevant coral communities is poorly documented. On Ningaloo Reef, Collins, Zhu,

Wyrwoll, and Eisenhauer (2003) showed that fore-reef zones, at present overlain by their Holocene counterparts, are made up of two distinct assemblages. Domal *Porites* and *Montastraea* pioneers, thinly encrusted by coralline algae, have changed upwards through an arborescent *Acropora*-rich assemblage into an algal pavement. This replacement may reflect an upward-shallowing succession accompanied by an increase in water agitation through time.

Huon Peninsula. Along the northern coast of Papua New Guinea, in the Huon Peninsula, a series of uplifted reef terraces of late Pleistocene age contain a well-preserved coral fauna (Chappell, 1974). Pandolfi (1996, 1999) analysed the coral composition and species richness of a sequence that includes nine reef generations ranging in age from 125 to 30 ka. Most of the major reef-building episodes showed well-preserved distinct environments, from the bottoms to the tops of terraces, identified as lower fore-reef, upper fore-reef, reef crest and reef-flat respectively (Figure 3.14). These zones contain a total of 122 coral species. The lower reef slope, estimated to have developed between 30 and 20 m deep, is typified by the dominance of *Diaseris* spp., *Diploastrea heliopora* and *Favia matthai*. The upper reef slope, ranging approximately from 20 m to surface, comprises mostly *Acropora palifera*, *Favia pallida*, *Montastrea annuligera*, *Goniastrea retiformis*, *Platygyra pini* and domal *Porites* spp., together with occasional *Symphyllia agaricia* and

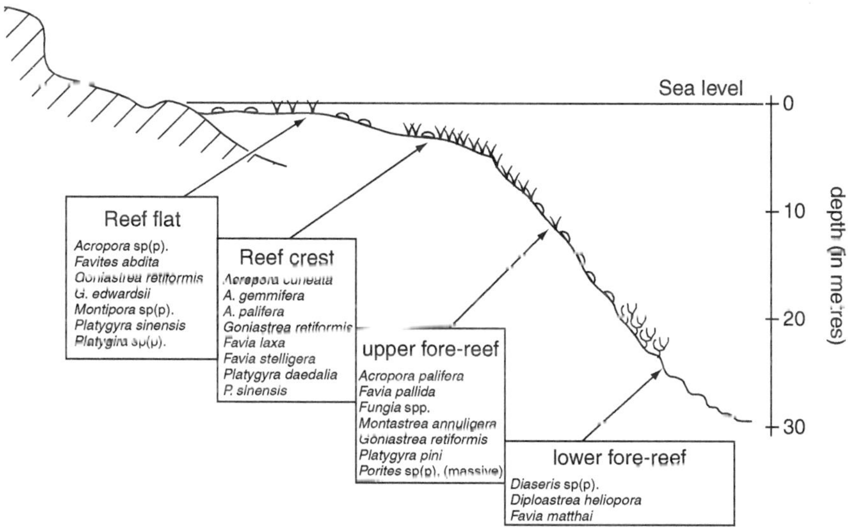

Figure 3.14 Reconstructed typical, coral reef zonation observed in the successive Pleistocene reef terraces, Huon Peninsula New Guinea. For each reef zone the most abundant coral taxa are indicated. Note overlap of some coral taxa may result from reworking and downslope displacement of coral colonies. Modified and redrawn from Pandolfi (1996).

S. recta. The reef crest is dominantly composed of *Acropora cuneata*, *A. gemmifera* (*humilis* group), *A. palifera*, *Goniastrea retiformis*, *Favia laxa*, *F. stelligera*, *Platygyra daedalea* and *P. sinensis*, locally with *Leptoria phrygia*, *Hydnophora microconos*, *Montastrea curta*, *Pocillopora* sp., *Acropora hyacinthus* and *Stylophora pistillata*. The reef flat is dominated by *Acropora* spp., *Favites abdita*, *Goniastrea retiformis*, *G. edwardsii*, *Montipora* spp., *Platygyra sinensis* and *Platygyra* spp. The compositions of these Pleistocene coral assemblages are similar to those described from the equivalent modern reef zones (Nakamori, Campbell, & Wallensky, 1995). Local differences in environmental parameters have played an important role in determining the composition of coral assemblages. These differences resulted in locally distinct populations. However, during the nine successive high stands over the 95-ka interval, populations display a clear constancy in both composition and species diversity especially in reef-crest and reef-slope settings. This constancy is considered due partly to the long-term prominence of similar few forms that continue to dominate in the nearby living reefs. The geographic range of the spatially prominent species appears to be no greater than that of the uncommon species that are also widespread.

The Ryukyus. The Ryukyu islands (southwestern Japan) also include sets of well-developed, raised Pleistocene reef limestones (so-called 'the Ryukyu Group') up to 50 m thick (Iryu, Yamada, Matsuda, & Odawara, 2006). Studies of the modern reef biota and associated sediments around the islands has allowed the assignment of different types of Ryukyu Group limestones to distinct depositional reef environments (Iryu et al., 1995; Nakamori, Iryu, & Yamada, 1995; and references herein). Sagawa, Nakamori, and Iryu (2001) analysed the compositions of coral assemblages from drillholes and outcrops on and off the small islands of Irabu-jima and Shimoji-jima. The limestones that constitute the core of both islands are shown to have been deposited during the early to middle Pleistocene (1.5–0.3 Ma) and represent a wide range of depositional environments, from shallow reef flat to deep fore-reef slope. Five coral assemblages have been identified; each typical of a particular reef zone (Figure 3.15):

(1) An assemblage dominated by foliaceous, encrusting and lamellar coral forms and including *Leptoseris yabei*, *L. hawaiiensis* and *L. papyracea* along with *Pachyseris speciosa*, *P. rugosa*, *Cycloseris* spp., *Diaseris* spp., *Zoopilus echinatus* and *Cyclarina lacrymalis*. This population is the analogue of the *Leptoseris scabra* community that is today observed along lower fore-reef zones at 30–50 m deep in the Ryukyus.
(2) An assemblage mainly composed of foliaceous, encrusting and laminar corals, such as *Oxypora* spp., *Pectinia* spp. and *Mycedium* spp. By analogy to the present-day *Oxypora lacera* community found in the same area,

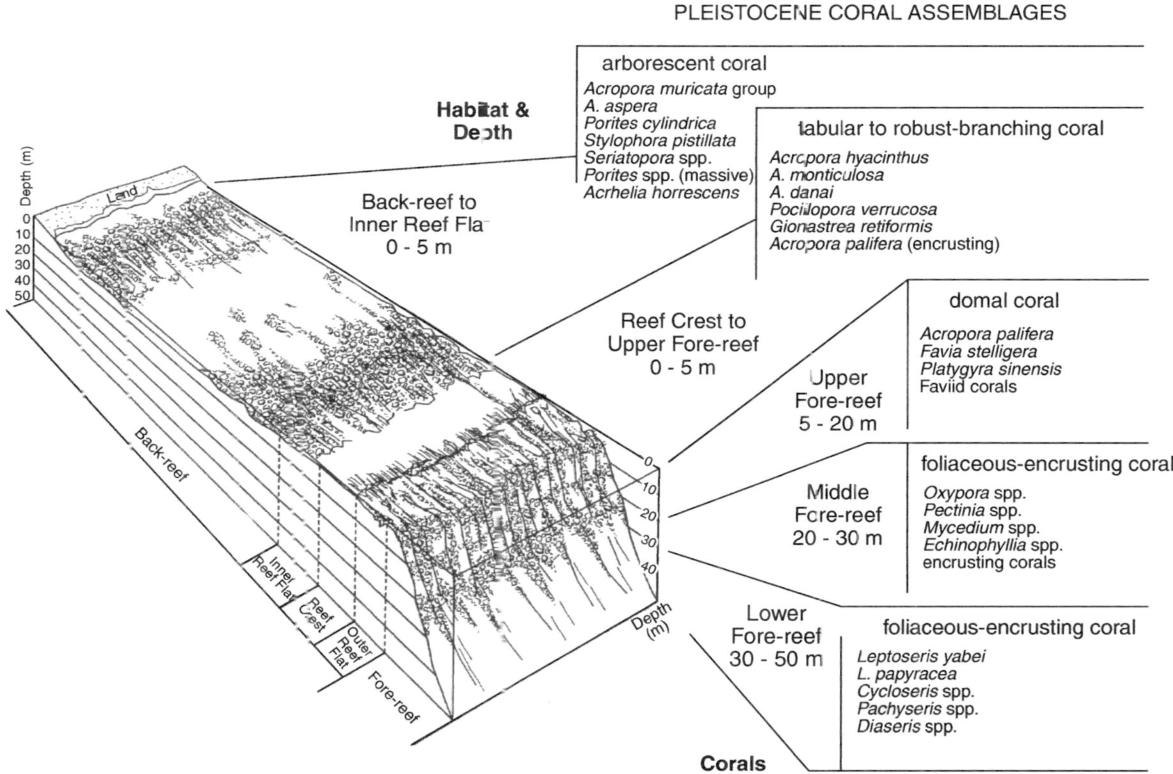

Figure 3.15 Reconstructed, typical coral reef zonation observed in the Pleistocene reef exposures from the Ryukyu Group, Japan. For each reef zone the composition of the coral assemblages with the most abundant coral taxa is indicated. Modified and redrawn from Sagawa et al. (2001).

this population is regarded as characteristic of middle fore-reef slopes at depths of 20–30 m.

(3) An assemblage consisting mostly of domal faviid corals (*Favia stelligera*, *Platygyra sinensis*, *P. ryukyuensis* and *Favites* spp.), together with *Acropora palifera*. The modern counterpart is represented by the *Favia stelligera* community, inhabiting upper fore-reef zones (0–10 m deep).

(4) An assemblage rich in tabular- and robust-branching corals including species of *Acropora hyacinthus* and *A. humilis* groups respectively. These are comparable to the modern *A. hyacinthus* and *A. aspera* communities encountered on reef-crest and upper fore-reef zones at depths of 0–5 m.

(5) An assemblage dominated by arborescent corals such as *Acropora muricata* (formerly *A. formosa*) and *A. aspera* groups. Associated species are *Stylophora pistillata* with *Porites* spp., and *Acrhelia* (*Galaxea*) *horrescens*. This population is analogous to the modern *Montipora digitata*, *Porites cylindrica*, *Porites nigrescens* and *Acropora aspera* communities that inhabit inner reef-flat to shallow back-reef zones (less than 5 m deep) in the Ryukyus.

Whatever the age of the coral limestone unit considered, stability in taxonomic composition and species diversity of the coral assemblages seem to have been maintained through a 1.2 Ma interval in spite of the repeated, numerous falls in sea level and sea surface temperature. But, given the lack of quantitative data, it is difficult to propose hypotheses concerning possible species extinctions or rarefactions.

Great Barrier Reef of Australia. In eastern Australia, the composition of coral assemblages during the Pleistocene development of the Great Barrier Reef (GBR) has been analysed by Webster and Davies (2003) on the basis of two cores extracted from outer- and inner-shelf reefs (Ribbon Reef 5 and Boulder Reef) respectively. Braithwaite et al. (2004) gave a simplified picture of the coral distribution in Core Ribbon Reef 5. Regarded as initiated at approximately 600 ka (Alexander et al., 2001; Braithwaite et al., 2004; Obrochta, 2004; Dubois, Kindler, Spezzaferri, & Coric, 2008), the GBR developed through at least five successive reef-building episodes separated by coralline alga-dominated (rhodolithic) depositional events. In Ribbon Reef 5 core (210 m long), the earliest coral-rich unit is found at around 130 m below the present reef surface (Figure 3.16). The base of the Boulder Reef core (86 m below reef surface), typified by the occurrence of coral-bearing beds, gives an age of 210 ka. In the reef units, three major coral assemblages were identified, each representing a distinct reef environment.

(1) An assemblage typified by the prominence of robust-branching forms including species of *Acropora humilis* (*A. monticulosa*) and *A. robusta* (*A. robusta*, *A. palmerae*) groups together with *A. palifera*, *Stylophora pistillata*, *Pocillopora damicornis* and *P. verrucosa*. Associated corals are faviids

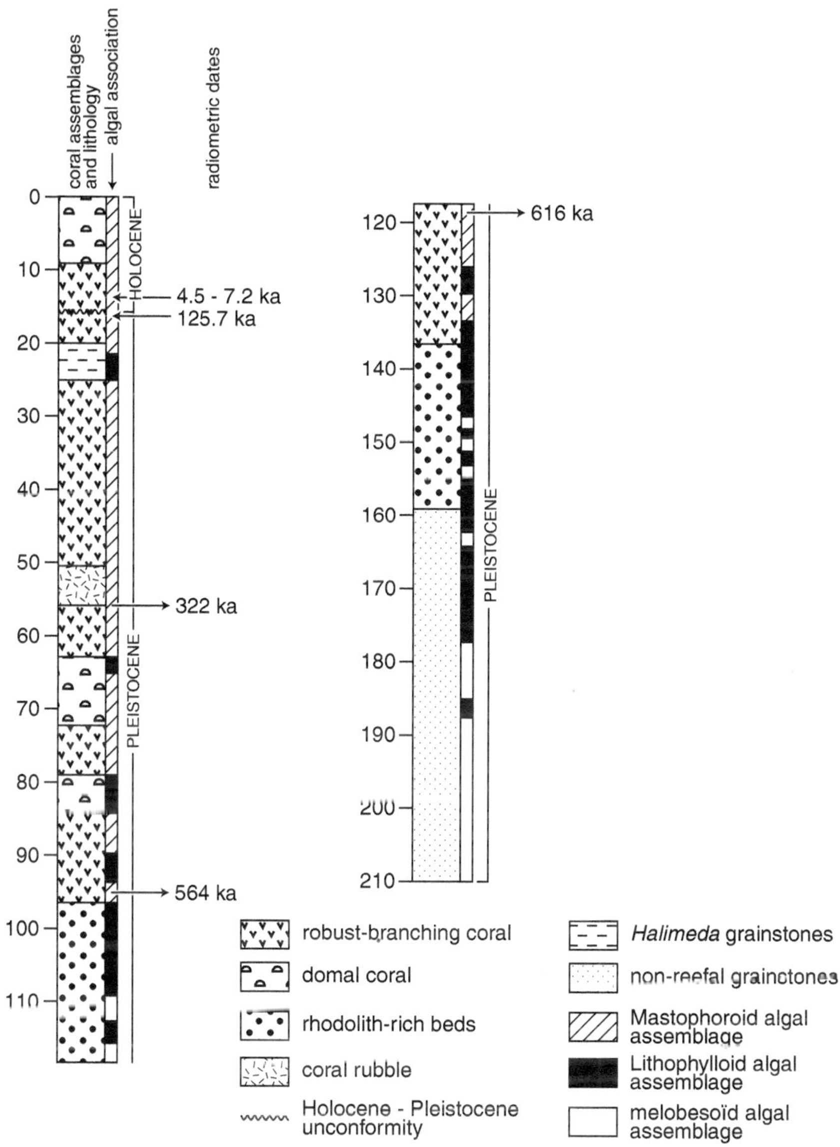

Figure 3.16 Log summarizing the lithology and composition of coralgal assemblages in the Ribbon Reef 5 core (Central Great Barrier Reef of Australia). The successive unconformities within the Pleistocene sequence are not shown. Radiometric dates are expressed in radiocarbon years for the Holocene sequence, and are derived from $^{234}U/^{238}U$ values for the Pleistocene sequence. Adapted from Webster and Davies (2003).

(*Goniastrea*). By reference to modern, northeastern Australian counterparts, this assemblage is considered to have inhabited high-energy, reef-crest to upper fore-reef zones (i.e. outer reef margin) at depths of less than 10 m and was found in the core from Ribbon Reef 5.

(2) An assemblage mostly composed of domal forms such as *Porites* (*P. cf. lutea*, *P. cf. solida*) and faviids (*Favia pallida*, *Favites flexuosa*, *F. chinensis*, *Leptoria phrygia*, *Leptastrea cf. purpurea*, *L. transversa*, *Platygyra daedalea*, *Cyphastrea* sp., *Echinopora pacificus*, *E. cf. lamellosa*, *E. cf. gemmacea*) with common encrusting (*Porites cf. lichen*, *Montipora* sp., *Pachyseris speciosa*, *P. rugosa*, *Hydnophora exesa*, *Pavona varians*, *P. venosa*) and occasional gracile branching forms (*Acropora horrida* group). This suggests that the relevant habitat was typified by lower water-energy and/or deeper conditions (i.e. inner reef margin on an outer-shelf reef) when compared to that of the robust-branching assemblage. These domal coral populations were present in the Ribbon Reef 5 core.

(3) An assemblage also dominated by domal poritids (*P. australiensis*, *P. cf. cylindrica*, *P. murrayensis*, *Goniopora*) but in association with faviids (*Cyphastrea microphthalma*, *Echinopora mammiformis*, *E. hirsutissima*), and devoid of encrusting forms. Subordinate forms include *Pocillopora verrucosa*, *Stylophora pistillata* and *S. hystrix*. The habitat is thought to reflect a lower-energy, higher-turbidity setting like a leeward reef-flat zone on an inner-shelf reef. This assemblage is typical of the Boulder Reef core.

On Ribbon Reef 5, temporal changes in the coral assemblages were expressed by transitions from the robust-branching assemblage to a domal assemblage and in reverse. The coralline algal assemblages in successive reef generations experienced similar variations in composition (Braga & Aguirre, 2004). Three major coralline algal assemblages were identified: a mastophoroid assemblage, typical of the shallowest reef environments, a lithophylloid assemblage, mainly occurring in deeper reef settings and a melobesoid assemblage, mainly occurring in open-shelf environments. The robust-branching coral-dominated communities are thickly encrusted by the mastophoroids *Hydrolithon onkodes* and *Neogoniolithon fosliei* with minor occurrences of *Lithophyllum* and *Sporolithon*, while the domal coral community includes the thinner thalli of the lithophylloids, the *Lithophyllum pustulatum* group, in association with a number of other *Lithophyllum*, *Mesophyllum* and *Sporolithon* species. The coralgal alternations may represent the response of reef growth to a variety of environmental constraints. The most conspicuous feature is the repeated occurrence of both assemblages downcore in the different reef units over approximately 600 ka beneath Ribbon Reef 5. This is interpreted as expressing the remarkable constancy in the taxonomic composition of the same coralgal assemblages since the initiation of the GBR, at least on the outer shelf, despite being affected by numerous cycles of severe climate perturbations. But, as in the Ryukyus, it is not possible to predict possible changes in the regional species pool.

Henderson Island. In southeast Polynesia, Henderson Island (Pitcairn Group) is one of the easternmost islands of the west Indo-Pacific province. It forms an uplifted atoll-like limestone island of late Pleistocene age on which fossil reef marginal and lagoonal coral populations, dated at about 330–630 ka (Stirling et al., 2001), have been well preserved. Paulay and Spencer (1988) have described the compositional traits of these populations. A total of 26 scleractinians were identified from the different fossil reef zones. At least eight of the fossil species found have not been reported in the modern reefs of the island. Three of the eight regarded as locally extinct forms (*Pocillopora damicornis, Favia stelligera, Fungia scutaria*) still live on nearby islands. Fossils of the genus *Leptoria* are present on Henderson but are apparently unknown in the Pitcairn Group. A total of 44% of fossil corals are not found living on Henderson. The conspicuous high turnover rate is interpreted as expressing long-term variations in species composition and a lack of stability in the site since the middle (?) Pleistocene. This may be due to the marginal biogeographic location of the island, making both the survival and repeated settlement of species difficult.

3.3.2.2. The latest Pleistocene to Holocene

Current knowledge of the composition of reef communities since the Last Glacial Maximum (LGM, i.e. the past 24 ka) is derived largely from drilling investigations (see Montaggioni, 2005, for review). In the Indo-Pacific, up to seven hundred subsurface boreholes have penetrated about 80 modern reefs and exceptionally through recently submerged reefs. Additional information on coral assemblages comes from scattered, uplifted Holocene reef sections (Figure 3.17). But, quantitative data are scarce and the collected biota has been identified at a variety of taxonomic levels, making inter-site comparisons of community composition difficult. However, three models of ecological successions may be defined from the analysis of cores and outcrops.

The first model relates to sections that exhibit a single coral assemblage from the initiation stage to the reef top (Figure 3.18). These sections are generally found beneath modern exposed or sheltered reef crest/flat and fore-reef environments that began to accrete 10–7 ka ago. This model is illustrated by the13-m thick sequence extracted from the Toliara barrier reef-flat (southwest of Madagascar) approximately dated 7.6 ka at base (Camoin, Montaggioni, & Braithwaite, 2004). The coral-dominated community is typical of high-energy reef-margin settings and consists of robust-branching forms (*Acropora robusta* group, mainly) associated with *A. humilis* group, *Pocillopora cf. verrucosa*, *P. eydouxi* and occasional domal

Figure 3.17 Close-up of a Holocene (about 6 ka) reef exposure showing a community dominated by tabular-branching *Acropora hyacinthus* group along with robust-branching *Acropora palifera* and faviids, Huon Peninsula, Papua New Guinea (photograph by L. Montaggioni).

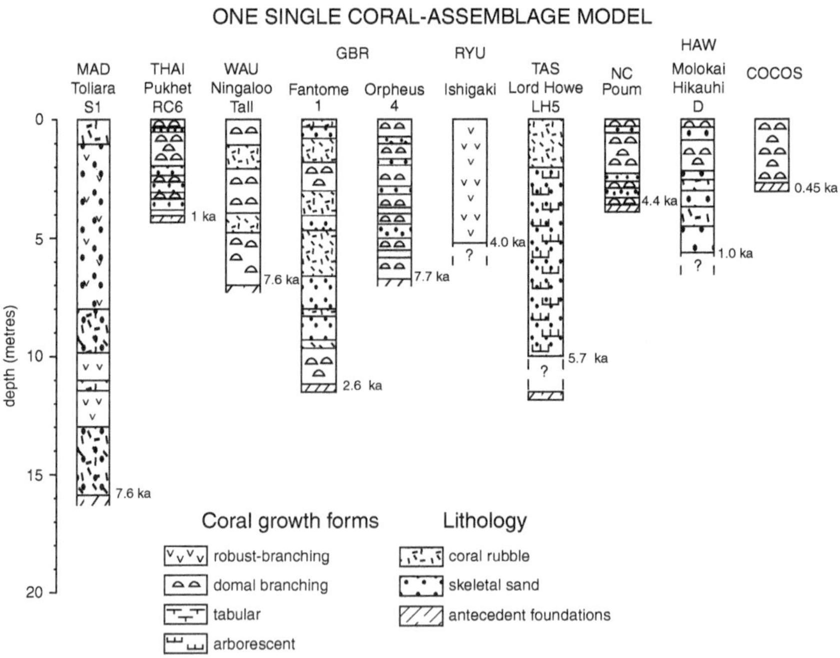

Figure 3.18 Core logs showing Holocene sequences composed of single coral assemblages from Indo-Pacific reefs. These reflect constant environmental conditions during vertical reef accretion. MAD = Madagascar; THAI = Thailand; WAU = Western Australia; GBR = Australian Great Barrier Reef; RYU = Ryukyu Islands; TAS = Tasmanian Sea; NC = New Caledonia; HAW = Hawaiian Islands. Numbers and letters refer to a specific core extracted from a given reef site. From Montaggioni (2005).

Favia cf. stelligera, *Diploastrea* and *Heliopora*. Colonies are thickly encrusted with coralline algae and foraminifera (Homotrematidae). Molluscs include the bivalves *Arca plicata*, *Ctenoides annulatus* and the gastropods *Turbo* sp., *Cypraea nucleus* and *Columbella turturina*, all characteristic of upper fore-reef environments. A single-community model has also been described from protected inshore areas like the Panwa Peninsula in south Thailand (Tudhope & Scoffin, 1994; Scoffin & Le Tissier, 1998). There, the coral assemblages form 2–4 m thick sequences deposited from 5.7 ka. Corals mostly include domal poritids and faviids (*Porites lutea*, *P. lobata*, *Goniastrea aspera*, *G. retiformis*, *G. favulus*, *Platygyra sinensis*, *P. daedalea*, *Favites abdita*, *F. chinensis* and *Favia* spp.) together with domal agaricidae (*Coeloseris mayeri*), various branching acroporidae (*Acropora nobilis*, *Montipora digitata*) and siderastreidae (*Psammocora digitata*). Bivalves such as *Pendum spondyloideum*, *Arca ventricosa* and *Barbatira helbingii* are common. This community is typical of back-reef zones subjected to heavy mud loading (15–20 mg l^{-1} of suspended sediment). The single-community model typifies many other reef localities such as Ningaloo Reef on the Western Australian margin (Collins et al., 2003), Fantome (Johnson & Risk, 1987) and Orpheus Islands on the Australian GBR (Hopley, Slocombe, Muir, & Grant, 1983), Lord Howe Island from western Pacific (Kennedy & Woodroffe, 2000), Ishigaki Island in the Ryukyus (Yamano, Kayanne, & Yonekura, 2001, 2003), Poom reef in New Caledonia (Cabioch, Montaggioni, & Faure, 1995), the Hawaiian archipelago (Grigg, 1998; Engels et al., 2004) and Cocos Island, eastern Pacific (Macintyre et al., 1992). Data from fore-reef-zones are rare. Collins et al. (2003) carried out drilling operations through the fore-reef slopes of Ningaloo Reef (Western Australia), demonstrating the continuous development of a 7.5 m thick domal *Porites*-dominated assemblage with thick coralline algal encrustations, over the past 7.6 ka at depths of 10–35 m relative to the present sea surface. Such homogeneous compositions of reef community within a given sequence probably reflect the persistence of ambient conditions from the earlier stages of colonization to upward coral growth at the stillstand. In open-sea-facing settings, the initial colonizers maintained pace with the rising sea level until stabilization. In areas affected by high mud input, only siltation-tolerant communities such as those dominated by poritids and faviids, could have grown.

The second model of ecological succession relates to the stacking-up of two distinct coral assemblages in a given ocean-facing sequence. In most instances, a deeper water, lower-energy, coral assemblage is replaced upwards by a shallower, higher-energy coral-dominated community. On modern margins in exposed or semiexposed sites, the bases of Holocene sequences consist of either domal poritids and faviids, or arborescent and tabular acroporid frameworks, representing the pioneering assemblages that started to grow at depths of from 10 to more than 20 m, as indicated the present thicknesses of the sequences. The overlying assemblage is usually

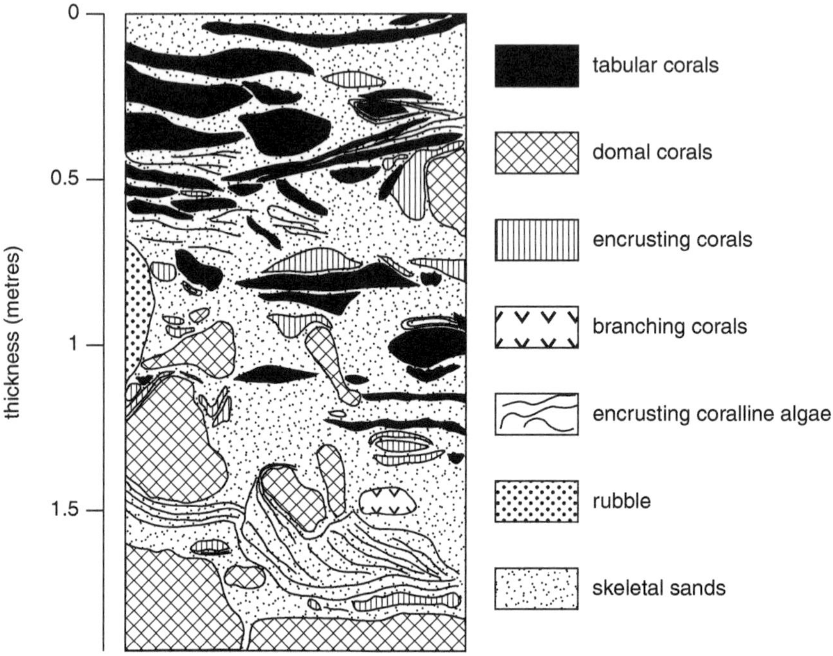

Figure 3.19 Composition of coral assemblages from the section of a Holocene reef exposure, Nakazato area, southwest of Kika-jima Island (Ryukyus, Japan). The domal coral forms mainly include *Porites* spp. and faviids. The tabular growth forms are dominated by *Acropora* spp. This succession is interpreted as shallowing-upwards. Modified and redrawn from Webster et al. (1998).

composed of robust acroporids and pocilloporids thickly encrusted by coralline algae (*Hydrolithon* mainly) or by algal pavements, representing both the shallowest and highest energy communities. In sequences from more protected reef rims and/or affected by high turbidity, the lower sections comprise domal poritids and faviids or foliaceous *Montipora*/agariciids originally settled at depths of not more than 15 m. The upper sections are dominated by domal poritid/faviid growths or tabular to arborescent acroporid assemblages. One of the best illustrations of this model is provided by Webster, Davies, and Konishi (1998) from the analysis of both boreholes and exposures on the raised fringing reefs of Kikai-jima Island (Central Ryukyu Islands, Japan) (Figure 3.19). These reefs form four distinct, step-like terraces around the island and developed at sea levels of 9–6, 6–3.4, 3.8–2.6 and 2.9–1.6 ka respectively. The modern reef started to grow approximately 1.6 ka ago. Four distinct upper fore-reef slope to shallow reef-flat and one deeper fore-reef coral assemblage were delineated in which a total of 30 coral genera and 70 species were identified. After comparison with modern assemblages determined from nearby reefs, their likely environmental settings were recognized. On the outermost sections of terraces located along the windward coast, the common distribution pattern consists of a significant upward decrease in domal colonies and a correlative increase in tabular- and robust-branching coral forms. The most important initial colonizers directly growing on the Pleistocene basement include domal poritids (*Porites lutea, P. lobata* or *P. australiensis*) and faviids (*Leptoria phrygia, Favia pallida, Goniastrea retiformis, Montastraea* sp., *Favites* sp., *Platygyra* spp., *Cyphastrea microphthalma, C. seralia*). The domal coral-dominated assemblage was regarded as settled at depths ranging from 5 to 10 m. These corals are overlain by an *Acropora*-dominated assemblage, mostly composed of *A. hyacinthus, A. humilis* groups, *A. palifera* and *A. monticulosa.* The taxonomic replacement not only represents a significant change in growth shapes but is accompanied by a decline in generic richness. The assemblage predominantly composed of tabular- and robust-branching corals was regarded as grown at depths of less than 5 m. This successional pattern, was therefore interpreted as a shallowing-upward sequence in a high-water-energy environment. Similarly, in areas subjected to high turbidity, a two-phase succession is observed locally. Domal poritids and faviids are again prominent in the lower unit, while the near-surface interval is dominated by domal and columnar coral forms (*Goniastrea retiformis, Favites* sp., *Montipora* sp., *Acropora* sp., *Millepora exaesa* and *Heliopora coerulea*). Similar upward-shallowing successions have also reported from Mayotte in the Comoro Islands (Camoin et al., 1997, 2004), Mahé in the Seychelles (Braithwaite et al., 2000), the Houtman Abrolhos Islands in southwestern Australia (Collins et al., 1993), Kume Island in the Ryukyus (Takahashi, Kobe, & Kan, 1988), Koror in the Palau Islands (Kayanne, Yamano, & Randall, 2002), Guam in the Marianas (Kayanne, Ishii, Matsumoto, & Yonekura, 1993), Mangaia in the Cook Islands (Yonekura et al., 1988), several outer- and mid-shelf reefs of the Australian GBR, among them Yonge, Myrmidon and Stanley Reefs (Hopley, Smithers, & Parnell, 2007), Ribbon 5 Reef (Webster & Davies, 2003) and One Three Reef (Marshall & Davies, 1982) (Figure 3.20). Coral successions typified by shallow-water assemblages overlain by deeper ones (upward-deepening sequences) can also be found locally. Cabioch et al. (2003) described the composition of coral-coralline alga-dominated assemblages in cores extracted from the uplifted reef terrace of Urélapa Island (Vanuatu, southwest Pacific). The core sequences record reef growth history from 23 to 6 ka. Two distinct coral assemblages have been recognized (Figure 3.21). Overlying the antecedent foundation at core depths ranging between 61 and greater than 90 m, the earlier assemblage developed from 23 to about 11.5 ka. It is composed principally of robust-branching *Acropora* spp. associated with scattered domal faviids intensively encrusted by the coralline algae *Hydrolithon cf. onkodes, Dermatolithon cf. tesselatum, Lithophyllum cf. molluccense* and *Neogoniolithon cf. fosliei.* This assemblage occurs preferentially in shallow, high-hydrodynamic-energy reef margin settings (0–6 m deep). An assemblage, consisting dominantly of domal *Porites* (*P. lutea, P. lobata*) together with

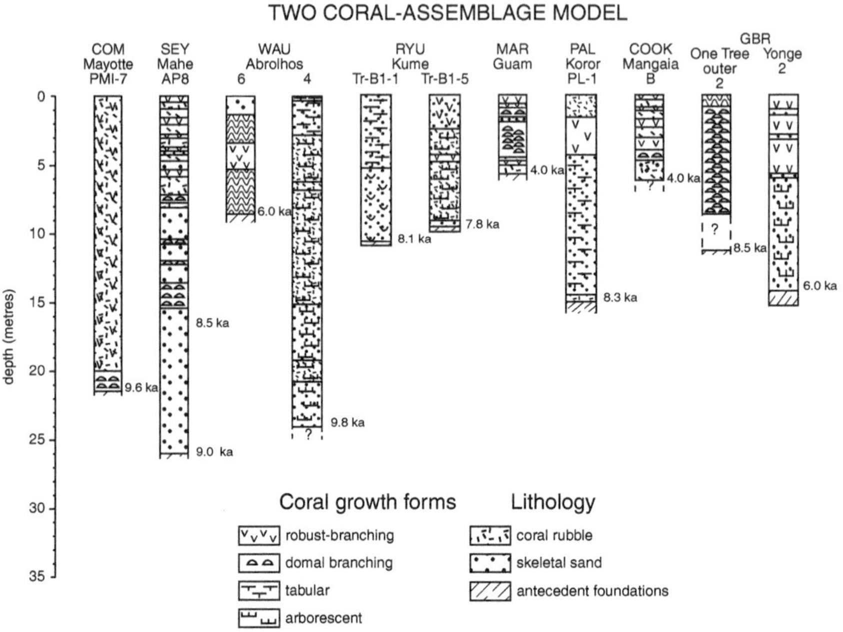

Figure 3.20 Core logs showing Holocene Indo-Pacific reef sequences composed of two different coral assemblages that reflect shallowing-upward successions. Note deeper, lower-energy coral assemblages are overlain by shallower, higher-energy corals. COM = Comoro Islands; SEY = Seychelles Islands; WAU = Western Australia; RYU = Ryukyu Islands; MAR = Mariana Islands; PAL = Palau Islands; GBR = Australian Great Barrier Reef. Numbers and letters refer to a specific core extracted from a given reef site. From Montaggioni (2005).

occasional branching acroporids and thin incrustations of *Lithophyllum* sp., *Mesophyllum* sp. and *H. cf. onkodes*, occupies the upper sections from about 70 to 61 m core depth. It was flourishing between around 11.5 and 6 ka, inhabiting a lower-energy environment 10–20 m deep. The upward replacement of shallower by deeper coral forms is interpreted as reflecting an abrupt deepening and subsequent decrease in wave energy, probably linked to a rapid jump in sea level.

The third model of ecological successions was reported from cores that exhibit recurrent alternations of shallower, higher-energy and deeper, lower-energy coral assemblages. Frequently, such composite successions are found beneath reef margins or reef flats that have developed over periods of about 10,000 years. For example, the reef crest pile from the outer barrier of Tahiti Island is composed of 1–10 m thick, alternating assemblages dominated by either shallower *Acropora robusta* group or deeper, lower-energy tabular *A. cytherea* group, arborescent *A. clathrata*, domal *Porites* spp. (*P. lobata*, *P. lutea*), encrusting *P. lichen* and arborescent *P. nigrescens* (Cabioch, Camoin, & Montaggioni, 1999). This type of succession is also

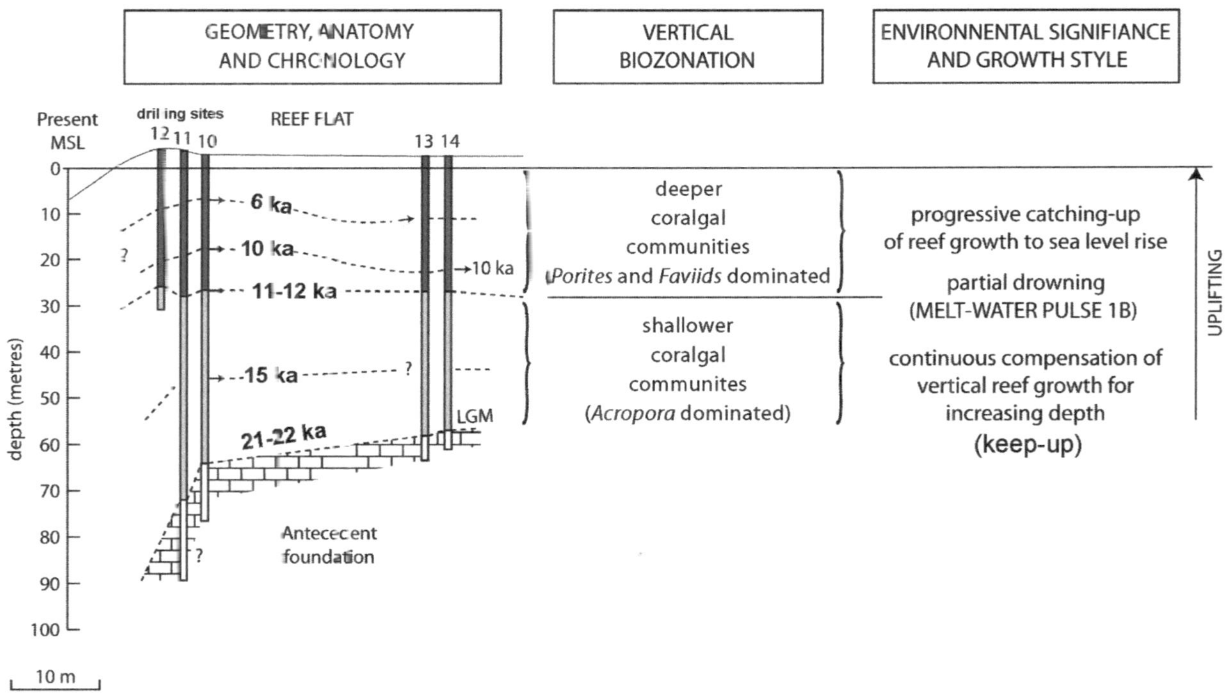

Figure 3.21 Core log showing a typical deepening-upward sequence from an exposed Holocene reef terrace, Urélapa Island, Vanuatu. The robust-branching coral-dominated assemblage in the lower section is overlain by a domal coral assemblage from about 70–60 m core depth to the top. Adapted from Cabioch et al. (2003).

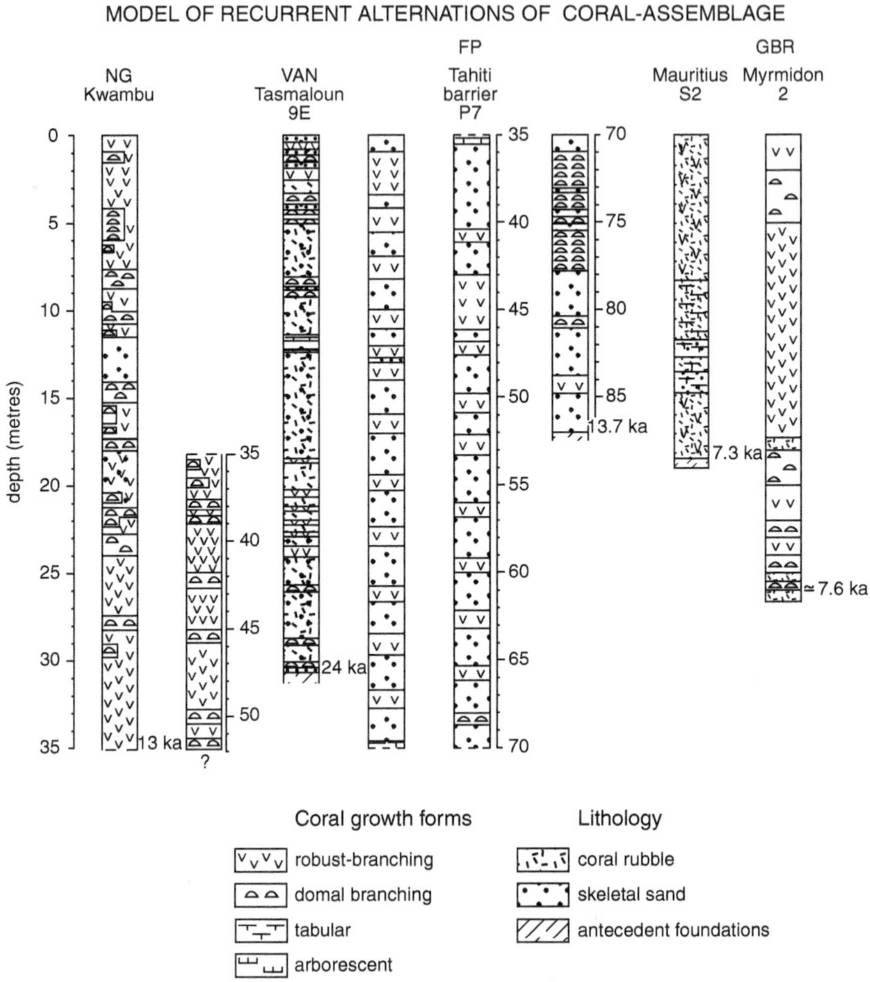

Figure 3.22 Core logs showing late Pleistocene to Holocene Indo-Pacific reef sequences composed of recurrent alternations of different coral assemblages. These successions reflect repetitive changes in environmental conditions during vertical reef accretion. NG = New Guinea; VAN = Vanuatu Islands; FP = French Polynesia; GBR = Australian Great Barrier Reef. Numbers and letters refer to a specific core extracted from a given reef site. From Montaggioni (2005).

exemplified by fringing reefs from Kwambu (Huon Peninsula in New Guinea; Chappell & Polach, 1991), Tasmaloun (Vanuatu Islands; Cabioch et al., 1998), Pointe-au-Sable (Mauritius Island; Montaggioni & Faure, 1997) and a number of mid- and outer-shelf reefs on the Australian GBR (among others are Stanley, Wheeler, Cockatoo, Myrmidon and Viper Reefs; Hopley et al., 2007) (Figure 3.22). The repeated abrupt replacement of coral assemblages during reef accretion is likely to reflect rapid lateral displacement of communities across the drilling sites in response to changes in ambient environmental conditions, like changes in water energy in relation to acceleration or slow-down in the rise of sea level or changes in turbidity.

3.3.2.3. The recent past

The effects of disturbing events on the structure of Indo-Pacific reef-coral communities are poorly documented from the subfossil record. However, although they are insufficient to explicitly test hypotheses about the changing scale of biotic turnover, they may demonstrate coral growth under changing disturbance conditions through time.

Pandolfi et al. (2006) presented a case study of coral mass mortality from the Holocene exposed fringing and barrier reefs of the Huon Peninsula (Papua New Guinea). This study has important implications for testing the potentiality of recent reefs to survive major disturbance. The survey of nine step-like reef terraces, ranging in age from 11 to 3.7 ka, indicates that several mass depletion episodes occurred at a frequency of less than 1 per 1.5 ka. The most severe destruction event, reflected in the simultaneous death of up to 90% of the corals, took place *ca.* 9.4–9.1 ka and was caused by the deposition of volcanic ash. A phase shift from coral- to algal-dominated assemblages immediately followed this event. But, the re-settlement of coral communities and subsequent reef growth were rapid, within less than a century interval. This means that rapid recolonization can quickly restore the functional abilities of reef communities following disturbance. However, the post-disruption reef communities display marked differences, compared to their pre-disruption analogues (Figure 3.23). *Acropora palifera* stands and arborescent *Acropora* predominated originally but decreased significantly in abundance after the volcanic event. By contrast, *A. hyacinthus* and *A. humilis* groups, formerly uncommon, became prominent after reef recovery. The coralline alga *Hydrolithon onkodes* was abundant before and during the disturbance, but poorly represented afterwards. The renewed community structure persisted in part for about 2 ka after reef rejuvenation.

On Réunion Island, most fringing reefs have suffered from eutrophication since the end of the eighties, resulting in rapid phase shifts from coral- to coralline alga-dominated communities. Subsequently, fleshy algae, and locally cyanobacterial films, increased opportunistically and became the dominant surface cover (Montaggioni, Cuet, & Naïm, 1993; Chazottes, Le Campion-Alsumard, Peyrot-Clausade, & Cuet, 2002). The core extracted from the Trou d'Eau (western Réunion Island) showed a continuous deposition of relatively well-preserved debris of arborescent *Acropora cf. muricata* (Montaggioni, 1977) (Figure 3.24). This suggests constancy of coral growth for at least the last 8 ka until it was interrupted, and supports the idea that the current disturbed state is unusual.

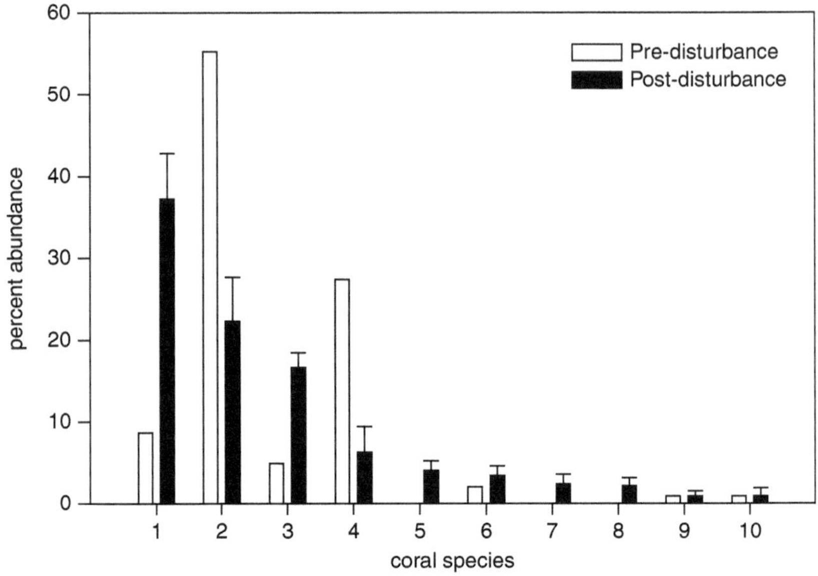

Figure 3.23 Percent abundance of coral species (relative to the total species in the coral assemblages) prior to and after the 8.5 ka mass-mortality event, Bonah River section, Huon Peninsula, Papua New Guinea. Composition of coral assemblages: 1 = *Acropora hyacinthus* group; 2 = arborescent *Acropora*; 3 = *A. humilis* group; 4 = *A. palifera*; 5 = *Stylophora*; 6 = *Pocillopora*; 7 = *A. robusta* group; 8 = *Heliopora*; 9 = *Porites* spp.: 10 = *Galaxaea*. Vertical bars refer to mean standard error. Modified and redrawn from Pandolfi et al. (2006).

In southern areas of the South China Sea, on the basis of high-precision U/Th dating (accuracy: up to ±1 to 2 yr) of coral colonies, Yu et al. (2006) demonstrated that at least six large-scale regional mass-mortality events have occurred over the past two centuries. Destruction especially affected *Porites*-dominated assemblages. Most of these events are postulated to have been caused by high-temperature bleaching during El Niño years (Figure 3.25). More recently, the bleaching event of 1998 was accompanied by marked changes in the composition of coral assemblages from shallow subtidal settings along the east coast of Bali Island, Indonesia (Piller & Riegl, 2003). The composition of the pre-disturbance coral assemblage was typified by the dominance of almost monospecific thickets of the arboresecent *Acropora cf. vaughani*. Most of the *Acropora* colonies died after bleaching and were overlain by foliaceous growths of almost monogeneric *Montipora*, which in turn were covered and rapidly settled by an encrusting *Montipora* species. The *Montipora* cover is now protecting the dead acroporid framework against mechanical and biological destruction. Thus, such a turnover event has the potential of being preserved in the fossil record.

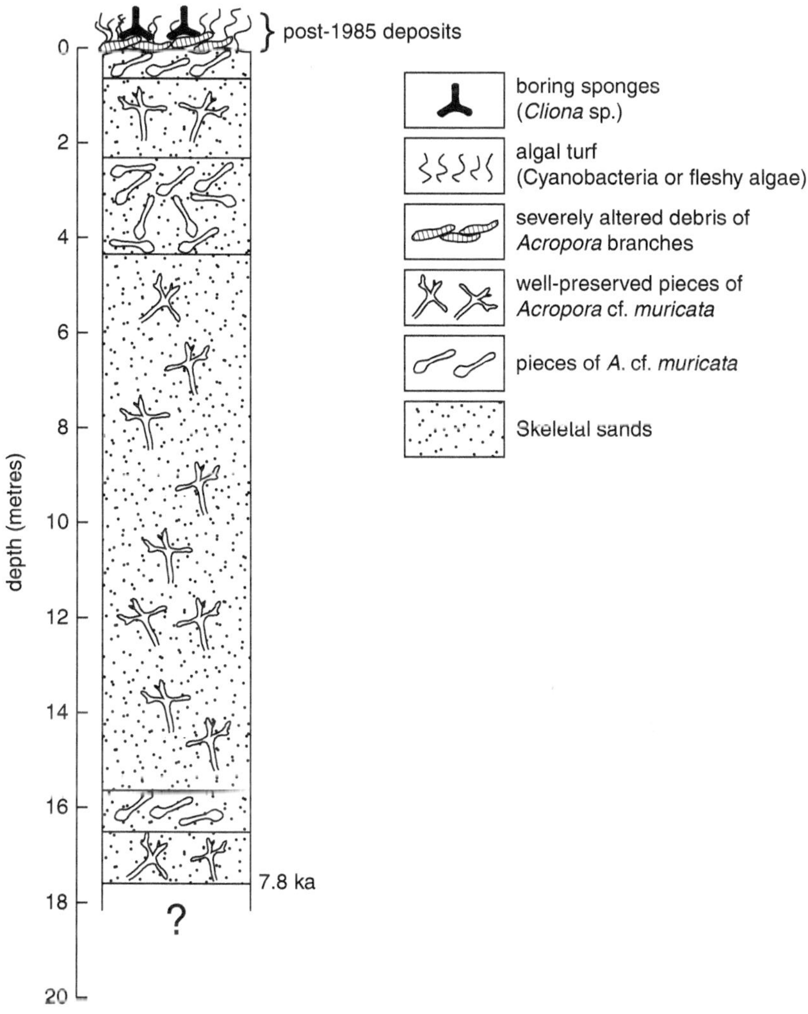

Figure 3.24 Recent changes in the structure of the inner reef flat community from Trou d'Eau fringing reef, Réunion Island (western Indian Ocean). From the middle of the 1980s, the original *Acropora cf. muricata*-dominated community was killed and replaced by non-calcifying organisms. Adapted from Montaggioni (1977) and Montaggioni et al. (1993).

Like that of the Caribbean, the face of Indo-Pacific coral reefs has changed locally and regionally significantly over the past three decades (see Wilkinson, 2004, for review). The major features are (1) the mortality of corals due to natural and anthropogenic disruptions, and subsequent reduction of coral cover with space opening for competition; and (2) invasion of the dead coral substrate by highly competitive fleshy, filamentous and even calcifying algae,

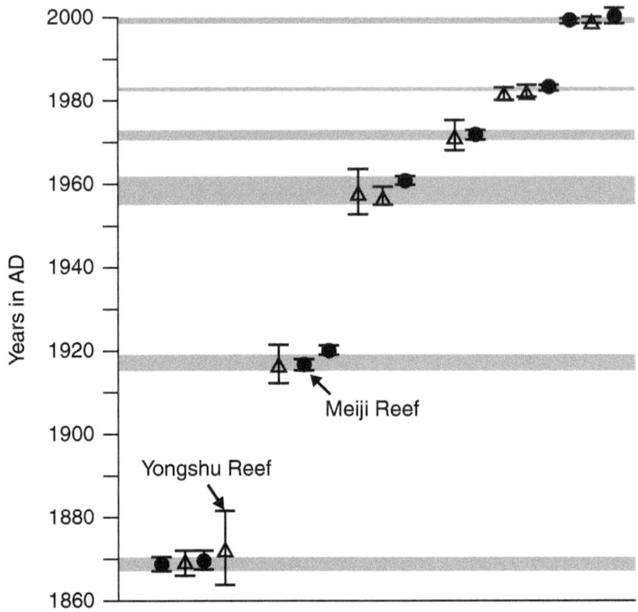

Figure 3.25 Ages of the mass-mortality events that affected domal *Porites* colonies from reef-flat zones at Yongshu and Meiji sites (South China Sea) during the past two centuries. From Yu et al. (2006).

thus limiting coral recruitment and the recovery of coral populations. As on Caribbean reefs, these features have ultimately blurred zonation patterns that were formerly characteristic of reefs in the region.

3.4. Dynamic Patterns of Reef Communities

Investigations of reef-community dynamics, i.e. the time over which the community remains stable, rebuilds without significant variation, or markedly changes in composition, have been carried out at palaeoecological to neoecological time scales, that is from tens or hundreds of thousands years to a few decades. It is important to note, however, that the dynamics of reef biotas reflects the evolution of the biota pool at regional to provincial scales and not that of the individual reef systems that have obviously been unstable during the Quaternary.

3.4.1. Reef-Community Stability

Changes in environmental conditions during the Quaternary have been regarded as responsible for having severely affected reef growth. Until the beginning of the 1990s, palaeoecologists were quick to point out the potentially detrimental effects of alternating episodes of cooling and warming and concomitant fall and rise in sea level on the stability of fossil reef communities (Stoddart, 1976; Taylor, 1978; Potts, 1983, 1984; Potts & Garthwaite, 1991; Paulay, 1990, 1991, 1994).

By contrast, most recent studies have provided strong evidence for the long-term maintenance of reef community structure and for striking similarities between Pleistocene, Holocene and modern reefs during at least the past 500 ka. Workers have shown almost without exception, that Pleistocene reef-coral assemblages display compositions and zonation similar to those of modern reefs at the same location, at least prior to the recent major biotic crisis. Such statements have been made on the basis of evidence in the western Atlantic and Indo-Pacific (in particular, Crame, 1980, 1981; Jackson, 1992; Hunter & Jones, 1996; Pandolfi, 1996, 1999, 2002; Pandolfi & Jackson, 1997, 2001, 2007; Greenstein, Curran, & Pandolfi, 1998; Pandolfi, Llewellyn, & Jackson, 1999; Aronson & Precht, 1997, 2001; Webster & Davies, 2003; DiMichele et al., 2004; and references herein). In the Caribbean, the structure of coral communities appears to have changed little since the Plio-Pleistocene turnover event and, thus, was thought to have experienced long-term stasis. Although reefs suffered repeated exposure during low sea-level stands, they reassembled to recurrently produce similar community patterns. *Acropora*-rich communities dominated almost continuously during interglacial high sea-level stands over the past few hundred thousand years with the persistence of *Acropora palmata* as the dominant reef-crest builder. The structure of Pleistocene reef-coral communities is therefore thought to be ordered and predictable to a high degree of confidence over broad spatial and temporal scales (Figure 3.26). As stressed by Hubbard et al. (2005), shallow-water coral assemblages dominated by branching acroporids might be regarded as the 'norm' and their occurrence or absence might be indicative of past reef health.

Additional arguments have come from the analysis of reef molluscan faunas. Gardiner (2001) investigated molluscan assemblages preserved in a late Pleistocene reef (San Salvador, Bahamas) depleted by a 1.1–1.5 ka sea-level fall during the last major interglacial stage (about 125 ka). The findings indicate that similar molluscan populations grew within two distinct episodes of reef building (at approximately 132–125 and 125–119 ka) despite their demise following a 5–6 m drop in sea level and subsequent community rearrangement. Based on a comparative analysis of fossil and modern molluscan faunas from Aldabra Atoll (Seychelles), Taylor (1978) stressed that the biota of the Indo-west Pacific reef province was relatively stable, despite the probable occurrence of changes in the tropical Indian Ocean species pool during the late Pleistocene. The thermal constancy over much of the region, the broad extent in latitude, the great variety of tectonic settings and thus habitat diversity have maintained the long-term

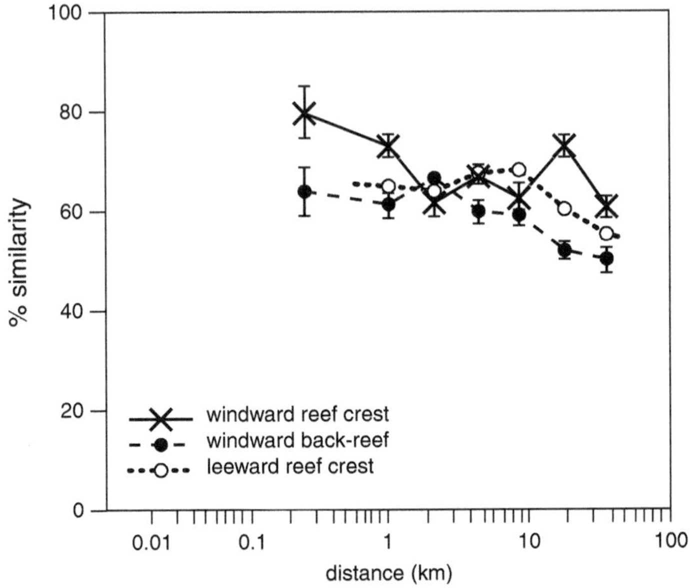

Figure 3.26 Degree of similarity in the species composition of late Pleistocene coral assemblages as a function of distance between different sites on Curaçao. Note that the species composition has remained relatively stable in the reef settings considered throughout the region. Modified and redrawn from Pandolfi and Jackson (1997).

stability of the faunal province. The Indo-Pacific species pool is speculated to have survived climatic deterioration during glacial episodes resulting in provincial retraction and changes in habitat. Species may have been capable of resettlement in areas from which they were excluded during low sea-level stands in spite of possible drastic changes in reef habitats. Kohn and Arua (1999) showed striking similarities between the composition of 1.8 Ma old gastropod faunas and that of the modern fauna (about 80% of species are still extant) from Viti Levu (Fiji, western Pacific), suggesting the persistence of a stable community structure over a considerable period through habitat stability.

Unfortunately, most of the data from fossil assemblages is not represented by qualitative inventories of the principal identifiable coral frame builders that are compared to their modern counterparts. Such inadequate sampling may result in artefacts. Consistent data, supporting the persistence of species composition, have been demonstrated using a species relative-abundance census. As a whole, uncommon taxa appear to respond in the same way to environmental constraints as the commonest (Pandolfi, 2002).

A number of ecological models of community structure have been invoked to explain the persistence of reef-coral community attributes over broad temporal and spatial scales (DiMichele et al., 2004). Thus, the co-occurrence of forms at any given time may reflect an association composed of species with mutual benefits or with similar responses and tolerance to environmental conditions, to an association of geographic isolates, or to an association dominated by species releasing abundant offspring. An alternative model relates to the so-called 'unified neutral theory of biodiversity and biogeography' by Hubbell (1997). This points out that the number of individuals remains constant within a given community and invasion by new individuals first requires space to be opened by the death of previous occupants. The community is considered neutral because all individuals are potentially subject to similar life and evolutionary constraints (death, reproduction and speciation) in a given time span. Hubbell's neutral model seems to account for the maintenance of coral community structure in the Pleistocene reefs of Papua New Guinea and Barbados demonstrated by Pandolfi (1996, 1999). The composition of both reef communities is typified by limitations in membership (i.e. the high level of community integration within reefs) of coral assemblages through space and time and appears to express predictable species assemblages rather than random groupings.

3.4.2. Reef-Community Variability

The reef-stability hypothesis is based almost exclusively on the census of well-preserved coral and molluscan taxa from the late Pleistocene. Little is known regarding the usually poorly preserved faunas from the middle to early Pleistocene. In addition, the dynamics of other major reef dwellers such as coralline algae and the green alga *Halimeda* have yet to be explored (Hillis, 2001). Nevertheless, Pandolfi (1996, 1999) emphasized the fact that Pleistocene reef-coral assemblages experienced spatial rather than temporal variability during repeated sea-level fluctuations. In addition, Precht and Miller (2007) pointed out that Pleistocene coral assemblages within the same environment appear to have varied more between contemporaneous reefs from different locations than between reefs of different ages at the same site. Local environmental disturbances had greater effects in controlling the taxonomic composition of coral populations than had global, climate-driven changes. These populations appear not to be dispersal-limited, meaning that local factors were critical in population dynamics. The marked role played by local factors on both local composition and regional species richness and by regional factors on local composition and richness suggests that both local and global environmental controls are of importance in disrupting the distribution patterns of coral species. Examples attesting to the fluctuating composition of coral species and other reef biota with changing environmental conditions over thousands to ten thousands of years have been described. These challenge the expectations of long-term stasis in reef communities.

Thus, extinction, geographic contraction and rarefaction events have been reported from both the reef provinces. The genus *Pocillopora* was reported by Geister (1977) from the latest Pleistocene of San Andrés Island, Aruba and Barbados, where it developed as extensive thickets. No identification at the specific level was possible because of the uncertainty of the linkages with the extant Indo-Pacific and the older Tertiary Caribbean pocilloporids. However, this form has affinities with *P. palmata*. *Pocillopora* became extinct between 62 ka and the late Holocene. Similarly, while *Dendrogyna cylindricus* was formerly common on Pleistocene reefs, it has become rare today on Grand Cayman and *Acropora prolifera* has not been recognized there either. By contrast, while the free-living coral *Manicina areolata* is present in the Pleistocene of Grand Cayman and in many present-day lagoons in the Caribbean, it is absent from the modern lagoons of the island. The hydrocoral *Millepora*, which commonly occurs on modern reefs in shallow water, seems to be missing from the Pleistocene of Grand Cayman. Despite the large number of Pleistocene reefs in the Caribbean, *Millepora* has only been reported from reefs in Barbados, San Salvador (Bahamas), Key Largo (Florida), San Andrés Island and the Dominican Republic. Its scarcity may have resulted from changes in community structure or may be an artefact of preservation (Hunter & Jones, 1996).

Two scleractinians, *Pocillopora cf. palmata* and the organ-pipe growth form (*M. nancyi*) of the *Montastraea annularis* species complex, which were common throughout the Caribbean islands during the 125 ka reef-building episode, disappeared between 82 and 10 ka. Pandolfi (1999) and Pandolfi et al. (2001) analysed the distribution patterns of both species throughout the Caribbean and, more particularly, from exposed reef terraces on Barbados, considered to range in age from 82 to more than 600 ka. Within the temporal distribution of the *Montastraea annularis* species complex, the number of organ-pipe *Montastraea nancyi* individuals appears to have increased from about 500 to 125 ka, whereas other growth forms (massive, columnar and lamellar) of the complex declined in abundance. However, none of these various forms became extinct in this period. Variations in species abundance among the members of the complex through time are believed to have resulted from a long-term competitive hierarchy, first promoting the organ-pipe form that later disappeared. The selective extinction of *M. nancyi* and *P. cf. palmata* as well may have been caused by instability in the community structure, possibly following the drop in sea level during the Last Glacial Maximum at around 24–19 ka. Depletion may have occurred in only a few thousands years. The species confined to oceanic islands with restricted shelf areas may have been affected by greater disturbance in their spatial distribution than those inhabiting larger continental shelves. When reduction in habitat area exceeded a critical threshold value, populations of widespread, highly competitive coral species probably suffered rapid extinction. After the extinction of *M. nancyi*, significant changes have affected the structure of metapopulations composed of the *M. annularis* species complex. The three surviving growth forms have experienced ecological and morphological changes. The disappearance of the competitively superior, faster-growing organ-pipe form released new space for slower-growing, columnar and domal *Montastraea* that had originally been excluded, at least partly, from shallow habitats. At the same time, the columns of *M. annularis* became narrower, resulting in an increase in their linear extension rate (Pandolfi et al., 2001). Similar extinction events may have occurred during previous glacial periods, but due to the lack of related fossil records, these cannot be demonstrated.

In the Red Sea, a few coral species (*Turbinaria peltata, Pavona minuta*) appear to have become locally extinct, while still living in the rest of the west Indo-Pacific (Veron, 2000). Comparing the late Pleistocene molluscan assemblages to their modern counterparts, Taviani (1997, 1998) listed the fossil taxa in three categories: totally extinct, locally extinct, rare or moved further south. For instance, the limpet *Diodora impedimentum* suffered complete extinction. Geographic eradication affected the gastropods *Rhinoclavis vertagus, Columbella turturina, Cerithium madreporicolum, Conus litteratus* and the bivalve *Cucullaea cucullata*. Rarefaction was experienced by the gastropods *Cypraea moneta, Olivia bulbosa* and the bivalve *Corbula taitensis*. These events are interpreted as reflecting a basin-wide turnover during the last glaciation. Biotic disturbances that affected coral reef ecosystems worldwide were amplified within the semi-enclosed Red Sea basin. A drop in sea level has disrupted the structure and internal organization of reefs and the hydrologic regime as well; water exchanges between the Red Sea and the Indian Ocean diminished and, subsequently resulted in hypersaline conditions. Similarly, on Aldabra Atoll (Seychelles) and along the Kenya coast, there have been major changes in the distribution of molluscan assemblages since the late Pleistocene (Crame, 1981, 1986). Of the five bivalve *Tridacna* species present in the Pleistocene limestones, only two (*T. maxima, T. squamosa*) are found today. The other species are now restricted to the Indonesian-west Pacific region (Taylor, 1978; Crame, 1980). Differences in the compositions of successive molluscan faunas on Aldabra are chiefly explained by changes in habitats through time, changes in random biotic factors (timing and order of recruitment) or changes in the species pool.

Grigg (1997) stressed that *Acropora* species were present throughout the entire Hawaiian island chain during the Pleistocene. However, *Acropora* is still living in the central islands of the archipelago, whereas it is missing in the high southwesterly volcanic islands. Paulay and Spencer (1988) pointed out that the coral fauna on Henderson Island (Southeast Polynesia) has experienced large-scale turnover from the late Pleistocene onwards. In addition, Paulay (1990, 1991) claimed that species compositions and geographic distribution of coral and bivalve assemblages present on the reefs of south Pacific islands were severely altered during the Pleistocene. Despite local to regional similarities between early Pleistocene and present-day gastropod assemblages in the western Pacific, Kohn and Arua (1999) considered that local extinctions may have occurred in some habitats, and it is not possible to infer the persistence of the entire species assembly on Fijdi throughout the Quaternary since comparison was based on only two faunas, 1.8 Ma apart in age, with a lack of information between. The sudden decline, geographic contraction or rarefaction of originally widespread Pleistocene reef faunas may have taken place as habitat reduction became critical.

Recent examples of coral responses to environmental disruptions strongly support the hypothesis of habitat limitation as the foremost trigger inducing rapid instability of coral populations. Since the 1980s, the prominent *Acropora cervicornis* and *A. palmata* species have suffered important reductions in population sizes almost everywhere in the Caribbean related to habitat destruction (Lewis, 1984; Hughes, 1994; Aronson & Precht, 1997). Similarly, Hillis (2001) suggested that the pattern of habitat availability for *Halimeda* species has significantly changed over the past 1–2 Ma, probably promoting local extinction or migration, as attested by their present-day distributions. Threshold effects in species extinction and spatial reduction, therefore, have to be considered in estimating reef responses to changing environmental conditions.

3.4.3. Reef-Community Stability Versus Variability: The Time-Scale Question

Data on the ecological dynamics of reef communities in both provinces are confusing when considering variable time scales. Some reefs appear to have undergone severe disruption in the distributional patterns of coral communities at both small temporal and spatial scales. Several workers (Tanner, Hughes, & Connell, 1994; Bak & Nieuwland, 1995; Connell, Hughes, & Wallace, 1997; Aronson & Precht, 1997; Aronson et al., 2002; Aronson, Macintyre, Wapnick, & O'Neill, 2004; Aronson et al., 2005; Piller & Riegl, 2003; Aronson & Ellner, 2007; Precht & Miller, 2007) have demonstrated that Pleistocene and recent reef communities have shown unpredictable variations in populations of coral species on millennial to decennial time scales. The structure of reef communities extending over areas smaller than 1 km can also vary greatly. For instance, when analysed over short time scales, the structure of Quaternary coral communities within habitats appears to have exhibited high variability (Jackson et al., 1996; Pandolfi & Jackson, 1997, 2007; Aronson et al., 2002; Aronson & Ellner, 2007; Macintyre, 2007). Hubbard et al. (2005) discussed the palaeoenvironmental significance, of the occurrence of widespread hiatuses in the *A. palmata* record throughout the western Atlantic in the past 7.5 ka. Whatever the cause of the gaps in *Acropora* occurrence (disease, bleaching or other events), the identification of contemporaneous acroporid devastation events at the regional scale indicates that spatial continuity cannot be significantly correlated with the temporal maintenance of a preserved species. During the recent reef crisis in the Caribbean, the transition from the former coral-dominated to the present macroalgal-dominated systems, which has operated on a decadal scale, reflects such a short-term variability (Wooley, 1992).

The trends in biotic shifts at present observed in living assemblages have proven difficult to predict from the fossil record since they do not have any clear geological precedent. The analysis of ecological successions from a historical point of view may provide clues with which to understand present-day biotic structures and predict future alterations in response to natural and/or anthropogenically driven disruptions. A comparison of temporal patterns of coral species dominance between different geographic areas may highlight the role of severe disturbances in scale-dependent variability of reef communities.

These observations shed light on important issues: What do the Pleistocene and Holocene reefs actually encapsulate regarding community variability or stability? How are events linked to temporary decline and recovery of coral taxa preserved in the fossil record, in cores as well as outcrops? What were the respective effects of changes in community dynamics and taphonomy on the fossil record? Answering these questions is crucial to understanding the patterns and controls of coral assemblage persistence at varying time scales.

Finally, the discordant results on reef dynamics obtained using neoecological and geological approaches raise the following question: Is the apparent long-term stability of Pleistocene reefs real or an artefact of preservation if short-term instability events occurred at too small scale to be recorded? The most striking feature of late Pleistocene coral assemblages is that they formed limited memberships, probably as a result of the evolutionary history of coral reef systems. Reef communities are open entities with a high degree of complexity and interconnectedness, especially in terms of trophic structure, contributing to the maintenance of an overall equilibrium (Wood, 1999). Limited community membership seems to favour reef-coral stability because only certain species are accepted and incumbent ones deter the settlement of new recruits (Jackson, 1992, 1994; Pandolfi, 1996, 1999).

3.5. Conclusions

A significant body of information regarding the patterns of reef community structure and coral species diversity has been generated by the study of both Pleistocene and Holocene reefs in the two geographic provinces. A few species largely dominate community composition in a variety of reef zones. This situation commonly faithfully reflects reef zonation patterns equivalent to those at present in most living reefs. The most striking zonal feature of coral assemblages is the generalized prominence of the genus *Acropora* in reef communities in time and space in both the Caribbean and Indo-Pacific provinces.

Despite the remarkable consistency in the spatial distribution of distinct coral forms, several coral assemblages may have occurred in a single zone. Conversely, a single assemblage may not necessarily relate to a distinct reef zone. Interpretations of reef-coral distributions therefore depend heavily on the scale at which assemblages are recognized. The larger the areal scale, the more assemblages and zones overlap. Therefore, smaller scales should be used in the definition of reef assemblages to faithfully compare outcrop and core data and to identify possible subzonal changes within a single depositional environment. Disregarding such subzonal changes may result in misinterpretations in palaeoenvironmental reconstructions, particularly from the analysis of reef cores.

Distributional patterns of coral species indicate that the Pleistocene reef communities consist of non-random populations, ordered and predictable over broad spatial scales. Local species preferences for particular environmental constraints are observed in the persistence of diversity in coral communities. However, the spatial continuity of a coral species in the Quaternary record cannot be systematically correlated with its temporal persistence. This fact must be kept in mind when changes in the community structure of present-day reefs at time scales of decades are compared with those observed in the Holocene and the Pleistocene over time scales of hundreds to thousands years or longer. The determination of how short-term changes are preserved in the fossil record is critical in relating the absence of a given taxon to its recent decline. Misreading such features may limit the use of the Quaternary record as a model for rapid changes in modern reefs.

Patterns of ecological succession are typified by the gradual incorporation of longer-lived species into communities, expressed by the replacement of most small, slow-growing, species by large, fast-growing colonies. The progressive rise to dominance of assemblages of larger species probably requires time intervals of several hundreds of years. Long-term dominance by slow-growing domal forms in the early stages of reef development is also predictable from relationships between temporal sequences of assemblages and spatial compositions of communities. Pioneering, mainly domal, coral forms are known to be well adapted to high-stress environments while species prominent in the later phases of succession are typical of low-stress situations. The biozonation observed on modern reefs may also express a potential temporal zonation since it results from an increasing environmental stress gradient; the zones grade from those subject to higher-energy, shallow-water conditions and bearing robust-branching acroporid assemblages (plus pocilloporids in the Indo-Pacific) rich in coralline algae, to those subject to lower-energy, deep-water conditions and dominantly exhibiting domal faviid or poritid assemblages. Where there is a decrease in accommodation space through time, domal coral assemblages can be replaced by robust-branching forms. An increase in accommodation leads to a reverse succession. Such long-term transitions have been convincingly demonstrated in most Quaternary sequences.

From the analysis of a limited number of reliable case studies, which mostly relate to the late Pleistocene sequences, it is concluded that the taxonomic composition and diversity of coral assemblages has remained remarkably constant through time, despite experiencing multiple cycles of global climate perturbation. The stability of coral populations is regarded as effective over at least the past 500 ka. This assertion may be extended in the past, by the 1.5 Ma long coral record from the Ryukyu Islands. It should be stressed that despite local and regional changes in species diversity due to significant range retractions and some extinctions, the overall composition and species diversity of reef-molluscan populations have maintained a remarkable constancy over the past hundreds of thousands years, particularly throughout the west Indo-Pacific. This assertion rests on the compositional attributes of the 1.8 Ma old molluscan fauna of Fiji. Unfortunately, the long-term, pool-stability hypothesis is difficult to demonstrate. Little is known about reef generations older than the last interglacial period. Any generalization regarding the long-term dynamics of reef-coral communities requires further census over broad spatial scales.

Although marked extinctions may have been occluded by time-lag effects, recolonization of inundated shelf substrates by reef faunas has obviously followed episodes of extreme environmental disruption. During low sea-level stands, relict populations survived in adequate refuges until conditions favourable for repopulation and expansion returned. This is supported by the fact that provincial species pools were able to continuously re-populate reef systems after devastation by repeated falls in sea level. The greater spatial variability of late Pleistocene reef communities over several glacial and interglacial episodes (as distinct from temporal variation) provides evidence that local environmental conditions had a greater influence in determining the abundance and diversity of reef faunas than global, climate-induced constraints.

In the Caribbean, the quantification of the community membership limitations of Pleistocene reefs indicates that less than one-third of the species present in the provincial coral pool were included in the communities. However, as emphasized above, palaeoecological studies have mostly been devoted to reefs of the last interglacial, thus restricting the applicability of the findings. In the Indo-Pacific, the question of reef limited membership is almost unexplored. Extensive work is therefore required to permit an initial comparison with modern communities and infer possible changes in community structure in the Quaternary.

Since the late 1970s, living populations of corals, especially acroporids, have experienced large-scale devastation and local to regional turnover in the Caribbean and in a number of localities in the Indo-Pacific. By contrast, coral populations during the Holocene appear to have been permanently luxuriant. Outside hurricane-swept areas, mass destruction events in the Holocene reefs were infrequent (about 1 disruption per 1–1.5 ka), placing these events in the frame of natural rates in healthy reef systems. As a whole, these destructional episodes are easily identifiable in the sediment record, at least at centennial resolution, and therefore allow the coupling of disturbance events with changes in community structure. Holocene instances demonstrate that the functional abilities of renewed coral communities were fully reacquired following rapid re-settlement within a time range shorter than 100 yr. Further studies of disturbances suffered by Holocene reefs in the Indo-Pacific province, coupled with the relatively well-documented Caribbean cases, will undoubtedly help place mortality events in a temporal frame, providing a better understanding of the nature of reef community structure (i.e. random or time-organized assemblages) and the long-term responses of reef systems to disruption. However, comparisons between changes in community structure in modern reefs on scales of decades and those affecting the fossil couterparts should only be made cautiously.

CHAPTER FOUR

Controls on the Development, Distribution and Preservation of Reefs

4.1. Introduction

Reef growth is influenced by a variety of biotic and abiotic factors (Figure 4.1). These act at varying temporal and spatial scales (for instance, see Buddemeier & Hopley, 1988; Karlson & Hurd, 1993; Smith & Buddemeier, 1992; Brown, 1997a,b; Kleypas, 1997; Hubbard, 1997; Kleypas, McManus, & Menez, 1999; Lough & Barnes, 2000; Knowlton & Jackson, 2001; Harriott & Banks, 2002) and are involved to varying degrees in the control of the daily and seasonal life histories of individual reef inhabitants and in their interactions within their communities. They determine patterns of reef distribution locally, regionally and globally, over periods ranging from decades, centuries and millennia to millions of years, and control the preservation of reefs in the geological record (Wood, 1999). A key concern is how shallow-water reef ecosystems have responded to these factors. Reefs throughout the Quaternary were formerly believed to have responded to changes in biotic constraints and physicochemical regimes in essentially the same way. However, recent analysis of present-day functioning of coral reefs has indicated that they consist of highly complex, disturbed and non-equilibrium communities that have responded to ambient ecological variation with great flexibility, although restricted by identifiable thresholds (the tolerance threshold concept, in the sense of Hopley, 1994).

Reef communities have generally not entered the geological record without suffering damage. The manner of their demise may differ, and fossil skeletons from reef-tract and lagoonal environments have, with time, passed through a filter that may alter the original ecological signals. Postmortem alterations include the selective destruction of individuals and age-classes, removal from the life habitat, and mixing of successive generation within habitat. Thus, the detection and identification of the alteration suffered by reef skeletal material during life and between death

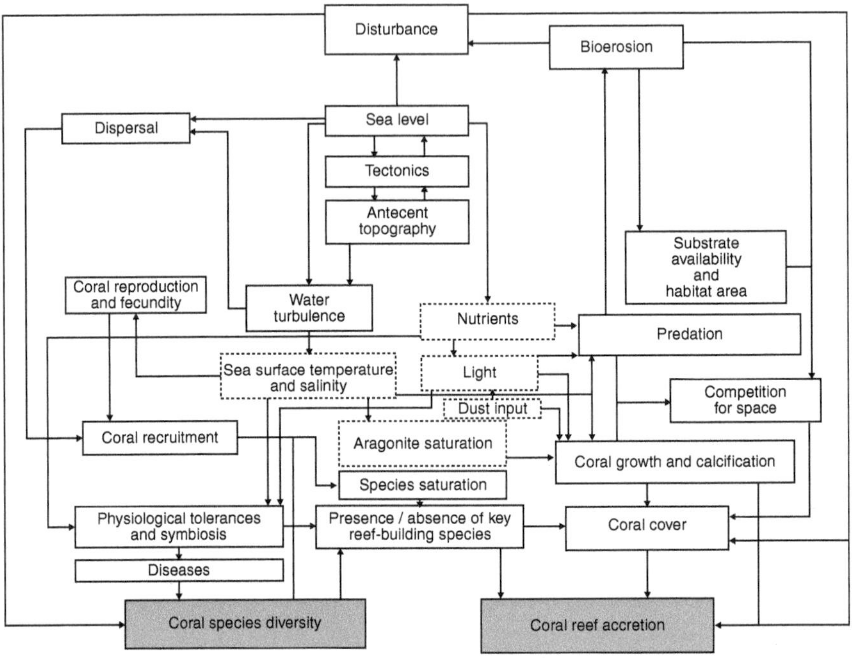

Figure 4.1 Summary of the main factors controlling the development and distribution of tropical coral reefs. Grey-coloured boxes refer to end-products. Modified and redrawn from Harriott and Banks (2002).

and fossilization, provide a challenge of paramount importance to any meaningful interpretation of reef communities and environments in the fossil record. It may be hampered by postmortem loss of information caused by poor or lack of preservation and by the transport of skeletal material from life sites (Scoffin, 1992). However, Quaternary reef-dwelling assemblages appear to preserve a high proportion of this critical information and there is apparently a high degree of fidelity between a given fossil community and the adjacent modern counterpart (Pandolfi & Greenstein, 1997; Edinger, Pandolfi, & Kelley, 2001; Pandolfi & Jackson, 2007; Greenstein, 2007).

The present chapter sets out to address the following questions:

(1) What roles did the various biotic and environmental factors play in the development and distribution of Quaternary reef communities?
(2) To what extent have postmortem processes altered the compositions of fossil communities and, as a corollary, what is the degree of similarity between modern and fossil assemblages?

4.2. Controls on Reef Development and Distribution

4.2.1. Biotic Controls: The Role of Recruitment, Species Saturation, Competition, Predation, Symbiosis and Disease

Sammarco (1996) and Pandolfi and Jackson (2007), among many others, have discussed the role of a variety of intrinsic processes in the maintenance of coral diversity in local reef communities over short-term time scales. The distribution and recruitment of coral larvae was considered to be a key influence on both the distribution and abundance of new assemblages (the recruitment limitation hypothesis). Sexual reproduction appears to be an important factor. Corals have two strategies for sexual reproduction, broadcast and brooding. Brooding species release male gametes into ambient waters. These fertilize the female gametes from a colony of the same species. The fertilized eggs develop into planulae that are released from the parent colony to become planktonic before eventually settling on a suitable substrate and growing to form new genetically distinct colonies. By contrast, broadcast spawning corals release both male and female gametes coevally into the water. The density and timing of the release is critical to maximize the probability of successful fertilization but planulae are already subject to the vagaries of ocean currents. Broadcast spawning is the dominant form of sexual reproduction in corals, but appears to be more common in Indo-Pacific species than in those in the Caribbean. The primary recolonizing corals of the Indo-Pacific region include both broadcasters (*Acropora*) and brooders (*Pocillopora*, *Seriatopora*), whereas common species in the Caribbean (*Agaricia*, *Porites*, and *Favia*) are brooders. Applied to the prediction of Pleistocene coral populations at the scale of individual islands in the western Caribbean, Pandolfi and Jackson (2007) concluded that the recruitment limitation hypothesis did not account for the overall dominance of *Acropora palmata* and *A. cervicornis* reputed to release spawn in reduced quantities. The dominant mode of reproduction for these branching corals is asexual; new colonies form by breaking off branches that continue to grow.

An additional factor that may have governed species compositions in fossil reef communities is the dispersal capabilities of the species. Do corals that have a broadcast mode of reproduction possess greater dispersal potential than those that are brooders? In the Indo-Pacific region, the ranges of both broadcasting and brooding corals are at present comparable. Coral species that release numerous larvae capable of long-distance displacement are believed to maintain constancy in community structure through time and space (Hubbell, 1997). However, some brooding species like *Pocillopora damicornis* are, and seem to have been, extremely widespread. In the Caribbean, by reference to their living analogues, the dominant taxa in Pleistocene reef assemblages are suggested to not have had a higher capacity of releasing more larvae than scarce species (Pandolfi & Jackson, 1997).

An alternative factor controlling reef community composition may have been the saturation state of coral species richness, as presented by local communities with respect to the regional species pool. The maintenance of community composition largely depends upon the size of the regional coral pool. Any community that has the potential to incorporate additional species is defined as undersaturated, and when it is no longer open to invasion from regional supplies, it is considered to be saturated. In the latter case, biotic exchanges and ambient environmental conditions play a major role in defining the structure of local populations. In the Pleistocene of the Caribbean, local coral species diversity may be regarded as having been undersaturated due to the dominance of a limited number of species, even though subordinate forms occupy any remaining niches. The particular composition of the regional species pool is more important than its overall diversity. As a result, colonies produced by a reduced number of the species present within the pool may dominate numerically over those of other species.

A fourth limiting condition is that of metapopulation dynamics in which interactions between coral dispersal and competitive abilities are thought to control species richness at any given reef site. Pandolfi and Jackson (2007) argued that the metapopulation dynamics hypothesis has to be taken into account in order to explain the stability of Pleistocene reef-coral populations in the Caribbean at broad temporal and spatial scales. For instance, Pandolfi (1999) suggested that the limited dispersal potentialities of coral larvae were responsible for the decline of some species during the Last Glacial Maximum (about 25–19 ka). This event may have been triggered by habitat fragmentation and reductions in area in response to the fall in sea level. Reduction of the surface area of critical substrates below a tolerance threshold probably has an inimical effect on coral growth and survival, insofar as the maintenance of local species richness requires continuity of larval exchanges among disparate reef communities. Habitat area is believed to be one of the major limiting environmental factors driving species richness in the western Pacific (Bellwood, Hughes, Connolly, & Tanner, 2005). The geometry and geographical characteristics of reef areas within a domain influence species richness at the regional scale, and richness tends to peak in the middle parts of the domain (the mid-domain effect in the sense of Bellwood et al., 2005).

Other critical limitations are linked to biological interactions. These govern the population density of coral species through competition, predation, symbiosis and disease. The competitive advantages of some coral growth forms may explain the prevalence of few species over others in a particular reef zone. In the southern Caribbean (the islands of Barbados, Curaçao and San Andrés), Pandolfi and Jackson (2007) found that the taxonomic composition of Pleistocene coral communities was controlled more by the ecological particularities of a few coral species than by distances separating the communities. Over distances of kilometres to tens of kilometres, within a given island, the similarity of coral species abundance remains markedly high, reflecting non-random species assemblages. However, at the scale of hundreds to thousands of kilometres from island to island, although there are marked discrepancies in some community structures (Pandolfi, 2002), the dominant frame-building species (*Acropora palmata*, the *Montastraea annularis* species complex, and *A. cervicornis*) remain constant, implying a high level of similarity among the different coral frameworks. It is likely, therefore, that biotic exchanges and interactions between habitat-dimensions and regional characteristics of the coral pool are responsible not only for the prevalence of the same coral species associations throughout the Pleistocene deposits of the southern Caribbean, but also for minor differences in their community compositions. Similarly, competition between corals and algae (Figure 4.2) is regarded as fundamental in determining the structure and composition of reef communities when macroalgae come to dominate reef corals (McCook, Jompa, & Diaz-Pulido, 2001, and references therein). However, large-scale replacement of corals by macroalgae may reflect coral mortality due to a

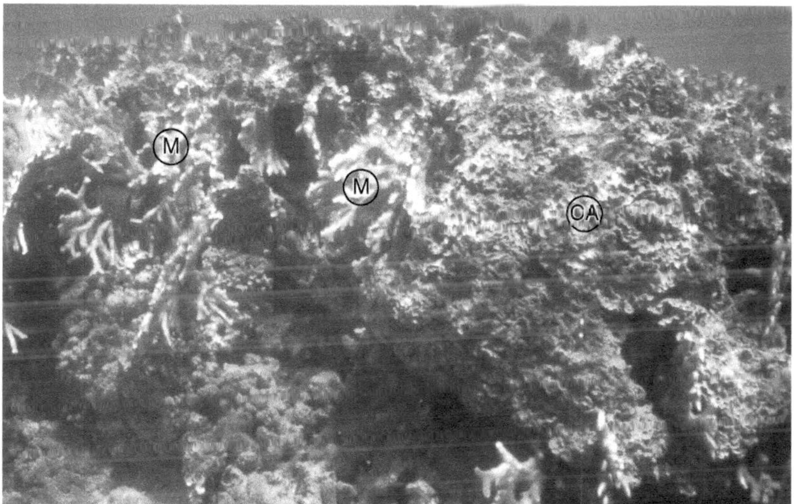

Figure 4.2 Competition for space between the hydrocoral *Millepora platyphylla* (M) and overgrowths of the coralline alga *Neogoniolithon* sp. (CA) on the inner reef flat, La Saline fringing reef, Réunion Island, western Indian Ocean (Photograph by L. Montaggioni).

variety of disturbances rather than competitive interaction. This may be accompanied by competitive inhibition of coral recruitment, resulting in ecological phase shifts (Done, 1992a). A variety of vertebrate and invertebrate predators and grazers are known to feed on living corals. These are involved in a complex of processes that may control the trajectory of reef development over time. Some have substantial effects on the structure of reef communities and can trigger transitions between different types of reef (Carpenter, 1997, and references therein). The symbiosis between coral and zooxanthellate algae was demonstrated to be accompanied by sensitivity to environmental stress, resulting in a disruption of the host–algal association and subsequent bleaching (Muller-Parker & D'Elia, 1997, and references therein). The main determinants of symbiotic disruption include high light, and particularly ultra-violet light levels, elevated temperature and high nutrient concentrations that may be coupled with high turbidity and low salinity.

The apparent increase in the incidence of coral diseases seen in many areas is most commonly attributed to three factors, including water pollution, elevation in sea surface temperature (SST) and overfishing (Rosenberg, Koren, Reshef, Efrony, & Zilber-Rosenberg, 2007). Microbially mediated diseases and syndromes have caused widespread mortality among corals and associated organisms in the Caribbean, in particular since the beginning of the 1970s (Precht & Miller, 2007, and references therein). Reef drilling investigations indicate that mass mortality of corals, especially acroporids in some Caribbean areas, is a relatively recent phenomenon, occurring within the last hundred or at most thousand years (Wapnick et al., 2004). To our knowledge, there is no evidence of any link between potential pathogenetic activity and deterioration of reef biotas in the Pleistocene. The taphonomic signatures associated with band disease events in corals are easily confused with some types of borings. The epidemic that devastated the echinoid *Diadema antillarum* communities throughout the Caribbean in the early 1980s did not produced any recognizable signal in the sedimentary record, strongly suggesting that evidence of such events is not reliably preserved in the geological record (Lessios, Robertson, & Cubit, 1984; Aronson et al., 2005).

Although there is no direct evidence of biotic interactions among organisms in fossil reefs, the predictable patterns of temporal and spatial maintenance in community structure observed in Pleistocene sequences suggests that biotic attributes have been important in influencing the compositions of coral populations. Crame (1980, 1981) noted that within the late Pleistocene reefs exposed along the Kenyan coast, patterns of coral species diversity and composition may have been governed primarily by intrinsic factors. Interspecific competition might have operated during the initiation of these communities, but diversity increased or decreased as conditions varied. Sharp increases in species richness were induced by species immigration and evenness or by spaces opening in paucispecific, arborescent acroporid stands. Conversely, declines in diversity were caused by local dominance by dense overgrowths of arborescent corals. Branching pocilloporids, domal poritids and faviids and *Galaxea fascicularis*, regarded as important pioneers, were outgrown by fast-growing, strongly competitive, new arrivals like acroporids. Thus, the major short-term successional trend in shallow-water sequences was rapid dominance of arborescent and tabular corals. In deeper waters, under lower light conditions, foliaceous and encrusting corals were metabolically more efficient than ramose and tabular-branching colonies and pre-empted space.

4.2.2. Abiotic Controls: The Role of Physical and Chemical Disturbances

Some physical and chemical factors controlling reef development are directly related to seawater properties (e.g. SST and salinity, nutrient levels, turbidity and hydrodynamic energy). These effects are presented and discussed in Chapter 7. Other controls include substrate availability, antecedent topography, tectonics, dust input and changes in atmospheric CO_2 and sea levels. These may operate synergistically in complex and contrasting ways over differing temporal and/or spatial scales.

4.2.2.1. Substrate availability and refuges

Veron (1995) demonstrated that the decline of coral species diversity eastwards across the Pacific is primarily controlled by substrate availability, but also by the survival and dispersal capacity of recruits. In the Indo-Pacific, most of the dominant reef species are broadcast spawners, releasing sperm and eggs simultaneously (Hughes et al., 1999). As already indicated above, patterns of recruitment differ among broadcasters and brooders depending upon stock size, larval survival, and settlement behaviour. Such disparities may explain differences in the rates of substrate recolonization during periods of sea-level rise, because reef sites may be sources or sinks for given coral groups. The inhibition of coral settlement may therefore reflect the contraction or absence of suitable nurseries for particular species during low stands. Shelf edges, banks and seamount summits provide the best candidates to serve as refuges and centres of dispersal of coral larvae. The recolonization of potential substrates during transgressive phases requires the establishment of oceanic circulation regimes suitable for larval transport. The insular marine biota of the Pacific reef domain is typified by wide-ranging species, suggesting that long-distance dispersal usually occurred after low sea-level stands because endemism is mainly restricted to the most remote island groups (Meyer, Geller, & Paulay, 2005).

By contrast, and based on molecular data, Benzie (1999) pointed out that gene flow patterns of reef species and corals in particular have been controlled principally by events related to global climate and sea-level changes since at least the beginning of the Pleistocene rather than by current regimes similar to those of the present-day ocean circulation. Infrequent pulses of long-distance dispersal best account for the patterns of genetic variability of a variety of modern reef dwellers. In the Indo-Pacific, spatial models of genetic divergence conform to predicted patterns of maintenance and survival of isolated populations within multiple refuges during low sea-level stands, followed by highly pulsed dispersal events corresponding with interglacial high sea-level stands. These population distributions persisted for several thousand years after their establishment, exhibiting unique sets of genotypes and little genetic differentiation. In this context, isolated populations may have evolved from a number of different refuges. Thus, the isolation and spatial limitation of most refuges within the Indian Ocean (the Red Sea, western Indian Ocean islands, the Maldives) and the Pacific (the Ryukyus and central Pacific islands) appear to have been incompatible with the maintenance of high-diversity biotas. However, the evidence suggests that marine speciation occurred regularly over small spatial scales, leading to localized endemism and high diversity. For example, Meyer et al. (2005) observed that turbinid gastropods form at least 30 geographically isolated clades in the Indo-West Pacific, separated by distances of less than 200 km.

4.2.2.2. Antecedent topography

Antecedent basement theories developed largely on the assumption that the physiography of Holocene and modern reefs has been governed to a great extent by shelf foundations (Daly, 1915; Hoffmeister & Ladd, 1944; Steers & Stoddart, 1977; Purdy, 1974; Guilcher, 1988, pp. 45–50; Purdy & Bertram, 1993; Grigg et al., 2002; Hopley, Graham, & Rasmussen, 1997, 2007, pp. 253–260). Large-scale topographic features such as the elevations of shelf breaks and atoll summits, the general distribution of topographic highs and the overall slopes of shelves are thought to directly constrain reef locations. In general, there are antecedent elements to the general physiography of fringing reefs. Hopley and Partain (1987) and Smithers, Hopley, and Parnell (2006) provided a classification for the northeastern Australian fringing reefs in part based on the nature of reef foundations. Choi and Holmes (1982) and Ginsburg and Choi (1983) accepted the idea of antecedent control on the development of the shelf-barrier reefs atolls of Belize, although they regarded it as reflecting the influence of fluvial siliciclastic distribution. Gischler and Hudson (2004) concluded, on the basis of borehole evidence, that variations in the date of establishment of different areas of the Belize Barrier Reef were a consequence of variations in elevation of the antecedent topography. Lidz, Shinn, Hine, and Locker (1997) presented seismic data to show that reef development in the Upper Florida Keys was driven by the antecedent Pleistocene bed-rock topography. Where the bed-rock is absent, reefs are absent. Kayanne et al. (2002), on the basis of drill cores, concluded that reef development on the Palau islands in the western Pacific Holocene is also 'primarily constrained at its foundation by the antecedent topography of the Pleistocene substratum'. However, Holocene reef growth can occur locally largely independent of antecedent topography. Present reef morphology may result simply from the interplay between the rate of postglacial sea-level rise, prevailing water-energy conditions, and the response of biological activity to these forcings, and consequently has no resemblance to that of the underlying palaeotopography (Walbran, 1994). Similarly, Adey (1978) suggested that rates of reef growth are 'sufficient to have produced virtually all refoid rims' and that no specific antecedent surface was required to generate 'the classical bioherm configuration'.

Examination of outcrops and cores from many modern atolls and barrier reefs demonstrates that their morphology, and in particular the saucershape, has been produced principally by the subaerial dissolution of the antecedent foundations. For instance, in the northwestern Tuamotus, central Pacific, Mataiva atoll appears to possess a reticulated lagoon, divided into numerous shallow pools by a network of shoals. The remains of an old, extensively calcitized reef tract, regarded as of Miocene age, crop out locally along the outer reef rim (Pirazzoli & Montaggioni, 1986). Boreholes made through pool-separating shoals revealed the structural attributes of the lagoon. Holocene sediments, composed of skeletal gravels to carbonate mud are interbedded with a few *Porites* corals, overlies a pre-Holocene, irregular karst palaeotopography, on rocks ranging from Miocene to late Pleistocene in age, locally overlain by phosphate deposits. The age of the base of the Holocene deposits ranges from 6.5 to 6.0 ka. This means that the pre-Holocene foundation has experienced severe, long-term subaerial erosion resulting in the development of a series of central basins, prior to deposition of phosphorites, Holocene inundation and sediment filling (Pirazzoli & Montaggioni, 1986). Observations, particularly from Fiji (Ferry et al., 1997), Mururoa (Buigues, 1997), the isolated carbonate platforms of Belize (Gischler & Lomando, 2000; Gischler, 2007) and mid-shelf reefs on the Great Barrier Reef (Hopley et al., 2007) confirm the major control of dissolution on atoll-like morphology. There is also subsurface evidence from Mayotte Island (Zinke, Reijmer, & Thomassin, 2001, 2003a,b), the Belize barrier systems (Gischler, 2007), outer-shelf reefs on the Great Barrier Reef (Hopley et al., 2007), and the Florida reef tracts (Lidz et al., 1997) indicating that barrier-reef like morphology is also dissolution driven.

Small-scale topographic features such as changes in slope, pre-existing mounds and channels, and substrate type, have probably had more influence in facilitating or preventing coral settlement than overall shelf architecture (Grossman & Fletcher, 2004; Montaggioni, 2005). Regional differences in the timing of reef initiation and in growth patterns can be explained in terms of substrate attributes. Reefs preferentially colonize karst surfaces of limestones and rough lava flows, whereas smooth-surfaced metasedimentary outcrops and unconsolidated sediments are apparently less suitable (Cabioch et al., 1995). The slope of the substrate may have a direct control on the growth forms of pioneering corals and on the compositions of the resulting assemblages (Webster, 1999). Encrusting and foliaceous forms are better adapted to growth on steep slopes ($>40^{\circ}$) than branching or domal forms.

4.2.2.3. Tectonics

Tectonic processes operate over varying timescales according to local geodynamic settings, but usually at rates markedly lower than those of changes in sea level. Given the magnitude and rates of Pleistocene and Holocene eustatic fluctuations during periods of rising sea level (2–50 times higher than those of local tectonic movements), tectonically induced changes are likely to have been overwhelmed by eustatic changes and consequently do not seem to have been a major control on reef growth over short time scales (from 1 ka to less than 20 ka). Hydroisostatic processes have played a significant role in the altitudinal distribution of reef tracts (see Chapter 9, Section 9.4). This is especially evident in the Holocene record (Dickinson, 2004). Mid-Holocene reefs have emerged on a number of tropical islands in response to global isostatic adjustment (Mitrovica & Milne, 2002). Mantle flowage to compensate for the transfer of mass from Pleistocene circumpolar ice-sheets to the global ocean, caused by deglaciation, resulted in a low-latitude drawdown in sea level during the late Holocene and subsequently in reef emergence.

Excluding hydroisostatic effects, the tectonic environment of Quaternary reef systems has been controlled principally by the complex interplay of vertical and or horizontal motions over periods of thousands to tens of thousands of years. Contrasting models of reef evolution throughout the Pleistocene to Holocene are regarded as expressing a response to differences in geodynamics (Scoffin & Dixon, 1983; Scott & Rotondo, 1983a,b; Montaggioni, 2000; Hopley et al., 2007, pp. 18–34).

On passive margins, differences in depth to the reef foundations across shelves can be explained primarily by long-term subsidence. According to Hopley et al. (2007, p. 275), faulting may account for the contrasting configurations of reefs on the eastern Australian margin. This area is thought to have been affected by subsidence pulses until at least the late Pleistocene; but movements appear to have continued up to the present. Based on the present stratigraphical position of the earlier reef generation found in the core extracted from Ribbon Reef 5 (Webster & Davies, 2003;

Braithwaite et al., 2004), a subsidence rate averaging 0.2 mm yr^{-1} is indicated for the northern central Great Barrier Reef area for the past 600 ka. This has had significant effects on reef growth since the middle Pleistocene with the development of reef sequences that show a substantial expansion compared to the nearby barrier reef system of New Caledonia (Cabioch, Montaggioni, Frank et al., 2008). By contrast, in the southern Great Barrier Reef, movements along active faults are suspected to have caused uplift of reef tracts within the past 6 ka (Kleypas & Hopley, 1993). Similar divergent growth histories can also be demonstrated for continental barrier reefs in the Caribbean. For example, Purdy et al. (2003) reported that the distribution of Recent reef deposits on the Belize platform was strongly driven by the position of near-surface folds and faults. The modern barrier reef results from a late Quaternary colonization of the edge of an underlying carbonate shelf margin by corals. Along the margins of the Red Sea, the location and orientation of Pleistocene to modern reef tracts clearly reflects the interplay of crustal rifting, differential tilting and uplift of tectonic blocks (Braithwaite, 1982; Plaziat et al., 1998; Dullo & Montaggioni, 1998).

On active margins, reef distribution and geometry are mainly controlled by regional tectonic history. At active junctions of lithospheric plates, high-intensity earthquakes cause metre-scale coseismic uplift. Successive reef generations from early Pleistocene to Holocene in age typically form emergent step-like terraces in the Caribbean (Mesolella, 1967; Geister, 1980; Radtke, Grün, & Schwarcz, 1988; Taylor & Mann, 1991; Mann, Prentice, Burr, Pena, & Taylor, 1998; Feuillet, Tapponnier, Manighetti, Villemant, & King, 2004) and in the Pacific (Chappell, Ota, & Berryman, 1993; Ota et al., 1993; Hantoro et al., 1994; Bard, Hamelin, Pirazzoli et al., 1996; Cabioch et al., 1998, 2003; Mann, Taylor, Lagoe, & Quarles, 1998; Taylor et al., 2005). Earthquakes have also resulted in large landslides that cut through the reef tracts. The areal extent of reefs affected by landslides depends upon tectonic uplift rates, the thickness and geometry of the reefs and the average gradient of the land surface. Ota, Chappell, Berryman, and Okamoto (1997) reported that landslides affected the Pleistocene reef terraces of the Huon Peninsula (Papua New Guinea) throughout the late Quaternary. A total of 26 landslide events were recorded within the past 120 ka. During the Holocene, two dated sets of landslides have been identified between 7.8–6.3 and 1.3–0.8 ka respectively.

On mid-plate volcanic islands, thermal subsidence is the dominant tectonic process. Subsidence rates range from 0.25 to 0.50 mm yr^{-1} on Tahiti (Bard et al., 1996) to up to 2 mm yr^{-1} on Hawaiian islands (Campbell, 1986; Ludwig, Szabo, Moore, & Simmons, 1991; Moore, Ingram, Ludwig, & Clague, 1996; Webster et al., 2006). In addition, around volcanic centres, due to the localized load of the island masses, the underlying oceanic crust is deformed, forming moat-like depressions around the islands. This results in the development of arches as flexural forebulges beyond the down-warped area elevated up 200 m above the average depth of the ocean crust. In the Pacific, carbonate high islands and atolls located near the crests of arches possess emergent Quaternary reef terraces at elevations ranging from a few metres to up to several tens of metres (Montaggioni, 1989; Nunn, 1994; Dickinson, 2004).

4.2.2.4. Dust input

Ice-core records from Antarctic sites indicate that in comparison to the present-day, dust fluxes have been 10–12 times larger during the glacial maxima of the past 400 ka and 27–30 times larger during the Last Glacial Maximum (Petit et al., 1999). In addition to their negative effect (cooling) on climate forcing (Bar-Or, Erlick, & Gildor, 2008), the large volumes of dust and atmospheric aerosols that would have been deposited in tropical surface waters may have enhanced the effects of inimical factors in reef growth during Pleistocene low sea-level stands. Thus, the decline in reef vitality in the Caribbean over the past 25 years is suspected to be partly linked to dust fluxes from African deserts (Shinn et al., 2000). The phytoplankton blooms that currently occur off Hawaii are also interpreted to be a result of increasing climatic desertification, promoting large dust storms in Asia and atmospheric transport of iron-rich particles eastwards (Chadwick, Derry, Vitousek, Huebert, & Hedin, 1999). Similar conditions probably operated during the Last Glacial Maximum, particularly in the southwestern Indian Ocean where dust fluxes were three to five times greater than those in the Holocene (Marcantonio et al., 2001) and also in the western Pacific (De Deckker, Tapper, & Van der Kaars, 2002). Colder and drier episodes at 24–18.5 and 12.8–11.6 ka seem to coincide with dust events and the spatial restriction of Indo-Pacific reefs (Montaggioni, 2005).

4.2.2.5. Atmospheric CO_2 and aragonite saturation

The saturation state of ocean surface waters with respect to aragonite is temperature-dependent (Kleypas, 1997). In modern seas, SST values and saturation state appear to be positively correlated. However, in equatorial areas (0–15° latitude) there is a relative depression of aragonite saturation caused by a significant decrease in the evaporation/precipitation ratio and enrichment of CO_2 in surface waters by local upwelling (Opdyke & Wilkinson, 1993). Aragonite saturation is usually at a minimum in higher latitudes (up to 35°). The calcification rates of many autotrophic organisms increase as a function of increasing carbonate saturation and decreasing pCO_2 (Gattuso, Frankignoulle, Bourge, Romaine, & Buddemeier, 1998; Brocker et al., 1999). Calcifying reef communities grow optimally in warm waters where aragonite is supersaturated within values ranging between 4.1 and 3.1 (Kleypas, Buddemeier, et al., 1999; Silverman, Lazar, & Erez, 2007). Under such conditions, photosymbiosis is regarded as promoting coral growth (Buddemeier, 1997).

During the Last Glacial Maximum, surface waters are assumed to have experienced higher values of aragonite saturation (probably around five to six times) compared to the present day (Buddemeier et al., 1998). During deglaciation, tropical waters are likely to have been affected by a marked decline in carbonate saturation to 4.4, in response to an increase in atmospheric pCO_2. This interpretation is supported by experimental manipulation of calcium concentrations. Variations in carbon dioxide levels seem to have been maintained within the tolerance thresholds for reef calcification throughout the past 18 ka (Buddemeier, Gattuso, & Kleypas, 1998; Gattuso et al., 1998). Revisiting the 'coral reef hypothesis' of Opdyke and Walker (1992), Vecsei and Berger (2004) stated that carbonate production by calcifying benthic populations inhabiting shallow-water settings, particularly reef environments, contributed significantly to the rise in atmospheric CO_2 levels during the last deglaciation. This agrees with the apparent synchroneity between variations in pCO_2 and reef growth phases over the past 20 ka as observed in the Indo-Pacific province (Montaggioni, 2005). Extending these findings to earlier Pleistocene deglacial periods suggests that shallow-water carbonate production has been promoted by increasing aragonite saturation (Figure 4.3) and thus has probably resulted in a strong positive feedback to the increase in CO_2 flux and subsequent warming within the last millennia preceding interglacial peaks.

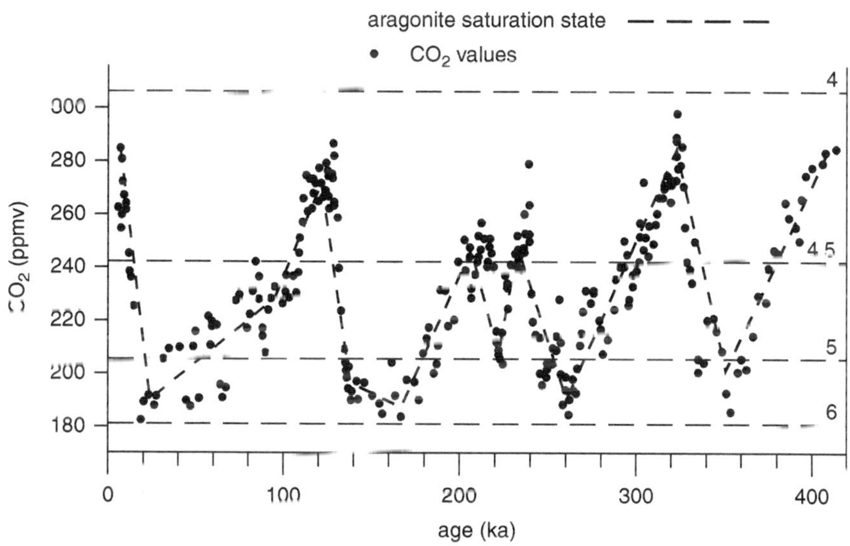

Figure 4.3 Relationship between changes in atmospheric CO_2 levels and aragonite saturation of tropical surface waters during the past 420 ka. The record of CO_2 measured on enclosed air bubbles from the ice core at the Vostok site (Antarctica) was extracted from Petit et al. (1999). The rates of aragonite saturation are those proposed by Buddemeier, Gattuso, and Kleypas (1998).

4.2.2.6. Sea level

Sea level is known to have fluctuated dramatically during the Quaternary at orbital timescales (10^4–10^5 years) in response to climate variability (for instance, see Waelbroeck et al., 2002; see Chapter 9, Section 9.4.2). Such fluctuations have led to substantial changes in tropical, shallow-water environments affecting the extent, nature and distribution of reef habitats. However, the way in which reef communities have responded to these movements is still debated. Potts (1984) suggested that the 10–100 ka frequency of sea-level cycles was insufficient to control the fate of reef corals in an evolutionary sense, especially in the Indo-Pacific province, but allowed that intraspecific variability may have been promoted. Similarly Veron (1995) argued that changes in sea level had little influence on the diversity of Indo-Pacific reef-building corals on the scale of the Quaternary. In this area, only acroporids appear to have been slightly affected by drops in sea level. By contrast, the Caribbean record indicates that corals have suffered severe disturbance over the past 1.8 ka, resulting in significant faunal turnovers (see Chapter 3, Section 3.3).

It has been assumed that reef coral communities re-settled after suffering major and repeated drops in sea level during glacial periods since at least the early Pleistocene (Pandolfi, 1999). Based on the analysis of the genetic structure of reef inhabitants, however, Benzie (1999) indicated that the resurgence of populations from the same refugia, in response to repeated isolation during Quaternary low sea-level stands, occurred progressively through successive, transgressive events (Figure 4.4). Either original populations were affected by disruption and retraction and then coalesced as gene flow was re-established, or they retained sufficient genetic variation such that gene flow between populations remained limited. Some genetic variations that limit gene exchange today may have developed long before the youngest isolating events of the Last Glacial Maximum. The genetic divergence of reef-associated groups or species (most within the Plio-Pleistocene) has operated at varying timescales. Some may have occurred rapidly in as little as a few centuries, whereas other populations may have successfully coalesced after multiple isolations. The confusion in coral taxonomy is probably a reflection, at least in part of spatial heterogeneity resulting from the vicariant isolation of populations during low sea levels followed by partial re-integration. The recent pattern of genetic divergence provides evidence of multiple intrabasinal centres of evolution in reef species, especially during low sea-level stands, and within-basin evolution rather than expansion from the Indo-West Pacific high-diversity locus.

Paulay (1996) studied the effects of sea-level fluctuations on the compositions of bivalve assemblages in the central Pacific islands. He concluded that changes in sea level during the Pleistocene influenced the likelihood of speciation on isolated islands by altering the maintenance of local communities. Species were affected differentially according to their

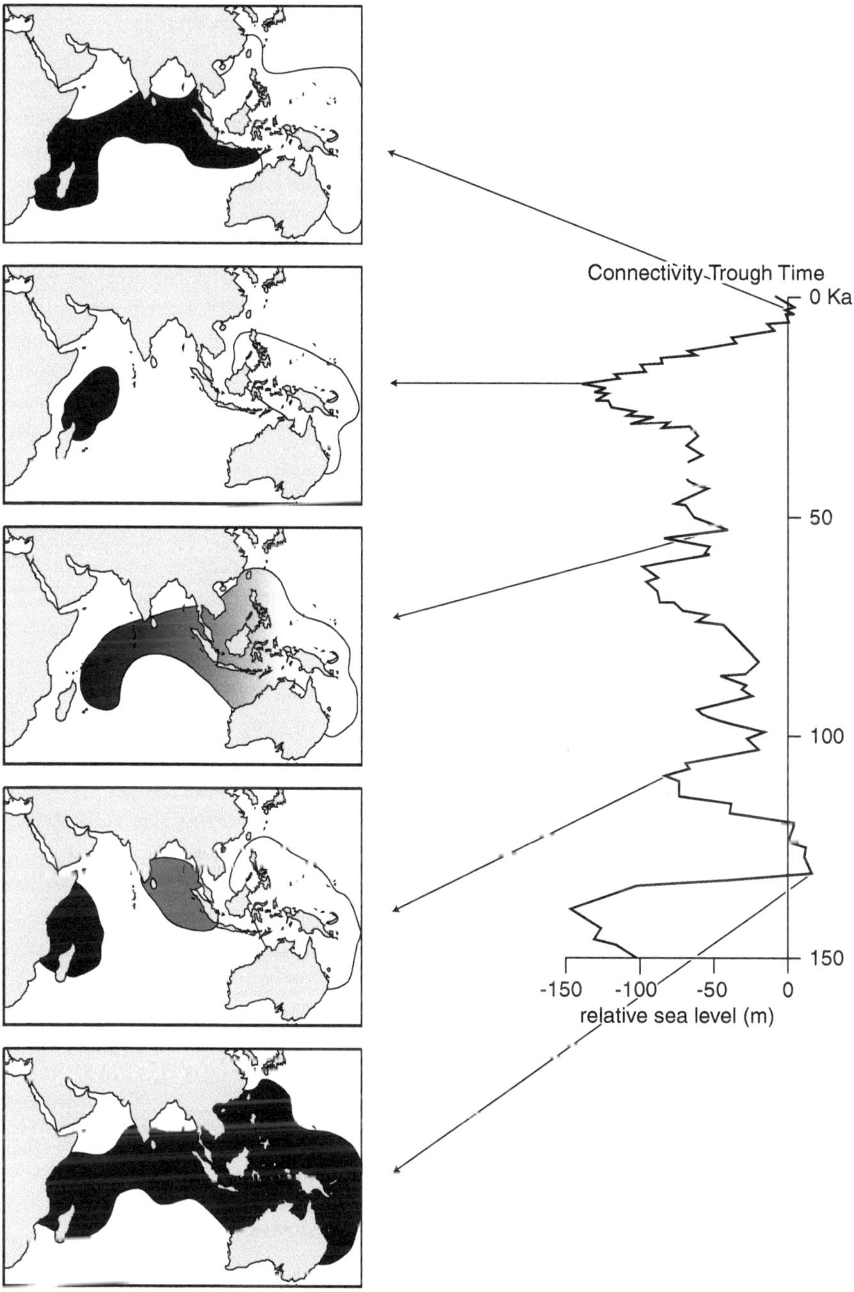

Figure 4.4 Hypothetical distribution of a coral taxon in response to repeated isolation over the past 150 ka of changes in sea level. During low sea-level stands, populations were probably separated from refuge to refuge. During high sea-level stands, populations may have exchanged genes and coalesced, or they may have incorporated sufficient genetic changes that gene flow between populations was severely restricted. The figure suggests that populations that have limited exchanges today were first affected by genetic changes long before the latest low sea level and subsequent isolation at around 20 ka. Modified and redrawn from Benzie (1999). The sea-level curve for the past 150 ka is adapted from Waelbroeck et al. (2002).

life habitats. During sea-level falls, inner-reef forms suffered local extinction, but re-settled and expanded back into the area during transgressive episodes. By contrast, metapopulations of outer-reef bivalves maintained a relative long-term stability over both low and high sea-level stands, promoting the appearance of endemic forms.

Global changes in sea level have combined with tectonics and antecedent topography to determine the location and geometry of reefs. Hubbard (1988) stressed that variations in tectonics and topography have led to different sea-level histories that are reflected in site-specific reef development patterns. Abrupt changes in sea level, at rates of up to 20 mm yr^{-1} demonstrated for the last deglaciation (e.g. the past 19 to about 8 ka, Bard et al., 1990; Blanchon & Shaw, 1995a,b; Siddall et al., 2003), probably also occurred during previous glacial cycles, at least since the Middle Pleistocene Climatic Transition (at about 1 Ma) with the emergence of high-amplitude glacial variability. The model of Kleypas (1997) assumes that during the last deglaciation a rate of sea-level rise exceeding 10 mm yr^{-1} resulted in coral reef drowning, thus restricting area-specific accretion rates by up to 5%. This model appears not to be of general value because coral communities would have been able to keep pace with sea level rising at rates averaging up to 20 mm yr^{-1} (Montaggioni, 2005). For example, reef crests on the barrier reef of Papeete (Tahiti, French Polynesia) have compensated for episodic jumps in sea level (Montaggioni et al., 1997). Similar compensational growth events are likely to have occurred throughout Pleistocene glacial cycles, promoting the persistence and areal expansion of reef systems in the course of sea-level change.

A lack of accommodation space is also thought to be a limiting driver of reef growth. The review of fringing reef development scenarios by Kennedy and Woodroffe (2002) shows how accommodation space controlled by sea-level position has exerted a commanding influence on reef physiography and architecture throughout the Holocene. From the insular shelf of south Oahu (Hawaii), Grossman, Barnhardt, Hart, Richmond, and Field (2006) demonstrated that a lack of accommodation space and frequent wave disturbance have been responsible for the restriction of framework development and vertical reef accretion below wave base during the Holocene. There was a clear transition from vertical to seaward reef development as sea level stabilized at around 5–3.5 ka. Sediment production and progradation increased significantly, filling shelf channels.

4.3. Controls on Reef Community Preservation: The Taphonomic Approach

The fate of organisms and their remains after death is the focus of taphonomy. On coral reefs, physical, chemical and biological taphonomic processes play major roles in determining the styles of framework and rubble preservation (Macintyre, 1984a; Hutchings, 1986). The effects of taphonomic processes on reef substrates may be either constructive or destructive (Scoffin, 1992; Perry & Hepburn, 2008). Constructional processes relate to both encrustation, that is deposition of additional biogenic carbonate material on and within the primary framework, and marine cementation. Destructional processes include both bioerosion, that is the degradation of hard substrates by biological processes prior to and after burial, and physical disruption (mainly by storms). The main purpose behind taphonomic research in coral reef systems is to detect bias affecting the fossil record and thus to test the assumption that the death assemblages of present reef biotas provide reasonable counterparts of fossil reef assemblages.

4.3.1. The Distribution of Taphonomic Signatures

The zone in which skeletal hard parts are most likely to suffer rapid postmortem degradation extends from the reef surface–water interface to several centimetres depth. This zone is referred to as the *taphonomically active zone* (TAZ in the sense of Davies, Powell, & Stanton, 1989). The degree of taphonomic degradation affecting corals is determined by the time during which skeletons remain within the TAZ (the residence time).

4.3.1.1. The modern and Holocene record

The physical, chemical and/or biologically controlled processes altering carbonate skeletons and/or sediments in recent reef environments have been reviewed by Macintyre (1984), Hutchings (1986), Bromley (1990) and Scoffin (1992). In addition, Perry and Hepburn (2008) described an array of specific, identifiable taphonomic signatures that are potentially diagnostic in terms of depositional environments.

Encrustation. Calcified encrusting organisms (epibionts) mainly grow on the surfaces of hard reefal substrates (Figure 4.5). The major taxa include non-geniculate coralline algae, foraminifera, bryozoans, serpulid worms and some bivalves and gastropods. Coralline algae play a key role as secondary reef builders, living on the substrate surface, and are typically photophilic or light-loving organisms. Foraminifera also serve as first-order binders growing on the undersides of coral rubble or within cavities (coelobites) and are dominated by the genera *Homotrema*, *Carpenteria*, *Gypsina*, *Planorbulina* and *Acervulina*. Bryozoans and serpulids are less common and mostly occupy cavity (cryptic) niches. Nevertheless, examination of cavity-dwellers (sciaphilic or shade-loving organisms) may contribute significantly to the reconstruction of sequences of coral deposition (Martindale, 1992;

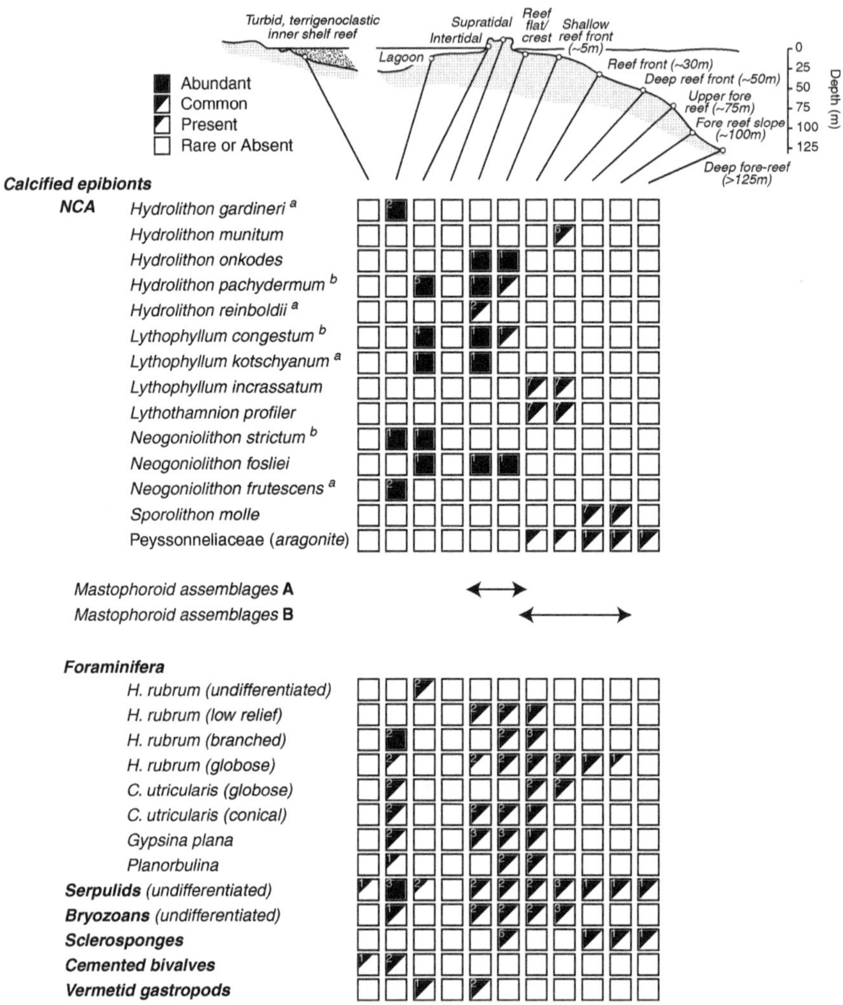

Figure 4.5 Sketch illustrating the distribution of the different types of calcified epibionts encrusting substrates across an idealized reef system. NCA = non-geniculate coralline algae; a = only reported from the Indo-Pacific reefs; b = only reported from the western Atlantic–Caribbean reefs; A = dominated by *Hydrolithon* and *Neogoniolithon* species; B = dominated by *Mesophyllum* and *Lithothamnium* species. Organisms referred to as: 1 = occur predominantly on exposed, high-illuminated substrates; 2 = occur predominantly on sheltered, low-illuminated substrates; 3 = occur on both exposed and cryptic niches; 4 = chiefly in mid-intertidal habitats; 5 = chiefly in high-intertidal habitats; 6 = associated with deep, cryptic niches in these habitats; 7 = reported from algal nodules (rhodoliths). Modified and redrawn from Perry and Hepburn (2008).

Perry, 2001). Their occurrence, diversity and sequential patterns within cryptic microhabitats have the potential to act as benchmarks for substrate residence times. The first substrate colonizers are represented by solitary forms (e.g. foraminifera and serpulids) that are later overgrown by colonial bryozoans or by encrusting algae.

The distribution patterns of calcified epibionts are driven by both biotic and abiotic factors. Like corals, the shapes of encrusters respond to prevailing environmental constraints, principally wave exposure. Depth and light, nutrient availability and biotic interactions (competition, predation) are also important determinants of the nature of encrusters and the development of successional stages. Widespread cover of the substrate by encrusters is typical of high-energy, outer reef settings. By contrast, encrustation is usually limited in low-energy, high-turbidity, back-reef environments. This has been assumed to result from the scarcity of suitable substrates for colonization in response to rapid burial due to stronger siltation rather than from lower light levels.

Boring. Organisms living within reef substrates are mostly bioeroders capable of penetrating carbonate rocks using chemical dissolution or mechanical abrasion (Hutchings, 1986). They include cyanobacteria, chlorophyte and rhodophyte algae, fungi, foraminifers, worms, sponges, bivalves and cirripedes. The contribution of boring organisms to framework alteration in Quaternary reefs can be inferred from trace fossils. The identification of the organisms responsible for traces is possible at least at the group level, in most cases. The taxonomy of fossil borers (ichnotaxa) is based on the morphological traits of their preserved traces (Figure 4.6). Two main groups of borers are identified from the size of their traces: macroborers and microborers.

Macroborers are typified by the production of boreholes larger than 1 mm diameter. Those operating in modern reefs are well known and include sponges (mostly Clionidae), bivalves (Lithophaginae and Gastrochenidae), sipunculids (Phascolosomatidae, Aspidosiphonidae), polychaete worms (Cirratulidae, Eunicidae, Fabriciinae, Spionidae) and barnacles (Perry & Bertling, 2000). On a reef system scale, sponges represent the most prominent infaunal eroders (as much as 75–90% of total macroborers). Bivalves are also efficient agents of coral bioerosion, producing typical vase- or funnel-shaped boreholes easily identifiable in the fossil record as the ichnogenus *Gastrochaenolites*. Polychaete worms are common initial colonizers of both living and dead coral substrates. Locally they can be responsible for as much as 35% of infaunal bioerosion in modern and Quaternary reefs. Scoffin and Bradshaw (2000) pointed out that macro-endoliths in coral framework can be separated into two categories, those inhabiting dead- and live-coral substrates respectively. Two styles of cavities were recognized. In dead coral substrates, cavities are sinuous and branched,

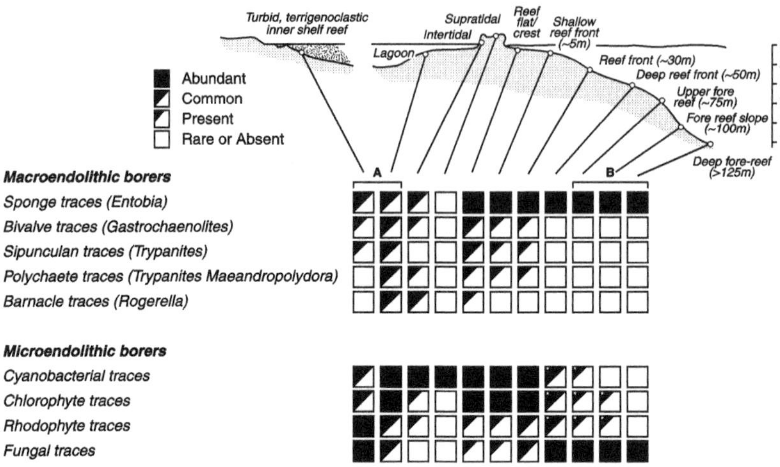

Figure 4.6 Sketch illustrating the distribution of different types of traces produced by macroendolithic and microendolithic borers through substrates, across an idealized reef system. Macroborers: A = macroboring exhibits a high variability in terms of borer abundance and assemblage diversity; B = macroboring operates through multiple phases of bioerosion, resulting in a high degree of substrate alteration. Microborers: 'white spot' symbol indicates that borings primarily develop parallel to substrate surface. Modified and redrawn from Perry and Hepburn (2008).

cross-cutting the original coral growth banding at random; these reflect active excavation by endolithic organisms (euendoliths) that include sipunculid worms, pholad and mytilid bivalves. By contrast, living coral skeletons are characterized by cavities created by passive endoliths (paraendoliths) that embed themselves in the living tissue and are entombed within the coral as the skeleton grows around them. The resulting traces tend to parallel the extensional direction of the coral. In addition, living corals may house a group of passive endoliths that occupy existing cavities within the skeleton (cryptoendoliths). Passive endoliths include pyrgomatid barnacles, spirobranch worms, some *Lithophaga* bivalves, gastropods, cryptochirid crabs and upogebiid and alpheid shrimps. The identification of both types of infestation of cavities in dead coral colonies provides an estimate of the elevation reached by the colonies above the surrounding soft sediment surface prior to burial. Fossil corals that occupied relatively elevated positions commonly exhibit extensive infestation by euendolithic borers and poor preservation of growth surfaces. By contrast, colonies that are infested mainly by para- and cryptoendoliths probably grew close to the soft sediment surface and were buried shortly after death. Thus, the taphonomy of macroborers can provide information on the environment in which fossil corals have grown.

As with encrusting species, a number of biotic and environmental factors have been invoked to explain both substrate colonization and the density of macroborers in reef environments (Perry & Bertling, 2000; Perry & Hepburn, 2008, and references therein). Due to the cryptic locations of most macroborers, light can be disregarded as a direct control. Wave energy appears to be a key determinant of macroborer distributions because it controls the rate of flow and thus transport of plankton through the reef structure. Additional but less efficient controlling factors include the rate of cover by encrusting forms, water depth and physical properties of corals (growth forms, colony sizes, density). Boring intensity may vary as a function of coral growth forms and the amount of dead basal area potentially offered to borer settlement. Massive coral colonies that do not limit borehole size are more commonly infested than branching forms. In addition to the form of the substrate coral, the type and extent of boring is influenced by substrate availability and the time spent within the TAZ. For example, in deep fore-reef environments, coral growth rates are lower and the residence times of dead corals are longer; and they are therefore, exposed to eroder recruitment for a greater time than in shallow-water settings. Variations in both skeletal density and structure exert additional controls on bioerosion by macroborers; in general, the higher the skeletal density, the lower the infestation. However, although it is assumed that boring is a direct function of skeletal density, the protection provided by higher-density corals against grazers (fish, sea-urchins and gastropods) may outweigh the increased energy cost of removing the higher-density skeletal material (Highsmith, 1981).

The extent of encrustation may play an important role in determining the intensity of macroboring. Generally, infestation decreases with increasing substrate overgrowth. The primary factors governing macroborer distribution have not been clearly differentiated and are presumably interactive, with their relative importance differing from site to site. Taxonomic uncertainties regarding a number of macroboring ichnospecies mean that it may be difficult to accurately determine the relationship between species and reef zones. As outlined by Perry and Hepburn (2008), this restricts the overall use of macroboring imprints as proxies for palaeoecological reconstruction, but the identification of local macroboring features may help to define the degree of taphonomic alteration.

Other boring features result from the activities of microborers. These produce boreholes that vary in diameter from about 1 to 100 μm. Microborers include phototrophic cyanobacteria, chlorophytes, rhodophytes and heterotrophic bacteria and fungi. The distribution of most of these organisms is primarily driven by variations in light penetration with depth, although some are able to operate within wide depth ranges. Relatively homogenous microendolithic assemblages are found throughout tropical reef regions at depth intervals related to specific light levels. Thus, and because of their light dependency, endolithic cyanobacteria and algae provide useful criteria for reconstructing palaeodepth (Vogel, Gektidis, Golubic, Kiene, & Radtke, 2000). Many microborers are taxonomically identified as ichnospecies on the basis of their trace attributes.

Burrowing. Burrowing organisms alter the sediments on which or in which they live through a variety of processes (mainly displacement and ingestion) that result in bioturbation, producing changes in grain size, sorting, texture and spatial rearrangement of layers (Bromley, 1990; Tudhope & Scoffin, 1984; Scoffin, 1992; Kosnik, Hua, Jacobsen, Kaufman, & Wust, 2007). Bioturbators are principally crustaceans (soldier crabs, alpheid and thalassinid shrimps). Generally, these groups occur in mutually exclusive zones, limited in distribution by the texture of the sediments (Suchanek, 1985; Bradshaw, 1997). Soldier crabs are found in sandy intertidal environments. Alpheid shrimps are found in intertidal and subtidal back-reef to reef-front zones depending on species. Thalassinid shrimps are mostly present in subtidal back-reef and fore-reef slopes. On a time scale of several millennia in the case of a prograding reef system, successive phases of seaward reef accretion will result in a shallowing-upward sequence from fore-reef through reef-flat and back-reef sediments to beach deposits. These will be characterized successively by thalassinid, alpheid and crab burrows (Bradshaw, 1997) (Figure 4.7).

4.3.1.2. The Pleistocene record

Studies of specific encrusting taxa remain patchy. The most detailed investigation of calcareous reef encrusters was carried out by Martindale (1992) on the Pleistocene rocks of Barbados. Assemblages of encrusters appear to be either of uniform or of mixed compositions. Uniform veneers developed on upward-facing surfaces of corals in back-reef settings. They consist of foliaceous coralline algae (*Mesophyllum*, *Neogoniolithon*, *Tenarea* or *Hydrolithon*) intermingled with the foraminifera *Gypsina plana* and *Planorbulina*, together with laminar and globose growth forms of *Homotrema rubrum* and isolated bryozoans and serpulids. Mixed veneers reflect changes in assemblage suites. Ranging from 80 to less than 10 mm in thickness, they are encountered on both sides of branches of reef-crest *Acropora palmata*. Upward-facing coral surfaces were first covered by thick layers of photophilic coralline species (*Hydrolithon*, *Lithophyllum*, *Tenarea*) overlain by thinner *Lithophyllum*, *Neogoniolithon* and *Mesophyllum* crusts. Sciaphilic encrusters such as the foraminifera *Planorbulina*, conical *Carpenteria utricularis* and *Homotrema* together with ascophoran bryozoans grew over the algal layers. The outermost part of the veneers consists of thin algal thalli,

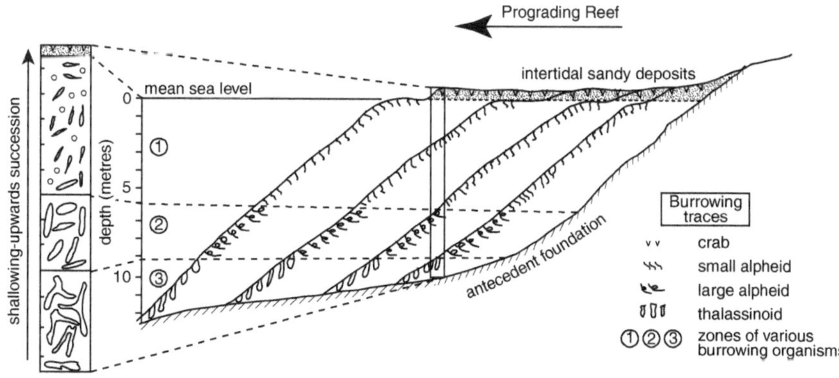

Figure 4.7 Idealized zonation of traces produced by burrowing organisms according to depth. Soldier crabs inhabit intertidal sandy sediments particularly in beaches and islets. Smaller alpheid shrimps live preferentially in intertidal to subtidal reef-front deposits, whereas larger forms are found downslope from about 6 to 9 m. Thalassinid shrimps mainly colonize deeper zones. In the case of a prograding reef system, the vertical sequence at a given location will consist of a shallowing-upwards succession from fore-reef to beach zones through shallow-water, subtidal to intertidal reef sediments. The diagnostic features are represented by off-reef thalassinid, near-reef alpheid and inner-reef soldier crab traces respectively. Modified and redrawn from Bradshaw (1997).

overgrown by branching *Homotrema* and globose *Carpenteria*, spirorbid worms and various bryozoans. By contrast, downward-facing coral surfaces carry millimetre-thick crusts devoid of the initial photophilic algal layer. The first stage of encrustation is represented by thin crusts of *Neogoniolithon* and *Lithophyllum*, covered in turn by *Mesophyllum* and *Lithothamnion*. The youngest laminae are composed of foraminifera (gracile branching *Homotrema rubrum*, globose *Carpenteria utricularis* and *Planorbulina*). The mixed assemblages are characteristic of high-energy, shallow reef settings. Both uniform and mixed encrusting assemblages are similar to those found in modern and Holocene Caribbean reefs.

Descriptions of boring activity are limited to a very few sites. Jones and Pemberton (1988) described the bivalves *Lithophaga* preserved in-place within their boring cavities and *Gastrochaenolites torpedo* in massive corals in a palaeolagoon on Grand Cayman in the Caribbean. Klein, Mokady, and Loya (1991) described macroboring traces in massive poritids from uplifted Pleistocene reefs in the northern Red Sea. Perry (2000) provided a detailed report on the compositions of macroborer assemblages that have infested a fringing reef complex in north Jamaica dating from the last interglacial. Twelve distinct types of boring traces were recognized and assigned to specific ichnotaxa. Most trace types were referred to the ichnogenus *Entobia* with morphological attributes similar to those of the sponge *Cliona* spp. found on modern coral reefs. Other types of traces were related to the ichnogenus *Gastrochaenolites* regarded as resulting from boring by a gastrochaenid bivalve, and to the ichnogenera *Maeandropolypora* and *Trypanites* interpreted as representing traces produced by worms. In addition to these distinct morphospecies, Perry (2000) found traces produced by indeterminate ichnogenera, probably produced by barnacles and sponges. The relative abundance of trace types varied between reef environments. At the scale of the entire fringing reef complex, the abundance of groups of borers, expressed as a percentage of the substrate released by boring, was as follows: sponges 64.7%, worms 25.8%, bivalves 8.2% and unidentified 1.3%.

4.3.2. Taphonomic Features as Criteria for Identifying Reef Sub-Environments and Depositional Events

4.3.2.1. Identification of reef sub-environments

Perry and Hepburn (2008) showed that taphonomic criteria may improve the interpretation of the depth zonation of *in situ* coral assemblages. Since taphonomic processes vary between cross-shelf reef systems with depth, wave energy and water clarity, their signatures provide an efficient tool to delineate sub-environments, using successions of taphonomic features and the relative abundance of key encrusters and/or borers. Each sub-environment appears to be typified by a distinctive taphonomic signal (Blanchon & Perry, 2004). This approach is especially useful when the identification of reef environments relies on cores extracted from reef tracts that may be dominated by a single or very few coral species and/or were periodically subjected to the influence of hurricanes. Because such sites are dominated by coral clasts that have been deposited homogeneously in a range of different environments, the interpretation of the internal structure of Holocene and older reefs in terms of depth zonation and reef development is particularly difficult and a taphonomic approach is therefore required.

The cross-shelf distribution of microboring varies significantly. Infestation is generally higher in back-reef and inshore environments than on the outer-shelf. High-energy, shallow reef-front zones (2–10 m below mean sea level) are typified by widespread, thick calcareous encrustation by photophilic organisms such as the coralline algae *Hydrolithon* (*H. onkodes* and *H. fosliei*), *Neogoniolithon* and *Lithophyllum* along with the foraminifer *Gypsina plana*. Sciaphilic assemblages mainly composed of globose *Carpenteria utricularis* and branched forms of Homotremidae, together with bryozoans, serpulids and sclerosponges, occur within cavities. Microendolithic traces are ubiquitous and produced mostly by cyanobacteria (*Scolecia filosa, Erygonum nodosum, Fascichnus frutex, Fascichnus dactylus, Rhopalia catenata*) and chlorophytes (*Ichnoreticulina elegans, Cavernula pediculata*). Macroboring traces result predominantly from the activities of sponges

(*Entobia* spp.), bivalves (*Gastrochaenolites*) and worms (*Trypanites/Maeandropolydora*). The amount of material removed averages 25% of the bulk volume. Reef-crest and reef-flat zones (0–2 m below mean sea level) display taphonomic features comparable with those described from the reef-front. The distribution density of epi- and endobionts is highest in areas where wave energy prevents sedimentation. Sponges, mainly clionids, are the most efficient macroborers, responsible for about 90% of substrate disintegration in reef-crest and shallow reef-front zones. Bioerosional traces are mainly referred to *Entobia* spp. (*E. convulata*, *E. ovula* and *E. glomerata*). Subordinate borers include bivalves (*Lithophaga*, *Gastrochaena*, *Petricola*), polychaetes, sipunculids and barnacles, all with a heterogeneous distribution varying from site to site. High-turbidity inner-reef areas, influenced by siliciclastic supply, are characterized by significant variation between substrates in terms of the extent of both encrustation and boring. Calcareous encrustation is usually limited to thin isolated crusts of coralline algae (*Lithophyllum*, *Lithothamnion*, *Neogoniolithon* or *Lithoporella*) and isolated foraminifera, serpulids and bryozoans. There is marked variability in the levels of bioerosion from clast to clast. Key species of microborers are similar to those found in clear waters, but they form compressed assemblages reflecting the lower light penetration; rhodophyte and fungal traces occur preferentially in shallower areas. Although macroboring assemblages may show a shift from dominance by clionids to dominance by lithophagid bivalves and worms (MacDonald & Perry, 2003), sponges commonly responsible for from 55% up to 75% of the substrate disintegration in lagoonal and nearshore settings. Deep reef-fronts (about 50 m depth) are characterized by limited encrustation, with only thin crusts of coralline algae (*Mesophyllum*, *Lithothamnion*, *Spongites*, *Lithoporella*), foraminifera (*Carpenteria utricularis*, *Acervulina* sp.), bryozoans and serpulids. Sponges remain the dominant borers in deeper environments (about 15 to >110 m) and are locally responsible for 98% of substrate infestation. Microendolithic processes are less effective. The most significant at these depths are fungal traces (*Orthogonum fusiferum*, *Saccomorpha spherula* and *Polyactina araneola*). In rubble-dominated deposits in inter- and supratidal zones, coral fragments show varying degrees of abrasion and rounding. Some clasts are thinly encrusted by layers of coralline algae. Bioerosion varies widely but is generally of low intensity. The percentage of material removed is the lowest of any reef environment, on average 10%. The common ichnofacies include *Entobia* isp., *Gastrochaenolites* isp. and *Trypanites* isp.

4.3.2.2. Identification of short-term depositional events

One other potentially viable application of reef taphonomic features relates to the recognition of short-term depositional events (Martindale, 1992;

Scoffin, 1992, 1993; Blanchon & Perry, 2004; Perry & Hepburn, 2008). The nature of taphonomic successions has been demonstrated to be useful to differentiate storm-controlled and non-storm deposition of coral detritus.

Under fair-weather conditions, the deposition of coral rubble is gradual, resulting in the progressive burial of rubble beds (Figure 4.8A). Clasts occupying the surface of the sediment pile experience high water agitation and high light irradiance, particularly in shallow reef-front and reef-flat settings. Thus, clasts are, to a large extent, overgrown by thick veneers of photophilic encrusting species. Periodically, the surface of the deposit is coated by freshly deposited clasts, leading to the relative displacement of earlier deposits downwards. Rubble formerly at the surface and now in the process of burial will be subject to decreasing water turbulence and light intensity, and the composition of encrustations will transform successively from photophilic-dominated assemblages through semi-cryptic assemblages to those composed mostly of sciaphilic organisms. Each successive assemblage overgrows its antecedents and the encrustation sequence reflects the gradual transition through different microhabitats. Under storm conditions, reefs have been assumed to record deposition controlled by a rapid pulse in which live coral material is transformed into a death assemblage (Greenstein, 2007, and references therein). The overall composition of encrusting assemblages is similar to that found in fair-weather sequences, but as a result of the instantaneous deposition of clasts, different encrusting sequences may develop (Figure 4.8B). Overprinting of successive encruster developments may depend on the pre-depositional history of the clasts reworked by the storm. Generally, there is little overprinting in a deposit assumed to be linked to a unique storm event. Each clast is coated by a single encrusting succession related to the depth of burial in the deposit. The base of a fresh storm deposit is typified by thin crusts of a sciaphilic assemblage. The intermediate parts of the deposit are colonized by thin semi-cryptic to photophilic assemblages, whereas the uppermost layers are overgrown with thick sequences of photophilic encrusters. The thickness of the crusts in these layers is governed by the periodicity of the storms generating the deposits. For instance, Perry (2001) applied observations in Holocene and modern, shallow-water reefs from the Caribbean to the Pleistocene of Barbados. He was able to differentiate two distinct *A. palmata* units resulting from *in situ* framework accretion and storm deposition respectively. Storm deposits preserved in the Pleistocene reef exposures formed repetitive successions of discrete (0.4–1 m thick) depositional units. Each unit is characterized by a vertical sequence from 1 to 2 mm thick of sciaphilic encruster-rich veneers at the base to photophilic encruster-rich veneers increasing gradually up to 20 mm thick at top. The upper surfaces of storm units are usually colonized by pioneering coral colonies (mainly *Agaricia agaricites*) the sizes of which prove growth over less than 10 years.

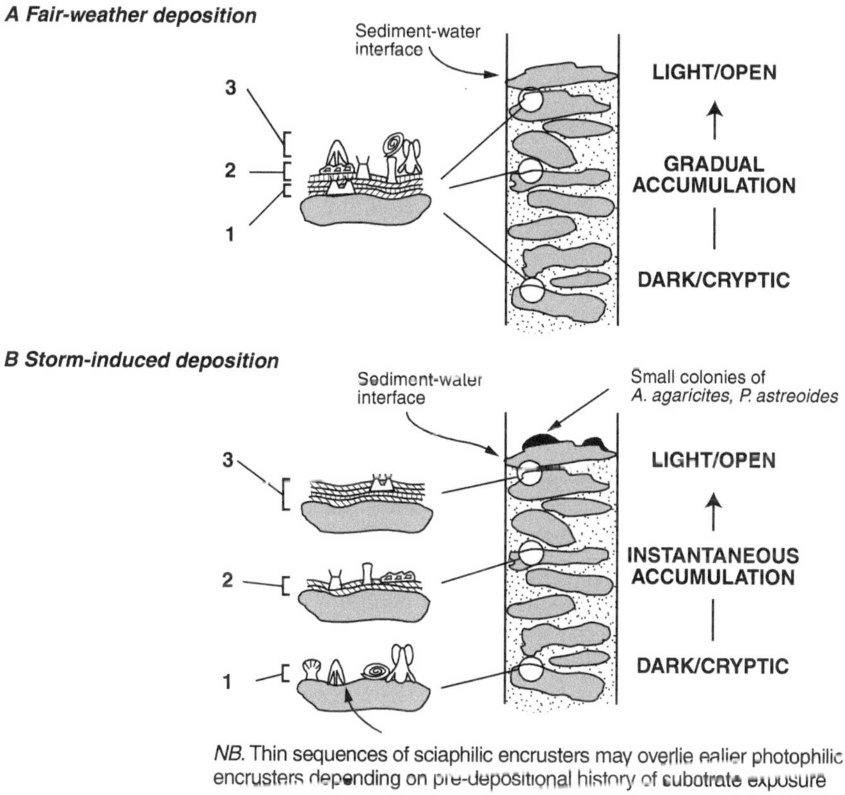

Figure 4.8 Sketch illustrating changes in the sequence of successive encrustation events during coral rubble deposition as a function of water energy. (A) Fair-weather deposition: stage 1 = encrustation by photophilic organisms including coralline algae (*Hydrolithon* sp., *Neogoniolithon* sp., *Lithophyllum* sp.) and foraminifers (*Gypsina plana*, low-relief *Homotrema rubrum*). The crust thickness is controlled by the duration of exposure close to the sediment–water interface; stage 2 = encrustation by semi-cryptic organisms, including corallines (*Hydrolithon* sp., *Lithophyllum* sp., *Neogoniolithon* sp., *Sporolithon* sp. and *Titanoderma* sp.), foraminifers (branched and globose *H. rubrum*, *G. plana*) and bryozoans; stage 3 = encrustation by sciaphilic organisms, including foraminifers (globose and conical *Carpenteria utricularis*), serpulids, cheilostome bryozoans and sclerosponges. (B) Storm-induced deposition: stage 1 = thin sequence of sciaphilic encrusters including serpulids, foraminifers (conical and globose *Carpenteria utricularis*), cheilostome bryozoans and sclerosponges; stage 2 = thinner sequence of photophilic and semi-cryptic encrusters, including corallines (*Hydrolithon* sp., *Lithophyllum* sp., *Neogoniolithon* sp., *Sporolithon* sp. and *Titanoderma* sp.), foraminifers (branched and globose *Homotrema rubrum*, *Gypsina plana*) and bryozoans; stage 3 = thick sequence of photophilic encrusters including corallines (*Hydrolithon* sp., *Neogoniolithon* sp., *Lithophyllum* sp.) and foraminifers (low-relief *H. rubrum*, *G. plana*). Modified and redrawn from Perry and Hepburn (2008).

Storms events in the sedimentary reef record may also be identified by the distribution of burrowing and related textural characteristics. The density of burrows and the degree of alteration decrease significantly in sediments overlain by storm beds. In areas subjected to frequent storm events, the original textural attributes of the sediment tend to be well preserved due to the short period during which infaunal animals have been reworking the sediment (Bradshaw, 1997).

Based on the detailed examination of Pleistocene reef assemblages in the Caribbean, Meyer, Bries, Greenstein, and Debrot (2003) noted that the proportion of corals preserved in the growth position increased in sites that today are subjected to a lower frequency of tropical storms. They suggested that the orientation of the coral colonies might be used as a reliable metric to identify past storm events from the Quaternary record. By contrast, Bishop and Greenstein (2001) remained sceptical that hurricane-related signals could be identified in Pleistocene coral accumulations, in part because it is virtually impossible to differentiate the sudden supply of living coral elements in death assemblages after fossilization.

Greenstein and Moffat (1996) attempted to determine the residence time of coral material on the sea floor using taphonomic features (encrustation and macroboring) shown by *Acropora palmata* and *A. cervicornis*. Comparing late Pleistocene exposures of the Bahamas with those of their modern coral counterparts nearby, they found that the modern coral material was markedly more degraded than that preserved in the Pleistocene. This suggested that the fossil coral specimens had been exposed to alteration on the sea floor for a shorter period than their modern equivalents. Sudden burial, related to storm events, was assumed to have occurred during a fall in sea level that killed acroporid-dominated communities but limited damage by encrustation and boring.

4.3.3. Taphonomic Controls on Modern and Fossil Reef Communities

4.3.3.1. Coral communities

As Greenstein (2007) pointed out, palaeontologists have paid little attention to the question of coral preservation in comparison to that of other marine calcifying invertebrates. There are several reasons for this. Reef-building corals are considered primarily to be more resistant to taphonomic alteration than most other benthic skeletal organisms living in temperate and tropical seas. In addition, the complex reef community includes a great variety of potentially fossilizable taxa displaying different growth forms and with different chemical and mineralogical attributes. As a result, the varied components follow different taphonomic pathways and the end-product is difficult to study comprehensively. Greenstein (2007) in a compilation of studies of modern to subfossil reef-coral taphonomy identified two key topics: the controls on the taphonomic bias suffered by living corals and the fidelity of coral death assemblages as counterparts of living communities.

Taphonomic bias. Our understanding of the effects of alteration on the postmortem evolution of coral communities is principally based on studies conducted in the Florida Keys, the Great Barrier Reef and the Bahamas. It has been assumed that an assemblage of dead corals deposited close to a living coral community provides a relatively faithful picture of the once-living community, and that the living assemblage provides a reasonable proxy for the fossilized community. Death assemblages are defined as in-place dead coral colonies and coral rubble deposited up to about 10 cm within the sediment in close proximity to the living assemblages. Clearly, these assemblages were trapped within the TAZ and thus affected by erosional processes.

Gardiner, Greenstein, and Pandolfi (1995) investigated the role that intrinsic (colony growth forms, taxonomic membership) and extrinsic (hydrodynamic energy, abrasion and dissolution rates) factors play in the degree of deterioration of dead coral assemblages from low-energy patch reefs and high-energy reef-tract zones in the Florida Keys. Corals from protected patch reefs exhibited significantly greater alteration than those deposited in wave-exposed reef-tract settings. A comparison of the taphonomic degradation of massive faviids (*Favia fragum*), branching acroporids (*Acropora cervicornis*) and encrusting (*Millepora alcicornis*) in the two reef zones indicates that there are only insignificant differences in deterioration from zone to zone or, in other words, there was no environmental control on taphonomic degradation of the different corals. Biotic factors appeared to be more effective than environmental conditions (exposure to waves) in controlling the level of alteration. Irrespective of the environment, faviid colonies are more extensively affected by boring (attack by bivalves and sponges) than acroporids and hydrocorals. Thus, the diversity and abundance patterns of coral growth forms are clearly important determinants of the degree of taphonomic alteration that affects a coral community.

Investigation of the relative importance of biotic and abiotic factors to the level of coral taphonomic degradation was extended to deep reef environments (20–30 m) of the Florida Keys by Greenstein and Pandolfi (2003). A comparative analysis of three different growth forms between shallower and deeper reef environments shows that there are significant variations and gradients in the intensity of coral alteration within the various reef habitats. Abrasion and dissolution were more severe in patch-reef and reef-crest environments. Variations in rates are assumed to result from the different energy regimes in which the corals accumulated. By contrast, invasion by borers and encrusters was higher in deeper settings (Figure 4.9). Variations in coverage by epi- and endobionts were likely to be controlled

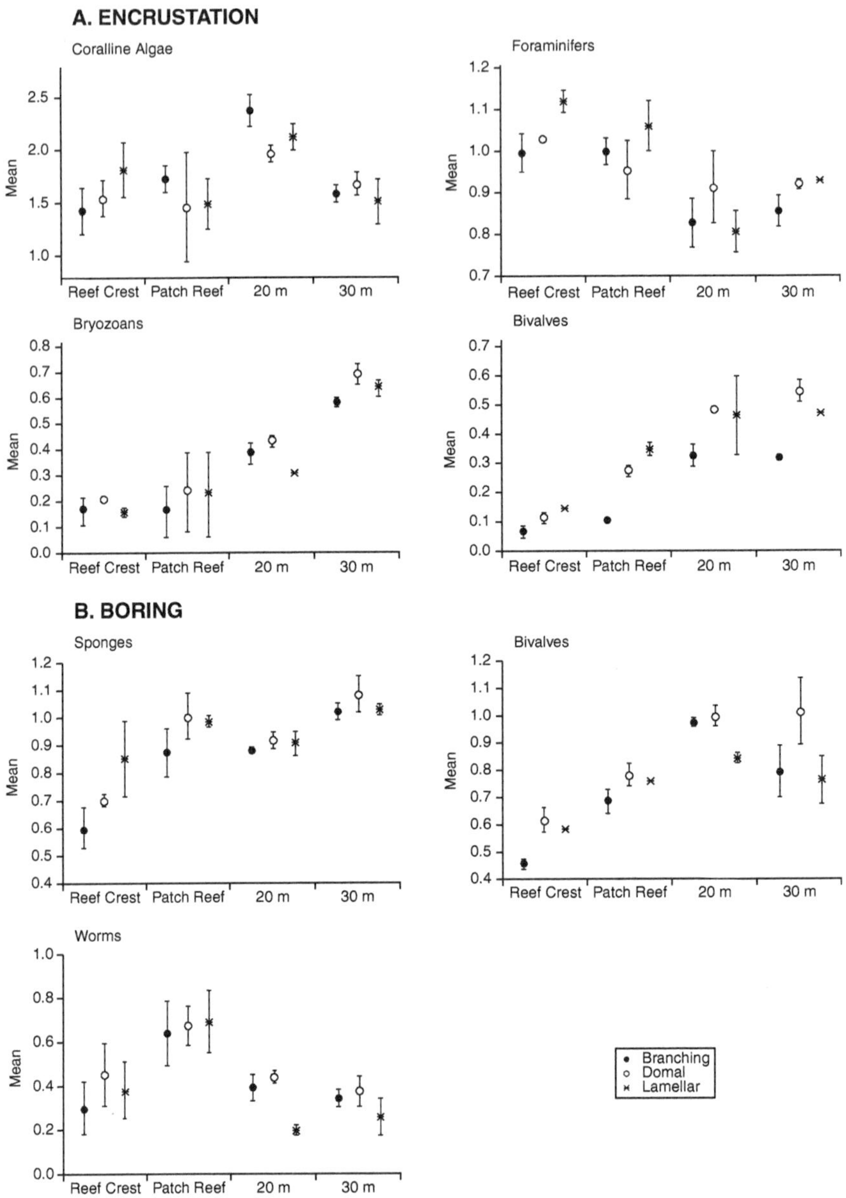

Figure 4.9 Estimates of biologically mediated alteration suffered by coral specimens of branching, domal or lamellar growth forms living in different environments (patch reef, reef crest, fore-reef). Measurement of the biological variables uses the percentage of surface area of a coral specimen covered by the encrusters or perforated by the borers. The coral is scored 0 if encrusters/borers were absent, 0.1–1 for 1–25% occurrence, 1.1–2 for 26–50% occurrence, 2.1–3 for 51–75% occurrence and 3.1–4 for 76–100% occurrence. Error bars refer to standard errors of the mean (95% confidence intervals). (A) Average occurrence by encrusters (coralline algae, foraminifers, bryozoans, bivalves). (B) Average occurrence by borers (sponges, bivalves and worms). Modified and redrawn from Greenstein and Pandolfi (2003).

by the residence time of the skeletal material in the TAZ and by nutrient availability, both of which increased in deeper-water environments. However, the responses of the specific growth forms to taphonomic degradation are similar regardless of the reef environment.

Pandolfi and Greenstein (1997a) earlier carried out a similar but taxonomy-independent study at Orpheus Island on the Great Barrier Reef, using the same methods of data collection and statistical analysis. Colony growth forms and invasion rates by boring and encrusting organisms were taken as intrinsic driving factors in combination with water depth, environmental energy and physicochemical factors. The respective roles of biotic and abiotic factors were estimated for massive, branching and free-living coral types in shallow (2–3 m) and deeper (6–7 m) settings on both leeward and windward parts of inner fringing reefs. In contrast to reefs of the Florida Keys, the taphonomic degradation of corals from the Great Barrier Reef varied between different growth forms. Statistical analysis of variables indicated that massive corals are markedly more affected by biological and physicochemical agents than branching and free-living forms. In addition, death assemblages encountered in deeper water and in sheltered settings were more altered than those in shallow-water or windward settings. Skeletal density, areal extent of colony coverage by living tissue and extensional rate were assumed to be the main biotic factors driving the intensity of postmortem deterioration. Paradoxically, although massive coral forms were more prone to suffer attack by borers, encrusters and significant dissolution, they were nevertheless able to maintain a higher skeletal integrity and therefore remain preserved for a longer period than other corals within the TAZ, irrespective of reef setting.

According to Greenstein (2007), wave energy may be negatively correlated with taphonomic preservation. Any coral growth form in a sheltered environment is able to resist alteration in the TAZ for a longer time than a comparable form in high-hydrodynamic-energy settings. This is because it escapes the reworking and subsequent partial disintegration by waves that can occur prior to burial. Hunter and Jones (1996) reached similar conclusions, observing that corals that grew on the patch reefs and in the reef tract of the late Pleistocene Ironshore Formation (Grand Cayman Islands) responded differentially to taphonomic processes as a function of local water agitation. Conversely, lower water energy and higher depositional rates may have favoured rapid burial and preservation. However, from the comparative analysis of modern and Pleistocene high-energy acroporid-dominated assemblages in the Bahamas, Greenstein and Moffat (1996) concluded that the degree of coral preservation was relatively independent of wave energy in areas subjected to high accumulation rates. Whatever the exposure to waves, the presence of well-preserved coral

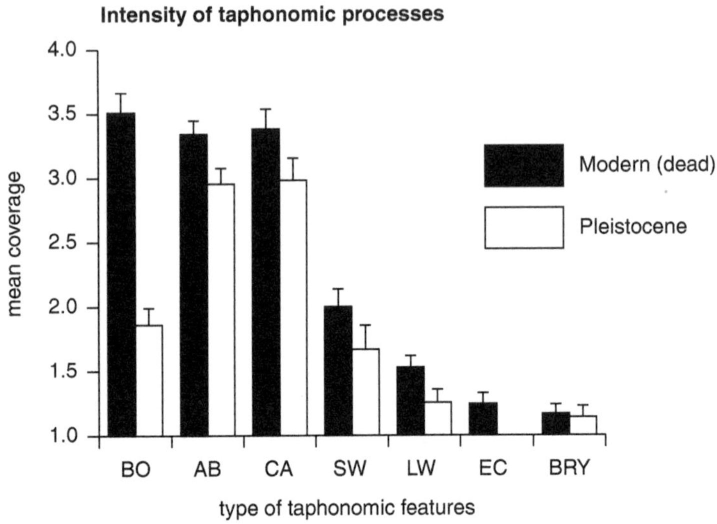

Figure 4.10 Estimates of taphonomic alteration suffered by modern (dead) and late Pleistocene colonies of *Acropora palmata* and *Acropora cervicornis* at San Salvador, Bahamas. Measurement of the biological variables uses the percentage of surface area of a coral specimen covered by the encrusters or perforated by the borers. The coral is scored 0 if encrusters/borers were absent, 1–1.9 for 1–25% occurrence, 2–2.9 for 25–50% occurrence, 3–3.9 for 51–75% occurrence and >4 for 76–100% occurrence. Error bars refer to standard errors of the mean (95% confidence intervals). Types of alteration: BO = boring (primarily sponges and lithophagid bivalves), AB = abrasion, CA = encrusting coralline algae, SW = encrusting worm tubes (diameter < 1 mm), LW = encrusting worm tubes (diameter > 1 mm), EC = encrusting corals and BRY = encrusting bryozoans. Modified and redrawn from Greenstein and Moffat (1996).

growth forms in the fossil record appears to require rapid entombment of both living and dead corals during reef accretion (Figure 4.10).

Fidelity of coral death assemblages. The ability of coral assemblages to reflect the once-living community appears to differ significantly between reef habitats (shallow versus deep-water settings, and high versus low water agitation) and between the western Atlantic–Caribbean and the Indo-Pacific province.

Pandolfi and Michin (1995) conducted a comparative taphonomic analysis of living coral communities and dead assemblages on fringing reefs in Madang Lagoon, Papua New Guinea, western Pacific (Figure 4.11). They were able to demonstrate that the original structure of the living coral communities was more faithfully reflected by their dead analogues in protected reef-crest zones than by those in high-energy reef-crest sites. In protected reef environments, death assemblages were likely to result from in-place accumulation, thus preserving most of the attributes of the

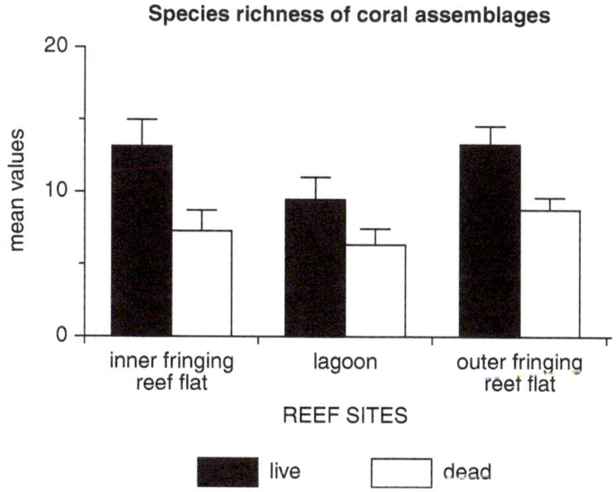

Figure 4.11 Species richness for living and dead coral assemblages at three different sites (protected, inner reef flat, protected lagoon and medium-energy, outer reef flat) from Madang Lagoon Area, Papua New Guinea. Error bars refer to standard errors of the mean (95% confidence intervals). Modified and redrawn from Pandolfi and Michin (1995).

once-living community, whereas dead corals deposited in strongly agitated sites may include reworked material and preserve little of the original community composition. The patterns of zonation observed in living reefs were not generally retained after death of the corals. Irrespective of wave exposure or depth, the living communities exhibit a greater richness than their dead counterparts. This may be due either to the longevity of some species, that exceeds the duration required to severely alter their skeletons in high-energy habitats, or to the selective preservation of some growth forms present in the original community. Estrada, Alvarez, Edinger, and Pandolfi (2004) later carried out a taphonomy experiment in Madang Lagoon, using live massive, branching and free-living corals in order to assess the effects of the various taphonomic variables (encrustation, boring and mechanical abrasion) on the intensity and patterns of alteration, between buried and exposed coral specimens from protected back-reef to high-energy patch reef settings. They observed (1) that the control of coral growth form on the taphonomic variables was very low: massive corals with the highest skeletal density were the most infested by macroborers, whereas branching forms with the lowest density were predominantly affected by abrasion; (2) that biological and mechanical alteration varied from site to site: the more exposed the site, the greater the alteration effects; (3) that burial had a strong control on taphonomy: exposed coral colonies suffered greater biological and mechanical alteration than buried ones; and (4) that the record of taphonomic evolution was strongly overprinted by high wave energy levels, and high-frequency burial–exhumation cycles promoted biological alteration.

Greenstein and Pandolfi (1997) compared the taxonomic compositions and diversity of living coral communities and adjacent death assemblages in reef-tract and patch-reef environments of Key Largo, Florida (Figure 4.12). Although the death assemblages differed in details of composition and diversity, they exactly reflected the zonation of living corals in the same environments. The disparity between the taxonomic compositions of living communities and death assemblages seems to have been caused by a significant growth form bias in the death assemblages in both reef-tract and patch-reef sites. Massive growth forms were prominent in the life assemblages but poorly represented in the dead analogues, whereas branching forms were of relatively low abundance in the living communities but predominated in the corresponding death assemblages. Finally, there were no significant differences in diversity between life and death assemblages.

Comparing the results from reefs in Madang Lagoon, Papua New Guinea, with those of shallow-water reefs in Florida (Pandolfi & Greenstein, 1997b), Greenstein and Pandolfi (1997) and Greenstein (2007) pointed out that there are both similarities and differences between the living coral communities and the corresponding death assemblages. In

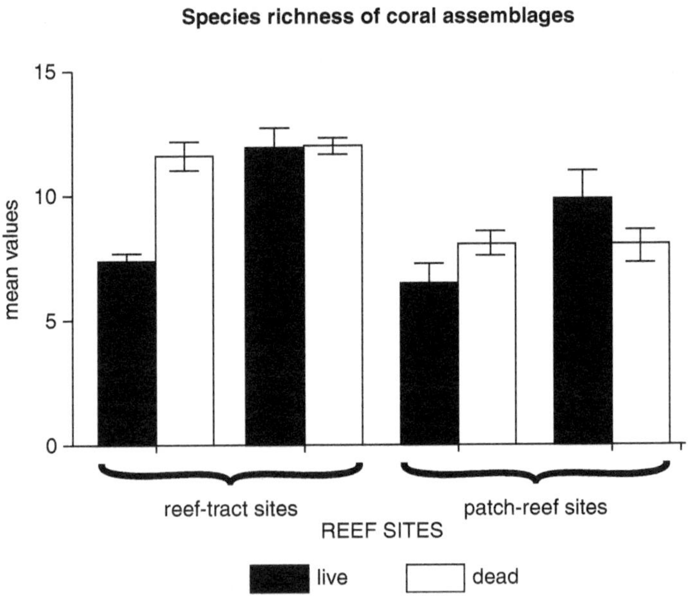

Figure 4.12 Species richness for coral life and death assemblages at two different types of reef sites (high-energy reef tract, lower-energy patch reef) in the Florida Keys. Error bars refer to standard errors of the mean (95% confidence intervals). Modified and redrawn from Greenstein and Pandolfi (1997).

neither area did the species diversity of the death assemblages faithfully reflect that of the living assemblages. However, there was a marked interprovincial contrast between the taxonomy of living and dead assemblages. Three kinds of differences were detected. First, and contrary to the Florida reefs, the fringing reefs of Madang Lagoon contain life assemblages with significantly higher diversity than those of the adjacent death assemblages. The differences between the two provinces may be linked primarily to differences in the diversity of the living communities. There is a significantly higher coral diversity, especially of branching species, in the Indo-Pacific province (up to 100 species, with *Acropora, Pocillopora* and *Stylophora* as the dominant genera). Thus, the precise identification of a set of branching growth forms present in coral debris is difficult or impossible, even if the corresponding species are present in an adjacent life assemblage. In addition, most gracile branching forms probably fail to resist disintegration. Second, the death assemblages of the Florida reefs are enriched in species not found in the life assemblages; only 57% of the dead species are encountered live, whereas in the Madang Lagoon 94% of the dead species are found in the living communities. Third, in the shallow-water reefs of Florida, most species present in death assemblages are absent from the adjacent living framework. This may be a consequence of the recent reef crisis in the Caribbean (see Chapter 3, Section 3.3.1). Many species are at present relict and are only found in death assemblages. Similar disparities are seen in deeper reef sites where death assemblages are again more diverse than in adjacent living communities.

To our knowledge, the only work devoted to the comparison between Holocene, living and death assemblages has been on the lagoonal reefs of Papua New Guinea (Edinger et al., 2001). These authors demonstrated that the composition of coral assemblages from uplifted Holocene reefs clearly reflects a mixture between that of the life and death assemblages found in modern reefs close to the fossil outcrops (Figure 4.13). This suggests that the Holocene assemblages consist of time-averaged coral cohorts that overlapped successively before uplift. As a result, species diversity appears to be higher in the subfossil deposits than in their modern analogues. In addition, and in contrast to the living assemblages, branching growth forms predominate over massive and lamellar forms in the subfossil accumulation as a result of their presumed higher growth and postmortem accumulation rates. Riegl (2001) showed that, in the fringing reefs at Dubai (Arabian Gulf), the composition of coral rubble did not directly reflect that of the living community and branching acroporids are by far the most common detrital constituent (Figure 4.14).

Studies comparing the compositions of modern life and death assemblages and their Pleistocene analogues by Greenstein and Curran (1997) were based on late Pleistocene coral deposits in the Bahamas and modern coral assemblages from the Florida Keys. They were able to show that the Bahamian fossil

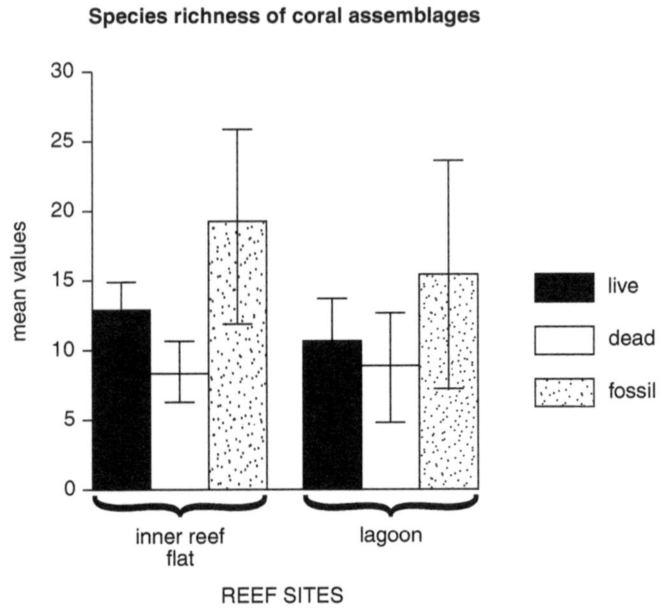

Figure 4.13 Species richness for living, dead and fossil (Holocene) coral assemblages from different modern sites (protected, inner reef flat, protected lagoon) and from the adjacent raised reef terraces respectively, at Madang Lagoon Area, Papua New Guinea. Error bars refer to standard errors of the mean (95% confidence intervals). Modified and redrawn from Edinger et al. (2001).

assemblage more faithfully reflected the structure and composition of the Floridian living communities than adjacent death assemblages. Despite its geographic distortion, this work strongly suggested that coral death accumulations may not represent reliable 'protofossil' assemblages. However, by restricting data collection to the Florida Keys area, Greenstein, Curran, et al. (1998) generally confirmed the conclusions of Greenstein and Curran (1997). The composition of coral assemblages in the late Pleistocene Key Largo Limestone is quite similar to that of nearby living communities.

However, these findings are not of general application. For example, on San Salvador (Bahamas), comparison of the living and dead assemblages from a mid-shelf patch-reef and Pleistocene assemblages from adjacent emergent patch reefs and reef tracts revealed that variability in composition between living and Pleistocene assemblages is markedly greater than that between the death assemblage and Pleistocene deposits (Greenstein, Harris, et al., 1998) (Figure 4.15). Although the environments of the modern and fossil assemblages were not exactly similar, it was assumed that the differences between the modern and ancient coral communities result from the recent demise of *Acropora cervicornis*-dominated communities and their replacement by *Porites porites*. However, despite a search for possible

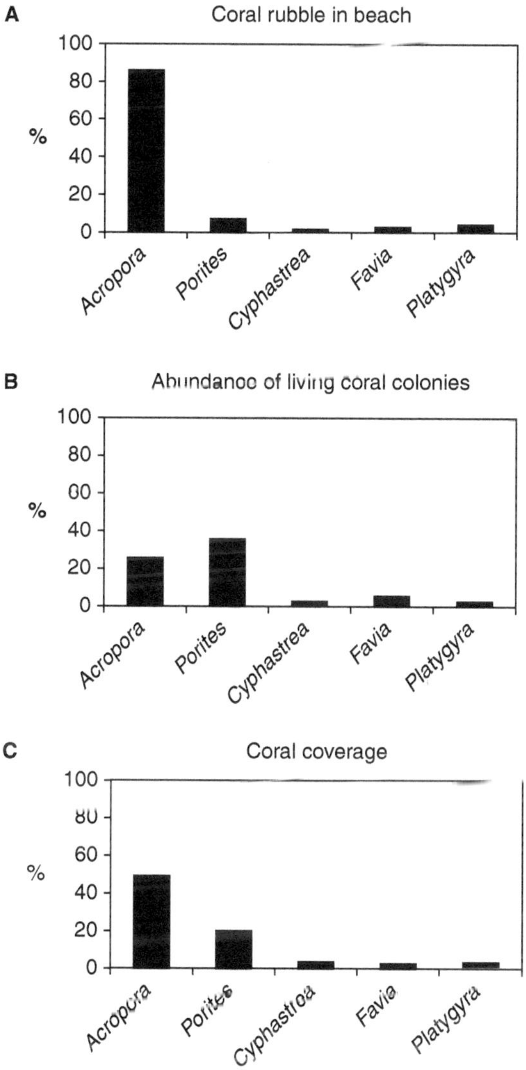

Figure 4.14 Comparison between the compositional attributes of living coral assemblages and coral rubble, on fringing reefs, Dubai (Arabian Gulf). (A) Generic composition of beach coral rubble. (B) Abundance of living coral colonies (surveyed along 50 m point count transects). (C) Relative generic coral coverage in the living assemblages. Modified from Riegl (2001).

transitions in the late Pleistocene coral reef communities in the same Bahamian sites, Rothfus and Greenstein (2001) concluded that there was no evidence of any mass mortality of any of the widespread coral species that could be comparable to the recent crisis.

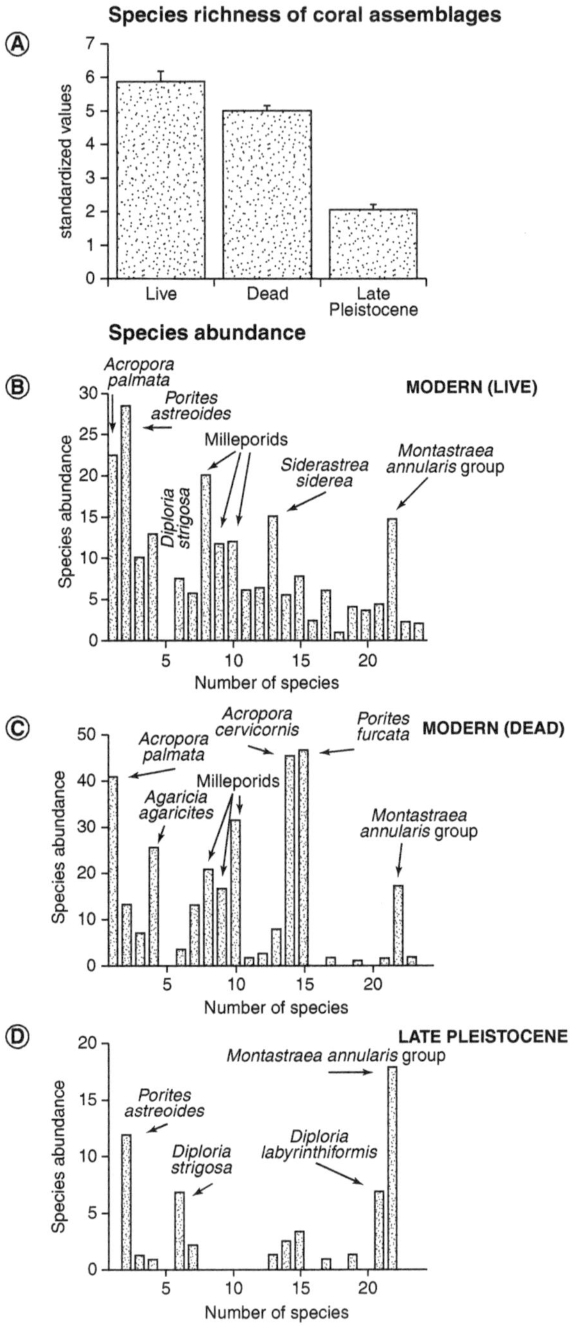

Figure 4.15 Species richness and abundance for living, dead and late Pleistocene coral assemblages from San Salvador, Bahamas. (A) Comparison of coral species diversity. Decrease in species richness results from the absence of milleporids and species that are relatively scarce in the living communities. (B–D) Frequency distribution of dominating coral species in living, dead and late Pleistocene assemblages respectively. Modified and redrawn from Greenstein, Harris and Curran (1998).

Finally, Greenstein (2007) concluded that dead coral assemblages cannot generally be regarded as reliable proxies for Pleistocene communities. The compositions of preserved fossil assemblages typically seem to reflect amalgamations of living reef corals and surrounding death assemblages.

4.3.3.2. Molluscan communities

Biotic and abiotic factors have been assumed to interact in controlling the degree of shell damage of molluscs (Lockwood & Work, 2006). Biotic factors include shell mineralogy, organic content and the size and thickness of the shell, whereas abiotic factors relate to life habitat, hydrodynamic energy and nutrient levels.

The most common alteration features that affect molluscan shell material are loss of the periostracum, surface and edge abrasion, fragmentation and disarticulation. Calcitic shells are commonly thickly encrusted in comparison to non-calcitic specimens. Organic-rich shells are commonly severely fragmented, and their edges are rapidly abraded. The levels to which other damage variables differ amongst biotic factors vary according to habitat type. Generally, epifaunal populations experience consistently more severe alteration than infaunal ones because they are usually deposited postmortem above the sediment–water interface. But locally, infaunal taxa may be subject to greater damage due to intensive internal abrasion.

A limited number of studies have focused on the degree of similarity between living and dead molluscan faunas in shallow-water coral reef environments. Zuschin and coworkers investigated the relationship between the compositions of living molluscan assemblages, hard substrate types and water depth and the fate of the relevant empty shells in reefs of the northern Red Sea (Zuschin, Hohenegger, & Steininger, 2000; Zuschin & Stachowitsch, 2007) and the Seychelles, western Indian Ocean (Zuschin & Oliver, 2003). As a whole, consistent differences in the composition, abundance and distribution patterns of living and death assemblages, mainly due to postmortem biases, were detected (Figure 4.16). The compositions of life and death assemblages appear to be more driven by the nature of the substrates than by water depth. From a taphonomic point of view three categories of hard substrate-associated molluscs can be differentiated. Those living in close relationships with corals (mainly byssate pteriomorph bivalves, *Pedum, Tridacna* and gastropods, *Coralliophila* and the encrusting vermetid *Dendropoma*) are easily incorporated into coral frameworks after death and are therefore preserved in the fossil record. Molluscs dominated by bivalves, the Chamoidea and Spondylidae that encrust hard substrates typically remain exposed to alteration for much longer periods and are affected by time-averaging; but may be rapidly buried at their habitat sites.

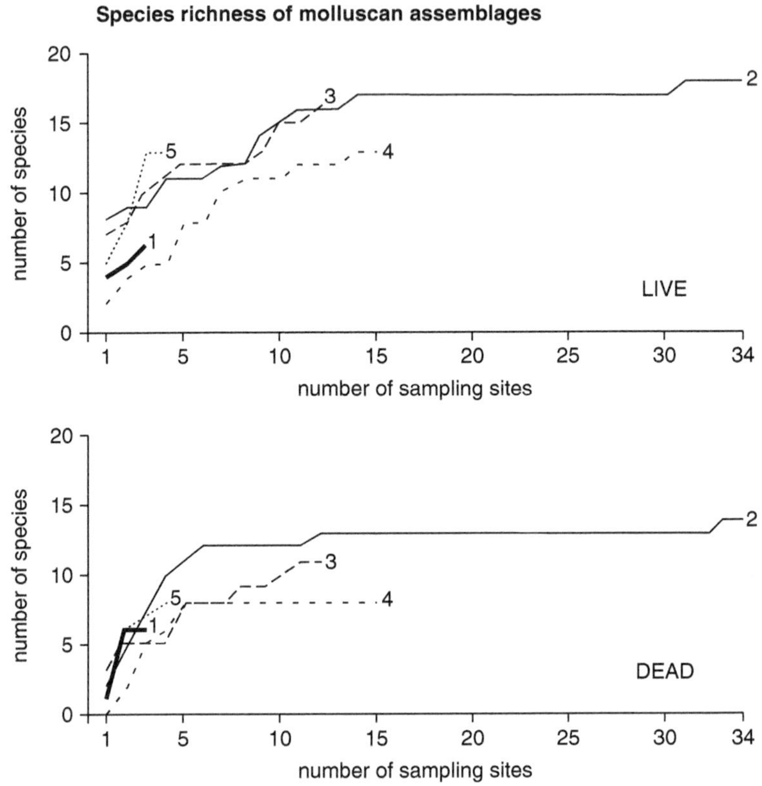

1. Reef flat zone from isolated patch reefs (*Stylophora*)
2. Reef flat zone from outer fringing reefs (*Porites*, Faviids)
3. Fore-reef zone (*Acropora, Porites, Millepora, Stylophora*)
4. Rocky grounds (lamellar corals)
5. Lagoonal coral patches (*Acropora, Stylophora*)

Figure 4.16 Cumulative curves of species diversity for living and dead molluscan assemblages on different hard substrate types from coral reefs, Safaga Bay, northern Red Sea. The corals cited refer to dominant forms. Modified and redrawn from Zuschin et al. (2000).

Parsons-Hubbard (2005) produced an inventory of the taphonomic characteristics of molluscan assemblages found in soft sediments in open-shelf, reef-tract and lagoonal environments on St Croix and Isla de Mona (Puerto Rico) in the northeastern Caribbean. This inventory indicates that there has been a loss of fidelity in the diversity and composition of modern death assemblages to a given environment, attributed mainly to postmortem transport and mixing of shells. These assemblages fail to define the distribution areas of their corresponding living communities or their taxonomic traits.

Two-year-long taphonomic experiments by Lescinsky, Edinger, and Risk (2002), investigating molluscan preservation in both eutrophic and mesotrophic reefs in the Java Sea (Indonesia), showed that (1) shell fragmentation was negligible; (2) encrustation was greater in offshore mesotrophic than in nearshore eutrophic sites, but for animal encrusters, rates and volumes of encrustation were greater in eutrophic settings; (3) rates of bioerosion appeared to be higher in eutrophic sites; (4) the intensities of encrustation and bierosion were strongly correlated with the productivity levels of ambient plankton, suggesting that shell alteration may serve as a proxy for reconstructing primary palaeoproductivity and thus nutrient supply.

Walker, Parsons-Hubbard, Powell, and Brett (2002) examined the role of predation on the fate of experimentally placed gastropod shells from fore-reef (15 m) to foreslope (262 m) environments off Lee Stocking Island (Bahamas). Shell breakage mainly occurred at shallow-shelf depths (<30 m). The potential for predation increases with time of exposure. Crabs, fish and stomatopods were responsible for most predatory alteration of the shells. However, gastropod shell damage appears not to be linked to any particular predator, but rather to a variety of potential predators including echinoderms and worms. Finally, molluscan remains were shown to suffer from higher rates of shell dissolution and bioerosion in carbonate sediments than in siliciclastic deposits. In reefal carbonate deposits (Kidwell, Best, & Kaufman, 2005), this probably results in greater taxonomic bias of preserved skeletal elements, but less time-averaging.

Limited information is available on the response of Pleistocene reef molluscan assemblages to taphonomic damage. In uplifted Pleistocene reef terraces along the Red Sea coasts, there is a marked reduction in the species abundance of molluscan assemblages, particularly of aragonitic forms (Taviani, 1997). This is assumed to have been primarily caused by severe diagenetic dissolution after emergence during low sea-level stands. Only thick walled shells of *Tridacna* and *Strombus* are still well preserved. However, rapid coral and coralline algal overgrowth, may potentially assist in preserving ecological information. As a result, encrusting *Dendropoma maxima* appears to be a common component in Pleistocene Red Sea corals.

4.3.3.3. Foraminiferal assemblages

Foraminiferal assemblages in sediments are a composite mixture of living and dead individuals. Like molluscan assemblages inhabiting soft-substrates, transport, deposition and reworking can severely alter foraminiferal associations. Mechanical abrasion and dissolution of small tests, and bioerosion of larger ones, play prominent roles in test damage (Yordanova & Hohenegger, 2002). Martin and Liddell (1991) devised experimental simulations of abrasion, dissolution and bioerosion for some common reef foraminifers, based on turbulence and water quality conditions in their respective natural settings. Test degradation appears to be relatively inhibited in carbonate environments, thus promoting time-averaging and the mixing of modern and relict assemblages.

In intertidal zones, the most effective taphonomic degradation is associated with dissolution. This results in differential preservation of calcareous and agglutinated tests; although the latter are highly susceptible to both oxic and anoxic conditions whereas calcareous tests are relatively well preserved in anoxic sediments (Berkeley, Perry, Smithers, Horton, & Taylor, 2007). As a result, in many shallow-water reef-tract environments, highly abraded calcareous tests belonging to a few larger forms (*Amphistegina*, *Archaias*, *Calcarina*, *Baculogypsina* and *Marginopora*) occur as lags that dominate foraminiferal assemblages to the point of erasing the distributional patterns of living species (Montaggioni, 1981; Martin, 1999).

The relative abilities of time-averaged foraminiferal assemblages to reflect reef zonation are mainly the result of water turbulence, cross-shelf and cross-reef topography, habitat-depth range, settling velocity of empty tests and/or intensity of in-place taphonomic alteration (Hottinger, 1983; Martin & Liddell, 1991; Glenn-Sullivan & Evans, 2001; Hohenegger & Yordanova, 2001; Yordanova & Hohenegger, 2002). In sites periodically subjected to tropical storms, mixing of foraminiferal assemblages is periodically enhanced (Figure 4.17). Tests of those inhabiting back-reef areas may be moved into fore-reef zones while fore-reef species are frequently swept into shallower reef environments. Poorly preserved tests indicate an allochthonous origin or reworking of relict assemblages. In protected regions, representative numbers of robust species living *in situ* are present as empty tests in time-averaged assemblages, whereas delicate forms are selectively destroyed, particularly in windward settings.

4.3.3.4. Echinoderm assemblages

The ability of skeletal detritus derived from reef echinoderms to reflect specific disturbance events has been intensively debated since the end of the 1970s. Reef workers have searched for evidence of either explosions echinoderm populations or mass mortalities from the recent past back to Holocene times, comparing living, dead and subfossil assemblages.

Over the past four decades, reef corals have suffered extensive mortality caused by population outbreaks of a coral predator, the asterid *Acanthaster planci*, on many Indo-Pacific reefs (see review by DeVantier & Done, 2007). Considerable controversy emerged regarding the factors responsible

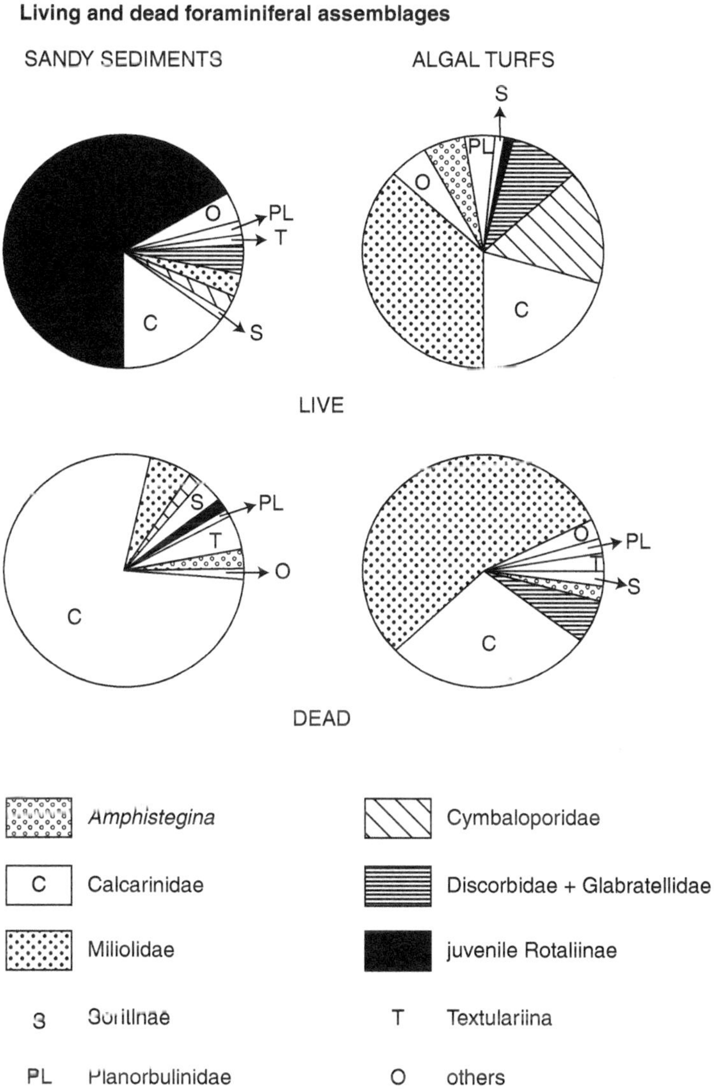

Figure 4.17 Compositions of living and dead assemblages of foraminifera in both algal turfs and adjacent sandy deposits, from the outer reef flat zone, Apo Reef, Mindoro, Philippines. Modified and redrawn from Glenn-Sullivan and Evans (2001).

for outbreak events and in particular the relative importance of natural versus human-induced causes. In order to resolve the dispute, attempts have been made to assess the potential of *Acanthaster*-derived skeletal particles in surface and subsurface sediments to record outbreak events in the geological past. However, field experiments have been inconclusive and suggest that previous assumptions were invalid. Greenstein, Pandolfi, and Moran (1995) found no identifiable signature of a post-outbreak mortality of *Acanthaster* in surrounding surface sediments as a result of intensive taphonomic disturbance (water turbulence, bioturbation). It seems that the only way to identify historical outbreak events is likely to be to use the typical traces of predation (feeding scars) left by *Acanthaster* on massive corals.

Taphonomic bias has drastically affected the fossil record of echinoids (Greenstein, 1993). However, the preservational patterns of regular, epifaunal forms (e.g. *Diadema, Echinometra, Tripneustes, Eucidaris*) and irregular, shallow-burrowing forms (*Mellita, Leodia, Meoma*) are quite different. Subfossil skeletal elements from regular echinoids reflect the distribution of once-living populations more faithfully than those of irregular species. Fragments of regular forms, although sparse, are strictly confined to areas colonized by living individuals. By contrast, elements produced by irregular taxa have experienced displacement and thus appear to be widespread compared to their living analogues. This is likely to be controlled by a low resistance of regular echinoid tests to displacement and reworking. The occurrence of preserved fragments thus indicates *in situ* deposition and hence may be used as a tool for palaeoenvironmental reconstruction. However, like that of *Acanthaster*, the sedimentary record of echinoids was demonstrated to be unable to preserve short-term events such as a massive and sudden mortality of populations caused by disease.

In the absence of rapid burial, crinoid skeletons appear to suffer virtually total disarticulation. The response of comatulid crinoids to postmortem damage was investigated by Meyer and Meyer (1986) on a fringing reef of the Australian Great Barrier Reef. They showed that there is no evidence that buoyant displacement plays a prominent role in postmortem dispersal of crinoidal skeletal elements. Fish predators may control taphonomic bias by selectively removing calyx parts from the living assemblages of the habitats. Surface sediments contain a time-averaged deposit of disarticulated, usually abraded crinoid ossicles, which results from both fair-weather and storm-related processes. However, particle sorting is minimal and the concentration of ossicles therefore directly reflects the abundance of adjacent live populations.

A census of fossil echinoids in late Pleistocene reefs at San Salvador Island (Bahamas) by Greenstein (1993) indicated that, in contrast to surveys of corals and molluscs, their preservation potential of was very low. However, the skeletal durability of irregular echinoids appeared to be higher than that of regular forms, as shown by comparison of living and dead populations. Fragments from regular forms are rare in reef-tract deposits whereas the remains of irregular echinoids are relatively common in sands interpreted as representing back-reef environments. These findings suggested that the poor Pleistocene record of regular taxa may result from taphonomic biasing. Similarly, good preservation of irregular echinoid tests is more likely to be controlled by environmental constraints than by biotic attributes (e.g. life habit, or degree of skeletal susceptibility to alteration).

4.4. Conclusions

A myriad of biotic and abiotic factors appears to participate in governing the attributes of coral reefs and maintaining diversity in reef communities over time scales varying from decades to tens of thousands years. These factors drive coral growth forms, the taxonomic compositions of reef communities, and the distribution, nature, geometry and post-mortem to long-term preservation of reef tracts.

Species diversity and population abundance are controlled by differing modes of reproduction, larval dispersal and patterns of recruitment to specific sites. However, the dominance of populations by a limited number of coral species with reduced larval supplies indicates that alternative or additional controls are operating at various spatial and temporal scales. Under conditions of saturation with respect to species richness, the compositions of reef communities depend upon biotic interactions (competition, predation and herbivory and disease) and on local abiotic parameters rather than the size of the regional species pool. Migration rates, dispersal and competitive abilities, and also habitat availability, can determine species abundance and survival patterns. Low rates of larval dispersal, limitation in recruitment, and habitat degradation or reduction below thresholds, may have inimical effects on coral reefs. In this view, the structure of reef communities is principally dependent on the ability of species to adapt to ambient constraints, and ecologically selected species are only incorporated into the community as 'limited members' (Pandolfi, 1996).

These are important factors when attempting to explain the persistence of Pleistocene reef communities through space and time. However, disturbance regimes are also major determinants in the degree to which biotic factors are able to influence the structure and distribution of reef communities; their frequency and duration, together with intensity control the resilience of local communities. Disturbance prevents communities reaching a climactic state, thus promoting rapid structural changes and reef population dynamics. Reef growth responds differentially to local and global disturbances and no single factor can satisfactorily account for changes in distributional and growth patterns on local to regional scales. These patterns are driven by the synergy of a variety of environmental parameters, especially during glacial intervals when nutrification levels seem to have been a major determinant of coral growth. Similarly, hydrodynamic effects, particularly those linked to cyclonic activity, have had important controls on the structure and taphonomic features of coral assemblages (see Chapter 7, Section 7.2.5). Substrate availability that indirectly governs larval recruitment is the most biogeographically limiting of all extrinsic factors. The distances between available substrates and centres of coral dispersal, and modification of circulation regimes on a provincial to regional scale, may partly explain variations in the taxonomic compositions of reef communities from one site to another. Based on data from the last glacial cycle, the influence of changes in dissolved CO_2 on aragonite saturation, and subsequently on reef building, appears unclear, probably because patterns of calcification have not changed significantly since the last glacial, despite a marked increase in atmospheric CO_2 levels. Other potential controls including tectonics, palaeotopography and atmospheric dust fluxes, together with SST, salinity and turbidity levels discussed in Chapter 7, have acted as modulators of the major controls, mainly at local scales.

Taphonomic alteration severely affects reef substrates and thus challenges their preservation. Taphonomic processes are typically dominated by biological encrustation (mainly coralline algae), macroboring (mainly sponges and bivalves), and microboring (mainly cyanophytes). The degree of preservation of the original reef community structure and the potential to record short-term changes in the structure of fossil reef communities are controlled by the attributes of organisms, including the types of growth forms, the original skeletal mineralogy and nature of live assemblages, wave exposure, rates and modes of burial and the intensity of encrustation and boring. High species diversity and high coral cover need not necessarily result in extensive in-place, lasting reef framework.

The relationship between living, dead and subfossil assemblages is more obvious for corals and molluscs than for echinoderms and foraminifera. However, caution is required in interpreting coral and molluscan death assemblages. A variety of factors alter the degree of fidelity of death assemblages: a greater degree of time-averaging, and drastic changes in life communities over short-term scales, making the ecological information recorded by adjacent death assemblages more representative of previous life generations than of the living assemblage; and a differential response of growth forms to taphonomic bias. The most informative coral assemblages are those that inhabited low-energy environments, whereas the best-preserved molluscan assemblages are those that lived close to hard substrates and corals and are overgrown by coralgal framework after death, or those consisting of species with thick-walled, calcitic shells.

The taphonomic attributes of Quaternary reef sequences may aid in the identification of temporal changes in depositional environments, as determined from coral, molluscan and foraminiferal assemblages or successional styles of encrustation development. Distinct sequences of taphonomic features have the potential to aid delineation of contrasting sub-environments of reefs. Quaternary reefs retain integrated information on reef communities encompassing ecological time, and allowing better detection of subtle anomalies in community structure than life assemblages. Thus, despite their substantial taphonomic alteration, Pleistocene and Holocene reefs provide a reliable long-term census of taxonomic composition and community structure. As emphasized in Chapter 3, the most appropriate use of Quaternary reefs as an environmental proxy is to assess long-term changes in community structure patterns of modern reefs, and in particular the relative abundance of common species, the eventual reduction in species diversity, and the disappearance of dominating taxa.

CHAPTER FIVE

Patterns of Carbonate Production and Deposition on Reefs

5.1. Introduction

Although global coral reef productivity has varied during the Quaternary in response to climate changes, reef systems have probably remained among the most important producers of calcium carbonate in the oceans even during lower sea-level stands (Kleypas, 1997). Estimates of mean reef carbonate production on a global scale have been extrapolated from studies of individual reef systems (Chave, Smith, & Roy, 1972; Vecsei, 2004). Today, calcification rates by coral reefs range between 6.8 and 8.3×10^{12} mol yr^{-1}. Most of the carbonate produced (about 7×10^{12} mol yr^{-1}) accumulates *in situ*, and the rest is washed into the oceans (Milliman, 1993; Milliman & Droxler, 1996; Schneider, Schulz, & Hensen, 1999). Thus, carbonate production by reefs is regarded as playing a major role in the global carbon cycle (Kleypas, Buddemeier, et al., 1999; Gattuso & Buddemeier, 2000; Suzuki & Kawahata, 2003; Vecsei & Berger, 2004) representing one-sixth of the carbonate produced yearly in the global ocean (Langer, Silk, & Lipps, 1997).

Sedimentologically speaking, coral reefs can be regarded as the end products of a variety of processes including construction (*in situ* framework accretion), destruction (sediment production through bioerosion and wave action) and sediment deposition (after transport and reworking within and on the periphery of areas of framework). Attempts have been made to incorporate all of these processes into an overall carbonate depositional model at the scale of a single reef system (Stearn & Scoffin, 1977; Smith & Kinsey, 1978; Land, 1979; Hubbard, Burke, & Gill, 1986; Hubbard, Miller, & Scatturo, 1990; Harney & Fletcher, 2003; Hart & Kench, 2007). Knowledge of the growth and/or carbonate production rates of frame builders, and associated reef dwellers and bioeroders is critical, because the sediments thus released represent significant volumes (Hubbard et al., 1990; Braithwaite et al., 2000; Hewins & Perry, 2006; Hart & Kench, 2007) contributing to the net calcium carbonate budget (Scoffin et al., 1980). Net production represents the amount of calcium carbonate remaining within the reef as framework and detritus following exports to adjacent oceanic waters.

Although Holocene reef accretion results for the most part from filling of framework cavities, back-reef and lagoonal areas by loose sediments

(Marshall & Davies, 1982; Tudhope, 1989; Braithwaite et al., 2000; Montaggioni, 2005; Purdy & Gischler, 2005), there have been few studies of the compositions of sandy sediments deposited subsurface (Cabioch, 1988; Colby & Boardman, 1989; Tudhope, 1989; Degauge-Michalski, 1990, 1993; Smithers, Woodroffe, McLean, & Wallensky, 1992). Coupled analyses of the compositions of surface and subsurface detritus are also scarce, and are restricted to the work of Colby and Boardman (1989), Smithers, Woodroffe, McLean, and Wallensky (1992) and Degauge-Michalski (1993).

Since the pioneering work of Thorp (1936), Emery, Tracey, and Ladd (1954), Ginsburg (1956, 1964), McKee, Chronic, and Leopald (1959), Maxwell, Day, and Fleming (1961), Folk and Robles (1964) and Lewis and Taylor (1966), efforts have been made in the last three decades to establish the relationship of surface sediment compositions to the adjacent reef community structure. These have included the potential value of skeletal constituents as indicators of reef facies and depositional environments in cross-shelf profiles. To quantify the spatial extent of sediment types on a large scale, tentative mapping investigations have been conducted in the last decade using traditional sediment sampling combined with acoustic surveys and multispectral satellite imagery (for instance, see Riegl, Halfar, Purkis, & Godinez-Orta, 2007). However, in the western Atlantic and the Indo-Pacific, detailed information on the compositions and distributions of carbonate sediment types remains restricted to a few individual reef systems.

The objectives of this chapter are to address the following: (1) What are the growth and carbonate production rates of reef builders and associated organisms, and what are the respective contributions of these organisms and relevant communities to total sediment production; (2) To what extent are the different reef sediment types reflections of the adjacent benthic communities and diagnostic in terms of depositional environments; (3) What are the differences in rates of deposition between differing sedimentary piles and their major controls?

5.2. Patterns of Reef Carbonate Production

The gross production of reef carbonates is highest on outer reef margins where corals and other calcifying organisms have high cover rates and water energy is high. Production tends to decline significantly in lower hydrodynamic energy back-reef and lagoonal settings where cover rates are lower.

5.2.1. Growth and Production Rates of Reef Dwellers

Estimates of growth and gross carbonate production rates by calcifying organisms on modern reefs (expressed in kg $CaCO_3$ m^2 yr^{-1}) rest mostly on census-based methods applied to a limited number of individuals from small areas over short periods and subsequently extrapolated over broader spatial and temporal scales. An alternative technique used to quantify reef-wide carbonate production is alkalinity reduction, a measure of daily changes in water chemistry.

Estimates of linear accumulation (i.e. one-dimensional mass accumulation) rates in the Holocene record, are core derived. Reef accretion rates represent net carbonate production over some period of time, assuming the reef surface represents present time. The values, expressed in millimetre per year (mm yr^{-1}) are computed from core records and radiometric dating by dividing the thickness of a given core interval by the time over which it accumulated. Accretion and production rates calculated from dating therefore represent approximate time-averaged values.

5.2.1.1. Corals

Estimates of coral growth rates are usually based on direct measurements of individual colonies using alizarin staining (measuring the vertical or lateral accretion between an introduced alizarin line and the living surface of the coral). Yearly extension rates are converted to carbonate production rates using skeletal densities of 1.4–1.8 g cm^3 according to coral growth forms, and the mean percent cover of each coral species or growth form. Corals typically produce two-thirds of total reef carbonate budgets (Payri, 1988) but may locally represent more than 90% (Hubbard et al., 1990).

Vecsei (2001), Dullo (2005) and Hart and Kench (2007) reviewed potential growth and/or calcification rates of modern scleractinian corals from the major reef provinces (Figure 5.1). Domal (massive) forms appear to be growing at rates averaging 10 mm yr^{-1} (range: 0.8–32 mm yr^{-1}) and have a gross carbonate production of from 3 up to 15 kg m^{-2} yr^{-1}. Robust branching corals have growth increments ranging from 33 to 130 mm yr^{-1}. Gracile branching (arborescent) colonies develop at rates averaging 100 mm yr^{-1}. Tabular forms grow at rates rarely exceeding 70 mm yr^{-1}. The carbonate production of both branching and tabular corals varies between about 1 and more than 25 kg m^2 yr^{-1} depending on species. In Florida Bay, growth and production rates of branching *Porites* were estimated to average 32 mm yr^{-1} and 0.014–1.17 kg $CaCO_3$ m^2 yr^{-1} respectively (Bosence, 1989). The lowest growth rates measured were from encrusting and foliaceous corals (0.8–24 mm yr^{-1}); the latter having production rates from about 3 up to 10 kg $CaCO_3$ m^2 yr^{-1}. However, there are no significant differences in growth and calcification rates of corals of similar growth forms living within similar environments in the Caribbean and Indo-Pacific. Shallow-water (<10 m) domal colonies are characterized by growth increments and gross calcification rates ranging from 5 to

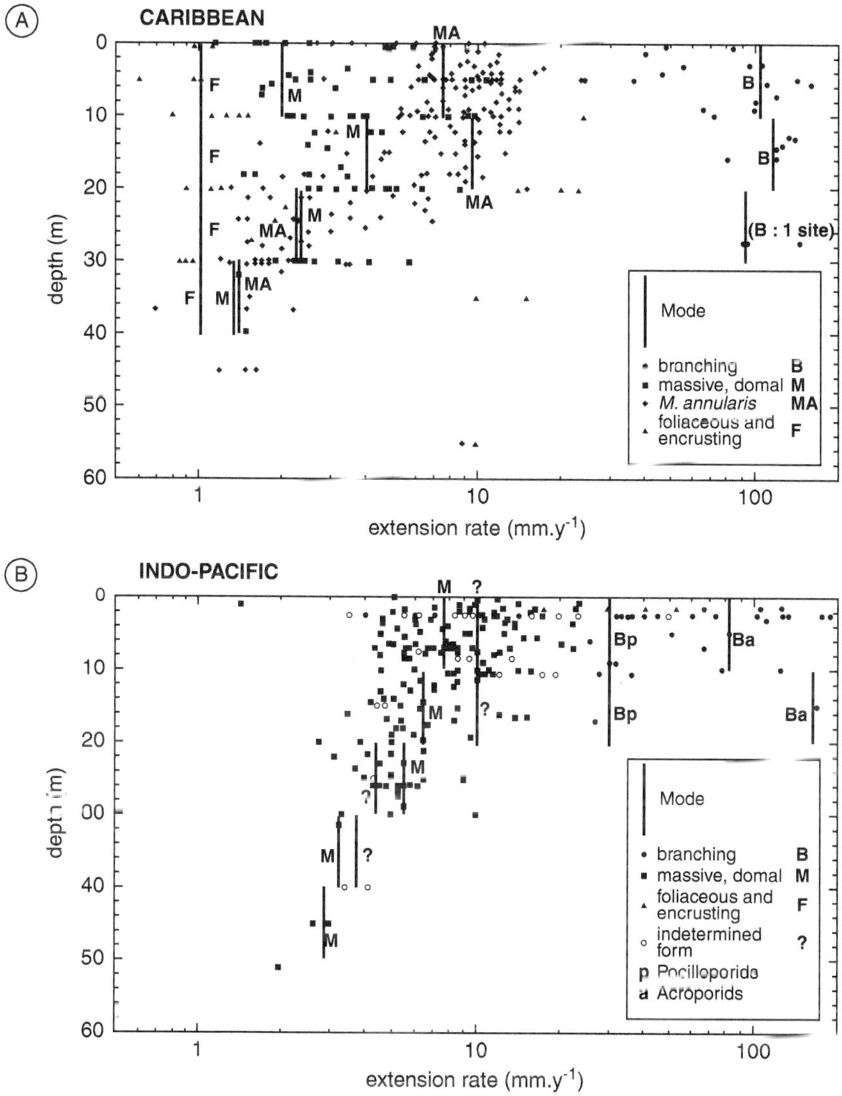

Figure 5.1 Potential linear extension rates of different reef-building coral growth forms in the Caribbean (A) and Indo-Pacific (B) provinces. Modified and redrawn from Vecsei (2001).

13.5 mm yr^{-1} and 5 kg m^2 yr^{-1} respectively. By contrast, the extension and calcification rates of any given coral species decrease significantly relative to increasing depth and decreasing light intensity (Bosscher & Schlager, 1992). Production rates of massive *Porites lutea* colonies from reefs in the

Gulf of Aqaba (northeastern Red Sea), were estimated to range from 8.4 to 3 mm yr^{-1} and 9 to 5 kg carbonate $m^2 yr^{-1}$ at depths between 0–5 m and greater than 30 m (Heiss, 1995). In addition, as demonstrated by Grigg (1982), growth and production rates decrease with increasing latitude. Measurements of shallow-water *Porites lobata* heads showed that these parameters vary between 12 and 3 mm yr^{-1} and 17 and 5 kg carbonate $m^2 yr^{-1}$ along a latitudinal gradient from about 19° to 28° north. The coral contribution is less than 20% of the total carbonate reef budget at the highest latitude.

Growth rates of Quaternary reef corals are poorly documented. In Indonesia, Crabbe, Wilson, and Smith (2006) compared radial growth rates from fossil massive *Porites* and *Favites* with those of their living counterparts in adjacent modern reefs. Values were of the same order of magnitude ranging from 15 to 10 mm yr^{-1} according to depth. Johnson and Pérez (2006) measured extensional rates of the massive genera *Porites*, *Monastraea* and *Goniopora*, ranging in age from late Oligocene to Pleistocene from across the Caribbean, and compared these values with records of modern coral growth rates (Figure 5.2). The results reveal that there were marked differences in linear extension rates among colonies of different ages in this area for the past 30 Ma. Apparently, rates were lower in the late Miocene and higher during the late Oligocene, the Pleistocene and Holocene. Given that calcification is known to be promoted by lower atmospheric CO_2 levels (Kleypas, Buddemeier, et al., 1999), higher growth rates in the late Oligocene and Recent times may have been triggered by decreasing levels of carbon dioxide.

5.2.1.2. Coralline algae

Growth rates of geniculate and non-geniculate coralline algae are usually expressed as vertical accretion of the thallus. In tropical regions, continuous growth ranges between <1–2 mm yr^{-1} and 5–20 mm yr^{-1} for encrusting and branching forms respectively (Adey & Vassar, 1975; Stearn et al., 1977; Agegian, 1981; Matsuda, 1989; Hubbard et al., 1990; Payri, 1997; Hart & Kench, 2007). The carbonate production of coralline algae tentatively inferred from growth rates, varies widely as a function of thallus shape, bulk skeletal density, cover rate, predation intensity and depth. Lower values are obtained from assemblages chiefly composed of encrusting forms subject to minimal light levels and range from 0.003 to 0.020 kg $CaCO_3$ $m^2 yr^{-1}$. Higher values are recorded from dense assemblages dominated by branching forms growing in shallow waters and experiencing low grazing pressure (0.17 to more than 2.5 kg $CaCO_3$ $m^2 yr^{-1}$). Locally coralline algae can contribute from about 1.5% to more than 40% of the total gross carbonate productivity of a reef system (Hubbard et al., 1990; Harney & Fletcher, 2003).

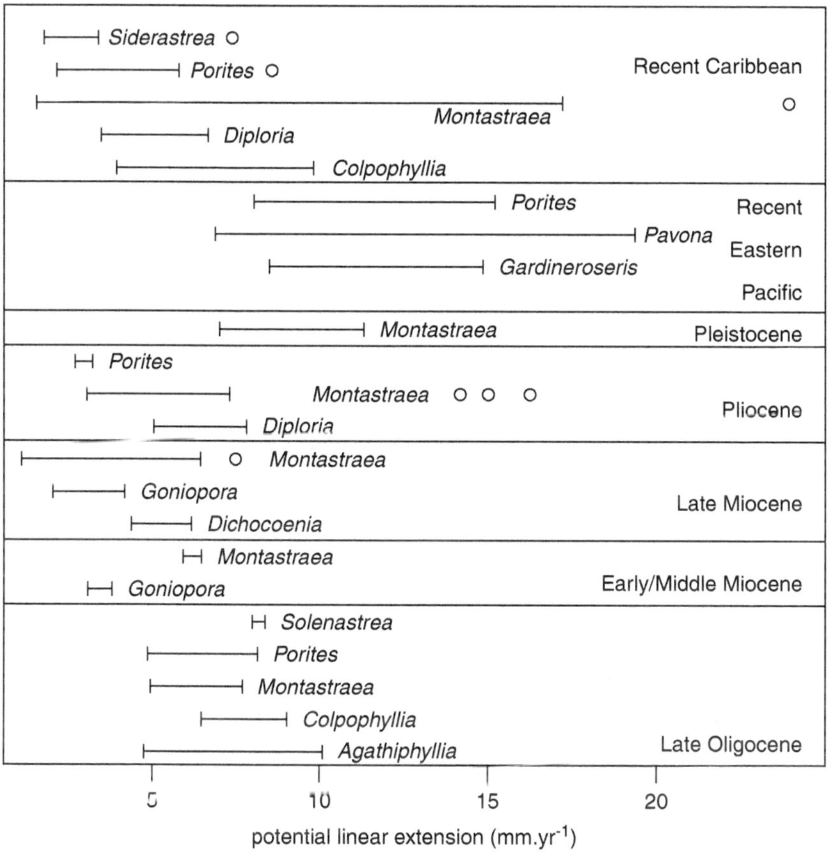

Figure 5.2 Estimated ranges of annual growth rates of some Cenozoic coral forms in the Caribbean. With comparative data from Eastern Pacific corals. Modified and redrawn from Johnson and Pérez (2006).

5.2.1.3. Rhodoliths

Estimates of growth rates of algal nodules (see Section 5.3.1 for description) revealed that those of tropical forms are up to an order of magnitude higher than those of temperate species (Bosence, 1983a). Similar contrasting results have been obtained from a number of reef areas and environments. In most reef systems, branching to columnar rhodoliths from reef-flat and back-reef environments appear to have developed at rates varying between 2.5 and 3 mm yr^{-1} (Adey & Vassar, 1975; Stearn et al., 1977; Montaggioni, 1978). However, in Bermuda and French Polynesia, Bosellini and Ginsburg (1971) and Payri (1997) found that the mean growth rates of shallow-water, columnar rhodoliths do not exceed 0.4 and 0.15–0.60 mm yr^{-1} respectively. Massive rhodoliths deposited at depths of from about 30 to more than 60 m appear to grow at rates substantially lower than most shallow-water nodules and range between 0.1–0.4 mm yr^{-1} (Vogel, 1970; Bosellini & Ginsburg, 1971; Montaggioni, 1978) and 0.01–0.09 mm yr^{-1} (Focke & Gebelein, 1978; Reid & Macintyre, 1988; Littler, Littler, & Hanisak, 1991). The nuclei of some of these deep-water forms give radiocarbon ages of 0.48 to about 1.5 ka (Focke & Gebelein, 1978; Montaggioni, 1978; Reid & Macintyre, 1988; Goldberg, 2006) that indicate that a proportion of living rhodoliths are in fact relic forms that have recently been recolonized. Data on carbonate production by rhodoliths show wide ranges according to nodule shape, reef setting and the method of estimation used. Values vary between 0.003 and 0.30 kg $CaCO_3$ $m^2\,yr^{-1}$ (Payri, 1997).

5.2.1.4. *Halimeda*

Carbonate production by all *Halimeda* species together has been estimated to contribute about 8% to the total world carbonate budget (Hillis, 1997) varying between about 0.028 and 2.2 kg $m^2\,yr^{-1}$ calcium carbonate on average (Van Tussenbroek & van Dijk, 2007). Estimates of the growth rates vary depending upon the methods used (Multer, 1988; Payri, 1988), but primarily upon a variety of biotic and environmental factors. For instance, soft-substrate (psammophytic) and hard-substrate (lithophytic) species seem to have production rates that differ by several orders of magnitude. Using data from the barrier reef complex of Moorea (French Polynesia), Payri (1988) demonstrated that the lithophytic species *H. opuntia* (about 0.975 kg calcium carbonate $m^2\,yr^{-1}$) has growth rates 13 times higher than those of the soft-bottom *H. incrassata* f. *ovata* (about 0.075 kg). By contrast, Harney and Fletcher (2003) calculated that on a windward Hawaiian reef, *H. opuntia* produced sediment at rates of 0.6–3 kg $CaCO_3$ $m^2\,yr^{-1}$, exceeding 6.5 kg in dense meadows. In Florida, the lagoons are particularly depauperate in *Halimeda* standing stocks, with a production of only 0.004–0.030 kg $CaCO_3$ $m^2\,yr^{-1}$ (Bach, 1979; Bosence, 1989). By contrast, in a similar environment in the Mexican Caribbean, *H. incrassata* was shown to be capable of releasing 0.815 kg $CaCO_3$ $m^2\,yr^{-1}$ (van Tussenbroek and van Dijk, 2007). The highest production rate (*for Halimeda incrassata*) was obtained from a Panamanian lagoon with up to 2.3 kg $m^2\,yr^{-1}$ (Freile & Hillis, 1997).

5.2.1.5. Molluscs

Although shelly molluscs provide a significant proportion of modern reef sediments, their contribution to the carbonate budget is poorly documented. Available data indicate that molluscan carbonate production varies greatly, depending on the size and density of living species and the environment (Bosence, 1989; Hart & Kench, 2007). Production ranges from less than 0.001 up to 0.60 kg $CaCO_3$ $m^2 yr^{-1}$. Higher values (>0.070 kg) have been obtained from dense micromolluscan assemblages living in sandy beds, while lower values (<0.005 kg) have been reported from isolated macromolluscs living on hard bottoms.

To our knowledge, the only attempt at estimating carbonate production of individual sediment types was made by Bosence (1989) from samples dominated by molluscan detritus, in Florida Bay. Molluscan–foraminiferal grainstones to wackestones and molluscan mudstones appear to accumulate about 0.33 kg $CaCO_3$ $m^2 yr^{-1}$ each, whereas molluscan–*Halimeda* wackestones/mudstones reach deposition rates of about 0.9 kg $CaCO_3$ $m^2 yr^{-1}$.

5.2.1.6. Benthic foraminifera

Foraminiferal production represents approximately 4.8% of the global carbonate reef budget and 0.76% of present-day production in the world ocean (Langer et al., 1997). At the scale of individual reef systems, the contribution of foraminifera, usually estimated from the number or volume of tests in sediments, appears to have been restricted to free-living or epiphytic, larger forms (mainly soritids, nummulitids, amphisteginids and rotalinids). As with other sediment producers, the foraminiferal contribution varies greatly, depending on the composition of the assemblages, environment and depth. Overall, values range from 0.0001 to 0.002 kg $CaCO_3$ $m^2 yr^{-1}$ (Bosence, 1989) up to 2.5 kg $m^2 yr^{-1}$ (Hart & Kench, 2007). Higher production rates (>0.20 kg on average) are recorded from reef flats, adjacent back-reefs and beach zones, while foraminifera in deep lagoons and along fore-reef slopes and shelves tend to have lower turnover rates, producing less than 0.15 kg $m^2 yr^{-1}$ on average (Hallock, 1981; Sakai & Nishihira, 1981; Langer et al., 1997; Yamano, Miyajima, & Koike, 2000; Harney & Fletcher, 2003).

However, the determination of carbonate production by non-encrusting foraminiferal populations should only proceed with caution, since turnover rates of the relevant remains are underestimated, and are generally assumed to be less than 100 years. However, radiometric dating of *Amphistegina* tests collected from the surface of a sandy beach on Hawaii gave ages of more than 1.5 ka (Resig, 2004). This means that any quantification of changes in carbonate production has to be based on biotic censuses rather than on the analysis of detrital fractions.

5.2.1.7. Calcareous epibionts

Calcifying encrusting organisms (e.g. coral recruits, crustose coralline algae, bivalves, gastropods, bryozoans, serpulid worms and foraminifera) clearly contribute carbonate to both the reef framework and to detritus.

Mallela (2007) demonstrated that total carbonate production by framework encrusters in northern Jamaican sites ranges from about 0.070 to 0.159 kg m^2 yr^{-1} in clear, high energy waters, falling to 0.003–0.03 kg m^2 yr^{-1} appproaching zones subject to high-turbidity and reduced wave energy. In Florida Bay, Nelsen and Ginsburg (1986) and Bosence (1989) showed that the volume of lime mud produced annually by red algae and serpulid epiphytes living on *Thalassia* leaves varied from 0.055 to about 1 kg m^2 yr^{-1}, a markedly lower volume than that reported from Barbados (2.5 kg). On Bermuda reefs, Pestana (1985) found significantly lower production rates by bryozoans and coralline algae that colonized the thalli of the brown alga *Sargassum* (0.0034–0.0082 kg carbonate m^2 yr^{-1}). Such differences may be attributed to differing densities of the meadows that control the ability of plants to dampen wave activity and to trap fine grains (Almasi, Hoskin, Reed, & Milo, 1987).

5.2.1.8. Bioeroders

The destruction of reefal carbonate substrates by bioeroding organisms is one of the most important processes in carbonate production (Kiene, 1985, 1988; Hutchings, 1986; Chazottes, Le Campion-Alsumard, and Peyrot-Clausade, 1995; Perry, 1999; Zubia & Peyrot-Clausade, 2001).

Cyanobacteria and bioeroding fungi are estimated to be responsible for about 0.35 kg $CaCO_3$ m^2 yr^{-1} of substrate disintegration (Kleemann, 2001). Chazottes, Le Campion-Alsumard, and Peyrot-Clausade (1995) estimated that cyanobacterial and chlorophyte microborers produce 0.6 kg $CaCO_3$ m^2 yr^{-1} from a French Polynesian reef. Boring sponges, dominated by clionids, attack reef substrates by both chemical and mechanical means. However, Zundelevich, Lazar, and Ilan (2007) demonstrated that sponges remove around three times more carbonate by chemical than by mechanical means. The total volumes of carbonate released by populations of sponges vary from about 0.2 up to 20 kg $CaCO_3$ m^2 yr^{-1} (Kiene & Hutchings, 1994; Schönberg, 2002). Polychaete worms have an intensive bioerosive activity, resulting in the production of from about 0.6 to more than 2 kg carbonate m^2 yr^{-1} (Chazottes et al., 1995; Kiene & Hutchings, 1994). Bioeroding molluscs, including bivalves, gastropods and chitons, play a respectable role in carbonate recycling on reefs. The bioerosive potential of all molluscan eroders together on a given reef averages 0.15 kg $CaCO_3$ m^2 yr^{-1} (Kiene & Hutchings, 1994), but may locally reach 9 kg $CaCO_3$ m^2 yr^{-1} (Kleemann, 2001). On One Tree Reef, a mid-shelf platform reef (southern Great Barrier Reef of Australia), chitons alone (*Acanthopleura*) contribute to bioerosion budgets at levels comparable with those of echinoids and fish, with erosion rates that average 0.16 kg $CaCO_3$ m^2 yr^{-1} (Barbosa, Byrne, & Kelaher, 2008). During feeding, regular echinoids, mostly from the genera *Diadema*, *Echinothrix* and *Echinometra*, may locally erode substrates at rates equal to or higher than gross carbonate-framework production (Bak, 1994). Rates vary widely between reef environments, depending on the densities of individuals (from 1 to more than 50 individuals m^2) and range from 0.050 kg to as high as 20 kg $CaCO_3$ $m^2 yr^{-1}$ (Bak, 1994; Peyrot-Clausade et al., 1996; Mokady, Lazar, & Loya, 1996; Peyrot-Clausade & Chazottes, 2000; Carreiro-Silva & McClanahan, 2001; Toro-Farmer, Cantera, Londoño-Cruz, Orozco, & Neira, 2004; Herrera-Escalante, Lopez-Pérez, & Levte-Morales, 2005). Scarid fish are responsible for erosion of from 0.2 up to 9 kg $CaCO_3$ $m^2 yr^{-1}$ (Ogden, 1977; Bak, 1994; Peyrot-Clausade et al., 1996; Peyrot-Clausade & Chazottes, 2000) (Figure 5.3). Carbonate production by crustaceans is generally quite low, averaging 0.008–0.015 kg $m^2 yr^{-1}$.

External bioerosion from grazing is regarded as the dominant erosional process on reefs, but varies widely in intensity between sites. It may locally account for 60–85% of total bioerosion, resulting in the removal of more than 2.5 kg $m^2 yr^{-1}$ of carbonate (Chazottes et al., 1995: Peyrot-Clausade et al., 1996).

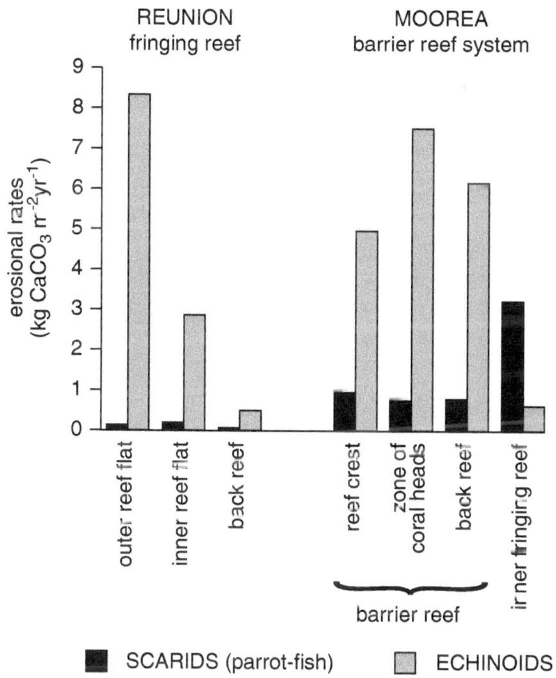

Figure 5.3 Erosion rates of scarid fish and ECHINOIDS from modern reef systems (the fringing reef of Réunion, western Indian Ocean; the barrier and fringing reef system of Moorea, French Polynesia, central Pacific). Modified and redrawn from Peyrot-Clausade and Chazottes (2000).

5.2.2. Carbonate Production at the Scale of Single Reef Systems

Estimates of contemporary production vary by several orders of magnitude between reef systems, within different reef environments, and between the zones of a single reef system, depending on the type of substrate and biota, and growth and cover rates. As shown by Vecsei (2001), framework reefs have higher productions than detritus reefs (Figure 5.4). Combining census-based and core-derived methods on the shelf-edge reef system at St. Croix (northeastern Caribbean), Hubbard et al. (1990) found that gross production at the scale of the entire reef averaged 1.21 kg m^2 yr^{-1}. Sediment export represented 0.30 kg $CaCO_3$ m^2 yr^{-1}, probably as a result of flushing by major storms. Vertical accretion rates ranged from 0.15 to 1.70 mm yr^{-1}, with a reef-wide average of 0.92 mm yr^{-1} over the past 2–3 ka. The derived net rates of carbonate production varied from 0.41 to 2.27 kg m^2 yr^{-1}, averaging 0.91 kg. These represent that part of the production preserved and stored within the reef system.

Similar results have been obtained from a reef-flat platform in northern Australia (Hart & Kench, 2007). Gross production was estimated at 1.66 kg $CaCO_3$ m^2 yr^{-1} on average. Present-day vertical accretion occurs at an average rate of 0.86 mm yr^{-1}, assuming a 25% erosion rate. Contrasting values were reported from the Caribbean and Indo-Pacific, using census-based studies of different reef environments (Figure 5.4). In general, carbonate production ranges from less than 1 to more than 10 kg m^2 yr^{-1}, averaging 4–5 kg, as a response to spatial variability in coverage by carbonate producers and differences in the compositions of assemblages living in any given zone (Chave et al., 1972; Stearn et al., 1977; Eakin, 1996; Scoffin, 1997; Harney, Grossman, Richmond, & Fletcher, 2000; Yamano et al., 2000; Harney & Fletcher, 2003). Studies based on alkalinity-reduction methods have provided results in close agreement with those derived from the census approach. Kinsey (1985), Kinsey and Hopley (1991) indicated that production rates vary between 0.5 and 10 kg m^2 yr^{-1} in lagoonal zones and on outer reef rims. On Moorea (French Polynesia), off-reef sediment export was estimated by comparing gross production rates calculated from specific dominant calcifiers (about 5 kg $CaCO_3$ m^2 yr^{-1}) with net production of 2.4 kg. This suggests that at least half of the production was exported to the ocean (Payri, 1988).

Rates of sediment deposition in the central Great Barrier Reef (GBR), a mixed carbonate/siliciclastic shelf system, have been estimated for the past 3 ka (Heap, Dickens, & Stewart, 2001). The deposition rate of the bulk sediment averaged from 0.60 up to 2.8 kg m^2 yr^{-1}. The carbonate component, consisting primarily of foraminiferal tests and molluscan grains, accumulates at rates ranging from 0.05 to 1.90 kg m^2 yr^{-1}. Siliciclastic accumulation rates are comparable to those of the skeletal sediment but

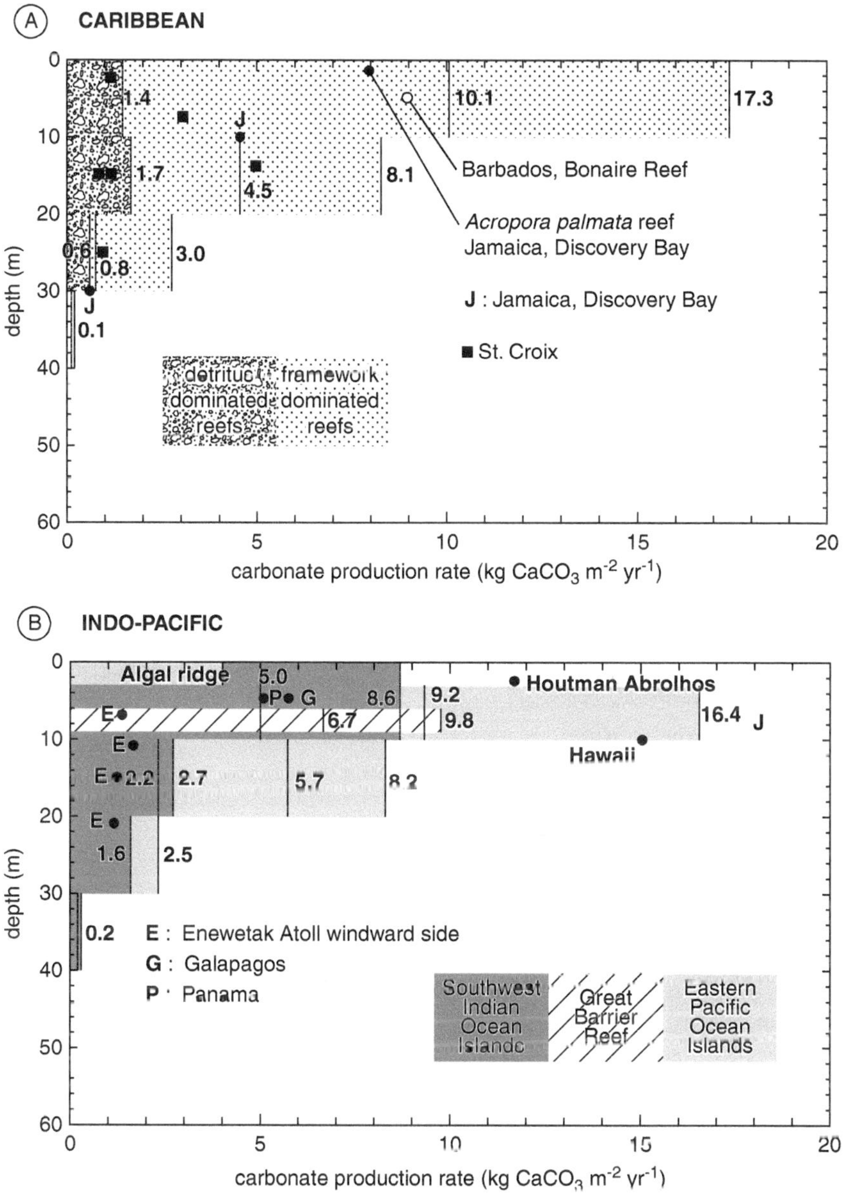

Figure 5.4 Estimated carbonate production of Caribbean (A) and Indo-Pacific (B) reef-crest and fore-reef zones, based on cover and growth rates of corals and associated biota and the amounts of early cements. Low and high values are estimated on the basis of 25% and 50% effective branching coral cover respectively. Total production appears similar in the two provinces and decreases exponentially with depth. (A) Caribbean: The production is markedly higher in framework-dominated reefs than in detritus-dominated ones. (B) Indo-Pacific: The production is comparable in reefs from continental and island areas. Modified from Vecsei (2001).

decreased markedly over time, reflecting the gradual impedance of the terrigenous supply by a laterally growing reef tract.

5.2.3. Reef Carbonate Production at Global and Provincial Scales

Estimates of carbonate production by shallow-water coral reefs at a global scale have been tentatively suggested by Milliman (1993), Kleypas (1997), Kleypas, Buddemeier, et al. (1999) and Vecsei (2004). Using measured environmental parameters from the modern tropics (sea surface temperature, salinity, nutrient levels, depth-attenuated level of photosynthetically available radiation, suitable reef-habitat area and topographic relief), Kleypas (1997) and Kleypas, Buddemeier, et al. (1999) calculated that the global production of modern coral reefs averages approximately 1.00×10^{12} kg yr^{-1} (1 Gt yr^{-1}), ranging from 0.9 to 1.68 Gt yr^{-1}. This estimate is close to values presented by Milliman (1993) but is substantially higher than those reported by Vecsei (2004) who used census-based measurements of biota, including fore-reef zones but excluding back-reef and lagoonal zones (approximately 0.75 Gt yr^{-1}, extrema: 0.65 and 0.83 Gt yr^{-1}). Reasoning at the provincial scale, Vecsei (2004) estimated that, according to the degree of fore-reef steepness, Caribbean reefs can produce about 0.9–2.7 kg $CaCO_3$ m^2 yr^{-1}, or 0.07–0.08 Gt yr^{-1}, whereas the total production for Indo-Pacific reefs ranges between 1.9 and 26 kg $CaCO_3$ m^2 yr^{-1} or 0.72 and 0.79 Gt yr^{-1} (Figure 5.5A).

Kleypas (1997) has also modelled reef carbonate production over the past 22 ka, since the Last Glacial Maximum, using appropriate data on sea level, temperature changes and shelf topography. The results indicate that areas available for reef growth were reduced to about 20% of those of the present day with carbonate production reduced to 27%, principally as a consequence of the reduction in space at the low sea-level stand (about 120 m below the present sea surface). At that time, global reef carbonate production is stated to have been less than 0.25–0.30 Gt yr^{-1}. Production appears to have increased rapidly from 11 to about 7–6 ka and then levelled off at about today's value, as sea level stabilized around its present position (Figure 5.5B).

5.3. Patterns of Reef Carbonate Deposition

5.3.1. The Nature and Distribution of Components in Superficial Sediments

The compositions and volumes of detrital sediments appear to be primarily controlled by their formative environment reflected in the nature of

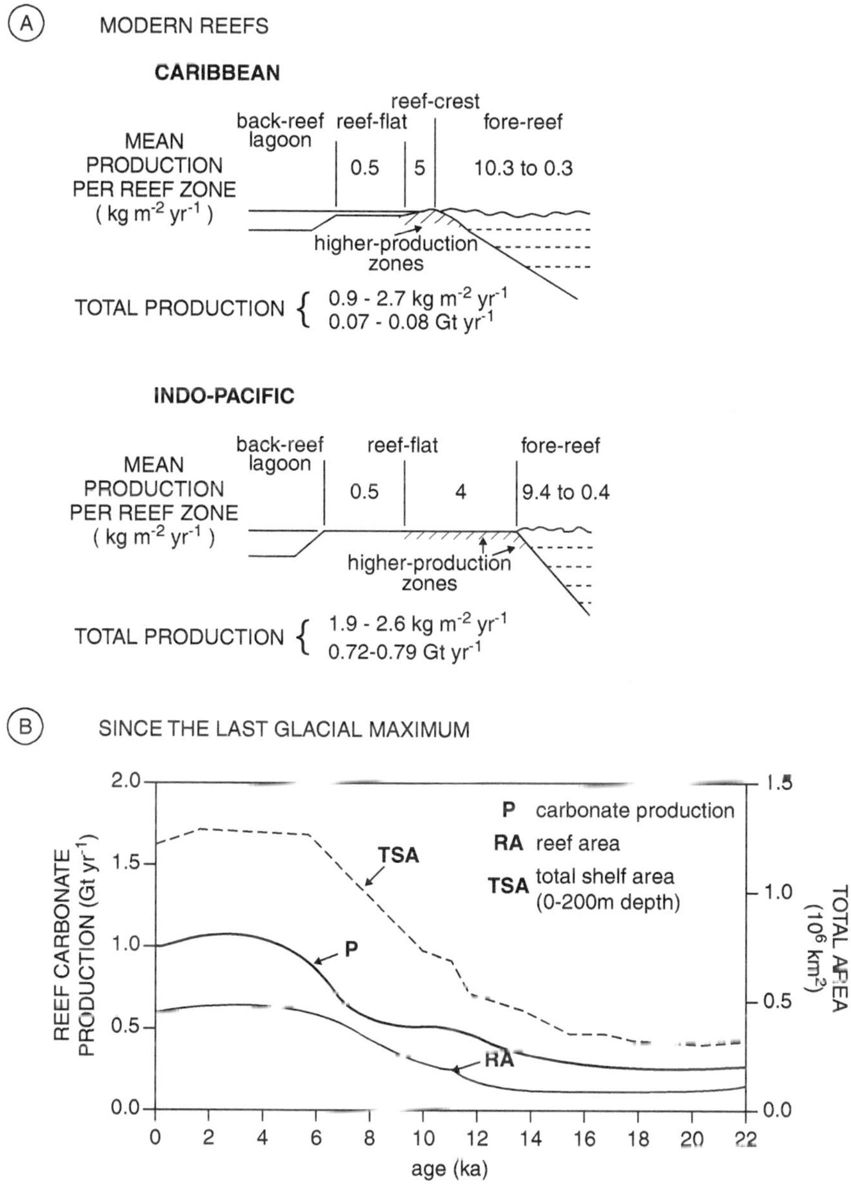

Figure 5.5 Estimates of global carbonate production rates of coral reefs. (A) Mean and total production of modern reefs in the Caribbean and Indo-Pacific provinces (simplified and redrawn from Vecsei, 2004). (B) Estimates of reef carbonate production, total shelf area (0–200 m depth) and total shallow-water coral reef area for the past 22 ka. The production increased proportionately as flooded shelf and reef areas increased (modified and redrawn from Kleypas, 1997).

benthic communities (constructors or eroders) and the hydrodynamic regime. Currents of removal are regarded as more active in the accumulation of skeletal sediments than currents of delivery (Orme, 1977b; Scoffin, 1987). As a result, the use of conventional textural analyses of skeletal deposits as a potential for interpreting the conditions of transport and deposition has proven difficult (Lewis, 1969; Braithwaite, 1982; Kench & McLean, 1997). Recognition of immobile versus mobile populations of skeletal deposits on the basis of hydraulic settling and threshold experiments have been suggested to have a greater potential for interpreting the role of physical and biological processes in reef sedimentation (Kench, 1997). Grain shape also varies and the hydrodynamic behaviour of rod-like and plate-like grains differs from that of equant particles (Maiklem, 1968a; Braithwaite, 1973). As a result of this behaviour, the remains of particular groups of organisms are commonly concentrated within specific size classes, and component analysis generates different results where different classes are analysed.

The overall composition of sediments may vary greatly between different reef environments and reef sites. The most important components are coral, coralline algae (especially, non-geniculate forms), green algae such as *Halimeda*, molluscs, and benthonic foraminifera (Figure 5.6). There are also significant differences in the contributions of sediment producers within specific size grades (gravel to mud, Scoffin, 1992). The gravelly to sandy fractions of deposits may contain additional components including minor skeletal contributors, non-skeletal grains of carbonate or siliciclastic origin. The finer-grained sediment fractions (<0.05 mm) consist predominantly of carbonate mud or clay-rich deposits. In fossil reefs, particularly those that have been subaerially exposed, the association of carbonate components may be a diagenetic artifact rather than a true reflection of the original biota. This reflects the differential susceptibility of the components to diagenesis; an original calcitic mineralogy confers a preservational advantage (see Chapter 8).

It is important to be aware that superficial sediments may result, at least in part, from long-term storage and supply from subfossil to fossil sediment reservoirs. The storage times of detrital material may locally be on a millennial scale (0.5–5 ka) as demonstrated by Harney et al. (2000). Sand-sized remains vary in age according to their production and turnover rates, the higher the turnover the younger the mean age of the components. Thus, the compositions of superficial sediments reflect the structures of former communities rather than those of adjacent living ones.

5.3.1.1. Corals

There is generally a marked variation in the proportions of coral detritus according to wave exposure (i.e. windward versus leeward) and/or substrate

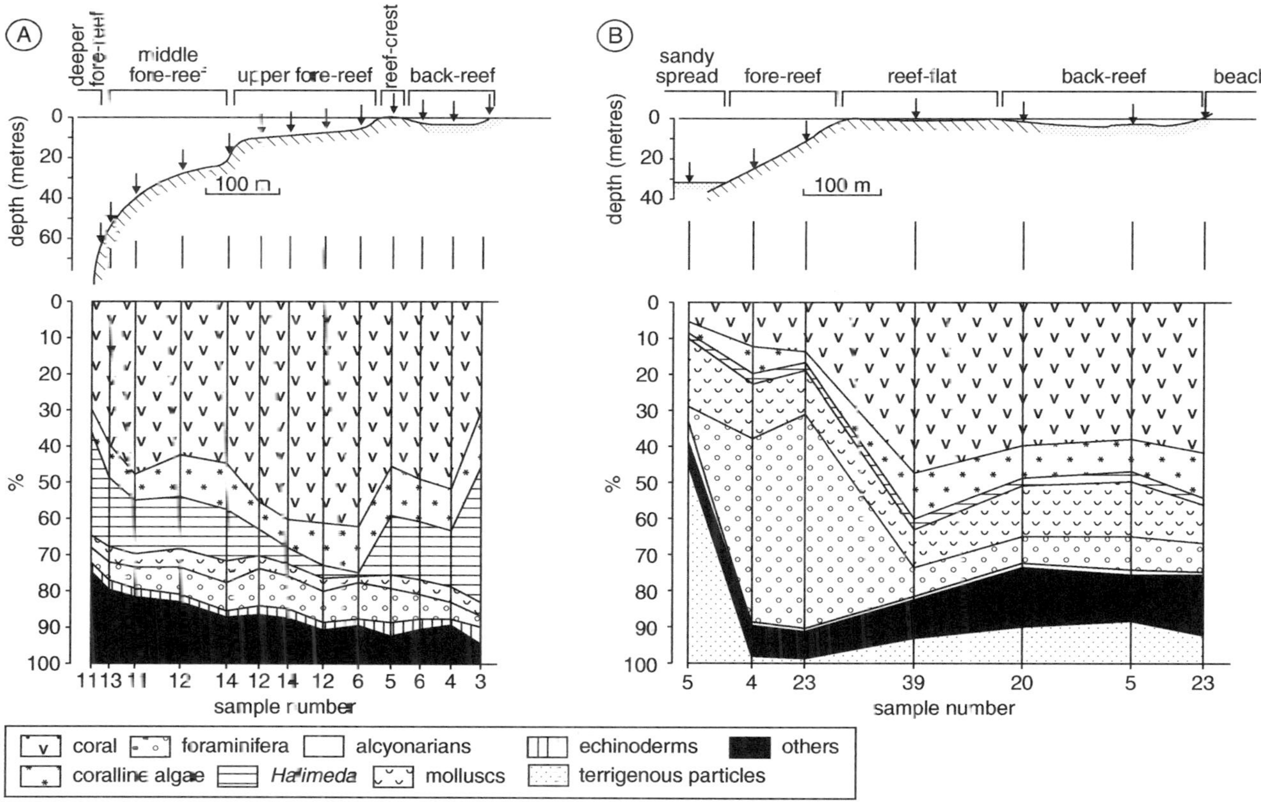

Figure 5.6 Gross physiography, location of sediment sample sites and relative percentage abundance (average values) of the major components in surficial sediments from two fringing reefs. (A) Northern coast of Jamaica, Caribbean (data from Boss & Liddell, 1987a). (B) Western coast of Réunion, western Indian Ocean (data from Montaggioni, 1978).

cover by coral assemblages. Coral colonies may be broken by storms and form gravel. The breakdown of these may later generate coarse sand (20–1 mm) and fine sand to silt fractions, around 0.25–0.025 mm (Orme, 1977b; Kench & McLean, 1997). The distinctive angular concave chips generated by clionid sponges lie in the size range of 35–45 µm (Goreau & Hartman, 1963).

Basic distinctions can be made within and between different environments in terms of the content of coral detritus as shown in the western Indian Ocean (Lewis, 1969; Masse, 1970; Braithwaite, 1982; Gabrié & Montaggioni, 1982a, 1982b; Montaggioni & Mahé, 1980; Montaggioni, Behairy, El Sayed, & Yusuf, 1986; Piller & Mansour, 1990), the Pacific (Maxwell, 1968; Weber & Woodhead, 1972a; Flood & Scoffin, 1978; Tudhope & Scoffin, 1985; Adjas, 1988; Spencer, 1989; Harney et al., 2000; Hewins & Perry, 2006) and the Caribbean (Boss & Liddell, 1987a; Macintyre et al., 1987). Fore-reef sites contain very variable amounts of coral fragments (from 2% to about 50% of the total sediment). By contrast, sediments of reef flats and proximal back-reef settings have coral content commonly approaching 60% and no lower than 20%. Generally, these values in part reflect the high cover rates of coral assemblages in the reef edge (30–80%). In most shallow lagoonal sand sheets and in adjacent deeper water areas of both barrier reefs and atolls, coral is commonly a secondary component forming from 3% to 15% of detritus on average (Weber & Woodhead, 1972a; Orme, 1977; Montaggioni, 1978; Tudhope, Scoffin, Stoddart, & Woodroffe, 1985; Chevillon & Clavier, 1988; Masse, Thomassin, & Acquaviva, 1989; Adjas, Masse, & Montaggioni, 1990; Smithers et al., 1992; Gischler, 1994; Chevillon 1996). The scarcity of coral detritus in these environments clearly indicates a local impoverishment of coral coverage (less than 10% of the substrate).

In reef sites at the southernmost limits of reef growth such as Lord Howe Island (31°33), Middleton and Elizabeth Reef (about 29°), the compositions of surface sediments appear to be relatively coral deficient, compared to most typical tropical fringing and mid-shelf reefs. Coral components are usually subordinate to coralline algae (Kennedy, 2003; Kennedy & Woodroffe, 2004). The proportions of coral in the sand-size fraction are on average less than 25%. Locally, and particularly in lagoonal areas, coral-derived fragments form from only 1% to about 20% of the sediment.

5.3.1.2. Coralline algae

Like corals, non-geniculate and, to a lesser extent, geniculate coralline algae are generally present in greater abundance in sediments deposited close to reef margins and coral patches. Their highest concentrations are usually encountered in very coarse to fine sands (2–0.15 mm) as a result of boring organisms.

The proportions of coralline fragments vary on average from <4% to 25% of the total sediment from deep fore-reef to back-reef zones, irrespective of exposure to waves (Lewis, 1969; Maiklem, 1970; Masse, 1970; Montaggioni, 1978; Gabrié & Montaggioni, 1982a; Delesalle, Galzin, & Salvat, 1985; Tudhope & Scoffin, 1985; Tudhope et al., 1985; Montaggioni et al., 1986; Boss & Liddell, 1987a; Macintyre, Graus, Reinthal, Littler, & Littler, 1987; Adjas, 1988; Flood & Scoffin, 1978; Masse et al., 1989; Spencer, 1989; Piller & Mansour, 1990; Chevillon, 1996; Gischler & Lomando, 1997; Hewins & Perry, 2006). But locally may exceed 40% (Jell & Flood, 1978). A local scarcity of coralline detritus probably reflects a low coverage of living coralline algae that compete unfavourably with fleshy macroalgae under conditions of low herbivory (Paulay, in Spencer, 1989).

Coralline algal detritus is widespread in subtropical environments and increases in abundance towards the southernmost limits of reef growth (Kennedy & Woodroffe, 2004). Thus, on Lord Howe Island, Kennedy (2003) argued that the overall dominance of coralline algae typical reflects a more subtropical rhodalgal assemblage rather than a tropical chlorozoan or chloralgal assemblage (in the sense of Carannante, Esteban, Milliman, & Simone, 1988). In this area the rapid increase in this important carbonate producer coincides with a general decline in coral extension rates.

5.3.1.3. Green algae *Halimeda*

Halimeda contributes selectively to detritus from coarse to very fine sands (1.5–0.1 mm) in a variety of reef settings (Orme, 1977b; Drew & Abel, 1985; Liddell, Ohlhorst, & Boss, 1988; Hillis, 1997). The distribution of *Halimeda* remains in surface sediments varies widely between reef sites and within and between reef zones as a response to ecological and hydrodynamical constraints. Due to its high buoyancy potential (Maiklem, 1968a; Braithwaite, 1973; Kench & McLean, 1997), *Halimeda* detritus can be easily dispersed throughout the different reef zones and preferentially accumulates in sheltered settings (in deeper fore-reefs, leeward reef flats, back-reefs and lagoons). Generally, the highest concentrations are found around and downstream from dense growths. Thus, *Halimeda* segments have occasionally been used as tracers for transport from the reef tract to adjacent basins (Johns & Moore, 1988). *Halimeda* grains may locally form substantial volumes in sand pockets, but be virtually absent from adjacent sediment pools within the same reef zone.

Halimeda debris varies considerably in local abundance in the Indo-Pacific region, ranging from 0% to 90% of the total sediment, irrespective of reef types (Chevalier et al., 1968a,b; Lewis, 1969; Gross, Milliman, Tracey, & Ladd, 1969; Maiklem, 1970; Masse, 1970; Maxwell, 1973; Milliman, 1974; Orme, 1977a,b; Flood & Scoffin, 1978; Orme & Flood,

1980; Braithwaite, 1982; Gabrié & Montaggioni, 1982b; Delesalle et al., 1985; Orme, 1985; Tudhope & Scoffin, 1985; Tudhope et al, 1985; Montaggioni et al., 1986; Adjas, 1988; Montaggioni, 1988b; Payri, 1988; Spencer, 1989; Smithers et al., 1992; Chevillon, 1996; Harney et al., 2000; Hewins & Perry, 2006). Hillis-Colinvaux (1980) indicated that the distribution of *Halimeda* species throughout the Indo-Pacific is controlled by biogeographical factors. *Halimeda* species are regarded as poor dispersalists with problems in moving to remote areas and difficulty growing at subtropical temperatures. Their low dispersal potential may account for their scarcity in the eastern Tuamotus and Henderson (Pitcairn) Island. A similar explanation can be invoked for their relatively low coverage in some areas of the western Indian Ocean (Montaggioni, 1978). The distribution of *Halimeda* in the high-latitude reefs of Middleton, Elizabeth and Lord Howe Islands off the Eastern Australian coast may be explained by inimical water temperatures; the alga decreases in abundance from north to south and is virtually absent from Lord Howe (Kennedy, 2003; Kennedy & Woodroffe, 2004). The low abundance of *Halimeda* on Midway and Kure atolls near the northwestern limit of the Hawaiian archipelago (around 28°N) may, like coral growth, also be temperature dependent (Grigg, 1982).

In the western tropical Atlantic, *Halimeda* is locally the most important sediment producer (Folk & Robles, 1964; Stoddart, 1964; Garret, Smith, Wilson, & Patriquin, 1971; Jordan, 1973; Milliman, 1973; Roberts, 1976; Wallace & Schafersman, 1977; Boss & Liddell, 1987a; Macintyre et al., 1987; Johns & Moore, 1988; Gischler & Lomando, 1999), but is totally absent from some areas (Milliman, 1967). A possible explanation for the lack of green algal production may be local nutrient limitations at variance with the ecological requirements of *Halimeda* species (Littler, Littler, & Lapointe, 1988).

5.3.1.4. Molluscs

Detrital molluscan shells and their derived grains commonly represent less than 10% of the total sediment components. But, bivalves and gastropods are locally by far the dominant sediment producers in lagoonal environments. They contribute mainly to sediment ranging from gravel to fine sand (20–0.15 mm). Broken bivalve shells are prominent in the larger size ranges, while microgastropods are characteristic of intermediate grades (1.5–1.0 mm). The distribution of molluscan remains is primarily controlled by the availability of living assemblages and only secondarily by the prevailing hydrodynamic regime. Generally, the boundaries of mollusc-dominated sediments coincide with those of the living assemblages (Piller & Mansour, 1990).

On most reefs of the Indo-Pacific, the proportions of bivalve and gastropod bioclasts average from 8% to approximately 26% of the skeletal material along fore-reef slopes and in reef-flat environments (Lewis, 1969; Stoddart, 1969a; Maiklem, 1970; Milliman, 1974; Orme, 1977b; Flood & Scoffin, 1978; Jell & Flood, 1978; Montaggioni, 1978; Montaggioni & Mahé, 1980; Braithwaite, 1982; Gabrié & Montaggioni, 1982a, 1982b; Delesalle et al., 1985; Tudhope & Scoffin, 1985; Montaggioni et al., 1986; Masse et al., 1989; Spencer, 1989; Chevillon, 1996; Harney et al., 2000; Kennedy, 2003; Kennedy & Woodroffe, 2004; Hewins & Perry, 2006).

Molluscan fragments are also ubiquitous in the Caribbean, amounting to from 8% to more than 30% of the sediment (Folk & Robles, 1964; Milliman 1974; Macintyre et al., 1987; Gischler & Lomando, 1999).

5.3.1.5. Foraminifera

Benthic foraminifera inhabiting reefs are among the most prolific sediment producers (Wantland, 1977; Hallock, 1981; Montaggioni, 1981; Tudhope & Scoffin, 1988; Langer et al., 1997). However, foraminiferal assemblages show important variations in distribution and state of preservation between different reef sites and environments. Like *Halimeda* segments, plate-like and subspheric tests are widely distributed throughout reef systems by virtue of their settling velocities and may locally form monospecific accumulations. Some may therefore be used as tracers of sediment transport across reef systems (Coulbourn & Resig, 1975; Montaggioni & Venec-Peyré, 1993; Li, Jones, & Blanchon, 1997).

On most Indo Pacific reefs, the proportions of foraminiferal grains vary dramatically from zone to zone. Foraminiferal detritus dominates on the fore-reef slope, representing from 15% up to 60% of the total sediment (Lewis, 1969; Masse, 1970; Montaggioni, 1978; Montaggioni & Mahé, 1980; Gabrié & Montaggioni, 1982a, 1982b; Montaggioni et al., 1986; Masse et al., 1989; Piller & Mansour, 1990). In reef-flat and proximal back-reef settings, the concentrations range from 1% to 15% (Harney et al., 2000, Kennedy & Woodroffe, 2004). Similar concentrations occur in many lagoons, as in Bikini and Enewetak Atolls (Milliman, 1974) and on isolated islands of the central Pacific (Spencer, 1989). By contrast, in the GBR region, Maiklem (1970), Maxwell (1973), Orme and Flood (1980), Flood and Scoffin (1978), Jell and Flood (1978) and Tudhope and Scoffin (1985, 1988) claimed that foraminiferal tests are the most abundant constituents, commonly forming approximately one third to one-half of all samples on reef rims, reef flats and inter-reef plains.

Generally, the contribution of foraminiferal tests to reef detritus in the Caribbean appears to be lower than that of most Indo-Pacific reefs, less than 15% (Milliman, 1974; Boss & Liddell, 1987a; Macintyre et al., 1987), although on the Belize-Yucatan platform (Gischler & Lomando, 1999), skeletal sediments locally consist of 50% foraminifera.

Pelagic foraminiferal tests are typically rare (less than 0.5% of the total sediment) but locally reach up to 5% in sandy spreads occurring at the base of some fore-reef zones and in inter-reef environments (Tudhope & Scoffin, 1985).

5.3.1.6. Other skeletal components

Bryozoan remains constitute a minor sediment component on most coral reefs worldwide. They are generally of low abundance, with values less than 1–2% of the total sediment (Masse, 1970; Gabrié & Montaggioni, 1982a, 1982b; Masse et al., 1989; Chevillon, 1996). Locally, however, they release detritus forming up to 6% of the total sediment (Lewis, 1969; Montaggioni, 1978; Braithwaite, 1982; Delesalle et al., 1985; Tudhope & Scoffin, 1985; Montaggioni et al., 1986) and occasionally reach maximum values of 15% (Hewins & Perry, 2006).

Aragonitic alcyonarian sclerites (spicules) contribute to sediments as indurated monospecific spiculites within cavities and as loose grains in the finer sand fractions of surficial detritus (Montaggioni, 1980; Konishi, 1981). Free spicules are present in very low concentrations, normally less than 1–3% of the total sediment (Masse, 1970; Braithwaite, 1982; Gabrié & Montaggioni, 1982a, 1982b; Tudhope & Scoffin, 1985; Tudhope et al., 1985; Masse et al., 1989; Smithers et al., 1992), but locally exceed 5–9% of the sediment (Montaggioni, 1978; Montaggioni & Mahé, 1980).

Echinoderms produce only a small fraction of identifiable sediment particles, generally representing 1–2% of the total sediment (Masse, 1970; Braithwaite, 1982; Delesalle et al., 1985; Tudhope & Scoffin, 1985; Tudhope et al., 1985; Montaggioni et al., 1986; Smithers et al., 1992; Chevillon, 1996; Hewins & Perry, 2006).

Crustacean shells (dominantly ostracods) and fragments range in abundance from 0.2% to approximately 5% of sediments (Lewis, 1969; Braithwaite, 1982; Gabrié & Montaggioni, 1982a, 1982b; Tudhope et al., 1985; Piller & Mansour, 1990; Smithers et al., 1992; Chevillon, 1996) but may rise above 10% in back-reef and lagoonal environments (Montaggioni, 1978; Montaggioni & Mahé, 1980; Piller & Mansour, 1990).

Fragments of serpulid crusts rarely rise above 2% of the total sediment (Lewis, 1969; Montaggioni, 1978; Gabrié & Montaggioni, 1982a, 1982b; Tudhope & Scoffin, 1985; Montaggioni et al., 1986).

Sponge spicules are confined principally to the deeper parts of fore-reef slopes and to back-reef and coastal zones that may locally carry relatively high coverages of siliceous sponges (Rützler & Macintyre, 1978; Naim, 1993). When present (mainly within the finer sandy fractions), they do not exceed 1–2% of the total sediment (Masse, 1970; Montaggioni & Mahé, 1980; Gabrié & Montaggioni, 1982b; Tudhope & Scoffin, 1985; Tudhope et al., 1985; Piller & Mansour, 1990).

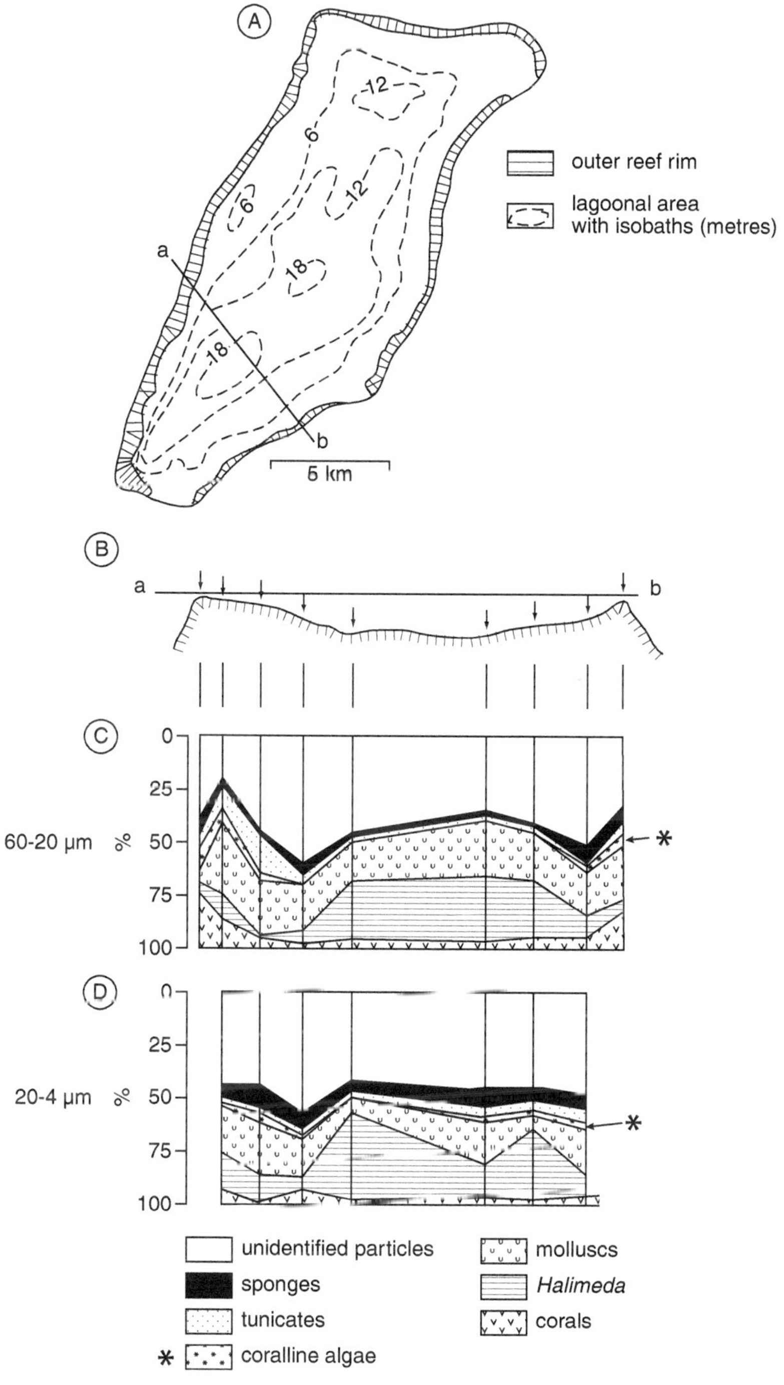
A
12
6
12
6
18
18
a
b
5 km
outer reef rim
lagoonal area
with isobaths (metres)
B
a
b
C
0
25
50
75
100
60-20 µm
%
*
D
0
25
50
75
100
20-4 µm
%
*
unidentified particles
sponges
tunicates
coralline algae
molluscs
Halimeda
corals

5.3.1.7. Non-skeletal and compound carbonate grains

These grains are heterogeneous, and include faecal pellets, aggregates and coated grains of varying origins (ooids, lumps and grapestones in the sense of Bathurst, 1975; Milliman, 1974) and probably also unidentifiable micritized bioclasts. They are generally of very low abundance (less than 2% of the sediment, or absent). Reef-related environments in the Caribbean appear to contain higher proportions of such grains than those of the Indo-Pacific province. The highest concentrations (from 12–76%) are found in lagoonal areas. The grains in these are mainly faecal pellets, and lumps and grapestones are rare (Milliman, 1973; Gischler & Lomando, 1999).

5.3.1.8. Unlithified carbonate mud

On present-day reef systems, carbonate muds (grains smaller than 63 μm in diameter) generally occur in the inner and/or deepest parts of lagoonal environments. The mud content is usually more than 50% of the sediment volume. Their origin was formerly extensively debated (see Milliman, 1974; Bathurst, 1975 for summaries). Most such material has been demonstrated to be of biogenic origin (Figure 5.7) resulting from the mechanical or bioerosional disintegration of original skeletal constituents (Pusey, 1975; Ellis & Milliman, 1985; Scoffin & Tudhope, 1985; Tudhope et al., 1985; Nelsen & Ginsburg, 1986; Tudhope & Scoffin, 1986; Adjas et al., 1990; Zinke et al., 2001; Gischler & Zingeler, 2002) or from the alteration (micritization) of skeletal grains (Reid, Macintyre, & Post, 1992; Reid & Macintyre, 1998). Chemically precipitated muds are largely restricted to lagoonal environments and to arid, subtidal, coastal flats (Purser, 1973). They probably form seasonally from waters supersaturated with respect to carbonate (Adjas et al., 1990; Macintyre & Aronson, 2006).

5.3.1.9. Free-living nodules

Mobile growths consisting predominantly of red algal rhodoliths are common components on modern reefs worldwide (Bosellini & Ginsburg, 1971; Adey & Macintyre, 1973; Konishi, 1975; Montaggioni, 1979a; Minoura & Nakamori, 1982; Bosence, 1983a, 1983b; Flood, 1983; Scoffin,

Figure 5.7 Composition of unlithified carbonate mud at Glovers Reef, a platform system offshore of Belize, Caribbean. (A) Physiography of Glovers Reef showing location of the sample transect (a–b). (B) Transect line with location of sampling sites. (C) Composition of the 62–20 μm fraction of the sediment. (D) Composition of the 20–4 μm fraction of the sediment. The mud composition was determined using point counting under a scanning electron microscope. Modified and redrawn from Gischler and Zingeler (2002).

Stoddart, Tudhope, & Woodroffe, 1985; Reid & Macintyre, 1988; Minnery, 1990; Tsuji, 1993; Piller & Rasser, 1996; Payri, 1997; Gischler & Pisera, 1999; Lund, Davies, & Braga, 2000; Foster, 2001; Rao, Montaggioni, et al., 2003; Perry, 2005). Free-living coralline algal, rhodoliths include both massive and branching nodules (Figure 5.8). The taxonomic compositions of rhodoliths differ between reef provinces, reef sites, and according to depth. Generally, the rhodoliths from shallower water environments (less than 5 m) consist predomnantly of the mastophoroids (*Neogoniolithon*, *Hydrolithon* and *Lithoporella*) together with the lithophylloids (*Lithophyllum*, *Dermatolithon*, *Tenarea*). In deeper water environments (greater than 10 m), the melobesioids (*Mesophyllum* and *Lithothamnion*), together with the sporolithacean *Sporolithon* are the most common. The peyssonnelid red

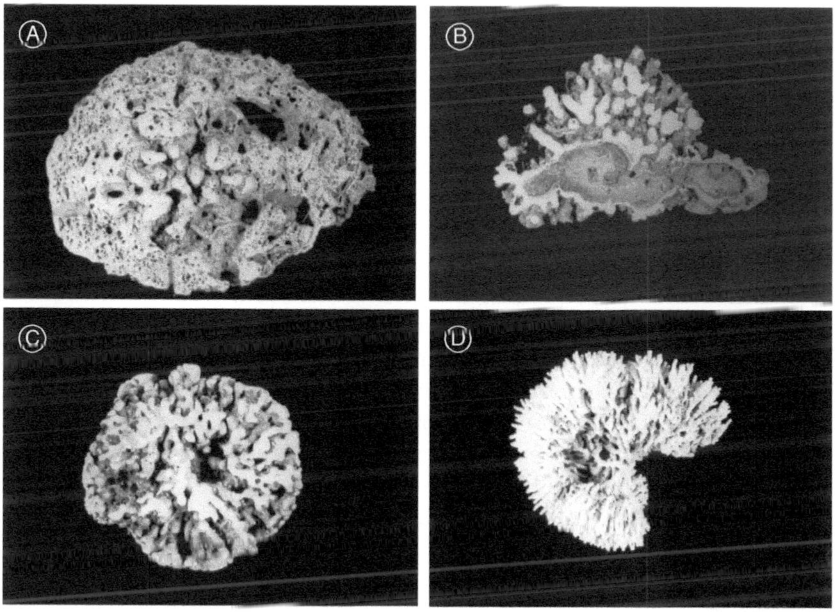

Figure 5.8 Form and internal structures of rhodoliths from different reef zones and sites, western Indian Ocean (photograph from L. Montaggioni). (A) Cross-section of an elliptical massive nodule composed of an algal nucleus of branching growth form, covered by laminar thalli (shorter diameter: 65 mm). Outer sandy spread, 60 m deep, west of Réunion. (B) Cross-section of an asymmetrical branching nodule composed of a coral nucleus and laminar, algal coatings. Height: 40 mm. Inner reef flat, fringing reef at La Saline, Réunion. (C) Cross-section of a sub-spheroidal nodule, monospecific in composition (*Lithophyllum*) showing a bumpy surface and a growth form of columnar type. Shorter diameter: approximately 80 mm. Inner back-reef zone, fringing reef, eastern coast of Mauritius. (D) Piece of a spheroidal gracile branching nodule, monospecific in composition (*Lithothamnion*). Diameter: 80 mm. Inner back-reef zone, fringing reef, eastern coast of Mauritius.

algae and encrusting foraminifera may locally contribute significantly to nodule growth in association with bryozoans, bivalves, serpulid worms and encrusting corals. On fore-reef terraces, shelf ridges and foreslopes, foraminifera may also contribute to rhodolith growth, equal in importance to coralline algae (Reid & Macintyre, 1988). Locally, branching types may be monospecific, resulting from the isotropic accretion of a single thallus (Montaggioni, 1979a; Piller & Rasser, 1996; Payri, 1997). The diameters of algal nodules ranges from less than 3 to about 15 cm, irrespective of shape, internal structure and habitat. Rhodoliths may be concentrated in particular environments; the number of individuals per square metre ranges from 1 to about 100 (Montaggioni, 1979a; Scoffin et al., 1985; Payri, 1997).

In addition to red algae, individual coral colonies locally develop in the form of free-rolling spheroidal balls (coralliths). These are known from the Indo-Pacific (Pichon, 1974; Scoffin et al., 1985; Riegl, Piller, & Rasser, 1996; Roff, 2008) and Caribbean (Glynn, 1974). Growth forms and taxa forming coralliths include massive *Porites*, *Cyphastrea* (*C. microphthalma*), *Siderastrea*, *Goniopora*, *Gardineroseris* and occasionally branching *Pocillopora* and *Pavona*. These nodules range from about 3 up to 25 cm in diameter.

The controls on nodule distribution are expected to lie along a continuum ranging from hydrodynamic energy and deposition to biological processes (mainly bioturbation). Movement by waves and currents and by browsing fish and crustaceans is considered to be necessary to maintain the globular growth form of free-living biogenic nodules. However, there is apparently no direct correlation between current velocities and the distributional pattern of such nodules (Scoffin et al., 1985). Generally, nodules are believed to encapsulate sensitive records of their formative and depositional conditions and thus to provide reliable palaeoenvironmental indicators (Bosellini & Ginsburg, 1971; Bosence, 1983b; Scoffin et al., 1985; Frantz, Kashgarian, Coale, & Foster, 2000; Halfar, Zack, Kronz, & Zachos, 2000).

5.3.1.10. Microbialites

These deposits result from trapping and binding of detrital material and/or mineral precipitation by benthic microbial communities (Burne & Moore, 1987; Golubic, 1991; Golubic, Seong-Joo, & Browne, 2000). Cyanobacteria-dominated deposits accrete subtidally to intertidally in a variety of environments from open marine to lagoonal, inner reef flat and beach settings and on substrates including loose sands, sea grass beds, algal turfs and crusts, consolidated sedimentary bottoms and living or dead coral surfaces (Rasmussen, Macintyre, & Prufert, 1993: Défarge, Trichet, Maurin, & Hucher, 1994; Reid, Macintyre, Browne, Steneck, & Miller, 1995; Macintyre et al., 1996; Steneck, Miller, Reid, & Macintyre, 1998; Webb, Jell, & Baker, 1999;

Reid et al., 2000; Sprachta, Camoin, Golubic, & Le Campion, 2001; Abed, Golubic, Garcia-Pichet, Camoin, & Sprachta, 2003; Gautret, Camoin, Golubic, & Sprachta, 2004; Gautret & Trichet, 2005; Pringault, de Wit, & Camoin, 2005). Some microbialites occupy cryptic niches, as observed in the GBR (Reitner, 1993; Webb et al., 1999) and in French Polynesia (Montaggioni & Camoin, 1993). Individual fabrics and structures are produced by filamentous cyanobacteria including *Phormidium*, *Symploca* and/or *Schizothrix*.

The proliferation of cyanobacterial mats and discrete microbialites in modern reef environments, particularly in lagoonal and coastal settings, is a recent phenomenon, first emerging at the beginning of the 1980s. This event was apparently coincident with a marked decline in the health of coral communities. Microbialites may compete with corals and other phototrophic builders that require similar high irradiance levels (Pringault et al., 2005). The settlement of microbialites on living colonies seems to cause corals to decline irreversibly.

The occurrence of lithified micritic crusts resembling microbialites has been reported from Quaternary reefs, mainly from deposits formed on deep fore-reefs slopes (Moore, Graham, & Land, 1976; James & Ginsburg, 1979a; Land & Moore, 1980; Brachert & Dullo, 1991; Dullo et al., 1998; Brachert, 1999; Cabioch et al., 2006; Camoin et al., 2006), in lagoonal and intertidal sites (Jones & Hunter, 1991) or in shallow-water caves (Macintyre, 1984b; Reitner, 1993; Zankl, 1993; Reitner, Gautret, Marin, & Neuweiler, 1995). Lithified micritic crusts have also been described by Macintyre and Marshall (1988), in Quaternary reef frameworks, but were not regarded as microbial. However, similar crusts associated with high-energy coral and coralline algal frameworks are present in cores penetrating the outer barrier reef of Tahiti and in adjacent lagoonal patch reefs (Figure 5.9), and these are interpreted as microbialites (Montaggioni & Camoin, 1993; Camoin, Gautret, Montaggioni, & Cabioch, 1999). Framework-associated microbialites that developed since the last deglaciation (in the past 19 ka) have been identified from a number of other reef sites in both shallow- and deep-water environments, including cryptic frameworks in the Caribbean (Zankl, 1993), the western Pacific (Australian Great Barrier Reef: Webb, 1996; Webb, Baker, & Jell, 1998; Vanuatu: Cabioch, Taylor, et al., 1999; Cabioch et al., 2006), the central Pacific (Camoin et al., 2006; Camoin, Iryu, McInroy, & the IODP Expedition 310 Scientists, 2006, 2007) and the Indian Ocean (Camoin et al., 1997). The presence of 'reefal microbialites' in shallow-water settings and 'slope microbialites' at depths of 10–20 m or greater than 100 m suggests differing histories of development and possibly also differing microbes. Reefal microbialites reflect a late stage of encrustation experienced largely by dead coral communities, while slope microbialites have usually been deposited as the ultimate stage of a biological succession indicating a deepening-upward

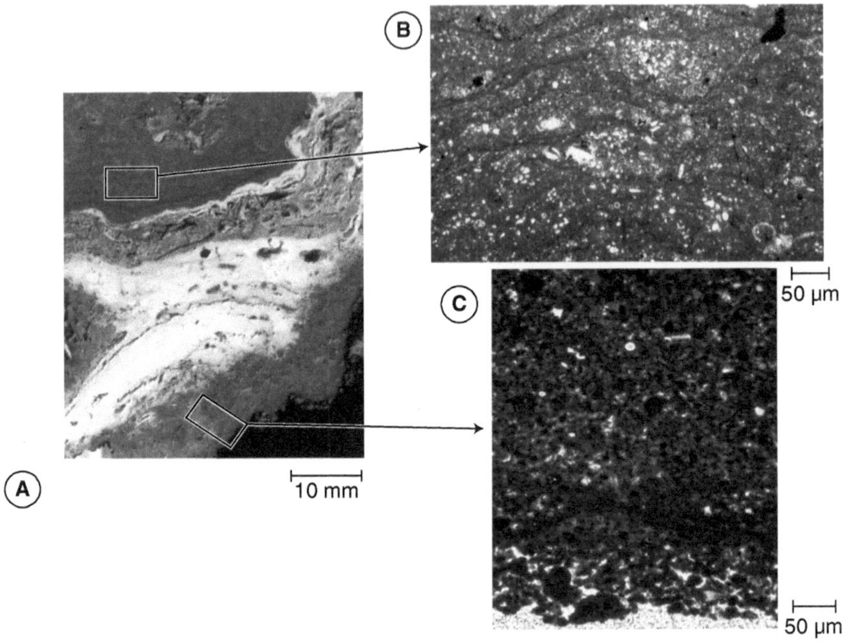

Figure 5.9 Holocene microbialites from a core extracted from the outer barrier reef pile, Tahiti, French Polynesia (Photographs by L. Montaggioni). (A) Core section (14.5 m below present reef surface) showing a coralgal assemblage (lamellar coral encrusted by thick coralline algal thalli) overgrown by thick microbialite layers. (B) At the top, thin section microphotograph of a laminated microbial crust. The size of the microbioclastic grains trapped in the micritic matrix averages 15 μm. (C) At the base, thin section microphotograph of a clotted, peloidal, micritic coating. In the central part of the picture, the diameter of darker peloids averages 20 μm.

sequence in which shallow-water corals and associated builders are replaced by deeper water assemblages. Both reefal and slope microbialites reflect changes in water quality, mainly indicating an increase in nutrients (terrestrial groundwater seepage, or upwelling during sea-level rise; Camoin et al., 2006).

5.3.1.11. Mixed carbonate–siliciclastic sediments

In reef settings close to terrigenous sources, siliciclastic material may contribute to sedimentation (see Doyle & Roberts, 1988 for a selection of case studies; and Perry & Larcombe, 2003; Macdonald, Perry, & Larcombe, 2005 for discussion). Sand- to silt-sized terrigenous grains of varying mineralogy (quartz, mafic grains and clay-minerals) may constitute significant volumes of the sediment. The mud fraction consists partly of clays minerals (metahalloysite, kaolinite, gibbsite and goethite) and amorphous silicates and represents 5–85% of reef sediments on volcanic islands

(Montaggioni, 1978; Zinke et al., 2001). The GBR is characterized by mixed carbonate/terrigenous deposits in a variety of settings (Maxwell & Swinchatt, 1970; Scoffin & Tudhope, 1985; Flood & Orme, 1988; Heap et al., 2001; Heap, Dickens, Stewart, & Woolfe, 2002).

In Holocene and Pleistocene successions, carbonate muds are found in a variety of reef zones, either as unconsolidated sandy to silty deposits or as indurated, wackestones to mudstones. Beneath reef flats, carbonate mud typically represents less than 10% of the total sediment (Johnson & Risk, 1987; Yamano, Kayanne, & Yonekura, 2001; Braithwaite et al., 2000; Kennedy & Woodroffe, 2000; Gischler, 2007). Sections from the deeper lagoons of barrier reefs and atolls retain higher mud contents, about 50–80% of the total sediment (Smith, Frankel, & Jell, 1998; Zinke et al., 2001; Zinke, Reijmer, et al., 2005; Gischler, 2003).

5.3.2. Classification of Sediment Types

Sediments may be differentiated using the major representative contributors and grain-size characteristics as descriptors. Conventionally, all types are named by reference to their lithified equivalents following the nomenclatures of Dunham (1962) and Embry and Klovan (1972). The use of these terms allows modern and fossil data to be compared. The most efficient method of classifying sediment types has proven to be multivariate analysis of component and grain-size data. This allows a meaningful differentiation of discrete sediment types, each of which is typified by a distinct grouping of major and secondary skeletal or non-skeletal components. Unfortunately, to date, there has only been a limited number of such statistical treatments (factor and cluster analyses) from either modern or fossil reef sediments in the literature (Figures 5.10 and 5.11).

5.3.2.1. Carbonate rudstone-dominated types

Coral-dominated rudstones. This sediment type consists of poorly sorted to unsorted, angular to rounded coral rubble together with clasts of bivalves, gastropods, coralline algae and a variety of sand-sized skeletal elements (Figure 5.12A). It forms a prominent component of most Holocene and Pleistocene sections, irrespective of ambient hydrodynamic energy conditions and zones. On modern reefs, coral rudstones are usually found in intertidal to subtidal storm generated gravel sheets deposited on the surfaces of reef flats and prograding into back-reef and lagoonal environments.

Coral rudstones may represent from 30% up to 60% of the total volume in sediment piles on exposed reef margins and in innermost back-reef zones (Tracey & Ladd, 1974; Macintyre & Glynn, 1976; Adey & Burke, 1977; Lighty et al., 1978; Fairbanks, 1989; Davies & Hopley, 1983; Johnson, Cuff, & Rhodes, 1984; Hubbard et al., 1986; Montaggioni, 1988b;

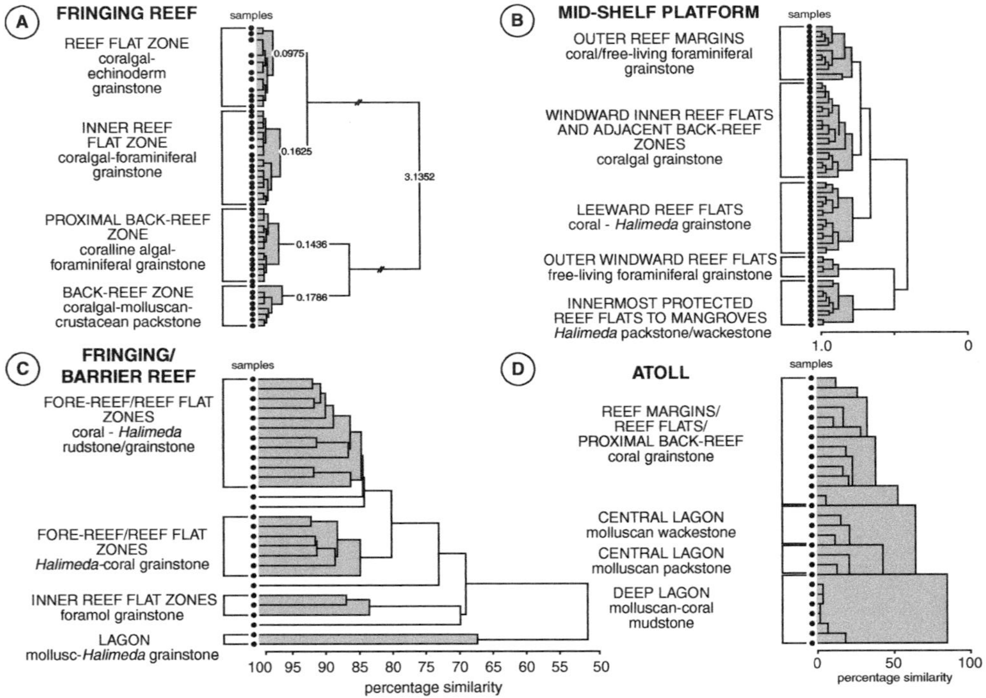

Figure 5.10 Differentiation of sediment types based on statistical analyses of component grain compositions. (A) La Saline fringing reef, Réunion, western Indian Ocean: cluster analysis performed using Ward's method (Euclidian distances on non-standardized variables). Euclidian distances are given for each sediment type (modified and redrawn from Chazottes et al., 2008). (B) Mid-shelf reef platforms of Low and Three Isles, northern Great Barrier Reef of Australia: Q-mode cluster analysis performed using Klovan and Imbrie's factor programmes (modified and redrawn from Flood & Scoffin, 1978). (C) Fringing-barrier reef system of Danjugan Island, Philippines, Pacific Ocean: cluster analysis performed using Renkonen similarity index (modified and redrawn from Hewins & Perry, 2006). (D) Rasdhoo Atoll, Maldives, Indian Ocean: cluster analysis performed using Euclidian distances on non-standardized variables (modified and redrawn from Gischler, 2007). Note the occurrence of coral, coralline algae, *Halimeda* and foraminifera dominated sediment types in fore-reef and reef-flat zones, while mollusc-dominated sediment types typify back-reef and lagoonal environments.

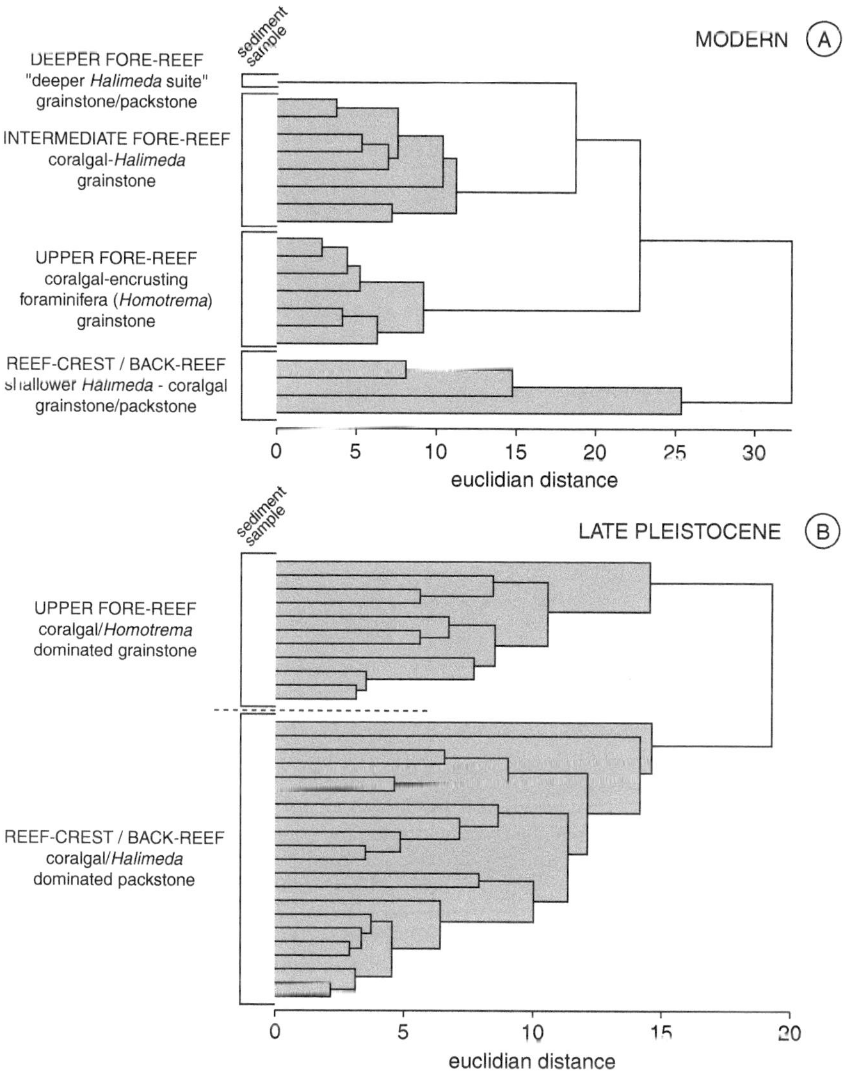

Figure 5.11 Comparison of component grain compositions between modern (A) and late Pleistocene (B) (Falmouth Formation) reef tracts in north Jamaica, Caribbean. The compositions of the sediment types in the fossil reef are similar to those of the upper fore-reef and reef-crest/back-reef zones respectively of the modern reef. Modified and redrawn from Boss and Liddell (1987a, 1987b).

Tudhope, 1989; Cortés et al., 1994; Blanchon, Jones, & Kalbfleisch, 1997; Montaggioni & Faure, 1997; Gischler & Hudson, 1998; Iryu, Nakamori, & Yamada, 1998; Braithwaite et al., 2000; Kennedy & Woodroffe, 2000; Collins et al., 2003; Sugihara, Nakamori, Iryu, Saski, & Blanchon et al.,

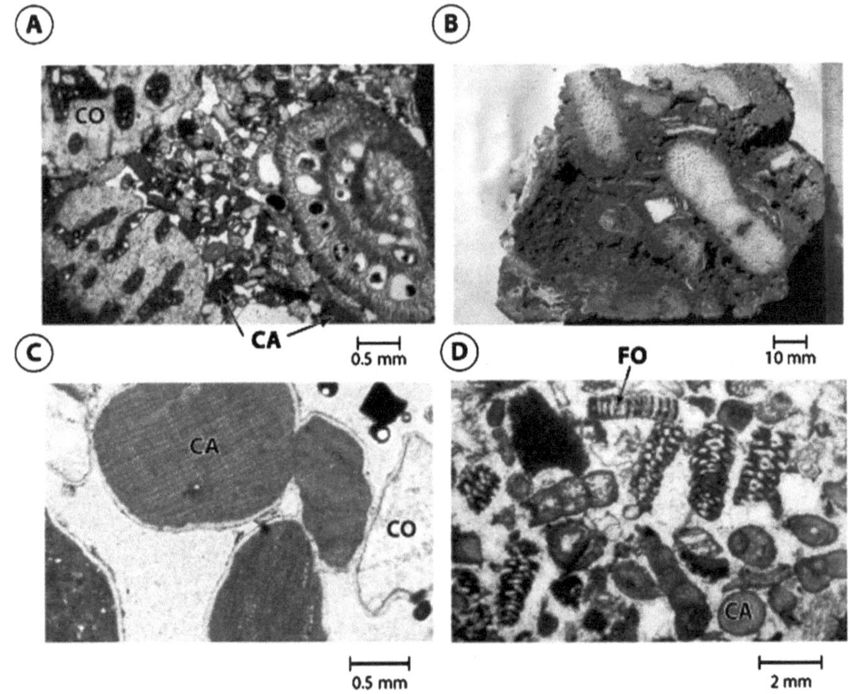

Figure 5.12 Thin-section photomicrographs of reef sediment types (from L. Montaggioni). (A) Poorly sorted, coralgal-foraminiferal rudstone from Holocene beach-rock, Tarama, the Ryukyus, Japan. The coarser fraction is composed of coral (CO), coralline algal (CA) grains and foraminiferal tests (Rotaliid). The finer sand-matrix fraction is dominated by coral and coralline debris. The shorter diameter of the rotaliid test is up to 2 mm. (B) Coral-dominated floatstone from a core section extracted from the outer barrier reef (10 m below present reef surface), Tahiti, French Polynesia. The coral fragments (about 1 cm thick) are derived from a *Pocillopora* colony. Associated grains are coralline algae and *Halimeda* plates. The matrix consists of microbioclasts, clay-rich mud and high-magnesian micritic cement. (C) Well-sorted, coralgal grainstone from internal sediments deposited in an intertidal reef flat, Moorea, French Polynesia. CO = coral; CA = coralline algae. The cement is an isopachous fringe of high-magnesian calcite. The sizes of grains range from approximately 0.5 to 1 mm. (D) Well-sorted, coralline algal-foraminiferal grainstone from an exposed, late Pleistocene reef flat, west coast of Mauritius, western Indian Ocean. The coralline fragments are mainly articulated *Amphiroa* (CA); the foraminiferal fragments (FO) are mainly of soritid tests. The average grain size is approximately 1 mm. The cement consists of blocky, low-magnesian calcite.

2003; Webster & Davies, 2003; Grossman & Fletcher, 2004; Blanchon & Perry, 2004; Hubbard et al., 2005). Fragments may be fresh, weakly encrusted or heavily encrusted by coralline algae and associated calcifiers reflecting differences in rates of deposition and burial (Perry & Hepburn, 2008).

Generally, gravel is supported by a sandy and/or muddy matrix (Figure 5.12B), locally giving way to a floatstone texture (Macintyre, 1977; Montaggioni & Camoin, 1993; Blanchon & Perry, 2004; Engels et al., 2004; Gischler, Hudson, & Pisera, 2008). In sediment accumulations in high-energy settings (exposed reef margins and flats), the matrix, where present may be fine-to-coarse sand consisting of typical reefal constituents including foraminiferal tests, micromolluscs, *Halimeda* plates, coral, coralline algae, echinoid and alcyonarian grains. In sediment piles from low-energy settings, the interclast matrix consists of fine sand, silt or mud. The mud may be carbonate or mixed with terrigenous clay components (Johnson & Risk, 1987; Smith et al., 1998; Yamano et al., 2000; Gischler & Zingeler, 2002).

Coralline algal-dominated rudstones. In Quaternary reefs and carbonate platforms, coralline algal rudstones are predominantly represented by rhodolith beds (Alexander et al., 2001; Webster et al., 2003, 2006; Payri & Cabioch, 2004; Braga & Aguirre, 2004; Kundal & Dharashivkar, 2005). In some cases, these have provided stabilized substrates for pioneering coral communities and predate reef initiation. The most striking examples of rhodolith limestones are described from the Pleistocene of New Caledonia and the Ryukyu Islands in the western Pacific. For example, Payri and Cabioch (2004) described an 8-m-thick rhodolith unit of mid-Pleistocene age (0.41–0.85 Ma; Cabioch, Montaggioni, Thouvery, et al., 2008) deposited at the base of a carbonate sequence in the southwestern New Caledonian barrier reef system directly overlying the bedrock. Based on its taxonomic composition, this deposit was interpreted as a suite of shallower (less than 10 m), high-to-moderate hydrodynamic energy and deeper (but less than 40 m), low-energy environments. In contrast to the photophilic coralline algae, little is known about the role of sciaphilic red algae (peyssonnelids) in the formation of algal rudstone, particularly from the Quaternary record. The only description to date is from the late Pleistocene of Grand Cayman Island in the Caribbean (Hills & Jones, 2000). Here, *Peyssonnelia rubra*, associated with coralline algae (mainly *Lithoporella*, *Lithophyllum*, *Hydrolithon* and *Neogoniolithon*) and other encrusters, has formed nodules up to 14 cm in diameter. The ages of these are estimated to range between about 250 and 600 years. The Grand Cayman rhodoliths are regarded as having grown in shallow waters (less than 14 m) surrounding back-reef coral patches.

5.3.2.2. Carbonate grainstone/packstone-dominated types

These consist of skeletal, coarse-grained (grainstone) to muddy (packstone) sands, unconsolidated or poorly lithified in modern, Holocene and late Pleistocene deposits and moderately to firmly cemented older rocks.

There have been few investigations of the biological compositions of the sandy detritus in Quaternary reefs. This is somewhat frustrating for two main reasons. First, because back-reef and proximal lagoonal accumulations consist predominantly of skeletal sands, occupying more than 80% of the total volume (Tudhope, 1989; Marshall & Davies, 1982; Davies & Hopley, 1983; Montaggioni, 1988b; Gray, Hein, Hausmann, & Radtke, 1992; McLean & Woodroffe, 1994; Cabioch, Camoin, & Montaggioni, 1999; Kennedy & Woodroffe, 2000; Zinke et al., 2001; Gischler, 2003). Sequences from reef margins, reef flats and patch reefs may contain continuous sand intervals up to 5 m thick, representing from 10% to 50% of the total rock volume (Davies, 1974; Henny, Mercer, & Zbur, 1974; Easton & Olson, 1976; Falkland & Woodroffe, 1997; Montaggioni & Faure, 1997; Webster et al., 1998; Cabioch, Camoin, et al., 1999, 2003; Yamano et al., 2001; Grossman & Fletcher, 2004; Woodroffe et al., 2004; Hubbard et al., 2005; Gischler, 2007, 2008). Second, there is a need to improve the databank on the compositions of sand piles because the proportions of the various components may reflect changes in environmental conditions influencing the structure of biological communities (Perry, 1996; Lidz & Hallock, 2000; Perry, Taylor, & Machent, 2006; Chazottes, Reijmer, & Cordier, 2008).

Coral and coralgal-dominated grainstones/packstones. As mentioned above (Section 5.3.1), coral and/or coral–coralline algal (coralgal) sands are usually restricted to upper fore-reef, reef-crest, reef-flat and adjacent back-reef zones (Figure 5.12C).

A number of subsidiary coralgal types have also been identified, based on their associated subordinate components. On eastern Red Sea reefs, a coral–octocoral (*Tubipora*) sediment is associated with typical coralgal sediments (Montaggioni et al., 1986). Coral-encrusting foraminifera and/or coral–bryozoan grainstones/packstones are also regarded as indicators of proximity to hard substrates (Mackenzie, Kulm, Cooley, & Barnhart, 1965; Wigley, 1977; Braithwaite, 1982; Reiss & Hottinger, 1984; Montaggioni & Venec-Peyré, 1993). In Jamaica, Boss and Liddell (1987a) indicated that the upper fore-reef zone differs from nearby back-reef and lower fore-reef areas in being characterized by the presence of a coral–*Homotrema rubrum* grainstone. The deep and middle fore-reef slopes are typified by the presence of coral–*Halimeda* and coralgal–*Halimeda* facies respectively. On Réunion, both coral–*Amphistegina* and coral–alcyonarian associations are recognized in the sandy accumulations spilling down fore-reef slopes. Locally, elevated proportions of foraminiferal tests and alcyonarian spicules reflect high densities of foraminiferal populations living upslope as epiphytes and soft corals inhabitating hard substrates in the vicinity (Gabrié & Montaggioni, 1982a).

Similar patterns are known from Holocene and Pleistocene reef sections. Coral and coralgal sediment types are ubiquitous, but the highest abundances of these components (up to 50% of total sand fractions) occur in sediments accumulated in fore-reef, reef-crest and outer reef-flat zones (Montaggioni, 1977, 1982; Webster et al., 1998; Kayanne et al., 2002; Cabioch, 2003; Collins et al., 2003; Yamano et al., 2003; Grossman & Fletcher, 2004; Gischler, 2003). For example, in emergent reef crests of late Pleistocene age on Mauritius (Indian Ocean), sand-sized coral and coralline algal grains represent 25–45% and 7–68% respectively of the total constituents (Montaggioni, 1982). Locally, foraminifera form a significant proportion of sediments (Figure 5.12D).

Sediments from the upper fore-reef zones of the late Pleistocene Falmouth Formation of north Jamaica are coralgal grainstones, consisting of 51–63% coral and 18–30% coralline algae (Boss & Liddell, 1987b). In the back-reef zones of the Falmouth Formation, a coral–*Halimeda* packstone has been identified, with a composition comparable to that of back-reef sediments in the adjacent modern fringing reef system (Figure 5.11).

In both Caribbean and Indo-Pacific reefs of Holocene or Pleistocene age, subordinate components in grainstones/packstones are derived mainly from benthic foraminifera and molluscs. Foraminiferal tests derived from encrusting groups (Homotremids mainly) and a variety of free-living forms dominated by amphisteginids, calcarinids, baculogypsinids, soritids and/or miliolids in the Indo-Pacific, and by asterigerinids, peneroplids, soritids and/or miliolids in the Caribbean. Relatively rare sand types, including alcyonarian (spiculite) grainstones have been described locally in various zones beneath reef flats and in shallower back-reef areas, (Montaggioni, 1980; Konishi, 1981; Johnson & Risk, 1987; Braithwaite et al., 2000).

***Halimeda*-dominated grainstones/packstones.** Where present in modern reefs, *Halimeda*-dominated sediments can be almost ubiquitous, but locally may serve as useful environmental markers. On mid-shelf reefs of the northern Australian Great Barrier, this type of sandy sediment is restricted to low-wooded islands, occurring in sheltered areas such as the lee of mangroves (Flood & Scoffin, 1978). Similar distributions have been described by Jell and Flood (1978) on reef platforms in the southern GBR where reef-flat detritus includes both typical chloralgal and chlorozoan facies (in the sense of Lees, 1975) dominated by *Halimeda* and coralline algae and by *Halimeda* and scleractinians respectively. Similarly, chlorozoan components dominate sediments from the innermost back-reef areas of Danjugan in the Philippines (Hewins & Perry, 2006). Based on the species composition and depth habitat of *Halimeda* suites, Boss and Liddell (1987a) distinguished two *Halimeda* sediment subtypes on Jamaican reefs: a shallow-water subtype (less than about 25 m) dominated by *H. opuntia* and *H. simulans*, and a deep-water subtype (greater than 25 m) rich in *H. copiosa* and *H. cryptica.*

In the northern barrier reef system of Belize, the proximal lagoonal zones are characterized by grainstones containing on average up to 40% *Halimeda* (Pusey, 1975).

As in modern reef sites, *Halimeda*-rich deposits are found in Holocene and Pleistocene sections from a variety of reef zones (Figure 5.13A, B).

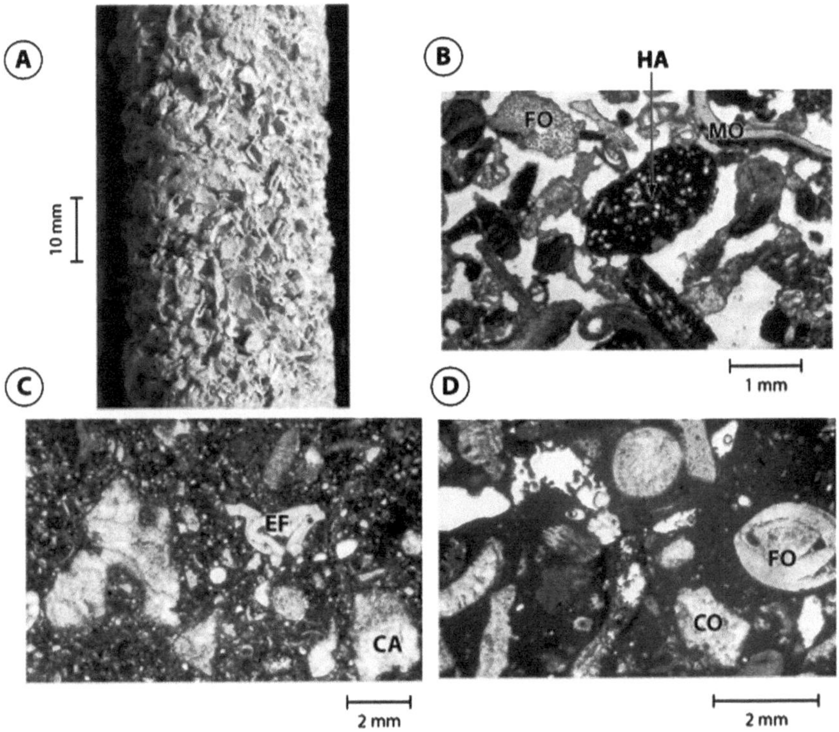

Figure 5.13 Thin-section photomicrographs of reef sediment types (from L. Montaggioni). (A) Twenty centimetres long core section composed of *Halimeda*-dominated grainstone, top of late Pleistocene sequence (11.80–12 m below present reef surface), Raine Island, northern Great Barrier Reef of Australia. (B) Well-sorted, *Halimeda*/mollusc-dominated grainstone deposited in an inner reef flat, Moorea, French Polynesia. HA = *Halimeda*; MO = molluscs; FO = foraminifera. The incipient cement is of grain contact or meniscus types. The central *Halimeda* plate is about 1 mm diameter. (C) Coral fragments in foraminiferal wackestone from an exposed, late Pleistocene, back-reef zone, westcoast of Mauritius, western Indian Ocean. CA = coral; EF = encrusting *Carpenteria* fragment. The matrix consists of microbioclasts (various skeletal debris, ostracods), clay-rich mud and low-magnesian calcite micrite. The larger skeletal grains range from 0.5 to upto 2 mm in diameter. (D) Coral-foraminiferal mudstone from a late Pleistocene core section (113 m below present reef surface) extracted from Ribbon Reef 5, Australian Great Barrier Reef. CO = coral; FO = *Amphistegina* test. The matrix consists of low-magnesian calcite mud. The diameter of the *Amphistegina* test is about 1.5 mm.

Halimeda plates are locally concentrated beneath inner reef flats, forming up to 35% of the total sand fractions (Marshall & Davies, 1982; Engels et al., 2004; Gischler et al., 2008), and beneath back-reef zones (Degauge-Michalski, 1990; Gischler & Lomando, 1999; Kayanne et al., 2002). Sequences from semi-exposed to protected environments may include *Halimeda* and/or chloralgal packstones, as reported from Holocene fringing reefs in New Caledonia (Cabioch, 1988), Pleistocene reef complexes from the Ryukyus, Japan (Nakamori et al., 1995), and from Barbuda, West Indies (Wigley, 1977). In core sections from west and central Pacific atolls, three different *Halimeda*-dominated sand types are recognized in lagoonal areas. From shallow, proximal to deeper, distal, areas these are successively coral–*Halimeda* grainstones, *Halimeda*–nummilitid–miliolid grainstones, and *Halimeda*–molluscan packstones (Yamano, Kayanne, Matsuda, & Tsujii, 2002). *Halimeda*-rich rudstones and packestones/wackestones of late Pleistocene to Holocene age have been described from deep fore-reef slopes of New Caledonian barrier reefs at depths of 85–250 m (Flamand et al., 2008) and off the Marquesas Islands at depths of 70–130 m (Cabioch, Montaggioni, Frank, et al., 2008). Two hypotheses were suggested to explain their occurrence at relatively great depth. These assemblages might have been deposited in place, representing a pause punctuating the postglacial sea-level rise, or have cascaded down the slope from reef margins.

Mollusc-dominated grainstones/packstones. Molluscan–coral, molluscan–coralline algal (Figure 5.14A) and molluscan–*Halimeda* sediments are typical of a number of inner back-reef zones from modern reefs. Examples have been described from the Indo-Pacific (Montaggioni & Mahé, 1980) and the Caribbean (Wigley, 1977; Macintyre & Toscano, 2004; Gischler, 2007). In addition, in both outer- and inner-reef environments, molluscan fragments may be mixed with substantial numbers of larger foraminiferal tests. For example, such an association, referred to the foramol facies of Lees (1975) and Wilson and Vecsei (2005), has been described from the shelf edge of the central GBR (Scoffin & Tudhope, 1985). In the Philippines, the foramol association occurs locally across the entire inner reef-flat zone, with two components (coral and Halimeda) forming up to 50% of the sediment (Hewins & Perry, 2006). On the Jordanian coast of the Gulf of Aqaba (Red Sea), the sediments from upper fore-reef slopes are of a molluscan–foraminiferal subtype, composed of about 50% coral and 33% foramol (Gabrić & Montaggioni, 1982b). In Florida Bay (Caribbean), Bosence (1989) described mollusc–foraminiferal grainstones to wackestones and mollusc–*Halimeda* grainstones to mudstones as the dominant sediment types.

As expected, in Holocene and Pleistocene lagoonal sequences, the limestones recovered are molluscan-dominated packstones (Perrin, 1989; Cabioch, Camoin, et al., 1999; Kennedy & Woodroffe, 2000; Zinke, Reijmer, Thomassin, & Dullo, 2003). Foramol facies have been described

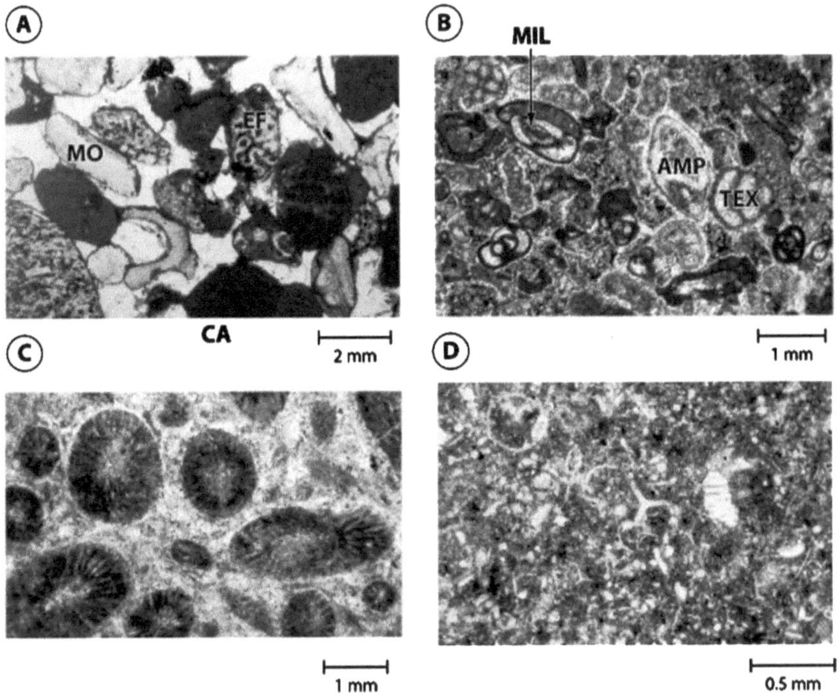

Figure 5.14 Thin-section photomicrographs of reef sediment types (from L. Montaggioni). (A) Well-sorted, molluscan-coralline algal grainstone from sediments deposited in a proximal back-reef zone, Mauritius, western Indian Ocean. CA = coralline algae; MO = molluscs; EF = encrusting foraminiferal. The grain size ranges between 1 and 2 mm. (B) Foraminiferal packstone from an exposed late Pleistocene back-reef zone, west coast of Mauritius, western Indian Ocean. The dominant foraminifera are miliolids and textulariids. MIL = miliolids; TEX = textulariids; AMP = amphisteginids. In the central part of the picture, the diameter of the *Amphistegina* test is about 2 mm. (C) Alcyonarian (spiculite) grainstone from an exposed late Pleistocene reef, Gulf of Aqaba, Red Sea. The diameter of the largest spicule sections is about 1.5 mm. (D) Fine-grained, sponge-rich wackestone from an exposed late Pleistocene back-reef zone, west coast of Mauritius, western Indian Ocean. The triactine spicule in the central part of the picture is about 0.2 mm diameter.

from mid-Pleistocene reefs in New Caledonia (Cabioch, Montaggioni, Thouveny, et al., 2008). On isolated carbonate platforms off Belize, Holocene deposits from lagoon shoals consist mainly of molluscan packstones–rudstones comprising, on average, 25% molluscan fragments. In the central lagoons of platforms, deposits are foramol wackestones (Gischler, 2003, 2007). Tebbutt (1975) and Gischler (2007) showed that the inner shelf lagoon deposits of the late Pleistocene reef systems of Belize, are also typified by a molluscan–*Halimeda* packstone rich in both bivalves and gastropods.

Foraminifera-dominated grainstones/packstones. As emphasized by Sugihara, Masunaga, & Fujita, (2006) the compositions of foraminiferal sediments in shallow-water reef environments, may change with latitude. Thus, a variety of foraminiferal types and subtypes, defined at family or genus levels, have been statistically defined in contrasting reef sites.

Numerous studies have focused on reef biozonation, based on the compositions of both living and dead foraminiferal associations from modern and fossil reef systems (review by Hallock & Glenn, 1986, and papers by Martin & Liddell, 1988, 1991; Vénec-Peyré, 1991; Hohenegger, Yordanova, Nakano, & Tatzreiter, 1999; Langer & Hottinger, 2000; Bicchi, Debenay, & Pagès, 2002; Yamano et al., 2002; Langer & Lipps, 2003; Fujita, Shimoji, & Nagai, 2006). For example, in the Gulf of Aqaba (Red Sea), four sediment types have been distinguished related to the depositional environments of the fringing reef system: a mixed encrusting *Acervulina*–free-living *Amphistegina* type in the deeper fore-reef zone; an encrusting Homotremid–*Acervulina* type typical of the upper fore-reef and reef-crest zones; a mixed Homotremid–*Amphistegina–Spirolina* type diagnostic of the reef-flat zone, and a Miliolid (*Triloculina, Quinqueloculina*)-Soritid (*Amphisorus, Sorites*) type characteristic of the back-reef zone (Gabrié & Montaggioni, 1982b). On western Pacific atolls, Yamano et al. (2002) identified three foraminiferal-dominated sediment types distributed from proximal, shallower to central, deeper, lagoonal areas and characterized by *Calcarina*, mixed *Calcarina–Heterostegina* and *Heterostegina* respectively. At Discovery Bay (Jamaica), *Archaias–Amphistegina–Asterigerina*-dominated grainstones are present across the entire reef system from fore-reef to back-reef zones (Martin & Liddell, 1988). In northern Belize, two distinct types have been identified: a peneroplid-grainstone and a miliolid-mudstone, derived respectively from the proximal and distal inner parts of the lagoon of the barrier reef system (Pusey, 1975).

The high abundance and dominance of foraminiferal tests is a common feature in Holocene and Pleistocene reef successions. Sediments dominated by encrusting foraminifera (Figure 5.13C) are very similar from ocean to ocean, with abundant *Homotrema* and/or *Carpentaria* (see Wigley, 1977; Montaggioni, 1982; Pandolfi et al., 1999). By contrast, free-living foraminiferal sediment types differ in composition within and between oceans, although some larger foraminifera such as *Amphistegina* are ubiquitous (Langer & Hottinger, 2000) (Figure 5.13D). In the Indian Ocean, reef-flat accumulations of Holocene and Pleistocene age are typified by the prevalence of *Amphistegina–Marginopora–Calcarina* grainstones (Figure 5.12D) and back-reef/lagoonal successions by Miliolid (*Triloculina–Quinqueloculina*)-*Textularia* packstones (Figure 5.14B) to mudstones (Montaggioni, 1978; Colonna, 1994; Braithwaite et al., 2000). In the western Pacific, Holocene sequences are characterized by a *Calcarina–Baculogypsina–Marginopora* association (Cabioch, 1988; Yamano et al., 2001, 2002; Kayanne et al.,

2002) that is also present in Pleistocene deposits (Lacroix, 2004). In the Pleistocene reef complexes of the Ryukyu Islands, Iryu et al. (1998) and Fujita et al. (2006) noted the occurrence of *Cycloclypeus–Operculina* and *Cycloclypeus–Heterostegina–Amphistegina* grainstones regarded as typical of deep fore-reef zones respectively. Wigley (1977) reported the presence of two distinct foraminiferal sediment types in early (?) Pleistocene reefal limestones of Barbuda in the West Indies: a *Homotrema*-coralline algal type and an *Amphistegina*-coralline algal type, both representing deposition in a reef tract.

Other reef-associated grainstones/packstones/wackestones. Atypical skeletal sediments may occur locally as a response to high cover rates by specific reef-dwelling communities. Alcyonarian grainstones are common in a few modern reef and fossil settings (Konishi, 1981), especially within reef-flat and back-reef deposits (Figure 5.14C). Sponge-rich wackestones occur in a number of deep fore-reef, back-reef and lagoonal environments (Land, 1976) (Figure 5.14D). Non-skeletal grainstones to packstones, consisting of ooids, pellets and/or grapestones, have been statistically differentiated in a few sites of different ages (Wigley, 1977; Piller, 1994). Such deposits are interpreted as originating in shallow waters, on the surfaces of unstable substrates and subsequently redeposited in lagoons as aeolian or storm sediments. In addition, grainstones to packstones rich in altered carbonate grains have been identified in several reef-associated sites, especially in enclosed or semi-enclosed lagoons, where they are locally the dominant sediment type. Pusey (1975) described a grainstone, composed of faecal pellets and micritized skeletal grains in the lagoon of northern Belize and a similar facies has been reported by Piller and Mansour (1990) in the northern Red Sea (Bay of Safaga).

A variety of composite terrigenous-skeletal grainstones to wackestones have been encountered locally. Terrigenous-coral grainstone types have been statistically differentiated in northern Red Sea reefs (Gabrié & Montaggioni, 1982b; Piller & Mansour, 1990; Piller, 1994) and on Réunion (Gabrié & Montaggioni, 1982a). In the northern barrier system of Belize, in distal parts of the lagoon, a mixed carbonate–terrigenous grainstone contains up to 47% skeletal grains, mostly molluscs and miliolids (Pusey, 1975).

5.3.2.3. Carbonate wackestone/mudstone-dominated sediments

Mud-rich sediments on modern reefs are typically dominated by molluscs, locally representing up to 50% of the skeletal components (Piller & Mansour, 1990; Gischler & Lomando, 1999; Zinke, Reijmer, Thomassin, & Dullo, 2003). Subtypes are locally rich in corals and/or free-living foraminifera (Figure 5.13D) and among the latter miliolids are the most abundant (Pusey, 1977; Wantland, 1977; Montaggioni, 1981; Yamano et al., 2002).

5.3.3. Temporal and Spatial Shifts in Skeletal Sediment Composition

The role of physical and ecological disruption events in controlling reef carbonate production and budgets has been discussed by Perry, Spencer, and Kench (2008). The potential for reefs to shift from a state characterized by communities dominated by carbonate-producing organisms to one in which communities consist mainly of soft thalloid algae is critical to the supply of detrital grains. Because the compositional patterns of detritus on reefs are primarily controlled by the nature of the living biological assemblages, shifts in reef community structure driven by natural or human-induced disturbance should be detectable from the analysis of the uppermost sediment layers. Attempts to detect such changes in the compositions of reef communities over time or space have been made in a few sites in the Caribbean (Perry, 1996; Lidz & Hallock, 2000; Perry et al., 2006; Greenstein, 2007; Precht & Miller, 2007) and in the Indo-Pacific (Chazottes, 1996; Chazottes, Le Campion-Alsumard, Peyrot-Clausade, & Cuet, 2002, 2008; Uthicke & Nobes, 2008; Schueth & Frank, 2008). Studies at Discovery Bay in north Jamaica by Perry et al. (2006) of temporal shifts, using cores from depths of 5–25 m, allowed a reconstruction of the history of reef lagoon sedimentation relative to bauxite contamination over the past 40 years. Abrupt changes in the composition of sediments in the core were reported at depths of 5–10 m. In the lower layers regarded as 'clean carbonates', constituents were dominated by corals (40% of the total components), molluscs (20–25%), coralline algae *Amphiroa* (10–15%) and *Halimeda* (10–15%). Near-surface and surficial sediments are composed primarily of *Halimeda* (20–30%) and *Amphiroa* (30–40%), while the proportions of corals and molluscs decline dramatically, expressing the lethal influence of bauxite input on the corresponding living communities (Figure 5.15). Lidz and Hallock (2000) compared the compositions of surficial sediments collected over a 37-year period from the Florida Reef Tract. The proportions of the major sediment producers (corals, *Halimeda* and molluscs) was shown to have changed markedly in the different reef zones through time. In the upper and the middle keys the proportions of molluscan and coral remains relative to *Halimeda* more than doubled for molluscs and tripled for corals. These changes are regarded as a response to ecological shifts in the reef communities, stimulated by both natural and anthropogenic disturbances, including cold water and nutrient inputs and disease. The increased production of molluscan and coral grains was promoted by accelerated bioerosion in response to a proliferation of boring organisms due to increased planktonic productivity. However, the potential for preserving such evidence

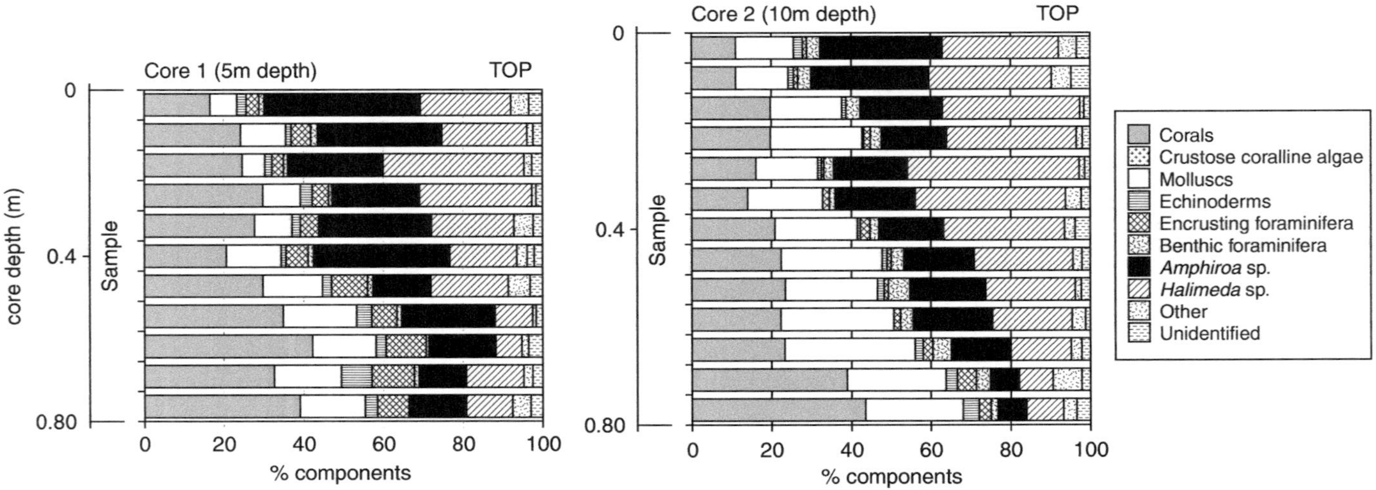

Figure 5.15 Core plots showing temporal shifts in the relative percentage abundances of skeletal components (>50 μm sediment fractions). Cores 1 and 2, 80 cm long, were extracted at depths of 5 and 10 m respectively, from the innermost part of Discovery Bay, north Jamaica, Caribbean. Note the relative decrease in the amounts of corals and encrusting coralline algal grains and the concomitant increase in *Halimeda* and coralline *Amphiroa* upcore. Modified from Perry et al. (2006).

of changing reef depositional patterns varies widely from site to site. Perry (1996) claimed that sands derived from fore-reef and reef-crest frameworks generally have a higher potential to record changes in sediment production patterns than those accumulated in back-reef environments. It was argued that this is because back-reef carbonates suffer extensive biogenic reworking, and dissolution, and are thus time-averaged deposits.

Renema and Troelstra (2001), together with Hallock, Lidz, Cockey-Burkhard, and Donnelly (2003) and Schueth and Frank (2008) have all suggested that foraminiferal assemblages, particularly those of the larger symbiont-bearing forms, are reliable indicators of changes in reef environmental conditions because they have water-quality requirements similar to corals. However, in comparison to corals that are long-lived organisms, relatively short-lived foraminifera offer the advantage of making it possible to identify suspected short-term stress events. Studies of spatial changes in community structure and sediment components in reef settings subjected to varying nutrient input were conducted experimentally on Réunion. Chazottes et al. (2008) demonstrated that, in areas where soft algal assemblages dominated over coral communities as a response to nutrification, there was a shift from coral to coralline algal-dominated detritus, together with the settlement of dense assemblages of boring sponges. This shift was accompanied by a decrease in sediment production and in the relative proportions of very fine sands to muds with increasing medium to fine sands, as a result of the decreasing activity of grazers. High proportions of coralline algal fragments and siliceous sponge spicules occurred in sediments from nutrient enriched areas in comparison to adjacent locations not subject to nutrification.

5.3.4. Depositional Rates of Reef Carbonates

An important dataset regarding the rates of Holocene reef deposition in a variety of geodynamic settings has accumulated (see Macintyre, 1988, 2007; Dullo, 2005, Montaggioni, 2005; Hopley et al., 2007, pp. 372–403 for reviews) (Figure 5.16). These rates have been shown, for the most part, to have been driven by changes in hydrodynamic energy in response to exposure and/or changing accommodation space (Blanchon & Jones, 1997; Blanchon et al., 1997; Hubbard, Burke, & Gill, 1998; Braithwaite et al., 2000; Montaggioni, 2005). Four types of reef-related accumulations can be delineated: growth frameworks forming reef edges (or margins); detritus-rich successions of sheltered inner-shelf reef edges and lagoonal piles; and *Halimeda* mounds.

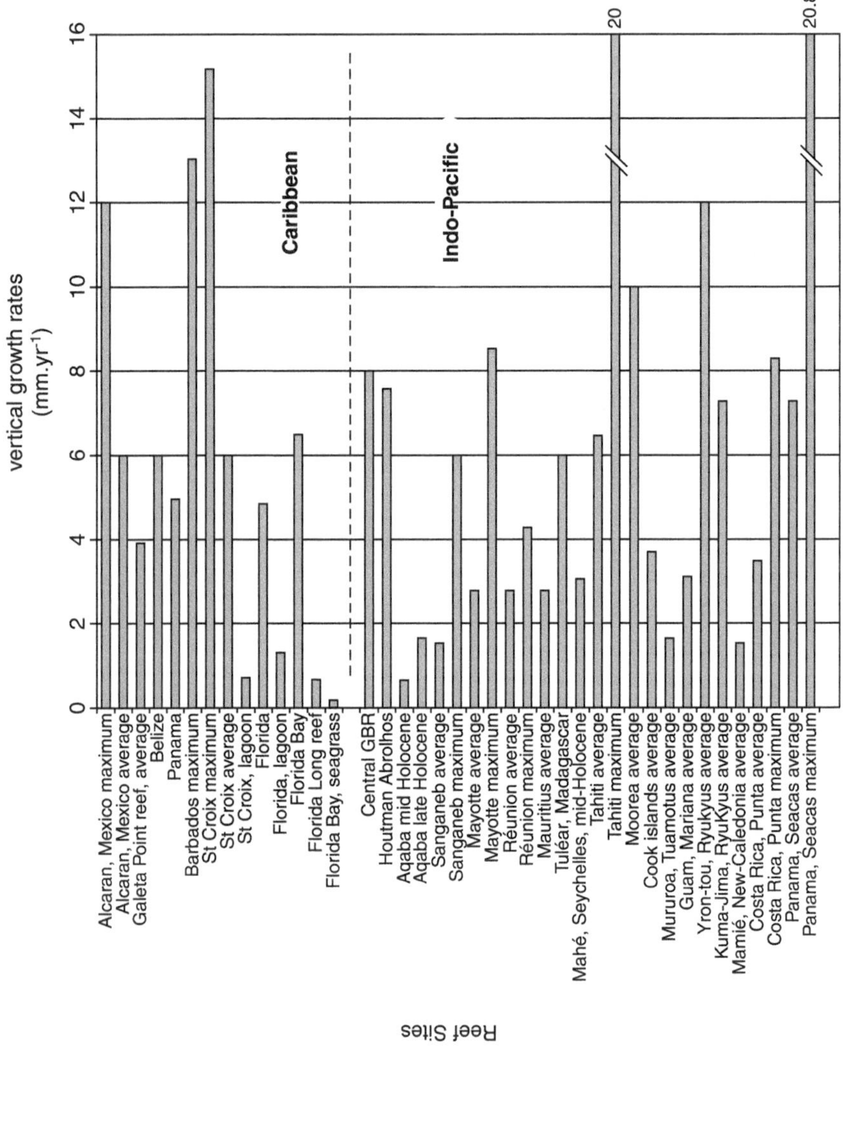

Figure 5.16 Vertical growth rates of selected reef systems in the Caribbean and Indo-Pacific regions during the Holocene. Data from Dullo (2005) and Montaggioni (2005).

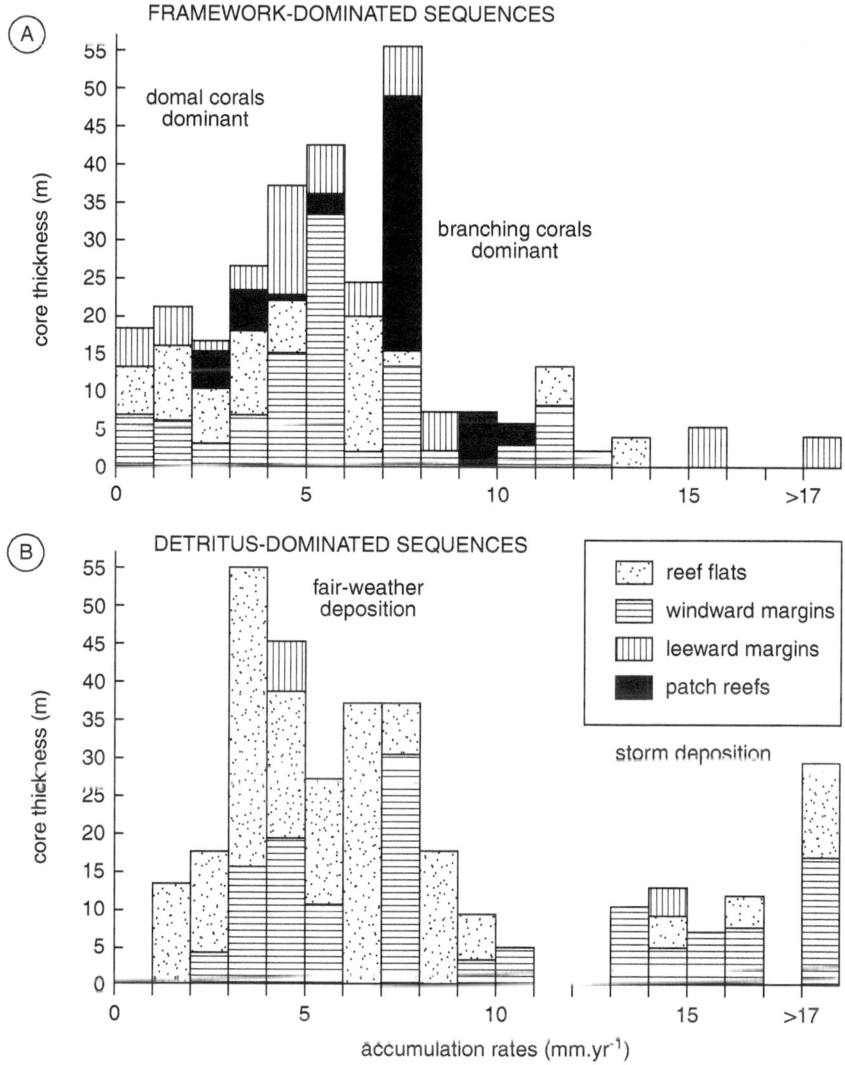

Figure 5.17 Vertical accumulation-rate ranges of framework-dominated (A) and detritus-dominated (B) reef sequences of Holocene age from the Australian Great Barrier Reef. The data are based on compositional analysis and dating of cores extracted from a total of 40 individual reefs. Core thicknesses represent the cumulative length of cored sections composed of either framework or detrital material. Adapted and redrawn from Hopley et al. (2007, Figure 11.4).

5.3.4.1. Reef-edge, framework-dominated aggregations

In reef accumulations dominated by growth framework, the total variation in vertical accretion rates ranges between <1 and about 30 mm yr^{-1} with a modal rate of 7–8 mm yr^{-1} (Figure 5.17A). The higher modal rates are generally recorded in cores containing a greater proportion of branching forms (Macintyre & Glynn, 1976; Davies & Montaggioni, 1985; Montaggioni, 1988b; Hubbard et al., 2005; Macintyre, 2007). The fabric of the coral framework has therefore been suggested to control accretion rates. Thus, when comparing the growth rates of different coral forms, higher accretion rates might be expected for reefs dominated by shallow-water branching corals than for those dominated by deeper-water, domal, foliaceous or encrusting colonies. However, this has proven a controversial concept and its validity has been questioned. Using findings from the Holocene development of the Belize barrier and atoll reefs, Gischler (2008) demonstrated that accretion rates appear to have increased with increasing palaeo-water depth. In addition, parts of the reef sequences dominated by massive corals, have apparently accreted slightly faster than those composed of branching acroporids. This can be explained by the higher resistance of massive corals to breakdown and the depth-habitat range (5–10 m) within which they are subject to lower disturbance and higher accommodation, whereas shallow-water (0–5 m) acroporids may repeatedly suffer disintegration and reworking during storms.

Gischler's (2008) conclusions in part agree with previous results from the Caribbean and Indo-Pacific provinces, but are not of general value. In framework-dominated sequences, high deposition rates are recorded from sections consisting of branching and domal coral communities. The vertical accretion rates of high-energy, robust coral frameworks locally reached 13–15 mm yr^{-1} (Glynn & Macintyre, 1977; Fairbanks, 1989; Montaggioni & Faure, 1997; Hubbard et al., 1998; Gischler et al., 2008). For comparison, low-energy, domal coral assemblages may have grown upwards at rates approaching 12–15 mm yr^{-1} (Cortés et al., 1994; Montaggioni et al., 1997; Camoin et al., 2004; Engels et al., 2004). However, although the vertical accretion rates of coral assemblages seem not to be governed directly by the growth habits of the corals, the highest rates measured (up to 20 mm yr^{-1}) coincide with the development of high-porosity frameworks laid down by tabular and arborescent acroporid assemblages. Rates of 20–30 mm yr^{-1} have been reported locally from arborescent acroporid-rich sections (Montaggioni et al., 1997; Kayanne et al., 2002). Notwithstanding these differences, domal coral frameworks are usually typified by average growth rates of 3–5 mm yr^{-1}, whereas the mean rates of branching forms are typically 5–8 mm yr^{-1}.

Vertical accumulation rates may vary within a single coral assemblage, depending on ambient conditions at the time of growth. Abrupt changes in the rates within a sequence may relate to changes in the composition of the coral assemblage in response to variations in accommodation space or hydrodynamic energy. The initial coral community is replaced by one better adapted to the new conditions. For example, at Rasdhoo Atoll (Maldives, Indian Ocean), a decrease in vertical deposition rates reported from the reef margin coincided with a change in the composition of the coral community from branching acroporids (growing at about 10 mm yr^{-1}) to domal poritids (growing at <4 mm yr^{-1}). This change occurred as the rate of sea level rise declined drastically and the reef top approached sea level within the 3 m depth range after 7–6 ka (Gischler et al., 2008). Thus, markedly higher rates of deposition are recorded from older Holocene reef sequences than from relatively contemporary deposits. A similar pattern has been described from the reef flat at Warraber Island (Torres Strait, northern Australia). This is at present emergent at mean low tide and accreted at a rate of about 4 mm yr^{-1} from 6.7 to 5.3 ka, but the present mean accretion rate is less than 1 mm yr^{-1} (Hart & Kench, 2007).

The rates of vertical deposition (aggradation) appear to be negatively correlated with those of lateral deposition (progradation). Vertical deposition efficiency decreases with increasing energy, while seaward accretion tends to be promoted by strong water agitation. In high-energy settings, the mean vertical accumulation rates average 5 mm yr^{-1} (extrema: 1.5 and 12 mm yr^{-1}). In these settings, lateral expansion rates may reach 300 mm yr^{-1} with a mode of 90 mm yr^{-1}. By contrast, in semi-exposed to protected reef margins, vertical accretion rates average 9 mm yr^{-1} (extrema: 1 and 25 mm yr^{-1}). These margins have developed laterally at maximum rates of about 85 mm yr^{-1} with a mode of about 50 mm yr^{-1} (see Montaggioni, 2005 for review). Two reasons may be invoked to explain why reef margins have developed vertically more slowly under higher energy conditions. First, the framework in these areas consists mostly of robust branching and massive forms that display lower growth potential rates compared to those of arborescent and tabular corals living preferentially in medium-to-low energy sites. Second, once reef tops have reached and are maintained within about 0–5 m water depth, high-wave energy probably inhibits framework development (Grigg, 1998; Grossman & Fletcher, 2004; Gischler, 2008) and promotes the displacement of detrital material downslope and backwards to the reef flat. Once the vertical accommodation space is filled, the dominant constructional margin process must change from aggradation to seawards progradation. Based on the analysis and dating of horizontal cores extracted from a steep, shelf-edge reef margin on St. Croix (Caribbean), Hubbard et al. (1986) demonstrated that lateral, seaward accretion at water depths of less than 30 m occurred at rates of 0.84–2.55 mm yr^{-1}, reflecting deposition of material slumped from the shallower parts of the reef front rather than in-place coral growth. On Buck Island, on the northeastern shelf of St. Croix, progradation rates of the fringing reef front range from 5 to 10 mm yr^{-1} (Hubbard et al., 2005). This explains how, in some instances, progradation rates may be higher than the growth rates of corals and associated calcifiers.

5.3.4.2. Reef-edge, detritus-dominated accumulations

In low-energy sites, reef tracts can best be described as detrital, sand-dominated, piles trapping scattered corals (Davies & Hopley, 1983; Montaggioni, 1988b; Hubbard et al., 1998; Kleypas & Hopley, 1993; Cabioch et al., 1995; Braithwaite et al., 2000; Yamano et al., 2001). In these accumulations vertical depositional rates are highly variable (Figure 5.17B). The pattern of detrital sedimentation defined by Davies and Hopley (1983) and revisited by Hopley et al. (2007, pp. 376–380) on the Australian Great Barrier Reef has been confirmed by most studies elsewhere (see, for instance, Hubbard et al., 1998; Braithwaite et al., 2000; Grossman & Fletcher, 2004; Montaggioni, 2005). Three accumulation rate ranges have been recognized, reflecting increases in the hydrodynamic energy gradient: low mean rates of about 5–6 mm yr^{-1} primarily represent rubble deposition during the early stages of reef settlement; intermediate mean rates of 5–10 mm yr^{-1} (extrema: 1 and >15 mm yr^{-1}) represent the steady filling of reef-flat, back-reef and lagoonal zones under fair-weather conditions; and higher mean rates up to 10–13 mm yr^{-1} (maximum >40 mm yr^{-1}) are related to rapid deposition of sand and rubble, presumably controlled by storms and cyclones. The highest rates of deposition have usually been reported from narrow reef systems such as fringing and platform reefs. In such sites, deposition promoted by low-frequency, high-energy events operates at rates 2–10 orders faster than those observed in large shelf-reef systems. On narrow reef systems, accommodation space may be filled rapidly compared to that available over wide-open barrier reefs where, in addition, debris may be washed away by strong currents.

Supratidal sandy deposits are common in reef systems where they have accreted in the form of ridges, cays or low islands, for the most part since the mid-Holocene. For example, based on radiometric dating, linear accretion rates of Warraber cay (Torres Strait, northern Australia) are inferred to have averaged 300 mm yr^{-1} over the past 3 ka as a result of the addition of approximately 1000 m^3 carbonate (Woodroffe, Samosorn, Hua, & Hart, 2007).

5.3.4.3. Lagoonal sediment accumulations

The rate of vertical deposition of lagoonal sediments varies between 0.1 and 15 mm yr^{-1} with mean rates of 4 mm yr^{-1} (Pirazzoli & Montaggioni, 1986; Smithers et al., 1992; Smithers, Woodroffe, McLean, & Wallensky, 1993; Cabioch, Montaggioni, Faure, & Ribaud-Laurenti, 1999; Zinke et al., 2001; Zinke, Reijmer, Thomassin, & Dullo, 2003; Yamano et al., 2002; Yang, Mazzullo, & Teal, 2004; see Montaggioni, 2005 for a review). Rates of deposition appear to decrease with increasing depth. Higher rates are recorded in shallower lagoons (1–6 mm yr^{-1} on average), and lower rates (from 0.3 to about 2 mm yr^{-1}) have been estimated for the deeper lagoons of wide barrier reefs and atolls. Such differences in rates may be attributed to the origin and type of the deposited material.

5.3.4.4. *Halimeda* mounds

The relief of *Halimeda* accumulations above the antecedent topography has been demonstrated using coring and seismic surveys to range from about 2 to more than 50 m (Davies & Marshall, 1985; Orme, 1985; Phipps, Davies, & Hopley, 1985; Orme & Salama, 1988; Marshall & Davies, 1988; Phipps & Roberts, 1988; Hine et al., 1988). Accumulation rates vary widely from site to site, ranging from less than 1 to more than 5 mm yr^{-1}. Assuming an initial porosity of about 50% and a mean density of 2.8 g cm^{-3} for *Halimeda* mounds, carbonate production is estimated to have varied between less than 1 and more than 4 kg $CaCO_3$ m^{-2} yr^{-1}.

5.3.5. Control of Reef Growth Styles on Rates of Deposition

During sea-level rise, the response of reef systems to increasing accommodation space was expressed in different ways, as demonstrated by deposits from the last deglaciation event (Davies, Marshall, & Hopley, 1985; Davies & Montaggioni, 1985; Neumann & Macintyre, 1985). Some systems developed vertically at rates balancing the rate of sea-level rise and maintained themselves within an appropriate shallow-water range throughout their accretion. This pattern is attributed to the "keep-up" growth style. Alternatively, reef growth was able to catch up with sea level before or after it stabilized ("catch-up" style) or ceased accretion soon after initiation ("give-up" style) (Figure 5.18).

Rates of vertical deposition are also seen to vary markedly with reef growth styles (Davies et al., 1985; Montaggioni, 2005; Hopley et al., 2007, pp. 383–385). Davies and Marshall (1979, 1980) were able to show that rates of reef deposition varied throughout the Holocene and can be represented by a sigmoidal curve. This S-shaped accretion pattern includes three phases of changing rate. The lower part of the curve relates to the early phase, with slow growth (less than 2 mm yr^{-1}), regarded as driven by inimical conditions during substrate colonization. The middle part expresses maximum rates of growth, ranging between 5 and 10 mm yr^{-1} in response to the establishment of optimal conditions. Finally, the uppermost part of the curve reflects a steady decline in aggradation rates (to less than 3–4 mm yr^{-1}) as the reef top approached the sea surface. This pattern is chiefly typical of reef piles that have aggraded following the 'catch-up' growth style. In 'keep-up' reef sequences, the early episode of slow growth is commonly missing, because aggradation was able to keep pace with rising sea level as soon as the substrate was inundated. The highest rates of

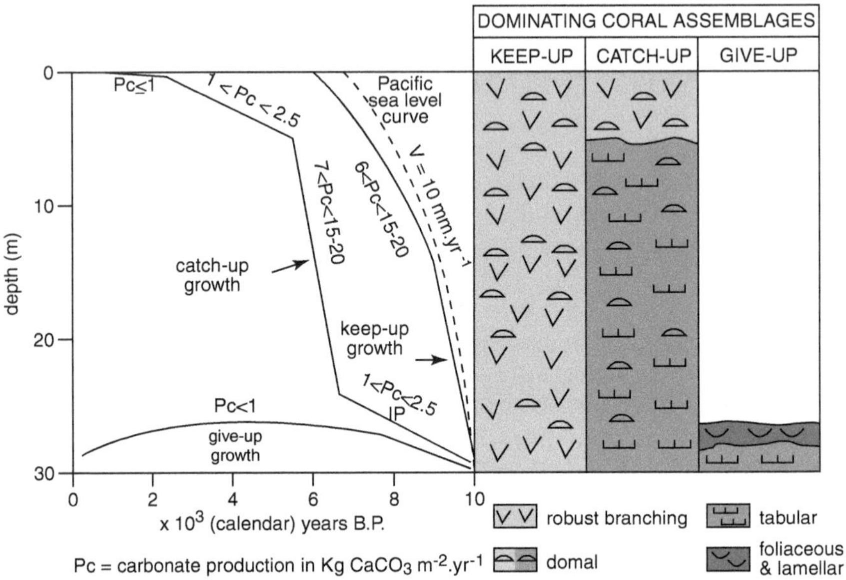

Figure 5.18 Reef growth styles, coral assemblages and carbonate production during the Holocene in response to sea level rise. Adapted from Neumann and Macintyre (1985), Hubbard (1997) and Masse and Montaggioni (2001).

aggradation were generally recorded at two distinct periods: first, during reef initiation, as the growth of pioneer coral assemblages were able to continuously compensate for the rapid increase in accommodation space; and second, during a period of accelerated growth when coral communities increased rates of vertical growth to escape drowning following an abrupt rise in sea level. Variations in deposition rates therefore appear to be tied partly to growth styles.

The rates and styles of reef growth are known to vary greatly within the same reef system, especially between different portions of a given reef-flat zone (Davies et al., 1985; Montaggioni, 1988b; Hubbard et al., 1998b, 2005; Grossman & Fletcher, 2004; Hopley et al., 2007, p. 380). Rarely can an entire reef system be categorized as a keep-up or catch-up system. For example, at Buck Island (on the northeastern shelf of St. Croix, Caribbean), the inner parts of the fringing reef flats grew vertically in a catch-up mode at rates of about 1.5–3 mm yr^{-1} throughout the reef-building phase, whereas the reef crest developed in a keep-up mode, at rates reaching about 9 mm yr^{-1} during the latest phase of growth (Hubbard et al., 2005). In the barrier reef of Palau, the keep-up has been typical of the development of the windward outer margins at mean rates of 6 mm yr^{-1}, whereas the leeward inner margins developed in catch-up style at rates of 3–4 mm yr^{-1} (Kayanne et al., 2002). However, growth styles may be independent of depositional processes and setting. Both 'keep-up' and 'catch-up' signatures pertain to framework and detritus as well, as in high-energy to sheltered reef flats. For instance, on the northwestern shelf of Tahiti (French Polynesia), the windward framework-dominated barrier margin and the detritus-dominated inner parts of the adjacent fringing reef flat developed in a 'keep-up' mode (Montaggioni, 1988b). It is clear that any curve of vertical accretion established from a single core strictly only reflects the behaviour of the coral communities at the core site and is not representative of the overall development history of the reef.

The net carbonate production of Holocene reefs during the period of vertical accretion can be estimated (Figure 5.18) taking into account a porosity value of 50% and a density value of 2.89 g cm^{-3} for the original framework and associated detritus (Smith, 1983). Reef sections dominated by faster-growing branching corals, with mean growth rates of about 10 mm yr^{-1}, have produced up to 10 kg $CaCO_3$ m^{-2} yr^{-1}. These estimates are close to values reported from active modern reef crests (Kinsey, 1983). Reef sections mainly composed of slower-growing branching to domal corals with growth rates of 5–8 mm yr^{-1} have released 8 kg $CaCO_3$ m^{-2} yr^{-1} on average. Reef sections with accretion rates of 1–4 mm yr^{-1} have produced less than 4 kg $CaCO_3$ m^{-2} yr^{-1}. Using these assumptions, the net rates of carbonate production can be estimated to have been as high as 15–20 kg $CaCO_3$ m^{-2} yr^{-1} during the period of faster growth, irrespective of the growth style (keep-up or catch-up style).

In areas of uplift, following high sea stand peaks and sea level stabilization, reef tracts have experienced a vertical movement to emergence. This has resulted in an apparent fall in sea level at rates of about 0.2–0.5 mm yr^{-1} over several thousand years. Reef deposition migrated downslope relative to antecedent reef tracts, such that younger corals have grown and associated sediments filled their flanks so that growth finally occupied lower locations. This process is referred to as the 'pack-up' growth mode according to Esat and Yokoyama (2006) and may result in polycyclic reef units (see Chapter 6, Section 6.6.4). It has operated throughout the Quaternary and is still functioning in a number of reef sites close to subducting plates.

5.3.6. Control of Latitude on Rates of Deposition

A number of workers have questioned the influence of latitude on reef deposition. Data relating to accretion rates are conflicting. Grigg (1982) showed that along the Hawaiian Island chain, the growth rates of seaward reefs gradually decrease with increasing latitude from about 11 to less than 1 mm yr^{-1}. The reefs were no longer been able to track rising sea level and began to drown as they have reached 29° north latitude, that is, the so-called 'Darwin Point'. In Florida, in areas more than 25° north, Holocene reefs form submerged terraces, having ceased growth between about 7 and 5 ka, presumably in response to a fall in sea surface temperature (Lighty et al., 1978; Macintyre, 1988). This suggests that accretion rates have slowed drastically, probably down to 1 mm yr^{-1}. By contrast, there appears to be little variation in depositional rates attributed to latitudinal differences on the Australian Great Barrier Reef that extends over a distance of about 2300 km (Davies & Hopley, 1983). Neither framework growth nor detrital accumulation rates are depressed with increasing latitude. Similarly, on Middleton and Elizabeth Reefs, and Lord Howe Island in the Tasman Sea, between 29°7′ and 33°30′ south, lagoonal deposition occurred at mean rates of 2–5 mm yr^{-1} during the mid-Holocene (Kennedy & Woodroffe, 2004; Woodroffe et al., 2004). In the highest latitude Japanese reefs, at 33°48′ north, vertical accretion rates of framework were more than 8 mm yr^{-1} (Yamano, Hori, Yamauchi, Yamagawa, & Ohmura, 2001).

5.4. Conclusions

Quantifying carbonate production of modern reefs at local, provincial, or global scales, and assessing changes in the cumulative production of Quaternary reefs, in response to changes in the environment over time, are difficult challenges. Calculations will suffer from large uncertainties irrespective of the methods used. Estimated values of present-day global production range between 0.65 and 1 Gt yr^{-1}. During the Last Glacial Maximum, primarily due to a reduction in the area of shelf substrates available for coral colonization, global production is estimated to have averaged 0.25–0.30 Gt yr^{-1}.

There are evident contrasts in the compositions and distributions of sediment types and of reef fabrics from zone to zone across reef systems, in response to differences in the processes of carbonate production and distribution. The major carbonate producers are limited to five biotic groups including corals, red coralline algae, molluscs, the green alga *Halimeda* and benthic foraminifera. The variations in the proportional abundance of these individual components within and between reef zones (fore-reef, reef-crest, reef-flat, back-reef zones and lagoons) are primarily controlled by the compositions of adjacent reef communities, the physiography of reef systems, proximity to terrigenous sediment sources, and ambient hydrodynamic regimes. In low-energy environments, the abundance of each component broadly coincides with the cover and/or production rates of the living producers. By contrast, in high-energy settings, the composition of the sediment only imperfectly reflects that of the adjacent communities due to sediment mixing in response to large-scale transport across the reef system. This emphasizes the complex relationships between the productivity of skeletal material by the relevant communities and the degrees of resistance to disintegration of individual skeletons. Coralgal rudstone/grainstones, in association with encrusting foraminifera-rich sediment types are typical of windward reef margins and outer reef flats. Coral–molluscan grainstones/packstones associated with foraminifera, dominated by *Amphistegina–Calcarina*, *Baculogypsina–Calcarina–Marginopora* or *Archaias-Asterigerina-Sorites* according to region and province, are transitional and are found mainly on inner reef flats and in adjacent, shallow lagoon areas. Molluscan–foraminiferal grainstones to wackestones, rich in *Heterostegina*, *Cycloclypeus*, *Operculina* and/or miliolids, extend into deeper reef settings along foreslopes or within lagoons. They can be covered locally by *Halimeda*-dominated sediment types.

The compositions of skeletal sediments and their potential short-term evolution relative to the structure of reef communities within contrasting Quaternary and modern environments provide additional evidence for the interpretation of the depositional history of Recent and Pleistocene, sequences particularly in cores or where *in situ* corals and other macrobiota are poorly preserved or lacking. Unfortunately, the means to deal with this important issue are still in their infancy, because there are few quantitative analyses of the biotic compositions of carbonate detritus in subfossil and fossil reefs. A full understanding of reef growth history during the Quaternary requires greater knowledge of skeletal deposition dynamics, particularly, of the sand fractions, from exposures and cores. Special attention must be paid to the development patterns of *Halimeda* bioherms that have apparently formed at the expense of coral reefs.

Changes in the growth rates of the biota, in hydrodynamic energy, and in accommodation space, are among the dominant factors governing sediment production and redistribution. They appear to result in contrasting styles of deposition, volumetric partitioning of sediments and large spatial and temporal variations in the development of the different biotic zones of a reef system. It is necessary in interpreting the thickness of a stratigraphic unit in terms of time of deposition, to assign different depositional rates to different reef fabrics. Although rates of change of accommodation space have varied during the Quaternary, the use of quantitative data as proxies for the duration of deposition remains viable, because the growth and production rates of individual organisms and reef fabrics are relatively well documented.

CHAPTER SIX

Reef Anatomy and Stratigraphy

6.1. Introduction

Early ideas of reef anatomy and stratigraphy were based on deductive reasoning. Certain processes could be seen or be inferred to occur, and from these a model could be built. For instance, Longman (1981, in Figure 10, p. 23) provided an idealized cross-section of a typical mature coral reef system showing the distribution of the morphozones at surface and within the pile. The internal structure was shown to perfectly reflect the reef-top zonation. In the figure caption, Longman, well aware of the poor reliability of such a scheme, wrote 'the reef framework is artistic, not realistic'.

The structure of Quaternary reefs, and especially that of Holocene reefs, is probably far more diverse than has sometimes been assumed. The development of light-weight drilling rigs (Macintyre, 1978; Thom, 1978) opened a new era of shallow coring worldwide from 1970s onwards. New data relating to the internal structure of Holocene reefs was obtained from a variety of tectonic settings. The following two decades were highly productive with reconstructions of the three-dimensional architecture of several recent reefs. In particular, encouraging results were obtained detailing the evolution of continental shelf reefs reflecting the drilling capabilities of Australian workers (see Hopley et al., 2007 for review). By contrast, there was less effective drilling through island reefs.

Although the number of attempts at accessing reef interiors has grown in recent years, the interpretations of coring results in terms of internal architecture have met with mixed success and elements of our knowledge remain uncertain for two related reasons. First, it has proved difficult to drill on reef edges swept by breaking waves. Second, although a reef framework may generate good recovery in cores, many reefs are dominated by coarse storm-derived detritus (see below) or sand. The former is commonly difficult to interpret and the latter difficult to recover. Nevertheless, Hubbard et al. (1990, 1998) demonstrated that interlocking, massive frameworks can be almost totally absent, and 40–90% of the core volume consists of detritus, mainly rubble and sand, and cavities. Similarly, from the study of fringing reef anatomy, Blanchon et al. (1997) claimed that the reef is 'storm rubble, not coral framework'. This explains why dated coral samples frequently indicate discrepancies in age-depth relationship in the two vertical succession attributed to recycling of older material over intervals of several decades to centuries. In general, the best results have been achieved by larger diameter drills, using casing or wire-line systems in which the walls of the boring are stabilized during drilling. There are advantages also in drilling loose materials if the rig can be adapted to vibrocoring, penetrating loose intervals with little disturbance. A corollary to this is that relatively powerful equipment is required and light-weight drills with small-diameter cores are unlikely to deliver the information required where the continuity or otherwise of accretion is at issue. Matters improve above 50 mm core diameter, but ideally cores should be even larger, giving greater information.

Surprisingly, there were deep boreholes through reefs long before any shallow investigations. As early as 1896, Funafuti Atoll in the southwestern Pacific was drilled in search of antecedent foundations. Similar land-based campaigns were conducted on a number of atolls and barrier reefs particularly in the Pacific. However, as pointed out by Steers and Stoddart (1977), while deep drilling significantly increased knowledge of Tertiary reef history, only rarely did it provide any substantive understanding of the stratigraphy and structure of Pleistocene reef sequences. This was probably due to limitations in the time constraints dictated by the different dating methods, particularly in the middle to early Pleistocene.

In other respects, in order to investigate the stratal patterns within reef depositional sequences and the nature and distribution of sequence boundaries, drilling has locally been coupled with seismic surveys. Unfortunately, not all of the reefs drilled were analysed in terms of seismic stratigraphy. Seismic profiles are available from only a limited number of reef sites in the Caribbean (Purdy et al., 2003; Toscano & Lundberg, 1999; Lidz, Reich, & Shinn, 2003), in the Indian Ocean (Purdy & Bertram, 1993; Collins, Zhu, & Wyrwell, 1998; Zinke et al., 2001) and in the Pacific (Harvey, 1986; Le Roy et al., 2008; Guille, Zhu, & Wyrwoll, 1996; Chardon et al., 2008). As matters stand, the development of computer modelling for reconstructing reef stratigraphy may perhaps be considered to partly compensate for the limited seismic data (Dalmasso, Montaggioni, Bosence, & Floquet, 2001).

The purpose of this chapter is to review current knowledge of Holocene to Pleistocene reef anatomy and stratigraphy. The topics considered are the morphology, architecture and composition of Holocene reef bodies, the structure and stratigraphy of Pleistocene barrier reefs, and atolls, and both emergent and submerged reef terraces. The final section discusses the use of computer modelling in reef studies.

6.2. Morphology and Anatomy of Holocene Reefs

Interpretations of the manner in which the coral biota can colonize an appropriate surface, and then develop vertically and laterally are based in part on speculation, but are in accord with data on anatomy gained from drilling operations in a variety of geodynamical settings (see Macintyre, 1988, 2007; Dullo, 2005, Montaggioni, 2005; Hopley et al., 2007, pp. 372–403, for reviews). Reef anatomy has been shown to a great extent have been driven by a limited number of environmental factors, primarily including antecedent topography (see Chapter 4, Section 4.2.2), high-hydrodynamic-energy events and terrigenous input (see Chapter 7, Section 7.2.5).

6.2.1. Nature and Compositions of Reefs

Although there is much variation in reef morphology (see Chapter 1, Section 1.2) since the earliest formal observations, three broadly defined zones have been recognized: the fore-reef, the reef edge (reef crest and reef flat) and the back-reef and/or lagoon. Primary (*in situ*) or secondary (reworked and subsequently recemented) framework-dominated accumulations usually dominate in exposed, ocean-facing reef margins. These contrast with the dominance of detritus-rich successions in sheltered, inner-shelf tracts and back-reef/lagoonal sequences, resulting in reef systems in which the interiors consist largely of rudstone- to mudstone.

6.2.1.1. The Fore-reef

The realization that coral reefs grew along their seaward margins led to the idea that they must grow over their own debris created by wave action. This inductive step was especially important in oceanic atolls where reefs apparently rise from great depth (Darwin, 1842). The notion of 'reef-talus' became an integral part of descriptions of fore-reefs to the point in some ancient examples that inclined banks of coarse debris alone were regarded as sufficient evidence of the former existence of a reef. The idea was pervasive, and numerous textbooks illustrate reefs in which the fore-reef comprises a steep debris slope. However, fore-reef slopes commonly appear to be covered by growing corals (for instance, see Braithwaite, 1971; Montaggioni & Faure, 1980; Hubbard, 1986). In fact, debris shed from the reef edge is more likely to be carried backwards by wave overwash and is not distributed across the fore-reef slope. As Blanchon et al. (1997) pointed out, overwash during hurricanes generates currents shoreward across reefs that reach speeds in excess of $5\ \mathrm{m\,s^{-1}}$. Storm surges in cyclones may be 2–15 m above predicted tide levels, and waves 8–10 m high can generate breakers of over 14 m (van Woesik, Ayling, & Mapstone, 1991). Thus, it seems unlikely that much material would be deposited along outer reef slopes.

Nevertheless, there are two situations in which sediment can accumulate in reef-front areas. As reef accretion reaches towards sea level, it becomes increasingly difficult for waves to propagate shorewards. It also becomes more likely that water piling on and draining off the reef during storms, will transport sand streams to accumulate below reef passes as fans at the foot of the slope or as turbidites in deeper water (Goreau & Land, 1974; Moore et al., 1976; Land & Moore, 1977). Such 'return currents' are weaker than those driving onshore transport as they are sustained only by gravity and the hydraulic head generated by the volume of water piled onto the reef, but they are significant. Hurricane-induced flows, such as those described by Hubbard (1992) on St Croix, US Virgin Islands, are largely responsible for flushing sand rather than coarser debris. Thus, following the hurricane of September 1989, net currents in the Salt River Canyon that were sustained for 4–6 h reached speeds of 2 m s^{-1} with additional oscillatory flows of 4 m s^{-1}. These removed an estimated 2 million kg of sand and scoured adjacent channel walls to bare rock. In the adjacent Cane Bay channel, some 336,000 kg of sand were removed in the same period, a volume estimated to represent the equivalent of 400–700 years of accumulation on the adjacent reef (Hubbard, 1992).

Where wave action is relatively benign and growth is relatively vigorous (in the absence of a debris slope) because the growth rate of corals colonizing the reef front increases towards the sea surface (particularly in turbid waters where light is more limited by depth), the growing surface will become progressively steeper. It therefore reaches a point where the weight of the mass will exceed the strength of the framework and blocks of metre dimensions will detach and slide to the foot of the slope. Blocks of this kind have been observed by Braithwaite (1971) in the Seychelles and by Hubbard, Burke, and Gill (1985, 1986) on St Croix. Fractured blocks that have been only minimally displaced may represent the source of the hurricane-derived megablocks described in areas like Mauritius (Montaggioni, 1978). However, neither off-reef flow nor slope failure will generate debris in the size ranges usually described as 'talus'. The missing part of this puzzle is to be found in the Pleistocene history of the reefs.

Coral growth on steep outer slopes is only possible on the shallower margins but it accumulates to sufficient thickness that the weight causes the attachment to the surface or the underlying substrate to fail. McIlreath and James (1984) figured platform margin deposits from Belize to illustrate deposition on a by-pass-type platform margin. Massive blocks and coarse debris generated by slope failure, bioclastic sand spilling from the platform edge, carbonate turbidites and hemipelagic ooze all intermingle to accumulate in extensive fans at the foot of cliffs in water in excess of 200 m deep. There may be discussion as to whether all of this material reflects deposition during contemporary reef growth, that is during the current or earlier high sea-level stands or is also a result of erosion during lowstands. But it is certainly cumulative. However, the heterogeneity of these deposits, and the intermittent deposition likely to lead to the generation of bedding, best fit the descriptions of talus on ancient reefs. It is arguable therefore whether such deposits are part of the reef assembly *sensu stricto* if they appear to be confined to this special platform margin case, or if indeed they are common because all ancient reef accumulations reflect cycles of sea-level change like those of the Pleistocene.

Shinn, Hudson, Halley, and Lidz (1977); and Shinn, Hudson, Robbin, and Lidz (1981) were able to map the upward transition of a reef-front spur. Kan, Hori, Nakashima, and Ichikawa (1995) and Kan, Hori, and Ichikawa (1997) presented a model describing the upward growth and lateral coalescence of framework pinnacles to establish a wall and were able to document the upward transition of an isolated reef-front patch that formed at about 3 m depth some 1.2 ka, growing seawards and reefwards to form a reef spur by 0.7–0.6 ka. However, and perhaps surprisingly, the evidence of growth was largely confined to directions normal to the reef front with little or no expansion across the adjacent grooves. As indicated in Chapter 5, framework areas are typically an order of magnitude more productive in terms of carbonate deposition than adjacent sand-dominated surfaces (for instance, see Chave et al., 1972; Stearn et al., 1977; Smith, 1983; Hubbard et al., 1990).

As previously emphasized by James and Ginsburg (1979a), the present-day architecture of the outer margins of atolls and barrier reefs is probably a reflection of both accretionary and erosional effects. Most possess subvertical drop-offs with a sudden break in slope. The occurrence of breaks in slope at different depths may reflect the elevation reached by reef growth during former sea levels or erosion during subaerial exposure. Variations in the depths at which slope breaks and submerged terraces occur are primarily a function of the local tectonic history. In the Tuamotu region, the boundary between the antecedent foundations and Holocene reef deposits lies between a few metres above present sea level and about 10 metres below. In the Society Islands, the equivalent surface occurs at depths ranging from about 30 to 90 m (Figure 6.1). The northwestern Tuamotu atolls have experienced significant uplift during the Quaternary, but with different amplitudes from site to site (Montaggioni, 1985). By contrast, the dominant tectonic process in the Society Islands was active subsidence (Montaggioni et al., 1987; Bard, Hamelin, Arnold, et al., 1996). Similarly on Niue, a high, atoll-shaped carbonate island, there is a submerged terrace at around 6 m depth (Wheeler & Aharon, 1997). Similar profiles have been observed in modern atoll margins (Montaggioni et al., 1987). On the outer margins of Mataiva and Mururoa atolls (French Polynesia), a sharp breaks in slope are found at 6 and between 8–11 m depth respectively above a 2–4 m high scarp. Boreholes on these sites indicate that the break in slope marks the position of a submerged terrace. This represents the top of a former, presumably Pleistocene or older reef rim, in part subaerially eroded and interpreted as an unconformity surface (Figure 6.1) on which the Holocene reef was initiated. A similar origin has been invoked for terraces at 6 and 12 m depth on Tikehau and Takapoto atolls (northwestern Tuamotus) and from Bikini and Enewetak atolls

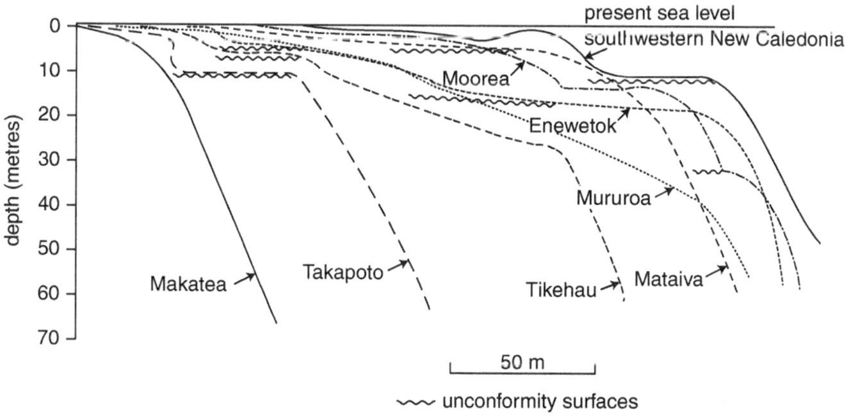

Figure 6.1 Representative profiles across the upper parts of the outer margins of a number of atolls (Takapoto, Tikehau, Mataiva, Mururoa, Enewetok) and barrier reefs (Moorea, New Caledonia). The bathymetric positions of the Holocene–Pleistocene unconformity surfaces are indicated. Modified from Montaggioni et al. (1987).

(Tracey & Ladd, 1974). In these instances, significant Holocene reef accretion appears to have been restricted to the inner parts of the terraces, forming a spur-and-groove system. By contrast, the outermost parts of the terraces and adjacent reef walls have been partly eroded, with relict relief overlain by coralgal veneers less than 1 m thick. As inclined boreholes on Mururoa have shown (Bard, Fairbanks, Arnold, & Hamelin, 1992; Buigues, 1997; Camoin et al., 2001), fore-reef deposits are young (14–17 ka) and are commonly relatively thin, probably because the antecedent slopes are steep, and at depth may include intercalations of 'talus' deposits, carbonate turbidites, a deep-water encrusting biota and lowstand reefs. Detailed bathymetric surveys have indicates that several breaks in slope are present to a depth of about 180 m. Inclined boreholes have demonstrated that these relates to unconformity surfaces marking the tops of Pleistocene reef generations. A similar morphological sequence scheme was recently described from the slopes of the New Caledonian barrier reef (Flamand et al., 2008).

6.2.1.2. Reef-edge, framework-dominated sequences

The upwards extension of frameworks results in an unbedded sequence, although progressive shallowing is reflected in the succession in the biota (see Chapter 3, Section 3.3). Successive growths and the interskeletal deposition of loose sediment mean that quite long pauses in growth activity, that may be reflected in the erosion of moribund colonies, will be only locally expressed, if at all. This apparent continuity has been taken as a *sine qua non* of reef framework of all ages. Hubbard et al. (1990) found no clear correlation between breaks in growth in successive boreholes on St Croix.

In reef edges, primary growth frame may generally represent about half of the total volume in cores (Macintyre & Glynn, 1976; Easton & Olson, 1976; Davies & Hopley, 1983; Montaggioni et al., 1997; Hubbard et al., 2005). The rest consists mainly of skeletal material (40%) and scattered empty cavities (less than 10% on average). In high-hydrodynamic-energy settings, especially in cyclone-swept regions, the internal structure of reef margins is characterized by a secondary framework. The total framework (primary and secondary) may comprise 30–60% of the cored material (for instance, see Blanchon et al., 1997; Montaggioni & Faure, 1997; Gischler et al., 2008).

The constructional potential of framework-dominated reef edges seems to depend on a variety of factors including coral community structure, coral cover and growth rate (Davies & Hopley, 1983; Kleypas, 1996; Van Woesik & Done, 1997), exposure to water agitation and depth (Shinn, 1963; James & Ginsburg, 1979b; Hubbard et al., 1986; Grigg, 1998; Montaggioni, 2005; Hongo & Kayanne, 2008). The attributes of the community, species composition and abundance, colony size frequency, and individual life expectancy and succession, have exerted controls on reef accretion throughout the Holocene. The resulting patterns are illustrated by near-shore coral structures from the southern Great Barrier Reef (Kleypas, 1996; Van Woesik & Done, 1997). In sites that experienced high population turnover, low settlement densities and high rates of skeletal disintegration combined with high turbidity, substantial coral growth was prevented. As soon as settlement occurred, the structure was dominated by encrusting and foliaceous coral forms, forming communities and incipient reefs and sediment accumulations lacking typical reef flats at sea level. By contrast, in sites that experienced high recruitment densities and long colony life expectancies and also escaped strong siltation, the development of luxuriant fringing reefs, mainly consisting of including branching acroporids and derived detrital deposits, was promoted. There is evidence of a correlation between wave agitation and depth, and the degree of reef development. The more extensive development of high-energy reef margins may have been controlled by factors that operated synergenetically. Wave-control, slope instability, triggered slumping of the upper reef face, the biology and ecology of corals and associated builders, including larval settlement, metabolism of the biota, food and nutrient supply, and siltation, (Yamano et al., 2003). All of these biotic and abiotic factors are encouraged in high-energy wind-wave fields, while lower hydrodynamic energy areas and particularly, inner-shelf settings, experience little slumping and extremes in temperature, salinity and turbidity.

6.2.1.3. Reef-edge, detritus-dominated sequences

Assuming that relative sea level remains stationary or is rising only slowly, as the growth frame extends towards the surface the severity of wave action increases and breakage and transport of coarser debris becomes more likely.

Wave-driven currents wash fan-like lobes of coralgal cobbles over the seaward margins of back-reef sediments. Close to the edge these are ultimately bound into the secondary frame, but within short distances may remain as loose coralgal surface sediment. Thus, in low-energy sites, reef edges can best be described as detrital, sand-dominated piles enclosing only scattered corals (Davies & Hopley, 1983; Montaggioni, 1988b; Hubbard et al., 1998; Kleypas & Hopley, 1993; Cabioch et al., 1995; Braithwaite et al., 2000; Yamano, Kayanne, et al., 2001).

In Holocene sequences, within a given zone, sandy detritus may locally remain similar in terms of composition and texture throughout the time interval of deposition. In particular, in high-hydrodynamic-energy settings, sands may be dominated by coralgal grainstones from base to top, as described from a number of Holocene reef sequences (Montaggioni, 1988b; Webster et al., 1998: Cabioch, 2003). However, in most cases, there is a marked change in constituent composition and grain size though time in response to the reduction in accommodation space and the subsequent increase in water agitation. For example, on a moderate water energy, outer fringing reef flat on Mahé (Seychelles), Braithwaite et al. (2000) described a 26-m-thick sequence composed of medium- to fine grained coral–molluscan–alcyonarian grainstones with nummulitids at the base, giving way upwards to coarse coral grainstone/rudstone rich in encrusting foraminifera and amphisteginids reflecting a shallowing-upward depositional model.

6.2.1.4 Back-reef/lagoonal sediment sequences

The flux of sediment from the reef edge to the back-reef zone is controlled largely by the frequency and severity of storms and varies seasonally and on decadal time scales. Up to 80% of skeletal sediment produced by windward reef flats can be regarded as available for transport into lagoons (Stoddart, 1969a,b). The volume of allochthonous sediments deposited in a lagoon is proportional to the area of the reef rim area at sea level (Tudhope, 1989).

The proximal parts of back-reef zones and lagoons are supplied dominantly with rubble and sandy sediments derived from the adjacent fore-reef and reef-edge zones. By contrast, coral is not generally a major component of sediments in the innermost parts of atoll and barrier reef lagoons and atolls. The primary source of sediments in these areas is *in situ* carbonate production, resulting mostly in the accumulation and biological disintegration of *Halimeda*, bivalve shells and foraminiferal tests with the contribution of allochthonous silty grains from the reef margins and locally, the mainland (Adjas et al., 1990; Smith et al., 1998; Yamano et al., 2002; Gischler & Zingeler, 2002; Yang et al., 2004).

As shown by Holocene sequences, lagoonal floors may remain under subtidal conditions, relative to present sea level for long periods, or fill up, passing from successive intertidal and supratidal deposits, according to their initial depth and width and sedimentation rates (Montaggioni, 2005). Following Ladd and Tracey (1949) and Ladd (1950) and Schlager (1981, 1993), Purdy and Gischler (2005) introduced the concept of the 'empty bucket model of reef sedimentation', based on the examination of back-reef sedimentation patterns. Contrary to the common view of the partly filled back-reef zones and lagoons of modern rimmed reef systems as an end-product, the authors demonstrate that the present-day condition of these environments is a transitional stage. As the reefs continue to develop the end point will probably be achieved by total filling. Accepting the growth of the reef edge as the limiting boundary, the degree to which the back-reef can be filled depends upon the net production of the system and the surface area required to be filled. On small barrier and platform reefs and atolls (less than 50 km^2 in area), the central zones originally at depths of less than 20–25 m below present sea level have commonly been almost completely filled. This nearly-full state is observed in a number of Indo-Pacific and Caribbean sites. On reef systems constructed on deeper antecedent substrates, the lagoons are only partly sediment-filled and display a typical basin topography, as emphasized by Zinke, Reijmer, and Thomassin (2003) and by Purdy and Gischler (2005). During phases of filling, transitions in sediment types have occurred within sequences. These may include abrupt changes in grain size and component composition. In most cases, the volume of coarse-grained deposits tends to increase up the sequence, with shallowing-upward successions of carbonate from transitions from mud, mudstone, to floatstone and wackestone–packstone and –rudstone. Such sequences formed as sea level was rising, probably in response to the rapid vertical accretion of adjacent outer reef margins. These increasingly supplied proximal lagoonal zones with skeletal detritus, while distal zones primarily generated detritus-enriched from *in situ* growing communities. Based on study of a small, enclosed platform reef from the Great Barrier Reef, Scoffin and Tudhope (1988) established a predictive model of complete lagoonal sedimentation. They showed that in this area the shallowing-upward sequences began with subtidal units consisting of *in situ* coral rudstones overlain by coral–molluscan floatstones/wackestones in which sorting, mean grain size and the proportions of margin-derived components increased near sea surface. The upper parts of the sequences are intertidal to supratidal units differentiated according to exposure to wave energy, coral rudstones to windward, and grainstones to leeward.

Although the facies and textural attributes of lagoonal sediments from postglacial to Holocene, remain poorly documented, data seem to indicate that this pattern is a common feature of a variety of system types, at least in subtidal sections, and is irrespective of the extent of the lagoon areas. These issues are of particular relevance to Pleistocene accumulations in which lowstands of sea level aborted the filling process, leaving an upstanding margin and lagoon floor subject to karst erosion. However, extrapolation to more ancient deposits suggests that rimmed platforms may have been more common in 'ice-house' periods.

There are, however, alternative models of lagoonal sedimentation. On Mayotte, in the Western Indian Ocean, Zinke et al. (2001) obtained a large set of high-resolution seismic profiles, backed by 20 gravity cores, from the present lagoon (10–65 m deep). They observed that the lagoonal sediments dated at around 11.6 ka at the base vary in depositional modes and in compositions (Figure 6.2). Transgressive and highstand systems tracts could be recognized locally, but have different ages and depths, that reflect variations in the underlying topography. A lowstand tract is represented by a palaeosol resting directly on the eroded surface of Pleistocene deposits that include volcaniclastic intervals. A transgressive systems tract onlaps or downlaps the unconformity and locally ranges in age from 11 to 7 ka. Three depositional systems have been identified in the lagoon: proximal transgressive layer; proximal and distal incised-valley fills; and a mid-lagoon transgressive layer. The highstand tract, representing the last 7 ka comprises a proximal terrigenous wedge, mid-lagoon and distal carbonate sands and muds and reefal carbonates. The proximal transgressive sequence consists from base to top, of onlapping to downlapping terrigenous deposits. The sediments grade upwards from molluscan mudstones to mixed carbonate–terrigenous foramol wackestones and coral–foramol packstones/rudstones. In the deeper, mid-lagoon areas, the onlapping succession consists of

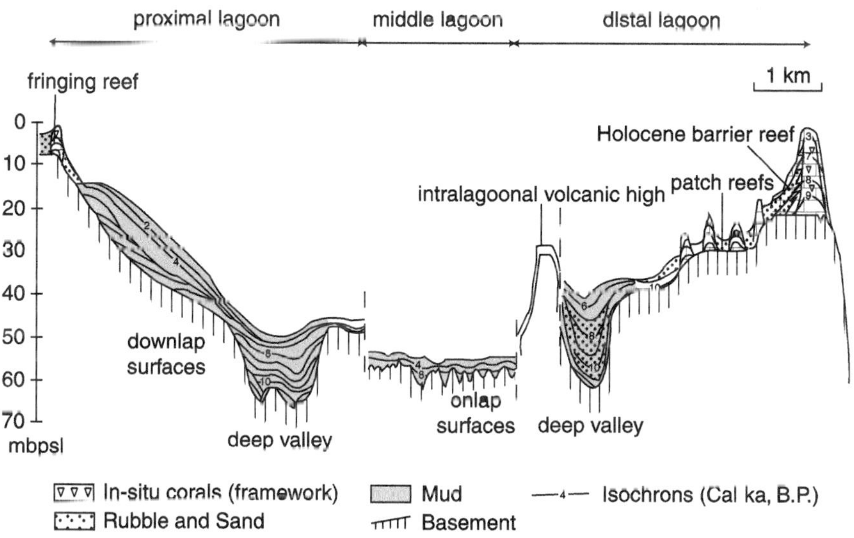

Figure 6.2 Interpretative seismic-based cross-section of the barrier reef and lagoon system on Mayotte, Western Indian Ocean. Basal unconformity, depositional units and time-constrained sequence boundaries are indicated. Modified from Zinke et al. (2001) in Montaggioni (2005).

terrigenous mud and molluscan–ostracod wackestones at the base, and mixed carbonate–terrigenous molluscan-echinoidal wackestones and foramol packstones at the top. The sequence in distal areas, the comprises foramol mudstones/packstones overlain by coral packstones/rudstones. Variations in the thicknesses of these units reflect the topography of the basement and proximity to the various sources, but the authors emphasize the steep relief created, noting that sediment production has not so far been enough to fill the lagoon. This study is one of the most detailed of its kind and illustrates the potential provided by seismic investigations.

6.2.2. The Thickness of Holocene Reef Sequences

The total vertical thicknesses of Holocene reef accumulations varies widely, according to the depth of the antecedent foundation relative to present sea level. In many sites, Holocene deposits are commonly less than 10 m thick (Easton & Olson, 1976; Macintyre & Glynn, 1977; Cabioch et al., 1995; Davies & Hopley, 1983). They are rarely older than 8 ka at the base. However, in the Caribbean, the thickest Holocene reef sequence extracted from Isla Perez is 33.5 m (Macintyre, Burke, & Stuckenrath, 1977). In the Pacific, continuous postglacial reef sequences more than 70-m thick have been described in the Huon Peninsula, Papua New Guinea (Chappell & Polach, 1991), the Tahiti barrier reef (Montaggioni et al., 1997) and Espiritu Santo Island in Vanuatu (Cabioch et al., 1998). These began to accrete vertically between approximately 22 and 13.8 ka.

6.2.3. Conceptual Models of Reef Deposition

Hubbard, Gill, Burke, and Morelock (1997) suggested that reef accumulation may be referred to four conceptual models. For slow rates of sea-level rise and high rates of production, the reef framework can be expected to respond by a progressive progradation. As the level of production falls, it may be taken up entirely in vertical aggradation with no forward accumulation. As it falls further, or as the rate of sea-level rise increases, the growth frame retreats upslope blanketing the surface and drowning rather than overgrowing early-formed sections. At the fastest rates of rise or lowest rates of production, the framework cannot be sustained and segments are abandoned with new growth following the retreating shoreline, and backstepping. Hubbard et al. (1997) recorded behaviour conforming to this last model from the shelf reefs on the southwestern coast of Puerto Rico, dated between 7.19 and 6.2 ka. Kennedy and Woodroffe (2002) proposed a classification of fringing reefs, with six types defined on the twin bases of their morphology and anatomy. Montaggioni (2005) revisited the reef growth models of both Hubbard et al. (1997) and Kennedy and Woodroffe (2002) and attempted to apply the dominant depositional patterns to fringing reefs, barrier reefs, and atolls. All can be accommodated in the following main four reef anatomy types (Figure 6.3).

1) Deposition within back-reef zones and lagoons took place at rates close to those of the adjacent reef edges, allowing the various parts of the reef system to reach sea level coevally and to form a nearly flat-topped surface across the entire profile (Figure 6.3A). Throughout the period of accumulation there was a sustained fine balance between growth of the outer reef rim and aggradation and back-reef accumulation through onlapping. This is equivalent to model A of Kennedy and Woodroffe (2002) and operated at Hanauma Reef, Oahu, Hawaii (Easton & Olson, 1976).
2) Relates to reef systems that have dominantly developed by lateral accretion of the reef margin. They started to grow in settings where vertical accommodation space was limited or absent (Figure 6.3B). In the Indo-Pacific, such Holocene reefs mostly formed close to the shore when sea level was around its present-day position. Reefs of this type can be related to models B and C defined by Kennedy and Woodroffe (2002), and are common in inner-shelf sites. The fringing reefs of Fantome (Johnson & Risk, 1987), Orpheus Island (Hopley et al., 1983) on the Great Barrier Reef, and Hikauhi, on Molokai, Hawaii (Engels et al., 2004), are typical of prograding systems. In the Caribbean, despite sea level only having reached its present position within the last centuries and the persistence of accommodation space until recently (see Chapter 9), drilling evidence indicates that progradation has been the dominant process on northern Buckland Island, US Virgin Islands (Hubbard et al., 2005). Lateral backward accretion may occur periodically. In this the reef develops through coalescence of offshore patches onto the fore-reef by framework growth and/or sediment infill (model D of Kennedy & Woodroffe, 2002). Such a pattern of progradation has been reported from the barrier reef at Toliara, Madagascar (Weydert, 1973).
3) In this type of reef anatomy (Figure 6.3C), reef initiation took place below present sea level, at depths where the rates of reef aggradation during the early Holocene were insufficient to fill available accommodation space. The reef body was therefore only able to "catch up" to sea level by successive accretion centres stepping backwards. Ultimately it consisted of a series of superimposed retrograding units. The fringing reefs at Hale O Lono, Molokai, Hawaii (Engels et al., 2004) mirrors this scenario. The landward migration of the accretion centres was assumed to result from the specific adaptation of coral communities to local environmental changes. Changes in the rates of the rising sea level may have contributed to the demise of former accretion centres by progressively modifying the local current regime and thus causing a

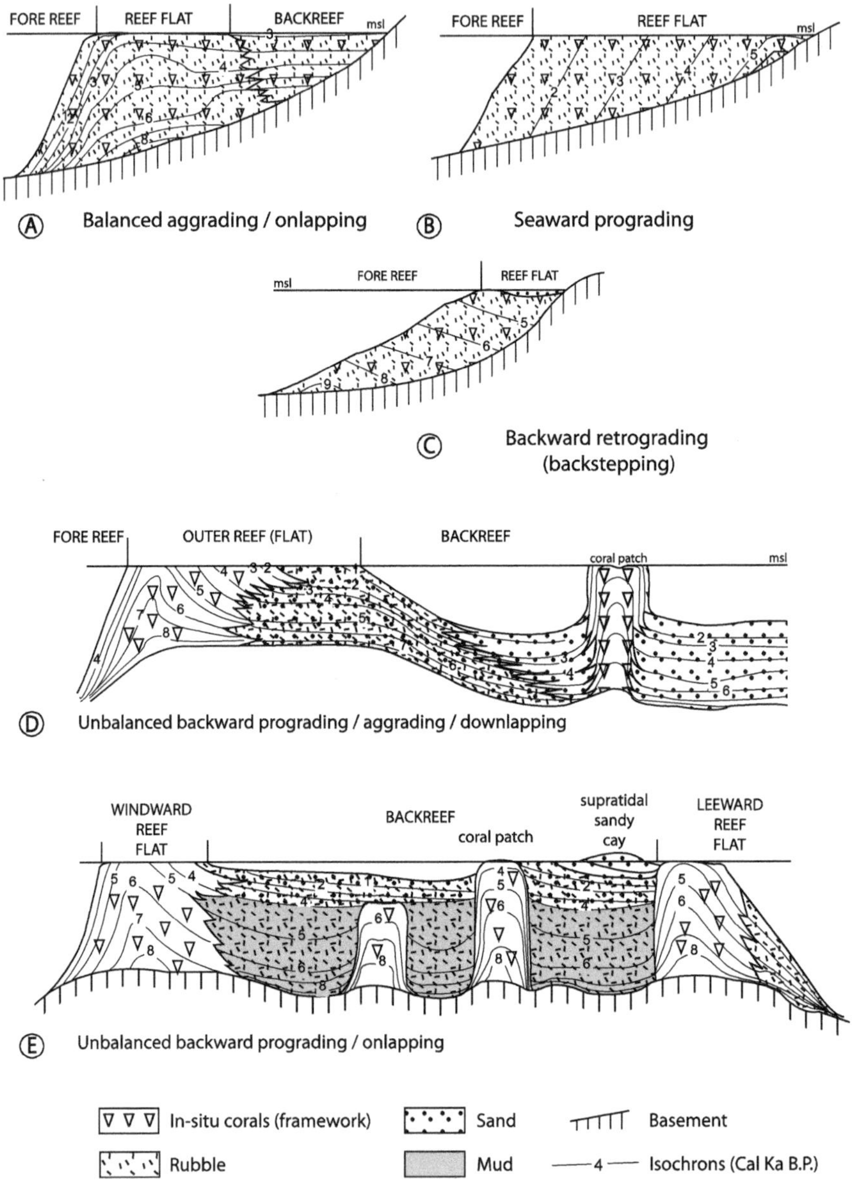

Figure 6.3 Generalized models of Holocene reef system anatomy based on dominant deposition: (A) balanced aggrading–onlapping model, (B) seaward prograding model, (C) backstepping model, (D) unbalanced aggrading–downlapping model and (E) unbalanced aggrading–onlapping model. Revisited from Kennedy and Woodroffe (2002) in Montaggioni (2005).

reduction of wave sheltering at the reef site. Coral communities were forced to migrate landward to recover habitats more suitable for their sustained growth (Engels et al., 2004).

4) This type of reef anatomy relates to reefs that comprise a well-developed outer rim enclosing a depressed back-reef area such reefs may form through multiple depositional processes. In back-reef settings, differences between stratigraphic sequences reflect the existence of local depositional modes, irrespective of the nature of sediment sources (carbonates or siliciclastics). As claimed by Smith et al. (1998), these differences are presumably dictated by the timing of reef rim growth and its relationship across the reef system to sea-level position in the course of sea level rise. Early during flooding, the back-reef zones were probably open systems in which sediment accumulation was continuously disturbed by water circulation. Except for the deepest depression incising the inner floors, the current regime promoted coral colonization and winnowing of finer particles and, as a result, deposition of coral gravel. As the reef margins developed vertically, water agitation within back-reef areas decreased progressively and the sedimentation style changed. This resulted finally in the grading of sediments into sand and/or muddy sands. As the transgressive phase ended, around 7 ka in the Pacific, 3 ka in the western Indian Ocean, and more recently in the western Atlantic (see Chapter 9, Section 9.4.2), most reef rims that grew through the keep-up mode and formed protecting walls thus favoured trapping of sediments within back-reef environments. The fact that individual back-reef zone responded with different growth modes and sedimentation rates to local environmental constraints such as size, depth and topography of the antecedent buckets or basins, proximity to sediment sources suggests that lagoonal deposits have encapsulated local rather than regional or global events. This explains why the main phase of back-reef filling started at around 7.6 ka in the Pacific, irrespective of reef types (Marshall & Davies, 1982; Davies et al., 1985; Pirazzoli & Montaggioni, 1986; Smith et al., 1998; Cabioch, Camoin, & Montaggioni, 1999; Yamano, Kayanne, et al., 2001; Kennedy & Woodroffe, 2002). Two distinct patterns of bed termination can be inferred in back-reef settings on the base of isochron distribution or seismic profiles. Sediment sheets may accumulate as downlapping clinoforms (Figure 6.3D) at the inflection of the slope between the rim-lagoon transitional slope and the lagoon floor. The isochrons delineate a series of talus cones deposited by by-passing of sediments from the windward margins or inshore siliciclastic environments towards the centre and distal portions of the lagoons or along the outer slopes of leeward reef margins. Back-reef deposition can also occur with onlapping beds generated by termination of gently dipping or nearly horizontal sheets against dipping antecedent substrate or the flanks of intralagoonal coral patches. The isochrons are parallel to the lagoon

floor surface (Figure 6.3E). Concurrent with back-reef filling, vertical accretion of the reef margin continued until it reached present sea-level. The reef rim prograded over early deposited back-reef sediments and partial to complete filling may occur. According to their initial depth, width and sedimentation rates, the back-reef floors remained under subtidal conditions, and either filled up completely or passed through intertidal to supratidal deposits. On reef platforms, barriers and atolls, the Holocene deposits of which commonly overlie deeper antecedent foundation, the lagoons are partially filled by sediments, particularly where they are partly or wholly enclosed, and exhibit the typical basin topography referred to by Zinke, Reijmer, and Thomassin (2003); Zinke, Reijmer, Thomassin, and Dullo (2003) and Purdy and Gischler (2005). The major depositional process has formed a downlapping sequence. This operates on many barriers and atolls, for instance, at Mayotte (Zinke et al., 2001), Palau (Kayanne et al., 2002), Middleton (Woodroffe, Kennedy, Hopley, Rasmuseen, & Smithers, 2000), Enewetak (Tracey & Ladd, 1974), Tarawa (Marshall & Jacobsen, 1985), Mururoa (Perrin, 1989) and on outer-shelf ribbon reefs (Davies & Hopley, 1983).

6.3. Structure and Pleistocene Stratigraphy of Barrier Reefs and Atolls

6.3.1. Barrier and Shelf Reefs

The reef systems of continental margins in particular are commonly only a part of more extensive carbonate platforms. Few of these have been investigated in any detail by boreholes, but there have been extensive seismic surveys aimed at providing conceptual models for comparison with ancient systems, principally for the oil industry. Notable among these is the work of Davies, Marshall, and Hopley (1985) tracking the evolution of the Great Barrier Reef, Queensland and Marion Plateaus off northeast Australia. These suggested carbonate deposition over the last 60 Ma with reefs recognized at a number of intervals. Brooks and Holmes (1989) identified nine seaward prograding clinoform sequences on the southern margin of the Florida Shelf using seismic data and limited coring. Depositional patterns are interpreted to have resulted from high-frequency sea-level fluctuations during the late Quaternary, with four sequences totalling some 330 m thickness deposited in the last 100 ka. However, these did not apparently include reefs. Finally, little is known of the anatomy and stratigraphy of barrier and shelf reefs and long cored sequences are rare.

To date, information is available from only a few sites in either the Caribbean or Indo-Pacific regions.

6.3.1.1. Case studies from the Caribbean

In Florida, at Key Largo in the Florida Keys, the analysis of 69 deep and shallow cores demonstrated that the Florida platform has evolved in three phases, covering the last 420 ka (Multer et al., 2002). The first phase is represented by units Q1 and Q2. Unit Q1 is typified by the abundance of siliciclastic grains and molluscan–foraminiferal grainstones deposited in relatively deep waters. Unit Q2 reflects a shallow carbonate environment containing coralline algae and branching corals. These two units are interpreted as deposited during the MIS 11 highstand. The second phase is represented by only one unit (Q3) that is assumed to have developed during the MIS 9 highstand. This consists of massive coral and coralline algal framework with mollucan–foraminiferal wackestones to packstones. In the third phase, units Q4 and Q5 formed during MISs 7 and 5 respectively. Unit Q4 is of very similar composition to Unit Q3, but the volume of coralgal framework is generally lower. Unit Q5 consists mainly of coralgal boundstones associated with large amounts of *Halimeda* plates. During the MIS 11 highstand, a shallow-water carbonate platform developed from a deeper environment. Then, during MIS 9 to 5, typical reef tracts were initiated and flourished. This stabilization phase characterized by reefs and leeward productive lagoons, was followed by lower sea levels, represented by a sequence of younger (MIS substages 5c, 5a) shelf-margin wedges, sediment veneers and outlier reefs. In the past 400 ka, there were at least four retrograding/prograding events, usually separated by episodes of subaerial exposure and karst development. The Florida scenario appears to differ from the evolution of western Pacific barrier reef complexes (see below). However, there are similarities in the timing and the processes of reef initiation. In both regions, coral reefs settled before MIS 9, overlying non-reefal, deeper deposits.

From the leeward margin of the Great Bahama Bank, two deep cores, up to 450 and 650 m-long respectively, were extracted, providing a detailed picture on the structure and depositional history of the Bahamian carbonate platform during the Pliocene and Pleistocene (Manfrino & Ginsburg, 2001). The architecture of the platform appears to have been controlled chiefly by reef growth in relation to changing sea level, and supports the contention that the pre-Pleistocene Bank was an atoll-like structure. The interpretation of depositional history is based on an integrated chronostratigraphy, including strontium isotope stratigraphy, magnetostratigraphy and nannofossil-biostratigraphy (Figure 6.4A). The upper *ca.* 125.9 m of the outermost core (Clino) was demonstrated to cover the entire Pleistocene, while in the core Unda, at the more landward site, the base of the

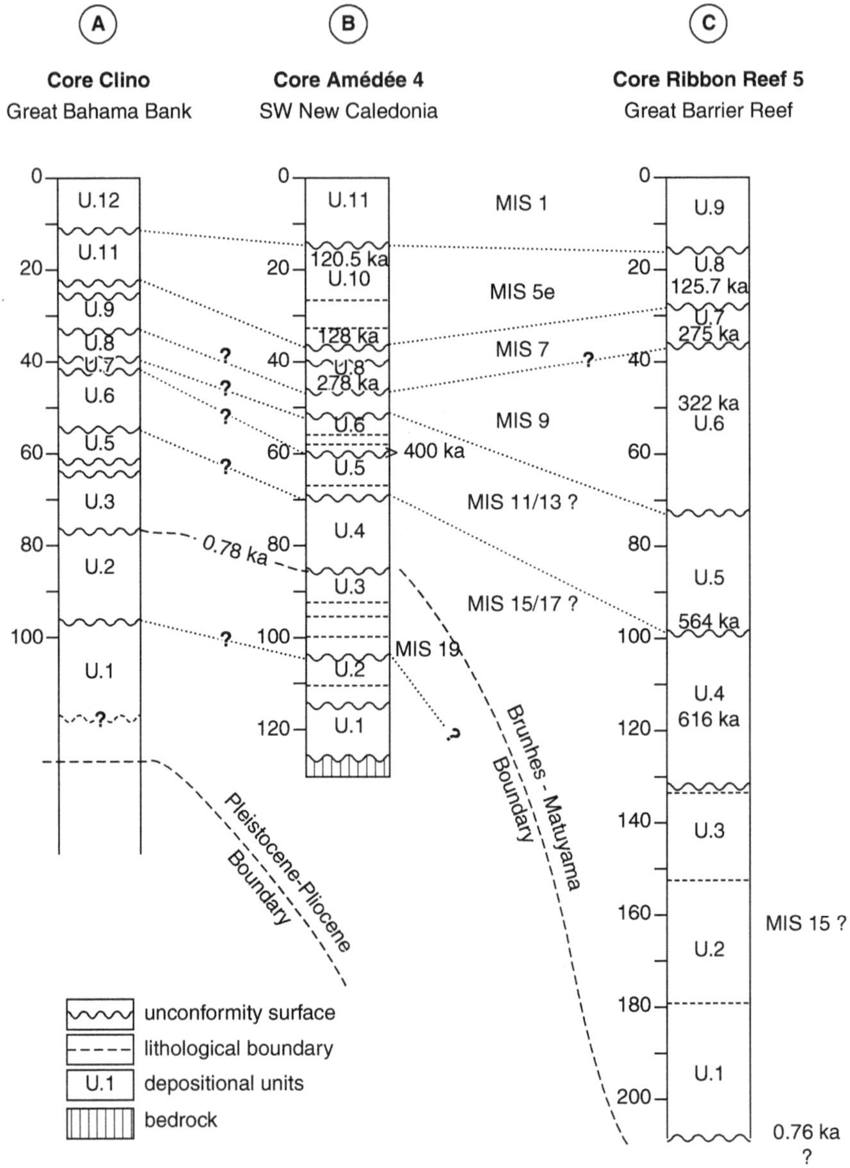

Figure 6.4 Chronostratigraphic correlation between cored sequences extracted from: (A) the Great Bahama Bank (Manfrino & Ginsburg, 2001; McNeill, Eberli, Lidz, Swart, & Kenter, 2001), (B) the southwestern New Caledonian Barrier Reef (Cabioch, Montaggioni, Frank, et al., 2008) and (C) Ribbon Reef 5, Great Barrier Reef (Alexander et al., 2001; Braithwaite et al., 2004). Age constraints are based on U-series dating, magnetostratigraphy, strontium isotope stratigraphy and/or nannofossil stratigraphy, and correlations with oxygen isotopic stages (MIS). The Brunhes–Matuyama chron boundary (at 0.78 ka) and the Pleistocene–Pliocene boundary are indicated locally. The different depositional units identified in each sequences (U.1 to U.12) are also indicated. Subunits are identified on the basis of abrupt changes in lithology.

Pleistocene appears to lie at a depth of about 97 m downcore. A rapid reef progradational phase occurred during the late Pliocene and early Pleistocene when the platform had accreted vertically to a subhorizontal surface. Ten glacial–interglacial cycles have been identified above the Bruhnes–Matuyama boundary, spanning the last 0.8 ka, and fit the number of sea-level fluctuations predicted by oxygen isotope records in the deep sea. The major phases of reef growth initiated during episodes of rising sea level from unconsolidated sediments provide evidence of upward shallowing through extensive aggradation. Seismic lines indicate that reflectors constitute surface unconformities of erosional origin, predominantly represent episodes of falls in sea level. The transition from a ramp-type platform to a flat-topped platform occurred in the mid-Pleistocene. In Clino, above the major reef-building phases, corals and skeletal sediments contributed dominantly to aggradation of the platform, with non-skeletal carbonates deposited only in the uppermost 20 m. In Unda, the post-reef units are composed of a series of peloidal and skeletal packstones to grainstones that are thought to be the counterparts of those accumulated during the Holocene.

In Belize, the extensive reef system includes a prominent barrier reef (James & Ginsburg, 1979a,b) 3–10 km wide and rarely deeper than 3 m. A wide lagoon lies behind the reef edge. The internal structure of the Holocene deposits within the Belize barrier reef system has been strongly controlled by tectonic history with subsurface topography, and in particular the position of near-surface folds and faults developed in both carbonates and terrigenous material (Purdy et al., 2003). The modern barrier platform started to prograde from a regional unconformity during the late Quaternary on the edge of an underlying carbonate shelf clinoform. The carbonate composition of the underlying Holocene deposits (Pleistocene reef limestones) has been confirmed by drilling operations through the modern barrier reef crest (Gischler & Lomando, 2000). However, seismic evidence indicates that the progradation of the carbonates is not a general feature; differential progradation or gravitational slumping, or both, may also have operated. In areas underlain by carbonates, karst dissolution has contributed to antecedent topography. Unfortunately, only relatively short cores were recovered from the Belize barrier reef (Gischler & Lomando, 2000; Gischler & Hudson, 2004; Gischler, 2007). These have provided information on the internal structure of reef tracts of the Holocene and the last interglacial (MIS 5). Older Pleistocene units are not yet chronologically constrained (Mazzullo, 2006).

6.3.1.2. Case studies from the Indian Ocean

The southwestern continental margin of Australia bears shelf-edge reef platforms (Houtman Abrolhos Groups). Seismic surveys and drilling

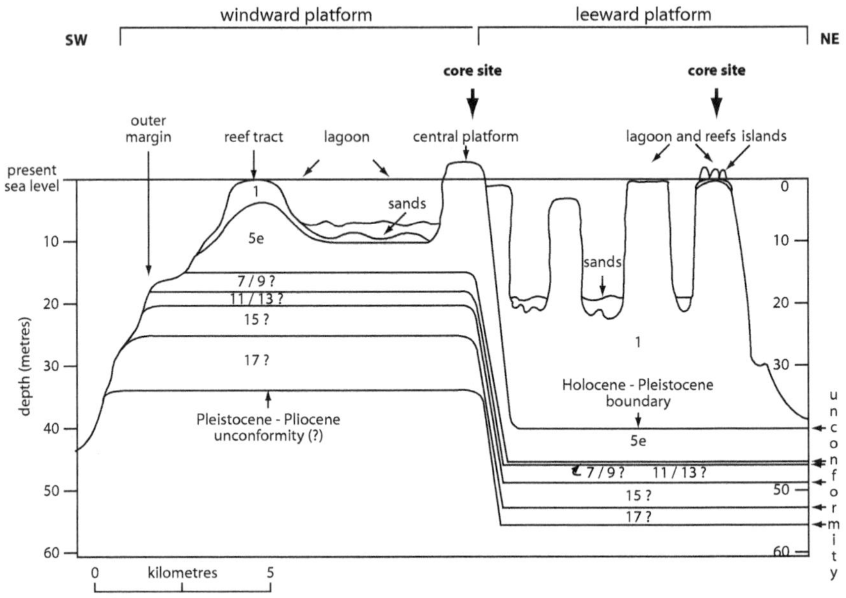

Figure 6.5 Idealized cross-section of the outer-shelf reef platform, Easter Group, Houtman Abrolhos, southwestern Australia. The cross-section is based on both seismic and drilling data. The position of expected MIS high sea-level units is indicated. Modified and redrawn from Collins et al. (1998).

operations by Collins et al. (1998) investigated the structure of coral reefs in the central part of the margin (Easter Group). The thickness of the Holocene varies from less than 5 m on the windward portion of the platform to a uniform 40 m on its leeward margin. Beneath the Holocene section, a total of six continuous reflectors have been identified and interpreted as subaerially exposed surfaces, consistent with drilling data (Figure 6.5). In windward settings, these surfaces were encountered at depths of approximately 4–9, 15, 18, 20, 25 and 34 m. In leeward settings, these discontinuities occur at depths of 40, 45–46, 48 and 56 m below sea level. The uppermost discontinuity is recognized as the last interglacial surface. The lowermost surface, lying between 34 and 56 m depth may be the Pleistocene–Pliocene boundary. The intermediate unconformities are likely to separate carbonate units ranging between 2 and 10 m in thickness and are expected to equate to Pleistocene high sea-level stands (MISs 5e, 7, 9/11, 13/15 and perhaps 17).

6.3.1.3. Case studies from the Pacific Ocean

In New Caledonia, the barrier reef system is typical of shallow rimmed shelves with marginal subvertical foreslopes, corresponding to the ‘escarpment margin' type of carbonate platform defined by Emery (1996). In this model, the vertical accretion of the platform dominates over its seaward accretion.

Seismic lines across the outer shelf to the lagoon, along the northeastern margin reveal the evolution and structure of the barrier reef system since the late Miocene (Chardon et al., 2008). Following extensional faulting in the late Miocene and severe fluvial erosion, a margin-forming clastic wedge was overlain by an aggradational fluvio–deltaic system, probably in the early Pliocene. This delta fan was partly buried by a transgressive carbonate sequence that aggraded to form a thick carbonate platform during the late Pliocene to early Pleistocene. This platform was overlain by the modern barrier system, probably during early-mid Pleistocene. The floor of the northeastern lagoon exhibits two successive incision/aggradation cycles related to late Pleistocene glacial–interglacial stages. The late Quaternary evolution of the architecture of the lagoon in northwest New Caledonia has been reconstructed from high-resolution seismic and multibeam bathymetry surveys (Le Roy et al., 2008). Two depositional sequences, capped by erosion surfaces have been identified. These unconformities delineate the floor of two incised-valley generations. Based on chronostratigraphic correlations with cores extracted from the nearby reefs, the deposition of the New Caledonian lagoon appears to have commenced around 200 ka. Two or three episodes of about 10-ka of low sea level during the late Pleistocene may account for the physiography observed today. During successive low sea level stands, erosional fluvial networks developed across the exposed lagoon. The outer part of the lagoon was connected to the foreslopes through passes continuous with deep valleys, presently about 80 m below sea level. The rivers have not incised the inner-shelf and proximal outer-shelf areas during low sea stands, but have produced an aggrading sedimentary wedge incised by several superficial channels (Le Roy et al., 2008). In combination with seismic surveys, recent drilling investigations have provided details of the structure and stratigraphy of the New Caledonian barrier reef tract (Chardon et al., 2008). Cores have been extracted from two islets (Kendec, Amédée), close to the western outer barrier reef, and range from about 129 to 149 m depth. On Amedée, the sequence consists of a total of 11 carbonate units, distributed into two contrasting lithological sequences (Figure 6.4B). Delineation of each successive unit is based on a combination of lithological, diagenetic and geochemical features, including abrupt changes in sedimentary facies, unconformity surfaces with subaerially generated deposits, and sharp changes in oxygen isotope stratigraphy. The upper sequence comprises units dominated by coralgal frameworks and derived skeletal material. The lower sequence contains detritus-dominated units with sparse coral debris. Carbonates appear to have begun to be deposited before 0.78 ka, according to magnetostratigraphic and nannofossil-based biostratigraphic results. These findings are in agreement with seismic records by Chardon et al. (2008). A magnetic field reversal was identified at around 117–122 m core depth, correlated with the unconformity between units 1 and 2 and attributed to the Matuyama–Jaramillo boundary. In addition, a second reversal, marking the Brunhes–Matuyama transition, found at approximately 80 m core depth is accompanied by a clear change in lithology. The lower carbonate units may relate to a ramp or a slope depositional prism rather than a typical reef platform. By contrast, the upper units, younger than 0.78 ka, are thought to be part of a well-differentiated reef system presumably representing inner reef-flat zones. In particular, the unit deposited during the last interglacial highstand (MIS 5 stage) consists of three distinct reef subunits separated by abrupt changes in lithology, each unit presumably related to the well-documented highstands between about 125 and 115 ka (see Chapter 9, Section 9.4.2). Accretion of the Holocene reef units is less than 14 m and did not begin before 8 ka. The evolution of the carbonate platform system is assumed to have been controlled by 100-ka orbital eccentricity cycles, at least in the relatively age-constrained upper units. The identification of subunits, based on changes in deposition styles, strongly support the influence of environmental disruptions operating within the 100-ka time span. By contrast, the subunits identified from the lower core sections older than 1–1.2 Ma, may have been deposited during 41-ka obliquity cycles. This assumption agrees with the work of Yamamoto et al. (2006) in the central Ryukyu Islands where changes in the mode of sedimentation were assumed to result from increased amplitude of sea-level cycles at the onset of the mid-Pleistocene climate transition (MPT). An alternative hypothesis to explain the occurrence of non-reefal deposits within the lower sequence suggests that it may consist of outer shelf to foreslope sediments accumulated during transitional, interstadial to stadial episodes rather than during high sea level stands.

There are significant differences in the structure and age of reef initiation on either side of the Coral Sea, between the New Caledonian and the northeastern Australian barrier complexes. On the Great Barrier Reef of Australia, a 210-m-long shelf-edge and foreslope sequence was recovered from Ribbon Reef 5 (Figure 6.4C). It is subdivided into three main sections containing a total of nine units (Alexander et al., 2001; Braithwaite et al., 2004). The lower section (units 1–2) consists of resedimented, mixed skeletal (benthonic and planktonic) material deposited on a foreslope. The intermediate section (units 3–4) includes laminated wackstones to grainstones interbedded with scleractinian corals and rhodoliths. The upper section (units 5–9) comprises a series of typical coral reef deposits separated by karst surfaces and palaeosol-draped unconformities. Magnetostratigraphy and strontium isotope dating indicate that reef initiation probably occurred between 0.88 and less than 0.5 Ma. The earlier limit fits the MPT, suggesting that the onset of reef growth may have been linked to that of the 100-ka, eccentricity-dominated sea-level cycles. One explanation for the retardation of reef initiation at the upper limit may be inimical conditions for extensive reef growth along the western border of the Coral Sea prior to 0.5 Ma. However, this assumption contrasts with the development of large reef complexes in areas from the eastern border of the Coral Sea to the far western Pacific.

Comparison between the Pleistocene sequences from the Bahamas, New Caledonia and the Australian Great Barrier Reef show some similarities in the records of depositional cycles over the past 0.78 Ma. In all three sites, 8–10 units have been recognized separated by sequence boundaries. However, the timing of reef initiation differs significantly. Whilst reef growth flourished as early as the late Pliocene in the Bahamas, the findings from New Caledonia and northeastern Australia tend to indicate that reef building was probably delayed until the mid-Pleistocene (Figure 6.4).

6.3.2. Atolls

When finally published, Darwin's ideas regarding subsidence were widely lauded and promoted but of course he had no proof. In London, the Royal Society commissioned the Funafuti borehole in order to test the hypothesis, and the data from this were described by Cullis (1904) and Finkh (1904) with more recent analysis by Ohde et al. (2002). The borehole encountered a substantial thickness of shallow-water limestones (339.5 m) and thus implied considerable subsidence, but did not reach the volcanic foundations required by Darwin's model. Up to 100-m-long carbonate sequences have since been extracted from a number of atolls: Kita-Daito-Jima Atoll, Japan (see Ohde & Elderfield, 1992), Bikini (Johnson, Todd, Post, Cole, & Wells, 1954) and Enewetak atolls (see Quinn, Lohmann, & Halliday, 1991) in the Marshall Islands, Mururoa Atoll in the Tuamotus (Deneufbourg, 1969; Repellin, 1977; Buigues, 1996; Camoin et al., 2001), Pukapuka and Rakahanga atolls, Cook Islands (Gray et al., 1992), Bougainville Guyot in Vanuatu (see Quinn, Taylor, Halliday, Collot, & Greene, 1991) and the Maldives (Gischler et al., 2008). However, the Pleistocene stratigraphy of most atolls is poorly constrained as a result of severe recrystallization, although many more drilling investigations and subsequent extensive stratigraphic and lithological studies have been carried out, particularly in the central Pacific (see cases studies in Vacher & Quinn, 1997). However, isotopic and fossil-based dating analyses have been tentatively applied to cored coral material from some sites. Strontium isotope ratios and, in some cases, U-series methods have proven able to provide substantially greater resolution than biostratigraphy.

At Enewetok Atoll, the Quaternary stratigraphy remains controversial. Strontium isotope chronology conflicts with biostratigraphy. According to strontium isotope data (Quinn, Lohmann, et al., 1991) and lithology (Wardlaw & Henry, 1986), the Quaternary sequence extracted beneath the lagoon is estimated to be approximately 100 m thick and to consist of 12 depositional units separated by subaerial unconformities. Beneath the outer rim, the sequence thickness is less than 70 m. The Holocene deposits referred to as Interval II are about 20 m thick in the lagoon, and consist of *Halimeda*-molluscan packstones and wackestones. The 10–12 m thick outer-rim Holocene section contains skeletal grainstones–packstones on the inner margin and coral rudstone to floatstone to seawards. The Pleistocene section (Interval II) is typified by coral floatstones and *Halimeda*-rich sand and mud in lagoonal cores, but is dominated by skeletal packstones and grainstones interlayered with coral boundstone towards the margins.

On Mururoa, seismic investigations, coupled with both vertical and inclined coring, have provided detailed information on the main architectural attributes of the reef pile. The thickness of the Quaternary carbonates inferred from magnetostratigraphy is about 63 m. The base of the Pleistocene sequence coincides with the Gilsa subchron at about 1.6 Ma, within the Matuyama reversal period (Buigues, 1997). The Pleistocene sequence averages 50 m thick beneath the atoll rim and the lagoon. Several subaerial exposure surfaces are inferred for the Pleistocene sequence, but have not been confidently identified beneath the outer reef rim and the lagoon except for a single karst surface at about 18 m core depth. This surface relates to the MIS 5–7 transition. However, 300 m long cores with seaward inclinations of 30–45° provided new constraints on the architecture of the outer margin of the atoll. This appears to consist of a series of aggrading to prograding, fringing tracts deposited during both highstands (MIS 13–3) and lowstands (MIS 12–4), as inferred from U-series dating (Perrin, 1990; Camoin et al., 2001). Thus, at least at its periphery, Mururoa is not erosional, but constructional in origin (Figure 6.6).

On Funafuti Atoll, as demonstrated by Ohde et al. (2002) using strontium isotope measurements, the most striking feature is the impressive thickness of the Quaternary sequence, that reaches 189–200 m, when compared to Enewetok (*ca.* 95 m) and Mururoa (*ca.* 63 m). This sequence was deposited during the last 1.5 Ma and is subdivided into at least five units. The uppermost unit relates to Holocene reef rim deposits, dated at about 8 ka at the base. The Pleistocene units, consisting largely of lagoonal deposits, are separated by karst unconformities at 26.4 m (Holocene–Pleistocene or MIS 1–5 boundary), 30 m (MIS 5–7 boundary), 65 m (MIS 7–9 or 11 boundary), 75–80 m (MIS 9/11–17 boundary) and 120 m core depth (age: 0.65–1.2 Ma). An additional hiatus in deposition is inferred at 170 m.

The complexity of atoll growth histories is well illustrated by reference to the Maldive Island Group in the central Indian Ocean. The island rims consist of ring-like reefs (atolls) tens or in several cases hundreds of kilometres across. Carbonate deposition in the Maldives began in the early Eocene, and there is evidence of exposure and local collapse within the preserved succession, but reef deposition did not commence until what is probably the Pliocene–Pleistocene (Purdy & Bertram, 1993). The depths of the larger lagoons and present relief correlate well with annual rainfall

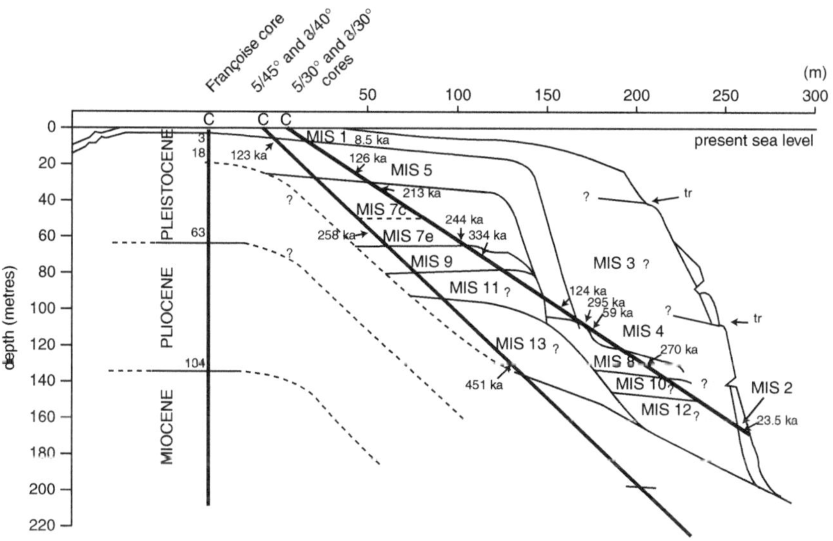

Figure 6.6 Internal structure and chronostratigraphy of the outer margin of Mururoa atoll, central Pacific, based on magnetostratigraphy and U-series dating of coral samples from vertical and inclined cores (C). The positions of the successive MIS highstand and lowstand stages are shown. Modified from Perrin (1990) and Camoin et al. (2001).

(Purdy, 1974; Purdy & Winterer, 2001). It is acknowledged that this may not have been identical during Pleistocene lowstands, but the present pattern of distribution appears to be close enough to provide a reasonable explanation. As suggested by seismic data, the most striking feature regarding stratigraphy is the contrast between a consistent earlier pattern of progradation towards the inner sea during the Miocene and the dominant aggradation during the Pliocene–Pleistocene. The thickness of the Quaternary sequence is difficult to estimate but is probably up to 200 m. This change in the depositional regime is likely to have been caused by drastic changes in sea-level behaviour. Lateral accretion is the response to a relatively stable sea level, while vertical accretion results from increases in the amplitude of sea level rise. The large scale reef passes, which are more common on the inner sides of macro-atolls, may reflect a regional drainage during lowstand periods (Purdy & Bertram, 1993). It is notable that such areas bear a strong resemblance to the reticulate banks described in lagoons of the Great Barrier Reef (Hopley, 1982) and Mataiva Atoll, French Polynesia (Delesalle, 1985; Pirazzoli & Montaggioni, 1986). The physiography of the Maldives appears to be a composite product of repeated glacial–interglacial cycles and erosion during exposure events has had more influence than subsequent reef growth. Perhaps the most significant conclusion is that the atolls of the Maldives are not the end members of a Darwinian progression from fringing reefs. Subsurface evidence shows clearly that the modern atoll-like reefs do not result from a preceding barrier reef stage of growth.

6.4. Stratigraphy of Emerged Reef Terraces

Locally, on coastal continental margins and on islands subjected to uplift, a series of emerged terraces are present that may reach several hundred metres in elevation. Terraces are commonly associated with Quaternary reefs, but there are problems in providing any coherent summary. This is in part because they are of two kinds: depositional terraces representing the net accumulation of a particular accretion period, and erosional terraces, the latter including both erosional benches and notches cut during periods of stable sea level. In addition, they reflect the interaction of two variables, eustatic and hydroisostatic variation in sea level, and tectonic or isostatic changes in the continental margin or island. The tradition has been to refer to these by their positions relative to present sea level. But factors influencing sea-level change vary from place to place, and although some can be regarded as 'global' they may nevertheless vary in magnitude. Chronological relationships provide a more robust basis for comparison, but as indicated in Chapter 7, diagenetic changes in the material available for dating mean that in some areas there are no dates and in others some may be unreliable. Thus, we are faced not only with the problem that the record is incomplete, but also the difficulty of providing accurate ages for those parts of it that are present.

6.4.1. The Huon Peninsula and Barbados Models

The terraces of the Huon Peninsula in Papua New Guinea are regarded as one of the most detailed records of the late Pleistocene (Figure 6.7). Described by Chappell (1974, 1986), Chappell et al. (1996), Ota and Chappell (1999) and Yokoyama, Esat, and Lambeck (2001), these developed over a distance of some 80 km. As a result of uplift on the leading edge of the Australian plate, they rise over 600 m, forming a stairway of more than 20 reef complexes that include at least two terraces below present sea level. Bloom, Broecker, Chappell, Matthews, and Mesolella (1974) were careful in their descriptions to differentiate the facies models expected in barrier and fringing reefs associated with submergence and emergence. These terraces largely record relative highstands of sea-level cycles as the margin was progressively uplifted and relative sea-level fell. Many of the reef profiles illustrated conform to models implying an offlap sequence of reef accretion

Figure 6.7 General view of exposed Pleistocene reef terraces on the Huon Peninsula, Papua New Guinea. Photograph by L. Montaggioni.

(model B of Montaggioni, 2005). Typically, the growth of each unit followed a rising relative sea level and is transgressive, but some are described as regressive and there are clearly erosional notches reflecting still stands during falling sea level. Reef edges are defined by the presence of *Acropora humilis* and an algal zone with *Acropora cuneata*. Beneath these are zones characterized by *Acropora palifera* and at greater palaeodepth, *Plesiastrea, Hydnophora* and *Echinophyllia*. Other facies include 'lagoonal' and 'fore-reef' sediments that are clearly differentiated in the lower terraces. There are also substantial intercalations of deltaic sands and gravels derived from local streams and used by Chappell (1974) in his interpretations. Radiocarbon dates of the lower terraces and U/Th dating of some of the older units have provided a reasonably well-constrained chronology. The ^{14}C data appeared in Polach, Chappell, and Lovering (1969) and the U/Th data in Veeh and Chappell (1970). The dated samples indicate sea-level maxima at 6.4, 29, 40, 60, 84, 106, 118, 137, 185, 218 and >250 ka. Chappell et al. (1996) were able to reconcile much of this sequence with the sea levels implied by deep-sea oxygen isotope data, but these results cannot be regarded as comprehensive. In his discussion, Chappell (1974) indicated that some reefs, formed during periods when rising sea level outstripped uplift, would show an age inversion, that would be deposited at a higher level, thereby reversing the 'normal' sequence of decreasing age with altitude. In addition, erosional notches (that reflect an important stasis) cannot be easily dated.

Like those of the Huon Peninsula, the raised reef terraces of Barbados are viewed worldwide as one of the major records of interglacial high sea stands during the late and middle Pleistocene and the sequence has been referred to as the 'Barbados Model' of sea-level change (Mesolella et al., 1970; Thompson & Goldstein, 2005). Recently, significant refinement of the Electron Spin Resonance (ESR) dating method allowed a revision of previously published stratigraphies (Schellmann & Radtke, 2004). The differentiation of reef units and associated erosional features was improved using mapping from air photograph interpretation. Up to 13 main terraces are identifiable (Figure 6.8). Some of these can be divided into subunits on the basis of separation by fossil cliffs, paleochannels, or paleolagoons. The uplifted reef terraces usually display a strong positive correlation between elevation and age. The youngest and lowest terrace developed at about 74 ka at the termination of the last interglacial highstand. It now lies about 2 m above sea level. The older and highest terrace, dated at about 410 ka, reflects deposition during MIS 11 and reaches about 120 m above sea level. The well-preserved terraces generally exhibit a steep seaward margin in the form of a relict cliff or slope. The fossil reef flat behind the outer margin typically have a low relief and change backwards into channels and lagoons. The lagoons are locally bounded by beach-rock lines, an erosional platform or a subvertical cliff. Some wave-cut platforms are 100–400 m wide.

Compared to the Huon Peninsula and Barbados, no other sequences provide this level of stratigraphic discrimination in relation to Pleistocene sea-level events. The stratigraphies that have been described elsewhere typically refer to only a few well-dated depositional units.

6.4.2. Other Reef Terraces Sequences

The most time-expanded sequences of uplifted reef terraces were described from north of Sumba Island (Indonesia) and from the Ryukyu Islands (Japan).

More than 11 terraces were recognized on Sumba Island, from present sea level to an elevation of 475 m (Pirazzoli et al., 1993). The oldest ESR age estimate (*ca.* 600 ka) was obtained for a terrace at present lying at approximately +250 m. The ages of the older and higher terraces were inferred tentatively from the interglacial stages of oxygen isotope chronology back to MIS 27 (*ca.* 0.99 Ma). However, large uncertainties may arise due to the intercalation of low sea-level reefs within the sequence.

Analysis of the carbonate-dominated Ryukyu Group revealed that reef initiation probably occurred in the early Pleistocene (Sagawa et al., 2001). The depositional patterns appear to have changed at around 0.8–1.0 Ma (Figure 6.9). Prior to this time, from 0.8 Ma to about 1.45–1.65 Ma, about 80 m of mixed carbonate-siliciclastics and carbonates (Unit 1) were deposited in fore-reef slope to shelf environments. These show no evidence of subaerial

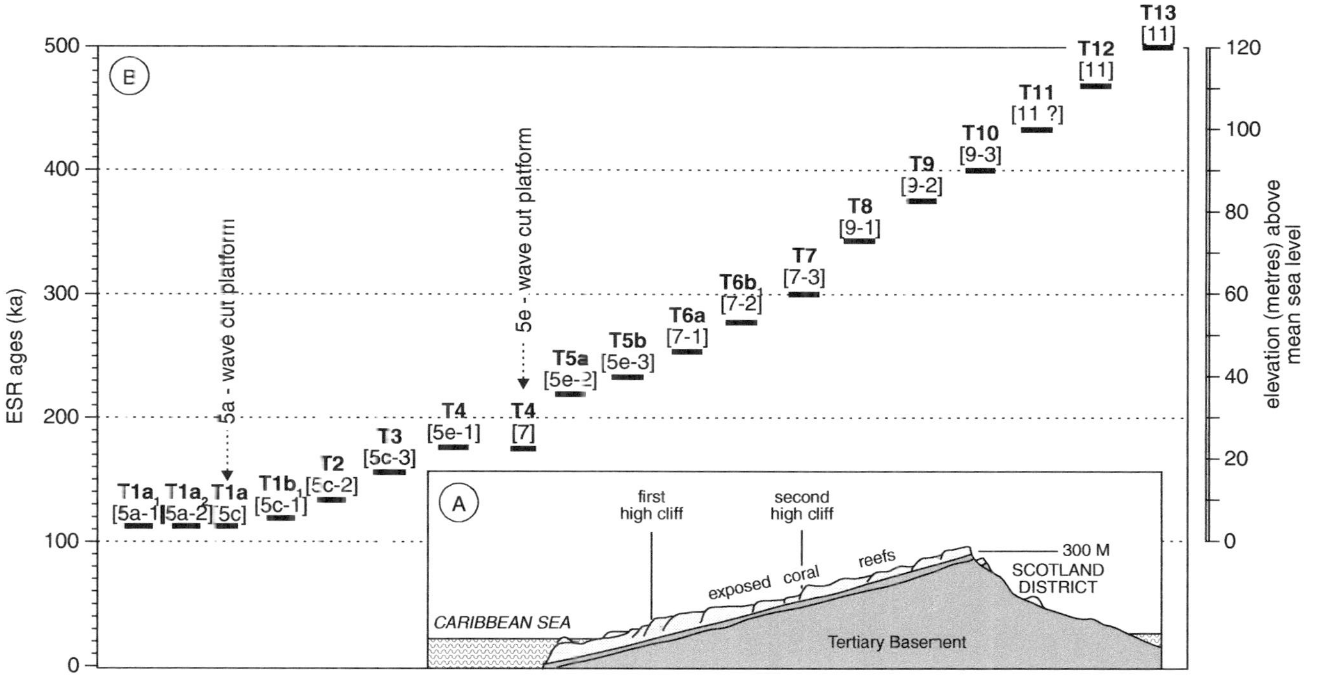

Figure 6.8 Morphostratigraphy of Quaternary reef terraces of Barbados, Carribean. (A) Geological section across the terrace succession and (B) ESR ages and elevations of terraces in southern Barbados. Each terrace is identified by combined letter–number nomenclature. The lowest and youngest terrace is referred to as T-1, while the highest and oldest is T-13. Each terrace is correlated to an oxygen isotope (MIS) stage and substage (number–letter nomenclature placed in brackets). Modified from Schellmann and Radtke (2004).

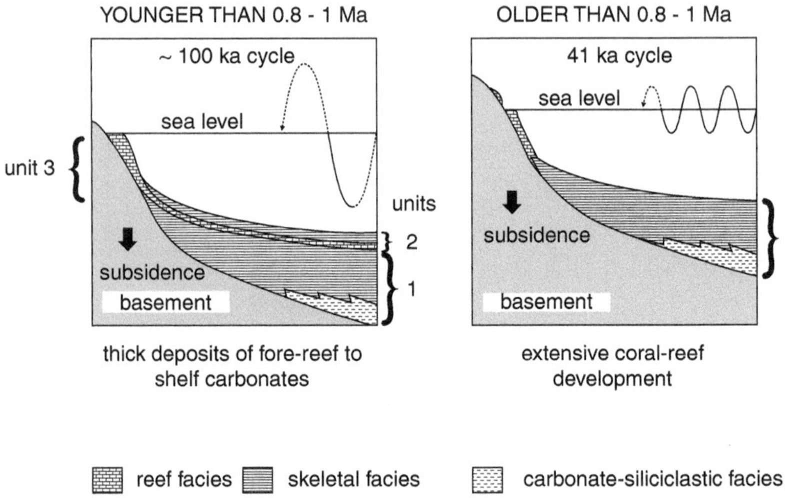

Figure 6.9 Idealized stratigraphic successions of the Pleistocene Reef Complex formation (Ryukyu Group), Okinawa-jima, central Ryukyus. Significant changes in depositional regimes occurred from the mid-Pleistocene climatic transition (at about 0.8–1.0 Ma) in response to changes in sea-level amplitude and frequency, from a 41–100 ka cyclicity. Modified and redrawn from Yamamoto et al. (2006).

exposure, suggesting that they were permanently under water (Yamamoto et al., 2006). They have not been overlain by reef deposits. The stratigraphic successions and configuration of lithofacies do not represent deepening or shallowing sequences but indicate repeated sea-level fluctuations. Thus, glacially controlled sea-level variations in the early Pleistocene are inferred to have been of relatively low amplitude (less than 50 m) and high frequency. Reaching less than 100 m in thickness, Unit 2 comprises two subunits. The basal subunit represents a coral reef complex that accumulated from 0.8 to 0.4 Ma. It is overlain by fore-reef to shelf carbonates rich in rhodoliths. Unit 3 consists of coral reef deposits younger than 0.4 Ma. The geometry and stratigraphy of units 2 and 3 reflect repeated sea-level oscillations with larger amplitude (60–100 m) but lower frequency, than with those inferred for the early Pleistocene. The timing of the change in the Quaternary depositional patterns may fit that of the MPT (Yamamoto et al., 2006) if it is considered to be centred at 0.9–0.8 Ma. This interval was typified by a change in the glacial–interglacial cyclicity from small-amplitude 41 ka to large-amplitude 100 ka periods (see Chapter 1, Section 1.4.2). The latest middle Pleistocene deposits consist of coral and benthic foraminiferal limestones capped with late Pleistocene to Holocene siliciclastics.

Raised reef terraces have been also encountered in locations along both sides of the Red Sea, (Gvirtzman, 1994; El-Asmar, 1997). In South Sinai, there is a series of five terraces, the ages of which range from the last interglacial (MIS 5e) to older than 300 ka, presumably MIS 11 (El-Asmar, 1997). Land, Mackenzie, and Gould (1967) found at least five terraces in the Pleistocene history of Bermuda. Braithwaite, Taylor, and Kennedy (1973) described Pleistocene sequences on Aldabra in the western Indian Ocean. Montaggioni (1982) documented the Pleistocene succession on Mauritius Island, and Braithwaite (1984) the sequence on the Kenya coast. Hantoro et al. (1994) recorded six major reef terraces rising to more than 700 m on the Kabola Peninsula of Alor Island, Indonesia. Additional works devoted to the developmental history of exposed reef terraces include the Ryukyus (Konishi et al., 1974; Radtke, Grün, & Mangini, 1997), Indonesian Islands (Chappell & Veeh, 1978; Pirazzoli et al., 1991; Hantoro et al., 1994; Bard, Hamelin, Pirazzoli, et al., 1996), Vanuatu (Cabioch, Corrège, et al., 1999), the French Antilles (Feuillet et al., 2004) and Haiti (Dodge, Fairbanks, Benninger, & Maurasse, 1983; Dumas, Hoang, & Raffy, 2006).

6.4.3. High-Carbonate Islands

There are a number of saucer-shaped high-carbonate islands. In the Indo-Pacific, the morphology of which resembles that of modern atoll lagoons. The present elevation of these islands, ranging from few metres to up to 100 m results from different generating processes linked to plate tectonics, including hotspots (e.g. Mangaia in the Cook island, Nauru and Niue in the southwest Pacific), volcanism at or near divergent plate boundaries (e.g. Makatea in the Tuamotus, the Cayman Islands in the northern Caribbean) or arc volcanism (e.g. Guam in the Marianas, the Tongatapu Island Group in Tonga, and the Loyalty Islands close to the Vanuatu Thrust). As the islands drowned due to crustal cooling, coral reefs developed around the volcanic cores and the volcanic pedestals were totally overgrown by reefal deposits. The time of deposition varies from site to site, but was usually between the early Miocene and mid-Quaternary. The time of uplift also differs from island to island, but was generally not before the early Quaternary. Uplift processes vary and reflect regional tectonics. On islands close to recent hotspot volcanoes (Tikehau, Anaa, the Makatea islands in the Tuamotus, Henderson Island in the Pitcairn Group), uplift may be explained by lithospheric flexure reflecting the overload of nearby volcanoes (e.g. Tahiti and Moorea, or Pitcairn) forming a bulge on the surrounding crust. By contrast, on Nauru (southwestern Pacific), emergence is attributed to the uplift of the seafloor carried up onto a mantle bump by plate motion. The Loyalty Islands and Guam were uplifted on the crest of the frontal arc formed at the convergence of the Pacific and Philippine plates.

The atoll-like morphology of most of these islands strongly suggests that it is inherited from dissolution during Pleistocene low sea levels (Montaggioni et al., 1987; Purdy & Winterer, 2001). Their modern coastal physiography and architecture mainly result from relative sea-level variations related to both tectonic movements and glacioeustatic sea level changes during the Quaternary. The cliffs surrounding these islands are typically rimmed by mid-late Pleistocene to Holocene fringing- or barrier-like reef tracts at a variety of elevations. For instance, on Makatea (Figure 6.10A), the Pleistocene deposits include two generations of reef terraces at 7 and about 30 m above present sea level that are connected to two upper notch lines (Montaggioni & Camoin, 1997). These terraces were tentatively dated by U/Th methods at 100–140 ka (MIS 5e) and 300–400 ka (MIS 9, 11 or 13) respectively. A Holocene-emergent reef terrace, 3.7–5.3 ka old, forms a peripheral fringing reef system, reaching an average elevation of 0.5 m above present mean sea level, overlying a pre-Holocene marine erosional platform. In nearby islands including Mataiva, Tikehau, Kaukura, Niau, Anaa and Rangiroa atolls, uplift was of markedly lower amplitude than Makatea. The remains of pre-Pleistocene carbonates culminate at approximately 1–4 m above present sea level, forming karstic, partly dolomitized 'feos', as they are referred to by Polynesian people (Figures 6.10B and 6.11). The Pleistocene to Holocene deposits consist predominantly of a few-metres-thick detrital carbonates with scattered, in-place coral colonies, locally covering phosphatic layers (Figure 6.11B). The stratigraphy of typically subsiding atolls like Mururoa does not include exposed relicts of pre-Pleistocene reefs (Figure 6.10C).

Grand Cayman Island in the Caribbean has a low-lying atoll-like platform around the margins, deposited during the last interglacial high sea level stand (Jones, Ng, & Hunter, 1997).

6.4.4. The Question of Multistage Terrace Development

Although it has been clear for some time that terraces can be generally correlated with MIS high sea levels, it has also become apparent that there are superimposed minor (suborbital) cycles. Intervals of deposition have not all been characterized by reef growth. Palaeosols and thick aeolian deposits are widely distributed in Australia (Brooke, 2001), in the Bahamas and Bermuda (Sayles, 1931; Vacher, Hearty, & Rowe, 1995) and are interbedded with reef-derived materials on Rodrigues and Aldabra (Western Indian Ocean) and the Kenya coast (Braithwaite, 1994, 1973, 1984). Such successions are best illustrated in the Pleistocene successions from Eleuthera in the Bahamas (Hearty, 1998). Much of this succession is of aeolian origin and only part of reef deposits, but nevertheless prolonged lowstands are marked by karst erosion surfaces and the deposition of palaeosols. Using aminostratigraphy to date the samples and comparing the results with marine oxygen isotope stages indicates deposition during stages 13 or 15, 11, 9 or 11, early and late 7, 5e, 5a and Holocene, separated by seven erosion surfaces bearing palaeosols. The MIS 7 and 5 highstands were found to be subdivided by intervals of

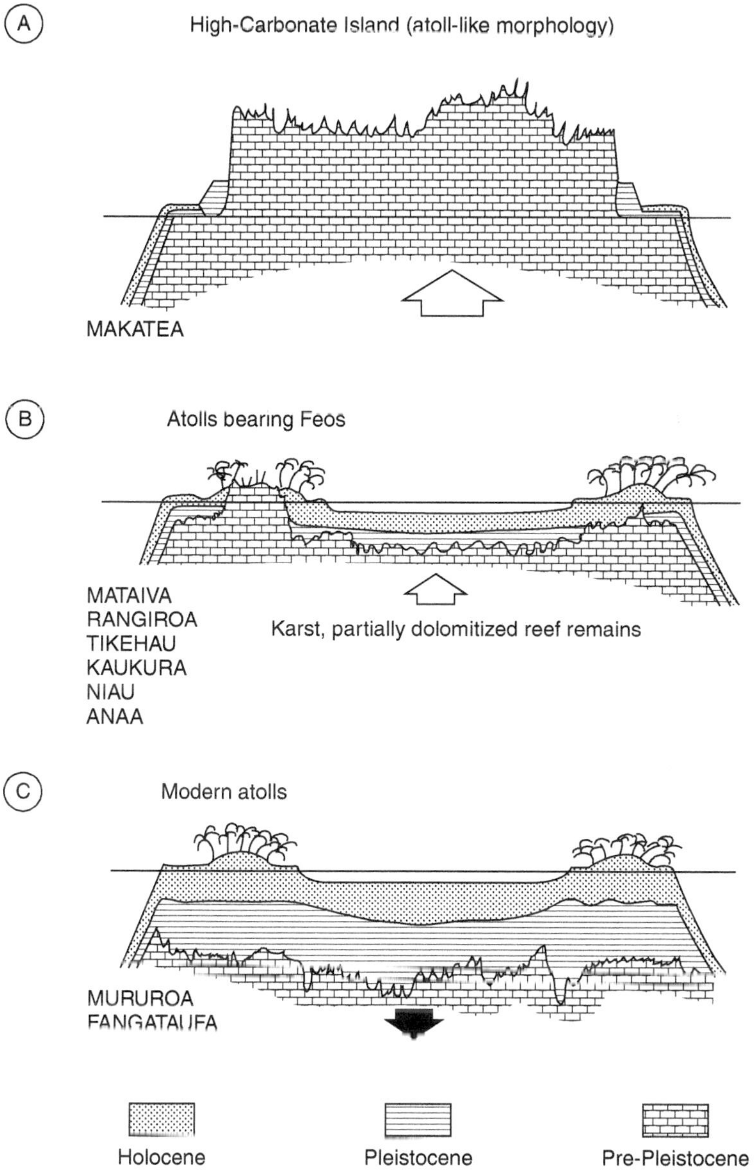

Figure 6.10 Structure and stratigraphy of atoll-like, high-carbonate islands (A), atolls with remains of pre-Pleistocene reefs (feos) (B) compared to those of modern atolls (C).

erosion, reflecting their polycyclic nature. As pointed out by Blanchon and Eisenhauer (2001), the reef architecture of some Barbados terraces is complex due to multistage reef growth in response to suborbital to millennial changes in sea level. A similar division of MIS 7 in a reef-deposited sequence is

Figure 6.11 Karst and dolomitized remains of an old coral reef, probably of Miocene age, Kaukura Atoll, northern Tuamatus, central Pacific. Photograph by P. Pirazzoli.

implied by the results from the Great Barrier Reef borehole (Braithwaite et al., 2004) and in boreholes on Mururoa (Camoin et al., 2001) and New Caledonia (Cabioch, Montaggioni, Thouveny, et al., 2008). The multistage origin of a Pleistocene reef terrace was also demonstrated in the Persian Gulf by Pirazzoli et al. (2004). On Kish Island, a terrace uplifted to about 32 m above present sea level originally represented two successive depositional events attributed to MIS 5e and MIS 7e high sea levels. This strongly suggests that slow uplift rates may cause polycyclic marine deposition with the location of successive interglacial reefs at similar elevations. A similar composite reef terrace was described from Great Inagua Island in the Bahamas (Kindler, Reyss, Cazala, & Plagnes, 2007). U-series dating and lithostratigraphy indicate that here the major building phase corresponds to MIS 5e. The lower parts of the terrace span MIS 7. These findings suggest that part of the island has undergone uplift and tilting in the last 100 ka.

6.5. Stratigraphy of Submerged Reef Terraces and Banks

Generally, coral reefs that developed during Pleistocene glacial–interglacial transitions (terminations) are at present submerged and usually overgrown by coralgal reefs and encrustations deposited during episodes of recolonization during the last deglacial period (Harris & Davies, 1989; Fletcher & Sherman, 1995; Toscano & Lundberg, 1999; Grigg et al., 2002). In the last 20 ka, progressive rise of sea level has been punctuated by phases of reef initiation and growth followed by reef demise. For example, within the last 26 ka, that is from the Last Glacial Maximum (LGM) to the late Holocene, four major phases of reef growth have been identified in the Indo-Pacific, locally interrupted by non-constructional or reef-drowning events (Montaggioni, 2000, 2005). In most cases, the successive reef generations are represented by distinct reef terraces and platforms, forming step-like topographic features along foreslopes. Such submerged terraces are well documented from Barbados (Fairbanks, 1989), western India (Rao, Montaggioni, et al., 2003), the Ryukyu Islands (Sasaki et al., 2006), the Great Barrier Reef (Harris & Davies, 1989; Beaman, Webster, & Wust, 2008), Hawaii (Webster et al., 2004, 2006), Tahiti (Camoin et al., 2007) and the Marquesas Islands (Cabioch et al., 2008). On cliff-like foreslopes, submerged coral buildups developed only as relatively thin veneers. These are exemplified in Mayotte, Comoros (Dullo et al., 1998) and on Mururoa (Bard et al., 1992; Buigues, 1997; Camoin et al., 2001). Unfortunately, little is known of the complex internal stratigraphies of drowned reefs worldwide due to the paucity of offshore boreholes and lack of seismic data.

6.5.1. Case Studies from Stable Areas

Submerged terraces have been reported from the foreslopes of several areas considered to be tectonically stable. These have been used locally as direct sea-level markers (see Chapter 9, Section 9.4.1).

On the continental slope of the Great Barrier Reef, submerged terraces have been described at depths of 40–70 m (Beaman et al., 2008). At this depth, the antecedent surfaces on which these rest would have been subject to several cycles of subaerial exposure, with the most recent coral growth additions relating to the period after the LGM. Reefs recently discovered in the Gulf of Carpentaria (Harris, Heap, Wassenberg, & Passlow, 2004; Harris, Heap, Marshall, and McCulloch, 2008) apparenty began their growth approximately 10.5 ka ago and persisted for some 2.0 ka, but by about 7 ka had virtually ceased to grow, and were unable to keep pace with rising sea level (Figure 6.12). These rest on an abraded surface of Pleistocene limestone at around 30 m depth.

Seismic investigations, together with dredging and/or coring in both the Pacific and Atlantic oceans, have shown that the calcareous alga *Halimeda* can form submerged banks and biohermal structures in tropical areas, representing deposits as thick and as extensive as the Holocene reefs themselves (Roberts & Macintyre, 1988). The thicknesses of *Halimeda* accumulations above the antecedent topography (usually interpreted as of Pleistocene age) range from about 2 to more than 50 m (Davies & Marshall, 1985; Orme, 1985; Phipps et al., 1985; Orme & Salama, 1988; Marshall & Davies, 1988; Phipps & Roberts, 1988; Hine et al., 1988). The results of radiocarbon dating indicate that the *Halimeda* mounds started to accumulate

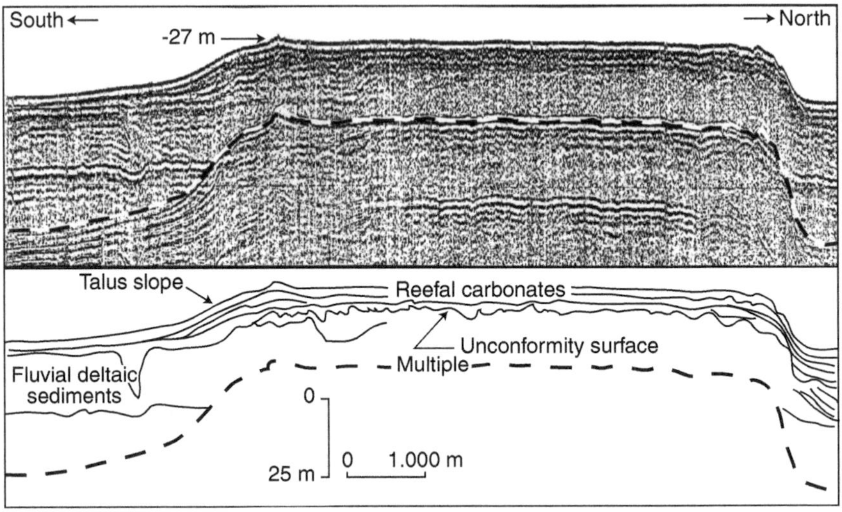

Figure 6.12 Seismic profile and interpretative cross-section of a submerged patch reef in the Gulf of Carpantaria, northern Australia. Harris et al. (2004).

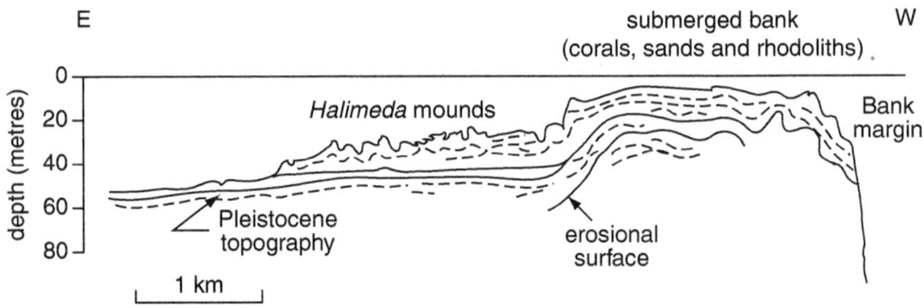

Figure 6.13 Interpretative seismic-based cross-section of *Halimeda* banks from the northern part of Kalukalukuang Bank, eastern Java Sea, Indonesia. Modified and redrawn from Phipps and Roberts (1988).

at different times throughout the Holocene from as early as 10 ka in the Great Barrier Reef (Marshall & Davies, 1988) to 3.5 ka in the eastern Java Sea (Phipps & Roberts, 1988). They have continued to grow actively to the present time. Lying in 20–100 m of water, these mounds consist of unconsolidated to firmly lithified packstones–grainstones dominated by unbroken, disarticulated *Halimeda* segments embedded into a poorly sorted sandy to muddy, skeletal to terrigenous matrix (Figure 6.13). Although *Halimeda* mounds are locally associated with coral reefs, they appear to be more extensively developed in areas where reefs are nearly absent. Hine et al. (1988), Marshall and Davies (1988) and Roberts, Aharon, and Phipps (1988) suggested that the remarkable development of *Halimeda* at the expense of reef-building corals occured in response to upwelling and nutrient overloading. The contribution of *Halimeda* mounds to shallow-water carbonate sedimentation and to carbon budgets on a global scale has probably been underestimated, particularly relative to the contribution of coral reefs. This assertion relies on studies from the northern Great Barrier Reef by Rees, Opdyke, Wilson, and Henstock (2007) that revealed that the outer-shelf *Halimeda* bioherms contain at least as much carbonate material as the nearby outer-shelf reefs, and probably up to four times more.

In the volcanic Marquesas Islands, widely regarded as relatively stable since the late Pleistocene, echosounding and dredging investigations on the reef foreslopes have revealed the occurrence of several distinct intervals of reef deposition (Cabioch, Montaggioni, Frank et al., 2008). Identification of the morphological and biological features of these combined with coral U-series dating, provide evidence of four successive reef generations between 26 and 9 ka, now ranging from 138 to 58 m depth. Shallow-water coralgal assemblages are replaced upwards by deep-water mollucan assemblages encrusted by microbialites and iron-rich hardgrounds. This replacement is interpreted as reflecting an upwards-deepening sequence.

On the margin of South Florida shelf, submerged reefs are found at 7–12 m below present sea level. Stratigraphically speaking, these consist of two units. The upper unit, 1.5–7 m thick, was deposited during the Holocene, from 8.9 to 5 ka. The lower unit, less than 10 m of which has been recovered is dated at 106–80 ka. It therefore records 26 ka of MIS substages 5c–5a. Locally, there is clear evidence of non-depositional episodes between MIS 5c/5b and 5b/5a. A distinct unconformity with evidence of subaerial exposure (caliche, paleosoils) separates the Holocene from the Pleistocene units. By contrast, there is no evidence of emergence within the Pleistocene unit (Toscano & Lundberg, 1999).

In stable areas, reef-drowning may reflect surges in sea-level rise during meltwater pulses linked to rapid deglaciation (see Chapter 9, Section 9.4). Blanchon and Shaw (1995a) examined the relative positions of a series of drowned reefs in the Atlantic–Caribbean province. These consist largely of *Acropora palmata*, typically dominant in waters less than 5 m deep. The reefs lie at depths of approximately 80, 50 and 15 m below present sea level. Drowning events were regarded as correlated with three separate catastrophic jumps in sea level. Subsequently, drilling off Grand Cayman by Blanchon, Jones, and Ford (2002) identified a 6-m jump in sea level approximately 7.5 ka ago, with the demise of the reef within a 160-year interval. Because the South Florida record covers one of the gaps (between 7.6 and 7.2 ka) recorded by Blanchon and Shaw (1995a,b) in the Caribbean, Toscano and Lundberg (1998, 1999) considered that the continuous record of reef growth precludes the occurrence of any rapid surge in sea-level rise. They interpreted the gaps in the early Holocene record as artefacts generated by limited sampling or core recovery.

6.5.2. Case Studies from Subsiding Areas

High-volcanic islands in mid-plate settings currently experience subsidence rates of 0.25–>1 mm yr^{-1}.

On Tahiti, drilling investigations conducted on the barrier reef complex (Montaggioni et al., 1997; Cabioch, Camoin, et al., 1999) and on adjacent fore-reef slopes (Camoin et al., 2007) have revealed the presence of two major lithological units: a postglacial carbonate sequence (Unit I) dated around 16 ka at the base, and a late Pleistocene sequence (Unit II). Beneath the modern outer barrier reef flat, the thicknesses of units I and II are about 87 m and 27 m respectively, overlying the volcanic basement. In the fore-reef zones, successive reef terraces of various lateral extents occur at 40–50, 60, 90 and 100 m depth. Beneath these structures, Unit I reaches 9–45 m thickness. The recovered thickest section of Unit II exceeds 65 m, but the volcanic foundation was not penetrated in this area.

The successions in Unit I primarily consist of coralgal–microbialite frameworks, commonly interlayered with loose skeletal sediments including coral and algal rubble and, occasionally, including interlayered volcaniclastics. Locally, coral assemblages are dominated by deep-water branching, domal, tabular and/or encrusting forms. Microbialites consist of dense laminated and thrombolitic fabrics; with the latter usually dominant. Corals are typically thinly encrusted by non-geniculate coralline algae, but at the base of the sequence the crusts are significantly thicker (up to 2 cm thick) and include vermetid gastropods. The latest stage of encrustation of coral colonies is represented by thick microbialite crusts dominated by massive laminated fabrics.

Detailed investigations suggest that 2–5 subunits can be recognized in Unit I based on internal structures and the nature of coral assemblages. Seven distinct coralgal assemblages have been identified. The most typical succession comprises a coralgal bindstone overlying a coralgal–microbialite framework. In the bindstone, the coral assemblage is dominated by deep-water encrusting and foliaceous colonies. Tabular colonies of *Acropora* occur locally in the lower part of the section, whereas in some areas the upper parts are locally capped by hardgrounds. The coralgal–microbialite frameworks dominantly consist of tabular acroporid colonies and branching poritids. Locally, however, the base of Unit I consists of poorly lithified skeletal grainstone rich in sand-sized volcanic lastic grains. The paleobathymetric interpretation of the coralgal assemblages at each drilling site, indicate that the fore-reef development reflects an upward-deepening sequence formed in response to a rapid rise in sea level. The contact between units I and II is generally an irregular unconformity with abundant dissolution cavities filled in part with unconsolidated skeletal and volcanic lastic sand. This surface probably formed during the penultimate deglaciation (Termination II).

Unit II, dated at around 32 ka (substage MIS 3a) at the top, primarily consists of coralgal framework interlayered at intervals with algal bindstone, microbialites, skeletal limestones and rubble. Basalt pebbles and lithoclasts occur throughout the unit. The coralgal frameworks are dominated by branching and encrusting *Porites*, locally associated with *Montipora*, *Pavona* and *Pocillopora*, and with widespread development of microbialite. The most complete sequence contains eight lithologically distinctive subunits. Several subaerially generated unconformities separate subunits in the upper sections of the Unit. These are probably linked to periods of sea-level fall during the late to middle Pleistocene.

Rapidly drowned reefs and associated deposits have also been described from Hawaii (Jones, 1995; Grigg et al., 2002; Webster et al., 2006). Based on high-resolution bathymetric and sidescan data, combined with remotely operated vehicle (ROV) and submersible observations with radiocarbon dating, the investigations of Webster et al. (2008) have focused on two shallower terraces at 150 and 230 m depth. The tops of these are covered by *in situ* deposits dominated by rhodoliths formed at from 30–60 m and on deep fore-reef slopes (60–120 m). These deposits indicate ages of 10.8–29.34 ka for the shallower terrace and 14.39–27.12 ka for the deeper, reflecting the transition into the LGM. A sedimentary model has been developed in relation to eustatic sea-level variations over the last 30 ka. Rhodoliths started to grow at 30–29 ka, following a fall in sea level of 50 m and an increase in bottom currents during the transition from MIS 3 to 2. The nodules continued to grow slowly throughout the LGM when sea level was relatively stable. Sea level rise early during deglaciation (17–16 ka) resulted in the complete drowning of the lower terraces. A more rapid rise between 14 and 15 ka caused the drowning of the upper terrace deposits and these were overlain in turn by deep-water *Halimeda* accumulations or oolites derived from shallow waters upslope.

Webster et al. (2004, 2008) also described 14 lowstand reefs from the Huon Gulf, Papua New Guinea, lying on the northern, rapidly subsiding margin of the Australian plate. The limestones recovered from the reef terraces include a variety of shallow-water coralgal assemblages identified by analogy with modern Indo-Pacific reefs. Higher energy communities were found at depths of 1,947, 2,121 and 2,393 m and provide ages of 344, 374 and 416 ka respectively. By contrast, lower-energy communities occur at depths of 1,113, 1,217, 1,612 and 823 m, with ages of 230, 250, 304, and 208 ka respectively. These reefs are typically multigenerational features composed of a series of terraces and benches. Such features reflect a complex history of growth, punctuated by drowning and minor backstepping in response to high-frequency climate and sea-level events. Radiometric dates indicate that the reefs have evolved over the last 500 ka as the Australian plate has subsided to more than 2,000 m depth. Data also suggest that the reefs not only drowned during major deglaciations, but also during abrupt and distinct climate change events such as interstadial–stadial transitions (Webster et al., 2008).

A similar scheme was established in the deep waters of Hawaii. Twelve reef terraces were identified and interpreted as deposited over the past 500–600 ka between MIS 5e/3 and MIS 15/14 (Webster et al., 2008). The deepest deposits, described by Jones (1995), lie at 1,650 m depth. Five sedimentary facies, including shallow coralgal reef, intermediate coralgal crust/nodule, coralline algal–foraminiferal crust, microbialites and hemi-pelagic/pelagic sediments are widespread and relate to a range of paleoenvironments from shallow reef to deep, open shelf. There are no seismic or drilling records describing the internal structures of any of these terraces. However, ROV observations of the MIS 6/7 reef face indicate the existence of several alternations of coral-dominated and algal-dominated communities. Further support for a complex stratigraphy comes from the MIS 10/11 reef sequence. ROV sampling showed at least six distinct reef units. These are typified by different *in situ* coral framework types that are separated by planar surfaces assumed to represent either thick coralline algal crusts and/or subaerial exposure surfaces.

On Mururoa (Buigues, 1997), ROV observations have revealed prominent terraces at 10, 20, 40, 55 and 65 m depth with a vertical wall between 110–20 and 200–230 m depth. Caves reported at 80–90 and 120–150 m depth. The terraces are regarded as the erosional tops of carbonate platforms developed during the Pliocene–Pleistocene but seem not to have been accurately dated, although the deeper caves have been referred to the LGM (>18 ka).

On Mayotte (Dullo et al., 1998; Camoin et al., 2004), direct observations by submersible have identified a number of coral-bearing terraces. Dates of 18–20 ka were obtained from reworked corals on the reef wall. Their presence is explained by the lack of accommodation space on the steep slope. Drowned reefs at 65–55 m depth have been identified locally. At their base these consist of shallow-water branching corals, but these are encrusted with thin veneers of deep-water, platy living forms. Samples consisting of shallow-water coralgal assemblages from a terrace at 110–115 m depth (uncorrected for isostatic sea level change and subsidence) provide a date of 17–18 ka. The cemented slope at depths ranging from 175 to 245 m depth is dated at 27.6–37.4 ka. The sediment also contains reworked shallow-water corals. Two karst systems have been identified, one at 150–155 m depth consists of caves up to 3 m diameter extending 2 m into the reef wall. The surfaces are covered in small-scale dissolution features but are overgrown by shallow-water corals, dated at 18.4–16.1 ka. A second karst horizon is present from 120–125 m depth. Caves here are up to 3 m deep and connected by solution pipes and channels.

The increasing refinement of chronologies of sea-level change, based on both the records of deep boreholes and sequences of raised limestones, has provided an accurate measure of the global pattern of subsidence first outlined by Darwin (1842) and referred to by Lyell (1853). Scott and Rotondo (1983a, 1983b) suggested that islands could be used as 'dipsticks' to map areas of upwelling and of lithospheric cooling. Subsidence is largely a reflection of cooling that occurs progressively as the plate moves towards its ultimate subduction, eventually reaching a point where the rate of submergence exceeds the rate at which coral growth is able to keep pace, thereafter the reef system is drowned. Grigg (1982) defined this appropriately as the 'Darwin Point'. Leaving aside the special cases of tectonic disturbance, the identification of regions of differential subsidence led to the realisation that in specific areas reef growth might or might not be able to keep pace with sea-level rise. The most spectacular 'give-up' example, the Darwin guyot north of the Marshall Islands, lies in 1,266 m of water but retains a rim with a central lagoon-like basin around 18 m deep. Ladd et al. (1974) were able to show that the rim here includes Cretaceous shallow-water rudists that imply subsidence for at least 60 Ma.

6.5.3. Case Studies from Uplifting Areas

The best documented example comes from the foreslopes of Barbados where Fairbanks (1989) described three distinct terraces at 10, 40 and about 70 m deep. The cored sequences range between 35 and about 70 m in thickness (see Chapter 3, Figure 3.9).

6.6. Reef Stratigraphy and Numerical Modelling

Computer simulations in which boundary conditions can be varied were initiated in the 1980s. Numerical modelling may help to better understand reef development in two ways. It provides a visual picture of the influence of the different controlling parameters on the stratigraphy and geometry of reef bodies and challenges empirical analysis and interpretation based on field observations or seismic records resulting from the application of conceptual models of reef growth.

Graus, Macintyre, and Herchenroder (1984, 1985) reconstructed the zonation and Holocene development of Caribbean reefs. Similar, attempts to assess the synergetic effects of sea-level fluctuations, subsidence, reef accretion rates, and marine, subaerial and fluvial erosion on reef island morphology and architecture were made by Paulay and McEdward (1990) using a computer-simulation model. Bice (1988) provided a two-dimensional forward model relating gross sediment production to eustatic changes in sea level superimposed on subsidence. Bosscher and Schlager (1992) attempted to model the effects of photosynthesis and light penetration related to water depth on coral growth. Model results were compared with the Alacran Reef, Mexico, two reefs on the Great Barrier Reef and reefs on the windward platform of St Croix. Modelling of the Alacran Reef (Isla Perez) showed generally good correspondence although the model suggested that growth did not begin until about 8 ka, later than the interpretation of Macintyre et al. (1977). The St Croix model also the reproduces drowning of the shelf-edge reefs as a result of the flow of turbid waters, and the continued growth of barrier and fringing reefs as described by Adey (1978). The growth curves simulated by the models seemed to provide acceptable comparisons with Bowl and Myrmidon reefs on the Great Barrier Reef described by Davies et al. (1985). Simulations of the Belize Reef margin also reproduced the general pattern of four consecutive downstepping reef increments that conformed to the hypothetical pattern suggested by James and Ginsburg (1979).

There have also been attempts to model reef terraces and variations in sea level (Turcotte & Bernthal, 1984). Assumptions regarding rates of uplift and harmonic variations in sea level with amplitude and period, but neglecting erosion, generate a sequence of terraces that compare reasonably well with raised terraces on the Huon Peninsula described by Chappell (1974). Interestingly, as the authors pointed out, a comparison by Bloom et al. (1974) of a sea-level curve inferred from oxygen isotope data superimposed on tectonic uplift can also generate a convincing terrace sequence.

During the last decade, an increasing number of studies using stratigraphic numerical modelling has emerged in attempts to quantify the major processes that drive reef growth. Most of these works focused on the growth history of Quaternary reef systems (Dalmasso et al., 2001; Warrlich, Waltham, & Bosence, 2002; Nakamura & Nakamori, 2007; Webster, Wallace, Clague, & Braga, 2007; Koelling et al., 2009). One- to three-dimensional growth models are able to simulate the internal structures of reefs in response to varying values of different parameters. Generally, the dominant controlling parameter is a user-definable sea-level curve. Other adjustable model parameters include the topography of the antecedent foundation, the maximum coral growth rate or carbonate production, depth-dependence and light attenuation, sediment transport and redeposition, subaerial and submarine erosion, and subsidence rate.

Dalmasso et al. (2001) compared a number of computing procedures used to simulate shallow-water carbonate platform development. It was argued that although geological data can provide accurate qualitative descriptions of the mechanisms involved in accretion, model simulations allow a quantitative evaluation of these and of the complex interactions between them. The principal comparison, between two models, CARBONATE and FUZZIM, reproduced the general stacking pattern and facies distribution of a Holocene reef section from Mauritius. However, the authors concluded that there was that there is no perfect system for modelling carbonate platforms.

Following works by Bosscher and Schlager (1992), Nakamura and Nakamori (2007) reconstructed the growth history of a fringing reef system over the last 5 ka using a 'geochemical' model, based on changes in diffusion-limited and length-enhanced calcification across the reef profile from the reef front shorewards.

Detailed numerical models of drowning mechanisms of atolls have been presented by Warrlich et al. (2002) and related to sequence stratigraphy. The coding used was an improved 3-D version of the CARBONATE programme. This was designed to simulate a variety of sedimentary processes and platform settings, including rimmed, shelved, isolated and attached platforms, carbonate ramps and mixed carbonate–siliciclastic systems. The results imply that although relative sea level is important, antecedent topography, environmental setting, and early diagenesis also have a profound influence. A distinction is made between sea-level behaviour during 'ice-house' and 'greenhouse' periods. Boundary parameters of the models include: carbonate production under several distinct scenarios (sediment dispersal, differentiating bed-load and suspended load, and distinctions between grains and matrix); subaerial dissolution; process-based facies; and eustatic sea-level change. Simulations were compared with an 'idealized' atoll described by Handford and Loucks (1993). Simulations reproduced offlap, transgressive tracts and aggrading highstand growth, the effects of rising sea level, moderated by greenhouse and ice-house variations, and those in which the controlling parameters reflect changes in rates of production or sea-level rise or fall. Successive scenarios were applied including drowning by accelerated sea-level rise, drowning by frequent subaerial exposure, and drowning by minimal environmental conditions. A rapid rise in sea level produced a bucket-shaped morphology that developed prior to drowning, since only the outer atoll rim can maintain pace with the rising sea level. Final drowning results in an abrupt transition from the atoll rim to deep, oceanic deposits. Drowning catalysed by reduced carbonate rim production, in response to unfavourable environmental constraints, was predicted to produce a partial filling of the lagoon prior to drowning. Eroded atoll rim sediments are transported preferentially lagoonwards. Episodic exposure of the atoll rim results in production shut down and subsequently in reduced sediment export to the lagoon. Repeated emergence and flooding of an atoll provide contrasting results. A model using an 80-m amplitude of 100-ka cycles does not predict drowning, but instead results in the development of a flat-topped mounded morphology. Increasing the subsidence rate from 0.1 to 0.3 mm yr^{-1} causes permanent drowning. However, the authors were careful to point out that there are limitations, not least because the rates of some processes are not known or the processes themselves are poorly understood. Fundamental errors in estimates of boundary conditions or failure to include driving factors will be exposed in model results.

Webster, Wallace, et al. (2007) applied numerical modelling to the evolution of two drowned lowstand Hawaiian reefs, today situated at 150 and 400 m depth and deposited during MIS 7/6 and MIS 5e/3. Comparing field observations and simulations allowed a reliable scenario to be constructed for the last two glacial cycles, that is the last 250 ka. Modelling of the internal reef stratigraphy suggests that the number and thickness of reef units, and the frequency and duration of emergence and drowning events, are primarily controlled by the frequency and amplitude of glacioeustatic sea-level fluctuations (Figure 6.14). Modelling parameters that best explain the field data include a subsidence rate of 2.5 mm yr^{-1}, vertical accretion rate of 2.5–2.85 mm yr^{-1} for the shallow-water coralgal assemblages and a variable topography for the antecedent foundations. The two reefs started to grow during MIS 7 and MIS 5e highstands respectively. Both experienced a long and complex growth history, developing periodically during about 90 ka. Precessional (23 ka) and higher frequency, suborbital sea-level oscillations mainly drove reef evolution and therefore explain the observed internal structure of each reef. The reefs have suffered repeated and rapid (<5–10 ka) subaerial exposure and drowning. These cycles produced a complex 'layer-cake' stratigraphy consisting of typical, shallow reef units separated by subaerial unconformities or deep-water (30–60 m) coralgal to open-shelf (>120 m) deposits. The final reef demise occurred during the penultimate and last deglaciations at about 133–134 and 12–14 ka respectively.

Using a spreadsheet-based reef growth model, Koelling et al. (2009) investigated the development of Pleistocene to Holocene reef systems in a variety of tectonic settings including subsiding islands and uplifting areas. The sea-level curve used for the model run covers the last 200 ka at orbital to suborbital time scales (Waelbroeck et al., 2002). The model allows the effects of a range of boundary conditions such as maximum reef growth, erosion rates, subsidence and uplift on both parameters, reef morphology and internal structure to be tested. Models may also help to identify favourable positions for boreholes in reef tracts that may provide critical information for reconstructing sea-level history. Modelling investigations of reef growth on the slowly subsiding (0.25 mm yr^{-1}) island of Tahiti over the last 200 ka showed the formation of a well-differentiated reef terrace at 35 m below present sea level, a drowned terrace at 65 m depth and a break in slope at 95 m depth (Figure 6.15), in part consistent with bathymetric surveys and drilling data from Tahiti fore-reef slopes (Cabioch, Camoin, et al., 1999; Camoin et al., 2006). Applied to the rapidly subsiding (2.5 mm yr^{-1}) island of Hawaiian, the model generated two major drowned reef terraces, at 140 m and 460 m depth, broadly in accordance with field data and with the numerical modelling shown in Figure 6.14. Modelling reef growth in a situation of rapid uplift such as that of the Huon Peninsula has proven to be more critical. There were pronounced discrepancies

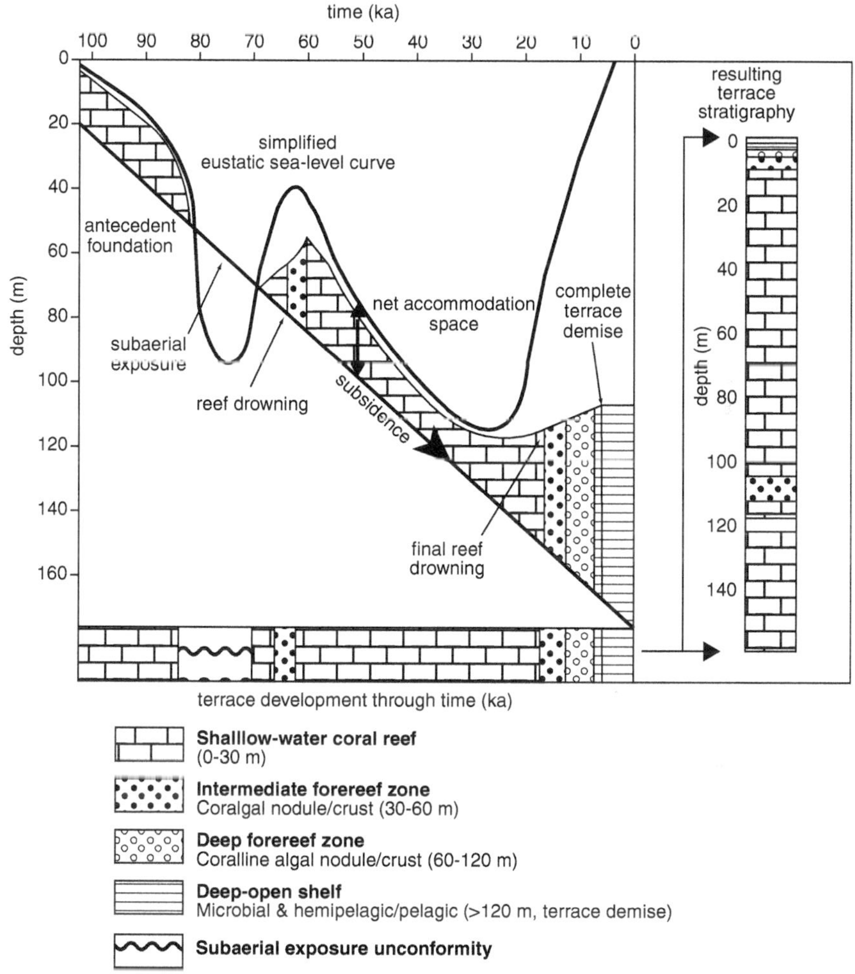

Figure 6.14 Hypothetical evolutionary model of a rapidly subsiding reef system over the last 100 ka. Rate of subsidence ranges from 2 to 3 mm yr^{-1}. The stratigraphy of the reef system presents a number of different depositional packages reflecting rapid sea-level oscillations. Shallow-water reef bodies are interbedded with deeper, fore-reef to open-shelf deposits and/or separated by subaerially exposed surfaces depending upon the rate and amplitude of changes in sea level. Modified and redrawn from Webster et al. (2007).

between the virtual models and field observations, depending on the range of uplift rates and frequency in sea-level fluctuations run in the models that proved particularly sensitive to sea-level forcing. Using an uplift rate of 1.9 mm yr^{-1}, an erosion rate of 0.1 mm yr^{-1}, an average vertical reef accretion of 10 mm yr^{-1} and the Waelbroeck's et al. (2002) sea-level curve as boundary conditions, the model is able to generate only five major

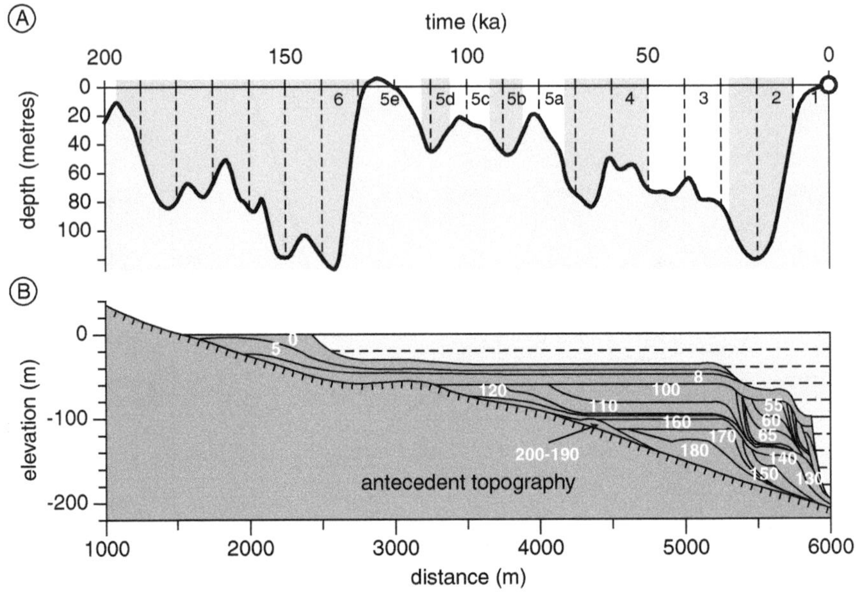

Figure 6.15 Numerical modelling of barrier reef growth on Tahiti, central Pacific, over the last 200 ka. (A) Eustatic sea-level curve inferred from oxygen isotopic chronology (Waelbroeck et al., 2002) with indication of MIS stages and (B) reconstruction of reef growth history using the following boundary conditions: sea-level curve in panel A; rate of subsidence (0.25 mm yr^{-1}); maximum vertical accretion rate (5.0 mm yr^{-1}); erosion rate (0.30 mm yr^{-1}); mean slope angle of the antecedent foundation (3.4°). The numbers within each depositional unit indicate the corresponding age in thousand years (ka). Adapted from Koelling et al. (2009).

terraces, ranging in age from the last deglaciation to MIS 6e uplifted by from 10–245 m above present sea level. The number of terraces predicted was significantly lower than that actually observed. Using a higher uplift rate (2.8 mm yr^{-1}) and higher frequency of sea-level changes (suborbital to millennial scales), the stratigraphic model reproduced almost all of the well-differentiated terraces of the Huon Peninsula reef succession in response to steeper gradients in sea-level rise and fall.

Numerical modelling of reef growth throughout the Quaternary may be improved if the isostasy corrections for different tropical latitudes taken into account. The relationship between water depth and reef accretion may be revisited. In particular, the depositional rates of deep fore-reef carbonate producers such as non-geniculate coralline algae, benthic foraminifers and the green alga *Halimeda* have to be incorporated in the models when fore-reef settings are gently dipping, thus promoting their preservation. Additional environmental factors usually regarded as little importance in controlling reef growth (mainly tidal sea-level fluctuations, and wind stress) may be also important determinants at very shallow depths.

6.7. Conclusions

Drilling operations, whether or not they are coupled with seismic surveys have revealed the composite nature of fringing reefs, barrier and shelf reefs, and atolls. The findings are consistent with data from studies on emerged reef terraces on tectonically active margins. On precessional (20–23 ka) and suborbital time scales, the occurrence and timing of reef growth events, broad-scale reef geometry, architecture and stratigraphy have been controlled by increasing paleo depths, associated with abrupt eustatic sea-level rise concurrent with decreasing light, decreasing sediment flux and increasing nutrients.

However, although no other sequences provide a higher level of discrimination of reef anatomy than exposed terraces, the stratigraphies that have been described typically relate to only a few depositional units. Local differences in uplift or subsidence, but also in patterns of erosion, mean that there is no single sequence that includes a representative record of all depositional patterns. Although it has been clear for some time that terraces can be generally correlated with Milankovich cycles, it has also become apparent that there are superimposed minor (suborbital) cycles. As emphasized by Montaggioni (2000) and Webster et al. (2007), drilling of reef structures in subsiding areas may provide greatly expanded stratigraphic sections compared with similar reefs in tectonically stable and uplifted regions. Thus, drowned coral reefs represent a unique archive of reef stratigraphy in response to sea-level and climate changes during the last six glacial cycles.

The timing and duration of Pleistocene reef growth episodes remain difficult to accurately estimate from reef stratigraphy for several reasons. First, the stratigraphic reef record is far from complete and, in the best cases, only represent some of the former interglacial stages. In addition, glacial low sea-level episodes are rarely preserved in exposures and rarely accessible by drilling. Second, the precise ages of individual reef units are difficult to acquire back to 400 ka, largely as a result of the instability of the mineral required for dating. Uranium-series dates have been successfully applied in some areas and also amino acid stratigraphy, magnetostratigraphy, ESR and strontium isotope measurement. These records, although by definition fragmented, may be compared with stable isotope data from deep-sea deposits. As would be expected, the preserved deposits are broadly correlated with periods of high sea-level stands during warm intervals, whereas gaps are interpreted to reflect lowstands during glacial periods. Significantly, in typical sequences, lowstands are not only represented by intervals of non-deposition, but also by periods of erosion that removed much or all of the preceding depositional increment and thus created some of the space for the subsequent deposition of the next increment.

The timing of reef deposition is therefore poorly constrained for the lower half of the mid-Pleistocene to early Pleistocene even if different dating methods are applied simultaneously to a same sequence. Third, as a result of the limited resolution of both isotopic and biostratigraphic dating, the duration of a sea stand during an interglacial or, still worse, a glacial stage may be miscalculated resulting in controversial interpretations of accretion rates and stratal thicknesses and subsequently of rates of sea-level oscillations and tectonic motions.

Several lines of evidence indicate that the depositional regimes of reefs have changed during the Quaternary. The timing of these changes may be equated with the timing of the MPT. From about 1.8 to 1.0–0.8 Ma, the mode of reef sedimentation was controlled by low-amplitude, high-frequency, 41-ka obliquity cycles. Over the past 1.0–0.8 Ma, reef development has dominantly responded to higher amplitude, lower frequency, 100-ka eccentricity cycles.

Assessing all of the parameters controlling reef development necessitates further examination of reef architecture and stratigraphy and a reappraisal of chronological frameworks. Numerical modelling has been shown to be a powerful tool to quantitatively assess the sensitivity of reef systems to sea-level changes and tectonic displacements. Because both the overall geometry of reefs and their internal age structure may be revealed for periods represented by oxygen isotope chronologies and related sea-level curves, predictive models may help in resolving temporal uncertainties over broad time scales. However, computer models have to be improved by incorporating additional variables such as wind stress, cyclonic effects, salinity and long-term changes in frequency and amplitude of climate cycles.

CHAPTER SEVEN

Reef Hydrogeology

7.1. Introduction

Water movements and water chemistry are not only critical to the initiation, growth and ecology of reef systems but also exert a commanding influence on chemical environments long after the reefs generated have been buried. The structure, palaeoecology and diagenesis of Quaternary reefs, as described in this volume, are direct reflections of varying flows throughout their entire formative cycle. Although, as will become obvious, there may be substantial overlap, the hydrogeology of reef systems can be divided into two main hydrological theatres: systems that control the ecology and growth of the living elements of the reef and their derived and dynamically distributed products, and those bearing on the subsequent diagenesis experienced by the reef assembly. The first have clear implications for the distribution of the varied bio- and lithofacies (and thus porosity) within the system and for the early beginnings of diagenesis. The second provide a record of the changing patterns of circulation related to variations in climate and sea level that are reflected in the character and distribution of cementation. Hearn, in his introduction to the special issue of *Coral Reefs* (2001) devoted to coral reef hydrodynamics, pointed to the wide range of scales of circulation that are now considered important, ranging from the hundreds of kilometres for ocean currents to the submillimetre scale of the diffusive boundary layer limiting nutrient uptake by plants. Here, we will examine water movements of different origin and at a range of scales for reef systems throughout the tropics, noting, however, that the Quaternary record of smaller scale movements can only be inferential.

From our present perspective, it is surprising to find that as recently as the 1940s, there was still speculation, following the subsidence-controlled theory of reef growth by Darwin (1842), that the cumulative growth of present reefs is simply a result of a large relative rise in sea level (Revelle, 1954). It is now clear that although subsidence is commonly, but not always, the cumulative result, additional glacioeustatic and hydroisostatic changes in sea level, discussed in Chapter 9, have superimposed a series of secondary oscillations. These Milankovitch cycles that show well-defined cyclicity of approximately 100 ka over the last 6–700 ka and 41 ka in the preceding 1.4 Ma provide the first-order controls on reef accumulation and subsequent diagenesis.

Second-order controls on reef growth and accumulation are a reflection of regional patterns of ocean currents. These fall into two groups, with dramatic contrasts between large barrier and shelf-reef systems like the Great Barrier Reef (GBR) and the Belize and Honduras Shelf, bordering continental margins, and the reefs and atolls of oceanic islands. From the perspective of Quaternary reefs, only regional scale circulation and gross environmental hydrodynamics are likely to generate a recognizable signature.

7.2. External Hydrology: Water Characteristics and Reef Responses to Waves and Currents

7.2.1. Sea Surface Temperatures

7.2.1.1. Temperatures and global limits to reef growth

Sea surface temperatures (SSTs) are a major determinant of the growth and distribution of modern coral reefs (Andrews & Gentien, 1982). The species richness of zooxanthellate corals correlates broadly with latitude and decreases with distance from the equator (Veron, 1995). Grigg (1982) showed that in the Pacific, rates of reef calcification also decline linearly as a function of latitude due to the decrease in mean SST towards the upper limits of the tropical belt. Although some reef-building corals are able to tolerate temperatures lower than 14–15°C or higher than 36°C for brief periods, most reef growth is limited to regions with SST between 18°C and 33–34°C for most of the year. About 50% of zooxanthellate species are able to occupy sites in which the average minimum temperature may be as low as 14°C, but these areas are unable to generate reefs. The net effect is that present coral reefs are broadly confined to the tropics, 23·5° north or south of the equator, but in detail, their world distribution is also strongly influenced by regional currents. These include the Somali Current, the Equatorial Current in Indonesia, the NE and NW Australian Currents, and the Caribbean Current. On the west coasts of South America and South Africa, cold Antarctic waters are driven north of the tropic and towards the equator by the Humboldt and Benguela Currents, limiting coral growth. By contrast, in west Australia, the cold north-flowing west Australian Current is deflecting away from the shore by the narrow (<100 km wide) Leeuwin Current, flowing south at speeds of 0.1–0.4 m s^{-1} (James et al., 1999). Hugging the shore, this has supported reef growth on Ningaloo reef and in the Houtman Abrolhos at 29°S since at least the mid-Pleistocene. Northern areas within this system are characterized by a high diversity of reef faunal and floral assemblages, but in winter (August) southern temperatures drop below 20°C, and biotas south of the Abrolhos are transitional to a cool-water association of coralline algae, molluscs and bryozoa (James et al., 1999). Off the eastern Australian coast, coral growth on Elisabeth and Middleton Atolls at about 29°S and Lord Howe Island at 31°S also owe their existence to a south-flowing warm current, the east Australian Current.

In the northern hemisphere, in Japan, the warm North Pacific Drift carries coral growth north of the corresponding tropic to Iki Island at 33°N (Yamano et al., 2001) and Miyake-jima at 34°N (Tribble & Randall, 1986). However, in this area, important periodical local temperature fluctuations result from the deflection of the Kuroshio Current and limit coral growth (Nadaoka et al., 2001). In the Atlantic Ocean, Bermuda, at 32°N is bathed in the warm waters of the Gulf Stream. Since Miocene times (see Chapter 2), two distinct coral provinces have been recognized, the Atlantic-Caribbean and the Indo-Pacific provinces, that reflect the relative hydrological isolation of these areas.

7.2.1.2. Intratropical temperature variations

Variations in water temperatures also influence coral distributions within the tropics. In northern areas of the Great Barrier Reef of Australia, SSTs average 24.2°C in winter (August) and 29.5°C in summer (December), within assumed limits for coral growth. By contrast, on the southernmost island consisting of coral, Lady Elliott Island (about 24°S), average winter temperatures fall below 20°C and there are no reefs further south within the GBR (Brandon, 1973). Lower temperatures in the area off Cooktown (latitudes 14–15°S) seem to be a reflection of a locally narrower shelf and relatively wide passages through the outer barrier, allowing the flow of a greater volume of Coral Sea water. Isern, McKenzie, and Feary (1996) showed that during the early Quaternary, temperatures in the western Coral Sea increased sharply from 22 to 23°C to modern values of 26–28°C. Similar increases were recorded in all ODP Leg 133 sites off NE Australia and appear to indicate the development of a warm water pool in the Coral Sea. It would be tempting to see in this the time of initiation of the central GBR, but because the dating of the latter is uncertain (Montaggioni & Venec-Peyré, 1993), it is not clear whether these events coincided. The north to south temperature gradient within the GBR is reflected in major variations in the percentages of coral cover and the relative abilities of the corals to recover from the results of disturbance.

Similar mesoscale thermally limited distributions are seen elsewhere. In the northern Red Sea, south of Suez, the lowest temperatures occur in February (winter) and average 17.4°C, increasing southwards to reach 22°C in the Red Sea proper (Edwards, 1987). However, cold northerly winds in the north may also drive overturning, with the result that surface waters are cooler and there is a wider fluctuation in mean values. In summer, northern areas experience a maximum SST of 26°C, increasing towards the Straits of Bab al Mandab to >32°C. On shallow reef flats in the Sudan, at low tide, corals locally apparently survive temperatures that briefly approach 40°C. This is significant because experiments by Berkelmans and Willis (1999) on the central GBR suggest that for some corals, at least, temperatures of this order are close to the 5-day 50% bleaching threshold, thought to reflect the upper thermal limit of growth. In areas subjected to heat stress, symbiont zooxanthellae suffer a reduction in photosynthesis due to depleted pigment concentrations, but under severe stress, they are lost from corals (coral bleaching). The corals can recover, but if the temperatures remain high, or if events are repeated too frequently, they die.

In the Caribbean, Smith (2001) described how both weather and hydrographic conditions combined to promote a thermal disturbance that resulted in a coral-bleaching event on Lee Stocking Island, Exuma Cays in the Bahamas in August of 1990. Although weather conditions at the time were commonly cloudy, waters on the shallow shelf had warmed over several weeks, apparently as a result of a reduction in evaporation and thus of evaporative cooling. Waters of about 31°C flowed off the shallow Banks, mixing with those around the Exuma reefs to reach temperatures of approximately 30°C.

Temperature differentials have been shown to act as drivers for currents within reef systems. Monismith, Genin, Reidenbach, Yahel, and Koseff (2006) showed that at the northern end of the Gulf of Eilat, shallow nearshore regions experience large temperature changes relative to deeper areas offshore subject to the same rate of heating or cooling. Where nearshore conditions are warmer, flow is offshore at the surface and onshore at depth, producing upwelling, whereas where they are cooler, the pattern is reversed. Both heating and cooling are observed during the summer and are reflected in a diurnally reversing exchange flow, but in the winter, there is little heating with the result that flow at the surface is predominantly onshore.

7.2.1.3. Historical changes in temperature limits

Reef coral populations respond to regional thermal changes by latitudinal shifts in diversity and abundance, and of geographical range boundaries (Precht & Aronson, 2004). There is good evidence that thermal controls on coral growth were important throughout the Holocene. During cold-water intervals, SST locally decreased to the currently accepted minimum of 18°C for reef growth, resulting in marked changes in the structure of coral communities (Figure 7.1). These included lower coral cover, lower generic diversity and smaller colony sizes relative to those of the present (Abram, Webster, Davies, & Dullo, 2001). By contrast, Veron (1992b) observed that

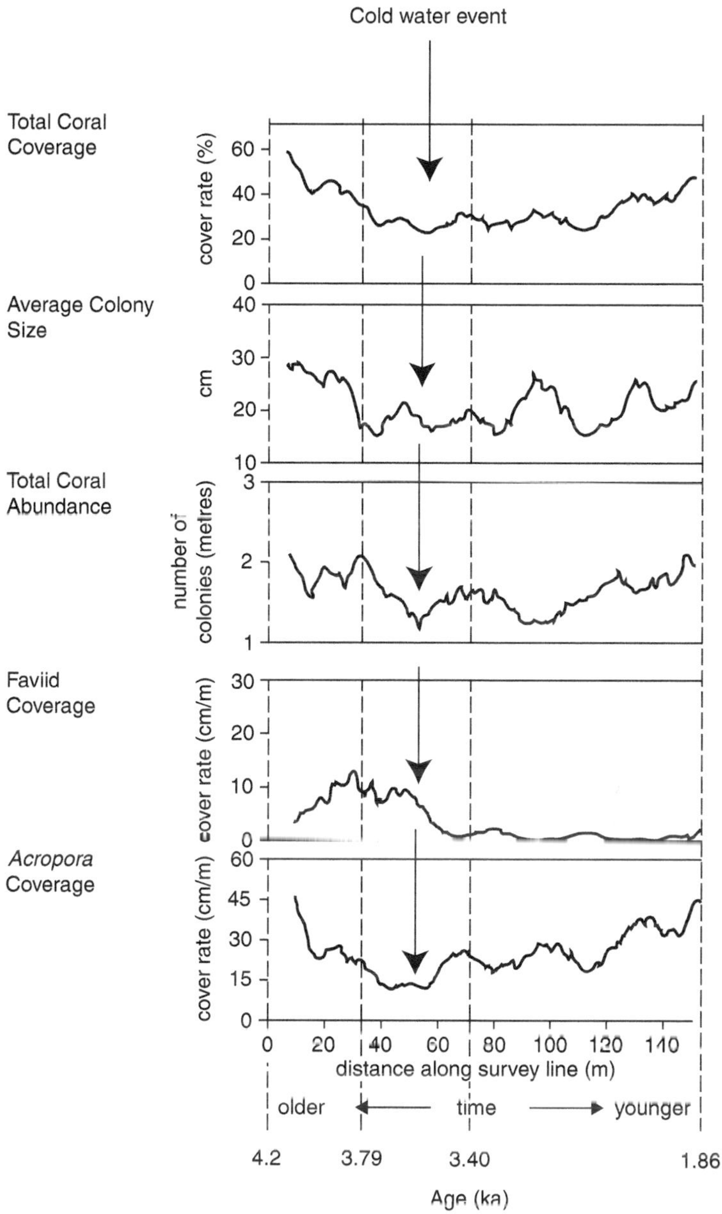

Figure 7.1 Relationship between changes in SST (recorded from the oxygen isotopic composition of domal *Porites*) and biological attributes of exposed Holocene reefs at Kadon (Kikai-jima Island, Ryukyus, Japan). The biological attributes (total coral coverage, average size of coral colonies, total coral abundance, coverage by faviids and *Acropora*) were measured using the line intercept transect method along a 150-m survey line across the Holocene reef terraces. The full curves represent a 10-m average values of the measured data. The age of these reefs ranges from 4.2 to 1.8 ka, from the mainland seawards. The cold water event that occurred at around 3.7 ka resulted in a drop in temperature below the currently accepted 18°C minimum for reef growth, and a subsequent disturbance in coral coverage, abundance and colony size. Modified and redrawn from Abram et al. (2001).

during the Holocene, an increase in SST of less than 2°C in the subtropical seas of Japan (at around 35°N) allowed the development of a 'high-latitude' coral community in a region that today is almost devoid of corals. Similarly, along the east coast of Florida (Precht & Aronson, 2004), warming during the Holocene climatic optimum (10–6 ka) enabled acroporid communities to expand further north. Greenstein and Pandolfi (2008) examined species distribution along the west Australian coast compared with that of Pleistocene species in the same area, concluding that there has been a significant range shift with the present distribution reflecting a retreat to the north. High-amplitude climate changes on a millennial scale, identified in the late Pleistocene 'Dansgaard–Oeschger cycles', culminating in cold episodes, have influenced reef development and distribution but the evidence is not always clear. In uplifted areas like Papua New Guinea, these cycles promoted the deposition of distinct reef terraces during brief warming phases (Chappell, 2002). Although they have not yet been identified in sequences older than late Pleistocene, such cycles may also have played a role in the sudden collapse and expansion of coral reefs during earlier glacial cycles.

The nature of the thermal response of tropical seas to Pleistocene glacial and deglacial episodes is still debated, and the records of tropical SSTs during the glacial maxima are particularly controversial. The contrast between the SST of the Last Glacial Maximum (24–19 ka) and the present is variously estimated to have ranged from 1.5°C to around 5°C depending on the proxies used (Colinvaux et al., 1996; Beck, Récy, Taylor, Edwards, & Cabioch, 1997; Songoni, Bard, & Rostek, 1998; Guilderson, Fairbanks, & Rubenstone, 2001; Gagan et al., 2000, 2004; Powers et al., 2005). For the last interglacial episode (120–125 ka), the surface temperatures of reef waters are relatively well constrained. In Western Australia and Papua New Guinea, the mean annual SST and seasonality were similar to modern patterns, with a summer maximum of 27–29°C. The only difference was in a winter minimum of 21°C, about 1°C cooler than the modern minimum (McCulloch & Esat, 2000). The eastern Pacific Ocean was as warm or warmer than at present at 120 ka (Seki, Ishiwatari, & Matsumoto, 2002; Muhs, 2006). Although the SST in the northern Red Sea was comparable to present-day values during the last interglacial, seasonal variation was significantly greater, 8.4°C at 122 ka, compared to the present-day average of 5°C (Felis et al., 2004).

Some records are available over longer periods. During the 80-ka interglacial episode, the SSTs of tropical eastern Pacific waters appear to have been markedly lower than at present, perhaps as a result of cooling by the flow of subartic waters along the western U.S. coast (Muhs, Simmons, Kennedy, Ludwig, & Groves, 2006). Kilbourne, Quinn, Taylor, Delcroix, & Gouriou (2004) demonstrated that during the glacial stage at about 350 ka, the SST in the western Pacific was 3.5°C lower than at present.

However, data indicate that in the western Coral Sea, SSTs have changed by only about 1.5°C or less in the past 800 ka (Lawrence & Herbert, 2005). Reconstructions of SST for the past 1.25 Ma, based on marine sediments from the eastern equatorial Pacific, partly support these findings, with glacial–interglacial temperature variability ranging from about 1°C to 4.5°C and SST estimated to have been about 1°C higher than average late Pleistocene temperatures (Liu & Herbert, 2004). Cooling of about 5°C would take the temperature below the survival limit of most reef-building coral communities (Crowley, 2000b), but reef growth potential was probably not diminished, and reef systems, now submerged, developed throughout the tropics during the Last Glacial Maximum (Montaggioni, 2005). During most Pleistocene interglacial intervals, the areal distribution of coral reefs may have been little different from that of the present day.

To place these findings in a wider context, Isern et al. (1996) were able to derive a detailed sea surface palaeotemperature curve for the entire period from the middle Miocene to the Holocene, using data from ODP site 811 on the Queensland Plateau. Throughout this period, temperatures in this area were on average greater than the 20°C required for reef growth. However, although growth was active in the early Miocene, by the late Miocene and early Pliocene, there were repeated intervals with temperatures between 18 and 20°C. After about 13 Ma, growth declined as temperatures decreased and parts of the plateau may have been emergent at 11 and 7 Ma. Reduced reef growth rates continued until the late Pliocene or Pleistocene even though SSTs were 22–23°C. At these levels, corals are stressed and growth rates reduced so that it is difficult for reef growth to keep pace with subsidence. It was not until 600–700 ka that temperatures rose above this limit as a result of the formation of the western Coral Sea warm pool (Isern et al., 1996). Braithwaite et al. (2004) regarded this period, from 600 to 700 ka, as corresponding with the time of establishment of the present GBR.

7.2.2. Water Quality and Nutrients

7.2.2.1. The modern record

Coral reefs have generally been regarded as nutrient-poor areas. In areas experiencing high nutrient concentrations (nitrogen $>2.0\,\mu mol\,l^{-1}$ and phosphate $\geq 0.2\,\mu mol\,l^{-1}$), the growth of macroalgae that compete with reef corals for space is enhanced and bioerosion is intensified (Hallock, 1988; Chazottes et al., 2002; Sanders & Baron-Szabo 2005) (Figure 7.2). However, substantial carbonate production by reef corals can persist under mesotrophic to eutrophic conditions (Atkinson, Carlson, & Crow, 1995), even in high-latitude, non-reef, coral communities (Halfar, Godinez-Orta, Riegl, Valdez-Holguin, & Borges, 2005).

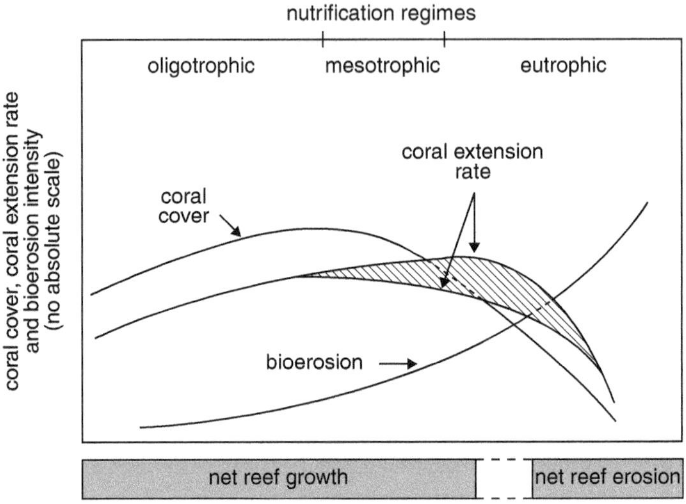

Figure 7.2 Controls of nutrification levels on reef growth characteristics (live coral cover, rates of coral extension, trend in intensity of bioerosion, and regime of net reef growth and net reef erosion). Note nutrification levels are regarded as related to both trophic regime and water turbidity. The cross-hatched area refers to variability in extension rates. Modified and redrawn from Sanders and Baron-Szabo (2005).

It has long been noted that regions of prolific growth on the GBR correspond to regions of increasing steepness in the continental slope (Maxwell & Swinchatt, 1970). From this and similar observations elsewhere, it has been argued that upwelling of deep water with an attendant renewal of nutrients is important to reef ecosystems, although the proportion of nutrients at any one time may not be large. Studies by Andrews and Gentien (1982) in the area of Myrmidon Reef, north of Townsville, provide evidence of the occurrence of upwelling. There are several different mechanisms including tides, island wakes and eddies and vortices that spin up in the shear zone where flow separates from the reef edge (Burrage, Steinberg, Skirving, & Kleypas, 1996). There are periodic (approximately 90 days) episodes of intensification of the east Australian Current, and measured oxygen and nitrate data indicate the contribution of upwelling during these intervals. However, there are also periods of downwelling with an exchange of outer shelf and slope waters. Subtropical lower water, at a depth of about 150 m, is characterized by a salinity of 35.7 psu and a mean temperature of 21.5°C and extends throughout the western South Pacific. During summer, upwelling isopleths slope strongly upwards and these waters intrude onto the shelf, producing density stratification. The nutrient content of upwelling waters is revealed in an increase in chlorophyll content, reflecting the growth of phytoplankton. By contrast, in winter, cool oxygenated waters cascade off the reef and down the slope. It seems that major current activity does not encourage nutrient-rich waters to penetrate far onto the shelf. However, there is evidence to the contrary. Andrews and Gentien (1982) showed that bottom intrusions of cool waters may extend across the entire width of the shelf. These surges of the order of $0.1 \pm 0.05\,\mathrm{m\,s^{-1}}$ are driven by onshore winds including both the NE monsoon and the SE trade winds. Large upwelling events that can occur as often as every summer, and represent an intrusion of waters from the Coral Sea, may displace as much as 33% of the water on the outer shelf and import two to six times more phosphate and nitrate that are typically present (Furnas & Mitchell, 1996). These are, however, to some degree, balanced by the export of nutrients in the waters that are displaced. Similar upwelling is reported from the Yucatan Peninsula (Furnas & Smayda, 1987).

The GBR illustrates one way in which upwelling might work. However, Kinsey and Davies (1979) showed that far from promoting reef growth, some nutrient increases might actually be detrimental. Their study of One Tree Reef suggested that the enrichment of waters in nitrogen and phosphate (to about 2 and $0.2\,\mu\mathrm{mol\,l^{-1}}$ respectively) would have adverse effects, and phosphate levels might be expected to cause at least a 50% suppression of calcification of corals and other organisms. There is, however, some evidence that increases in nitrogen may encourage calcification (Crossland & Barnes, 1974 in Kinsey & Davies, 1979), but Marubini and Davies (1996) subsequently showed that although increased nitrate levels do increase the density of zooxanthellate populations in corals, they simultaneously reduce rates of skeletogenesis. Montaggioni, Le Cornec, Corrège, and Cabioch (2006) used Sr/Ca and Ba/Ca ratios as proxies for palaeothermometry and nutrient load in order to track upwelling activity in New Caledonia over the past 6 ka. As might be expected, data from modern *Porites* reflect upwelling in response to the southeast trade winds that dominate summer periods. Mid-Holocene corals reflect an SST little different from that at present (+1°C), but upwelling activity was significantly greater, probably in response to increased wind activity. This is thought to have been responsible for the relatively late establishment of fringing reef growth along the southwest coast of New Caledonia at about 5.5 ka.

Relatively few data reflect the fate of the organic carbon generated within reef systems (Schrimm et al., 2002). Some is certainly trapped within the reef framework (Hubbard et al., 1990) and a proportion is released as carbon dioxide from reef waters (Gattuso, Frankignoulle, Smith, Ware, & Wollast, 1996), but there is also a net export to the open ocean. Schrimm et al. (2002) monitored seasonal variations in this flux in front of a reef pass in Moorea, French Polynesia. Although the net output of carbonate sediment was of the order of $60–100\,\mathrm{g\,m^{-2}\,d^{-1}}$, the contribution of organic carbon was only a few grams per square metre per day but was marginally higher during the summer season. The overall control on flux lies with local hydrodynamic conditions and their influence on the residence time and outward flow of water from the lagoon.

7.2.2.2. The Holocene–Pleistocene record

The settlement of coral communities and reef expansion or reduction during the last glacial cycle is assumed to have been governed, to a large extent, by nutrification rates (Buddemeier & Hopley, 1988; Marshall, 1988; Montaggioni, 1988b; Hopley et al., 1997). Changes in nutrient levels are also suspected to have been important to reef coral evolution and turnover during the Cenozoic (Hallock & Schlager, 1986; Kauffman & Fagerström, 1993; Wood, 1993; see Chapter 2).

During the Last Glacial Maximum, productivity in some ocean basins probably increased substantially, reflecting shallowing of the thermocline and increased upwelling and nitrate supply to surface waters (Muller & Opdyke, 2000). The western Equatorial Pacific, in particular, experienced increases in nitrate levels and productivity that were 1.5–2 times greater than at present. During the last deglaciation, the rapid rise in sea level resulted in a gradual reorganization of ocean circulation and a significant enhancement of large-scale upwelling activity. In many tropical regions, this led to the nutrient enrichment of waters of shelves and coastal areas (Marshall, 1988). In areas where the transfer of nutrients from southern high latitudes to the equator was disturbed during the last full glacial, the re-establishment of the modern thermocline regime occurred early during the deglaciation (Loubere, 2001). Additional indirect evidence of the influence of upwelling on tropical shelves during the past 14 ka is provided by the deposition of laminar microbialites on foreslopes and within incipient reef coral frameworks. These appeared during phases of rapid sea-level rise and are interpreted as reflecting substantial increases in nutrient levels (Brachert & Dullo, 1991; Montaggioni & Camoin, 1993; Camoin et al., 1999, 2006; Cabioch et al., 2006). Upwelling and nutrient input have also been invoked to explain the presence of expanded *Halimeda* bioherms found at depths of 20–90 m on a number of tropical shelf margins (Roberts & Macintyre, 1988; Hopley et al., 2007, pp. 183–190). Similar conditions were probably present during the late Pleistocene (about 120 ka) because the basement rocks beneath many Holocene reefs in the northern GBR consist of similar dense, lightly cemented, *Halimeda* accumulations (Hopley et al., 2007, p. 403).

7.2.3. Salinity

7.2.3.1. The modern record

Corals are generally regarded as limited to areas of 'normal' marine salinity (35–36 psu), but are relatively tolerant to variations. They typically grow within a range from 30 to 38 psu, but some species can survive at levels as high as 46 psu. In the tropics, sea surface salinity (SSS) is believed to be governed by a combination of atmospheric convection, evaporation and ocean dynamics; significant variability in SSS is today associated with El Niño/Southern Oscillation (ENSO) cycles.

On the GBR, salinity varies seasonally, with lower values during February to May. The maximum salinity in northern regions, probably as a result of evaporation, is about 36.15 psu and the minimum is about 33.6 psu. The shelf between 16° and 18° shows a much greater range. Rainfall in this area may total 3800 mm, most falling between January and March, and minimum salinity averages drop below 32 psu. South of latitude 20°S variations in salinity decrease, with averages between 35 and 36 psu, reaching their lowest values in April (Brandon, 1973). In spring and early summer, when salinities reach their maximum, there is commonly a well-defined gradient with the highest values near shore, decreasing seawards. However, in northern regions, at the beginning of the NW monsoon, there is a reversal with a gradual decline in values until low-salinity waters near shore establish a positive gradient offshore.

In the Red Sea, salinity at the southern end is close to that of the Gulf of Aden (about 36.5 psu), but towards the end of the summer, outflowing surface waters reach salinities of 38–39 psu. Paradoxically, reflecting their lower density as a result of higher temperatures, these float on top of the inflowing Gulf of Aden water. High rates of evaporation in northern areas as in the Gulf of Suez result in salinity that typically exceeds 40 psu. Before the construction of the Suez Canal, the Great Bitter Lakes had a salinity of at least 70 psu (Edwards, 1987) and would thus have formed a significant barrier to coral migration between the Red Sea and the Mediterranean, even assuming some interconnection at higher sea levels.

In the Persian Gulf, the limited water interchange through the narrow Straits of Hormuz mean that salinities in the central parts of the Gulf average 37–40 psu and on shallow parts of the Trucial coast may exceed 60 psu, again reflecting high rates of evaporation (Purser & Siebold, 1973; Mubarak & Kubryakov, 2000). Over wide areas, salinity increases with depth as dense, more saline, inshore waters sink downslope and flow outwards below less dense surface layers (Figure 7.3). Thus, the salinity window available for coral growth is narrow and some organisms, notably *Halimeda*, are excluded.

Although the Amazon and Orinoco rivers are major contributors to the north-flowing Equatorial Current that forms the bulk of the Caribbean Current, they seem to have little effect on the characteristics of water in the Caribbean. Salinities in the well-mixed surface layer range from 35.7 to 36.1 psu, whereas temperatures to a depth of around 50 m are between 27 and 28°C. Caribbean deep water is substantially cooler (10–8°C) but is only found below about 200 m (James & Ginsburg, 1979a). During the winter (October to January), strong northerly winds may bring brief cold spells with heavy rain, and in southern areas, the annual rainfall is up to 700 mm yr^{-1}.

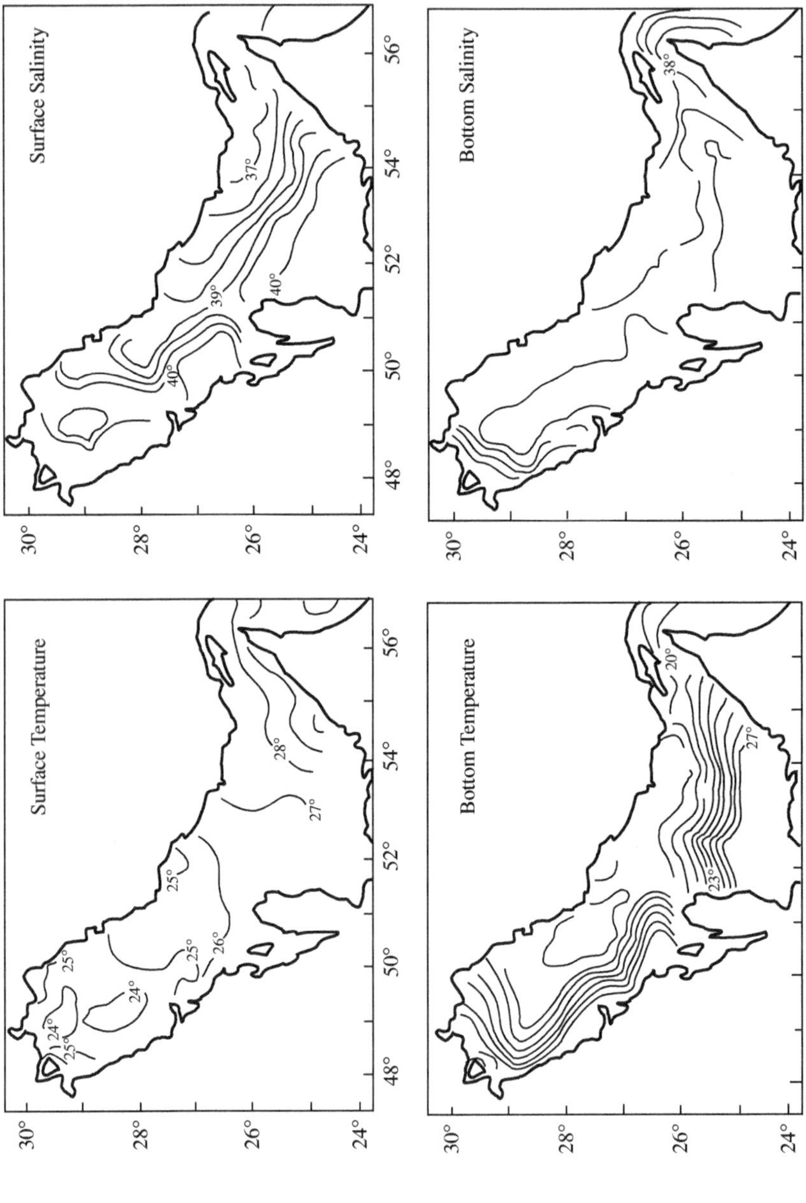

Figure 7.3 Summer temperature and salinity on the surface and on the bottom of the Persian Gulf (modified and redrawn from Mubarak & Kubryakov, 2000).

7.2.3.2. The Holocene–Pleistocene record

Little is known of the variability of tropical SSS during the Pleistocene. Brown (1997b) considered that the salinity ranges during glacial–interglacial intervals in the last 150 ka were comparable to those during the last glacial–Holocene interval, and reflected a freshening of surface waters during interglacials. By contrast, Kilbourne, Quinn, Taylor, Delcroix, et al. (2004), using oxygen isotope and Sr/Ca records from a 350-year-old coral from Vanuatu, showed that during the same lowstand glacial episode, surface waters in the southwestern Pacific were probably up to 2 psu fresher than at present. Assuming that the palaeosalinity anomaly has not exceeded ±2–4 psu relative to that at present since the middle Pleistocene, SSS changes cannot be thought to have been a limiting factor in coral and reef growth at a regional scale.

7.2.4. Water Turbidity

7.2.4.1. The modern record

Sediment transport during both storms and fair-weather periods plays a significant role in water quality. High sediment input (up to 100 mg l^{-1}) is detrimental to the growth of reef framework (Dodge, Aller, & Thomson, 1974; Cortés & Risk, 1985; Rogers, 1990; Fabricius & Wolansky, 2000; Sanders & Baron-Szabo, 2005) and results in changes to reef zonation (Acevedo, Morelock, & Olivieri, 1989). In part, these reflect a decrease in light, critical to coral growth (Bosscher & Schlager, 1992). However, as demonstrated in nearshore environments on the Australian Great Barrier Reef (Woolfe & Larcombe, 1999; Larcombe, Costen, & Woolfe, 2001) and in Jamaica (Mallela & Perry, 2007), coral communities can safely live within a wide range of turbidity and light levels, and form sites of relatively high diversity and coral cover under highly turbid conditions (Figure 7.4A).

Sediment loading exercises an important control on coral distribution and reef development and greatly influences the presence or absence of coral species, growth forms and growth rates (Anthony & Connolly, 2004). Corals vary in their ability to trap and remove sediment. A number of studies have investigated the response of individual corals. Among these, Stafford-Smith and Ormond (1992) examined the responses of 42 species with wide Indo-Pacific distributions. Major, fast-growing framework builders such as branching *Acropora* and *Pocillopora* are usually weakly tolerant of chronic turbidity and tend to be replaced by siltation-resistant species dominated by slow-growing, domal, foliaceous and encrusting forms (Kan, Hori, Nakashima, & Ichikawa, 1995; Kleypas, 1996; Van Woesik & Done, 1997) and solitary species. The tolerance of corals to sediment loading is chiefly defined by calice width. Generally, corals with small polyps are more susceptible to high sediment input than those with

Figure 7.4 Coral communities adapted to high-turbidity environments, fringing reef at Magnetic Island, Australian Great Barrier Reef (photographs by L. Montaggioni). (A) General view of the microatoll-dominated coral community at low tide. (B) Close-up of a microatoll (*Platygyra sinensis*) partly smothered with mud. The height of the colony above the water level is about 20 cm.

large ones. There is a positive correlation with the sizes of calices. All species with calices of >10 mm diameter are able to reject loads of up to 50 mg cm^{-2} (Figure 7.4B). Those with calices of less than 2.5 mm diameter are poor at rejection and intermediate forms (2.5–10 mm) have more variable abilities. However, *Porites* with small polyps can be very effective in removing clays and silts (Bak & Elgershuizen, 1976) and may dominate shallow, turbid-water communities (Tudhope & Scoffin, 1994). Where stress is prolonged, the coral assembly may be dominated by more active species with meandrinoid calices (Dodge & Vaisnys, 1977). Thus, many corals are able to survive and indeed thrive in waters of moderate to high turbidity. However, the ability of individual corals and assemblages to survive coating by sediment depends not only on their direct action but also on active water motion. Branching *Acropora* may not recover well where it is dependent on a self-clearing mechanism but will clear rapidly and continue to grow in a hydrodynamically active environment that aids removal. In general, stressed environments subject to heavy sedimentation show progressively fewer species, less live cover, and a greater abundance of branching forms (Rogers, 1990).

Heterotrophy is favoured under persistent high turbidity, while autotrophy is more appropriate to coral growth under lower turbidity pressure (Anthony & Fabricius, 2000). The phenotypic plasticity of coral colonies provides a major means of regulating light capture during growth, resulting in near-optimal internal irradiance across a large spectrum of environmental light conditions (Anthony, Hoogenboom, & Connolly, 2005). The minimum irradiance limit for reef development averages 6–8% of surface light (Cooper, Uthicke, Humphrey, & Fabricius, 2007). As light intensity declines, depth-related coral successions become more compressed and the depth at which reefs disappear decreases (Hallock & Schlager, 1986). On large continental shelves such as the GBR, although reef corals can grow at depths as great as about 100 m in outer-shelf settings, in mid-shelf sites, the lower limit of light saturation compatible with growth is about 20 m and is less than 3–4 m in inshore areas (Hopley, 1994; Van Woesik & Done, 1997). By contrast, on mid-ocean atolls, the minimum irradiance for coral growth extends to 150–160 m.

Increased terrestrial runoff has been a cause for concern in the Abrolhos reefs off the coast of Brazil (Segal et al., 2008). Although the reefs cover an area of some 6000 km^2, the coral community is characterized by relatively low coverage and high endemism. Sedimentation rates in nearshore reefs reach 225 $mg\,cm^{-2}\,d^{-1}$ in the winter, but detailed monitoring has shown that although rivers are important, wind-driven resedimentation due to Polar Front activity, characterized by intense surface winds reaching peaks of 8 $m\,s^{-1}$, has by far the greater influence (Segal et al., 2008). In Kenya, reefs have been in decline since the 1970s, apparently in response to sediment loading (Van Katwijk et al., 1993). This is greatest in the north where reefs in the Malindi–Watamu area lie south of the Sabaki river that drains a watershed of approximately 40,000 km^2 with a discharge of 5000 $m^3\,s^{-1}$ during the wet season. Sediment flux has risen steadily since the 1960s and is currently between 7.5 and 14.3 $Mt\,yr^{-1}$.

7.2.4.2. The Holocene–Pleistocene record

Little is known of the turbidity of waters to which early Holocene or Pleistocene reef systems might have been exposed. Cores through Holocene reefs in nearshore turbid zones of the GBR reveal an internal structure that typically lacks a contiguous framework. Corals are scattered and embedded in mud or muddy sands (Smithers & Larcombe, 2003; Smithers et al., 2006) forming 'coral marls' in the sense of Sanders and Baron-Szabo (2005), that is, muddy to sandy units containing potentially reef-building corals. Kleypas (1996) suggested that the demise of 8–6 ka inner reefs in the southern Great Barrier resulted, at least in part, from an enhanced siltation stress. This was caused by an increase in their relative confinement following the catch-up growth of reefs on the outer shelf to reach sea level. In the Indo-Pacific Warm Pool, the termination of late Pleistocene reefs may be explained by a substantial increase in rainfall as monsoonal conditions were re-established at about 14–13 ka, resulting in an increased supply of terrigenous silts from rivers (De Deckker et al., 2002). Differences in the structure of coral populations in the various environments of the late Pleistocene (the Ironshore Formation) reefs on Grand Cayman (Hunter & Jones, 1996) probably reflect the relative abilities of corals to reject sediment. The dominance of fast-growing, branching *Acropora* species on the reef tract, and their virtual absence from lagoonal patch reefs, is interpreted as driven by the turbidity levels of ambient waters.

On a more general level, Potts and Jacobs (2002) suggested that during the Pliocene and early Pleistocene, as a result of sea-level oscillations of 20–30 m with a period of about 10 ka, water depths over shelves rarely exceeded 20–50 m. During low sea-level stands, large areas of tropical shelves were probably bounded landwards by extensive flat plains with low-lying wetlands, mangroves and seagrass beds. In their view, terrigenous sediment and nutrient inputs were likely to be high during low sea-level stands, and shallow shelf and inshore sediments may have been frequently resuspended, limiting reef growth. However, it has also been argued (Alexander, 1996; Braithwaite et al., 2004) that results from ODP borehole 823A indicate that during low sea-level stands, much of this sediment would have been trapped in the wetlands and mangroves, leaving the reefs in clearer water. Increases in terrigenous material are correlated with reworking during high sea-level stands.

7.2.5. Hydrodynamics: The Effects of Tides, Currents, Waves, Tropical Storms and Tsunamis

7.2.5.1. Tides and regional currents

Hydrodynamics including tidal currents and wave energy exert the primary controls on the ecological zonation of shallow-water reef crests (See Chapter 3, section 3.2).

Case studies from the Indo-Pacific. The effects of wave and tidal-controlled flows on the sediment flux across different reef types are relatively well documented (Nelson, 1996; Storlazzi, Ogston, Bothner, Field, & Presto, 2004; Kench & Brander, 2006; Presto, Ogston, Storlazzi, & Field, 2006; and references therein). The Australian Great Barrier Reef has been used as the outstanding model of a continental platform-margin reef, particularly in relation to water circulation. Winds and tides are regarded as the main mechanisms driving water movements (Wolanski, 2001; Hopley et al., 2007; Lambrechts et al., 2008) (Figure 7.5). In winter, southeasterly winds that rarely exceed 15 m s^{-1} dominate, and in summer northerly winds average only 7.5 m s^{-1}. In the north, mean annual rainfall is 1100–2100 mm, but to the south, it decreases to around 1000–1200 mm. Intense rainfall events contribute significantly to these high totals and bursts of 800 mm in a day are a potential source of stress to reef corals when they correspond with low tides (Hopley et al., 2007). A mixture of diurnal and semi-diurnal tides is controlled by two amphidromic nodes, one near New Zealand and the other in the Coral Sea. Tides are amplified as they cross the shelf, and the range is 2.3 m near the southern margin, 10.3 m between latitudes 21° and 22°S (Broad Sound) and an average of 3 m in the remainder of the province (Maxwell, 1968). The eastward extension of the Barrier Reef shelf in the Broad Sound region accentuates the effect as the shallow shelf retards the tidal flow, raising water levels long before the tide reaches narrower shelf areas (Maxwell, 1968). Tides exert an important control on the extent and duration of exposure of the reefs during low water, and the semi-diurnal flow on and off the Platform on which the GBR rests entrains approximately 15% of the total water volume accommodated by the shelf. The net flow associated with tides provides a significant transport system for both larval distribution and sediment transport, together with a pump for subsurface flow of water through the reef margin (see below). Large-scale circulation is dominated by the south-flowing waters of the East Australian Current, trapped against the coast in near-surface waters above the permanent thermocline. Meanders and mesoscale eddies are formed as the current crosses the Tasman Sea, with protuding mushroom jets off Cairns and Townsville (Burrage et al., 1996). Modelling by King and Wolanski (1992) confirmed the presence of a south-flowing current on the mid- and outer shelf, with a wind-driven north-flowing current on the inner shelf and along the shoreline. Reef density has a significant effect and where reefs are numerous, currents are slowed and plankton (including larvae) is retained in contact with reefs for longer periods. In areas of high reef density, tidal currents may be 3–4 m s^{-1}.

At 450 km long and with lagoons covering an area of about 24,000 km^2, the New Caledonian barrier reef is the second largest in the world. The islands lie within a large-scale anticyclonic gyre centred at about 15°S.

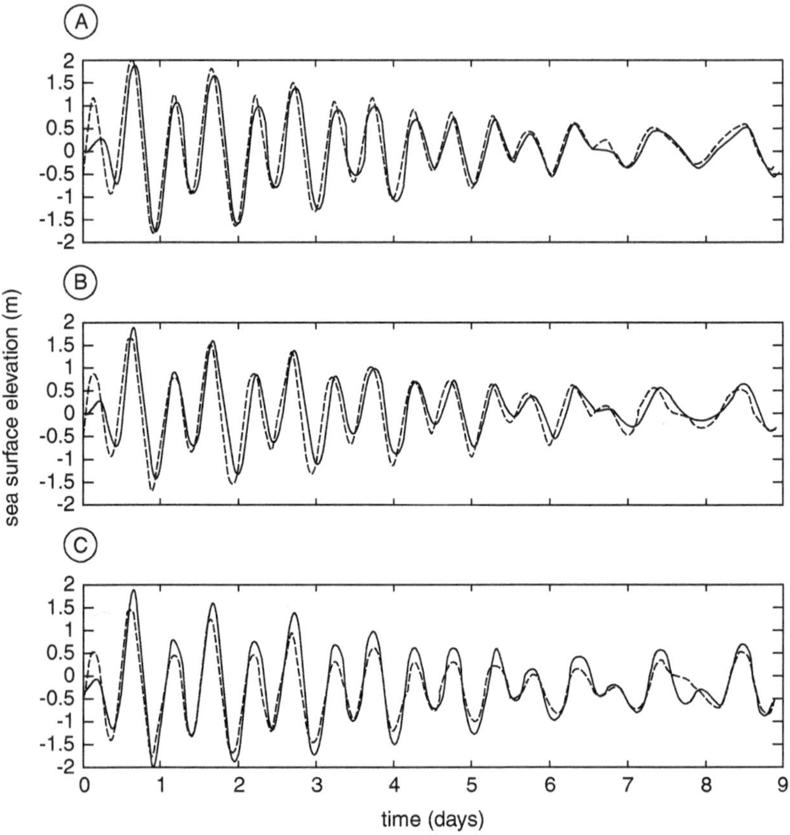

Figure 7.5 Tidal water elevations over the Australian Great Barrier Reef (GBR). Dashed lines represent measured sea surface elevations, while the full lines refer to numerical predictions. (A) Northern GBR (Pelican Island). (B) Central GBR (Stanley Reef). (C) Southern GBR (Gannet Cay). Modified and redrawn from Lambrechts et al. (2008).

The semi-diurnal tidal range is typically small, 1.7 m, generating currents flowing from southeast to northwest within the principal lagoon and locally reaching speeds of $0.3\,\mathrm{m\,s^{-1}}$ (Douillet, 1998). However, winds dominated by the SE trade winds at average speeds of $8\,\mathrm{m\,s^{-1}}$ (Douillet, Ouillon, & Cordier, 2001) have a greater influence on water and sediment movements, even where the tidal flow is at a peak, with the effects most marked in shallow water. Severe cyclonic winds are experienced in the area once or twice a year. For most Pacific reefs, the tidal range is typically less than a metre.

Along the southeastern African coast, in Kenya, with a tidal range of 3.2 m at spring tides, wave-driven inward circulation reaches 2000–$4000\,\mathrm{m^3\,s^{-1}}$ and provides an important barrier to the flow of polluted waters from tidal creeks (Angwenyi & Rudberg, 2005). In the central and northern Indian Ocean, the tidal range reduces from 1.5 to (locally) 0.3 m.

The Red Sea with a narrow connection to the Indian Ocean has only local small-amplitude oscillatory tides (Edwards, 1987) with the maximum range of 0.6 m near the mouth of the Gulf of Suez, and 0.9 m between the Dahlak Archipelago and Kamaran. Tidally driven water movements apparently generate few currents.

Although oceanic islands are considerably smaller, patterns of circulation may be equally complex. Tidally driven circulation in Cocos (Keeling) Atoll where structure controls both lagoonal circulation and energy distribution was described by Kench (1998). The lagoon is approximately 102 km^2 with a relatively broad pass to the south, numerous channels to the east and an open margin to the north and west. Cocos lies within the westward-flowing Equatorial Current but during November to March falls within the Equatorial counter current. The tidal range is 0.5–1.1 m with a maximum of 1.3 m. Currents in the southern and eastern passes increase for about 1.5 h^{-1} after low tide and reach peaks of 0.5 and 0.68 $m\,s^{-1}$ around 2 h before high tide. In the shallow lagoon, currents are typically less than 0.2 $m\,s^{-1}$ but reach 0.3 $m\,s^{-1}$ on falling spring tides.

Case Studies from the Caribbean. The Belize barrier reef in the western Caribbean, at about 170 km long, ranks fourth in the world. The tidal range in this region varies from as little as a few centimetres up to a maximum of around 30 cm. The principal current sweeping the area, the Caribbean Current, filters through the islands of the Antilles, generating numerous eddies and gyres before flowing westwards along the north coast of South America and swinging northwards along the Belize coast (Murphy, Hurlburt, & O'Brian, 1999). Wust (1964) mapped rates of flow, indicating the average speed for the entire system, by the time the current reaches Florida, flow is estimated to be of the order of 0.3–0.4 $m\,s^{-1}$. The prevailing winds are from the east and average 4.1 $m\,s^{-1}$ (James & Ginsburg, 1979a) driving onshore waves.

Tides in the Bahamas vary from 0.4 m on the banks to 0.8–1 m on open coasts. As a consequence, there is little tidally induced mixing of salt and freshwaters. A prominent feature of the topography and hydrology of the islands is the presence of large-scale dissolution features. These are identified at the surface as 'Blue holes' and appear both on the shallow shelf and inland where they are referred to as 'cenotes'. These vertical shafts are more generally 50–100 m deep. Some connect to extensive horizontal passages that provide routes for enhanced circulation. More than 14 km of surveyed caves form the Lucayan Caverns on Grand Bahama. These connect to passages of 2–3 m diameter that appear to have developed preferentially around island margins. Tidally driven water-table fluctuations increase with the depth of the hole (Whitaker & Smart, 1997) and provide a graphic illustration of connectivity. In a 90-m-deep cenote in North Andros, 18 km from the coast, there is a tidal lag of 216 minutes, whereas in an adjacent borehole 34 m deep, the lag is 277 minutes. Offshore, Blue Holes and connecting passages are commonly characterized by strong reversing tidal currents.

7.2.5.2. Winds and waves

It seems self evident that the impact of waves has a profound effect on the morphology, structure and ecology of the reef (for instance, see Kench, Brander, Parnell, & McLean, 2006). Conversely, reef geometry contributes to the determination of wave energy on reefs (Brander, Kench, & Hart, 2004). What is less obvious is the distribution of wave energy. It might be supposed that although hydrodynamic energy will vary within the system, any increases or decreases will be proportionately distributed. This is not so. Madin et al. (2006) showed that reefs on the exposed SE margin of Lizard Island in the GBR have little or no energy gradient when waves are small, but as size increases, the gradient steepens and becomes localized at the reef crest. For wave heights of over 3 m, during most tide conditions, most of the wave height that is lost over a 300 m distance of propagation (95%) is dissipated in the first 50 m. Equally surprising is the fact that a 6-m wave passing shorewards over the reef crest will be of similar size to a 2 m wave by the time it reaches the back of the reef. Madin et al. (2006) suggested that two zones can be recognized, the reef crest, represented by the first 50 m of wave contact with intense and relatively frequent motion episodes, and the reef flat where water movements are relatively uniform and benign. Similar observations were made by Lugo-Fernandez, Roberts, and Suhayda (1998) on Tague Reef, St Croix. Here, there is a net energy decrease of 78–88%. Tidal changes in depth, of about 0.3 m, result in a further 20% reduction between the fore-reef and back-reef. Most energy is lost in the process of the wave breaking, by spilling or plunging, rather than in bottom friction. These observations are important where disturbance is seen as a key factor in generating space for colonization. They also suggest that the more severe effects of storms and hurricanes will be confined within much smaller areas than has commonly been supposed.

The major wind system affecting the GBR for about 9 months of the year is that of the SE trade winds. The most vigorous reef growth occurs in areas open to the swell generated by the dominant SE trade winds, but behind the main barrier, tidal currents locally provide sufficient nutrient renewal for active growth, as in the Pompey Reef complex in the southern region, and on the margins of the Whitsunday Channel. In the central and northern regions, tidal scour behind the outer barrier has produced a sediment-free zone. Where currents pass through reef breaks or along channels, mounds of coarse debris may form, locally consisting almost entirely of *Halimeda* and *Marginopora* debris. Rees et al. (2007) suggested that in this area, *Halimeda* bioherms on the outer shelf contain at least as much, and possibly four times more, carbonate sediment than is present in the ribbon reefs to the east. In the nearshore zone, the net transport of water and sediment is from south to north. However, in addition to these broadly defined movements, the proliferation of small pinnacle reefs results in a large number of local gyres and areas of turbulence, and these may also promote local upwelling (Hopley et al., 2007). Circulation within the GBR system has been mapped by Luick, Mason, Hardy, and Furnas (2007). Neutrally buoyant tracers released into the central lagoon during the north-flowing current season (January to August) take 50–150 days to exit at the northern end. In the northern areas, speeds of north–south depth-averaged reversing currents range from about 0.2 to $-0.2\,\mathrm{m\,s^{-1}}$ (Figure 7.6). By contrast, most particles (70%) introduced to south-flowing currents (August to December) cross the lagoon in 20–330 days to enter the outer-shelf reefs, with the more rapid transit times reflecting favourable winds. An important conclusion is that the residence times are such that planktonic larvae and nutrients may be held for periods of 1 month to 1 year, with the implication that they transform long before they reach the outer reefs. Studies of reef lagoons on the GBR indicate that currents are generally less than $0.2\,\mathrm{m\,s^{-1}}$ but are larger where there are openings in the reef crest. As a result, flushing times vary from 12 h to 6 days (Hopley et al., 2007). However, wind-driven or wave-driven circulation may dominate at times, depending on position and local morphology.

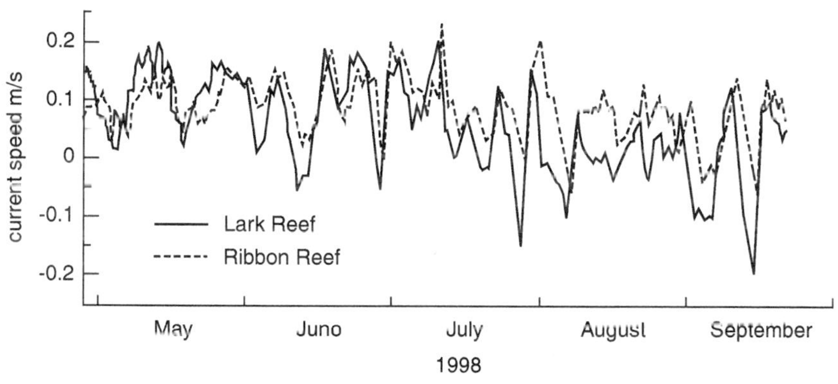

Figure 7.6 Speeds of north–south components of depth-averaged currents recorded over 6 months (May to September 1998) near Lark Reef (water depth: 51 m) and Ribbon Reef (depth: 30 m), northern GBR. Currents were measured using a bottom-mounted acoustic Doppler current metres. Analysed tides were removed from the data. A 35-hour-cut-off period as a pass filter was applied prior to plotting. Modified and redrawn from Luick et al. (2007).

Although, as indicated, there are important regional currents in the Caribbean, locally, greater effects are attributed to waves overtopping the reef edge (Roberts, Wilson, & Lugo-Fernandez, 1992; Symonds, Black, & Young, 1995). Coronado, Inglesias-Prieto, Sheinbaum, Lopez, and Ocampo-Torres (2007) showed that on the Puerto Morelos fringing reef of Yucatan, with a microtidal regime, normal circulation is generated by wave-induced flow (0.02–0.03 $m\,s^{-1}$) entering the lagoon over a shallow reef flat and driving strong flows (0.2 $m\,s^{-1}$) and leaving through the northern and southern channels. Superimposed on wind-generated wave activity, the geostrophic response of the Yucatan Current may locally raise sea level and increase wave power. Under normal conditions, the residence time for water within the lagoon is about 3 h but measurements during Hurricane Ivan in 2004 indicated a reduction to 0.35 h. By contrast, during summer, when wave heights are low and sea level reduced as a result of intensification of the offshore current (from 0.9 to around 2 $m\,s^{-1}$), water exchange in the lagoon is also significantly reduced.

Wind- and wave-driven circulation has also been described by Jouon, Lefebvre, Douillet, Ouillon, and Schmied (2009) in the southwestern semi-enclosed lagoon of New Caledonia (Figure 7.7). The local wind has a fetch of only a few tens of kilometres at maximum. The wave field is fetch limited with a short, mean wind wave period (<5 s). On Miyako Island at the southern end of the Ryukyu Group, south of Japan, the tidal range is about 1.5 m. Water is driven into the lagoon over the shallow western rim of the reef flat and then flows out through a channel on the eastern margin. Residence times range from as little as 1.5 to 9.3 h^{-1}, and it is argued that it is these that enable the reef to support a high level of productivity.

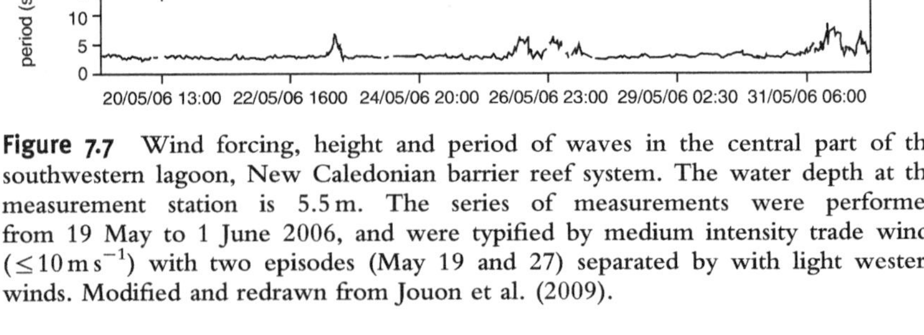

Figure 7.7 Wind forcing, height and period of waves in the central part of the southwestern lagoon, New Caledonian barrier reef system. The water depth at the measurement station is 5.5 m. The series of measurements were performed from 19 May to 1 June 2006, and were typified by medium intensity trade winds ($\leq 10\,m\,s^{-1}$) with two episodes (May 19 and 27) separated by with light western winds. Modified and redrawn from Jouon et al. (2009).

Currents are typically unidirectional throughout the tidal cycle but may pause at low tide when the reef flat is emergent. There is also a slack at high tide as a result of the increasing depth at which waves break. The short residence times imply that most of the mobile organic production on the reef flat probably leaves the system without decomposing (Kraines et al., 1998).

Work by Munk and Sargent (1954) showed that at Bikini, the distribution of marginal spurs and grooves was correlated with the mean annual distribution of wave power. Similar observations have been repeated many times elsewhere (e.g. Yamano et al., 2003) and are reflected in the division of oceanic reef systems into windward and leeward groups. This division underlines both the dynamic effects of wave power in controlling reef structure and the importance of water movement in delivering nutrients.

Atkinson, Smith, and Stroup (1981) produced a detailed study of circulation in Enewetak lagoon (Figure 7.8). Warm surface waters in the surrounding ocean extend to about 70 m depth, and as the deepest pass is only 56.7 m, these are normally the waters entering the lagoon, mostly along the eastern windward margin, surf driven over the reef perimeter by the prevailing trade winds at speeds of 0.1–1.5 m s^{-1}. By contrast, on the leeward margin, waters move back and forth across the reef but have only a small net outflow, equivalent to about 6% of the water entering the lagoon from the east. South Channel with a maximum depth of 22 m has a nearly continuous outflow with current speeds of 0.08–0.3 m s^{-1}. The southwest passage is less than 2 m deep with a reversing flow incorporated into that of the flanking reef. The waters of the lagoon, 64 m deep, are apparently fully mixed and of uniform density with little variation in either salinity or temperature. Surface flow to the southwest, about 2% of wind speed, is 5–15 m deep with a reverse flow between 10 and 30 m depth moving at about half of the surface speed (2–4 cm s^{-1}). Below 30 m, the flow moves southwards at 1–2 cm s^{-1}. The residence time of water in the lagoon appears to be about 4 months.

One of the key features of smaller islands is the development of eddies. Eddies around Guam were described by Wolanski, Richmond, Davis., Deleersnijder, and Leben (2003). They result from interference with the North Equatorial Current flowing at speeds of 0.1–0.2 m s^{-1} that dominates regional circulation from June to December. Eddies vary in scale from a 200 km diameter cyclonic eddy, with a current speed of up to 0.5 m^{-1} advected eastwards in August 2000, to small-scale structures shed from the tips of the island and other local topographic features. It was suggested that collectively these may be sufficiently energetic to return eggs and larvae to the island and thus self-seed the reefs from which they originated. Eddies were also identified in a 2003 survey of West Maui, Hawaii (Storlazzi, McManus, Logan, & McLaughlin, 2006).

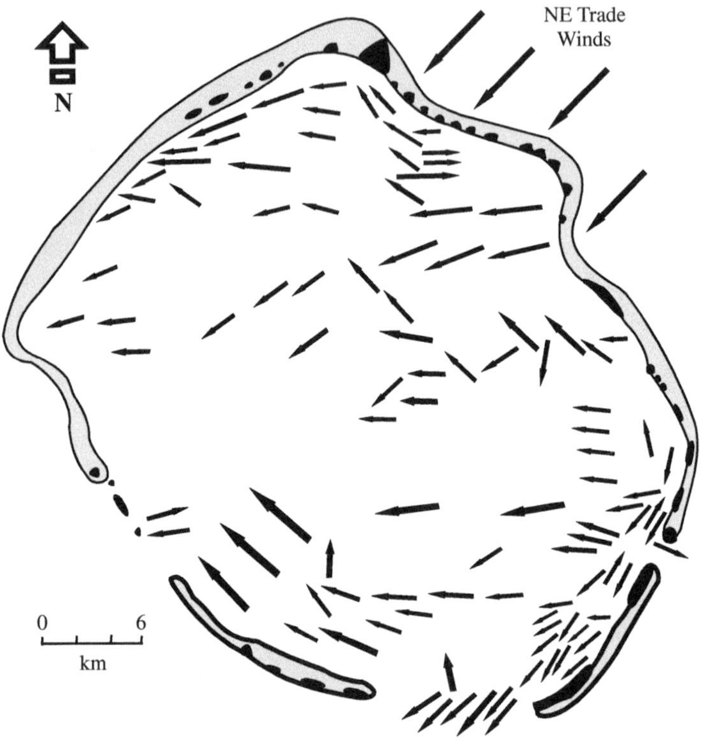

Figure 7.8 Wind-driven, surface currents in the lagoon of Enewetok Atoll (redrawn from Atkinson et al., 1981).

7.2.5.3. Hydrodynamics and coral morphology

A number of studies have investigated the hydrodynamic effects of coral morphology and in particular the resistance of corals to waves and currents and the influence it has on ecology (Chamberlain & Graus 1975; Graus, Chamberlain, & Boker, 1977; Graus, MacIntyre, & Herchenroder, 1984). Wave resistance can be related to reductions in branch size, degree of branching and branch orientation relative to water movements. However, branches have the effect of increasing the surface area of the colony and thus in part depend for their efficacy on the existence of adequate water movement. Massel and Done (1993) used analyses of shear, compressional and tensional forces generated by waves to conclude that, so long as corals are firmly attached, they can resist all waves regardless of colony size or shape or cyclone intensity. Vulnerability increases with age, new recruits are small and therefore offer relatively little resistance to flow and are less likely to be broken or dislodged. However, it is evident that many corals are detached by storms. With current predictions of increasing storminess linked to climate change, Madin and Connolly (2006) have attempted to forecast the level of disturbance of reef corals during such events, incorporating oceanographic models, engineering theory and shape factors. Calculations of shape factors of colonies, the strength of the substrate to which these are attached and the attenuation of waves across the reef platform, permit predictions of changes in size, structure and species composition of coral assemblages caused by recent and likely future hydrodynamic disturbances. Results accord well with field tests on Lizard Island in the GBR.

7.2.5.4. Storms, cyclones, hurricanes and typhoons

The modern record. Storms, cyclones, hurricanes or typhoons are important limiting factors in coral colonization and growth and in sediment generation and deposition (Geister, 1977, 1980; Harmelin-Vivien, 1994; Grigg, 1998; Grossman & Fletcher, 2004; see also Chapters 3 and 4) (Figure 7.9).

Winds in excess of 17 m s^{-1} are referred to as 'tropical storms' but when they reach 33 m s^{-1}, they become cyclones in the SW Pacific and Indian Oceans, hurricanes in the North Atlantic and NE Pacific Oceans, and typhoons in the NW Pacific. In general, they occur in belts from 7° to 25° north and south of the equator and thus are broadly similar in their distribution to the major reef systems of the world. Frequency varies widely, and in some areas, it is tied to El Niño cycles that are characterized by an increase in hurricane activity in the eastern Pacific and a decrease in

Figure 7.9 Typical cyclonic deposits on reef tracts: storm ridge (about 1 m above sea level) mainly composed of pieces of arborescent *Acropora* growth forms, inner-shelf reef at Molle Island, Whitsunday Group, Australian Great Barrier Reef (photograph from L. Montaggioni).

activity in the Atlantic and Caribbean. In recent years, a global increase in frequency has been putatively linked to global warming. The western Pacific experienced 85 major storms in the period 1975–1989 but 116 from 1990 to 2004. Water driven across reefs during severe storm surges may be more than 10 m above normal levels and may sweep along or across the shore, leaving a trail of destruction.

The effects of hurricanes and other storms have been reviewed by Massel and Done (1993), Scoffin (1993), Harmelin-Vivien (1994) and Morton, Richmond, Jaffe, and Gelfenbaum (2006). Catastrophic disturbances have been widely recognized as parameters controlling community structure, generating changes by mechanical destruction, by reforming surface morphology and by clearing surfaces for recolonization. They may locally modify sedimentation and increase turbidity, and also cause changes in salinity and sea level, promoting local flooding. Done (1992b) examined the effects of cyclone Ivor in 19–20 March 1990, on the ecology of the Ribbon Reef area of the GBR. Ivor was ranked level 2 on the Saffir-Simpson scale. Relatively severe damage, locally denuding tens to hundreds of metres of reef slope to depths of 20 m, was focused within 50 km of the storm path. In some sheltered areas, corals were depleted by breakage, dislodgement or local avalanches, but patches remained apparently unaffected. As might be expected, the vulnerability of particular species is in part a reflection of their cross-sectional area, and hence drag, but also their adhesion to, and the relative stability of, their substrate. Areas dominated by branching forms are less at risk when these are small, but become increasingly vulnerable with time. Done (1992b) described the ecology of the denuded areas as 'reset' with a new communities developing in response to recruitment and succession. By contrast, where damage is reduced, survivors of the existing community may recover and re-establish the earlier community structure.

The effects of Hurricane Hugo on St Croix in the US Virgin Islands in September 1989 were documented by Hubbard (1992). Sustained winds were $>204\,\mathrm{km\,h^{-1}}$ with gusts to $306\,\mathrm{km\,h^{-1}}$, generating waves of 6–7 m height along the more exposed south coast. Large volumes of sand were flushed from shallow-water areas, estimated at 336,000 to 2×10^6 kg. Significantly, the calculated transport rates were eleven orders of magnitude higher than those measured during fair weather.

A detailed study by Blanchon and Jones (1997) examined exposure-related coral morphology and shelf-edge architecture around Grand Cayman. The island is rimmed by an 87-km-long shelf-edge reef consisting of coral-armoured buttresses about 100 m long and 10 m wide that rise to within 12 m of mean sea level and are separated by narrow sediment-floored canyons. Pruning of buttresses and flushing of canyons during storms provides the principle control on the initiation and maintenance of shelf-edge architecture. Reef accretion proceeds until a depth is reached at which hurricane-induced wave action removes weaker corals at the same rate that they are replaced. Blanchon and Jones (1997) called this the 'hurricane accretion threshold' and on Grand Cayman it is proportional to exposure.

Harmelin-Vivien and Laboute (1986) were able to compare the effects of six hurricanes that swept through French Polynesia between December 1982 and April 1983. The degree of damage varied, depending on the width of the fore-reef terrace and the angle of the slope, influencing wave attenuation. Coral destruction varied with depth, but the shallow community, better adapted to vigorous wave motion, fared better than that in the deeper zones. Coral destruction was severe (60–80%) at depths of 15–30 m and nearly 100% in areas deeper than 35 m. The outer slope had become a rubble zone with remnant coral patches aligned perpendicular to the reef front. On low angle slopes as observed at Moorea, material was transported onshore and thrown up as a boulder rampart, but on the steep slopes of Tikehau Atoll, broken corals rolled downslope, creating trails of destruction.

Larcombe and Carter (2004) showed that in addition to their hydrodynamic effects, cyclone systems are responsible for large-scale water movements that overprint what can be regarded as normal circulation on the GBR. During these storms, significant volumes of water are driven onshore generating a strong north-flowing current (cyclone corridor) behind the barrier formed by shelf-edge reefs (Figure 7.10). Sediment transport and surface scouring within this zone are dominated by aperiodic currents flowing in excess of 1–3 m s^{-1}. Gagan, Chivas, and Herczeg (1990) recorded the effects of cyclone Winifred (a 50–70-year storm) in the area east of Innisfail in the central GBR in February 1986. A mixed terrigenous carbonate, normally graded storm layer, locally more than 11 cm thick, extended up to 30 km offshore in waters up to 43 m deep. There was an increase in carbonate nearshore following transport of at least 15 km, and a loss of mud from the mid-shelf. A year after the storm, the inner-shelf sediment still formed a recognizable depositional unit, whereas that offshore had been bioturbated and had effectively lost its identity. Following cyclone Sadie in 1994, flood plumes from rivers spread across the shelf, but after cyclone Violet in 1995, SE winds again confined turbid waters to a narrow coastal zone. Similar winds following the Burdekin flood of 1981, and floods from the Johnstone River after cyclone Winifred in 1986 (Gagan, Sandstrom, & Chivas, 1987), held low-salinity (30 psu) water close to the shore. During the latter event, most terrestrial plant debris was deposited within 2 km of the shore and none moved more than 15 km offshore. Gagan et al. (1987) concluded that, excluding exceptional Burdekin River floods, terrestrial runoff has not reached the active reef in historical times, but that material deposited nearshore may be resuspended during cyclones and thus reach growing corals during secondary transport.

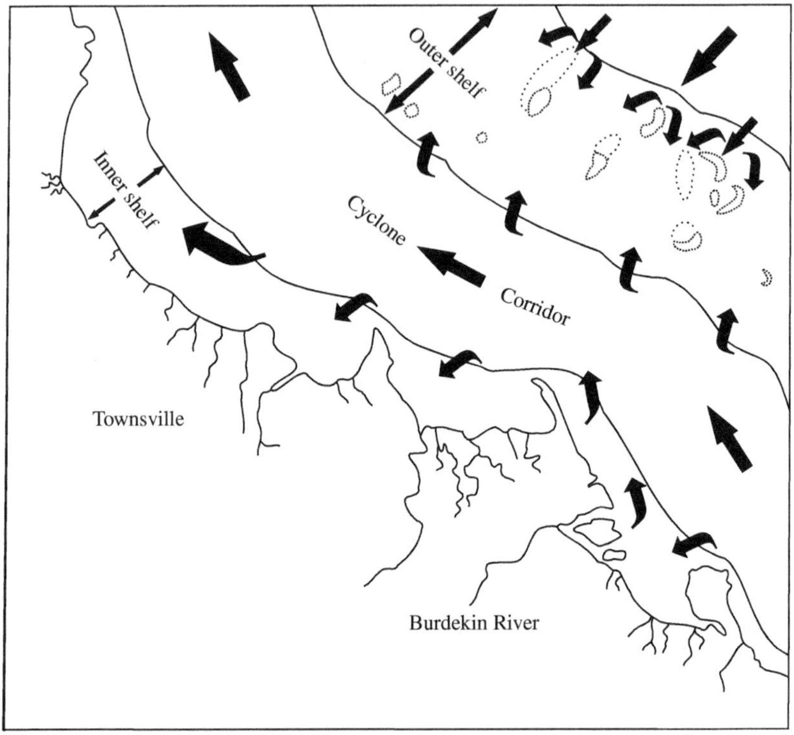

Figure 7.10 Cyclone-generated, north-flowing current along the inner part of the Australian Great Barrier Reef shelf (cyclone corridor) (redrawn from Larcombe & Carter, 2004).

An additional effect of storms has been reported by Nadaoka, Nihei, Kumano, Yokobori, and Wakaki (2001) from a reef on Ishigaki in the Ryuku Islands. Hydrodynamic, thermal, salinity and turbidity fluctuations were measured during passage of a typhoon. There was an abrupt decrease in water temperature of about 1°C, coupled with an increase in salinity as the storm passed. The temperature change seems to have been related to an increase in thickness of the surface mixing layer and relative upwelling, with a north-flowing current driving relatively turbid river-derived water away. Such effects can be expected to be of increasing importance towards the limits of coral growth but are unlikely to be significant at low latitudes.

The historical and geological record. The intensity, frequency and distributional patterns of cyclones are primarily driven by changes in SST. Simulations show (Knutson, Tuleya, & Kurihasa, 1998) that an increase of about 2°C will result in a 5–10% enhancement of cyclone activity. One conclusion that might be drawn from this is that, as a result of colder temperatures, cyclonic activity during maximum glacial phases was probably lower than at present. In the following deglacial episodes, the temperature-driven increase in the severity of cyclones may locally have prevented reefs from maintaining pace or catching-up with rising sea levels, thereby favouring reef drowning. Nott and Hayne (2001) suggested that over the past 5 ka, the GBR has experienced a series of supercyclones with air pressures of >920 hPa (wind speeds of $>250\,\mathrm{km\,h^{-1}}$) every 200–300 years and that these were a dominant influence in shaping reef communities. From a geological perspective, the significant effect of storms of all kinds is the breakage of coral and redistribution of debris. Reworked ridges and bars consisting of blocks that may range in size from a few centimetres to metres are a significant feature of Low Isles, Pipon Reef, Three Isles and elsewhere in the GBR (Scoffin, 1993). Stoddart (1963, 1971) suggested that severe hurricanes have demolished spurs on reef fronts, leaving smooth truncated reef surfaces. Ball, Shinn, and Stockman (1967) estimated that between 160,000 and 320,000 hurricanes struck the Florida Keys region in the course of the Pleistocene.

High-intensity storms appear to stimulate larger phytoplanktonic blooms (Babin, Carton, Dickey, & Wiggert, 2004) and such storms may also therefore have indirectly contributed to the disturbance of coral growth during deglaciation. Conversely, long-term storm activity may have had a positive influence. It may locally have increased rates of deposition by instantaneously incorporating large coral clasts into the reef pile. On the GBR anatomy and ecological successions have varied as a function of swell strength and/or cyclone frequency and intensity. In many cores from Holocene reefs, the material directly deposited on reef foundations is predominantly unconsolidated detrital sediment, commonly coral rubble, whereas the upper, younger, parts of the sections are dominated by reef framework. This suggests that the pioneering stages of reef accretion were commonly disturbed by framework reworking within the so-called 'cyclone corridor' that is regarded as having formed shorewards of the incipient reefs as sea level was rising (Larcombe & Carter, 2004). By contrast, the sediment pumping effect of cyclones, particularly in cyclone corridors, may have contributed to rapid winnowing of finer particles deposited on inner shelves, and thus reduced the duration of high turbidity events that hampered reef colonization. Such events probably occurred several times in the history of Pleistocene reef growth

Similarly, the 'high-energy window' concept, originally applied to mid-Holocene reef growth history (Hopley, 1984), can also be extended to the Pleistocene record. During postglacial, ice-melting episodes, in areas where vertical reef growth was unable to compensate for the rapid rise in sea level, an interval of higher wave energy may have been interpolated before reef growth was again able to 'keep-up'. The high hydrodynamic energy window may have remained open from the time of inundation of the foundations until the reef tops reached sea level and formed a protecting barrier. High-energy events have probably been most effective in areas where reef building started from deeper foundations (>20 m), because waves operating over incipient reefs require depths greater than 5 m to be dynamically efficient. The efficiency of waves is thought to have increased in sites where sea level has risen rapidly (at around 10 mm yr^{-1}). As sea level stabilized, reef growth rates commonly declined to around 1.5 mm yr^{-1} as reef tops approached the sea surface. In such instances, high-energy conditions have been maintained for longer. The role of high-energy windows is likely to have been enhanced during the supercyclone events of the past 5 ka reported by Nott and Hayne (2001) from the Australian Great Barrier Reef. Similar exceptional events probably also occurred during interglacial, high sea-level stands.

Historical records of cyclone effects on reefs over the last three centuries indicate that there may be markedly different patterns of disturbance within the same reef system (Precht & Miller, 2007; Zhao, Neil, Feng, Yu, & Pandolfi, 2009; and references therein). Recovery may be rapid and can be completed in less than 5 years. Storms may promote the dispersal and expansion of dominant coral species by clonal reproduction of storm-derived colony fragments. This suggests that the history of damage has a significant influence on reef development and resilience. The 'intermediate disturbance hypothesis' of Connell (1978) assumes that species diversity is highest in areas subjected to intermediate disturbance levels, in terms of frequency and intensity. Although the specific influence of storms on coral diversity is difficult to prove, application of the intermediate disturbance hypothesis to cyclones suggests that such events tend to increase diversity in reef sites dominated by branching or foliaceous corals, whereas in sites dominated by more resistant growth forms, no significant changes in diversity have been recorded (Rogers, 1993). Pandolfi and Jackson (2001) observed in the Pleistocene reefs of Curaçao that variability in coral community structure within a given habitat was much lower than that between different habitats. There were only limited within-habitat changes due to the lack of severe hurricane activity in this area. This reinforces the idea that disturbance cannot explain high levels of within-habitat order in coral community composition. Pandolfi and Jackson (2007) compared the effects of water turbulence on the compositions of coral assemblages in late Pleistocene shallow-water reef environments on the leeward sides of three southern Caribbean islands. Reef crests from Barbados are likely to have experienced the highest wave energy in relation to severe hurricane activity, those from Curaçao were severely washed by open-sea swells, while those from San Andrés, outside the hurricane belt and partly protected from open-sea influences, were subjected to much lower water agitation. Their results show that coral assemblages are easily predictable and vary in species richness according to wave exposure.

7.2.5.5. Tsunamis

In addition to major storm events, the positions of many reef systems close to tectonic plate margins mean that they will also be swept periodically by earthquake-induced tsunamis. In many areas, tsunamis may also result from volcanic eruptions and major landslips (Morton et al., 2006). In contrast to hurricanes and tropical storms, the effects of tsunamis on coral reefs have been poorly documented (Hughes, 1993). In general, effects are considered similar to those of severe storms, but they may also exceed them with high impact waves breaking on the reef margin and large volumes of water, resembling a storm surge, swept shorewards (Madin & Connolly, 2006). Stoddart (2007) filled a gap in our knowledge of the devastation suffered by reefs during tsunamis, editing a special issue of *Atoll Research Bulletin* devoted to the effects of the earthquake and tsunami that struck Indian Ocean reefs in 26 December 2004. Surveys indicated a highly variable degree of coral damage and mortality and of shoreline erosion as a function of the distance from the Sumatran epicentre. In northern Sumatra, close to the epicentre on the eastern margin of the ocean, only about 15% of reef areas were severely devastated, although there were significant sectors in which the seafloor had been uplifted or submerged and *in situ* corals were exposed or carried to inappropriate depths by earthquake activity (Foster et al., 2006). In NE Sri Lanka (NARA, 2005), early reports noted widespread *in situ* breakage of branching corals including *Acropora*, *Montipora* and *Pocillopora* in exposed sites. Other areas were overrun by rubble that included trees and coral blocks of metre dimensions. In Thailand, of the areas surveyed (Turak, Veron, & Sanpanich, 2005), only 14% of reefs had suffered severe damage with some 50% showing moderate damage and the remainder unaffected. However, reports suggest that recovery is already underway and that it will be substantial within a decade. In the Seychelles, towards the western margin of the basin, levels of damage varied according to location and substrate type. The most impacted reef sites relate to the high islands on which direct coral damage exceeded 25% of the surveyed substrates.

It is difficult to differentiate the coral rubble deposited by storms and that resulting from tsunami-generated waves (Scoffin, 1993). Coral ridges, boulders of corals or blocks of reef framework ranging from a metre to several metres diameter (Figure 7.11A and B) may be found on coastlines or perched on the seaward margins or within reef flats in many Indo-Pacific (Montaggioni, 1978; Bourrouilh-Le Jan & Talandier, 1985; Scoffin, 1993; Nott, 2003; Kench, Nichol, Smithers, McLean, & Brander, 2008; Scheffers, Scheffers, Kelletat, & Bryant, 2008; Yu, Zhao, Shi, & Meng, 2009) and Caribbean reefs (Scheffers et al., 2006; Scheffers, Haviser, Browne, & Scheffers, 2009). The largest blocks (up to $50\,m^3$) are of Holocene age (about 6–0.50 ka) and were deposited at elevations ranging from a few metres to tens of metres. Given their size and weight, they have generally been assumed to have been lifted by waves generated by tsunamis rather

Figure 7.11 Catastrophic deposits (photographs from L. Montaggioni). (A) Large (about 6 m^3) coral block deposited at the surface of an emergent, Holocene storm ridge, northwest coast, Moorea Island, French Polynesia. (B) Huge reef block (up to 100 m^3) interpreted as transported by tsunami-induced waves. It is composed of seaward-dipping grainstone–rudstone beds (probably, beach-rock), and culminates at about 4 m above present mean sea level on the backshore. A coral fragment collected at the upper part of the block yielded a U/Th age of 6.2 ka. Tamarin village, western coast, Mauritius Island, Indian Ocean.

than cyclones (Scoffin, 1993; Radtke et al., 2003). Their age rules out suggestions that they are shoreline accumulations formed during interglacials and later uplifted (McMurtry et al., 2007; Webster, Clague, & Braga, 2007). Damage can be severe as Scheffers et al. (2006) reported on the destruction of Holocene reefs on the northeast coast of the island of Bonaire. Here, debris totalling some 6×10^6 t and including blocks, the largest of which were estimated at 50–260 t, were dated by radiocarbon and electron spin resonance methods. The ages of most clustered between 3.1 and 4.1 ka, although some were younger, and suggest discrete depositional events at 3.1, 1.5 and 0.50 ka. Offshore terraces consist of bare late Pleistocene (stage MIS 5e) limestones. The authors concluded that these deposits represent the remains of a Holocene reef that had been completely destroyed by a tsunami, with the debris subsequently redistributed by hurricane events. The reef has apparently been unable to regenerate in the past 3.1 ka. On the southwestern coast of Mauritius, at Choisy, Montaggioni (1978) observed a 120-ka-dated reef block of about 200 m^3, 7 m above present sea level. The block consisted of encrusting and domal corals, thickly encrusted by coralline algae, suggesting that it was derived from the framework of the reef edge. Given its size, it was interpreted as having been cast up on the shore during a late Pleistocene tsunami.

Morton et al. (2006), in a comprehensive review, listed criteria by which the deposits of extreme storms and tsunamis might be differentiated. Much of the emphasis is on the transport mechanism of megablocks with the conclusion that long-period tsunami waves have a more sustained effect and therefore have a higher transport capacity. However, for deposits of smaller grain size, it is difficult to differentiate transport mechanisms. Satapoomin, Phongsuwan, and Brown (2006) described deposits attributed to the 2004 tsunami in Thailand. Structures of the resulting sand sheet included surface bedforms, parallel laminae, landward- and seaward-inclined laminae and dipping laminae attributed to antidune formation, with the ensemble interpreted as the deposits of supercritical flow during inflow. Clay drapes on seaward-dipping laminae were attributed to outflow. Most deposits show normal grading but there is localized reverse grading. Related descriptions are available in Morton et al. (2007, and references therein).

7.3. Groundwater Hydrology

7.3.1. Characteristics of the Reef Hydrological System

Reefs form close to the sea surface and even small changes in sea level may leave them exposed to the atmosphere and subject to attack by surface waters. It might be supposed that because reefs grow in seawater, they would also be saturated with seawater. This is true for contemporary oceanic reefs that lack associated islands, but even small islands may generate freshwater lenses. In reefs fringing continental areas, hydraulic head may drive freshwater from these considerable distances beneath the reef

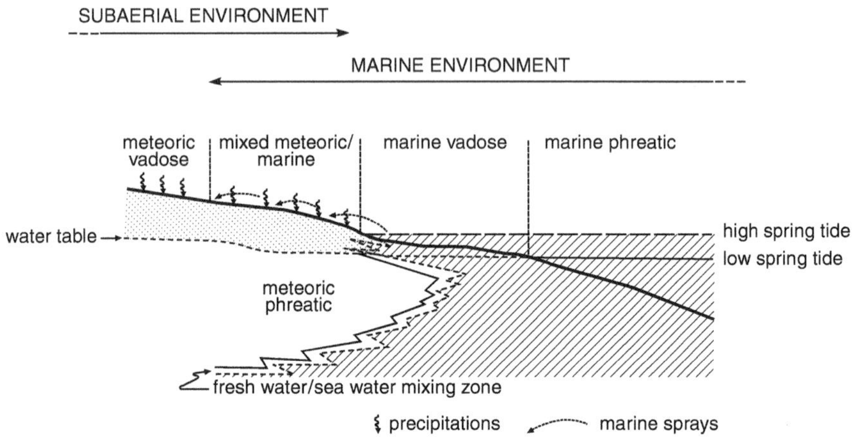

Figure 7.12 Distribution of the major nearshore diagenetic environments as defined on the basis of tidal cycles and of chemistry of interstitial waters (modified and redrawn from Coudray & Montaggioni, 1986).

(Figure 7.12). Where freshwaters are involved, the interstitial water body is stratified, with individual units characterized by distinctive chemistries. The thicknesses of these units and the boundaries between them depend upon local relief, the porosity and permeability (hydraulic conductivity) of the rocks and on ambient rainfall, but in general, two groundwater zones may be recognized. The water table defines the surface of the saturated phreatic zone in which water chemistry may vary widely. In porous limestone islands, it is seldom more than a few metres above sea level, subject to local stratigraphical and structural controls, and cementation (porosity reduction) related to these factors may result in the development of perched water bodies. On the Pacific island of Niue, 71 m high and with a high rainfall (>2 m), the water table is only 60–140 cm above sea level (Wheeler & Aharon, 1991).

Above the water table, the vadose zone contains both air and water within pores, the proportion of the latter depending upon rainfall. Flow is intermittent and water is typically present as grain-surface films, held between grains by capillary action and in coarser sediments as pendent droplets. In larger pores close to the water table, the water surface defines the limit of active crystal growth. This is a chemically active region near the surface, fresh meteoric waters derived from surface rainfall contain carbon dioxide (CO_2) dissolved from the atmosphere and from soils, together with humic and other acids. These waters are chemically aggressive and are responsible for dissolving carbonates. At greater depth, and extending into the phreatic zone, CO_2 degassing and evaporation increase saturation and drive crystallization of pore-filling cements.

The contrast in density between fresh and salt water allows the freshwater to float on top, and the interface is depressed below sea level so that the fresh water forms a Ghyben–Herzberg lens. An island of approximately 200 m diameter and receiving sufficient rainfall is required to maintain a freshwater lens in Holocene sediments (Budd & Vacher, 1991) and some 2 km diameter is necessary to maintain one in Pleistocene limestones (Cant & Weech, 1986). However, the detailed form of this lens depends both on rainfall (recharge) and the nature of porosity, together with topography. As a 'rule of thumb', the height of the water table above sea level and the depth of the marine water below have traditionally been seen as governed by a ratio of approximately 1:40. However, Budd and Vacher (1991) examined means of predicting the thicknesses of freshwater lenses beneath limestone islands, and concluded that while the Ghyben–Herzberg model may be satisfactorily applied to contemporary systems, it is unreliable when used to estimate palaeo-water table or palaeo-sea-level positions in Quaternary sequences. They suggested as an alternative that the thickness of the freshwater lens is better approximated as about 1% of the palaeo-island width and that this is more easily estimated. With sufficient rainfall, topographic relief of only a few metres can force the boundary between fresh and salt waters substantially below sea level. Given sufficient hydraulic head (reflecting high rainfall and/or a larger land mass), fresh waters may be driven laterally within a confined aquifer to lie beneath saline pore waters (Johnson, 1983). This is important because it means that the effects of freshwater diagenesis do not necessarily reflect exposure, and long-term submergence below sea level does not guarantee a lack of dissolution or alteration. The lower boundary of the vadose zone migrates up or down, depending upon recharge (rainfall), and the chemically active zone of migration of the fresh-saltwater boundary forms the 'mixing zone', the thickness of which depends upon the amplitude of the surface fluctuations. Buddemeier and Oberdorfer (1986) suggested that on Enewetak Atoll, the freshwater body is probably only a few metres thick, whereas the mixing zone is likely to be tens of metres.

The boundaries of the water table and the mixing zone are not static. They vary with the weather and in response to seasonal cycles, and locally may also show significant diurnal variations, reflecting tidal movements. The amplitude of the latter movements is correlated with tidal range but depends upon local poroperm characteristics and may be substantially damped.

There is considerable variation in the hydraulic conductivity of both Holocene reef materials and Pleistocene limestones. The latter present particular problems because, in addition to lithologic variation, they typically include a series of karst erosion surfaces, reflecting low stands of sea level, and these may be characterized by early cementation, fractures and both grain scale and metre-scale dissolution. Buddemeier and Oberdorfer

(1986) attempted to provide guiding values. They suggested that coral heads, algal frameworks and fine-grained lagoonal sediments have conductivities in the range 0.001–1 m d^{-1}, whereas coarse sand and reef rubble are likely to have values between 10 and 1000 m d^{-1}. Larger scale dissolution cavities and fractures may generate conductivities of the order of 2000 m d^{-1}.

The chemistry and distribution of groundwaters provide important controls on the subsurface features generated. Mylroie et al. (2001) supplied a general description of what Vacher and Mylroie (2002) referred to as 'epigenetic karst' developed on recently deposited carbonates, following the terminology of Choquette and Pray (1970). These are summarized in Jenson et al. (2006). Where the island is wholly carbonate, waters are autogenic, that is are derived from meteoric precipitation falling directly on the limestones. In such areas, two cave types may be developed, reflecting the distribution of chemically aggressive waters (Smart et al., 2006). Pit caves are vertical pipes extending from a few metres up to 50 m depth. The upper boundary of the freshwater lens is marked by the formation of water-table caves, but other caves may also form in the mixing zone at the bottom of the lens. The presence of a body of non-carbonate rock may modify the system. Vadose water may be intercepted above sea level and form contact caves, and streams in these may mechanically erode and transport material. Allogenic waters, originating from non-carbonate terrain are generally chemically aggressive, and inflow from these is likely to form large contact caves.

7.3.2. Flow in Holocene Reefs

7.3.2.1. The influence of permeability and conductivity

A series of works have investigated groundwater flow within Holocene reefs. Small-scale movements within the fringing reef flat of Orpheus Island at the northern end of the Great Barrier Reef, Queensland, were investigated by Parnell (1986). Using fluorescent dye tracers, he was able to show a net seaward movement in the order of 40 m d^{-1} and, long term, a considerable vertical mixing within the zone examined. Downward mixing is more rapid than upward mixing. When the reef flat is covered at high water, the hydraulic gradient within the sequence below is likely to be low. Under these conditions, diffusion is the most important process, and movement within the sediment is largely dispersive. By contrast, at low water, dispersive flow or advection dominates, predominantly seawards, although movements are likely to reverse in concert with tidal cycles. Buddemeier and Oberdorfer (1986) and Oberdorfer and Buddemeier (1986) investigated flow within Davies Reef in the central GBR that was driven by differences generated by currents, winds or tides. As might be expected in such a heterogeneous system, flow rates were quite variable, reflecting wide variations in vertical and horizontal permeability that in the high-permeability zones were of the order of 10 m d^{-1}. Exchange occurred both laterally and vertically. This is important for two reasons: interstitial waters have been shown to be nutrient rich and oxygen depleted (Andrews & Muller, 1983; Buddemeier & Oberdorfer, 1986; Sansone, Andrews, Buddemeier, & Tribble, 1988) and their low pH is significant in early diagenesis.

Unconsolidated Holocene sediments, ranging from fine sand to coarse coral rubble have hydraulic conductivities varying from 10 to 2000 m d^{-1} (Buddemeier & Oberdorfer 1986). Florida Bay muds consisting of micron-size aragonite needles with a bulk porosity of 45–85% have conductivities of 10^{-1} to 10^{-3} m d^{-1} (Swart & Kramer, 1997). Consolidated deposits, the nature of which was not defined, have conductivities varying from 0.1 to 2 m d^{-1}. There are major vertical heterogeneities, varying over four orders of magnitude. Somewhat surprisingly, underlying Pleistocene rocks, in which a series of exposure surfaces are associated with dissolution and extensive cementation, show a similar range of hydraulic conductivities, but there are also clearly high-permeability zones of the order of thousands of m d^{-1}. Buddemeier and Oberdorfer (1986) calculated that for a head difference of 5 cm across a reef 300 m wide, it would take 3 months for a particular body of water to move through the reef. They also noted, however, that if the gradient changes or flow reverses then residence times will be significantly greater. With intercalated high-permeability units such as are commonly present in Pleistocene sequences, vertical flow could reduce the residence time to 2.5 days. High permeabilities may extend to considerable depth. Ladd and Schlanger (1960) reported tidally driven variations in wells on Enewetak Atoll that indicate substantial flows at depths of more than 600 m. However, high permeabilities at shallow depths mean that the signal from tidal motions is readily transmitted, providing a mechanism for mixing fresh surface waters with the underlying saline water body. Data from Kwajelein Atoll (Hunt & Peterson, 1980) suggest that mixing is more closely related to depth than to distance from the shoreline and results in an extensive transition zone. In this location, the total volume of freshwater was calculated to have varied by more than 20% in a single year of observation.

7.3.2.2. Chemical and nutrient gradients

Andrié et al. (1998) described circulation within the Tahiti barrier reef, basing their observations on analyses at 12-depth intervals within a 150 m borehole drilled. The upper section of the hole to 88 m deep penetrated the Holocene/postglacial reef. Below, and beneath a karst surface, Pleistocene limestones extend to 115 m and below these, the boreholes passed through basalt to a total depth of 150 m. Samples were analysed for temperature, salinity, silica, ^{18}O, ^{3}He, CFC-12 and tritium. Results showed that at all levels, interstitial waters differ significantly from those of the surrounding ocean. Within the basalt basement and the Pleistocene reef, compositions appear to reflect the slow intrusion of deep Pacific waters (approximately 350 m) but there is a clear overprint of ^{3}He and silica originating within the basalt. There is a marked chemical discontinuity at 88 m where the transition to the postglacial sequence reflects both a change in the hydraulic regime and the disappearance of the geothermal flux outside the basalt. The composition of the water still reflects the addition of components from below, but also the intrusion of surface-derived water driven by turbulence and wave surges and local contamination by brackish groundwater of meteoric origin.

A number of studies have attempted to determine the nutrient contribution, if any, of groundwaters emerging at or near the reef surface. Umezawa, Miyajima, Kayanne, and Koike (2002) used two methods to estimate the nitrogen input to two reefs on Ishigaki Island, southwest of Japan. Measurements of dissolved inorganic nitrogen in groundwaters and estimates based on land use within their respective catchments agreed within a factor of two. Differences were expressed in terms of a denser growth of seagrass (*Thalassia*) on Shiraho reef, receiving the highest nitrogen input. Investigations on Hawaii (Street, Knee, Grossman, & Paytan, 2008) determined the origins and nutrient contribution of submarine groundwater discharge across the leeward shores of three islands. Nutrient additions were found to be greatest with measured nutrient fluxes of 0.04–40 mmol N m^{-2} d^{-1} and 0.1–1.6 mmol P m^{-2} d^{-1}, where there was a substantial contribution of meteoric groundwaters, but recirculation of seawater may also provide a means of transferring terrestrially derived nutrients. Taniguchi, Burnett, Cable, and Turner (2002) catalogued the widespread occurrence of freshwater springs in the littoral zone but also a more pervasive seepage unmarked by any obvious spring. This was recognized as a potentially significant route by which dissolved compounds reach the oceans, whether these are nutrients or pollutants. Estimates of volume vary from $<1\%$ of surface runoff up to about 10% (Garrels & MacKenzie, 1971; Zekster, Ivanov, & Meskheteli, 1973) to an optimistic 31% of total water flux (Lvovich, 1979), but some local studies have suggested values as high as 40% of surface runoff (Moore, 1996).

Quantitative results cannot be extrapolated to all reefs, each of which will have its own boundary conditions, but are important in indicating the critical controls on the chemistry of the circulating system and thus limits on cementation or dissolution that may take place. These include the initial permeability distribution, the geometry of the water bodies and surface recharge, defining surface and subsurface hydrology. The chemistry of the system is apparently dominated by oxidation of organic matter and is basically anoxic. Nitrogen and phosphorous concentrations are one to two orders of magnitude higher than in the overlying seawater, but pH is lower reflecting an increase in dissolved CO_2. Alkalinity is variable and may locally reflect either dissolution or cementation. Results obtained by Morse et al. (1985) in the Bahamas suggest that interstitial waters are close to equilibrium with respect to Mg-calcite, slightly supersaturated with respect to calcite and aragonite and substantially supersaturated with respect to dolomite. However, the initial growth of cements may be followed by dissolution as pCO_2 increases in response to the oxidation of organic matter.

Boreholes on Tikehau Atoll at the northern end of the Tuamotu archipelago (Rougerie, Fichez, & Dejardin, 1997; Rougerie, Jehl, & Trichet, 1997) showed wide variations in chemistry reflecting their positions relative to islets (motus). In two boreholes on reef flats, waters with salinities of 35.9–35.7 psu are oxic to approximately 27 m depth and suboxic to 33 m, where salinity is lower. In two holes situated in a reef pass adjacent (100 m) to a large island, water from 0 to 10 m is related to the freshwater lens of the island, and a marked anoxic zone between 10 and 20 m reflects the seaward extension of the mixing zone with salinities of 30–34 psu. Below about 30 m, the salinity is 35.5 psu. A single borehole on a lagoonal pinnacle is anoxic from top to bottom. The local deep penetration of the oxic zone in the face of interstitial decomposition of organic matter is attributed to wave-driven penetration of ocean waters, but $\delta\ ^3He$ values in interstitial waters imply an upward flow within the reef framework. Two boreholes through the barrier reef protecting Tahiti harbour (Rougerie, Fichez, et al., 1997) to 50 and 150 m depth respectively showed the Holocene/postglacial to extend to 50 m depth and the base of the carbonates to lie at 110 m depth. Waters are oxic to around 20 m depth, apparently as a result of lateral penetration of oceanic water. Below this, however, the redox potential is negative (−120 to −130 mV). The pH varies from 7.9 at the surface to 7.6 at 50 m. Salinity in the shallower hole is 35.7 psu.

7.3.2.3. Geothermal gradient and convective circulation

In contrast to wells drilled in continental sites, where temperatures typically increase with depth, a number of studies of oceanic islands have reported very different geothermal gradients. Temperatures were measured in both the Bikini and Enewetak boreholes as part of the 1952 investigations of these sites (Swartz, 1958). On Enewetak, there is a rapid temperature decrease to depths between 300 and 420 m, decreasing more slowly to 6.4°C at 954–1014 m. A closely similar profile is seen in surrounding ocean waters, and in one borehole, water levels followed those in the adjacent ocean in both phase and amplitude, attesting to the high permeability of the rocks penetrated. Similar results were obtained from repeated measurements and two further boreholes. Holes drilled on Bikini in 1948, and measured in 1952, showed a profile similar to that in the shallower parts of the Enewetak holes, decreasing from 28.7°C at the surface to 15.7°C at 430 m. Swartz (1958) also recorded translations of work by Sugiyama (1934, 1936) on Kita-Daito-jima that parallel profiles for Enewetak and Bikini with temperatures ranging from 22.5°C at the surface to 8.75°C at 431 m depth. Leclerc, Jean-Baptiste, and Texier (1999) presented evidence from boreholes through Mururoa Atoll that also shows a temperature decrease from approximately 26°C at the surface, to about 21°C at around 350 m beneath the atoll rim before increasing to around 35.0°C at 750 m beneath the lagoon (Figure 7.13A). Thus, notwithstanding an initial cooling, as in continental areas, the geothermal gradient here increases within the volcanic rocks on which the atoll rests. Simulations indicate that temperatures decrease downwards with increasing hydraulic conductivity of the karst layers (Figure 7.13B).

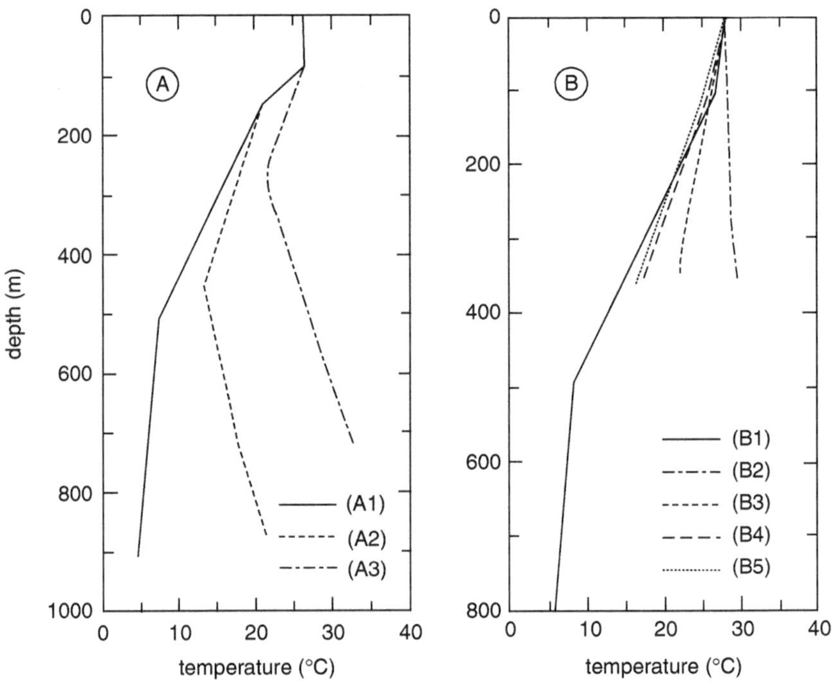

Figure 7.13 Vertical temperature (T) profiles from Mururoa Atoll. (A) Measures from the surrounding ocean (A1), from boreholes beneath the atoll outer rim (A2) and beneath the lagoon (A3). (B) Simulated T for different values of the hydraulic conductivity at the karstic layer (K), assuming a carbonate hydraulic conductivity of $10^{-4}\,m\,s^{-1}$. (B1) = oceanic T profiles used in the simulations; (B2) = simulated T profile for a homogeneous carbonate pile; (B3) = simulated T profile with a K value of $10^{-3}\,m\,s^{-1}$; (B4) with a K value of $10^{-2}\,m\,s^{-1}$; (B5) with a K value of $10^{-1}\,ms^{-1}$. Modified and redrawn from Buigues (1997) and Leclerc et al. (1999).

Partly reflecting this latter observation, it has been suggested that convective circulation in islands, driven by the geothermal gradient within subjacent volcanic piles, might be responsible for the delivery of additional nutrients to reef biotas at the surface. The fact that lagoon waters are generally less depleted in nutrients than surrounding surface waters led Rougerie and Wauthy (1986, 1993); Rougerie, Fichez, et al. (1997) and Rougerie, Jehl, et al. (1997) to suggest that convective circulation in the rocks below, which they referred to as 'endo-upwelling', might be responsible (Figure 7.14). However, although the circulation is real, its efficacy in this regard is doubtful. Both observation and modelling (Leclerc et al., 1999) indicate that nutrient flux can only provide a minor contribution. Leclerc et al. (1999) suggested that both lateral and vertical circulation of ocean waters should also occur within isolated islands and platforms. In order to replicate the negative temperature gradient indicated by the field data described above, it was necessary for their models to include a karst horizon. Where a barrier of this kind is near the surface, flows are directed laterally and a significant proportion of the net flow never reaches the surface. However, Umezawa et al. (2002) described nutrient

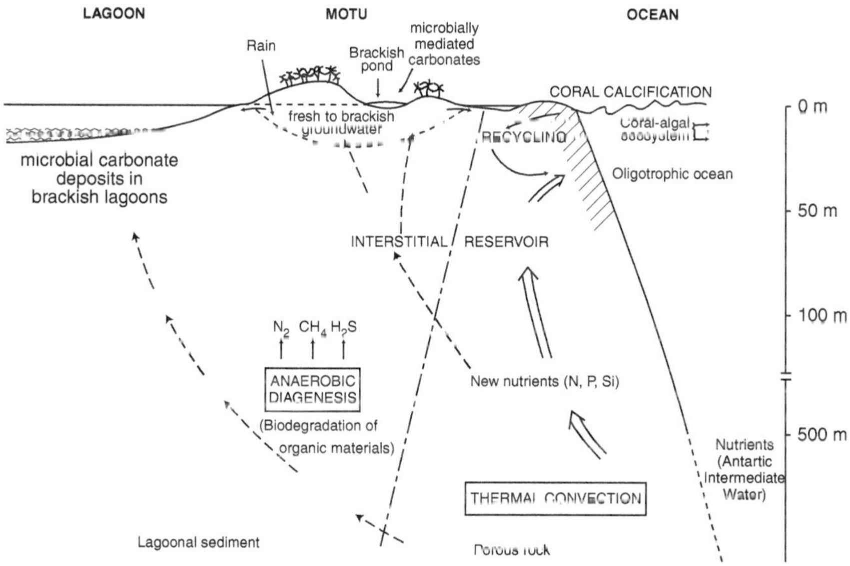

Figure 7.14 Conceptual model of interstitial water circulation by thermal convection (endo-upwelling) in Pacific atolls. Nutrients (phosphorus, nitrogen and silicium) are supplied from deep oceanic waters (Antarctic intermediate waters) and from recycling within the reef pile, thus maintaining high nutrient concentrations. Brackish waters may occupy ponds of the emergent reef rim and enclosed lagoons, promoting the development of microbial mats in which carbonate precipitation occurs. Modified and redrawn from Rougerie, Jehl, et al. (1997).

distribution in groundwater discharge through Shiraho and Kabira reefs on Ishigaki Island in the Ryukyu Group. With some variation between the two sites, the waters in both contain significant nitrogen that directly augments the reef nitrogen budget. This may be derived from land use in the adjacent watershed, but it has a potential control on biomass and illustrates the efficacy of groundwater flow in delivering nutrients to the reef.

Density-driven circulation in atolls and platform margins is clearly important, both in terms of assertions regarding nutrient supply and, more generally, statements relating to diagenesis. Modelling by Jean-Baptiste and Leclerc (2000) showed, using relatively simple arguments, that there is a negative feedback between convective flow and heat transfer such that, irrespective of the hydraulic conductivity of the medium, there is an upper limit to geothermally driven fluxes. Calculations indicate that although convective circulation is compatible with dolomitization models, it is one to two orders of magnitude too low to be significant in nutrient transfer.

An additional, albeit uncommon, aspect of interstitial waters in Holocene reefs is provided by Pilcher and Dix (1996) from Ambitle Island, Papua New Guinea. Here, boiling waters (about 100°C) emerge from discrete ports in the reef 10–15 cm in diameter, and as diffuse bubble streams at depths of 5–10 m. Flow rates in the larger vents are estimated to be 300–400 $l\,min^{-1}$. Precipitates include aragonite, ferroan Mg-calcite and Fe oxyhydroxides. Isotopic data indicate that the waters are of low salinity (<5 psu) and are of meteoric origin. In this context, it is important to note the association of shallow, cold water, coral banks with methane seepage described by Hovland (1990).

7.3.3. Flow in Pleistocene Reefs

Quaternary reefs and their associated carbonate platforms are typically characterized by an incremental stratigraphy in which periods of relative highstand of sea level are reflected in accumulation, whereas lowstands generate erosion surfaces that are commonly of karst form and draped with palaeosols. In many areas, there are also intercalations of aeolianites. The slopes surrounding reefs may bear lowstand wedges that in fringing reefs can include siliciclastic deposits, as in the GBR (Dunbar & Dickens, 2003; Braithwaite et al., 2004). Against this background of discontinuously stratified and widely varying permeabilities, there is no simple model to describe groundwater circulation and several significant driving forces have been identified. Generally, flow rates decrease as depth increases, but where the Pleistocene surface is shallow, hydrology may be dominated by karst permeability. It is important to note that many of the global figures quoted for flow are derived from non-carbonate rocks. Groundwater movements have been investigated in both large-scale carbonate platforms and oceanic islands.

7.3.3.1. Case studies from the Caribbean

Data from the Bahamas (Kohout, 1965) and the Florida shelf (Kohout, Henry, & Banks, 1977) formed the basis of large-scale circulation models, later referred to as 'Kohout circulation'. Investigations indicated that groundwater does indeed circulate and may be driven by several mechanisms, head-driven flow, geothermal heating and by density contrasts relating to salinity. A well-defined halocline develops beneath such platforms, and it has been shown that the geothermal heat flux beneath this is sufficient to drive a large-scale flow beneath the shelf and the adjacent ocean floor. Fanning et al. (1981) found evidence of circulation to depths of 500–1000 m, generating warm (40°C) springs emerging from the sea bed 45 km SSW of Ft. Myers, in Florida, and later observations by Paull, Chanton, Neumann, Coston, and Martens (1992) demonstrated that circulating waters also emerge from the Florida shelf margin escarpment as cold seeps associated with the escape of methane.

On a smaller scale, the islands of the Florida Keys consist of Pleistocene limestones formed during the last interglacial. Those of the Upper Keys (the so-called Key Largo Limestone) represent a reef tract, whereas the Lower Keys (Miami Limestone) are largely oolites deposited in tidal bars. Groundwaters in the Upper Keys are typically brackish, whereas the larger islands of the Lower Keys support small freshwater or slightly brackish lenses (Halley, Vacher, & Shinn, 1997). The water table tends to fluctuate with the tides and the range decreases inland. Rainfall varies dramatically over short distances and although the recharge response is rapid, the decay is also rapid, reflecting high permeability, and boundaries between facies units correspond with boundaries of conductivity.

The hydrogeology of the Bahamas islands, in which only a small proportion of the underlying rocks are of reef origin was reviewed by Whitaker and Smart (1997) who also emphasized the heterogeneity of permeability. Holocene deposits are unconsolidated or partly consolidated, and are generally thin bioclastic or oolitic sands derived offshore. Cementation varies regionally, and there is a partially cemented zone at and below the water table as a result of CO_2 degassing. Variations in cementation are in part a reflection of the higher annual rainfall in the northwest (1550 mm), double that in the southeast (690 mm). There is a consistent relationship between island area, the volume of the fresh water lens and mean annual rainfall (Figure 7.16). Potential evaporation is of the order of 1200 mm in the north, but is probably higher in the south, although no estimates are available. However, because of the high permeability of the surface, water is typically added to subsurface systems rather than lost to evaporation. Surprisingly, the correlation between lens volume and mean annual recharge appears poor. Total porosities may be 40–50% but with a relatively low transmissivity. Much of the Pleistocene sequence consists of the Lucayan Limestone, poorly cemented packstones and wackestones, with high hydraulic conductivity. This forms the major freshwater aquifers on Bahamian Islands, but these are never large. The areas overlying freshwater are typically four to six times greater in islands consisting of Holocene deposits than in those made up of Pleistocene limestones, reflecting the greater conductivity of the latter. The stratigraphical sequences are hydrologically partitioned by a series of subaerial (karst) exposure surfaces, on average one every 3 m, described by Beach (1995). As in Florida, the water table shows a rapid response to rainfall and there is a clear seasonal rise during the wet summer months followed by a slow decline in the dry winter season (Whitaker & Smart, 1997).

Whitaker and Smart (1990) described active circulation beneath the northern Great Bahama Bank (Figure 7.15). On Andros, expansion of the lens is paralleled by a decrease in salinity, but in areas where there are numerous tidal creeks or cavernous porosity, the lens is itself relatively thin, typically less than 35 m. Waters with salinities of 38–42 psu are generated by evaporation on the shallow surface of the Bank. These are driven eastwards beneath Andros Island and mix with waters of 36–37 psu and temperatures of 19–20°C drawn in from Tongue of the Ocean to the east. Two flow differentials have been identified. Water levels on the western margin of the banks are higher than those on the east, as a result of flow from the Florida Current. The resulting hydraulic head may be sufficient to drive water eastwards through the Banks. Oceanic Blue Holes on Andros, connecting with extensive horizontal caves along the eastern margin of the island, show

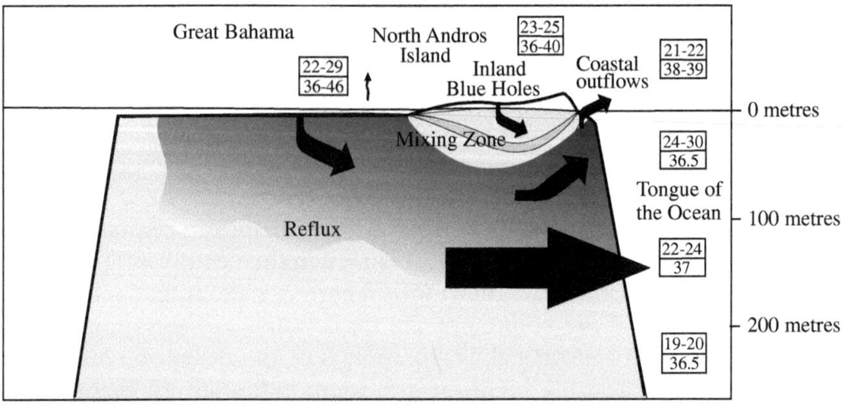

Figure 7.15 Large-scale circulation model for the Bahamian carbonate platform (Kohout circulation). The numbers given in the boxes refer to salinity of groundwater (top) and seawater (base) (redrawn from Whitaker & Smart, 1990).

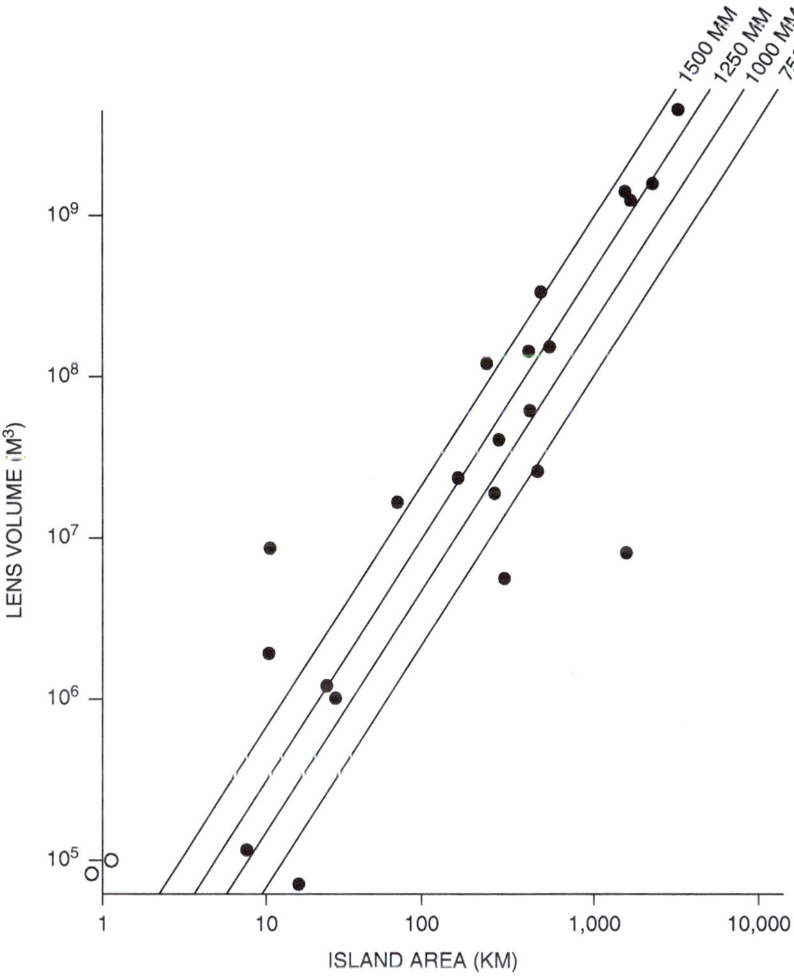

Figure 7.16 Relationship between island area, volume of freshwater lens and mean annual rainfall in the Bahamian Islands. Open circles represent islands exclusively composed of Holocene deposits. Modified and redrawn from Whitaker and Smart (1997).

strong currents that reverse semi-diurnally. These augment any eastward flow during low tides, and at these times currents are significantly stronger, providing a net discharge that is up to 4 psu more saline than water drawn in from the ocean during high tides. Two density-driven mechanisms are also important. Where the Bank emerges, the circulation within the freshwater lens drives water in the mixing zone seawards. The resulting discharge is compensated by an inflow of saline water at depth. Whitaker and Smart (1990) referred to this as buoyant circulation. By contrast, where the water at the surface of the banks becomes more saline as a result of evaporation, the increase in density drives a downward flow (reflux) displacing the less dense waters beneath. Although this latter mechanism was originally regarded as a reflection of hypersaline waters (Adams & Rhodes, 1960), it has since been shown theoretically (Simms, 1984) to be possible with only a slight elevation in salinity (of around 37–42 psu). Finally, geothermal heating may cause less dense waters to rise within the Bank and to flow laterally, with cold oceanic waters drawn in at depth to replace them. Circulation patterns based on these mechanisms have been extensively modelled by Whitaker, Smart, and Jones (2004) over distances ranging from a few tens to hundreds of kilometres.

Hanshaw and Back (1980) investigated the extensive carbonate platform of Quintana Roo, Yucatan Peninsula, Mexico. Quarries and low sea cliffs along the Caribbean coast reveal Pleistocene limestones, including reef deposits and aeolianites. In the northern part of the platform, annual rainfall averages slightly more than 1000 mm yr^{-1} and is either lost to evapotranspiration or percolates rapidly to join subsurface flow. The freshwater lens is about 70 m thick, floating on salt water of nearly oceanic composition. The high transmissivity of the system is demonstrated by the fact that around 80 km from the coast, where the land surface is about 30 m high, the water table is only 1.5 m above mean sea level. High production wells generate little drawdown and ^{14}C analyses indicate that the water is generally modern. The mean annual recharge is approximately 150 mm and the average discharge to the sea is about 8.6×10^6 m^3 yr^{-1} for each kilometre of coastline. Virtually, all water samples are in equilibrium with calcite and calculations suggest this state equates to the dissolution of 37.5 Mt (2.5 mmol l^{-1}) of calcite per year for each square kilometre. The mixing zone is about 3 m thick and extends inland for about 1 km, and within this, there is the potential for dissolution of a further 1.2 mmol l^{-1} yr^{-1}. However, local conditions may be influenced (sometimes detrimentally) by the uncontrolled extraction of water from wells. Smart et al. (2006) recorded the development of an extensive anastomosing network of flooded caves on the east coast of the peninsula. These extend 8–12 km inland and apparently formed relative to the present fresh/saline-water mixing zone. The coast is incised into the subhorizontal late Pleistocene (MIS 5) shelf margin, comprising reef and back-reef deposits at least 12 m thick, extending up to 10 km inland. Further inland, the surface consists of Pliocene and Miocene limestones resting on some 300 m of carbonate-rich breccias associated with the Chicxulub meteorite impact structure. An arc of cenotes (Blue Holes) connects to caves marking the fractured outer rim of the impact area. Contrasts in hydraulic conductivity within this zone promoted cave formation (Hildebrand et al., 1995). Coastal discharge was greater than at present when the caves formed, and phases of cave development can be related to glacioeustatic changes in sea level. Significant cave development occurred only when passages were occupied by mixing zone waters with active flow removing lowstand sediment fill. A characteristic feature of the system is an almost ubiquitous roof failure. Breakdown piles are common throughout the caves, reaching heights of up to 20 m. The frequency of mechanical failure reflects the relative weakness of these young rocks and their lack of cementation (Smart et al., 2006).

Beddows, Smart, Whitaker, and Smith (2004) found, in the Yucatan Peninsula and on Andros in the Bahamas, that aquifers in inland blue holes are stratified and were able to measure both temperature and salinity profiles to depths greater than 100 m below the water table. Earlier studies in Yucatan had reported high temperatures within the mixing zone and had interpreted them (Stoessell, Coke, & Easley, 2002) as evidence of rising geothermal waters. However, Beddows et al. (2004) showed that because of the contrast in density between the fresh and saline waters, stratification in many holes is stable. In addition, as a result of their greater specific heat capacity, the mixing-zone waters heat up in response to surface insolation, reaching temperatures ranging from 0.1 to $>10°C$ higher than those of either the overlying fresh water or the more saline waters beneath. They referred to this process as 'heliothermic' heating but did not rule out the possibility that geothermal convection might also occur. Many of the models applied to circulation patterns assume convective circulation.

7.3.3.2. Case studies from the Pacific

Investigations of Pacific islands have also produced data on the variability of hydrological conditions.

Niue with a surface area of $259\,km^2$ is one of the largest carbonate islands in the south Pacific (Aharon, Rasbury, & Murgulet, 2006). The core of the island consists of Miocene and Pliocene limestones with two dolomite intervals, and with the volcanic basement beneath the south-western area. A complex of Pleistocene fringing reefs forms a narrow terrace rising to 23 m around the island, terminating in steep sea cliffs. The island has been subaerially exposed at intervals since the early Pleistocene as a result of crustal warping (Wheeler, Aharon, & Ferrell, 1999). There is a narrow (about 100 m) modern fringing reef. No surface streams are present and the vadose zone is 30–70 m thick, but, like Barbados, the water table is currently only 1.83 m above sea level. The freshwater lens is approximately 40 m thick in the centre of the island, thickening to about 150 m before tapering out near the coast (Wheeler & Aharon, 1997). The freshwater–saltwater mixing zone is at least 40 m thick. Tidal fluctuations, albeit dampened, are transmitted to the water table because of the high permeability of the rocks. The mixing zone discharges at sea level in numerous springs (Aharon et al., 2006). Steeply dipping (30–40°) terraced caves occur predominantly in the Pleistocene rocks on the leeward side of the island. These formed along the seaward margin of a migrating freshwater lens, and their leeward distribution presumably reflects a reduced seawater flushing. Inland interconnecting flat-roofed water-table caves contain features reflecting alternating vadose and phreatic conditions. Stalagmites in the flanking caves grew about 1.5 times faster than those in the interior, but are only active in the lowest caves. Analyses of the dolomite in the Tertiary deposits indicate relatively high concentrations of Fe, Cu, Zn and Mn that imply that the waters responsible for its formation passed through the underlying volcanic rocks. Hovland (1990) suggested that the water might have included exhalative fluids. The discharge of meteoric-derived waters to the coastal margin is well established in some areas, reaching as much as 10 m yr^{-1} (Taniguchi et al., 2002). Average rates appear to be below 0.1 m d^{-1} although they can be higher than local rainfall.

On Mururoa and Fangataufa Atolls, Pleistocene limestones overlie Tertiary carbonates that are again partly dolomitized and rest on volcanic basements. Geothermal heating within the volcanic rocks causes convection in which waters rise centrally and are replaced by cooler ocean waters drawn in laterally (Buigues, 1997).

The island of Makatea in the Tuamotu archipelago also rests on a volcanic cone covered by a thick Neogene succession in which there is similar extensive dolomitization and also superficial phosphatization (Montaggioni & Camoin, 1997). Most of the island is surrounded by high-energy fringing reefs, up to 100 m wide, principally along the southern coast. The Pleistocene succession occupies only a limited area, comprising two terraces dated at 100–140 ka and 400 $\pm$ 100 ka. Like Barbados, Makatea was subject to Pleistocene tectonic uplift and the island is almost entirely surrounded by steep cliffs. During glacial lowstands, extensive dissolution by meteoric waters caused the enlargement of karst cavities in both Pleistocene and Neogene deposits, with some potholes up to 75 m deep. Notches in the cliffs occur at four levels, associated with open caves and galleries. During the last sea-level highstand, limestones of the 120 ka reef terrace were dolomitized as a result of percolation of magnesium-rich waters through the Miocene dolomites.

Enewetak Atoll in the Marshall Islands consists of about 40 low-relief islets surrounding a lagoon 40 km long and 32 km wide (Buddemeier & Oberdorfer, 1997). The present reefs and islands grew up on a basement consisting mostly of Pleistocene material, deposited during the last interglacial. The eroded surface of this occurs at depths of 8–10 m beneath the islands. Unusually for an ocean island, the tidal range on Enewetak is relatively large. Rainfall on the atoll averages 1470 mm y^{-1}. Groundwater characteristics have been investigated on 10 of the atoll islands. Residence times for meteoric water range from 4.5 to 6 yr^{-1} with one that may be as low as 2 yr^{-1}. Only Enewetak has a freshwater body that is sufficiently extensive and persistent to be considered as a water resource; but the residence time is uncertain and may be less than 5.6 yr^{-1}. Extensive pit sampling indicates that there are substantial variations in head, tidal response and water quality over distances of tens of metres (Buddemeier & Oberdorfer, 1997). Boreholes on Enjebi Island (Enewetak) show a linear salinity distribution, increasing with depth (Figure 7.17). The permeability of the Pleistocene succession is thought to exceed that of the Holocene by

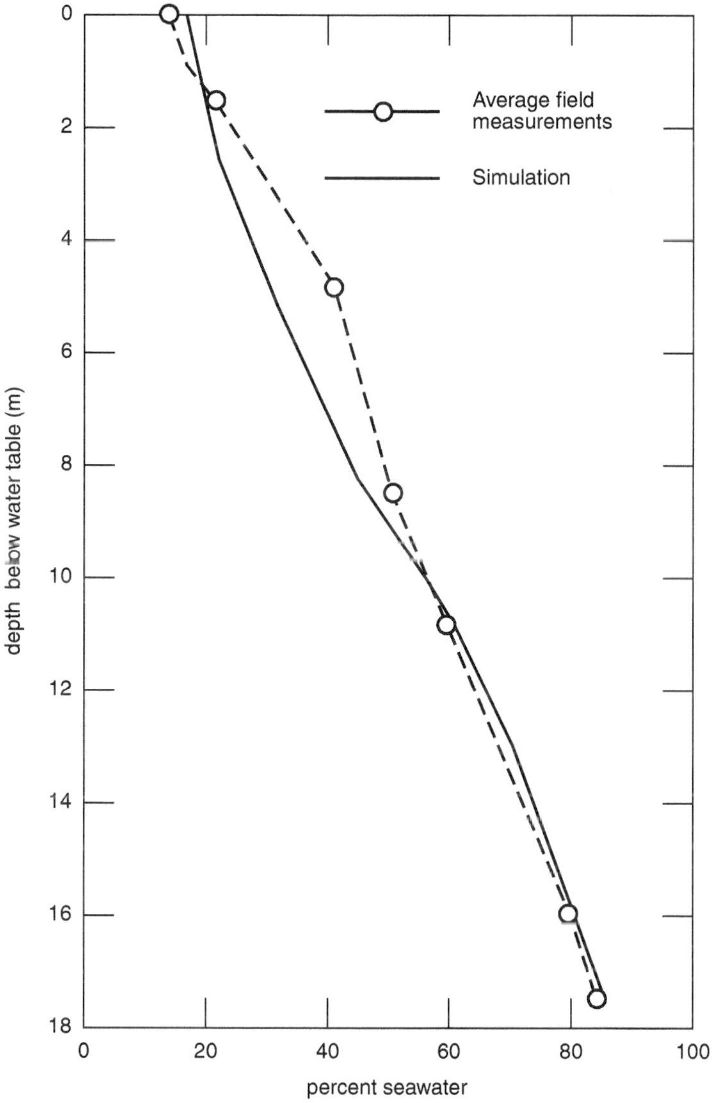

Figure 7.17 Measured and simulated, averaged salinity profiles from groundwater at Enjebi Island (Enewetak Atoll). Salinity is expressed as percentage of seawater into groundwater (from Buddemeier & Oberdorfer, 1997).

one or two orders of magnitude (Buddemeier & Oberdorfer, 1997) and as a result, the tidal lag is locally as low as $0.25\,h^{-1}$. Simulations suggest that mean flow directions reverse with the tide, forming inward flows during high tide and outward flows during low tide, and with this oscillation the key factor in mixing. On Enewetak, pits showed head varying from 0.2 to 0.44 m and tidal lag from 1.7 to $3.7\,h^{-1}$ with no coherent pattern. These differences are attributed in part to short-range variations in the surface of the Pleistocene. Boreholes typically show a strong stratigraphical influence with particular emphasis at dissolution unconformities. Mixing with the underlying seawater appears to be the major means by which freshwater is lost. Waves driven across the reef are ponded in the lagoon, with the resulting head driving a seaward flow, principally through the Pleistocene succession. There is no unequivocal evidence of where water emerges but the presumption is that it is from subsea springs.

These are described from many coastal karst terrains and have been observed on the reef platform on the Kenya coast (Braithwaite, 2005).

7.4. Conclusions

Rapid changes in bathymetry and water agitation have been major forcing factors controlling coral population dynamics and zonation. Waves, and wind-driven flows, are the primary influence on the shallower portions of reefs, whereas tidal currents dominate the outer regions and the island shelf. However, these independent systems are commonly opposed, and generate a zone of cross-shore horizontal shear and a front along which turbid lower salinity water is held inshore of clear higher salinity water. Convergent flows may also generate eddies and, in the absence of large waves, accumulate sediment. Turbid areas tend to be correlated with regions of poor coral health.

Water turbulence, especially that resulting from extreme events (cyclones, hurricanes and tsunamis), control the morphology and structure of coral assemblages and the compositions of detrital sediments, breaking and remodelling the reef surface and also providing surfaces for recolonization that contribute to species diversity, reef preservation and depositional patterns discussed in Chapters 3, 4 and 5 respectively. The extent to which variations in SST modulate coral growth and reef development is not well constrained, largely because it is difficult to separate thermal effects from those reflecting changes in nutrient supply and turbidity. During glacial intervals, decreasing temperatures do not seem to have markedly affected reef building, although falls in sea level restricted its distribution. There is compelling evidence of substantial reef development during several Pleistocene glacial intervals, and particularly during the Last Glacial Maximum. By contrast, nutrification levels seem to have been a major determinant of coral growth. Following the demise and burial of the reef system, promoted both by tectonic and eustatic changes in sea level, groundwater circulation has continued to influence chemical environments controlling the character and distribution of cements discussed in Chapter 8.

The hydrogeology of islands has an important control on their diagenesis. Large-scale circulation within carbonate platforms may be driven by hydraulic head, geothermal effects and so-called Kohout circulation, and similar effects are reported on a smaller scale in many oceanic islands. In these, as a result of the high permeability of Pleistocene sequences, hydrology is commonly separated into two or more layered systems giving the dual-aquifer concept common among reef islands. Rainfall is an important first-order control on the volume and characteristics to the freshwater body. In small islands, the base of the freshwater body is effectively truncated as a result of the relatively rapid flow of saltwater beneath. The tidal lag within islands decreases with distance from the shore but is primarily a response to local variations in permeability.

The compositions of meteoric waters vary as a result of their passage through rocks. They may be oversaturated or undersaturated with respect to calcite and aragonite, but simultaneously saturated with respect to dolomite, particularly when they are mixing with seawater (Ward & Halley, 1985). Thus, locally, groundwaters may be responsible for dissolution, neomorphism or dolomitization of the rocks through which they pass. Dolomitization of Pleistocene limestones seems to be less common than in the Tertiary limestones beneath. These and other diagenetic issues are discussed in Chapter 8.

CHAPTER EIGHT

Reef Diagenesis

8.1. Introduction

The term diagenesis relates to the physical and chemical changes that occur in sedimentary deposits and which result in the formation of rocks. These include physical compaction, dissolution, *in situ* mineralogical transformations and the growth of new minerals as cements. There is a long history of investigation of these processes, beginning with work such as that of Cullis (1904) but driven on more recently by the interest of the Oil Industry in porosity development in reef structures. A selection of more recent contributions can be seen in publications edited by Bricker (1971), Schneidermann and Harris (1985), Schroeder and Purser (1986), Gautier (1986) and McIlreath and Morrow (1990), with the review by Macintyre and Marshall (1988).

It can reasonably be argued that the transformation of the sediments and growth frameworks of reefs to the porous limestones, familiar on many tropical coastlines, begins while deposition is still in progress. However, far larger changes occur following burial and in particular during uplift. The history of Quaternary reef systems has been punctuated by repeated sea-level changes that have been reflected in profound changes in water chemistry and, as a consequence, changes also in the mineralogy and petrographic textures of the resulting rocks. This transformation depends on two factors: the inherent instability of the component minerals and the long-term instability of the environment in which they have been deposited.

The organisms that are the principle sediment formers in reefs secrete skeletons that consist of three principle minerals, aragonite, low- and high-magnesium calcite (Mg-calcite). The proportions of these vary in different phyla and their initial distribution within the reef system is therefore controlled *pro rata* by the ecology of those organisms The mineralogy of organisms has apparently varied through geological time and one of the reasons put forward to explain this is the fact that the chemistry of the oceans has also varied. Throughout the Phanerozoic, composition has fluctuated between periods when aragonite has been the common inorganic precipitate and periods when the dominant form has been calcite (Sandberg, 1983). It has been argued that such secular changes may be driven by changes in the Mg/Ca ratio of the water, controlled in turn by changes in seafloor spreading rates (Stanley & Hardie, 1998). However, the extent to which such changes are able to influence the compositions of minerals secreted by organisms is arguable and recent work by Kiessling, Aberhan, and Villier (2008) has suggested that there is no correlation, arguing instead that long-term changes in skeletal mineralogy have been predominantly regulated by mass extinctions.

All three minerals (aragonite, high- and low-magnesium calcite) are also represented in organic and inorganic precipitates that locally make significant additional contributions to the reef mass. Post-deposition, two additional minerals may be added, dolomite and phosphates, both reflecting the flow of waters through the rock mass. The mineralogical changes that take place in Pleistocene limestones reflect a progressive, but commonly incomplete, stabilization of the mineral assembly in response to the influence of the various water chemistries described in Chapter 7.

8.2. Mineralogy of Sediment Components

Of the principal organism groups, reef-building scleractinian corals consist almost exclusively of aragonite, although traces of strontianite (Greegor, Pingitore, & Lytle, 1997) and even calcite have been recorded (Houck, Buddemeier, & Chave, 1975; refuted by Macintyre & Towe, 1976; and Gladfelter, 1983). However, there is some evidence (Stolarski, Meibom, Przenioslo, & Mazur, 2007) that in the past scleractinians did secrete calcite and interesting experiments by Ries, Stanley, and Hardie (2006) show that at least three living species are able to secrete calcite when grown in seawater of inferred Cretaceous composition. Among the octocorals (Alcyonaria) the so-called 'precious coral' *Corallium rubrum* (Weinbauer, Brandstatter, & Velimirov, 2000) and *Sinularia* (Rahman & Oomori, 2008) secrete sclerites of Mg-calcite, and some gorgonians also contain sclerites of aragonite, Mg-calcite and amorphous hydroxyapatite (Bayer & Macintyre, 2001) that are dispersed when the colony dies, but most are unmineralized. Calcareous algae fall into two groups. The red algae that include encrusting species such as *Porolithon*, *Hydrolithon* and *Lithothamnion*, as well as articulated branching forms such as *Amphiroa*, *Jania* and *Corallina*, common in reefs, secrete Mg-calcite crystals in their cell walls, forming a rigid structure. The green algae including *Halimeda* and *Penecillus*, form branching and sometimes segmented plants, but secrete loosely bound needle-like crystals of aragonite in their cell walls. The latter are of particular interest because the crystals closely resemble those precipitated inorganically from seawater and carbonate production by these algae has been implicated in the generation of large volumes of carbonate 'mud' (see Chapter 5, Section 5.3.1). Among other encrusting species, calcified bryozoans primarily secrete calcite, although some include secondary growths of aragonite. The foraminifera also include important encrusting species that, together with benthic and planktic forms, secrete skeletons of Mg-calcite. Reefs support a diverse molluscan assemblage, the skeletons of which include aragonite and calcite, and sometimes both minerals together, with calcite showing a range of magnesium compositions. Echinoderm skeletons typically consist of Mg-calcite (up to 20 mol% Mg), characteristically forming porous crystals that are exceptionally large in comparison to those of other groups. Echinoid spines, for example, may be centimetres in length and yet effectively consist of single crystals. With the exception of barnacles, that have calcite skeletons, arthropods are generally poorly mineralized and do not form significant sediment components. Among the annelids, serpulids may secrete aragonite, calcite, or both and are locally important encrusters. Finally, microbial activity, principally of cyanobacteria, is responsible for the growth of encrusting stromatolites that include precipitated calcite (Defarge et al., 1994, 1996; Camoin et al., 1999). Both aragonite and Mg-calcite are metastable. Over time — but particularly in the presence of undersaturated waters that may reflect lower temperatures, the addition of CO_2 or dilution by meteoric water — they are subject to alteration. It is important to note, however, that relative stability is not only controlled by mineralogy but also by structure, the ways in which the mineral elements are incorporated into the various skeletons and their relationship to the organic components.

The processes responsible for the transformation of these skeletal remains into sediment have a direct effect on the sizes of the grains generated. The net result is that both the reef framework and associated sediments include varying proportions of unstable components. The rates of production of grains and the general controls on their distribution are discussed in Chapter 5, Section 5.3.1.

8.3. Cements in Quaternary Reef Limestones

Cements represent the inorganic and organically mediated crystallization of new mineral phases, aragonite, calcite (and sometimes also dolomite or phosphates) within inter- and intragranular pores of the sediment, in response to changes in the physico-chemical environment, with the result that this becomes a rock.

Cements in Quaternary limestones fall into two groups: synsedimentary, formed during or soon after deposition under marine conditions, and those crystallizing from fresh (meteoric) or mixing-zone waters following sea-level change. In saturated solutions random collisions of ions must overcome kinetic barriers to form stable clusters and these nuclei then enlarge to become crystals. In carbonate sediments, the crystals forming the grains act as critical nuclei, thereby avoiding this hurdle, but additional independent nucleation is also possible. The ultimate sizes of crystals depend on the rate of formation of new nuclei relative to their collective rate of growth. Crystal habit may be modified by differing rates of growth. In calcite, it may be energetically easier for ions to be added to dislocations exposed on the c-axis faces of crystals than to relatively perfect prism faces, and as growth rates increase crystals become increasingly elongate.

The general nature of these changes is well known, and it can be shown experimentally that similar crystal forms may be generated by a variety of driving mechanisms.

8.3.1. Controls on Cement Morphology

8.3.1.1. Contamination

Numerous 'foreign' components added to crystallizing solutions have long been known to influence crystal growth (Buckley, 1951). Variations in form described by Grigor'ev (1965) included a range from 'split crystals', to bow-ties, wheatsheafs and spherules that were attributed to sectoral adsorption of 'impurities' on specific growing faces. Split crystals (Figure 8.1) result from the division and divergence of molecular scale elements of the growing crystal with the degree of divergence reflected in this cognate series. Split crystals are common in calcites in Mururoa Atoll (Braithwaite & Camoin, unpublished). No specific driving agent has yet been identified but a variety of organic compounds might be considered.

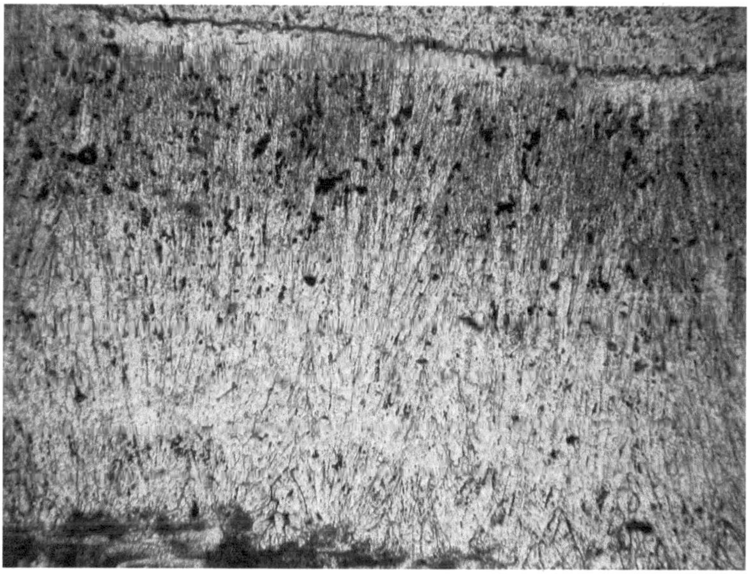

Figure 8.1 Radial-fibrous calcite. Note centres of nucleation and divergent fibres, split crystals, increasing in number outwards (crossed polars). Outer reef rim, Mururoa Atoll, 94.5 m core depth, late Pleistocene. Image width = 0.8 mm. Photograph by C. Braithwaite.

8.3.1.2. Growth rates and reactant supply

Given and Wilkinson (1985) were among the first to demonstrate a systematic relationship between crystal form and mineralogy and rates of reactant supply. Low-magnesium calcite typically forms equant crystals under meteoric conditions, but also in deep and/or cold marine settings and following deep burial. These are environments where saturation and/or rates of fluid flow are low and waters are likely to be relatively acidic, containing more dissolved CO_2. At high levels of supersaturation and during CO_2 evasion, typically in vadose settings, crystals are sometimes acicular and may form fibres or whiskers. A large proportion of the radial-fibrous crusts, common in many Quaternary reefs, reflect growth in the vadose zone. In warm shallow marine environments, precipitation is commonly driven by decreases in the solubility of CO_2 as a result of heating. In such environments in recent sediments, acicular crystals of high-magnesium calcite are common but radial-fibrous crusts of high-magnesium calcite are also found in the deeper parts of the Mururoa core (Aïssaoui, Buiges, & Purser, 1986) and are interpreted as of marine origin.

8.3.1.3. Changes in water chemistry

The composition of the water may have similar controls, with the proportion of magnesium in particular influencing crystal shape. Morse, Wang, and Tsio (1997) showed how both mineralogy and morphology can be influenced by subtle changes in water chemistry. The Mg:Ca ratio at which precipitation of carbonate switches from calcite to aragonite is strongly controlled by temperature, producing a range of 'split crystal' forms in both calcite and aragonite. Ricketts (1980) and Fernández-Díaz, Putnis, Prieto, and Putnis (1996) showed how the morphology of growing calcite and aragonite crystals may be controlled by the supersaturation state of the solution. In a pure $CaCO_3$ system, calcite crystallizes at lower supersaturation than aragonite (Ca^{2+} and CO_3^{2-} of $<20\,\text{mmole}\,l^{-1}$), forming simple rhombohedra, commonly with stepped faces and skeletal 'hopper' or dendritic forms typical of rapid growth (Sunagawa, 1982). Aragonite grows at higher supersaturations, forming spherulites, bundles of needle-like crystallites (split crystals) with orthorhombic prism faces. When small concentrations of magnesium are added to the solution, nucleation is initially inhibited, but when it resumes calcite crystallizes as rough spheres, dumbbells and wheatsheaf (bow-tie) forms also typical of split crystals. The greatest division and the most nearly spherical shapes occur in calcites with the highest magnesium content. Spherular (botryoidal) cements are a common feature of many reef sequences (Figure 8.2). In some of these, diverging elements within crystals may have clear boundaries but in others can only be recognized in a sweeping radial extinction. Aragonite crystals

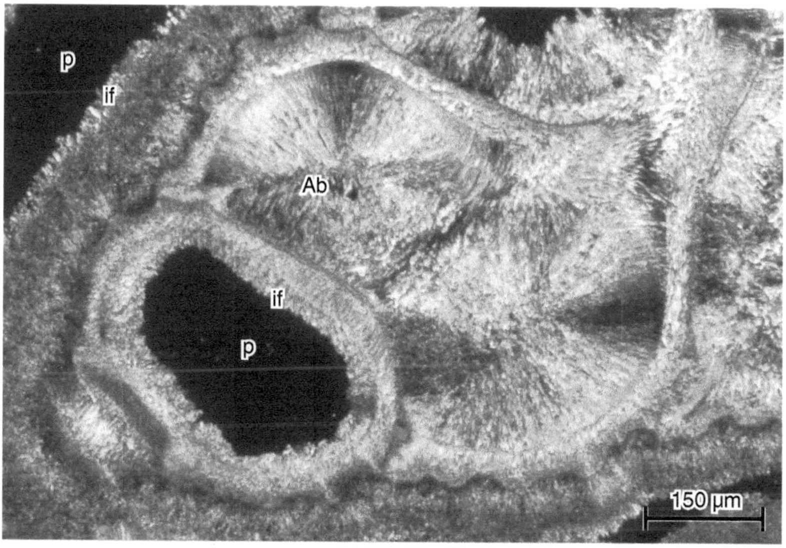

Figure 8.2 Isopachous epitaxial overgrowths (if) of acicular aragonite and spherular aragonite (Ab) around and within a gastropod shell respectively, outer reef-flat zone, Réunion Island (crossed polars). p = primary pore. Photograph by L. Montaggioni.

formed at the highest supersaturations are identical to those resulting from growth without magnesium.

8.3.1.4. Rates of fluid flow

The simultaneous influence of both fluid flow and composition on crystal growth has been examined in experiments by González, Carpenter, and Lohmann (1990). Waters with lower than six times supersaturation form flattened rhombohedral crystals but with increasing saturation terminations become more acute. These relationships can explain the transitions seen in many areas between 'nail-head' and 'dog-tooth' terminations in blocky calcites (Figure 8.3). Fibrous arrays and equant crystals are produced from fluids with overlapping saturations and concentrations but with differing Mg:Ca ratios. González et al., 1990 showed that, contrary to the observations of Given and Wilkinson (1985), rhombs formed in pools at saturations of 4.6 and Mg:Ca ratios of 0.03, whereas elongate crystals developed at supersaturations as low as 1.3 and Mg:Ca ratios of 2.83.

Frisia, Borsato, Fairchild, and McDermott (2000) described the controls of fluid flow and composition on the morphology of aragonite speleothems. In these, aragonite forms both needle crystals and radiating prismatic clusters (baguettes) consisting of parallel subcrystals. These are locally intergrown by rhombohedral calcite associated with fibrous calcite

Figure 8.3 Dog-tooth terminations of low-magnesium calcite crystals, late Pleistocene reef-associated deposits, Mauritius. SEM photograph by L. Montaggioni.

comprising radiating clusters of crystals with ragged interpenetrant boundaries. Calcite locally also replaces fan-arrays of former acicular aragonite crystals. The primary controls on the morphology of the aragonite apparently include degassing (CO_2 evasion), evaporation, and the fluid Mg:Ca ratio. Degassing and evaporation generate needle crystals whereas fan-like arrays form at low flow rates when the Mg:Ca ratio is >1.1.

8.3.1.5. Microbial control

In addition to the abiotic influences on crystal growth, many workers have identified, and in some cases also reproduced in experiments, a microbial control (see Chafetz, 1986 and references therein). Peloidal cements, consisting of rhombic microcrystals of Mg-calcite (micrite) have been widely reported in reefs and their interpretation has been controversial. Macintyre (1985) concluded that some could be formed as a result of abiotic chemical processes. However, similar micropeloids and clotted fabrics have also been attributed to bacterial precipitation (Chafetz, 1986;

Buczynski & Chafetz, 1991) with similar fabrics described by Camoin and Montaggioni (1994) and Camoin et al. (1999, 2006) from Tahiti reefs. Buczynski and Chafetz (1991) also described structures that superficially resemble spherules but lack the radial structure consistent with a split-crystal mode of growth.

Although a microbial influence is not necessarily expressed in distinctive structures, it appears to act in the formation of a variety of marine cements. Neumeier (1999) designed experiments to reproduce the cementation of beach rock. Simulated conditions included tidal movements, temperature, evaporation and water composition, controlling both salinity and saturation. A proportion of the experiments were inoculated with undetermined microbes and observed for 3–4 months. A variety of cements were generated including bladed and micritic high-magnesium calcite, and micritic and acicular aragonite, and the crystals generated were similar to those in typical beach rocks (Bricker, 1971).

González-Muñoz, Ben Chekroun, Ben Aboud, Arias, and Rodriguez-Gallego (2000) inoculated agar–agar gels with cultures of *Myxococcus xanthus*, a common soil bacterium, and produced nucleation and crystal growth, apparently as a result of local supersaturation from diffusing microbial metabolites. Mg-calcite crystallized with forms including spherules, dumbbells and wheatsheaf clusters consisting of bundles of subcrystals resembling in all respects those generated abiotically. Verrecchia et al. (1995) described the formation of calcite spherules within the mucilaginous sheaths of cyanobacterial cells in laminar crusts in Pleistocene calcretes. Braithwaite and Camoin (unpublished) have described fibrous calcite cements apparently nucleated on microbial filaments (Figure 8.4). Although experiments are able to reproduce the morphologies adopted by crystals, they unfortunately demonstrate that similar crystal forms can be generated within a range of physico-chemical conditions. Thus, although morphological variants attest to the fact of environmental change, with few exceptions they offer no guidance as to the precise nature of that change. It is likely that most crystal growth reflects the interaction of a number of factors rather than a single specific cause.

8.3.2. Textures of Cements

Crystalline cements may be difficult to differentiate from coarse neomorphic crystal growths. Grigor'ev (1965) described the 'geometric selection' that results from the growth of cement. Following random nucleation on the substrate, crystals with their fastest-growing axes normal to the surface are able to grow rapidly into the lumen of the pore whereas those with more nearly tangential orientations rapidly find their growth blocked and are overgrown. In carbonate sediments where a large proportion of nucleation is provided by existing surfaces (the crystals forming bioclastic grains),

Figure 8.4 Calcite cement nucleated on matted microbial filaments. Outer reef rim, Mururoa Atoll, 45.12 m core depth, late Pleistocene (plane polarized light). Image width = 0.8 mm. Photograph by C. Braithwaite.

cement crystals commonly preserve the orientation of these within the original grain surface. Acicular epitaxial overgrowths of aragonite are common within corals (Figure 8.5). Epitaxial overgrowths are also observed from calcitic bioclasts (Figure 8.6). However, as a result of geometric selection, cement can in principle be recognized by three criteria outlined by Bathurst (1958). From the margins to the centre of the pore, crystals decrease in numbers because of overgrowth, increase in size and become more strongly oriented, although the last of these may not be obvious where the crystals filling the remaining space may be rooted on any of the surrounding pore walls.

An important point regarding both primary grains and cements is their subtle variation in composition. In ancient rocks, differences in composition may be revealed by staining (e.g. Dickson, 1966) or by the use of cathodoluminescence (e.g. Meyers, 1991). Although staining remains practical, cathodoluminescence seems to be generally ineffective in Quaternary deposits. However, bioclasts in recent deposits commonly retain their original organic matrices and crystals growing in environments rich in organic matter (fulvic acids in particular) incorporate these into their structure with the result that they fluoresce in ultraviolet (UV) light (Ramseyer et al., 1997) with brighter zones reflecting greater incorporation of organic compounds. UV fluorescence commonly reveals growth zones within corals for example (Scoffin, Tudhope, & Brown, 1989b) and within

Figure 8.5 Epitaxial overgrowths of aragonite in a coral acroporid skeleton. pA = skeletal (primary) aragonite; sA = epitaxial, inorganic (secondary) aragonite. SEM photograph by A. Ribaud-Laurenti.

Figure 8.6 Echinoid spine with epitaxial overgrowth. Note, however, that these do not show the common optical continuity but are 'split crystals' with divergent molecular structure reflected in radial extinction (crossed polars). Outer reef rim, Mururoa Atoll, 94.5 m core depth, late Pleistocene. Image width = 0.8 mm. Photograph by C. Braithwaite.

abiotic crystals (Braithwaite & Montaggioni, 2009). In the latter, the occurrence of such compounds has been related to the intensity of surface run-off from the adjacent land mass.

8.3.2.1. Marine cements

On reef systems, there is a relationship between the degree of early cementation, the proximity to the open ocean, and the textural attributes of the sediments affected. The closer the sediments are to areas of wave agitation (driving water through the pores) and the coarser the deposits, the greater their potential for early marine cementation (Aïssaoui & Purser, 1985) (Figure 8.7).

A variety of crystal forms has been described in marine cements. Aragonite forms acicular crystals that may grow as isopachous fringes (Figure 8.8) or as epitaxial growths on the surfaces of aragonitic bioclasts (Figure 8.2). Radial clusters of prismatic crystals also occur locally together with spherular (Figure 8.2) or botryoidal growths (Figure 8.9). Calcite is typically magnesium-rich. It also forms radial-fibrous fringes (Figure 8.1) and isopachous crusts of needle-like crystals (Figure 8.10) that may be difficult to differentiate from those of aragonite. The dominant fabrics of high-magnesium calcite are structureless and peloidal micrites (Figure 8.11).

Surface seawater is generally supersaturated with respect to both aragonite and calcite. Broecker (1974) suggested that in the tropics, it may locally reach

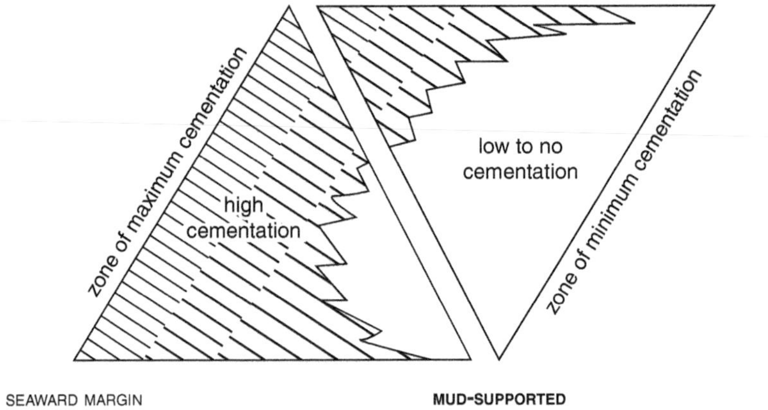

Figure 8.7 Relationships between the location and textural attributes of reef sediments and the potentiality of early marine cementation. Modified and redrawn from Aïssaoui and Purser (1985).

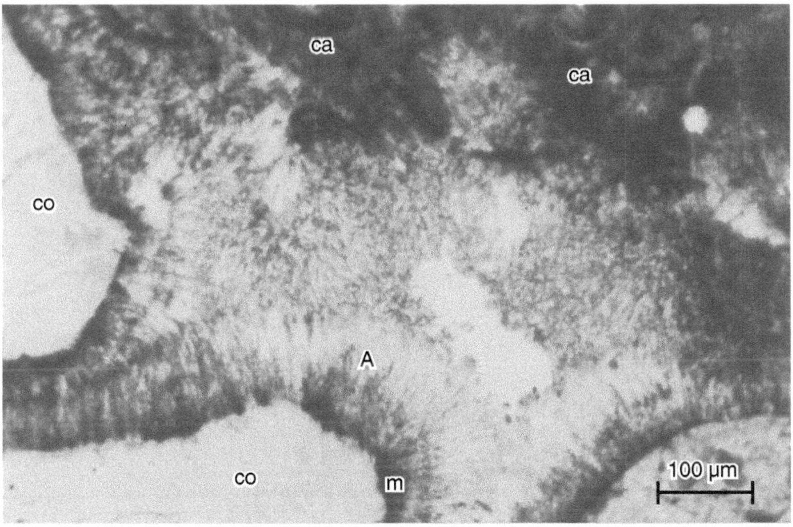

Figure 8.8 Isopachous fringes of acicular aragonite surrounding skeletal grains, outer reef-flat zone, Réunion (plane polarized light). A = aragonite; m = micritic envelope; ca = coralline algae; co = coral. Photograph by L. Montaggioni.

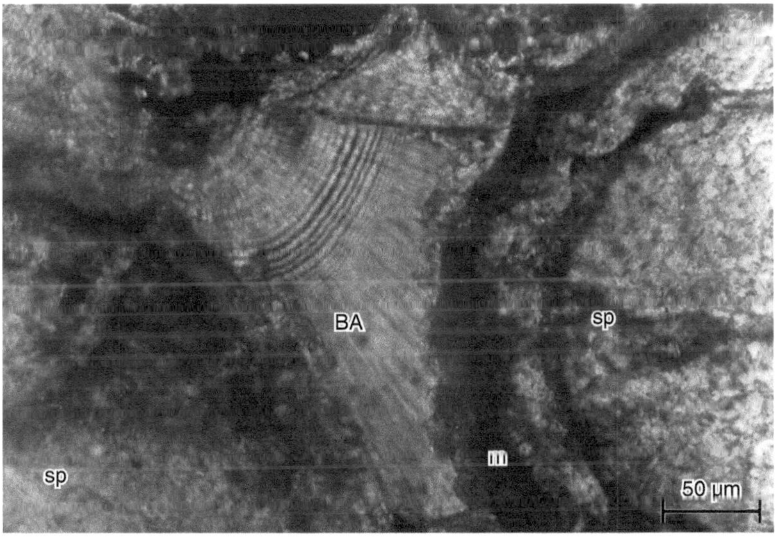

Figure 8.9 Botryoidal aragonite cement (BA). The cement fills an intergranular pore from an alcyonarian spicule, modern reèf flat, Réunion (crossed polars). sp = spicule; m = micrite envelope. Photograph by L. Montaggioni.

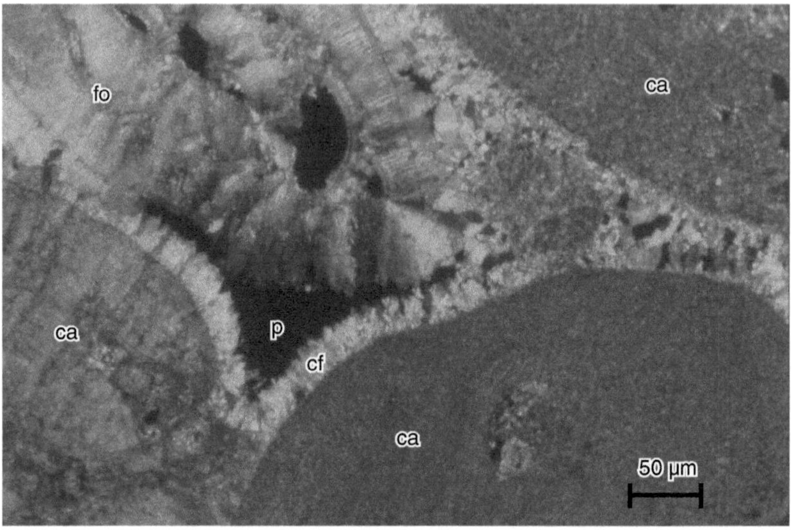

Figure 8.10 Isopachous fringes of prismatic, high-magnesium calcite (cf) surrounding skeletal grains (ca = coralline alga, fo = foraminiferan), modern beach rock, Curieuse, Seychelles (crossed polars). p = residual pore. Photograph by L. Montaggioni.

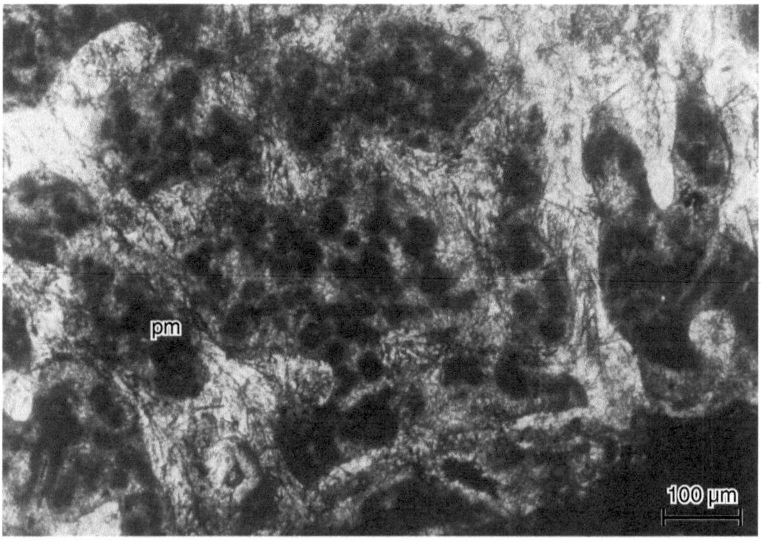

Figure 8.11 Micrite peloids (pm) filling intraskeletal cavities in coral, modern reef crest, Réunion (plane polarized light). Photograph by L. Montaggioni.

seven times saturation but it is subject to a wide range of modifying influences and varies with time. Studies by Sanyal, Hemming, Hanson, and Broecker (1995) suggest that the pH of the deep Atlantic and Pacific oceans was higher during the last glacial interval, decreasing at the beginning of the Holocene. By contrast, Langdon et al. (2000) modelled the effects of the present increase in atmospheric CO_2, predicting that by the middle of the 21st century it will reduce the concentration of CO_3^{2-} of the ocean surface by 30% relative to the pre-industrial level, reducing the calcium carbonate saturation by an equal percentage. However, for the present, surface waters are capable of precipitating carbonates and the question is not so much why modern carbonate sediments are cemented but why they are not. The answer to this problem seems to lie in rates, partly in rates of sedimentation, low rates allowing grain surfaces to remain in contact with seawater for longer periods, but also in rates of flow. An effective pump is essential and it is in this latter area that coral reefs excel. Waves, tides and other currents, together with density contrasts promoted by thermal and salinity variations, are able to drive large volumes of seawater through the porous reef mass (see Chapter 7). Given the amount of carbonate typically present in solution, the volume of water that must pass through the sediment in order that it can become fully cemented has been variously estimated at between 30,000 (Berner, 1971) and 10,000 (Bathurst, 1983) pore volumes. Rates of flow decrease with depth below the sediment–water interface and thus rapidly cemented zones are typically thin crusts, slab-like plates or nodules. Fine-grained sediments and coarse materials at greater depth commonly remain uncemented for protracted periods.

Earlier views that rapid lithification strictly reflects intertidal and shallow phreatic zones influenced by meteoric waters changed dramatically in the mid-1960s with the discovery of widespread early submarine cementation. An outstanding example of this was work by Shinn (1969) who described cements including both acicular aragonite and microcrystalline Mg-calcite forming in depths of up to 30 m in the Arabian (Persian) Gulf within the last 8 ka and some more recently. Temperatures, derived from stable isotope analyses indicate a range of 15–26°C and there is no evidence of either freshwater or hypersalinity. Land and Goreau (1970) reported widespread Mg-calcite cementation in reef-crest and fore-reef sediments in waters up to 70 m deep off the north coast of Jamaica. Most cement here is found within 1 m of the reef surface and radiometric dating indicates ages of up to 8.4 ka. Stable isotope analyses again excluded any influence of non-marine waters. Rapidly formed cements have also been described from the Great Barrier Reef (Marshall & Davies, 1981; Marshall, 1983), the Red Sea (Friedman, Amiel, & Schneidermann, 1974), Belize (Ginsburg & James, 1976), Barbados (Macintyre, Mountjoy, & D'Anglejan, 1968), Panama (Macintyre, 1977) and French Polynesia (Montaggioni & Pirazzoli, 1984). Macintyre and Marshall (1988) provided a comprehensive review of shallow marine cements.

A variety of crystal forms have been reported and those presented by Schroeder (1973) from Bermuda reefs are among the most diverse. Aragonite crystals included needles, laths and spherules, whereas Mg-calcite formed blocky, palisade and scale-like crystals together with micrite. However, these showed no systematic distribution and Schroeder (*ibid*) identified three groups of factors believed to influence development: the morphology and composition of the substrate, the direct and indirect influence of organisms present in the pores, and the microchemical environment, where the sizes of pores and rates of flow of waters through them control crystallization. Sequences of contrasting cement forms result from changes in the microenvironment during growth. Similar variations were described by Macintyre and Marshall (1988).

Extensive spherular growths of botryoidal aragonite up to 5 cm thick have been described from the deep fore-reef (65–120 m) of Belize (Ginsburg & James, 1976). These have an age of about 12.6 ka. Strontium values are 8120–8460 ppm, typical of normal seawater. Similar aragonite botryoids were described by Aïssaoui & Purser (1985) from Pleistocene deposits in New Caledonia and the Red Sea.

James, Ginsburg, Marszalek, and Choquette (1976) showed that in Belize, the first cement to form in the shallow subtidal reef surface is of Mg-calcite and includes bladed crystals and micrite, some of which is in the form of micropeloids. Aragonite is much less common. Cementation is sporadic seawards, but selectively developed within seaward-facing spurs where, presumably, the pump action of waves and tides is greatest. Lighty (1985) described cements from a buried reef, dated at around 7.15–9.44 ka on the Florida shelf. The greatest proportion of cement is found in reef facies from high-energy environments or areas where sedimentation rates were low. Only minor amounts of aragonite are present as epitaxial overgrowths on corals but locally pores are lined with prismatic Mg-calcite.

Thick crusts of Mg-calcite have been described from Mururoa Atoll (Aïssaoui, Buiges, & Purser, 1988; Braithwaite & Camoin, unpublished). These are typically radially fibrous and commonly show growth banding (Figure 8.1). Bundles of crystals show a sweeping extinction with elongate inclusions marking the boundaries between fibrous subcrystals. In some examples, few new nuclei are present but in others irregular fan-like arrays form cascades of new nuclei from layer to layer. Where sediment layers are thick, new nuclei formed on the resulting surface as growth resumed. Changes in growth style may follow each other within the same cement sequence. Similar cements are seen throughout the entire section sampled.

An exceptional example of aragonite cementation in reefs was described by Pichler and Dix (1996) and Pichler and Veizer (2004) from Ambitle Island, Papua New-Guinea. Here the fringing reef is exposed to the discharge of hot hydrothermal fluids and streams of CO_2 gas bubbles. Water depths are 5–10 m and fluid temperatures 89–98°C. However, dilution by seawater (1:9) is reflected in lower temperatures, calculated from isotopes ($\delta^{13}C = 1.9–2.2‰$ PDB and $\delta^{18}O = 14.2–14.7‰$ PDB). The aragonite forms in two distinct habits, euhedral pseudohexagonal crystals up to 2 cm long and later microcrystalline 'dendrites' locally almost occluding primary pores. Precipitation typically occurs within a metre of vents but there are minor seepages (and precipitation) throughout the bay. Although unusual, these observations may be considered in the light of precipitation from cool-water seeps described by Hovland (1990) in the North Sea and the Gulf of Mexico.

8.3.2.2. Subaerial cements and associated deposits

When sea level falls, the metastable components of the reef, aragonite and Mg-calcite grains, are typically subject to diagenetic alteration by meteoric water, first in the vadose zone and subsequently also in the phreatic zone. As the water passes through increasing volumes of rock, saturation rises and it becomes possible for low-magnesium calcite to precipitate. The cements are characterized by a non-uniform distribution. This was first indicated by experiments by Thorstenson, Mackenzie, and Ristvet (1972). Crystal forms may again vary, with blocky or prismatic cements ranging between flattened rhombohedral ('nail-head') and high-spired scalenohedral ('dog-tooth') forms. Radially fibrous cements are common in some areas. Crystal terminations provide an effective guide to position relative to the water table. Crystals in flooded pores, that is the phreatic zone below the water table, grow freely and generate normal terminations (Figure 8.12). In the vadose zone, the only water available for growth is held in surface films coating grains. Crystals may continue to grow within these films but their extension is starved against the surface of the film so that their collective growth forms a concordant flattened surface. Crystals in half-filled pores may have angular terminations beneath the water surface and flattened terminations above, the boundary marking the position of the water surface. Water droplets trapped within pore throats generate so-called meniscus cements that block the throats and form concave drapes on adjacent walls, rounding pores (Figure 8.13). By contrast, pendent water drops precipitate cement that thickens from the top to the bottom of the grain (Figure 8.13).

The patchy distribution of early meteoric cements has been replicated in experiments by Thorstenson et al. (1972) and Badiozamani, Mackenzie, and Thorstenson (1977), the last indicating the effects on crystal form of variations in sectoral growth rates dependent on temperature, saturation and fluid compositions.

Surfaces that have been exposed are typically characterized by karst erosion features that may extend many metres beneath the surface. The rocks beneath such surfaces may be extensively altered with pervasive

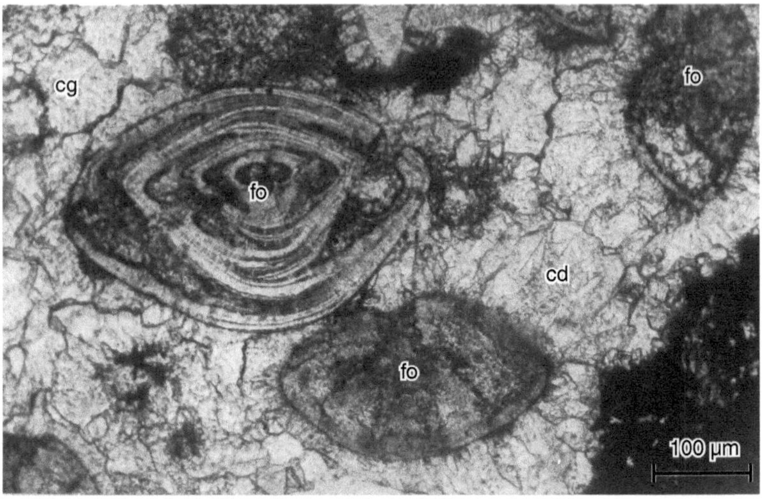

Figure 8.12 Prismatic (cd) and blocky (cg) cements of low-magnesium calcite formed within a meteoric phreatic zone, late Pleistocene reef-associated deposits, Mauritius (plane polarized light). fo = foraminiferal tests. Photograph by L. Montaggioni.

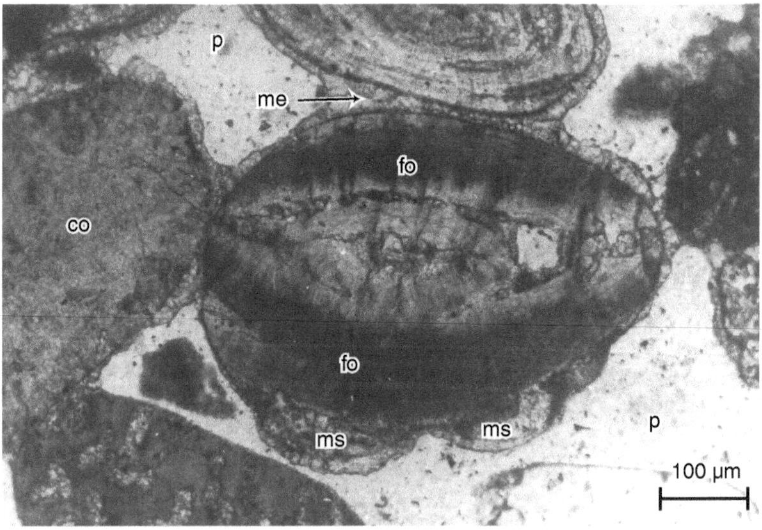

Figure 8.13 Meniscus (me) and pendent (microstalactitic) (ms) cements of low-magnesium calcite, late Pleistocene reef-associated deposits, Mauritius (plane polarized light). p = pore; co = coral; fo = foraminifera. Photograph by L. Montaggioni.

vuggy porosity reflecting dissolution. Erosion surfaces commonly carry a range of distinctive sediments. These include laminar, oncoidal and brecciated fabrics resembling those described from calcretes (Braithwaite, 1975, Aldabra; Braithwaite, 1984, Kenya; Coniglio & Harrison, 1983, Florida; Harrison, 1977, Barbados; Jones, 1991, Cayman Islands; Multer & Hoffmeister, 1968, Florida; James, 1972, Barbados; Read, 1974, Shark Bay). Muddy sediments characterized by alveolar septal fabrics have been described from a number of areas. These have an open spongy texture (Figure 8.14) and are morphologically similar to textures described by Esteban Cerdà (1974). They appear to form within polysaccharide films generated by microbial activity and are commonly associated with matted microbial filaments and clusters of needle fibres (James 1972; Jones & Kahle 1993) together with preserved rootlets. Alveolar fabrics have been widely interpreted as of subaerial (palaeosol) origin (Braithwaite, 1975, 1984; Wright, 1994), although Hillgärtner, Dupraz, and Hug (2001) regarded similar fabrics as formed under shallow marine conditions. Structureless micritic micropeloids are present in many cavities and groups may be incorporated within other cements. They are associated both with brownish laminated encrustations (cutans) and glaebules characteristic of palaeosols (Fitzpatrick, 1993). Some micropeloids present resemble those described by Macintyre (1985) and later attributed to bacterial precipitation by Chafetz (1986) and Buczynski and Chafetz (1991). However, Rossinsky,

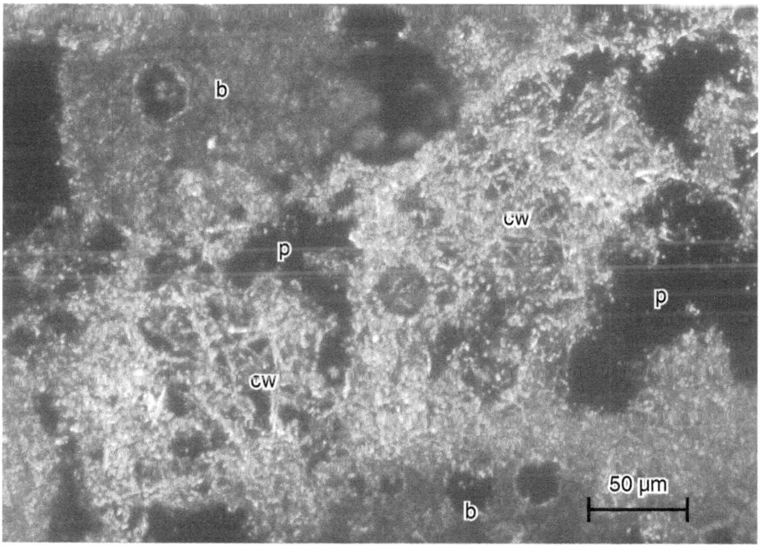

Figure 8.14 Alveolar septal fabrics showing needle fibres of low-magnesium calcite (cw), late Pleistocene calcretes, Rodrigues, Western Indian Ocean (crossed polars). p = pore; b = residual bioclasts. Photograph by L. Montaggioni .

Wanless, and Swart (1992) described laminated fabrics on the Islands of Providenciales and West Caicos in the Caribbean. These were interpreted as penetrative structures developed beneath subaerial surfaces rather than at a specific surface and, as the authors point out, they could easily be misinterpreted as reflecting individual exposure events.

8.3.3. Geochemistry of Cements

Given the difficulties of discriminating petrographically between cements of different origin, workers have for some years relied on geochemistry and on isotopic analyses in particular. In recent years more emphasis has been placed on stable isotope pairs, $^{12}C/^{13}C$ and $^{16}O/^{18}O$ have been shown to provide sensitive indices of environment. Some of the earliest works in this field were by Matthews (1968) and Allan and Matthews (1982) on Barbados. A detailed analysis of many of the issues involved is provided by Morse and Mackenzie (1990). Hudson (1977) in a seminal paper produced cross-plots of $\delta^{13}C$ against $\delta^{18}O$, where absolute values are expressed relative to standard PDB values. In these, rocks of differing origin can be placed in characteristic relatively well-defined fields reflecting their diagenetic history. Similar plots were expanded by Nelson and Smith (1996) (Figure 8.15). Similarly, James and Choquette (1984, 1990) showed that shallow marine sediments and marine cements are typically characterized by low positive values for both $\delta^{13}C$ and $\delta^{18}O$ (Figure 8.16). Meteoric groundwaters are relatively enriched in the lighter ^{12}C and, when they react with marine sediments, they take on more negative $\delta^{13}C$ values that become increasingly negative where soil waters or methane are involved. Plant tissues are depleted in ^{13}C as a result of metabolic processes and soil gasses are therefore significantly depleted in the heavier isotope, giving values of $\delta^{13}C$ of −20 to −35‰ PDB. In methanogenic environments these may fall below −40‰. The oxygen in the atmosphere, and thus in meteoric waters, dominates surface alteration products, giving rise to increasingly negative $\delta^{18}O$ values in comparison with those of unaltered marine sediments. Whereas $\delta^{13}C$ values for limestones that have reacted with meteoric waters may be −5 to −10‰ PDB, values for $\delta^{18}O$ are of the order of −3 to −8‰ PDB. Evaporation and an increase in temperature may drive $\delta^{18}O$ to more positive values as a result of fractionation. Mixing-zone waters have more positive $\delta^{13}C$ (+1‰) and negative $\delta^{18}O$ (−1‰) values.

Saller and Moore (1991) investigated the effects of climatic variation on the geochemistry of meteoric calcite cements in boreholes on Enewetak Atoll, Cat Island in the Bahamas and northeastern Yucatan. Oxygen isotope values are relatively consistent between the three areas, whereas mean $\delta^{13}C$ values for sparry calcite were −5.2‰ PDB for Enewetak, −3.4‰ for Cat Island and −6.0‰ for northeast Yucatan. Columnar calcites

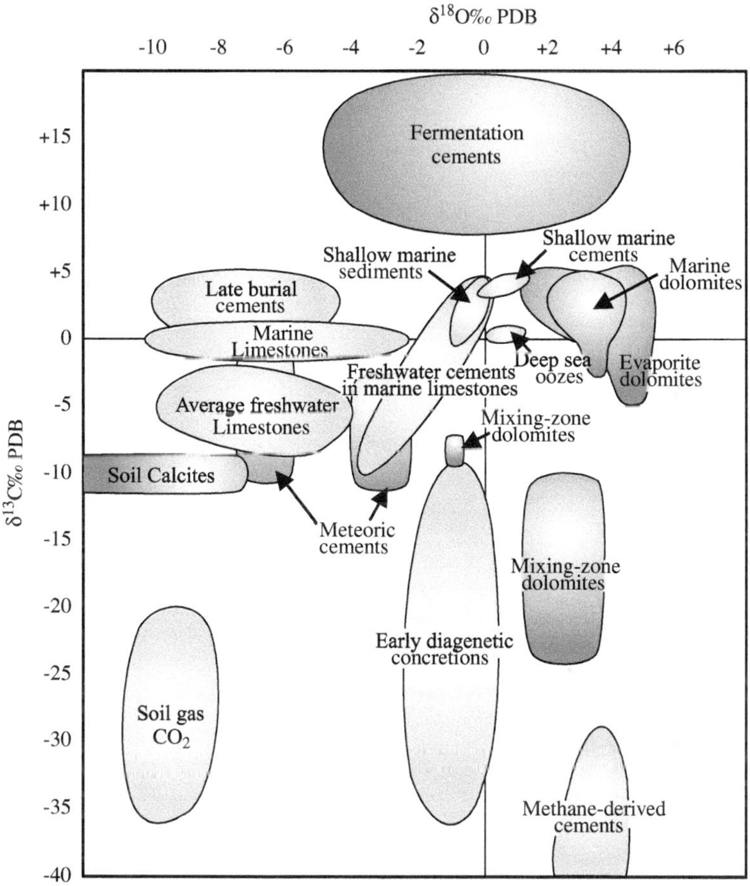

Figure 8.15 Cross-plots of stable oxygen and carbon isotope fields in carbonates indicating general environmental interpretations. Incorporating data from Hudson (1977) and Nelson and Smith (1996).

were −5.6‰ for Enewetak, −3.4‰ for Cat Island and −7.9‰ for Yucatan, with comparable variations in micritic and equant cements. Mean $\delta^{18}O$ values for sparry cements ranged from −6.5‰ for Enewetak, −4.1‰ for Cat Island and −5.7‰ for Yucatan. Columnar cements gave mean values of −6.9‰ for Enewetak, −3.5‰ for Cat island and −6.4‰ for Yucatan. These variations are not directly related to latitude. The $\delta^{13}C$ values in particular are significantly lower than those of the sediments when first deposited and reflect the addition of soil-derived CO_2 and organic matter, unrelated to distance below the exposure surface. Concentrations of strontium and magnesium also vary within and between the three localities. The mean strontium concentrations in sparry cements are 620 ppm for Enewetak,

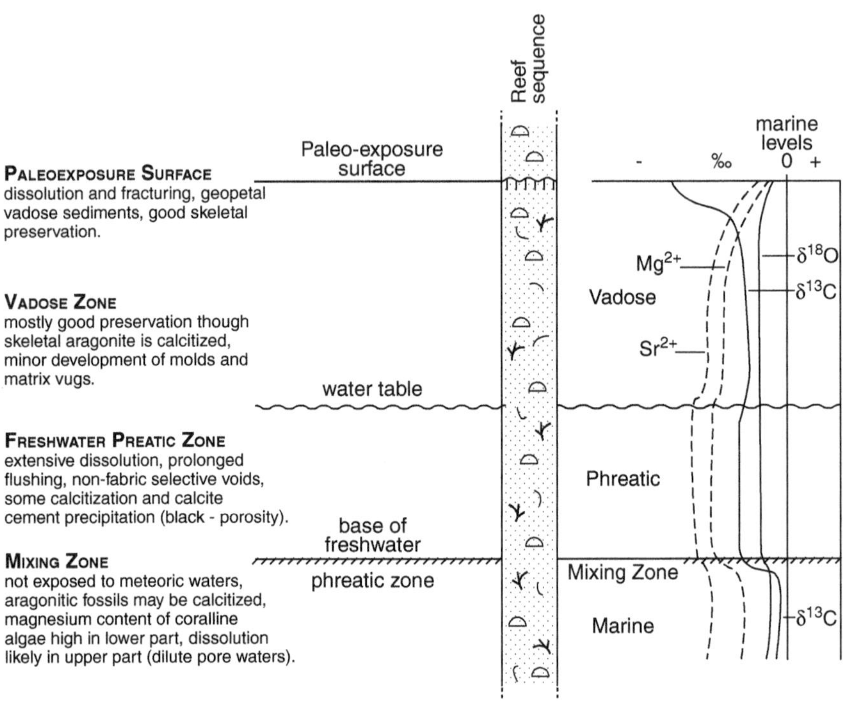

Figure 8.16 Schematic diagram illustrating the effects of meteoric (freshwater) diagenesis affecting a hypothetical core extracted from a reef limestone in a warm, moderately humid climate. The geochemical curves express results obtained from analysis of bulk samples. Modified and redrawn from James and Choquette (1984) in Morse and Mackenzie (1990).

1200 ppm for Cat Island and 700 ppm for Yucatan. The high values for the Bahamian samples are believed to reflect a higher strontium content of the original sediments. In all three areas, depositional Mg-calcite was rare. Magnesium was detected in cements but concentrations varied, with a mean of 1.0 mole% $MgCO_3$ for meteoric cement on Enewetak, 0.84 mole% for Cat Island and 2.2 mole% in Yucatan, the last probably reflecting the influence of mixing-zone waters. These results are interesting, making an important general point regarding the climatic setting of cements, but they are averaged over periods of up to 2.1 Ma and are likely to conceal significant downhole variation related to exposure surfaces.

Cements deposited by meteoric waters are typically low-magnesium calcites but, for reasons outlined above, may have morphologies identical to those of marine Mg-calcites. Kinsman (1969) showed that where aragonite altered to calcite in a closed system, the calcite was characterized by relatively high strontium (Sr) contents (700–1000 ppm) inherited from the original aragonite. In open systems, where larger volumes of water are involved, the Sr is flushed through and is typically lower than 350 ppm. However, in areas of low water–rock interaction, as in the vadose zone, concentrations may remain relatively high. In the same areas, magnesium lost from Mg-calcites may also be relatively high. Chung and Swart (1990) showed how the concentration of uranium could be used to differentiate vadose and freshwater phreatic cements from South Joulters Cay in the Bahamas. The uranium in vadose cements was less than 0.28 ± 0.11 ppm and those of the freshwater phreatic zone 1.80 ± 0.75 ppm. These values are all relatively easy to obtain and provide a useful guide to origins.

The radial-fibrous cements on Mururoa Atoll (Braithwaite & Camoin, unpublished) form two isotopically distinct groups. At three intervals towards the surface, brown radial-fibrous crusts are characterized by negative values for both oxygen and carbon isotopes. These range from −3.43 to −5.73‰ PDB (average −4.75) for $\delta^{18}O$ and +1.16 to −7.55‰ (average −2.74) for $\delta^{13}C$. At depth, colourless but apparently similar radial crystals have positive isotope values, ranging from −0.05 to +1.67‰ for $\delta^{18}O$ (average +0.93‰) and from +2 to +4.39 for $\delta^{13}C$ (average +3.46‰). Thus, whereas the brown crusts carry a freshwater signal the colourless Mg-calcite cements are marine.

Variations in strontium content of diagenetic calcites have been referred to above. Experimental studies by Malone and Baker (1999) have examined aragonite-to-calcite and dolomite-to-calcite transformations at temperatures ranging from 40 to 200°C. These are important because of the emphasis placed on Sr content in both diagenesis and palaeoceanography. They demonstrated a clear temperature dependence related to growth rates. Crystals that have grown rapidly tend to have high Sr contents whereas those formed by alteration over longer periods and with increasing burial depth become relatively depleted.

8.4. Replacement and Dissolution

All carbonate minerals are soluble, to some degree, but there is a recognized scale from the most soluble, aragonite, through high-magnesium calcite and low-magnesium calcite to dolomite (Chave, Deffeyes, Weyl, Garrels, & Thompson, 1962). Biogenic polymorphs also show varying solubilities that reflect differences in structure. Dissolution may occur where the rocks come into contact with cooler marine waters or where marine waters are diluted or are replaced by fresh meteroic waters. Two contrasting styles of replacement have been recognized, neomorphism, in which it is apparently a 'solid-state' process that preserved details of structure, and a more pervasive dissolution that may remove any trace of the original mineralogy and structure.

8.4.1. Neomorphism and Early Diagenesis of corals

Throughout the Quaternary, aragonite has been the common phase in both organic and inorganic marine carbonate precipitation. In many phyla with carbonate skeletons, the earliest biogenic mineral precipitate is an amorphous calcium carbonate (ACC, Weiner, Levi-Kalisman, Raz, Weiss, & Addadi, 2002). In most cases, the ACC is replaced (neomorphosed) by aragonite (Nassif et al., 2005). This in turn is also commonly replaced. The term 'neomorphism' was coined by Folk (1965) to describe a change in which aragonite typically dissolves along a molecular scale dissolution front and is immediately replaced by precipitation of calcite, with the dissolution surface marked by a narrow chalky interval of partial dissolution (Pingitore, 1976, 1982). Some remnants of the original aragonite may be incorporated into the growing calcite (Sandberg, Schneiderman, & Wunder, 1973; Sandberg & Hudson, 1983) (Figure 8.17). As the gap between altered and unaltered crystals is only of molecular dimensions, structural components and even colour may be incorporated into the newly grown crystals creating the illusion that this is a 'solid-state' process (Maliva, Missimer, & Dickson, 2000).

The diagenetic alteration of corals in particular has been studied in some detail, largely because of its effects in compromising both the accuracy of dating Quaternary reef deposits (see Chapter 1, Section 1.7), and proxies

Figure 8.17 Neomorphic replacement of coral, preserving details of trabecular structure. Note also the preservation of epitaxial crystals of aragonite within calcite overgrowths. Pleistocene reef unit, Ribbon Reef, Great Barrier Reef core (plane polarized light). Image width = about 1.2 mm. Photograph from C. Braithwaite.

used in reconstructing paleoclimates (see Chapter 9, Section 9.2). Cuif and Sorauf (2001) and Perrin (2003, 2004) showed that the fibres forming the skeleton are composite structures, each comprising a series of discrete micron-scale growth increments formed epitaxially on preceding growths (Figure 8.18). Early diagenesis results in the thickening and subdivision of the subcrystals within each increment towards the outer margins of the layer, while maintaining optical continuity. At a later stage, pervasive dissolution may also affect some growth increments. James (1974) indicated that

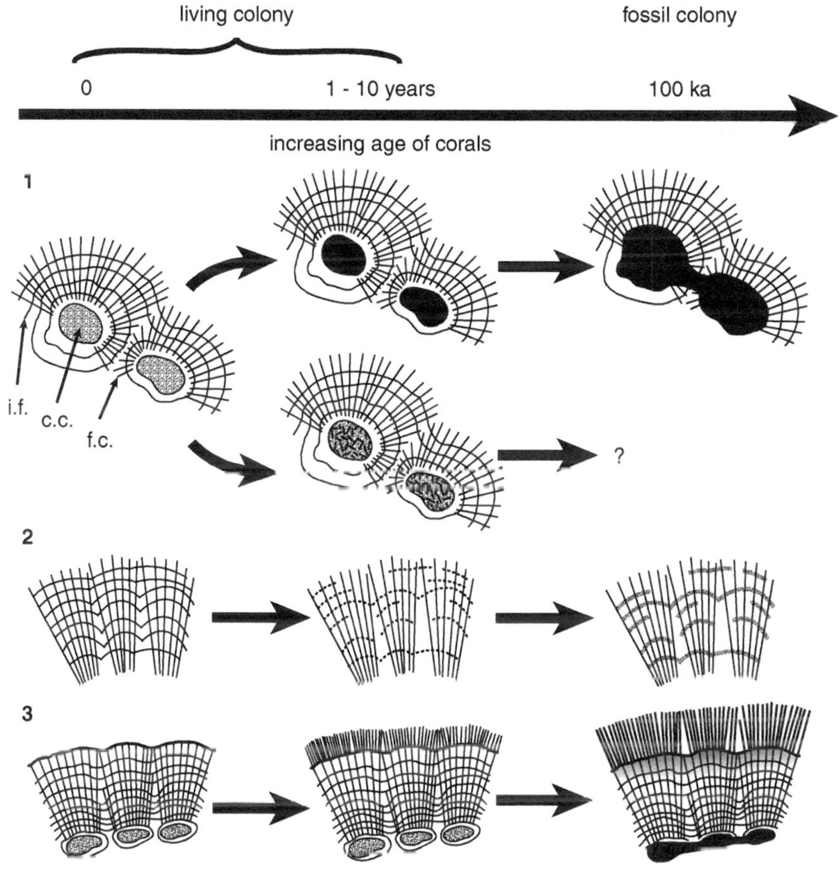

Figure 8.18 Diagenetic alteration of scleractinian coral skeletons through time. 1: Selective dissolution of calcification centres (c.c.) and later, local dissolution of growth increments (i.f.) in coral fibres (f.c.), with development of cryptocrystals in the calcification centres. 2: Alteration of the incremental zonation within networks of the coral fibres, resulting in a discontinuous distribution of the increments. Finally, microinclusions are deposited. 3: Syntaxial overgrowth and thickening of coral fibres composed of primary (biologically mediated) aragonite. The thickening process relates to the recrystallization of the fibres to secondary (inorganic) aragonite. Modified and redrawn from Perrin (2004).

Figure 8.19 Incipient dissolution (d) of aragonite needles in a coral skeleton dated at 9.56 ka, outer barrier reef zone, 38.5 m core depth, Tahiti. SEM Photograph by A. Laurenti_Ribaud.

dissolution of aragonite begins at the original centres of calcification, comprising micron-scale equant crystallites of aragonite, with the resulting microporosity allowing passage of reactive fluids (Figures 8.18 and 8.19). The centres may acquire epitaxial overgrowths of aragonite but may equally be overgrown by calcite. In the latter case, the aragonite may survive for some time in rocks that are otherwise substantially altered (Braithwaite & Montaggioni, in press). There are large variations in structures between species and thus in their responses to diagenetic change (Perrin, 2003).

Bar-Matthews, Wasserberg, and Chen (1993) were able to show qualitatively that in corals with varying degrees of visible alteration but no calcite, there were progressive increases in $^{234}U/^{238}U$ and parallel decreases in Na, S and Mg attributed to the inorganic growth of secondary aragonite. The source of the excess uranium was not known but it was clearly not in seawater. In addition, among others, Quinn and Taylor (2006) found that in living *Porites* with visible alteration, the original geochemical signatures are modified, with significant increases of $\delta^{18}O$ and Sr/Ca and decreases of Mg/Ca, resulting in a significant (but spurious) interpretation of sea surface cooling.

The confinement of the reactive surfaces, and the preservation of structural and sometimes colour features, led to the belief that neomorphic replacement takes place within a closed system. However, Martin, Wilkinson, and Lohman (1986) argued that there had to be an exchange with surrounding waters. They showed that in the Pleistocene Key Largo Limestone, variations in skeletal porosity, by moderating the access of fluids, exerted an important control. Maliva (1998) was subsequently able to apply a computer model that supported the lack of isolation. On a more mundane level, because aragonite has a denser structure than calcite, some 8% of its volume must be released in order that it may be replaced volume-for-volume and it is thus able to contribute to the saturation of surrounding fluids and to the growth of calcite cement elsewhere in the rock. As transformation occurs on a migrating front, along which the solubility of the dissolving components varies, largely as a result of differences in grain orientation, the boundaries between the two phases are typically irregular. Replacive crystals are commonly coarse, and irregular boundaries with re-entrants contradict the rules of boundary-free energy (Bathurst, 1983). While not all neomorphic calcite conforms to this pattern, where present, these features provide a convenient means of differentiating coarse neomorphic crystal growths from cement. An important deviation was recorded by Zhu, Marshall, and Chappell (1988) in Quaternary raised limestones on the Huon Peninsula, New Guinea. In these, aragonite fibres in corals become coarser, with length-to-width ratios decreasing to 5:1 while apparently retaining their original mineralogy. In many instances neomorphic crystals may continue to grow into adjacent pores as cement.

8.4.2. Neomorphism of Magnesian Calcite

As Chave, Deffeyes, Weyl, Garrels, and Thompson (1962) showed, Mg-calcite is unstable under normal atmospheric conditions and the solubility of the calcite increases with increasing magnesium content. In theory, Mg-calcite should be more soluble than aragonite at 15.5–16 mole % $MgCO_3$. In practice, the magnesium is commonly lost relatively rapidly, and the calcite assumes the low-magnesium form while preserving minute details of internal structure. Wollast and Reinhard-Derie (1977) showed that contrary to appearances, skeletal high-magnesium calcite is not generally homogeneous but consists of micron-sized domains that contain differing proportions of magnesium. Dissolution is therefore incongruent. The areas with the highest magnesium content dissolve first and are replaced on a molecular scale by calcite containing less magnesium that crystallizes syntaxially on adjacent lower magnesium surfaces. The magnesium released is carried away in solution, but the process continues by instalments until the entire grain has been replaced and is in equilibrium with the lower magnesium fluid. As a result, the process again resembles a 'solid-state' replacement, but because it is not limited to a single advancing front it does not generate the coarse crystals typical of aragonite replacement. However, Mg-calcite bioclasts including calcareous algae and echinoderm plates may also show pervasive dissolution of cell walls and stereome (Braithwaite & Camoin, unpublished).

8.4.3. Wholesale Dissolution of Carbonate Minerals

Wholesale pervasive dissolution results in the formation of pores that are visible in thin section or hand sample. Depending on the saturation state it may involve the selective removal of grains or growth generations of cements of specific compositions, or indiscriminate dissolution of bioclasts and cement. In a few instances, dissolution results in the complete disappearance of the bioclast but more commonly traces remain (Figure 8.20). Bathurst (1964, 1966) described how the surfaces of bioclastic grains are commonly infested with endolithic algae, fungae and cyanobacteria. These bore into the grain surface and many of the resulting holes are subsequently filled with either microcrystalline acicular aragonite (Lloyd, 1971) or Mg-calcite (Winland, 1968). Bathurst (1966) called the effect 'micritization' as it results in the replacement of the margins of the grains by microcrystalline 'mud' (micrite of Folk, 1959). In some examples, the whole grain may be replaced forming a structureless peloid. Where the

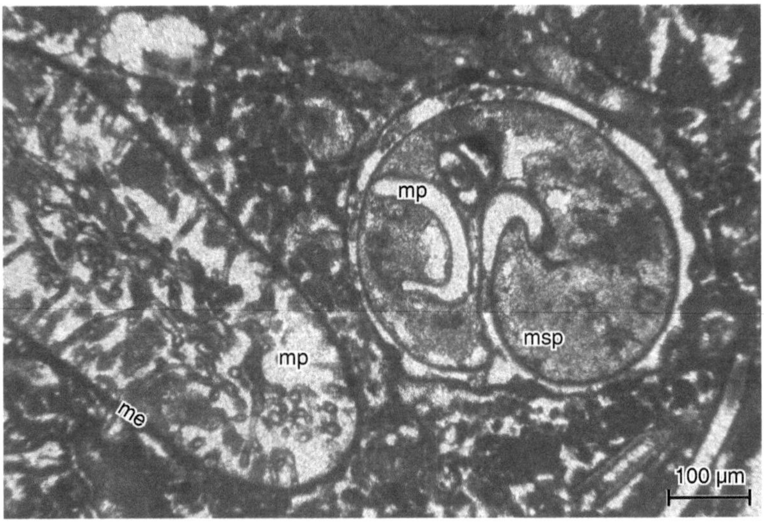

Figure 8.20 Dissolution of aragonite skeletal grains (green alga *Halimeda* right, gastropod shell left) resulting in moldic pores (mp). The *Halimeda* plate is surrounded by a micrite envelope (me) and contains the remains of utricles, while the gastropod shell is filled with low-magnesium microspar cements (msp) (plane polarized light). Late Pleistocene reef unit, core Amédée 4, southwestern barrier reef system, New Caledonia. Photograph by G. Cabioch.

'micrite' is Mg-calcite and the bioclasts are of aragonite, there is a marked contrast in solubility. Thus, as the bioclast dissolves the margin remains as a 'micrite envelope' (Figure 8.20). This preserves the shape of the grain, although it may subsequently collapse during compaction, and the intragranular pores are later filled by cement that is low-magnesium calcite. However, the term 'micritization' has also more recently come to be applied to general processes by which cryptocrystalline fabrics are formed within bioclasts by recrystallization (Reid & Macintyre, 1998, 2000). It might be thought that because they grow in equilibrium with surrounding waters, once formed cements would be stable. This is generally true, but crystal surfaces may be dissolved and specific compositional zones selectively removed, as demonstrated in Pleistocene limestones from the Great Barrier Reef (Braithwaite & Montaggioni, in press). In some areas such dissolution may lead to the collapse of the remainder of the crystal under compaction. Where waters are strongly undersaturated (typically meteoric waters) even low-magnesium calcite will dissolve. Where rocks have already achieved stability, there is no longer any significant differential in rates of dissolution of the grains present and the process is commonly indiscriminate, generating irregular vugs and channels, the distribution of which may bear no relationship to that of the original grains. Such effects are particularly prominent at or below subaerially eroded surfaces. Lowstand deposits are a special case. Because they are deposited when sea level is at its lowest within the cycle, they should remain immersed in seawater for some considerable time. From this, it might be supposed that they would remain in equilibrium and would therefore be expected to escape diagenetic alteration. However, Schlager and James (1978) found seawater-generated effects, including the dissolution of aragonite, that resembled those of freshwater diagenesis, in rocks of the deep slope of Tongue of the Ocean in the Bahamas. Dissolution may begin at a very early stage driven by aerobic oxidation, promoting acidity, or by anaerobic bacterial sulphate reduction (Walter & Burton, 1990; Perry & Taylor, 2006). Finally, although the preservation of aragonitic bioclasts generally implies that the rocks have not been exposed since deposition, aragonite may survive protracted exposure in a relatively arid climate (Brachert et al., 2006).

8.4.4. The Effects of Compaction

Frameworks weakened by the selective dissolution of bioclasts or of cement zones may show local collapse and breakage of micrite envelopes. Zoned growths of high- and low-magnesium calcites and calcian dolomites in Pleistocene limestones (Ward & Halley, 1985) show similar selective dissolution. Knox (1977) described concavo–convex contacts between grains in bioclastic aeolianites at Saldanha Bay on the west coast of South Africa. These were attributed to dissolution by freshwater films in the vadose zone, and referred to as reflecting 'vadose compaction'. Although they are likely to be present locally, such features have not so far been recorded in reef-related rocks.

8.5. Hydrological Control of Flow Rates

The key regional influence on diagenetic reactions is the hydrological environment. The detailed mechanisms driving flows are discussed in Chapter 7 and in case studies presented in Vacher and Quinn (1997). These vary from the regional effects of ocean circulation to local wave and tide-driven flows, and head-driven flows beneath emergent sequences. Examples are presented here to explain how their influence is applied.

8.5.1. Flow in Seawater

Carter, Simms, Moore, Roberts, and Lugo-Fernandez (1989) described the effect of increasing wave height on reef margins in St. Croix, U.S. Virgin Islands. This is responsible for generating a hydraulic head that provides for flow through the reef. Earlier work on the island by Zankl and Multer (1977) had shown that there are significant chemical gradients between interstitial waters and those of the open sea. Typical values of hydraulic conductivity were of the order of 100 darcies. The total flux was about $0.5\,m^{-3}\,m^{-2}\,d^{-1}$. Sampling indicated that within a metre of the reef surface, pore waters in the reef crest and fore-reef were suboxic whereas those in the back-reef zone were anoxic. There was a decline in pH from about 8.1 in ocean waters to less than 7.6 in pore waters. Hydrogen sulphide with its effect on increasing alkalinity, was measurable in the back-reef and reef crest. There was a progressive decline in saturation states to near saturation. Carbonate cementation predominantly occurred in the shallow fore-reef and reef crest. Dark iron sulphides are common in the back-reef and in the deepest sampled reef crest. Iron, and less commonly manganese, oxides appear in the shallowest back-reef and are abundant in the upper reef crest and fore-reef. Together these provide a guide to the oxidation states of the water.

The diagenetic effects of groundwater circulation have also been investigated on Enewetak Atoll (Quinn & Saller, 1997). Hydraulic conductivity of Pleistocene rocks on Engebi Island, estimated from model calibration and optimization range from 600 to $1000\,md^{-1}$ (Buddemeier & Oberdorfer, 1997). Shallow borehole measurements imply a dual circulation with wave-driven flow inwards through Holocene deposits and head-driven flow outwards from the lagoon through the underlying Pleistocene with vertical mixing. Temperatures within both boreholes decrease with depth together with carbonate saturation. Three zones were recognized. Near the surface an aragonite and Mg-calcite zone is characterized by the precipitation of cements. There is no systematic dissolution. The lower limit of this zone corresponds to the aragonite undersaturation depth of approximately 300 m. At greater depth to approximately 1000 m, that is the calcite undersaturation depth, aragonite dissolves while magnesium is lost from Mg-calcite and radiaxial low-Mg calcite is precipitated. Carbon and oxygen isotopes indicate precipitation from seawater at 13–26°C. Below 1000 m, the rocks are characterized by the dissolution of calcite and sporadic precipitation of dolomite, apparently in response to the circulation of deep seawater through the atoll margin. At Discovery Bay, Jamaica, dynamic movements of seawater through the reef have caused crystallization of cements that locally overprint earlier fabrics (McCulloch & Land, 1992). There is little effect in reef-flat and distal fore-reef facies where water movements are relatively slow and fabrics are dominated by vadose features generated during Pleistocene lowstands. By contrast, in proximal fore-reef and reef-crest deposits, high flow rates have resulted in the deposition of extensive Holocene high-magnesium calcite cements. Data from a recording current meter in a borehole indicate that the driving force includes both tidal and wave-driven components and may also incorporate other current flow (McCulloch & Land, 1992).

8.5.2. Flow Rates in Meteoric Waters

Investigation of the waters beneath the northern islands of the Bahamas by Whitaker and Smart (2007a) have shown the evolution of the chemistry of meteoric waters and its importance to diagenesis. Measurements involved analysis of major ions, dissolved oxygen and organic carbon, together with pCO_2 of soil gas. There is a relatively rapid equilibration with aragonite present, but reactions are water-controlled and thus promote dissolution rather than precipitation of calcite. Dissolution is mediated by organic reactions and is irregularly distributed as a result of variations in the thickness of the soil cover and the shallow depth of the vadose zone. Rainfall is intermittent and much of it occurs in brief heavy storms that rapidly recharge the vadose zone by surface run-off. However, waters are initially equilibrated with atmospheric pCO_2 but reach higher values ($[7.4 \pm 3.7] \times 10^{-3}$ bars) as a result of their interaction with soil gas. These waters have been found to have already dissolved $52 \pm 19\,\mathrm{mg\,l^{-1}}$ Ca, more than half of that within the freshwater lens below ($93 \pm 18\,\mathrm{mg\,l^{-1}}$). Within the freshwater lens the pCO_2 is 16 (± 8.30) $\times 10^{-3}$ bars and this additional increase is thought to reflect the oxidation of organic carbon derived from the surface and from within the lens. As a result, water within the lens is suboxic and sulphate reducing and reactions may locally be either aerobic or anaerobic. Towards the top of the lens, pCO_2 is lower as a result of degassing. Here and at the surface, some precipitation may be driven either by degassing or by evaporation.

8.6. Rates of Reef Diagenesis

8.6.1. Rates of Diagenesis in Marine Environments

Cuif and Sorauf (2001) and Perrin (2004) in their study of coral microarchitecture noted that diagenetic changes in coral skeletons may begin when the colony is still alive, underlining the fact that the living cellular surface has no influence on the dead core of the colony. The loss of soluble matrix proteins and sugars may begin at a very early stage, and insoluble components may also degrade. Experiments by Gautret (2001) established that, for the corals examined, there was a decrease in a variety of amino acids from the surface to depths representing more than 400 years of growth. Perrin and Cuif (2001) concluded that diagenetic changes may begin in as little as 10–20 years.

Marine cements are commonly the first to form and Quaternary reefs were among the first places in which rapid marine cementation was recognized. The most recent observations suggest that the time required for the earliest phases of cement generation may be as little as a year (Friedman, 1998). Grammer, Crescini, McNeil, and Taylor (1999) reported the results of experiments suggesting that the initial cementation by fibrous aragonite occurs within about 8 months in water depths of up to 60 m in the warm waters of the Bahamas Banks. Shinn (1969) recorded the growth of aragonite in the Persian Gulf within about 20 years.

8.6.2. Rates of Diagenesis in Freshwater Environments

The rate at which changes in original carbonate mineralogy occur in areas open to meteoric waters varies widely. Observations on Eleuthera in the Bahamas by Dravis (1996) indicated that aragonitic sediments may be case-hardened by precipitation of low-magnesium calcite in less than 10 years. This is important as it applies equally to reef systems and suggests that, because sediment is rapidly bound and is not available for transport, little is shed when reefs are exposed during subsequent low sea-level events. Case-hardening may also protect much of the rock mass from dissolution, with water diverted through restricted channels so that bioclasts in protected areas are able to retain both colour and mineralogy.

A comparison of terraces on Barbados (Matthews, 1968; Harris & Matthews, 1968) shows a marked decrease in the proportions of both aragonite and Mg-calcite with increasing age, with even the youngest exposed successions containing very little Mg-calcite (Steinen & Matthews, 1973). Mg-calcite and aragonite can be preserved within the vadose zone where there is little rainfall. In the phreatic zone, where pores are completely filled with water, stabilization of aragonite and Mg-calcite is very much faster and on Barbados (Matthews, 1974; Humphrey, Ransom, & Matthews, 1986) complete stabilization, that is, conversion to low-magnesium calcite, may occur in as little as 5 ka where flow rates are high. However, rates of change reflect local sea-level variations and hydrology and it would be unwise to regard these as absolute limits.

An important influence on the path of meteoric diagenesis is the time available for reactions to occur. This was graphically illustrated by Zhu et al. (1988). The focus of the investigation was a series of raised late Quaternary limestones in the Huon Peninsula, New Guinea. These reach elevations of over 700 m. At the southeastern end of the section, even Holocene reefs are 24 m above present sea level. Only reefs related to the last two glacial cycles had been dated at the time. Annual rainfall along the coast varies from 1000 to 4500 mm. The Holocene corals sampled are predominantly aragonite and many contain epitaxial aragonite cements, although some show evidence of neomorphism and others partial replacement by calcite. In low uplift areas where the rainfall is high, most Pleistocene corals have transformed to calcite. By contrast, in high uplift areas where rainfall is lower, corals commonly preserve aragonite. However, and somewhat anomalously, near the centre of the area, corals in younger terraces are typically altered whereas those older than about 120 ka retain aragonite. These effects are quite locally distributed, reflecting local hydrological conditions. Dissolution is widespread and is also patchily distributed. Four distinct zones are recognized: marine phreatic, characterized by aragonite and Mg-calcite cements; marine vadose, affected by seawater but with aragonite vadose structures and extensive bioerosion; meteoric phreatic with extensive secondary pores and calcite cements; and meteoric vadose with calcite cements showing vadose fabrics and widespread dissolution and neomorphism. These features were said to occur in the same thin sections and even in the same corals, but did not form recognizable sequences.

Comparable results were reported by Humphrey (1997) from the raised reef terraces in Barbados. Here mineral stabilization to low-Mg calcite proceeded rapidly in the phreatic zone and was effectively complete within 5 ka in high flow settings. Grains that were originally Mg-calcite are neomorphosed and aragonite grains completely dissolved. Coarse blocky calcite occludes both primary and secondary pores. Although present groundwater zones are easily recognized using geochemical techniques, it can be difficult to relate these zones to petrographic features. It is important to note here that on the west coast of Barbados (Harris, 1971 reported in Humphrey, 1997), only about 1% of phreatic waters were found to be derived from the vadose zone above. Oxygen isotope values are remarkably uniform. Budd (1988) suggested from observations on ooid-sand islands at Schooner Keys in the Bahamas that the complete transformation of the sediment to a rock consisting only of calcite could be completed in 4.7–115.5 ka in the freshwater phreatic zone and within 8.7–60 ka in the upper part of the mixing zone, depending on local climate and hydrology.

Further light on grain stabilization and the locus of mineral reactions was shed by McClain et al. (1992) in a study of meteoric diagenesis of biogenic carbonates in the Exuma Cays, Bahamas. Comparison of data from 27 wells, measuring pH, alkalinity, Cl^-, Ca^{2+}, Mg^{2+}, Sr^{2+}, $SO_4{}^{2-}$ and $\delta^{13}C$ allowed calculations of mineral saturation levels and rates of mineral transformations. These showed that compared to the aragonite-dominated Holocene systems described elsewhere (Budd & Land, 1990), there was relatively little calcite cementation and less aragonite dissolution, perhaps as a reflection of the greater dissolution of Mg-calcite. Mineral stabilization generally appears to have been significantly faster (106 cm^3 m^{-3} yr^{-1} as compared to 49 cm^3 m^{-3} yr^{-1} at Schooner Cays, Vacher et al., 1990) but transformation reactions have been less efficient and are confined to the vadose zone and upper metre of the phreatic zone. The distribution of CO_2 from soil gas and oxidation of organic matter is thought to have been a major factor controlling the distribution of reactions.

8.7. Diagenetic Sequences

8.7.1. The Control of Sea Level and Climate

There is a clear stratification of water chemistry with depth beneath exposed reef surfaces. This can be interpreted in a one-dimensional model such as that described by Matthews and Frohlich (1987) and provides a static model for the resulting carbonate diagenesis that describes a sequence of distinctive diagenetic facies assemblages related to positions relative to the water table and mixing zone (Figure 8.16). Thus, where the exposed surface lies above sea level, a diagenetic sequence is developed beneath it that can be summarized as: vadose calcite, early phreatic calcite, mixing zone dolomite and unaltered marine carbonates. However, the Quaternary period was characterized by subsidence and glacio-eustatic changes in relative sea level (discussed in Chapters 6 and 9) with repeated important effects on diagenesis.

Climate is also significant and changes are more pervasive and more widely distributed in areas of higher rainfall. Finally, the magnitude and rate of change of sea level are significant. High amplitude changes leave more rocks exposed and variations in the duration of exposure determine the degree to which the mineral assembly is able to achieve equilibrium. Gvirtzman and Friedman (1977), in a study of both modern and emergent Pleistocene reefs in the northern Red Sea, defined a sequence of progressive diagenesis, beginning with the initial fabrics of living corals and completed through the leaching of aragonite components and the precipitation of low-magnesium calcite from meteoric waters. Changes in sea level might be assumed to result in a paragenesis in which successive events simply overprint cement sequences. However, although forward modelling (Matthews & Frohlich, 1987) yields the lithological assemblies predicted, the result of varying exposure to the influence of widely diverse water chemistries means that the relationship of features to particular exposure surfaces is no longer systematic and is commonly obscured by overprinting.

A similar one-dimensional model was used by Quinn and Matthews (1990) to investigate the post-Miocene diagenetic history of Enewetak Atoll. The period was dominated by high-frequency fluctuations in sea level reflected in multiple episodes of meteoric phreatic diagenesis. The model indicated that subaerial unconformities are likely to be preserved, because surfaces are eroded and reduced during periods of exposure and are therefore likely to be flooded and buried during the subsequent sea-level rise. Without this erosion, changes in sea level will be recorded, if at all, only in the passage of the palaeophreatic lens. Modelling results suggest that less than 10% of lapsed time is recorded by sediment deposition. Although multiple diagenetic episodes are predicted, no evidence is presented that identifies the individual effects of these.

The results of simulations using a coupled two-dimensional model, presented by Whitaker, Smart, Hague, Waltham, and Bosence (1997) suggested that starting from a similar one-dimensional sequence, varying hydrological zones and glacio-eustatic sea-level fluctuations generate a stacked sequence of diagenetic zones showing lateral continuity and trends from the platform interior to the outer margin. As Matthews and Frohlich (1987) pointed out, the association implied by the model, between diagenetic zones and unconformity (exposure) surfaces, is commonly misleading. Whitaker et al. (1997) again concluded that, as a result of overprinting, diagenetic history 'could not be unravelled by traditional stratigraphic and sedimentological methods'.

Observations aimed at systematically testing these views have generated conflicting results. Melim (1996) examined data from three Bahamian and South Floridan cores. The conclusion reached from these was that there was no phreatic diagenesis that could be attributed to large-scale low sea-level stands. There is evidence of repeated exposure of shallow-water deposits, but meteoric lenses are restricted to a zone within 60 m or less of the exposed surface. Features associated with phreatic conditions found deeper in the cores appear to have developed during sea-level highstands, not lowstands. This was explained as a result of the greater percolation distance that allowed waters to reach saturation before entering the lens. However, this assertion is questionable because the greater relief during lowstands would offer a significantly larger hydraulic head to drive waters to greater relative depth. Melim, Westphal, Swart, Eberli, and Munnecke (2002) subsequently questioned the common assumption that exposure to meteoric waters offers the principle means of reducing polymineralic sediments to low-magnesium calcite. Boreholes on the western margin of the Great Bahama Bank provided an opportunity to examine aragonite-rich sediments that were originally deposited in relatively deep water, beyond the reach of meteoric waters that are here apparently restricted to around 10 m of the exposed surface. Cores drilled in the platform top to depths of 100–150 m had been altered in meteoric or mixing-zone waters, but all others had apparently been exclusively subject to marine pore fluids. The fabrics developed include moldic porosity, neomorphic replacement of aragonite and precipitation of blocky calcite and calcite microspar. However, these are all characterized by $\delta^{18}O$ values (average +1‰) commensurate with marine pore fluids. These results suggest that large-scale sea-level lowstands may not create environments capable of supporting chemically active meteoric lenses.

Examination of core from an inclined borehole on Mururoa Atoll (Braithwaite & Camoin, unpublished) has suggested a relatively systematic alternation of intervals of pervasive change related to exposure surfaces separated by intervals in which a single generation of cement reflects relative stability. Within the phreatic zone, precipitation of low-magnesium calcite cements occludes pores. Thus, once these are in place, the rock mass becomes relatively stable. There may be no pore space available for the crystallization of additional cements and any neomorphic changes may be difficult to differentiate. By contrast, in the vadose zone indiscriminate dissolution creates and maintains pores that are thus open to the deposition of new cement generations. On Mururoa, a low rainfall permitted only a relatively shallow freshwater lens that remained stable for relatively long periods. A characteristic feature of the Mururoa cements is the occurrence of widespread radial-fibrous calcite (Aïssaoui et al., 1988). This commonly shows multiple growth zones, although there is no consistent relationship between depth (age) and number.

Similar contrasting zones have been recorded from Enewetak by Quinn and Saller (1997) and Saller and Moore (1989) where slightly modified aragonite-rich intervals alternate with altered calcite-rich zones throughout much of the upper Miocene, Pliocene and Pleistocene. Saller and Moore (1989) noted that some zones of intense alteration occur immediately below subaerial exposure surfaces, although others seem to have no such association. Where they are related, the zone of intrafabric dissolution of aragonite is thought to lie below the associated freshwater lens. Examination of a diagenetic assembly from a core from the Great Barrier Reef (Braithwaite & Montaggioni, 2009) reveals a lack of any consistent relationship between cement complexity defined as number and variety of cements, and recognized exposure surfaces. The contrast is interpreted as reflecting differences in rainfall and in the catchment. The relatively high rainfall in the Great Barrier Reef and larger volume of water from the catchment flushes the system so that even within the phreatic zone waters remain relatively undersaturated. Radial-fibrous calcite described as widespread on Mururoa (Aïssaoui et al., 1988) is missing from the Great Barrier Reef core.

8.7.2. The Control of Porosity

Geochemical modelling of surface and groundwaters allowed Whitaker and Smart (2007b) to provide quantitative estimates of meteoric diagenesis in the Pleistocene limestones of the northern Bahamas. The rates of many processes can be determined indirectly and the authors point out that generation of porosity at a rate of 0.1% ka^{-1} representing $10^{-6}\,m^{-3}\,yr^{-1}$ for every cubic metre of rock per year and with a groundwater flux of $0.3\,m^{-3}\,yr^{-1}$ would provide $3.6\,mg\,l^{-1}$ of dissolved calcium. Whereas traditional observations are commonly limited in their scope, such methods allow large-scale distribution of processes to be analysed. The northern Bahamas are relatively wet with rainfall varying from 1175 mm on southern Andros to 1496 mm on Grand Bahama. Potential maximum evaporation on New Providence is estimated at 1610 mm yr^{-1}. The average residence time of the water in the areas investigated ranges from 6 to 9 years. Meteoric waters in the Bahamas equilibrate with aragonite, but reactions are water- rather than mineral-controlled (Whitaker & Smart, 2007b). Dissolution of the bedrock surface is greatest within soil-filled hollows, much of it a result of soil-derived CO_2 that continues to influence dissolution throughout the vadose zone. In total, more than 60% of the calcium in groundwaters is accounted for by these processes, although additional calcium may result from dissolution by CO_2 resulting from the oxidation of organic matter within the Ghyben-Herzberg lens. In the vadose zone, porosity forms at 1.6–3.25 ka^{-1} on average. Within the phreatic zone, dissolution is more modest with averages of 1.4–2.8 ka^{-1}. At deeper levels, oxidation of any remaining organic matter and reduced sulphur species from anaerobic processes may also increase porosity but at reduced rates up to 0.06% ka^{-1}. However, precipitation driven by evaporation from the surface of the water table, is estimated to be in the order of 4.0–0.65 ka^{-1}. The net result is a so-called 'porosity inversion' in which primary porosity is occluded by the growth of new cements while secondary moldic pores, channels and vugs are created.

8.8. Dolomite and Reefs

Dolomite (dolostone) has been widely reported in tropical oceanic islands from rocks ranging from Miocene to Pleistocene and Holocene

(Aharon, Socki, & Chan, 1987; Trichet et al., 1984; Montaggioni, 1985; Hein, Gray, Richmond, & White, 1992; Budd, 1997; Montaggioni & Camoin, 1997). The origins of dolomite have been extensively debated (review by Machel, 2004) and remain a matter of contention. The range of possibilities is relatively large for large-scale carbonate platforms and various works have tested large-scale numerical models, but are more restricted in oceanic islands. There are few examples were dolomite can be regarded as contemporary with the deposits in which it is found and for the most part it is clearly replacive with a relatively small contribution from cement.

8.8.1. Penecontemporaneous Dolomite

The descriptions by Alderman and Skinner (1957) and von der Borch (1965) of the lime muds in the Coorong Lagoon in SE Australia, were among the earliest to suggest direct contemporary precipitation of dolomite. The waters of the lagoon are hypersaline and deposit dolomite that is calcium-rich and non-stoichiometric with a poorly ordered structure. Discussion centred on whether this 'protodolomite' was a direct result of precipitation or reflected contemporary alteration of calcite. However, recent analysis has suggested that microbial activity, principally by mediating sulphate reduction, may be responsible. Wright and Wacey (2004) imply that this mechanism may drive large-scale precipitation but this remains a point of contention.

Although dolomite is apparently forming contemporaneously, there are few examples where this is unequivocally demonstrated in reef environments. Rao, Kessarkar, Krumbein, Krajewski, and Schneider (2003) described late Pleistocene dolomite crusts forming on the carbonate platform off western India in some 64 m depth. The organic components are progressively overgrown by dolomite crystals that ultimately obliterate their structure. Stable isotopes indicate hypersaline and anoxic conditions, and sulphide reduction has also been important with both pyrrhotite and marcasite present. The dolomite is characterized by $\delta^{18}O$ values averaging 4.1‰, similar to values (4.3‰) for microbially mediated dolomites formed in a Brazilian hypersaline lagoon (Vasconselos & McKenzie, 1997). The $\delta^{13}C$ values average −0.95‰ and are more positive than those of dolomites formed offshore in the same region, probably reflecting formation under hypersaline conditions. Teal et al. (2000) described dolomite cements less than 6.4 ka-old occurring in organic-rich sediments over an area of 15 km^2 of the Cangrejo Shoals of northern Belize. There is no evidence of recrystallization and mean $\delta^{18}O$ values of 2.1‰ suggest precipitation from normal seawater. However, $\delta^{13}C$ values ranging from −5.2 to +11.6‰ suggest that formation was promoted by sulphate reduction and methanogenesis in an anoxic environment.

Along the upper seaward slope of the fringing reef at Discovery Bay in Jamaica, Mitchell, Land, and Miser (1987) described minor quantities of relatively calcic and poorly ordered dolomite forming in a subtidal hardground at about 10 m depth. Rather than strictly contemporary, ^{14}C ages range between 1.2 and 7.63 ka. Euhedral crystals of dolomite of about 5 μm form a cement lining skeletal pores in red calcareous algae, rim Mg-calcite peloids and form overgrowths on echinoderm plates. In each case, they appear to have formed syntaxially on existing Mg-calcite crystals. Typically the dolomite is replacive with euhedral (idiotopic, Sibley & Gregg, 1987) or anhedral (hypidiotopic) crystals, but may vary from small crystals that mimic primary fabrics to coarser growths in which original structures are only outlined by inclusions or are not visible at all. There is no megascopic evidence of dissolution or neomorphism of either aragonite or Mg-calcite. Cements and replacive crystals are commonly characterized by zoned growth, indicated by clouds of inclusions under cathodoluminescence or in backscattered electron images. These reflect variations in the calcium and magnesium content of the crystals and the varying incorporation of components such as iron or manganese. There seems to be no doubt, given its situation, that this dolomite formed from seawater and it is unlikely that meteoric waters were channeled to the area. Slightly depleted $\delta^{13}C$ compared to the bulk sediment was interpreted to suggest that respiration of organic matter and perhaps also sulphate reduction in near-normal seawater may have been responsible.

8.8.2. Conceptual Models of Reef Dolomitization

Dolomite is typically replacive, but a number of conceptual models have been invoked to explain its presence in limestone sequences (reviewed by Morrow, 1982; Hardie, 1987; Braithwaite, 1991; Machel, 2004). Numerical models have been developed to simulate many of these (e.g. Whitaker et al., 2004) that for the most part rely on mechanisms that drive large volumes of seawater through existing carbonates, although some require evaporative or mixing-zone waters in which dolomite precipitation is chemically favoured over that of calcite or aragonite. It is generally agreed that ultimately seawater provides the only practical large-scale source for the magnesium required. It is important to realize that as dolomite is replacive, the margins of any dolomite body commonly cut across boundaries defined by lithology or time lines, and that it may be significantly younger than the host rocks.

Case studies presented in Vacher and Quinn (1997) describe the occurrence of dolomite on a variety of carbonate islands. Wheeler and Aharon (1997) reviewed attempts to apply a suite of models to dolomitization on Niue. Machel (1988) suggested a model in which trace-element trends might be used to qualitatively track the direction of flow. This would allow some differentiation of the various mechanisms proposed but relies on relatively large numbers of samples along a well-defined traverse or traverses and for this reason has not been widely applied.

Large-scale circulation in the Florida Platform described by Kohout (1965) became the basis for the mechanism, now referred to as Kohout circulation, proposed by Kohout et al. (1977) to explain the formation of dolomite. Waters in the mixing zone have long been held to be capable of causing dolomitization, but as they are likely to be undersaturated with respect to both aragonite and calcite they may also cause corrosion of the caverns in which they flow (Smart, Dawans, & Whitaker, 1988).

8.8.2.1. Evaporation and mixing-zone dolomites

The diverse explanations for the formation of dolomite are reflected in differences in interpretation of reef-associated Quaternary deposits. Dolomitization by seepage reflux, first proposed by Adams and Rhodes (1960) requires a barred basin in which brines are concentrated by surface evaporation. The increase in density generates a hydrological gradient in which gravity drives brines downwards to effect the replacement of existing carbonates. Mazzullo, Bischoff, and Teal (1995) described shallow-subtidal dolomitization of Holocene deposits on the Cangrejo shoals of northern Belize. The dominant sediments are sandy muds consisting mostly of Mg-calcite (72%) with 25% aragonite and minor low-magnesium calcite. The tidal range is less than 0.5 m. Average velocities in channels are of the order of $0.25\,m\,s^{-1}$ in dolomitic areas, but less than $0.08\,m\,s^{-1}$ in channels elsewhere. The salinity of the water is 38 psu and the temperature 26–30°C. The maximum age of the dolomites is estimated to be 5.6 ka. Up to 5% of dolomite is found in the central area of the shoals forming an interparticle cement on grains of all compositions and there is no obvious textural relationship. The crystals are mostly about 5 μm, poorly ordered and calcic. Mean $\delta^{18}O$ is +2‰ PDB but the $\delta^{13}C$ varies widely from −0.6 to +8.6‰. Analysis of pore fluids suggest that in the key areas, they do not differ significantly from seawater. The critical factor in the formation of the dolomite appears to be the high tidal flow rates, pumping large volumes of water through the sediments.

Calcium-rich dolomites have also been described from Mutalau lagoon, Niue (Schlanger, 1965). This area of restricted circulation is nearly completely enclosed by the surrounding reef. Schlanger (1965) suggested that a slight fall in sea level and evaporation of the lagoon to gypsum saturation, would promote a downward flow of magnesium-rich waters through the lagoon floor. Rodgers, Easton, and Downes (1982) provided more detailed chemistry and rejected this model in favour of a mixing-zone model. First proposed by Hanshaw, Back, and Deike (1971), this is based on the premise that substantial dilution (about 10:1) of marine waters by freshwater generates a fluid that may be undersaturated with respect to calcite but oversaturated with respect to dolomite. Rogers et al. (1982) argued that the lagoon at Mutalau is too small to provide the volume of magnesium required and that outward flow would begin long before the system became hypersaline. In partial confirmation of this view, no traces of evaporite minerals have been found. Analyses showed that the sodium content of the dolomites is less than 1000 ppm indicating a probable brackish-water origin. Rodgers et al. (1982) suggested that waters in the mixing zone with 3–37% seawater and 2–3 atm pCO_2, would be capable of dissolving calcite and precipitating dolomite.

In the Yucatan Peninsula, mixing-zone waters are oversaturated with respect to dolomite. Pleistocene limestones within the flow path, representing platform margin reefs, are locally dolomitized (Ward & Halley 1985). The delicate balance between saturation and undersaturation is graphically illustrated by the selective dissolution of some compositional zones within the dolomite (Ward & Halley, 1985, Figure 5C and D).

Dolomitized Pleistocene reef and reef-associated limestones (216-ka old) forming a terrace in southeastern Barbados have been described by Humphrey (1988). The dolomites are again calcium-rich and poorly ordered, and selectively replace Mg-calcite bioclasts mimetically, carbonate muds and aragonite bioclasts, the latter commonly with loss of the original structures. At some intervals dolomite and calcite cements alternate. Analyses of $\delta^{18}O$ and $\delta^{13}C$ provide average values of +2.6 and −14.9‰ respectively. Humphrey (1988) drew attention to the depleted carbon values that may reflect the incorporation of soil gasses or reducing conditions, ruling out the latter. Meteoric water was regarded as the dominant influence because these isotopic values preclude the mixing of more than 5% seawater. Dolomitization is thought to have occurred immediately following the sea-level highstand of MIS 7.3 at around 230 ka and to have taken place in an interval of less than 5 ka. Burns and Rossinsky (1989) contested the interpretation of these dolomites as of mixing-zone origin, but although they produced cogent arguments to support this view they did not offer any clear alternative. Humphrey's (1989) rebuttal centred on the geometry of the dolomitized sections that appear as isolated lenses within otherwise unaffected lithologies. Humphrey and Quinn (1989) described the occurrence of dolomite in all three Pleistocene terraces on Barbados at 216, 194 and 122 ka, typically within reef-crest facies. Data were used to generate a numerical model that simulated repeated dolomitization events attributed to mixing zone processes and related to sea-level fluctuations, sedimentation and subsidence. From this, it was argued that similar processes involving mixing-zone waters might dolomitize large volumes of platform-margin carbonates.

Dolomites from the Little Bahama Bank, Andros, New Providence, San Salvador, Great Iguana and Mayaguana in the Bahamas, and Jamaica have been described by Vahrenkamp, Swart, and Ruiz (1991). Five phases of dolomitization have been recognized in the area, ranging from early Miocene through late Miocene, late Pliocene, latest Pliocene–early Pleistocene and late Pleistocene, dated using $^{87}Sr/^{86}Sr$ ratios. In most localities, age increases with depth. The best-constrained episode (late Pliocene) is calculated to have lasted only 6 ka. Geochemical evidence and the geometry of the dolomite body support an origin from seawater in which circulation was driven by an overlying mixing zone. The timing of the events in the Bahamas, the Caribbean and the Pacific suggests that there is a link between the preferred times of dolomitization and the evolution of the Sr-isotope ratios in seawater.

Differences in the alteration of late Pleistocene (120 ka) limestones on the north coast of Jamaica were reported by Land (1973). Those within the present vadose zone are relatively unaffected, but in the phreatic zone aragonite has commonly been dissolved and Mg-calcite is rare, reflecting the loss of magnesium. Both low-magnesium calcite and calcian dolomite are precipitated in the rocks below as isopachous cements, replacing calcite. Based on isotopic values ($\delta^{13}C$ values = −8.4‰; $\delta^{18}O$ = −0.9 to −1.6‰.), the principle component of the hydrological system held to be responsible is meteoric water, but with small amounts of magnesium derived from mixing with seawater. Analyses of the composition of the water suggest that as little as 3–4% seawater may be added. The trigger for crystallization is thought to be CO_2 degassing at the water table.

8.8.2.2. Thermal convection and large-scale circulation of seawater

Later, Aharon et al. (1987) rejected the mixing-zone model for dolomitization of oceanic carbonate islands on the basis of stable isotope analyses. Values for $\delta^{18}O$ and $\delta^{13}C$ were +1.9 to +3.6‰ and +1.1 to +2.6‰ respectively, with strontium concentrations of 213–231 ppm, consistent with precipitation from normal seawater at 20–25°C. Seawater was believed to be driven through the rocks by thermal convection promoted by the residual heat within the volcanic pile beneath. Some support for this view was provided by the distribution gradients of Fe, Mn, Cu and Zn and a relatively high uranium content. A similar geothermal convection was described by Saller (1984) and Rougerie and Wauthy (1993). Saller (1984) invoked geothermal circulation to explain dolomitization on Enewetak Atoll. The presence of extensive dolomitized zones beneath a number of atolls (e.g. Mururoa, Purser & Aissaoui, 1985; Niue, Aharon et al., 1987; Resolution Guyot, Flood et al., 1996) is taken as *prima facie* evidence of active circulation as it is generally agreed that the process of dolomitization requires the passage of large volumes of ocean water. However, Wheeler and Aharon (1997) pointed out that deeper boreholes on Niue have since identified two discrete dolomite bodies separated by an interval of unaltered limestone. The lack of connection between these bodies therefore rules out a deep convective origin. Stable isotopes in both intervals are consistent with a seawater origin, but there seem to be several mechanisms that might have brought about dolomitization. Because alteration of the lower unit seems to have occurred at greater depth, it is suggested that Kohout circulation could account for this. Flow is driven by the contrast between warm platform waters flowing outwards, and cooler ocean waters drawn in at depth to replace them. However, the scale of circulation at 2–3 km depth, argues against an application to small ocean islands like Niue.

Suzuki et al. (2006) described dolomites from Kita-daito-jima in the NW Philippines. These extend from the surface to a depth of approximately 100 m. Based on strontium isotope stratigraphy, subsurface units are shown to have dolomitized at 5.5 and 2.0 Ma whereas surface deposits formed between 1.6 and 2.0 Ma. A variety of textural types with varying Ca and Mg contents were differentiated. However, oxygen isotope analyses suggest that all were formed in normal seawater, although this was probably significantly cooler than present values, suggesting that dolomitization may have been associated with eustatic low sea-level stands.

Large-scale circulation in carbonate platforms, where reefs typically form the rims, is able to supply the volumes of water necessary to provide the magnesium required for dolomitization. The circulation model in the Florida Platform (Kohout, 1965) became the basis for this mechanism. However, at surface temperatures there are kinetic limitations to the process that would generally mean that dolomite should not form. Work by Wilson, Sanford, Whitaker, and Smart (2001) has modelled geothermal convection in a platform 40 km wide and 2 km thick. In this, temperatures of greater than 50°C overcome kinetic issues and allow the formation of dolomite with complete replacement occurring in about 60 Ma. Where variations in permeability are introduced, the focusing of flow may locally reduce this time by at least 50%.

Dolomite has also been reported in shallow boreholes beneath the lagoon of Aitutaki in the southern Cook Islands (Hein et al., 1992). It is most extensively developed at depths of 36–70 m, in rocks of Pleistocene age interpreted as part of an outer reef crest. Aitutaki is described as an 'almost atoll' with a relatively large volcanic island in the northern part of the lagoon. Although little is known of its hydrology, this has undoubtedly played an important role in directing meteoric flow. The dolomite mimetically replaces bioclastic grains and also meteoric calcite cements, and forms void-filling cement. Crystals are poorly ordered protodolomite, with 3–10 mole% excess $CaCO_3$. Isotopic analysis of the dolomite (Hein et al., 1992, 2006) indicates consistent $\delta^{18}O$ values of +2.6‰ and $\delta^{13}C$ values of +2.7‰ PDB that imply precipitation predominantly from seawater. However, analysis of fluid inclusions suggests that some mixing-zone water may also be involved.

8.8.3. Cycles of Dolomitization

In a number of carbonate islands, dolomite occurs in the form of isolated lenticular masses embedded within carbonate sequences. Makatea in the Tuamotu Island Group (Southwestern Pacific) is characterized by five separate dolomitic intervals (Montaggioni & Camoin, 1997) (Figure 8.21). These follow two distinct paragenetic sequences. In one, marine cementation by aragonite and Mg-calcite is followed by extensive dolomitization and subsequently by dissolution. Alternatively, marine cementation is followed by a phase of dissolution and cementation by meteoric waters, forming low-magnesium calcite, followed by cementation by dolomite (Figure 8.22) and finally by extensive dissolution. The various dolomite bodies are broadly lenticular with the inference that they are related to the positions of freshwater lenses. Isotopic analyses provide $\delta^{18}O$ values of +2.2

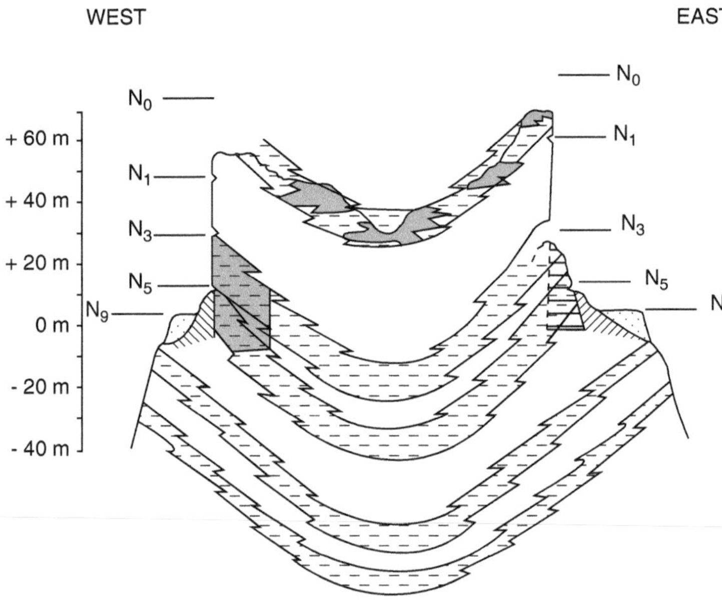

Figure 8.21 Schematic location of paleo-freshwater–seawater mixing zones in Makatea Island, northern Tuamotus, central Pacific. These zones are denoted by the dashed-line pattern. The positions of a number of former sea levels, based on notches and reef terraces (N_0–N_5), are indicated with their expected age: N_0 (2–1 Ma or 0.70 Ma), N_1 (1 Ma–0.70 Ma), N_3 (0.50–0.30 Ma), N_5 (last interglacial). N_9 is Holocene (6–1.5 ka). Differences in elevations of N_0 and N_1 on either side of the island may result from an early differential tilting. The position of the Holocene reefs is given by the light stipple pattern, while the positions of Pleistocene reefs are given by horizontally ruled (250–300 ka) and diagonally ruled (125 ka) patterns respectively. Dolomite distribution is indicated by dark areas. Modified from Montaggioni and Camoin (1997).

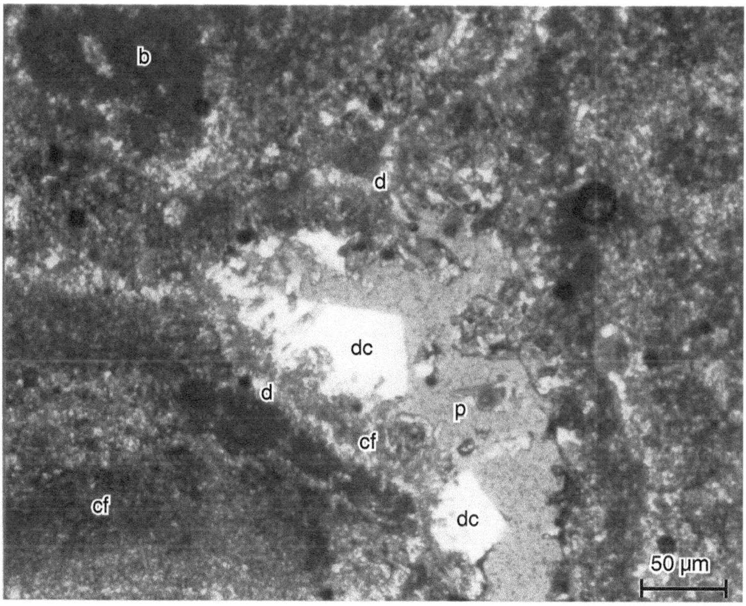

Figure 8.22 Dolomite rhombs as cements (dc) growing into a cavity created by dissolution (p). Locally, dolomite (d) replaces initial calcite fabrics (cf). b = preserved bioclast. Low-magnesium calcite appears dark-stained by alizarine red, while dolomite is white (unstained) (plane polarized light). Late Pleistocene reef unit, Makatea. Photograph by L. Montaggioni.

to +3.0‰ and $\delta^{13}C$ of +2.4 to +3.5‰ PDB that imply water with a composition similar to that of seawater. However, they cannot rule out a contribution of up to 40% freshwater. The dolomitized intervals are thought to reflect periods of stabilized sea level, extending downwards from about 70, 55, 45, 35 and 25 m. These elevations are loosely correlated with dated highstands, but because the dates obtained for the rocks are dates of deposition, when sea level was high, and the dolomite formed later, presumably when sea level was relatively low, the correlation between these events is uncertain.

On Niue, Wheeler et al. (1999) were able to demonstrate well-resolved repeated cycles of dolomite in Pliocene depositional cycles, alternating between mimetic and microsucrosic dolomites with transitions. By contrast, dolomitization in Miocene deposits was commonly incomplete and largely reflected in non-mimetic crystals. The mimetic dolomite is Ca-, Sr- and Na-rich (41 mole%, 234 and 386 ppm respectively) with relatively light $\delta^{18}O$ values (+2.8 ± 0.9‰). The sucrosic dolomite is relatively poor in Ca, Sr and Na (45 mole%, 142 and 241 ppm respectively) with heavier $\delta^{18}O$ (+3.6 ± 0.3‰). Mixing of these phases has resulted in linear geochemical trends. The non-mimetic phase is relatively uniform with enrichment in

Ca, Sr and Na (41 mole%, 293 and 339 ppm respectively) and heavy $\delta^{18}O$ values (+4.2±0.3‰). All three phases appear to have formed from normal seawater, but those in the upper unit probably reflect generation in the marine phreatic zone beneath the marine–meteoric mixing zone. The relatively heavy $\delta^{18}O$ values of the lower dolomite unit may reflect formation from cooler seawater at greater depth, although this raises the question of how quickly the process might have been able to proceed under such conditions.

8.8.4. Dissolution and Alteration of Dolomites

Dissolution of dolomite has been recorded in Quaternary rocks. Ward and Halley (1985) described the selective dissolution of zones in calcian dolomites in the Pleistocene of Yucatan. However, indiscriminate dissolution without regard to any compositional variation, is graphically illustrated by Machel (2004, Figure 9).

The process of replacement was first described by Von Morlot (1848) as 'dedolomitization' and this term was adopted by Shearman et al. (1961). However, it is clear that more than one process may be involved. Dissolution, as described by Jones and Luth (2002), allows the subsequent growth of calcite cement in the resulting pores, whereas 'calcitization' behaves as a 'solid state' replacement of dolomite by calcite, preserving the outlines of dolomite crystals and zones and sometimes also their precursor calcitic fabrics. Appreciable recrystallization (30%) of dolomites (neomorphism) has been achieved experimentally at temperatures of 50°C in as little as one year (Malone, Baker, & Burns, 1996). Gregg, Howard, and Mazzullo (1992) described a progressive 'recrystallization' of Holocene (younger than 3 ka) dolomite in peritidal crusts on Ambergris Cay, Belize. This is demonstrated by an increase in crystal size, decreasing $\delta^{13}C$ values (−0.9 to −5.5‰) and increasing $\delta^{18}O$ (+1.3 to +2.6‰). However, the process of replacement invoked is not the solid-state process usually implied. Instead the authors prefer a surface-energy driven dissolution–precipitation. Smaller crystals dissolve with the increasing saturation driving additional growth on the surfaces of larger crystals.

8.9. Phosphorites

The major sedimentary phosphorites of the world are associated with limestones. In Quaternary deposits, they are related particularly to exposure surfaces and with the percolation through these of phosphatic fluids. Discussions of the general characteristics of these deposits are to be found in Bentor (1980), Slansky (1986) and Notholt and Jarvis (1990). The primary accumulation of phosphate-rich materials on oceanic islands and on continental margins is a result of ocean upwelling. Upwelling is not necessarily a flux of nutrient-rich waters, but simply a system of constant renewal, sustaining the plankton and ultimately fish stocks on which the birds responsible for the initial deposits of guano feed.

Phosphates have been reported from a large number of reef-associated islands including Nauru (Aharon & Veeh, 1984), some in the western Indian Ocean (Hutchinson, 1950; Braithwaite, 1968, 1980, Remire, Aldabra), Amatuku (Funafuti, Rodgers, 1989), Aruba (Netherlands Antilles, Stienstra, 1985), Clipperton Atoll (Bourrouilh-Le Jan et al., 1985), Lady Elliott Island (Queensland, Australia, Chivas, Chappell, Polach, Pillans, & Flood, 1986), Raine Island (Great Barrier Reef, Australia, Baker, Jell, Hacker, & Baublys, 1998), Fanning Island (Roy & Smith, 1971) and Makatea (Montaggioni & Camoin, 1997).

Two groups with contrasting origins, have been identified.

8.9.1. Origins

8.9.1.1. Avian guano

As Hutchinson (1950) noted, guano deposits are most likely to form where seabirds in particular feed over wide areas and return to islands to rest or breed. Fresh avian guano consists largely of uric acid and contains about 4% P_2O_5 (McKelvey, 1973). Early diagenesis primarily involves the removal of the more soluble components leaving the relatively less soluble calcium phosphates. Rainfall is therefore important; islands with guano deposits are typically in arid areas. The residues from these early processes are ultimately relocated to react with the rocks beneath, forming phosphates. Most examples appear to be of Pleistocene age or older. The principal minerals are carbonatian-hydroxy-fluorapatite (francolite) with transitions to fluorapatite and dahllite, but on volcanic islands (Christmas Island; Trueman, 1965) iron and aluminium are incorporated to give barrandite, crandallite–millsite and a range of other phosphatic minerals. The phosphates appear in three forms: as replacements of carbonates, sometimes mimetic, but also as relatively structureless cryptocrystalline peloids; as a phosphatic residue following the dissolution of residual carbonates from replaced materials; and as a phosphate cement. The cements are typically multilaminar with 'colloform' textures that have been attributed to precursor deposits of a colloidal microbial gel. Some layers have a radially fibrous structure (Braithwaite, 1968). Deposits of this kind may form in much older rocks in caves and vugs many metres below the exposed surface (Braithwaite, 1980).

An isotopic study by Aharon and Veeh (1984) of material from Ebon Island, Nauru, Ocean Island and Makatea confirmed avian guano as a primary source of the phosphate. However, the relatively high fluorine content of the deposits on Ebon Island revealed that seawater had been a key medium of transmission, because neither fresh guano nor reef limestone provide, sufficient quantities of this element. Lighter $\delta^{18}O$ and $\delta^{13}C$ values for Nauru phosphates imply an origin involving meteoric waters, presumably older than 200 ka when the island was even more elevated than it is today. Aharon and Veeh (1984) argued in addition that these waters reflect monsoonal rainfall depleted in $\delta^{18}O$ in what may have been a drier climate in the tropics during ice ages.

8.9.1.2. Microbial mediation

Although the best-known island phosphorites are formed via guano deposition, there are deposits for which an alternative origin is proposed. Montaggioni (1985) described phosphatic rocks from Makatea that include oolitic grainstones, intraclast packstones and peloidal–oolitic wackestones with mammilated or botryoidal cements, but including isolated patches of unaltered carbonate (Figure 8.23). The phosphates rest on a distinctive karrenfeld surface and fill caves and dissolution pipes in the rocks beneath. Some accumulations form crude alluvial fans, indicating redistribution from earlier deposits. However, Montaggioni (1985) drew attention to earlier suggestions that the primary deposition was the result not of the accumulation of guano, but of marine organic matter on a submerged platform beneath relatively anoxic waters. Mechanisms of this kind have also been used to explain the occurrence of large-scale commercial

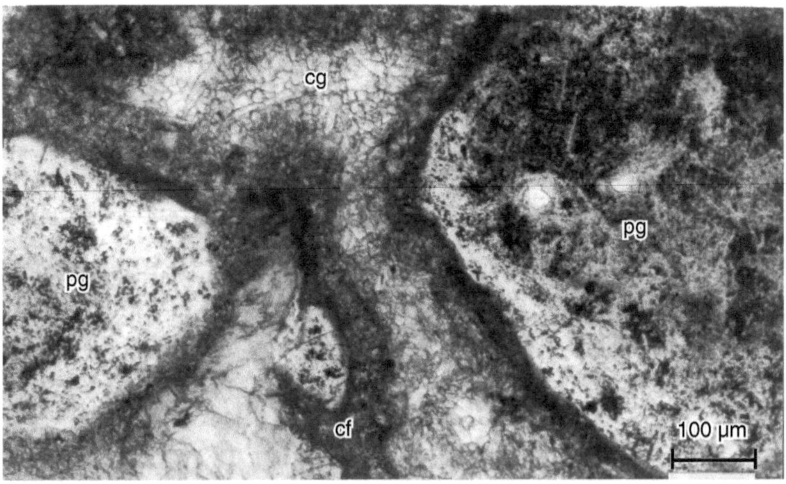

Figure 8.23 Phosphatic intraclasts (pg) successively cemented by isopachous rims (cf) and blocky mosaics (cg) of low-magnesium calcite, Makatea (plane polarized light). Photograph by L. Montaggioni.

phosphorites in the Middle East. Dahanayake and Krumbein (1985) suggested that microbes that included fungi were responsible for the formation of both microbial mats and coated grains in insular phosphorites. Bourruoilh-Le Jan et al. (1985) described lagoonal sediments that include phosphates, magnesite and kutnahorite (manganese-rich dolomite) from Clipperton Atoll off the west coast of Mexico. The lagoon on Clipperton is entirely enclosed, with the only overwash occurring during storm events. Surface waters are of low salinity and are responsible for the dissolution of marginal cliffs. Embayments on the northeastern and southeastern margins of the lagoon are characterized by littoral deposition of accretionary banks of phosphatic sediment. These are only 30–40 cm thick and rest on unaltered carbonate sediments. However, the nature of deposition remains in doubt. Clipperton was the site for a relatively ancient phosphate accumulation (formerly exploited commercially) and the lagoonal deposits may represent reworking of material derived from these. Jehl and Rougerie (1995) explained the concentration of organic matter as a result of the growth of cyanobacterial mats. These are initially preserved by local anoxic conditions but later degrade as the environment becomes more oxidizing, releasing PO_3^2 ions that react with underlying carbonates forming phosphatic crusts as on Tuvalu. Microanalysis and a comparative study of hydrocarbons extracted from the microbial mats and phosphates supports a genetic link. A study by Rougerie, Jehl, et al. (1997) of ponds on the rims of Tuamotu Atolls showed that microbial crusts, referred to as kopara, locally form thick stromatolites. Within these, the phosphate content increases with age and is related to the dissolved phosphate in interstitial waters. Work by Trichet and Fikri (1997) reinforced this view of the importance of microbial (cyanobacterial) activity in making the phosphorous of organic matter in lagoonal environments available for relocation. The critical part played in this by the relative anoxia of bottom waters has been underlined by an investigation of African island (part of the Seychelles group) by Gorshkov, Baturin, Bogdanova, and Magazina (2000). One of the most common non-phosphatic components of both island and ocean phosphorites that are unweathered is iron sulphides. These form minute pyrite framboids (FeS_2), consisting of clusters of crystals of about 1 μm diameter, as well as nanomicron-sized colloidal particles of troilite ($FeS.nH_2O$) attributed to microbial reduction of sulphates.

8.9.2. Age of deposition

Insular phosphates vary widely in age and, like the dolomites, phosphate genesis has occurred on more than one occasion. There seem to be few direct ages available for these deposits and inferential methods, based on the age of the material on which they rest, are unreliable. First, because these have a range of ages from Miocene to Holocene and second, because there was no fixed interval before phosphate deposition began. Veeh and Burnett (1978) reported uranium-series dates for phosphorites from Ebon Island in the Marshall Islands. The primary nature and entry of the uranium into the phosphates was determined by fission tracks. The upper age limit of the associated coral limestones was about 5 ka with radiocarbon ages of 2.8–3.4 ka. The uranium-series age determined for the apatite was around 4 ka. This seems reasonable if the 5 ka age is accepted, but overlooks the fact that the coral limestone would have been deposited during a high sea-level stand whereas the phosphates would have required a lowstand. Roe, Burnett, and Lee (1983) reported attempts at uranium disequilibrium dating of phosphate deposits from Vanua Vatu and Tuvuca in the Lau Group, Fiji. The limestones on which these rest range in age from lower Miocene to Pliocene, but the phosphorites provided ages of 111 ± 15 ka and older than 300 ka. Similar ages were reported from Mataiva Atoll in the Tuamotu Group (Roe et al., 1983). At the time these were the only records of deposition within this low sea-level interval. Uranium-series dates on Nauru phosphates, formerly one of the largest island phosphate deposits in the world provided ages of 140 and 300 ka (Roe & Burnett, 1985) and electron spin resonance dates (Chen, Brumby, Jacobson, Beckwith, & Polach, 1991) indicated ages ranging from 80 to 220 ka.

Chivas et al. (1986) reported phosphate development in a series of shingle ridges on Lady Elliott Island in the Great Barrier Reef. Radiocarbon ages suggested that phosphatization may have begun soon after these were deposited, at around 3.2 ka and continued intermittently until at least AD 1950. Baker et al. (1998) recorded similar phosphates on Raine Island in the northern Great Barrier Reef. These were precipitated in reef rubble deposited since about 4.78 ka, probably less than 1.2 ka. Isotopic evidence ($\delta^{18}O$) indicates that deposition was from meteoric waters with no influence of either evaporation or seawater mixing. By contrast, Aharon and Veeh (1984) recorded U/Th ages of >200 ka for apatite from Nauru and around 4 ka for samples from Ebon Island in the Pacific. Stienstra (1985) recorded Plio-Pleistocene phosphorites on Ceru Colorado, Aruba, in the Dutch Antilles that he regarded as being formed from guano more than 500 ka ago.

8.10. Conclusions

When first deposited, Quaternary carbonate sediments consist largely of aragonite and Mg-calcite, representing the skeletal remains of a variety of reef-dwelling organisms. The same minerals may also be added to the sediment during early marine cementation. If sea level falls, these unstable components are gradually replaced, either by neomorphism or by wholesale dissolution. Low-magnesium calcite cements form under the influence of meteoric waters and with time these dominate the system, with the loss of all unstable components. In some areas, exposure to waters rich in magnesium, among which seawater is the most significant, results in the replacement of parts of the assembly by dolomite, with the addition of dolomite as a cement. In arid areas close to sites of ocean upwelling, deposition of guano on exposed surfaces may lead to replacement by phosphates and deposition of phosphatic cements, largely under the influence of meteoric waters. In other island sites, phosphate deposition is likely to have been under the control of microbial activity and is not tied to exposure. The linkage between all of these mineral phases and water chemistry implies that the mineral assembly is sensitive to sea-level change. However, the results of this depend upon local rainfall and the extent and duration of the exposure episode. Although in a few cases, mineral assemblies can be related to specific exposure surfaces, more typically, because long-term changes represent an oscillation, parts of assemblies are overprinted. This means that although the diagenetic environment may be recognized, it may not be attributable to a specific exposure event.

CHAPTER NINE

Corals and Coral Reefs as Records of Climatic Change

9.1. Introduction

Because coral reefs occupy an environment at the interface of the ocean, land and atmosphere, they are ideally situated to monitor climatic events that operate at local and global scales throughout the tropics. They are able to incorporate into the skeletons of their individual coral colonies and into their structural framework and associated deposits as a whole, records of most of the environmental parameters that drive their growth.

Domal coral skeletons are considered to be first-order proxy indicators of climatic change in the tropics because: (1) they grow within specific geographical locations and habitats, forming the most common benthic communities in tropical seas; (2) they possess a mode of growth typified by the deposition of distinct annual bands that offer good chronological constraints; (3) they grow continuously and may live for more than 400 years, permitting continuous, relatively long-term records; (4) they grow at rates averaging 10–15 mm yr^{-1}, permitting weekly to monthly resolution; (5) they incorporate a large array of chemical tracers as well as morphometric characteristics significant in terms of environmental conditions; (6) they can be dated relatively accurately using radiometric methods for periods spanning the past 350–400 ka. The utility of geochemistry applied to reef-building scleractinian corals for the reconstruction of past climates has been supported by a large number of calibration studies. Pioneering work by Weber and Woodhead (1972b), Shen, Boyle, and Lea (1987) and Lea, Shen, and Boyle (1989) demonstrated that the variability of chemical element compositions in coral skeletons quantitatively reflects past surface chemistry of the oceans. However, the significance of geochemical signals in coral aragonite, in terms of climate variability, may be altered by both biological and diagenetic processes. During precipitation of skeletal aragonite, coral metabolic activity ('vital effects') may cause large compositional deviations from thermodynamic equilibrium and compromise precise calibrations of climate tracers (Rollion-Bard, Chaussidon, & France-Lanord, 2003; Sinclair & Risk, 2006; Meibom et al., 2007). Early and late diagenesis affecting skeletal aragonite in ambient seawater or after emergence may be difficult to detect and can also result in marked bias in both climate reconstruction (Enmar et al., 2000; McGregor & Gagan, 2003; Muller, Gagan, & Lough, 2004; Allison, Finch, Webster, & Clague, 2007) and age determination using radiometric methods (Siddall, Chappell, & Potter, 2006; Scholz & Mangini, 2007).

Reef systems have also revealed, in sedimentological and chronological data, how their spatial distributions, anatomy and compositions reflect sea-level variation over time. Sea level has significant implications for global climate (Shackleton, 1987) because many large- and small-scale changes in sea level broadly arise from changing climate. However, sea-level variability is constrained not only by climate, but also by a complex spatial pattern of interactions among tectonic and isostatic forcings that respond at different time scales (Lambeck, 2002; Milne, Long, & Bassett, 2005). The record of Quaternary sea-level change based on reef data has been intensively studied for the last two decades in both the western Atlantic and Indo-Pacific provinces (see reviews by Pirazzoli, 1991, 1996; Woodroffe & Horton, 2005; Montaggioni, 2005; Siddall et al., 2006; Hearty, Hollin, Neumann, O'Leary, & McCulloch, 2007; Hopley et al., 2007).

The main objectives of this chapter are to describe how changes in environmental conditions are monitored in the internal structures of reef tracts and in the physical and chemical compositions of coral skeletons. Four issues are considered: (1) the nature and climatic significance of coral geochemical proxies and the major climatic modes recorded by coral skeletons; (2) palaeoclimate reconstructions from the late Pleistocene to recent decades; (3) reef-associated structures and deposits diagnostic of sea level; (4) the reconstruction of sea-level changes from the late Pleistocene to recent decades.

9.2. Individual Coral Colonies as Records of Climate

Palaeoclimate reconstruction from individual coral colonies exploits a significant number of climatic signature patterns, including those of sea surface temperature (SST), precipitation, sea surface salinity (SSS), wind regime and/or ocean circulation.

9.2.1. Growth Mode of Banded Coral Skeletons and its Environmental Control

Domal corals grow at rates ranging from a few millimetres to about 3 cm yr^{-1}. As they grow, they generally produce a distinct seasonal pattern of alternating density bands (sclerobands) each year in their skeletons

(Buddemeier, Maragos, & Knutson, 1974). Variations in the width and density of bands are controlled by an array of environmental variables including latitudinal location, SST and salinity, light, season length, hydrodynamic energy and siltation rate (Scoffin, Tudhope, & Brown, 1989a, and references therein). Other controlling factors relate to coral metabolism (symbiotic exchanges, reproductive activity and nutrient availability). Corals therefore deposit sclerobands under varying conditions, leading to difficulty in the interpretation of density banding in terms of environmental changes. During growth, the aragonite of coral skeletons also incorporates various isotopes, minor and trace elements and some organic compounds. The variations in these are regarded as reflecting a moving average of environmental factors expressed in ambient waters during the time over which corals were growing. These factors include SST, hydrological balance (precipitation, evaporation, runoff and river discharge), light and nutrient levels, and ocean circulation.

The density-banding pattern is usually identified using X-radiography. The method is applied to slices cut through colonies or cores extracted from individual coral colonies along their major axes of growth. X-ray images are commonly used to select optimal sampling tracks for geochemical analysis and to define a precise chronology for the analytical record. The chronology of colonies, referred as sclerochronology, is generally based on counting the annual density-band pairs along their major axis of growth. The counting starts from the outermost layer at the colony top downwards. In the case of a living coral, defining the age is relatively easy because the date of collection is generally known and the banding pattern is usually clear. The age models may be improved using the seasonal cyclicity of some physical parameters, for example, linear extension and calcification rates, density (Lough & Barnes, 2000) and luminescent lines (Isdale, Stewart, Tickle, & Lough, 1998; Hendy, Gagan, & Lough, 2003), or using geochemical components in the coral skeletons, particularly those expressing the seasonality of temperature (Corrège, 2006) and light (Aharon, 1991) variation. Defining age models for fossil corals is substantially more difficult, especially for those extracted from reef sequences by coring. The specimens are commonly incomplete and may provide a biased record because the original colonies were not cored through the major axis of growth. In addition, the physicochemical attributes of the skeletons may have been compromised by postmortem diagenesis. Massive *Porites* and *Montastraea* are the most commonly used corals in palaeoenvironmental reconstructions in the Indo-Pacific and Caribbean provinces respectively. Occasionally, *Diploastrea heliopora* in the western Pacific (Watanabe, Gagan, Corrège, Scott-Gagan, & Hantoro, 2003; Corrège et al., 2004) and *Diploria strigosa* and *D. labyrinthiformis* in the Caribbean have also served as climate archives (Cohen, Smith, McCartney, & van Etten, 2004; Hetzinger et al., 2008).

9.2.2. Environmental Variables and their Proxies in Corals

9.2.2.1. Sea surface temperature

SST is by far the most significant variable for climate reconstruction and a number of palaeothermometers, both physical and chemical in nature, are embodied in massive corals.

Annual growth characteristics. The density-band patterns of massive coral skeletons allow the quantification of three growth attributes on an annual time scale: density, extension rate and calcification rate. The calcification rate is the product of the linear extension rate and the average density at which the skeleton was built during that extension. Changes in density, extension and calcification rates are controlled by environmental factors, including SST, light levels, water quality and sediment supply. However, given its dependence on oceanographic parameters, in most cases the coral extension rate is not easy to interpret in terms of climate change (Felis & Pätzold, 2004). Studies in the Indo-Pacific have nevertheless indicated that extension and calcification rates in massive poritid colonies can provide robust tracers for SST. Comparing findings obtained on the Australian Great Barrier Reef with data from other Indo-Pacific reef sites, Lough and Barnes (2000) concluded that: (1) the extension rate of domal poritids is negatively correlated with skeletal density and positively correlated to calcification rate and (2) the extension rate and calcification rate respond linearly to average annual SST (Figure 9.1). Average calcification increases by about $0.3\,g\,cm^{-2}\,yr^{-1}$ and the average extension rate increases by about $3\,mm\,yr^{-1}$ for each 1°C rise in SST. The high sensitivity of coral calcification to changes in SST has been supported by work in the central Pacific by Bessat and Buigues (2001), who showed that a 1°C rise in SST over the past two centuries has increased the average calcification rate of poritids by about 4.5%. Furthermore, based on spectral analysis, the annual calcification rate appeared to be significantly related to biennial (about 2.5 years), ENSO (about 4–7 years) and decadal-scale variability (about 21.9 years) frequency bands. In addition, Cohen et al. (2004) showed that X-ray intensity ratios, used as an indicator of variations in skeletal density, faithfully reflected SST variability in the subtropical North Atlantic gyre.

Oxygen isotopes. The ratio of oxygen isotopes, expressed as $\delta^{18}O$ [$\delta^{18}O$ = per mil deviation of the ratio of $^{18}O/^{16}O$ relative to the Peedee Belemnite standard (‰ vs. PDB)] incorporated in coral skeletons, is the most commonly used proxy for reconstructing high-resolution (monthly to near-weekly) SST. Pioneering work on the use of coral skeletal $\delta^{18}O$ as a palaeothermometer was by Weber and Woodhead (1972b) and Fairbanks and Dodge (1979). As the SST increases, the skeletal $\delta^{18}O$ decreases as a result of temperature-dependent kinetic fractionation effects (Figure 9.2A). However,

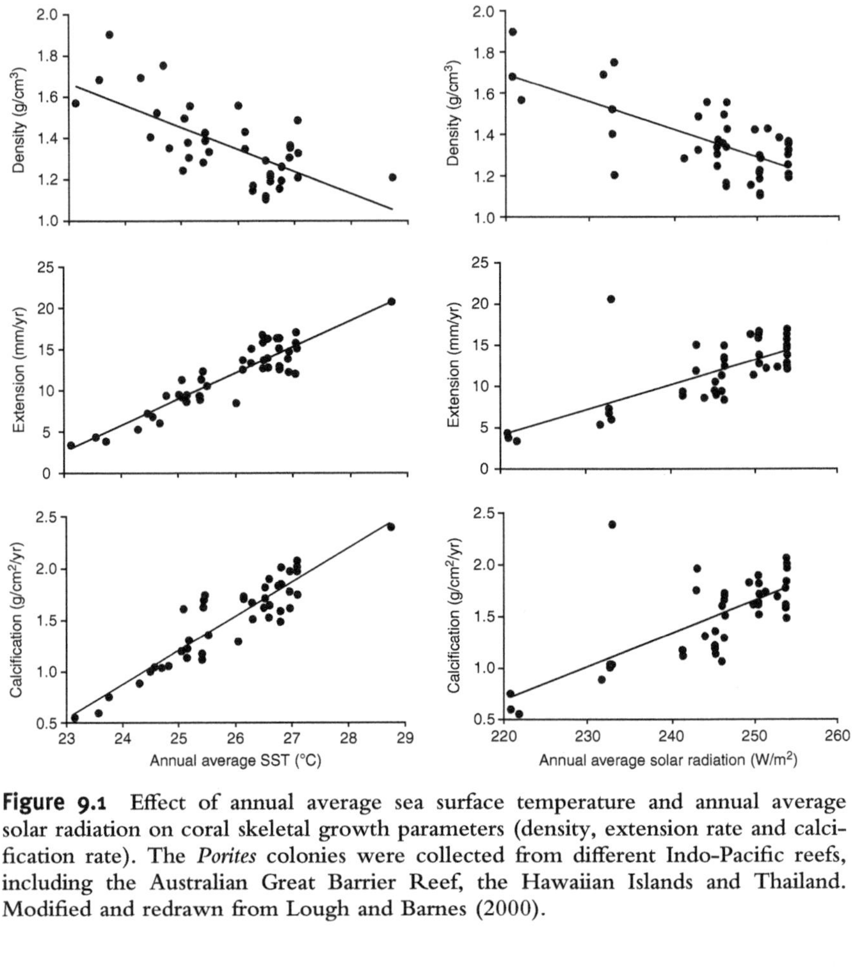

Figure 9.1 Effect of annual average sea surface temperature and annual average solar radiation on coral skeletal growth parameters (density, extension rate and calcification rate). The *Porites* colonies were collected from different Indo-Pacific reefs, including the Australian Great Barrier Reef, the Hawaiian Islands and Thailand. Modified and redrawn from Lough and Barnes (2000).

the $\delta^{18}O$ of the coral reflects a combination of the temperature and the $\delta^{18}O$ of the ambient seawater in which the skeletal aragonite precipitated. Skeletal $\delta^{18}O$ decreases as seawater $\delta^{18}O$ decreases. The seawater $\delta^{18}O$ is directly related to the hydrological balance. Evaporation results in an enrichment in ^{18}O, while precipitation, runoff and river discharge produce an enrichment in ^{16}O. Thus, coral $\delta^{18}O$ also provides information on changes in salinity (SSS) and the SST signal can be affected by the SSS component. In areas where the natural variability in SSS is low, variations in skeletal $\delta^{18}O$ primarily reflect variations in SST, whereas in areas where variability in rainfall, runoff and river discharge is high, coral $\delta^{18}O$ is more likely a reflection of changes in SSS. An additional constraint is provided by the fact that coral skeletons precipitate in isotopic disequilibrium with the ambient water; they are depleted in ^{18}O with respect to the isotopic composition of the water. This

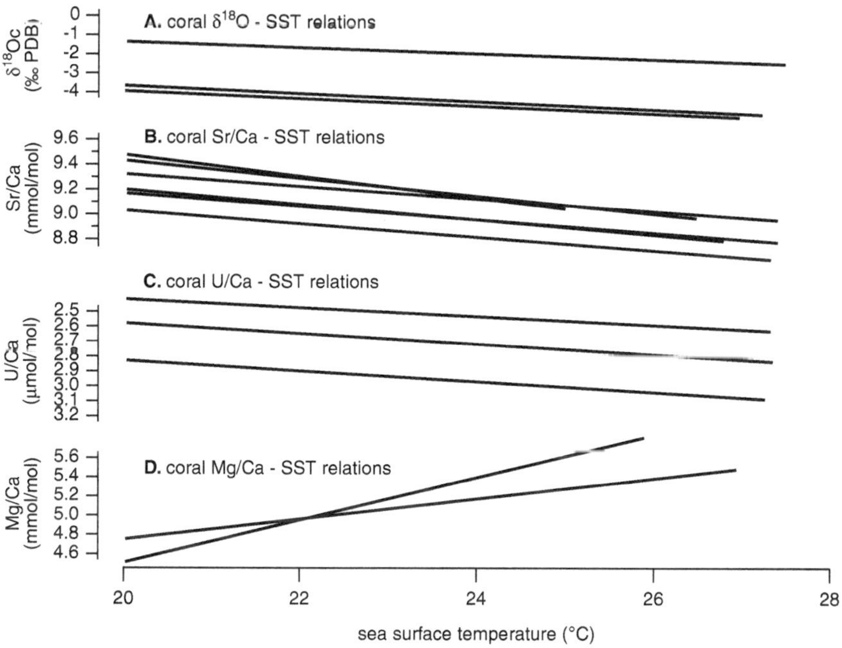

Figure 9.2 Linear regression between a variety of geochemical proxies in *Porites* corals collected in New Caledonia (western Pacific Ocean) and sea surface temperature (SST). (A) Coral $\delta^{18}O$–SST relationship; (B) coral Sr/Ca–SST relationship; (C) coral U/Ca–SST relationship; (D) coral Mg/Ca–SST relationship. Modified and redrawn from Ourbak et al. (2006).

disequilibrium is mediated during skeletogenesis by the metabolic activity of the symbiotic algae inhabiting the coral tissues through respiration, the so-called 'vital effect' (McConnaughey, 1989; Allemand et al., 2004). Apart from SST and the hydrological balance, the other parameters potentially driving coral $\delta^{18}O$ include extension rates, light level, productivity, food availability and variations in pH close to the calcification sites in the skeleton (Corrège, 2006). The actual effect of variations in extension rate is controversial, and has been regarded as both 'pronounced' (Felis, Pätzold, & Loya, 2003; Maier, Felis, Pätzold, & Bak, 2004) and 'insignificant' (Watanabe et al., 2003, and references therein). Similarly, there is controversy regarding the dependence of $\delta^{18}O$ on light intensity, feeding habits, metabolism and local flow regime. Some authors have claimed that low light levels tend to promote $\delta^{18}O$ depletion (Reynaud-Vaganay, Juillet-Leclerc, Jaubert, & Gattuso, 2001) and that well-nourished corals have lower $\delta^{18}O$ than starved ones, resulting in apparently higher SST signatures (Reynaud et al., 2002). Suzuki, Nakamura, Yamasaki, Minoshima, and Kawahata (2008) suggested that isotopic variations within an individual colony may be governed by local current intensity.

Others concluded that the oxygen isotopic signal is not altered by irradiance (Fairbanks & Dodge, 1979), food availability (Grottoli & Wellington, 1999) or metabolic activity (Juillet-Leclerc, Gattuso, Montaggioni, & Pichon, 1997).

Despite the complexity of the parameters affecting coral $\delta^{18}O$, observations on inorganically precipitated aragonite suggest that it does respond to variations in temperature. As SST increases, the coral $\delta^{18}O$ decreases. Near-weekly resolution records of variability of $\delta^{18}O$ in massive corals indicate that a 1°C rise in SST corresponds to a decrease of about 0.18–0.22‰ (Evans, Kaplan, & Cane, 2000; Grottoli, 2001; Corrège, 2006). However, the $\delta^{18}O$ of coral skeletons should be regarded as a relative rather than an absolute tracer of SST.

Strontium/calcium ratio. The Sr/Ca ratio of coral skeletons is regarded as a robust tracer of SST (see Beck et al., 1992; Gagan et al., 1998; de Villiers, 1999; Corrège, 2006). Compared to $\delta^{18}O$, the Sr/Ca ratio is more stable in seawater through time. There is an increase in Sr/Ca values as temperature decreases and Sr^{2+} probably substitutes for Ca^{2+} in coral aragonite when the temperature falls (Figure 9.2B). The coral Sr/Ca relationship does not seem to vary markedly for individuals inhabiting the same site, but the compilation of Sr/Ca–SST calibrations for the genus *Porites* by Corrège (2006) clearly show discrepancies in temperature calibrations between *Porites* colonies from different areas. The values of coral Sr/Ca calibrations express a temperature dependency ranging from about 0.0597 to 0.062 mmol mol^{-1} per 1°C (Gagan et al., 2000; Marshall & McCulloch, 2002; Felis & Pätzold, 2004).

The Sr/Ca ratios of coral skeletons may potentially be affected by factors other than SST. Controversial results suggest the influence of changes in growth rate during skeletogenesis (Swart, Elderfield, & Greaves, 2002; Allison & Finch, 2004; Felis et al., 2003; Cohen & Hart, 2004; Corrège et al., 2004; Reynaud et al., 2007). The amounts of both strontium and calcium ions in the ambient seawater also influence the Sr/Ca ratio in corals. The residence times of Sr and Ca in the ocean are known to be as long as millions of years. Thus, the Sr/Ca ratios of seawater has been believed to have been constant over time and space, and particularly since the Last Glacial Maximum (LGM; Guilderson, Fairbanks, & Rubenstone, 1994). However, de Villiers, Shen, and Nelson (1994) and Shen et al. (1996) pointed out that these ratios can vary markedly in the modern ocean either between locations, as a function of upwelling activity, or in the same location over the year. Moreover, during low sea-level stands, seawater Sr/Ca ratios may be altered through the release of strontium from the weathering of Sr-rich carbonates exposed on the shelves (Stoll & Schrag, 1998). The effect of changes in seawater Sr/Ca on reconstructed SST varies from 0.2 to 2°C (Corrège, 2006).

Other SST proxies. In addition to oxygen isotopes and Sr/Ca, other temperature-dependent elements present in coral skeletons are uranium (U/Ca), magnesium (Mg/Ca) and, to a lesser extent, boron (B/Ca). Due to the complexity of parameters governing the incorporation of uranium into corals, the use of U/Ca ratios as a thermometer remains in its infancy, although some significant results have been obtained locally (Min et al., 1995; Shen & Dunbar, 1995; Corrège et al., 2000; Fallon, McCulloch, & Alibert, 2003; Ourbak et al., 2006) (Figure 9.2C). The significance of coral Mg/Ca ratios in terms of temperature is still debated (Corrège, 2006, and references therein). To date, the only robust Mg/Ca–SST calibrations were by Mitsuguchi, Matsumoto, Abe, Uchida, and Isdale (1996), although additional calibrations were performed recently (Figure 9.2D). The potential sensitivity of B/Ca to SST was first tested by Hart and Cohen (1996) and by Sinclair, Kinsley, and McCulloch (1998) but further work is needed to assess the reliability of the boron palaeo thermometer. Other potential SST proxies in corals include Ca ($^{44}Ca/^{40}Ca$ and $^{44}Ca/^{42}Ca$) isotopes (Böhm et al., 2006) but are yet to be rigorously assessed.

9.2.2.2. Sea surface salinity

SSS is considered to be a paramount variable in climate reconstruction. It is known to control the thermohaline circulation and to greatly influence the ENSO phenomenon. In addition, the determination of changes in salinity at a given site may provide information on the hydrological cycle and on atmospheric variability, through changes in precipitation and evaporation patterns.

Oxygen isotopic composition has been successfully used in the past as a SSS monitor, alone or combined with the Sr/Ca in coral skeletons. Precipitation is depleted in ^{18}O relative to seawater and thus produces lower $\delta^{18}O$ values, whereas evaporation promotes the removal of lighter (^{16}O-rich) oxygen atoms from the ocean, resulting in higher $\delta^{18}O$ values in surface seawater. Since the precipitation/evaporation balance appears to be linked to changes in SSS, there is a strong relationship between the oxygen isotopic composition of seawater and SSS (see Corrège, 2006, and references therein). However, the seawater $\delta^{18}O$–SSS relationship also depends on the latitude and the area considered. For instance, in the tropical Pacific Ocean, the regression slope between seawater $\delta^{18}O$ and SSS varies from 0.27 to 0.42‰ psu^{-1} (Morimoto et al., 2002). Variability of this order may introduce an error of 0.1–2 psu in past SSS values derived from seawater $\delta^{18}O$ (Benway & Mix, 2004). As indicated, the oxygen isotopic composition of coral aragonite reflects both the local SST and seawater $\delta^{18}O$ components. Past variations in SSS can therefore theoretically be inferred from the $\delta^{18}O$ of the coral if the local seawater $\delta^{18}O$–SSS relationship is known as well as the SST component of the coral $\delta^{18}O$. In sites where annual to interannual SST variability is small and constant, although SSS variability is strong, the coral $\delta^{18}O$ values are usually regarded as directly reflecting changes in seawater $\delta^{18}O$. By filtering the $\delta^{18}O$ signal in the coral, the SST component can be eliminated and the residual $\delta^{18}O$ values are believed to strictly express SSS variations (Gagan et al., 2000; Le Bec, Juillet-Leclerc, Corrège, Blamart, & Delcroix, 2000; Corrège, 2006). In areas where variations in SST are large, a double-proxy method of coupled $\delta^{18}O$ and Sr/Ca measurements has proven to be a robust palaeosalinometer (McCulloch, Gagan, Mortimer, Chivas, & Isdale, 1994; Gagan et al., 1998; see Gagan et al., 2000 for a review). Assuming the relationship between the oxygen isotopic composition of the seawater and SSS is constant through time, the SSS component can be extracted from the coral $\delta^{18}O$ variations by derivation of SST values recorded in the Sr/Ca ratios. Similarly, coupled U/Ca and Sr/Ca analyses could potentially serve as a tracer of palaeo-SSS (Ourbak et al., 2006).

9.2.2.3. Precipitation

Barium/calcium. In marine waters and sediments, barium is associated with a variety of solid phases and is considered to be a powerful tracer of coastal and open sea processes (Prakash Babu, Brumsack, Schnetger, & Böttchert, 2002; Sinclair & McCulloch, 2004; Gonneea & Paytan, 2006). Barium is incorporated into coral skeletons by substituting for calcium at concentrations that closely reflect the Ba/Ca ratio of the surrounding waters. However, incorporation may also occur in the form of particulate barium organically bound at specific sites (Tudhope, Lea, Shimmield, Chilcott, & Head, 1996). Corals living in inshore settings are found to have higher Ba levels than those growing in reefs facing open sea. Shen and Boyle (1988) first suggested that this feature is a reflection of the barium supply from land to near-shore waters through terrestrial runoff. This idea was supported by further research and Ba/Ca ratios have been successfully applied to corals to reconstruct rainfall, land runoff and riverine input (Shen & Sanford, 1990; Tudhope et al., 1997; McCulloch et al., 2003; Sinclair & McCulloch, 2004; Montaggioni et al., 2006).

Stable oxygen isotopes. The $\delta^{18}O$ of coral skeletons can be also regarded as a qualitative proxy for changes in rainfall, as it is usually strongly linked to seawater $\delta^{18}O$. Heavy precipitation promotes a decrease in both seawater $\delta^{18}O$ and SSS. In areas where SSS varies significantly, or is even the dominant parameter in the skeletal $\delta^{18}O$ signal, changes in $\delta^{18}O$ values may reflect changes in rainfall (e.g. see Tudhope et al., 1996, 1997; Urban, Cole, & Overpeck, 2000).

Luminescence banding. Luminescent (or UV fluorescent) bands in massive coral skeletons were first detected by Isdale (1984). They were interpreted as resulting from changes in the relative proportions of terrestrial humic compounds incorporated into the coral skeleton during river flood events (Figure 9.3) generated by heavy precipitation (Isdale, 1984; Scoffin et al., 1989; Smith, Hudson, Robblee, Powell, & Isdale, 1989; Matthews, Jones, Theodorou, & Tudhope, 1996; Isdale et al., 1998; Lough, Barnes, & McAllister, 2002; Nyberg, 2002; Barnes, Taylor, & Lough, 2003; Hendy et al., 2003). Coral fluorescent bands may therefore serve as proxies for reconstructing variations in precipitation. However, controversial results have been obtained from other reef sites. Fluorescent bands may also be produced during the dry season (Scoffin, Tudhope, Brown, Chansang, & Cheeney, 1992) and may be present in corals living far from any river discharge, as a result of seasonal planktonic blooms (Tudhope et al., 1996). An alternative explanation for the fluorescence was proposed by Barnes and Taylor (2001) who suggested that variations in intensity may reflect variations in coral microstructure. In their view fluorescent bands are associated with low-density skeletal sites, and the reduction in calcification can be attributed to low salinity conditions reflecting terrestrial runoff.

Rare earth elements. The use of rare earth elements (REE) in corals, as potential tracers of marine chemistry, has received little attention (Sholkovitz & Shen, 1995; Fallon, White, & McCulloch, 2002; Wyndham, McCulloch, Fallon, & Alibert, 2004). REE are incorporated into the coral aragonite lattice in amounts closely reflecting their concentrations in ambient seawaters and are thus regarded as suitable for tracing environmental changes. The tracers commonly used are ratios of light and heavy REE (neodymium/ytterbium) and anomalies of cerium, the latter expressed as $3(\text{cerium}/\text{cerium}_{\text{shale}})/[2(\text{lanthanum}/\text{lanthanum}_{\text{shale}}) + (\text{neodymium}/\text{neodymium}_{\text{shale}})]$. REE compositions are expressed in normalized values relative to the composition of shale. Comparison of Nd/Yb and Ce anomalies between inshore and mid-shelf corals in the Australian Great Barrier Reef indicated that corals close to the coast were characterized both by higher REE concentrations (greater than 10 times) and by an enrichment in light REE (Wyndham et al., 2004). These differences were suggested to have been driven by seasonal changes in terrestrial runoff and river discharge.

9.2.2.4. Solar radiation

The environmental significance of variations in the concentrations of stable carbon isotopes incorporated into coral skeletons (expressed as $\delta^{13}C$, per mil deviation of $^{13}C/^{12}C$ ratio relative to the PDB standard) is difficult to interpret due to interactions between kinetic and metabolic processes during

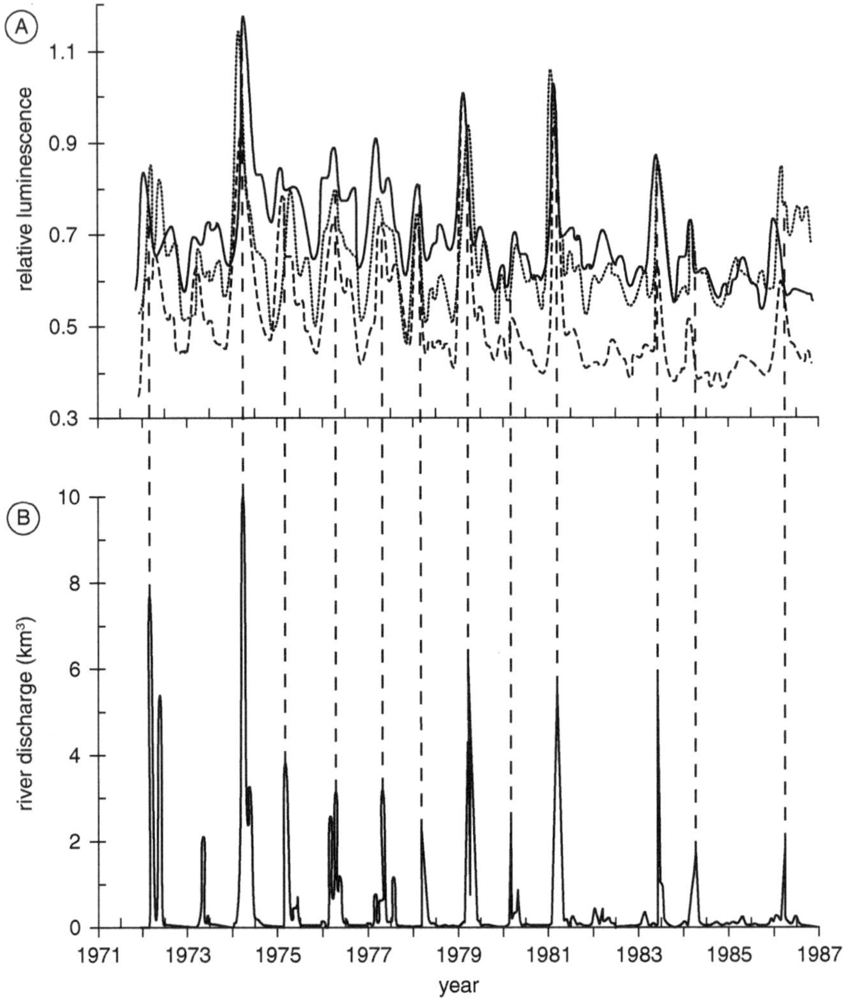

Figure 9.3 Comparison between dated luminescence profiles (A) through three *Porites* corals collected at Pandora Reef (central Great Barrier Reef of Australia) and the volume of fresh water discharge from the nearby Burdekin River (B) for the period 1972–1986. Luminescence lines were obtained from near-vertical tracks across slices from the coral colonies. Note the strong correlation between the luminescence records and riverine flooding episodes. Modified and redrawn from Barnes et al. (2003).

isotopic fractionation (Grottoli, 2000). On an interannual time scale, the variation in skeletal $\delta^{13}C$ is believed to be governed predominantly by algal photosynthesis within the coral polyp (Swart, 1983; McConnaughey, 1989; Juillet-Leclerc et al., 1997; Grottoli & Wellington, 1999) and to be related particularly to seasonal changes in light intensity, cloud cover and transparency of surface waters (Fairbanks & Dodge, 1979; McConnaughey, 1989;

Grotolli, 2002; Asami et al., 2004). Skeletal $\delta^{13}C$ increases with enhanced photosynthetic activity, and decreases in photosynthesis are accompanied by decreases in $\delta^{13}C$. The $\delta^{13}C$ of coral skeletons is also controlled by food uptake but in the absence of large heterotrophic variations in input, interannual cyclicity can be interpreted in terms of local changes in solar radiation resulting from seasonal differences in cloudiness (Juillet-Leclerc, 1999; Grotolli, 2002).

9.2.2.5. Atmospheric and oceanic circulation

The movements of the atmosphere are generated by differential heating of the earth's surface (see Chapter 1, Section 1.4). Oceanic waters around the world are driven by wind forcing and by density gradients. Surface oceanic currents move primarily owing to surface winds and the Coriolis effect forming large circular cells (gyres). Deep ocean circulation is driven by differences in the density of seawater (thermohaline circulation).

Cadmium/calcium and barium/calcium. Cadmium and barium are the most reliable proxies for upwelling. The behaviour and distribution of cadmium in the water column is very similar to that of phosphorous, a major macronutrient (Shen et al., 1987). The Cd concentration is low at the sea surface due to biological uptake and tends to increase with depth as a result of dissolution of sinking organic detritus. Upwelling, cold, deep waters are driven to the surface, and enrich the waters surrounding coral reefs in Cd (Figure 9.4). The Cd/Ca ratios of corals therefore appear to provide not only reliable records of variation in cadmium concentrations over time (e.g. see Reuer, Boyle, & Cole, 2003, and references therein), but also their relative intensity (Shen & Sanford, 1990).

The use of the Ba/Ca ratio as a record of palaeoupwelling in reef corals was pioneered by Lea et al. (1989) and discussed by Ourbak et al. (2006). The value of Ba as a proxy for upwelling or lateral advection has been demonstrated by Shen et al. (1992a,b), Tudhope et al. (1996), Anderegg, Dodge, Swart, and Fisher (1997), Fallon, McCulloch, van Woesik, and Sinclair (1999), Reuer et al. (2003) and Montaggioni et al. (2006) (Figure 9.5). The Ba signals appear to document a seasonal nutrient- and barium-rich upwelling to the sea surface. However, Cd was shown to be a better indicator of upwelling than Ba due its higher concentrations in the deep waters and lower input from the land.

Manganese/calcium. Manganese has been used tentatively as an indicator of upwelling. The most comprehensive study devoted to its use as an oceanographic proxy was by Shen et al. (1991). Mn appears to be a common element in both coastal and open seawaters, transported from land via river discharge and to the open sea by atmospheric and oceanographic fluxes. The

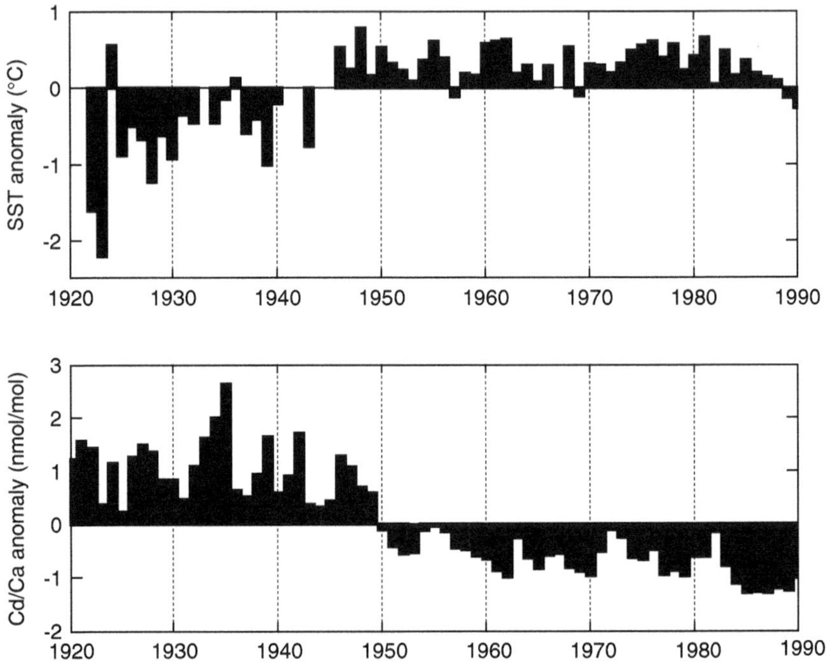

Figure 9.4 Comparison between mean annual sea surface temperature (SST) anomalies and Cd/Ca anomalies in the Tortuga area (Cariaco Basin, southern Caribbean). SST anomalies were calculated with respect to the 1920–1992 interval. Cd/Ca ratios were measured from a *Montastraea* coral at Isla Tortuga. Note the strong negative correlation between SST and the Cd/Ca time series. Modified and redrawn from Reuer et al. (2003).

conditions of inclusion of Mn into coral skeletons are still under discussion. It is generally supposed to be lattice-bound in coral aragonite at 10–50% of its water concentration. Other models of incorporation include trapping of discrete particles and adsorption in the form of an oxide or organic phase. In the eastern equatorial Pacific Ocean, high-frequency, interannual changes in Mn/Ca values have been inferred to have been controlled primarily by seasonal upwelling cycles. The periodicities shown by Mn/Ca are the reverse of those displayed by the Cd/Ca ratio, consistent with the mirror image distributions of Mn and Cd in the upper layers of eastern Pacific waters.

Lead. Lead is considered to be an excellent tracer of anthropogenic pollution resulting from industrial activities. It may be detected using the stable isotopes ^{206}Pb, ^{207}Pb and ^{208}Pb present in seawaters and sediments. It has been used as an indicator of air mass dynamics in the North Atlantic, where the dominant winds transport contaminant aerosols with specific

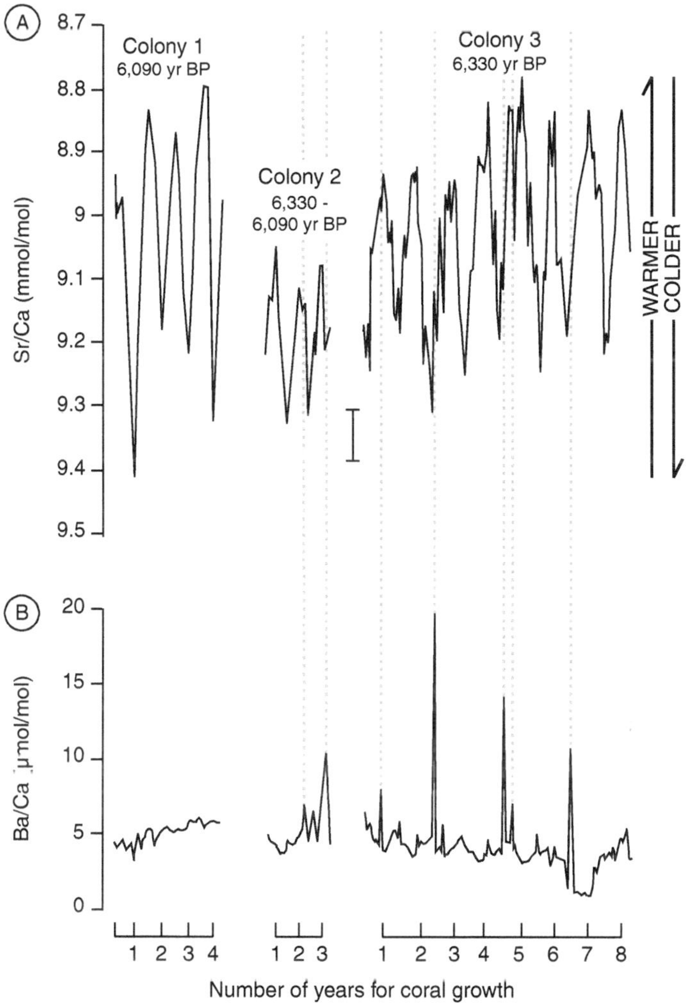

Figure 9.5 Comparison between raw Sr/Ca and Ba/Ca ratios plotted as time series for three mid-Holocene *Porites* corals extracted from the outer fringing reef flat at Vata-Ricaudy (southwestern New Caledonia). (a) Variations of Sr/Ca ratios. The yearly chronology is based on the observed Sr/Ca cycles, assuming the maximum and minimum values indicate the coldest (winter) and the warmest (summer) months for each year of growth. Colonies 1, 2 and 3 encompass 4, 3 and 8 growth-years respectively. The calibrated radiocarbon age of each coral is given. (b) Variations of Ba/Ca ratios. The vertical dashed lines show the link between Ba/Ca peaks and seasonal Sr/Ca peaks. Modified and redrawn from Montaggioni et al. (2006).

isotopic signatures. Ratios of $^{206}Pb/^{207}Pb$ range from 1.14–1.16 for trade wind easterlies to 1.19–1.22 for North American westerlies (Hamelin, Ferrand, Alleman, Nicolas, & Véron, 1997). Coral skeletons can also trap submicrometre-size particulate lead, transported by air and in ocean water masses and thus provide robust archives of atmosphere-derived pollution inputs to surface waters. Dodge and Gilbert (1984) pioneered the use of lead for reconstructing the history of atmospheric contamination over the past 150 years from coral cores collected in a variety of reef sites worldwide. Further investigations carried out by Shen and Boyle (1987, 1988), Reuer et al. (2003) and Desenfant, Veron, Camoin, and Nyberg (2006) focused on the North Atlantic region.

Radiocarbon. Changes in radiocarbon concentrations in corals provide reliable records of water mass dynamics on interannual to decadal time scales. The application of coral radiocarbon records in resolving ocean circulation patterns was reviewed by Grottoli and Eakin (2007). Radiocarbon (^{14}C) is originally derived from the stratosphere where it is naturally generated, and from nuclear bomb experiments conducted in the atmosphere from the end of the 1940s. At the sea surface, radiocarbon forming part of the atmospheric carbon dioxide, diffuses into the water and becomes part of the dissolved inorganic carbon (DIC). During coral skeletogenesis, the DIC present in the ambient seawater and including ^{14}C is incorporated into corals. Measurements of the radiocarbon in coral aragonite are reported as $\Delta^{14}C$ (the per mil deviation of $^{14}C/^{12}C$ of the sample relative to that of the 95% oxalic acid-1 standard) and corrected for fractionation to a $\delta^{13}C$ of −25‰. The $\Delta^{14}C$ reflects the radiocarbon content of the DIC. Variations in $\Delta^{14}C$ values in waters and in coral aragonite are controlled by the Suess effect (the dilution of the atmospheric concentrations in ^{13}C and ^{14}C by the admixture of large amounts of fossil fuel-derived CO_2) and reflect seawater movements, including vertical mixing and horizontal advection, and the supply of bomb-derived $\Delta^{14}C$. Prior to the testing of nuclear weapons, over the first half of the 20th century, coral $\Delta^{14}C$ values decreased due to the Suess effect. Direct CO_2 exchange between the atmosphere and deep ocean layers is quite limited, and this results in lower ^{14}C in deep waters as ^{14}C decays, and in lower $\Delta^{14}C$ in the DIC. The $\Delta^{14}C$ value of the DIC increases with increasing time spent by water masses at the sea surface until it reaches equilibrium with atmospheric $\Delta^{14}C$. Upwelling and vertical mixing supply low $\Delta^{14}C$-waters to the surface. The $\Delta^{14}C$ of surface waters is considered to be a robust, passive tracer for horizontal advection because the rates of biological processes affecting the $\Delta^{14}C$ of the DIC and radioactive decay are limited compared to those of surface water dynamics and the time scales studied (Guilderson et al., 2000; Felis & Pätzold, 2004). Changes in coral $\Delta^{14}C$ may also result from changes in the depth of the mixing layer or thermocline

(Toggweiler, Dixon, & Broecker, 1991; Rodgers et al., 2004). The strength and timing of upwelling events has been monitored using changes in the $\Delta^{14}C$ signal in a variety of reef sites in the Pacific and Indian Oceans (Druffel, 1981; Guilderson & Schrag, 1998; Druffel & Griffin, 1993, 1999; Grottoli, Gille, Druffel, & Dunbar, 2003; Grumet et al., 2004).

Other proxies of oceanic and atmospheric dynamics. The $\delta^{18}O$ records in corals may also provide a proxy for hurricane activity owing to their sensitivity to SST and seawater $\delta^{18}O$, both strongly related to rainfall. In areas subject to frequent and severe hurricane impacts, dramatic interannual to multidecadal changes in SST and precipitation can be attributed to long-term hurricane variability (Hetzinger et al., 2008).

The variability of skeletal $\delta^{13}C$ in corals is controlled by both photosynthesis and trophic regimes. Seasonal events of dense plankton blooms may prompt corals to partly modify their feeding practices, changing from autotrophy to heterotrophy. Increased ingestion of ^{12}C-rich zooplankton during periods of high plankton availability results in a drastic relative depletion in ^{13}C, with a mean $\delta^{13}C$ in corals of −15 to −21‰ PDB (Felis, Pätzold, Loya, & Wefer, 1998; Grottoli & Wellington, 1999; Grotolli, 2002). Because large plankton blooms are commonly triggered by deep vertical water mass mixing, strong negative $\delta^{13}C$ anomalies in coral skeletons provide a promising proxy for upwelling and vertical water mixing. Luminescence banding may be used as a monitor of trade wind variability in areas where wind speeds correspond with precipitation elevated rates (Nyberg, 2002).

9.3. Climate Reconstruction based on Individual Coral Colonies

9.3.1. The Record of the Last Decades and Centuries

Reef-building corals have a key role to play in climate reconstructions in historical times, given the limited length and number of instrumental records in tropical regions (Carriquiry, Risk, & Schwarcz, 1994; Corrège, 2006). Unfortunately, most corals are younger than 300–400 years and thus time series from individual modern corals rarely extend beyond the mid-16th century. Nevertheless, coral-based reconstruction of SST in the tropics clearly indicates the current global warming (Figure 9.6).

9.3.1.1. The Pacific Ocean

The tropical Pacific is covered by a relatively dense network of coral climate records, reflecting the key role of Pacific-centred ENSO in global climate

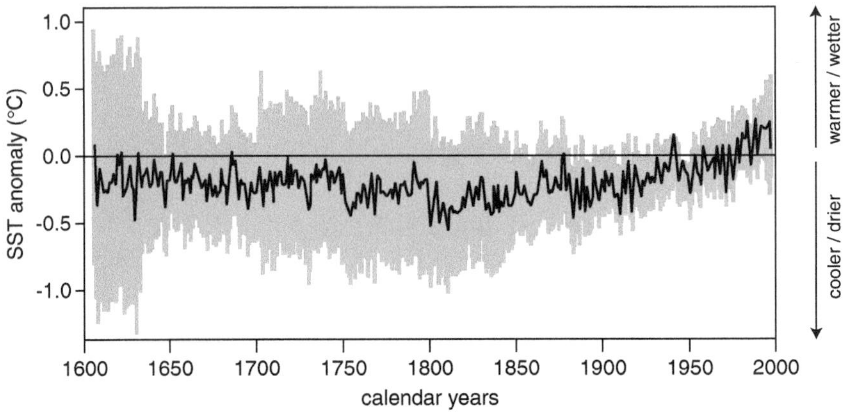

Figure 9.6 Coral-based reconstruction of sea surface temperature anomalies (with two sigma error bars) for the last 400 years in the tropics. SST series are standardized with respect to the interval AD 1961–1990. The mean reconstruction curve (thick black line) is developed from 14 coral multiproxy records obtained in the Indian and Pacific Oceans. Modified and redrawn from Wilson et al. (2006).

variability (Gagan et al., 2000; Grottoli & Eakin, 2007). Climate reconstructions indicate a long-term trend of decreasing $\delta^{18}O$ of about 0.17‰ PDB, corresponding to a warming of about 0.8°C from the middle of the 19th century. This appears to have started throughout the tropics at the end of the Little Ice Age in the mid-18th century (Wilson et al., 2006) and continues throughout the entire tropical Pacific, from the westernmost (Sun et al., 2004) to the easternmost areas (Linsley, Messier, & Dunbar, 1999). Given that the coral $\delta^{18}O$ signal is a composite of SST and SSS, the decrease in $\delta^{18}O$ may also incorporate a seawater freshening signal (Cole, 1996).

There have been abrupt spatial shifts in climate at interdecadal and decadal time scales across the oceans during historical times. Using composite coral records, Cobb, Charles, Cheng, and Edwards (2003) analysed the climate variability in the central tropical Pacific with a monthly resolution over the last millennium from time windows at AD 928–961, 1149–1220, 1317–1464, 1635–1703 and 1886–1998. Their findings reveal that throughout the 12th, 14–15th, 17th and early 20th centuries, SST varied within a relatively narrow range of about 0.6°C (Figure 9.7). Cooling periods seem to have started as early as the 10–12th centuries. These are expected to have witnessed the coolest and/or driest weather in the region for the last 1,100 years, comparable to modern La Niña conditions. In contrast, the climate of the 17th century was warm and humid, similar to that of modern El Niño conditions. The last decades of the 20th century were the warmest and wettest of the last millennium. The timing and structure of the warming trend in the central Pacific differs from Northern Hemisphere patterns during the early 20th century

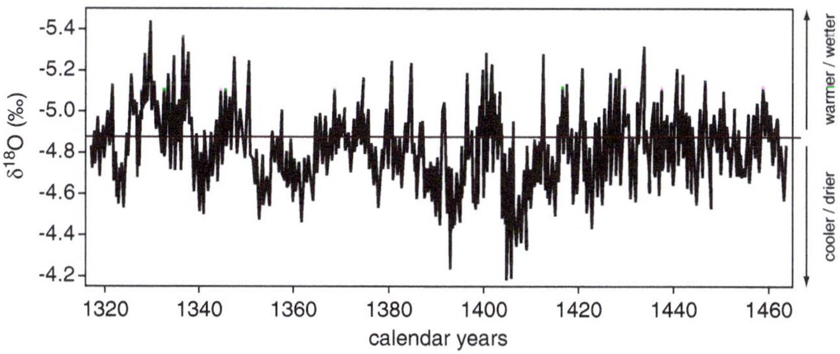

Figure 9.7 Overlapping oxygen isotope records at monthly resolution from corals collected at Palmyra Island (central Pacific Ocean). The record covers the AD 1320–1460 interval. The black horizontal line represents the average $\delta^{18}O$ values measured from a modern Palmyra coral for the 1886–1975 interval. Modified and redrawn from Cobb, Charles, Cheng, and Edwards (2003).

(Cobb, Charles, Cheng, and Edwards, 2003). Whereas the Northern Hemisphere SST showed a 0.4°C increase, the SST of the tropical Pacific was relatively stable. ENSO dynamics have varied greatly in the tropical Pacific from decade to decade throughout the last centuries and in some instances within single 10-year time intervals. ENSO frequencies and intensities were higher during some periods of the last millennium than those found during the last century. ENSO varied little during the 12th and 14th centuries, but variability was greater during the 17th century, relative to today. Most ENSO variance over the last millennium has probably been regulated by mechanisms internal to the ENSO system itself.

In addition, coral records from the South China Sea bear witness to a significant cooling event at about AD 490, followed by a warming episode at about AD 540 with SST comparable to the present day (Yu, Zhao, Wei, Cheng, Chen, et al., 2005). Distinctive cooling episodes on interannual time scales occurred in the western Pacific during the early 18th and early 19th centuries, probably in relation to active volcanism (Quinn et al., 1998). Within the West Pacific Warm Pool, in the Indonesian region, reconstructed SST anomalies since about AD 1780 are synchronous with Asian monsoon drought cycles, especially during major warm ENSO phases (D'Arrigo et al., 2006). In the southwestern Pacific, warming was accompanied by a severe freshening, particularly from AD 1870, as indicated by $\delta^{18}O$ data from Vanuatu (Kilbourne, Quinn, Taylor, Delcroix, et al., 2004) and the Great Barrier Reef (Hendy et al., 2002). By contrast, in the western part of the Coral Sea, off eastern Australia, SSS seems to have remained relatively stable over the last two centuries, following a freshening of surface waters that culminated around AD 1800 (Calvo et al., 2007). In the western tropical Pacific, SST and salinity have shown interdecadal variations for the last two centuries, in relation to variations in the position of the South Pacific Convergence Zone (SPCZ) and ENSO events (Le Bec et al., 2000; Bagnato, Linsley, & Howe, 2005; Juillet-Leclerc et al., 2006).

Since nuclear testing began coral $\Delta^{14}C$ records in the central Pacific show latitudinal trends that reflect differences in source waters. For example, $\Delta^{14}C$ values increased more rapidly along western boundary (Australia) than along the eastern boundary (Galapagos), because the currents on the eastern boundary incorporate older, deeper, ^{14}C-poor upwelling waters. Coral radiocarbon records also contribute to our understanding of the relationships between changes in ocean circulation patterns and climatic modes. In the eastern Pacific, for example, the appearance of upwelling waters apparently coincided with a shift from a negative to a positive Pacific Decadal Oscillation (PDO) phase and upwelling activity varies on interannual time scales (Grottoli & Eakin, 2007, and references therein).

9.3.1.2. The Indian Ocean

The Indian Ocean is also relatively well documented in terms of climate variability based on coral records. As in the tropical Pacific, coral $\delta^{18}O$ records at annual resolution reflect a general warming (of about 0.7°C elevation) and freshening trend since the 18–19th centuries extending from the east (Kuhnert et al., 1999) to the west of the Ocean (Pfeiffer, Timm, Dullo, & Podlech, 2004; Zinke, Dullo, Heiss, & Eisenhauer, 2004). In addition, Zinke et al. (2004) and Zinke, Pfeiffer, Timm, Dullo, and Davies (2005) provided a network of coral $\delta^{18}O$ and/or Sr/Ca data from the western to central Indian Ocean, locally spanning the last three centuries. Although ENSO events are centred on the tropical Pacific, their impact on both SST and rainfall in the Indian Ocean has been clearly identified. The impact of ENSO on both SST and atmospheric circulation in the southwestern Indian Ocean was particularly strong during the 18th century (Zinke et al., 2004). Work by Charles, Hunter, and Fairbanks (1997), Cole, Dunbar, McClanahan, and Mithiga (2000) and Pfeiffer and Dullo (2006) showed that in the Seychelles Islands (western equatorial Indian Ocean, 4°S), variations in coral $\delta^{18}O$-derived SSTs are driven by ENSO variance on interannual to decadal time scales. The ENSO linkages within the region have been statistically significant throughout the last 150 years (Figure 9.8) and are intensified during periods of high ENSO variability.

The climate of the Indian Ocean north of 10°S is in part controlled by the Asian monsoon. The monsoon occurs in the boreal summer and is typified by a seasonal reversal of surface winds and by changes in rainfall and evaporation. Cooling is caused by wind-induced mixing and evaporation. Recent studies indicate the great potential of coral geochemistry for resolving the interconnections between Asian monsoon and ENSO,

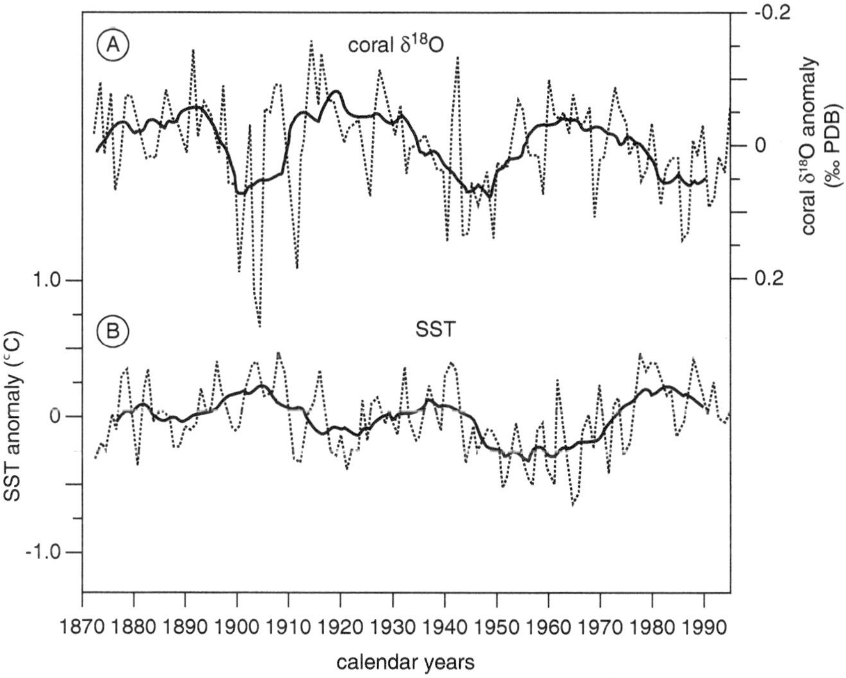

Figure 9.8 Comparison between the oxygen isotope record from a *Porites* coral drilled on Réunion (western Indian Ocean) and regional sea surface temperature. Coral $\delta^{18}O$ and SST anomalies are calculated relative to the mean for the interval AD 1871–1995. (A) Annual mean coral $\delta^{18}O$ (dashed line) and 11-point running average (thick black line). The coral isotope signal is composite, reflecting both changes in SST and in the evaporation/precipitation ratio, in relation to ENSO events. (B) Annual mean SST from the GISST (Global Ocean) 2.3 data set (dashed line) and 11-point running average (thick black line). Modified and redrawn from Pfeiffer et al. (2004) and Zinke et al. (2004).

as reflected in SST variations and precipitation anomalies (Grottoli & Eakin, 2007). Seychelles corals have captured records of cooling events during the boreal summer. In the Arabian Sea (southern Oman), variations in coral $\delta^{18}O$ record changes in both SST and precipitation (Tudhope et al., 1996). In particular, during the NE monsoon, oxygen isotope values are positively correlated with annual precipitation anomalies in India, whereas during the SW monsoon they reflect changes in upwelling intensity along the Oman coast.

Palaeoclimatic evidence for a teleconnection between the Indian Ocean and the PDO was provided by Crueger, Zinke, and Pfeiffer (2008) based on coral $\delta^{18}O$ data from Réunion and western Madagascar. The response of the oxygen isotope signals was strongly linked to the coupled SST/sea-level pressure of the PDO.

Precipitation anomalies related to the Indian Ocean Dipole variability have been detected in coral skeletons using Sr/Ca ratios or measurements of luminescence intensity and oxygen isotope analysis. Analysing a 24-year-old coral collected at Christmas Island, Marshall and McCulloch (2001) provided evidence of both warm and cool SST anomalies in the Sr/Ca signals, related to abnormal oceanographic conditions caused by the IOD in the eastern Indian Ocean. Similarly, negative and positive $\delta^{18}O$ values and changes in luminescence levels in a 20-year-old coral from Kenya were strongly correlated with periods of high and low rainfall related respectively to positive and negative IOD phases in the western Indian Ocean (Kayanne et al., 2006).

9.3.1.3. The Red Sea

To date, few coral-based climate reconstructions are available from the Red Sea. Oxygen isotope data from the southern Red Sea (approximately 15°N) were used by Klein et al. (1997) to estimate variations in the intensity of the Asian and African monsoons in relation to ENSO variability from the 1930s to 1990s. It appeared that coral $\delta^{18}O$ variations were predominantly driven by variations in the strength of surface water flows from the Indian Ocean to the Red Sea during the winter NE monsoon. The decadal variability in the $\delta^{18}O$ is strongly correlated with both the Indian Ocean SST and Pacific-based ENSO patterns. Felis et al. (2000) analysed a 245-year coral time series from the northern part of the Red Sea (about 28°N) at bimonthly resolution (Figure 9.9). A slight warming trend can be detected from approximately AD 1870 to the present, corresponding to a mean decrease in $\delta^{18}O$ of about 0.15‰ PDB. Interannual to interdecadal variability is closely correlated with variations in the North Atlantic Oscillation (NAO), ENSO and the North Pacific climate. These modes have played a significant role in the climate variability of the Middle East since at least the mid-18th century at a dominant frequency of 5–7 years. The influence of ENSO patterns on the regional climate has been confirmed by Rimbu, Lohmann, Felis, and Pätzold (2003). Oscillations with frequencies of about 70 and 22–23 years were identified in coral records (Felis & Pätzold, 2004) and were interpreted as probably linked to variations in the North Atlantic thermohaline circulation. Shifts in the teleconnection between the central Pacific and northern Red Sea occurred on interdecadal or longer time scales. In addition, the winter time series reveals connections with the Arctic Oscillation (AO), providing information on variations in winter circulation during the last 250 years (Rimbu, Lohmann, Felis, & Pätzold, 2004). Colder and drier conditions in the northern Red Sea have been linked to higher intensities of the AO/NAO and conversely warmer and wetter condition relate to lower intensities of the AO/NAO.

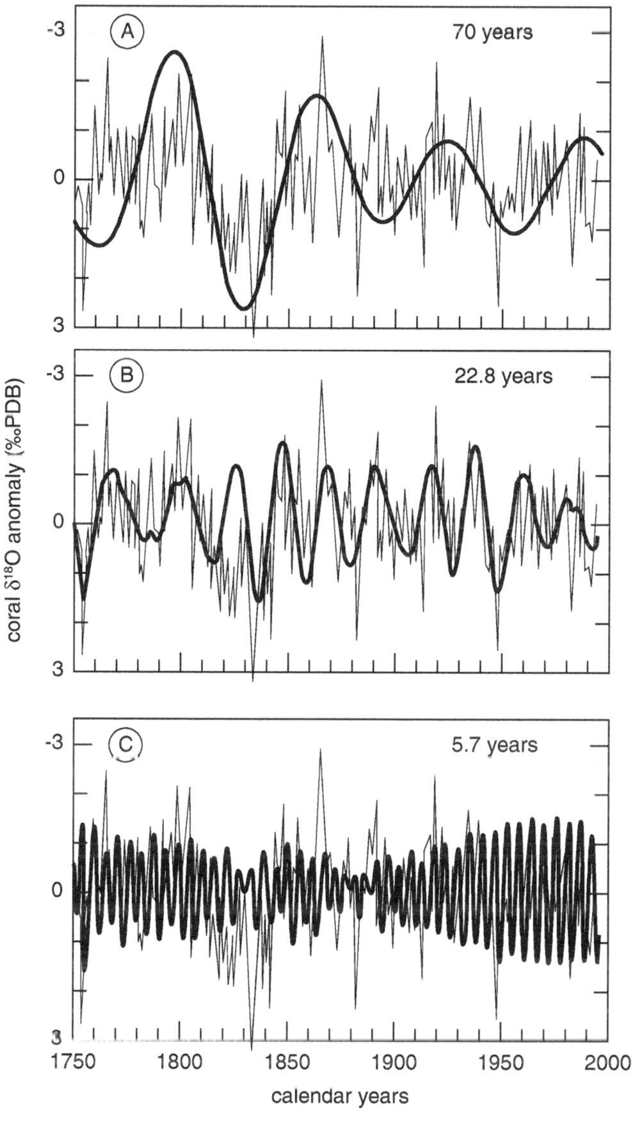

Figure 9.9 Results of spectral analysis performed on a coral oxygen isotope time series (AD 1750–1995) from the northern Red Sea (Ras Umm Sidd, Egypt) at bimonthly resolution. The thin black lines represent normalized mean annual coral $\delta^{18}O$ records calculated from the seasonal anomaly record. The most significant oscillatory modes (thick black lines) and their frequency are given. (A) The 70-year oscillation dominates the coral record and is probably of North Atlantic origin (NAO). (B) The 22.8-year oscillation reflects the influence of the Mediterranean Oscillation. (C) The 5.7-year oscillation denotes the control of ENSO events. Modified and redrawn from Felis et al. (2000).

9.3.1.4. The western Atlantic

Coral-based climate reconstructions are less well documented in the tropical Atlantic than those in the Pacific and Indian Oceans. The data available are summarized as follows.

The cooling period referred as the Little Ice Age has been detected in both the northern Caribbean and Bermuda using $\delta^{18}O$, Mg/Ca, Sr/Ca and/or density measurements. From the 18th century to the early portion of the 19th century, annual average SSTs in the region were about 1.5°C cooler than at present (Figure 9.10), and SSS showed large seasonal changes (Draschba, Pätzold, & Wefer, 2000; Winter, Ishioroshi, Watanabe, & Oba, 2000; Watanabe et al., 2003; Goodkin, Hughen, Cohen, & Smith, 2005).

In South Florida, Swart, Dodge, and Hudson (1996) interpreted the variability of coral $\delta^{18}O$ signals in terms of rainfall. Extending back to the mid-18th century, the coral record suggests that most of the 18th and 19th centuries were markedly drier than the second half of the 18th and early 20th centuries. On Belize, in the western Caribbean, a composite coral $\delta^{18}O$ record showed an annual variability of 0.6–0.8‰ PDB, representing changes in the monthly average SST of 3.4°C from AD 1815 to the present. There was a slight warming or freshening trend reflected by a decrease in $\delta^{18}O$ of 0.15‰ over the last two centuries (Gischler & Oschmann, 2005).

Analyses of coral cadmium and barium from Venezuelan tropical surface waters provide evidence for a reduction in coastal upwelling from the mid-20th century (Reuer et al., 2003). Potential controls are believed to be multiple and complex, expressing a non-linear climate system. One possible forcing mechanism may be, at least in part, a reduction in trade wind

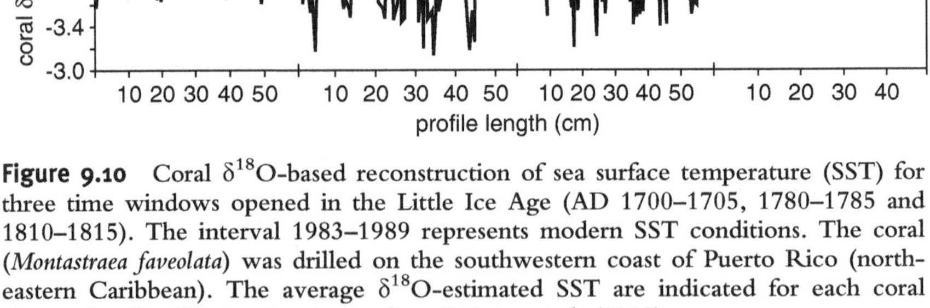

Figure 9.10 Coral $\delta^{18}O$-based reconstruction of sea surface temperature (SST) for three time windows opened in the Little Ice Age (AD 1700–1705, 1780–1785 and 1810–1815). The interval 1983–1989 represents modern SST conditions. The coral (*Montastraea faveolata*) was drilled on the southwestern coast of Puerto Rico (northeastern Caribbean). The average $\delta^{18}O$-estimated SST are indicated for each coral time series. Modified and redrawn from Winter et al. (2000).

intensity and an increase in tropical North Atlantic SST in response to a northward shift in the Atlantic Intertropical Convergence Zone (ITCZ) during El Niño events.

Atmospheric and oceanic circulation modes in the North Atlantic have been reconstructed using lead concentrations in corals from Mona Island in the northern Caribbean, within the ITCZ (Desenfant et al., 2006). Comparison of variations in the lead isotope ratios ($^{206}Pb/^{207}Pb$) of the corals with variations in the NAO indicates that during positive NAO phases and in spite of vigorous easterly wind flows, lead contaminants are actively moved from the continental United States to the northern Caribbean. During negative NAO phases, distinctively less radiogenic $^{206}Pb/^{207}Pb$ is transported by easterly trade winds as flow intensity decreases. These results provide evidence of transport pathways. Although easterly trade winds are weak during these negative events, they still promote lead transport to the northern Caribbean. The flow of strong easterlies during positive NAO events is obstructed to the south.

9.3.1.5. The eastern Atlantic

The climate dynamics of the Gulf of Guinea (eastern Atlantic) have been reconstructed using coral oxygen isotope analyses by Swart, White, Enfield, Dodge, and Milne (1998) who found a close correlation between average precipitation in sub-Saharan Africa and coral $\delta^{18}O$ values; higher rainfall is correlated with lower $\delta^{18}O$ values. This is probably due to the synchroneity of higher precipitation with flooding of the Niger and Congo rivers that affects SSS and coral $\delta^{18}O$ in the Gulf. Precipitation patterns in the region appear to be controlled by the magnitude of the Atlantic Dipole and the latitudinal position of the ITCZ.

9.3.2. The Holocene Record

Coral proxy records reveal that the Holocene, and particularly the mid- to late Holocene (7–1.5 ka), was punctuated by abrupt climatic shifts, larger than those documented from the previous millennium. In the tropical Pacific, there was a substantial and rapid increase in SST during the early Holocene, from about 10 to 8.9 ka (Beck et al., 1997). In the core area of the Indo-Pacific Warm Pool (IPWP), the rapid deglacial increase in SST to modern temperatures occurred at around 9 ka (Gagan et al., 2004). But SST subsequently became cooler. Between 8.9 and 7.4 ka, the IPWP was characterized by SSTs 1–3°C lower than today, presumably in response to the changing patterns of the ocean–atmosphere circulation (McCulloch et al., 1996). Similarly, in the southwestern Pacific, SSTs were about 1°C cooler from about 8.9 to 7.5 ka (Beck et al., 1997). Low SSTs have also been recorded from southern China Sea sites at approximately 7.5–7.0 ka.

The winter Sr/Ca-derived SST averaged 16°C with interannual to decadal winter anomalies of about 10.7°C. This variation is likely to have been related to a stronger Asian monsoon, resulting in a wider seasonality than today (Yu et al., 2004). Coral records are semi-continuous over the 8.9–7 ka interval and thus span the abrupt global drop in temperature at 8.2 ka (Alley & Ágústsdóttir, 2005).

The Holocene Climatic Optimum, a warm period roughly centred around 7–6 ka, has been identified in the western Pacific and particularly in the South China Sea (Yu, Zhao, Wei, Cheng, & Wang, 2005), in the Great Barrier Reef (Gagan et al., 1998) and New Caledonia (Montaggioni et al., 2006), on the basis of coral Sr/Ca and/or $\delta^{18}O$ data. Sea surface waters were probably 0.5–1°C warmer and more salty than those at present. The Indian Ocean and the Caribbean also seem to have been affected by higher SST and higher salinity at approximately 6.5 ka (Abram et al., 2007) and 7 ka (Gischler & Oschmann, 2005).

A number of cooling events have occurred in the Pacific in the past 6 ka. In the South China Sea, Yu, Zhao, Wei, Cheng, and Wang (2005) demonstrated an overall decreasing trend in SST from about 6.8 to 1.5 ka, with values depressed by 2.5–1.5°C, before reaching modern values (Figure 9.11). This decline was accompanied by a decrease in monsoon moisture transported from the South China Sea, consistent with a weakening of the Asian summer monsoon in response to a continuous reduction in insolation. In southern tropical Japan, coral $\delta^{18}O$ signals reflect cooling events of similar amplitude between about 3.8 and 3.4 ka, and during these periods SST may have occasionally been close to or below the currently accepted 18°C minimum temperature for reef growth (Abram

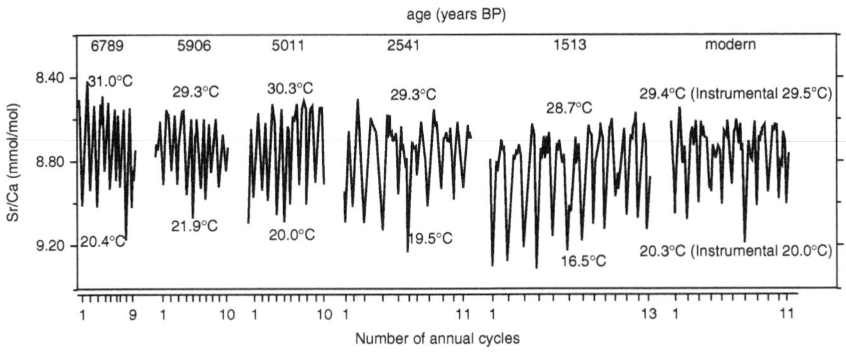

Figure 9.11 Coral Sr/Ca-based reconstruction of sea surface temperature (SST) for five time windows opened in the middle to late Holocene. The time series were analysed from *Porites* corals collected at Leizhou Peninsula, northern coast of the South China Sea. The average winter and summer Sr/Ca-derived SST and the U/Th ages (ka) of the studied coral colonies are given. Instrumental SST denote the present thermal conditions. Modified and redrawn from Yu, Zhao, Wei, Cheng, and Wang (2005).

et al., 2001). Several strong interannual or decadal cooling episodes of 1°C or more have been recorded in the southwestern Pacific during the mid-Holocene. In New Caledonia, coupled coral Ba/Ca and Sr/Ca records showed that coastal upwelling activity was stronger than that at present at 6.3–6 ka and was linked to cooling events (Montaggioni et al., 2006). In Vanuatu, a 1°C drop in SST occurred at around 4.15 ka. This suggests large-scale fluctuations in the depth of the thermocline and in the associated geostrophic circulation, resulting in phase shifts in the ENSO mode (Corrège et al., 2000). The tropical Indian Ocean has also experienced strong surface cooling episodes, accompanied by severe droughts during individual Indian Ocean Dipole events over the past 6.5 ka. These episodes have been interpreted as caused by strong cross-equatorial winds controlled by an enhanced Asian monsoon (Abram et al., 2007).

Long-term mid-Holocene changes in ENSO-related SST and SSS were reconstructed in the tropical Pacific using $\delta^{18}O$ and/or Sr/Ca data from corals (Corrège et al., 2000; Tudhope et al., 2001; Woodroffe, Beech, & Gagan, 2003; Gagan et al., 2004; McGregor & Gagan, 2004; Sun et al., 2005). The amplitude of ENSO events was shown to have been substantially reduced compared to those of the present between 9 and 6 ka by an estimated 60% (Tudhope et al., 2001) or 15% (Gagan et al., 2004; McGregor & Gagan, 2004) of average El Niño events (Figure 9.12). In addition, data from the Australian Great Barrier Reef suggest that SST variability and rainfall variability during ENSO periods were reduced by 20% and 70% respectively (Gagan et al., 2004). The frequency of heavy precipitation (warm El Niño phase) events appears to have changed from more than 15 years prior to 7 ka towards a present-day frequency of 3–7 years after 5 ka, indicating the onset of modern ENSO variability between 7 and 5 ka. From about 3 ka, there was an abrupt increase in ENSO amplitude. In southeast Asia, the influence of ENSO is likely to have been established by about 4.4 ka (Sun et al., 2005). The cause of the differences in ENSO behaviour during the early–middle Holocene is likely to have been differences in the earth's orbital configuration (Cane, 2005). In the central equatorial Pacific, interannual SST and SSS variability during ENSO periods was lower between 3.8 and 2.8 ka, but increased at about 2 ka; this is consistent with precessional changes in solar radiation seasonality, but also implies stronger teleconnection between ENSO and the ITCZ (Woodroffe et al., 2003). In the northern Caribbean, decadal to multidecadal variations in stable oxygen and carbon isotope ratios in corals from 7.2 to 5.2 ka-old were interpreted as reflecting local rainfall and/or freshwater flooding patterns. These patterns may have been controlled by the migration of the ITCZ and/or hurricanes and tropical storms during the mid-Holocene (Greer & Swart, 2006).

The variability of the Asian monsoon in the South China Sea during the mid-Holocene, about 4.4 ka, has been estimated from coral $\delta^{18}O$ (Sun

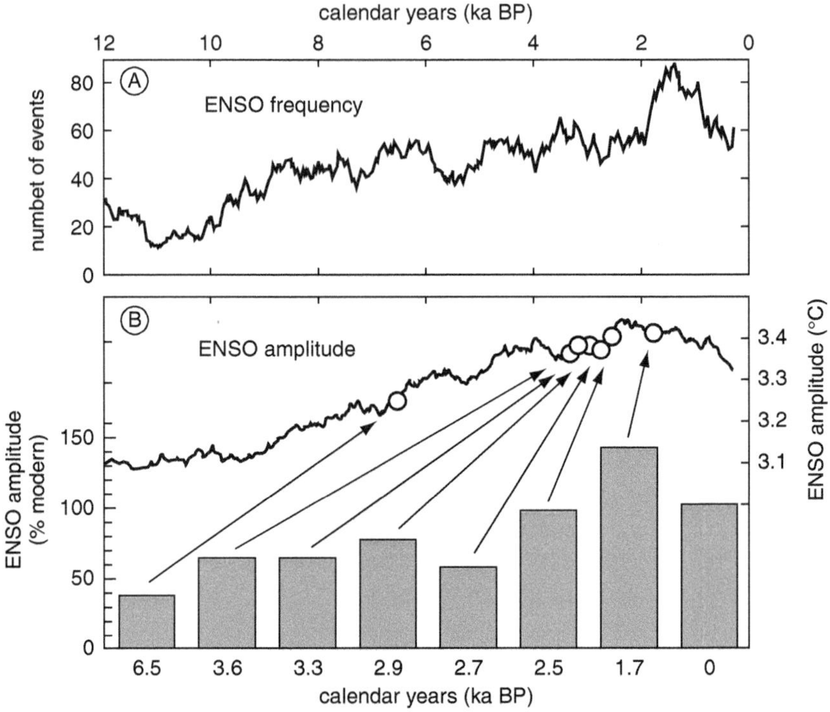

Figure 9.12 Holocene evolution of ENSO patterns in the equatorial Pacific. (A) Modelled number of El Niño events defined as mean December–February SST anomalies exceeding 3°C in the eastern equatorial Pacific in five 100-year overlapping windows. (B) Comparison between modelled amplitude of El Niño events with SST anomalies exceeding 3°C and relative amplitude of $\delta^{18}O$ variability in the 2–7-year ENSO band for coral records from Huon Peninsula, Papua New Guinea and Christmas (Kiritimati) Island, central equatorial Pacific. Modified and redrawn from Gagan et al. (2004).

et al., 2005). The annual $\delta^{18}O$ cycle was amplified by about 9%, compared to the present. This indicates that the interannual amplitude of both SST and SSS variability was stronger at that time. The ^{18}O-enrichment was probably driven by greater advection of moisture towards the Asian continent, and increased evaporation and vertical mixing in response to a strengthened mid-Holocene monsoon.

In the northern Red Sea, between about 5.8 and 4.5 ka, the amplitude of the seasonal variation of $\delta^{18}O$ in corals was greater than that at present. This has been interpreted as reflecting a larger seasonal contrast in SST and significant changes in the evaporation and precipitation regime. Summer rainfall of the African monsoon reached the northern end of the region, implying a northward migration of the monsoon. A change in climate may have occurred around 4.9–4.6 ka, corresponding to a reduction of moisture transport from the Indian Ocean (Moustafa, Pätzold, Loya, & Weber, 2000).

9.3.3. The Last Glacial Maximum to Early Deglacial Record

This period, defined here as ranging from about 25 to 10 ka, is poorly documented in terms of coral proxy climate reconstructions, largely because contemporary reefs are mostly either submerged or restricted to a few locally uplifted terraces.

During the LGM, SSTs appear to have been substantially lower throughout the tropics (Figure 9.13). In the IPWP, SSTs during the LGM were about 3°C lower than those at present (Gagan et al., 2004), whereas in other Pacific locations, Sr/Ca-derived SSTs range around 4–6°C (Beck et al., 1997). Coral $\delta^{18}O$ time series from the western Atlantic reveal a regional SST during the LGM that was depressed by about 4.5°C and a seasonality broadly similar to that of the present from 24 to 19 ka (Guilderson et al., 2001).

Throughout the tropics, changes in SST during the LGM may have been driven by changes in the radiative balance of these areas relative to the redistribution of energy towards higher latitudes (Guilderson et al., 2001). The synergy between low and high latitude climate is a direct consequence of the fact that in the northernmost and westernmost areas of the IPWP, shifts in SST were coeval with variations in the Northern Hemisphere summer insolation (Gagan et al., 2004). The climate linkages between low and high latitudes during the LGM are supported by simulations of ENSO mode by An et al. (2004). The results suggest large-amplitude, self-sustained interannual ENSO variability driven by a progressive shallowing of the thermocline in the equatorial Pacific as well as extra equatorial climate conditions.

During deglaciation, at around 13.7–13.1 ka, SST on Tahiti (central Pacific) were probably 0.5–1.5°C cooler than those at the present (Figure 9.13) with no marked difference in seasonality (Cohen & Hart, 2004). Estimates of postglacial to early Holocene temperatures from the southwestern Pacific indicate anomalies of 1–3°C below modern SST values (Gagan et al., 2004). These are supported by SST records from Vanuatu where coral $\delta^{18}O$ and Sr/Ca analyses indicate an anomaly averaging 4.5 ± 1.3°C below the present SST during the Younger Dryas interval at about 12 ka (Corrège et al., 2004). While periods with relatively warmer SSTs had annual amplitudes of about 3°C, comparable with modern ones, cooler periods were affected by larger amplitude variations of 5–6°C. These data reflect shallowing of the thermocline and suggest that the cooling of the Younger Dryas period was triggered by a contraction of tropical waters towards the equator. The SPCZ appears not to have been active during this event.

It is remarkable that most coral-derived SST results conflict with those based on microfossils — or modelling of SST reconstructions, that suggest a maximum cooling range of 1.2–3°C relative to modern temperatures during the past 25 ka (see Montaggioni, 2005, and references therein). Overestimation of the extent of LGM/Holocene cooling in coral records

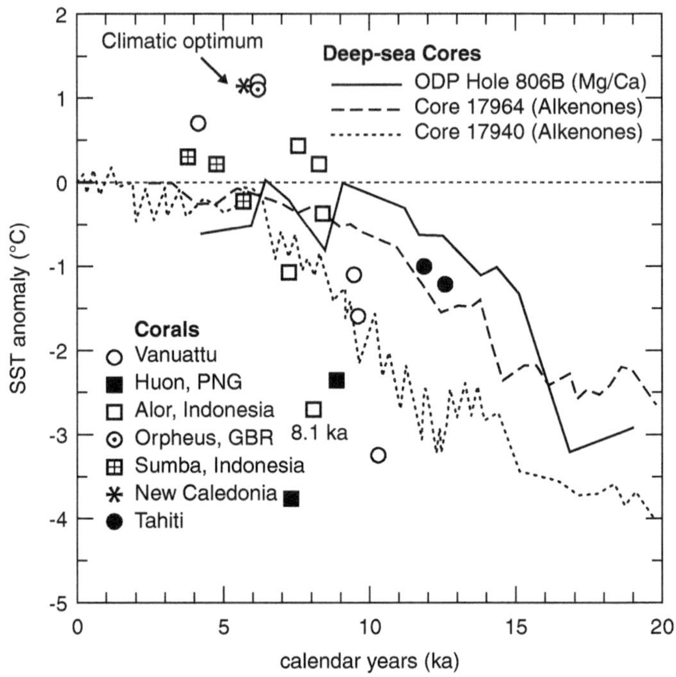

Figure 9.13 Reconstruction of sea surface temperature (SST) anomalies in the western equatorial Pacific during the last 20,000 years. Additional SST estimates are provided from the central Pacific. SSTs anomalies are calculated relative to late 20th century values. Reconstructed SSTs are derived from foraminiferal Mg/Ca, alkenone and coral Sr/Ca thermometry respectively. In ODP Hole 806B (Ontong Java Plateau), SSTs were estimated from surface-dwelling planktonic foraminifera. In cores 17694 and 17940 drilled in the southern China Sea, SSTs were reconstructed using measurements of alkenone ratios in calcareous nannoplankton. Coral Sr/Ca-estimated SSTs are based on fossil *Porites* samples from Espiritu Santo (Vanuatu), Huon Peninsula (Papua New Guinea), Alor and Sumba Islands (Indonesia), Orpheus Island (central Great Barrier Reef), Vata-Ricaudy Reef (southwestern New Caledonia) and Tahiti (Society Islands). Calendar ages were determined by U/Th TIMS or calibrated radiocarbon dating. Discrepancies between coral- and microfossil-based SST reconstructions are due probably to early diagenetic alteration of coral material. Modified and redrawn from Gagan et al. (2004).

may result from glacial–interglacial changes in oceanic Sr/Ca ratios (Felis & Pätzold, 2004) and/or early marine diagenetic alteration.

9.3.4. The Pleistocene Record

The network of coral-based climate data from Pleistocene deposits older than the LGM is still very sparse. This is mainly due to the scarcity of well-preserved, long-lived fossil corals in uplifted reef terraces or in cores.

9.3.4.1. The last interglacial

Coral records from the western Pacific and eastern Indian Oceans, based on Sr/Ca ratios, suggest that around 130 and 120 ka, mean annual SSTs were 29 ± 1 and 24 ± 1°C respectively (McCulloch & Esat, 2000). Similarly, other proxy and palaeontological data at seem to indicate that SSTs during the last interglacial were either very close to present-day temperatures or approximately 1°C higher (Tudhope et al., 2001; Muhs, Simmons, & Steinke, 2002; Winter et al., 2003) (Figure 9.14). Tudhope et al. (2001) investigated interannual ENSO-like variability in eastern Papua New Guinea on orbital time scales over the last glacial–interglacial cycle. The ENSO modes were found to have been repeated for at least the last 130 ka, even during glacial intervals. However, ENSO patterns have changed markedly, with a high deglacial/interglacial variability similar to that at present, probably modulated by both ENSO dampening during cool episodes and precessional forcing. ENSO events may have been weaker than those at present in the past 150 ka, especially under glacial conditions. These results are supported by data from Indonesia, at least for the last interglacial stage. During this period, ENSO patterns were broadly similar to those at present in terms of interannual variability in rainfall and SST (Hughen, Schrag, Jacobsen, & Hantoro, 1999). Variations in $\delta^{18}O$ and $\delta^{13}C$ time series in a 127-ka-old coral from the northwestern Pacific (Ruykyu Islands, Japan) were interpreted as reflecting an increased seasonality during the last interglacial. In particular, there was an enhanced evaporation of seawater compared to modern regional records. This may have been caused by an intensified seasonal insolation in the Northern Hemisphere, related to variations in orbital parameters (Suzuki et al., 2001). In the northeastern Caribbean, the seasonal SST variation during the last interglacial stage was also 1–2°C greater than it is today, primarily in response to winter cooling. The bias towards colder winters may be attributed to variations in low-latitude insolation induced by altered orbital parameters and modulated by atmospheric pCO_2 levels that were lower than they are now (Winter et al., 2003). Similarly, in the northern Red Sea, the last glacial period was typified by a larger amplitude SST seasonality. The latter is believed to have been primarily controlled by a more pronounced Arctic/NAO than that during the Holocene and at present, resulting in colder winters in the Middle East (Felis et al., 2004). Based on an analysis of coral luminescent banding, Klein, Loya, Gvirtzman, Isdale, and Susic (1990) established that the climate in the northern Red Sea during late Pleistocene interglacial periods (120 to older than 250 ka) was significantly wetter than today and typified by a possible summer rainfall regime.

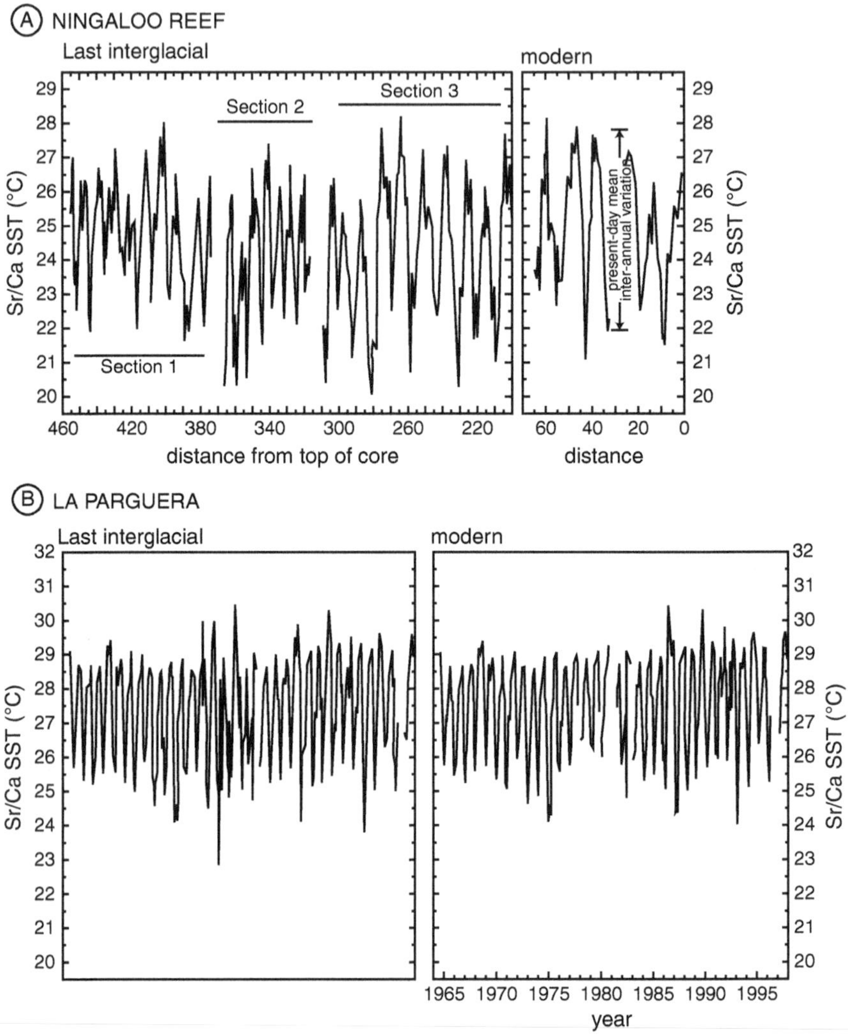

Figure 9.14 Coral Sr/Ca-derived sea surface palaeotemperatures (SSTs) from the last interglacial interval. (A) Palaeo-SST based on a fossil *Porites* collected from Ningaloo Reef (western Australia). SSTs from a Ningaloo modern coral are given for comparison (modified and redrawn from McCulloch & Esat, 2000). (B) Palaeo-SST based on a fossil *Montastraea* collected at La Parguera (Puerto Rico, northeastern Caribbean). SSTs from La Parguera modern coral are given for comparison (modified and redrawn from Winter et al., 2003). Note the mean annual temperatures during the last interglacial were comparable to the modern SST, but the seasonal range in SSTs was 1–2°C larger than at present, primarily due to colder winters.

The $\Delta^{14}C$ of coral skeletons from Papua New Guinea has been used to estimate variations in atmospheric radiocarbon levels during late Pleistocene intervals (Yokoyama & Esat, 2004). Peak values were coincident with sea-level rise events and with accelerated reef growth, but also with episodes of extreme cooling in the North Atlantic (Heinrich events). The connectivity between these widely separated events was suggested to reflect the interruption of the North Atlantic thermohaline circulation triggered by episodic partial disruption of the North American (Laurentide) ice sheet.

9.3.4.2. The penultimate deglaciation

During the penultimate deglaciation at about 128–132 ka when global sea level was 60–80 m lower in comparison to the present, the western equatorial Pacific, around eastern Papua New Guinea, is considered to have experienced a major cooling event, with SST about 6°C lower than today (McCulloch et al., 1999). By contrast, using pristine areas of coral skeletons rather than bulk samples, Allison et al. (2005) showed that in Papua New Guinea, SST depression during the 130-ka deglacial period was less than about 1°C compared to modern temperatures. As emphasized by Felis and Pätzold (2004), these findings again raise the question of whether the tropical zones were affected by marked cooling during glacial periods or whether coral-derived SST signatures were the result of diagenetic alteration.

9.3.4.3. Older interglacial–glacial periods

The nature of SST seasonality during the interglacial period from about 340 to 300 ka has been addressed by Ayling et al. (2006) on Henderson Island (southeast Pacific). Based on coral Sr/Ca time series, the amplitude of SST seasonal cycles was found to have been about 4.7 ± 0.75°C, exceeding the modern value. The more marked seasonality was attributed to an enhanced seasonality of insolation at the time of coral growth. The oldest period for which the tropical climate has been reconstructed from coral proxy records is that of the deglaciation at about 350 ka (Kilbourne, Quinn, & Taylor, 2004). Using a fossil coral from Vanuatu (southwestern Pacific), the authors found that Sr/Ca and $\delta^{18}O$ values account for an SST 2°C cooler and salinity 0–2 psu fresher than that at present. Seasonal SST variability seems to have been very similar to modern ranges while seasonal variations in salinity were reduced. These results are consistent with the migration of the SPCZ southwards during austral winters. In addition, they suggest that an ENSO-like mode operated about 350 ka ago.

9.4. Coral Reefs as Records of Sea-Level Change

Variations in relative sea level are the product of changes in the volume of seawater in the ocean basins and of vertical motion of the ocean floor. Large-scale ice–ocean mass redistribution, driven by the transition from glacial to interglacial periods, has resulted in both dramatic increases in ocean water volume and strong isostatic responses of the solid earth (Peltier, Farrell, & Clark, 1978; Lambeck, 2002; Milne, 2002). This mass redistribution and the consequent earth deformation referred as glacial isostatic adjustment produces the strongest signal in relative sea level in areas close to land-based ice sheets (near-field regions), as a response to rapid ice unloading. By contrast, the ice-induced component of the signal reduces in magnitude drastically in areas far from the major centres of glaciation (far-field regions) and the meltwater, eustatic, signal becomes dominant. Thus, far-field tropical regions appear to be ideal places to measure the eustatic component in the signal of relative sea level (Milne et al., 2005). However, far-field locations are also affected by a variety of other processes including meltwater loading of the seafloor, tectonic instability, the thermal expansion of ocean water, and changes in Antarctic mass balance, all of which may interact with the eustatic component.

The best recorders of changes in sea level in the far-field tropics are coral reef systems. The reconstruction of Quaternary sea-level changes is based on the analysis of reef-associated features (Lighty, Macintyre, & Stuckenrath, 1982; Davies & Montaggioni, 1985; Coudray & Montaggioni, 1986; Hopley, 1986a, 1986b; Pirazzoli, 1986, 1996; Cabioch, Montaggioni, et al., 1999; Dickinson, 2001; Hopley et al., 2007; Hearty et al., 2007).

9.4.1. Reef Evidence of Sea-Level Position

Coral reef systems preserve within their frameworks and associated deposits strong signatures of sea-level response. These can be defined as indicative, related to processes occurring directly at the sea surface, or directional related to processes occurring within varying depth or elevation ranges. Indicative features may be either depositional (reef flats, reef crests and microatolls) or erosional (marine notches and abrasion surfaces), and dominantly encapsulate a stillstand signature. Directional features are chiefly depositional, including the composition of coralgal communities and other reef dwellers in the growth framework, the geometry and diagenetic products of associated deposits, and the spatial arrangement of morphostratigraphic units. In addition, some bioerosional traces may delineate a clear depth zonation. Directional features may encapsulate either transgressive and/or regressive signatures (Davies & Montaggioni, 1985).

Sea-level indicators related to coral reefs have been reviewed by Pirazzoli (1991, 1996), Hopley (1986a, 1986b) and Hopley et al. (2007, pp. 61–72).

9.4.1.1. Reef flats and associated growth frameworks

The elevation of modern reef-flat surfaces is typically an indication of mean low-water spring tide levels. Common constructional features associated with reef flats include algal ridges and microatolls.

Algal ridges are typical reef-crest features on mid-Pacific atolls and on some high-energy Indian Ocean and Atlantic reefs. They develop subtidally within a 2–6 m depth range and can reach a maximum elevation of +2 m above mean sea level, directly related to wave energy. Algal ridges predominantly consist of crustose coralline algae including *Hydrolithon onkodes* and *Hydrolithon gardineri* in the Indo-Pacific, and *Porolithon pachydermum* and *Lithophyllum congestum* in the Caribbean (Adey, 1986). The precise definition of the position of a former sea level relative to relict algal ridges is by reference to measurements of the heights of their modern counterparts in similar sites.

Microatolls consist of individual subcircular colonies developed from hemispherical massive forms, mainly *Porites*, *Goniopora*, *Goniastrea* and *Montastraea* (Figure 9.15A), the vertical growth of which has been limited by exposure at a given tide level (Stoddart & Scoffin, 1979). For instance, on the Great Barrier Reef, characterized by meso- to macrotidal regimes (range: 2.5–6 m), the uppermost living surfaces of microatolls are assumed to approximate mean low-water spring tides, irrespective of the tidal range. Variation of up to 0.25 m is in relation to interannual changes in tidal amplitude (Hopley, 1986a; Hopley et al., 2007). Where water remains trapped in pools behind storm ridges as the tide level falls, water levels may remain permanently or temporarily above that of the open sea (moating effect). As a result, the uppermost level of coral growth in such moats is higher than that in reef-flat areas drained to the ocean, and microatolls in such areas overestimate the elevation of low-tide waters at levels up to mean high water neaps. Thus, reconstruction of former sea level based on fossil microatolls first requires the identification of the environment in which individual coral colonies grew when they were alive.

In the Quaternary record, reef flats are mainly preserved in the form of subhorizontal limestone terraces. Past sea levels are derived from the present-day elevations of fossil reef flats and associated biological structures and successions. However, there are limitations to this method. As emphasized by Davies and Montaggioni (1985) and Hearty et al. (2007), individual coral reef terraces are reliable recorders of sea-level stability (i.e. stillstands), but poor monitors of sea-level variation.

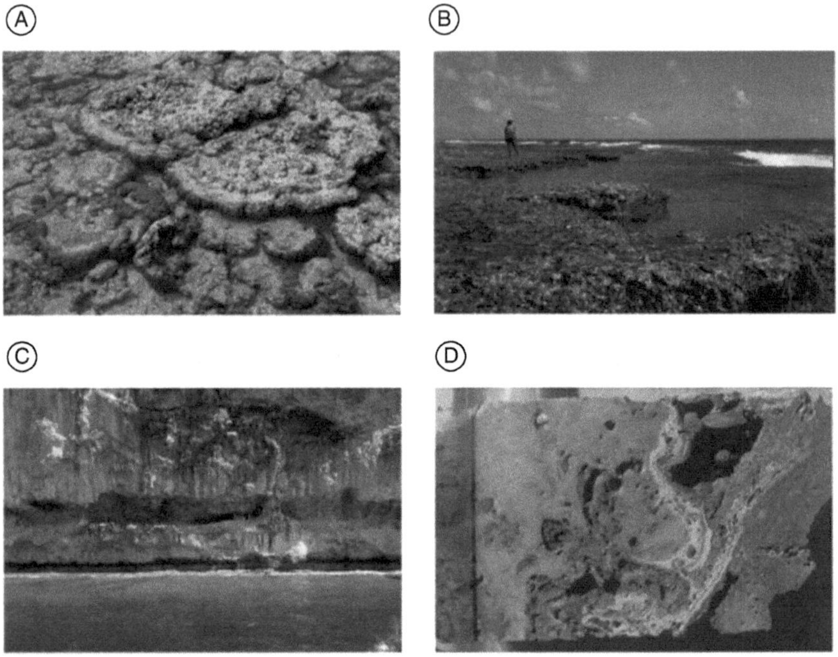

Figure 9.15 Typical reef-related sea-level indicators. (A) Reef flat dominated by coalescing *Porites* microatolls on Raine Island, northern Australian Great Barrier Reef (photograph by L. Montaggioni). (B) Emerged reef flat at elevation of 0.40 m above the modern reef flat, northwestern coast of Makatea, Tuamotu Archipelago, central Pacific. This terrace indicates the position of a highstand at around 4.5–5.3 ka (photograph by L. Montaggioni). (C) Two successive Wave-cut notches along the southwestern cliff of Makatea. The lower one is related to the mid-Holocene highstand, while the higher notch, at about 6 m above present mean low tide, indicates the relative position of the last interglacial highstand (photograph by L. Montaggioni). (D) Coralgal association of robust branching *Acropora robusta* group colony thickly encrusted by shallow-water coralline algae and vermetid gastropods. The later cavity-filling deposits on the right consist of laminated microbialites (photograph by L. Montaggioni).

Emergent reef terraces. In areas that are considered to be tectonically stable, or minimally displaced either up or down, the present-day elevation of reef terraces is assumed to represent the position of the sea surface at the time of their formation (Figure 9.15B). There are few descriptions of emergent algal ridges from fossil reef terraces. This apparent omission may reflect their low preservation potential in extreme energy settings (Davies & Montaggioni, 1985). On the northeastern side of Suwarrow Atoll (Cook Islands, South Pacific), fossil algal ridges dated at around 4.2–3.4 ka indicate up to 1 m of emergence (Woodroffe, Stoddart, Spencer, Scoffin, & Tudhope, 1990). From coring of algal ridges on Atol das Rocas (western Atlantic), Gherardi and Bosence (2005) demonstrated that sea level has oscillated over the past 4 ka.

Locally, emergent limestone terraces may display relatively well-preserved microatolls that document palaeo-low-tide levels. In the Indo-Pacific region, microatolls have played a significant role in deciphering the pattern of sea-level changes over the mid- to late Holocene (Chappell, 1983; Woodroffe et al., 2000; Hopley et al., 2007, pp. 64–66; Lewis, Wüst, Webster, & Shields, 2008). For instance, in the Society and Tuamotu Islands, many reef flats expose relict microatolls, the surfaces of which reach elevations of from 0.2 to about 0.8 m above the upper limit of living corals. These indicate a palaeo-sea level less than 1 m above the present position between 1 and about 5 ka (Pirazzoli, 1985; Pirazzoli & Montaggioni, 1988b).

In areas subject to rapid uplift, Holocene to Pleistocene reef flats are usually preserved in the form of a series of step-like subhorizontal terraces in which the geometry, stratigraphy and biozonation are still easily identifiable (see Chapter 6, Section 6.4). When the rate of uplift is known, these features allow the relative position of sea level to be accurately determined, as demonstrated by numerous works.

Submerged reef terraces. The discovery of relict submerged reefs along insular and continental foreslopes in various tropical areas (see Chapter 6, Section 6.5) has resulted in increasing information on the occurrence and growth history of reefs in relation to sea level. Macintyre (1988, 2007) and Montaggioni (2005) provided reviews for the past 25 ka of drowned reefs in the western Atlantic and Indo-Pacific respectively. As for emerged reef terraces, submerged counterparts appear to be relatively faithful markers of the course of sea-level change.

9.4.1.2. Erosional features

Cliffed limestone coasts are commonly incised by shoreline notches (Figure 9.15C). Pirazzoli (1986) provided a comprehensive review of destructional features affecting emergent limestones, particularly in reef sites, and providing a useful tool for the interpretation of past sea levels. Such features are prominent and readily identifiable palaeoshoreline markers on Pacific islands (Dickinson, 2001). As a generalization, the value of these features for precisely defining the elevation of former sea levels depends on the degree of exposure to wave agitation, the tidal range and the declivity of the limestone outcrops. Thus, the value increases from high to low hydrodynamic energy regimes, from macrotidal to microtidal environments and from gently sloping outcrops to vertical cliffs. The most reliable sea-level indicators are provided by tidal notches. The latter are typically midlittoral, formed by mechanical or bioerosional processes, with recumbent V- or U-shaped profiles, in which the lower parts of floors usually approximate to low spring tide level and the retreat point lies close to mean sea level. Notches form at average rates of 0.50–2 mm yr^{-1} (Hearty et al., 2007). Palaeoshoreline notches have been used to constrain Quaternary stillstands that may or may not be associated with intense karst formation, particularly in the tropical Pacific (Montaggioni, 1985; Montaggioni et al., 1985; Pirazzoli & Veeh, 1987; Woodroffe et al., 1990; Pirazzoli & Salvat, 1992; Hantoro et al., 1994; Dickinson, 2000; Berdin, Siringan, & Maeda, 2004).

9.4.1.3. Compositions of coralgal communities

There are very few coral species in either the Caribbean or the Indo-Pacific provinces that inhabit a sufficiently restricted depth interval to serve as robust sea-level indicators. In the western Atlantic, robust branching *Acropora palmata* living preferentially within 0–5 m depth in high-hydrodynamic energy sites has successfully been used to constrain sea-level curves from 17 ka to the Holocene (Lighty et al., 1982; Fairbanks, 1989; Toscano & MacIntyre, 2003; Gischler & Hudson, 2004; Hubbard et al., 2005).

In regions of the Indo-Pacific, robust branching acroporids (*Acropora robusta* group and *A. humilis* group) together with *Pocillopora verrucosa* can be regarded as the ecological counterparts of *A. palmata*. Their relatively narrow habitat range at depths of less than 10 m in agitated waters makes them potentially valuable sea-level markers (Faure, 1982; Pirazzoli & Montaggioni, 1988a). The association of the *A. robusta* group, *A. humilis* group species, *P. verrucosa* and/or *Goniastrea retiformis* with thick encrustations of *H. onkodes* and/or *H. gardineri*, encrusting foraminifera (*Homotrema* and *Carpenteria*), bryozoans and vermetid gastropods, develops at depths not exceeding 5–6 m (Montaggioni & Faure, 1997; Cabioch, Montaggioni, et al., 1999) (Figure 9.15D). The succession is very similar to that observed in cores from Caribbean reef fronts and crests (Perry & Hepburn, 2008) and has been applied as a sea-level recorder to a number of cores through modern reef crests, outer reef flats and fore-reef zones in the Indo-Pacific (Pirazzoli & Montaggioni, 1988a; Bard, Hamelin, Arnold, et al., 1996; Montaggioni & Faure, 1997; Montaggioni et al., 1997; Cabioch, Camoin, & Montaggioni, 1999; Cabioch, Montaggioni, et al., 1999; Cabioch, Montaggioni, Frank, et al., 2008; Camoin et al., 2001, 2004, 2007; Sasaki, Omura, Murakami, Sagawa, & Nakamori, 2004).

9.4.1.4. Other reef dwellers

Only a few molluscs can be regarded as significant in terms of sea-level position. In both Indo-Pacific and Caribbean reefs, the most robust sea-level indicators are encrusting vermetid gastropods that participate in the construction of reef-crest and outer reef flat growth frameworks. In French Polynesia, Pirazzoli and Montaggioni (1988a) described typical outer reef flat assemblages of *Serpulorbis colibrinus*, *S. annulatus* and *Dendropoma maximum*, in close association with *A. robusta* colonies and thick crusts of *H. onkodes*. In the Caribbean, the two main reef-building vermetids are referred to the genera *Dendropoma* and *Petaloconchus*. On back-reef shores, in both the Caribbean and Indo-Pacific regions, vermetids live at depths below mean sea level related to local water agitation. In the best conditions, the error range on estimates of sea level position is ± 0.10 m (Laborel, 1986). Additional fixed biological sea-level indicators on tropical coasts include oyster beds, barnacles and tube-worms restricted to the intertidal zone (see Baker, Haworth, & Flood, 2001).

The taphonomic signatures of reef frameworks can potentially also be useful tools to help in the identification of sea-level changes (Perry & Hepburn, 2008; see Chapter 4, Section 4.3.2). The use of taphonomic features may locally improve the resolution of interpretations initially based only on the compositions of coral assemblages. The signature provided by taphonomy is strictly directional. A reef framework that responds to sea-level rise according to the 'keep-up' growth mode exhibits limited vertical compositional change through the sequence, resulting in a relatively uniform taphonomic signature. Given that the reef top accretes upwards close to the sea surface, photophilic encrusters and typical shallow-water macro- and microboring traces dominate. In the fossil record, although an increasing number of sciaphilic species will colonize the progressively buried, deeper part of the framework, the taphonomic signature will remain near-surface in character (Perry & Hepburn, 2008).

9.4.1.5. Geometry of subtidal to supratidal sedimentary deposits

Modern reef flats and coastal environments are commonly characterized by skeletal sand and rubble, locally forming sandy islets (cays), beaches and storm ramparts deposited within the uppermost subtidal to intertidal or supratidal zones. On emergent palaeoreef systems, a relatively precise measure of former sea level position is given by the boundary between subtidal and intertidal sedimentary structures (Davies & Montaggioni, 1985; Hearty et al., 2007). For instance, in the Bahamas (New Providence Island), Hearty and Kindler (1997) were able to identify the transition between sandy beds deposited subtidally to intertidally during the last interglacial period on the basis of distinct internal structures. The former position of sea level can usually be defined within a decimetre-thick band around the subtidal–intertidal boundary.

Similarly, coral shingle ramparts have also been considered to provide sea-level criteria (Figure 9.16A). Modern deposits form asymmetric ridges with a gentle seaward slope resulting from storm and cyclonic activity (Scoffin, 1993). Relict analogues, termed rampart-rocks or coral conglomerates, may locally reach heights of up to 2 m above present mean spring tide levels. Typically the tops form a subhorizontal platform (Scoffin & McLean, 1978; McLean, Stoddart, Hopley, & Polach, 1978; Montaggioni & Pirazzoli, 1984; Collins, Zhao, & Freeman, 2006). McLean et al. (1978) concluded that on the Australian Great Barrier Reef, the upper platform formed at a time when sea level was about 1 m higher than today in the period 4.5–3.0 ka.

Holocene to Pleistocene beach-rocks, assumed to be cemented within the intertidal marine vadose zone, may also be used to define the former position of sea level. Where possible, the most efficient approach is to compare the elevations of modern and relict beach-rocks at the same site. Despite problems in the definition of the upper limit of marine lithification and in the interpretation of radiometric dates from skeletal beach detritus (Hopley, 1986b; Hopley et al., 2007, pp. 67–68; Neumeier, Bernier, Dalongeville, & Oberlin, 2000), there has been a number of attempts to infer mid- to late Holocene sea-level changes from beach-rocks throughout the tropics (e.g. McLean et al., 1978; Montaggioni, 1979b; Vieira & De Ros, 2006).

9.4.1.6. Fabrics and distributional patterns of cements

Longman (1980) and Coudray and Montaggioni (1986) claimed that diagenetic textures of limestones are diagnostic in terms of sea-level change. Among a variety of diagenetic features, cements, that is the binding precipitates within frameworks and around grains, may exhibit fabrics and mineralogy regarded as significant in terms of subaerial exposure and submergence (see Chapter 8 for review). Regional cement sequences have been described in a variety of reef environments as illustrating their reliability in the recognition of sea-level position. Montaggioni and Pirazzoli (1984) identified distinct suites of cements in French Polynesian rampart-rocks (Figure 9.16B). The diagenetic boundary between the intertidally to supratidally precipitated cements (zone of exclusively marine vadose cementation) and the subtidally to intertidally precipitated cements (zone of mainly marine phreatic cementation) represents the position of a particular water table level, closely linked to the former mean low-tide level. These rampart-rocks, deposited within the last 6-ka interval, provided evidence of a former stillstand, varying between about 0.45 and 0.8 m above present mean sea level from island to island. Similar results have been obtained by Gischler and Lomando (1997) and Blanchon and Perry (2004) at Belize and in the Gulf of Mexico respectively.

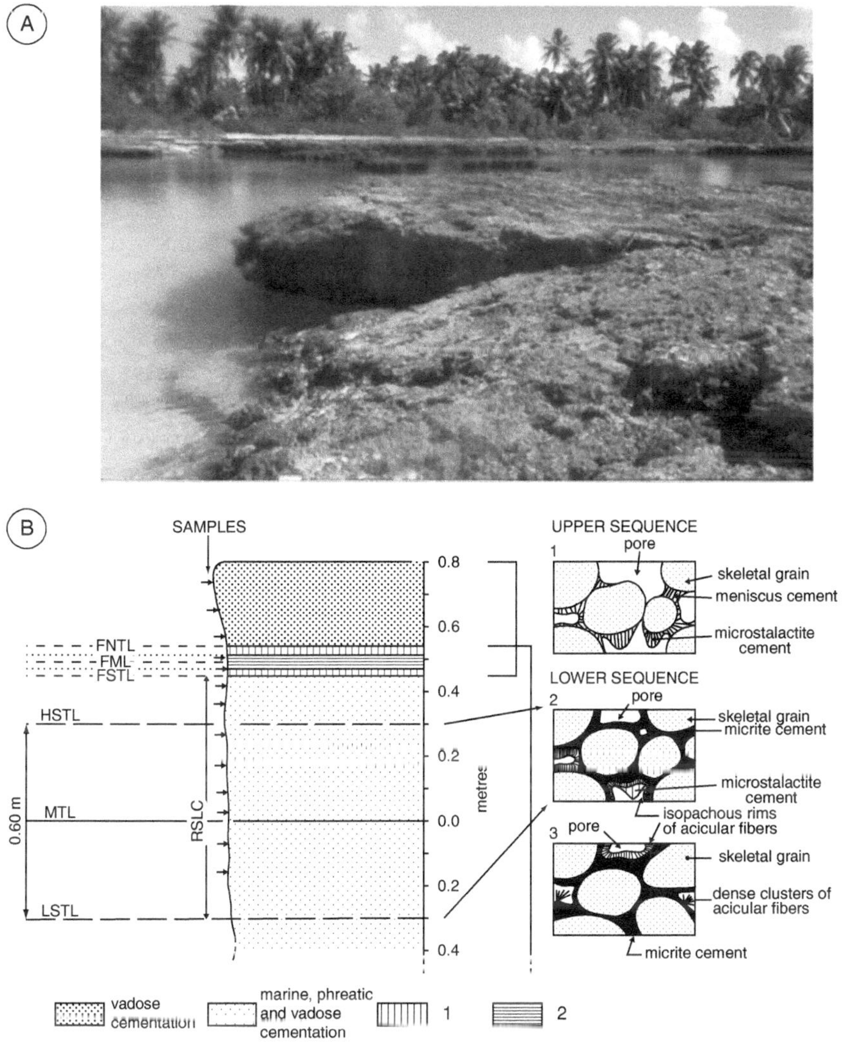

Figure 9.16 Exposed shingle ramparts as indicators of former sea levels. (A) General view of a rampart exposure, Mataiva Atoll, northwestern Tuamotus, central Pacific. Note the subhorizontal surface at the top (photograph by L. Montaggioni). (B) Left panel: Idealized cross-section of a rampart exposure showing the inter-relationship between the former water table level, the corresponding former low-tide levels and the present tide levels (assuming the tidal range has remained constant since lithification). 1 = range of uncertainty for the position of the former water table level as inferred from petrological criteria; 2 = range of uncertainty for the position of the former water table within the corresponding former range of low tides (FNTL = former neap low-tide level; FML = former mean low-tide level; FSTL = former spring low-tide level). The position of the present tide levels is also indicated — LSTL = present spring low-tide level; MTL = present mean-tide level; HSTL = present spring high-tide level. The distance between FSTL and LSTL represents the relative change in sea level (RSLC) since the mid-Holocene highstand. Right panel: Distribution of typical marine cements in

9.4.1.7. Stratigraphy of stacked reef sequences in cores

On subsiding mid-oceanic islands and passive margins, the reef generations formed in relation to rising relative sea level over successive glacial–interglacial cycles tend to pile up in the same place. In most cases, reef growth occurs during higher sea-level stands. Falls in sea-level produce a subaerial exposure surface. The high–low sea stand cyclicity results in a series of superimposed reef units. A robust inference of sea-level histories can be derived from physical and/or chronological stratigraphic relationships established in vertical profiles via reef drilling (see Chapter 6, Section 6.3). There are other potentially useful applications of reef stratigraphy to reconstruction of sea-level changes (Wheeler & Aharon, 1991). Thus, stratigraphic variations in the stable oxygen and carbon isotopes of reef carbonates can be used as a complementary tool. Shifts in the $\delta^{18}O$ and $\delta^{13}C$ signals can reveal the locations of both exposure surfaces and palaeo-water tables with great confidence, as a result of the specific isotopic signature of freshwater diagenesis (see Chapter 8).

9.4.1.8. Numerical modelling of reef growth

Changes in the frequency and amplitude of sea-level fluctuations result in changes in the sedimentary and stratigraphic attributes of reef systems over time. Computer modelling has been used to simulate shallow-water carbonate sedimentation and stratigraphy in order to better understand the complex interactions of depositional controls, especially the role of relative sea-level change (see Chapter 6, Section 6.6).

9.4.2. Reconstruction of Sea-Level Changes over Time

9.4.2.1. The middle to late Holocene

A large body of research conducted in the tropics in recent years has focused on sea-level history during the past 7 ka.

rampart exposures, according to diagenetic environments. The upper diagenetic sequence (1) strictly linked to the vadose zone, is typified by the occurrence of meniscus and microstalactitic cements, reflecting precipitation within a water-undersaturated environment (former intertidal zone). The lower diagenetic sequence exhibits a transitional zone at the top (2) containing an early cement generation with dense micrite and isopachous rim cements, and a second generation composed of pendant cements (microstalactites). This association results from lithification occurring successively within a water-saturated zone (the former subtidal zone), and an undersaturated zone (the present intertidal zone). The change from a subtidal to an intertidal zone reflects a drop in sea level through time. The lower diagenetic sequence at the base (3) exhibits only dense micrite, isopachous rims and dense clusters of acicular cement reflecting lithification under permanent phreatic conditions. Modified and redrawn from Montaggioni and Pirazzoli (1984).

The last millennium. Reef records of relative sea-level changes over the last millennium are based on measurements of changes in the elevations of living microatolls. The chronological framework in which these changes have occurred is defined using the annual skeletal density bands of colonies.

In the central Indian Ocean, Woodroffe and McLean (1990) showed from a series of sites throughout the Maldives, a net increase in sea level of approximately 6 mm between the 1960s and 1970s. In the eastern Indian Ocean, data gathered by Smithers and Woodroffe (2001) indicated that over the last century, sea level has risen at an average rate of less than 0.35 mm yr^{-1}, a rate markedly lower than that measured from the global tide-gauge network. In the central Pacific Ocean, sea level reached its lowest position since the beginning of the last millennium by the late 18th and 19th centuries at 0.2 m below the present level (Goodwin & Harvey, 2008). However, as in the Indian Ocean, relative sea level rose during the 20th century. From AD 1000 to 950 and during the last 500 years, sea level has oscillated with a sustained high multidecadal variability against the background of a sustained lowering due to the glacial isostatic readjustment. The discrepancy between the rates of sea-level lowering inferred from field observations and those calculated from geophysical modelling may result from severe ENSO climate anomalies around the Pacific basin (Goodwin & Harvey, 2008).

The 1–7 ka interval. In tectonically stable regions, the signal of relative sea level is usually characterized by a mid-Holocene sea-level highstand when the ice melting flux to the oceans decreased (Milne et al., 2005). The following fall in relative sea level to the present position was driven by glacial isostatic adjustment. Locally, the sea-level fall may reflect water loading of the continental shelves (hydroisostasy) that causes a concomitant uplift of the coastline (continental levering of Clark, Farrell, & Peltier, 1978) or from water flows that moved from the equator towards the collapsing forebulges of mid- and high latitudes (equatorial ocean siphoning of Mitrovica & Milne, 2002). All of these processes are recorded in the far-field tropics but not before the mid-Holocene due to the dominant imprint of the eustatic signature (Milne et al., 2005).

Comprehensive reviews of mid- to late Holocene sea-level variability have been provided by Pirazzoli (1991, 1996), Grossman, Fletcher, and Richmond (1998), Dickinson (2001), Woodroffe and Horton (2005), Angulo, Lessa, and de Souza (2006) and Lewis et al. (2008). The major question relates to the evidence of smooth or oscillating sea-level histories, particularly since sea level crossed over its present position (Figure 9.17).

In the western Atlantic, reef-derived sea-level reconstructions have been obtained from central South America and the Caribbean. On the eastern Brazilian coast, relative sea level crossed its present position by around 6.8–6.5 ka. A highstand at positions averaging 2–3 m above the present datum

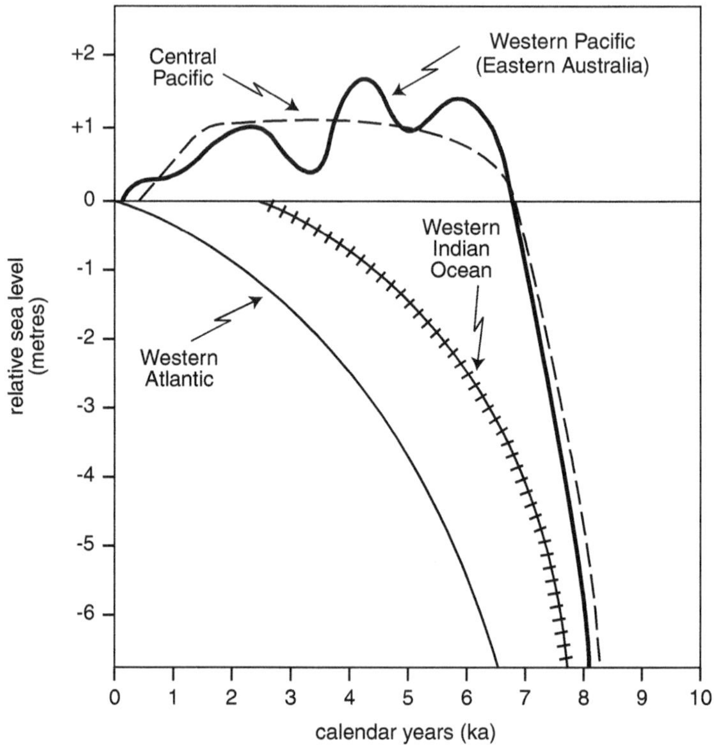

Figure 9.17 Selected sea-level curves for the past 8 ka based on reef features and associated coastal material. Modified and redrawn from Lewis et al. (2008): western Pacific (eastern Australia); Pirazzoli and Montaggioni (1988): central Pacific, French Polynesia; Camoin et al. (2004): western Pacific Ocean; Toscano and Macintyre (2003): western Atlantic.

has been recognized in the interval from 5.8 to 5 ka (Angulo et al., 2006). The geophysical model of Milne et al. (2005) suggests that the peak highstand was at 7 ka, with the level decreasing in elevation thereafter from +4 to +2.5 m along a north–south gradient. A period of sea-level stabilization occurred between 7 and 5 ka, followed by a steady decline to the present position. By contrast, there is no evidence for a mid–late Holocene highstand above present sea level in the Caribbean. Sea level does not appear to have reached its present position before the last millennium (Lighty et al., 1982; Toscano & Macintyre, 2003). This is attributed to subsidence of the Caribbean seafloor caused by the added water load reflecting melting of the North American ice sheets (Milne et al., 2005).

Throughout the Indian Ocean, the evidence for a mid-Holocene highstand varies from site to site (Camoin et al., 2004; Woodroffe, 2005).

Two highstands at +3.5 and +1.5 m above the present datum were identified in South Africa at about 4.7 and 1.6 ka respectively (Ramsay & Cooper, 2002). In southwestern Madagascar, emergent, in-place coral colonies dated at about 3–2 ka occur at elevations of 0.3–2.5 m above present sea level. Coral deposits about 3.7 ka-old were encountered at similar elevations on some non-granitic Seychelles Islands (Camoin et al., 1997, 2004). On Cocos Keeling, a single highstand younger than 3 ka is found at +0.5 m (Woodroffe & McLean, 1990). By contrast, in the Maldive, Laccadive and Chagos Islands, evidence for a sea level higher than that at present is poorly constrained although emerged reef conglomerates occur locally (Woodroffe, 2005). No remains of a high stillstand have been found on the volcanic islands of Réunion, Mauritius and Mayotte (Montaggioni, 1979b; Colonna, Casanova, Dullo, & Camoin, 1996; Camoin et al., 1997, 2004). Relative sea level in the western Indian Ocean has been assumed to have risen to its present position by about 3–2.5 ka and since then to have remained relatively stable. The eastern Indian coastline, extending over 150 km, exhibits two highstands, both culminating at approximately 3 m above present sea level, and dated at 7.3 and 4.3–2.5 ka (Banerjee, 2000). In the Houtman Abrolhos Islands, southwest Australia, Collins et al. (2006) described a 6.8-ka-old highstand culminating at 1.6–2 m above present sea level.

In the Pacific, observations indicate that relative sea level had crossed over its present position by approximately 7 ka. It was higher from about 7 to 6.5 ka, reaching elevations of +1 to +2.5 m prior to its fall to the present-day position. However, there were differences across the ocean in the timing and magnitude of the mid-Holocene highstands and in the nature of the late Holocene sea-level fall. On western and central Pacific islands, in areas unaffected by tectonic deformation, the dominant pattern of relative sea-level change was for an early Holocene rise in eustatic sea level, to be followed successively by a mid-Holocene highstand and a late Holocene fall in sea level, driven by glacio-hydroisostasy. However, the relative positions of the high sea-level stands inferred from biological indicators differ from those predicted by rheological models (Nakada, 1986; Mitrovica & Peltier, 1991; Nunn & Peltier, 2001; Mitrovica & Milne, 2002). In some island groups, the mid-Holocene highstand positions were disturbed by local uplift or subsidence to varying degrees (Dickinson, 2001). The most important discrepancy between field measurements and predicted estimates for the magnitude of the mid-Holocene highstand lies in the Society and Tuamotu islands and, to a lesser extent, in the southern Cook Islands. The explanation for this may lie in the thermal anomaly that characterizes the region (the South Pacific Superswell) and on which these island groups have risen. Standard mantle models used for global hydroisostatic predictions cannot account for local rheological properties (Dickinson, 2001). In the central Pacific, the highstand is thought to have come to an end by the first millennium AD (Dickinson, 2003) presumably in response to the decrease in the Antarctic meltwater input (Lambeck, 2002). This is consistent with field evidence presented by Goodwin and Harvey (2008) from the southern Cook Islands. The mid-Holocene highstand came to an end before 1.5 ka, probably between 2.5 and 2 ka. Relative sea level then declined from +1.3 to +0.45 m at around 1000 AD at a rate of 0.5 mm yr^{-1} (Goodwin & Harvey, 2008).

In eastern Australia, reconstructions suggest that sea level reached maximum elevations of +1.0 to +1.5 m at 7–6 ka and probably experienced two centennial-scale oscillations around 4.6 and 2.8 ka, before dropping to its present position from about 2 ka (Lewis et al., 2008). These data contrast with hydroisostatic models that predict a smoothly falling sea level (Lambeck, 2002). The timing of sea-level fluctuations in eastern Australia is in accordance with data from western Australia (Baker et al., 2001; Collins et al., 2006) and from other reef and non-reef sites in the western Pacific including Fiji (Nunn & Peltier, 2001), Malaysia (Tija, 1996), the South China Sea (Ma et al., 2003), Japan (Kato, Fukusawa, & Yasuda, 2003), the Philippines (Maeda et al., 2004) and Singapore (Bird et al., 2007). A similar sea-level behaviour is reported from South Africa, central South America and eastern India, suggesting that broadly similar hydroisostatic adjustments may have operated throughout the Southern Hemisphere (Angulo et al., 2006).

The rapid oscillations of sea level since the mid-Holocene may have been triggered on regional to global scales by a variety of climatic factors, including thermal contraction and the expansion of upper water masses (Mitrovica & Milne, 2002), changes in wind strength in relation to ENSO phases (Goodwin, 1998), and cycles of freshwater input controlled by repeated ice-sheet construction and ablation (Bond et al., 2001; Alley, Clark, Huybrechts, & Joughin, 2005). Differences in relative sea-level fluctuations over the last millennium, with respect to geophysical models, may indicate Pacific ocean-wide climate variability (Goodwin & Harvey, 2008).

9.4.2.2. The last deglaciation

The LGM lasted about 6,000 years, probably from around 26 ka (Peltier & Fairbanks, 2006; Cabioch, Montaggioni, Frank, et al., 2008), although Lambeck, Yokoyama, and Purcell (2002) defined the onset of the LGM at about 30 ka, the time when sea level first approached its lowest position (Figure 9.18). There is a consensus on the position maintained by global eustatic sea level at that time and data from both reef and non-reef sites indicate that it was at approximately −120 to −130 m compared to the present level (Colonna et al., 1996; Fleming et al., 1998; Yokoyama, Lambeck, De Dekkar, Johnston, & Fifield, 2000; Lambeck & Chappell,

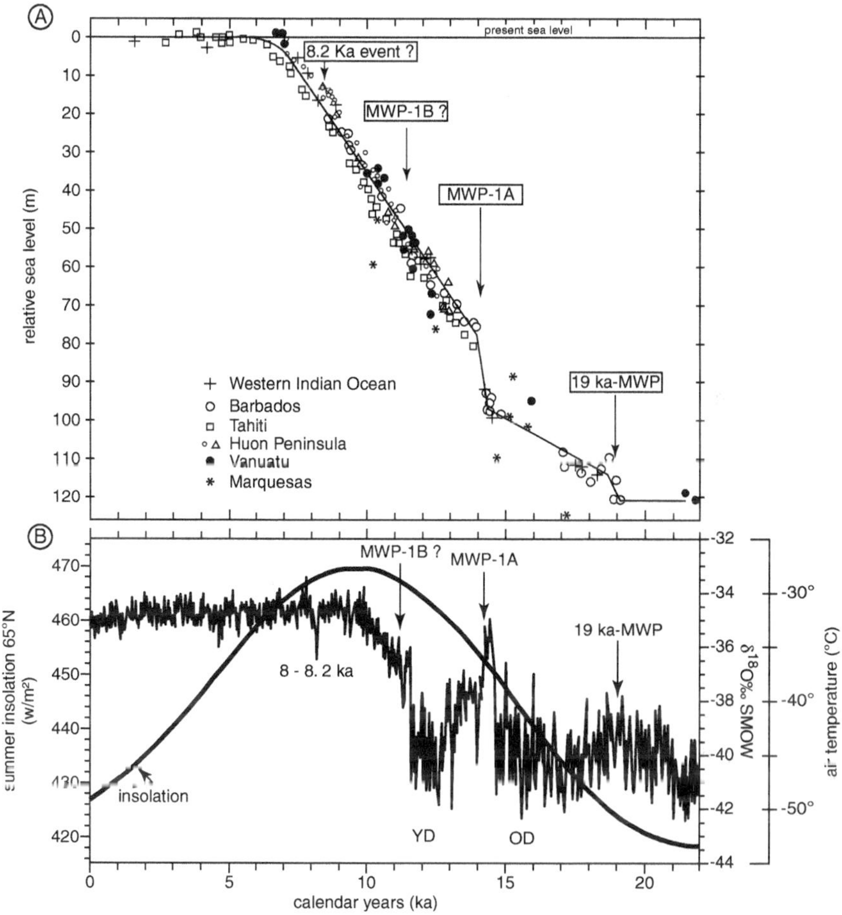

Figure 9.18 Relationship between sea-level changes and climate patterns during the last deglaciation. (A) Composite global sea-level curve for the last 22 ka based on the coral record. The timing of the expected meltwater pulses is indicated. From works by Camoin et al. (2004) in the western Indian Ocean; Fairbanks (1989) and Bard et al. (1990) in Barbados; Bard, Hamelin, Arnold, et al. (1996) and Montaggioni et al. (1997) in Tahiti; Chappell and Polach (1991) in Huon Peninsula, Papua New Guinea; Cabioch et al. (2003) in Vanuatu; and Cabioch, Montaggioni, Frank, et al. (2008) in the Marquesas. (B) Climate changes recorded by variations in oxygen isotope composition measured in the GRIP (Greenland) ice core. The summer insolation curve at 65°N is derived from Berger's (1979) work. The expected meltwater pulses are located in the $\delta^{18}O$ curve. YD = Young Dryas; OD = Older Dryas. Modified and redrawn from Bard et al. (1996).

2001; Clark & Mix, 2002; Peltier & Fairbanks, 2006; Cabioch, Montaggioni, Frank, et al., 2008).

A limited number of high-resolution sea-level records, based on cored reef sequences, have been used to approximate the global glacio-eustatic signal during the deglaciation following the LGM (Barbados: Fairbanks, 1989; Papua New Guinea: Chappell & Polach, 1991 and Edwards et al., 1993; Tahiti: Bard, Hamelin, Arnold, et al., 1996 and Camoin et al., 2007; Vanuatu: Cabioch et al., 2003; Caribbean: Toscano & Macintyre, 2003). Additional data have been obtained from submerged reef terraces (Colonna et al., 1996; Toscano & Lundberg, 1998; Webster et al., 2006; Cabioch, Montaggioni, Frank, et al., 2008). However, Fairbanks (1989) was the first to demonstrate, from the analysis of coral assemblages on Barbados, that the rise in sea level during deglaciation did not occur smoothly, but was punctuated by rapid jumps (Figure 9.18). These were interpreted as caused by dramatic meltwater pulses occurring at 13.7–14.2 and 11.5 ka, and termed meltwater pulses MWP-1A and MWP-1B respectively. An earlier high-magnitude meltwater pulse, referred as the LGM-Terminal MWP or 19-kyr MWP, was identified at about 19 ka in the Bonaparte Gulf (northwest Australia) by Yokoyama et al. (2000). Although this latter event remains a point of dispute, the fact that the demise of reef tracts in the Marquesas Islands (French Polynesia) dated at 26.6–25.3 ka occurred just at the onset of deglaciation strongly supports an abrupt rise in sea level (Cabioch, Montaggioni, Frank, et al., 2008). A later, low-magnitude meltwater pulse (MWP-2) at around 7.5–7.6 ka has been postulated by Blanchon and Shaw (1995a,b) and Blanchon et al. (2002) from a reef record on Grand Cayman (Caribbean) and by Bird et al. (2007) from date on Singapore. This event was regarded as linked to the climatic shift that occurred at 8.2 ka, and resulted in a rapid rise in global sea level due to the discharge of freshwater lakes from the Laurentide ice sheets (Clark, Marshall, Clarke, Licciardi, & Teller, 2001; Alley & Ágústsdóttir, 2005). However, the magnitude of the rise in sea-level was less than 0.50 m (Clarke, Leverington, Teller, & Dyke, 2004).

The existence of the MWP-1A has been confirmed by Hanebuth, Stattegger, and Grootes (2000), Camoin et al. (2007) and Cabioch, Montaggioni, Frank, et al. (2008). This event is thought to have coincided with the Older Dryas interval and not with the sharp Bølling-Allerød warming (Stanford et al., 2006). By contrast, the existence of both MWP-1B and MWP-2 are still matters of considerable debate (Bard, Hamelin, Arnold, et al., 1996; Clark & Mix, 2002; Zinke, Reijmer, Thomassin, & Dullo, 2003; Clarke et al., 2004; Bird et al., 2007; Cabioch, Montaggioni, Frank, et al., 2008). From the early deglacial phase (about 19 ka) to the mid-Holocene (about 7–6 ka), the rate of sea-level rise averaged 10 mm yr^{-1}. During the MWP-1A event, a rise of about 20 m in sea level probably occurred in less than 500 years at a rate of about 40 mm yr^{-1}. The rate of sea-level rise during MWP-1B has been postulated to not exceed 30 mm yr^{-1} (Blanchon & Shaw, 1995a) with comparable rates for the LGM-Terminal MWP (Yokoyama et al., 2000).

The controversies concerning the existence of some meltwater pulses are mainly due to significant and systematic discrepancies between reef records from Barbados, New Guinea, Tahiti and Vanuatu, and non-reef tropical records from the Sunda Shelf, Indonesia (Hanebuth et al., 2000) and northwest Australia (Yokoyama et al., 2000) particularly for the late glacial period (about 14–8 ka). They may be explained, at least in part, by the responses of the different coral communities to dramatic rises in sea level. Corals living close to the sea surface may have survived rapid jumps, the magnitude of which was insufficient to displace them from their habitat depth (less than 6 m) at the reef crest. MWP-1B and MWP-2 were characterized by smaller magnitude changes than MWP-1A and may therefore not have affected vertical reef development in the same way. Reefs were able to recover, leaving no resolvable framework records of these events.

The reef records from Barbados, Papua New Guinea and/or Tahiti have been used in a number of geophysical models to estimate the volumes of the ice sheets at the LGM (Milne, Mitrovica, & Schrag, 2002) and of the freshwater masses subsequently delivered by ice melt (Fleming et al., 1998; Lambeck, 2004). The geographical source of the water flux from the ice melt responsible for a particular sea-level jump continues to be debated (Lambeck, Yokoyama, Johnston, & Purcell, 2000; Clark, Mitrovica, Milne, & Tamisiea, 2002; Peltier, 2005). A number of models created with special reference to MWP-1A has been designed on the assumption that the freshwater supply came predominantly from the Northern Hemisphere. Applying a glacial isostatic adjustment model in which the Antarctic ice sheet contributes significantly to the relative sea-level rise generated by the MWP-1A event, Bassett, Milne, Mitrovica, and Clark (2005) resolved the discrepancies between the reef records. However, this model did not support the existence of a meltwater pulse at 11.5 ka.

9.4.2.3. The last interstadial period

Reconstructions of variations in sea level during the last glacial–interglacial transition (30–110 ka, MIS 3–5d) are derived mainly from the study of emergent reef terraces and occasionally of stacked reef sequences in cores (Figures 9.19 and 9.20A, B).

Chappell (2002), using a computer model of reef development controlled by sea level, identified a number of sea-level cycles between 30 and 65 ka (MIS 3a, 3b, 3c and 4) from reef terraces on the Huon Peninsula, Papua New Guinea. Each cycle lasted 6,000–7,000 years with a long episode of falling sea level issuing in a 1,000–2,000-year-long rise of

10–15 m. The timing of each reef generation was defined by U-series dating. The cycles peaked at 33, 38, 44.5, 52 and 58–60 ka. Sea-level elevations ranged approximately from 40 to 75 m below the present level. Fluctuations in sea level are associated with periodic, intense climate changes including Bond cycles (1,500-year frequency; Bond et al., 1997) and Heinrich events (release of ice-rafted detritus into the North Atlantic, indicating major ice breakouts from the Laurentide icecaps; Heinrich, 1988). Each rise in sea level was shown to have been coincident with a phase of deposition of ice-rafted detritus in seafloor sediments. However, as suggested by Esat and Yokoyama (2006), due to the rapidity of sea-level changes, corals on some terraces may have been distributed randomly without any sequential temporal ordering, with younger corals occupying lower positions. Locally, the random distribution of corals is due to irregularities in the depositional surface. Such distributions might locally alter the validity of sea-level curves based on age–height relationships. Nevertheless, the main features of sea-level behaviour during MIS 3 have been confirmed by work on other Pacific sites. On Malakula (Vanuatu), investigations of emerged terraces indicate that sea levels at 45–50 ka (MIS 3) reached elevations similar to those inferred from the Huon Peninsula, no deeper than 60 m relative to present (Cabioch & Ayliffe, 2001). On the uplifted island of Kikai (Central Ryukyus), three transgressive hemicycles have been recognized dated at about 52, 62 and 66 ka. The corresponding highstands can be correlated with the two older sea-level peaks from the Huon Peninsula (Sasaki et al., 2004). On Mururoa, MIS 4 has been identified in cored sections dated at 59 ± 0.2 to 69 ± 3 ka. The inferred palaeo-sea level ranges from 76 to 91 m below the present level, slightly deeper than that on the Huon Peninsula (Camoin et al., 2001).

On the Huon Peninsula, the uplifted reef terraces representing substages MIS 5a, 5c and 5d have been dated at approximately 85, 104 and 110 ka respectively. The corresponding sea-level positions were at about 20, 25 and deeper than 50 m below the present level (Chappell & Shackleton, 1986; Chappell et al., 1996). Based on electron spin resonance and/or U/Th dating, MIS 5a and 5c have also been recognized from raised reef terraces in Barbados (Schellman & Radtke, 2004; Potter et al., 2004). The mean ages of these terraces are centred around 76.7 ± 1.0, 84.3 ± 0.7, 101 ± 0.3 and 104 ± 0.9 ka. Assuming a constant rate of uplift of 0.27 mm yr^{-1}, sea-level highstands during MIS 5a have been estimated as reaching elevations of -19 and -21 m relative to the present position at 76 and 84 ka respectively, and an elevation of around 10 m at 104 ka. During MIS 5c, the heights of sea-level highstands ranged between approximately -13 and -25 m. Sea-level elevations for this period inferred from Barbados terraces are consistent with those from the Huon Peninsula and data from Haiti (northern Caribbean) are in partial agreement. MIS 5a and 5c each appear also to be typified by three sea-level oscillations, but there are

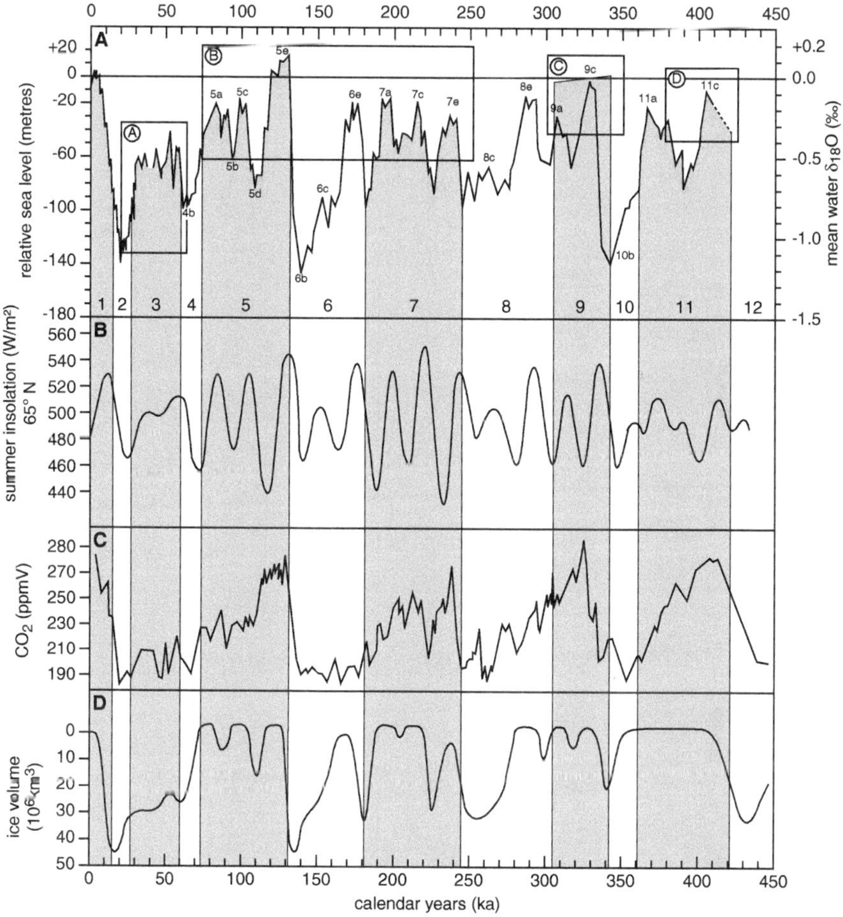

Figure 9.19 Relationships between relative sea-level changes and a variety of climate parameters (summer insolation at 65°N, atmospheric CO_2 levels and ice volume) for the past 450 ka. (A) Composite sea-level curve derived from the benthic foraminiferal oxygen isotopic ratios in the North Atlantic (cores NA-87-22 and NA-87-25) and equatorial Pacific (core V 19-30) (Waelbroeck et al., 2002). The boxes A, B, C and D refer to the subparts A, B, C and D of Figure 9.20 respectively. (B) Summer insolation curve (Berger, 1979). (C) Atmospheric CO_2 curve derived from Vostrok Ice Core, Antarctica (Petit et al., 1999) (D) Modelled ice volume curve (Loutre, 2003).

significant differences with regard to the elevations of highstands (Dumas et al., 2006). Similar discrepancies are observed on the southwestern Florida margin (Toscano & Lundberg, 1999), the Grand Cayman Islands (Coyne, Jones, & Ford, 2007), the Bahamas (Hearty & Kaufman, 2000), Bermuda (Wehmiller et al., 2004) and a number of Atlantic coastal locations north of

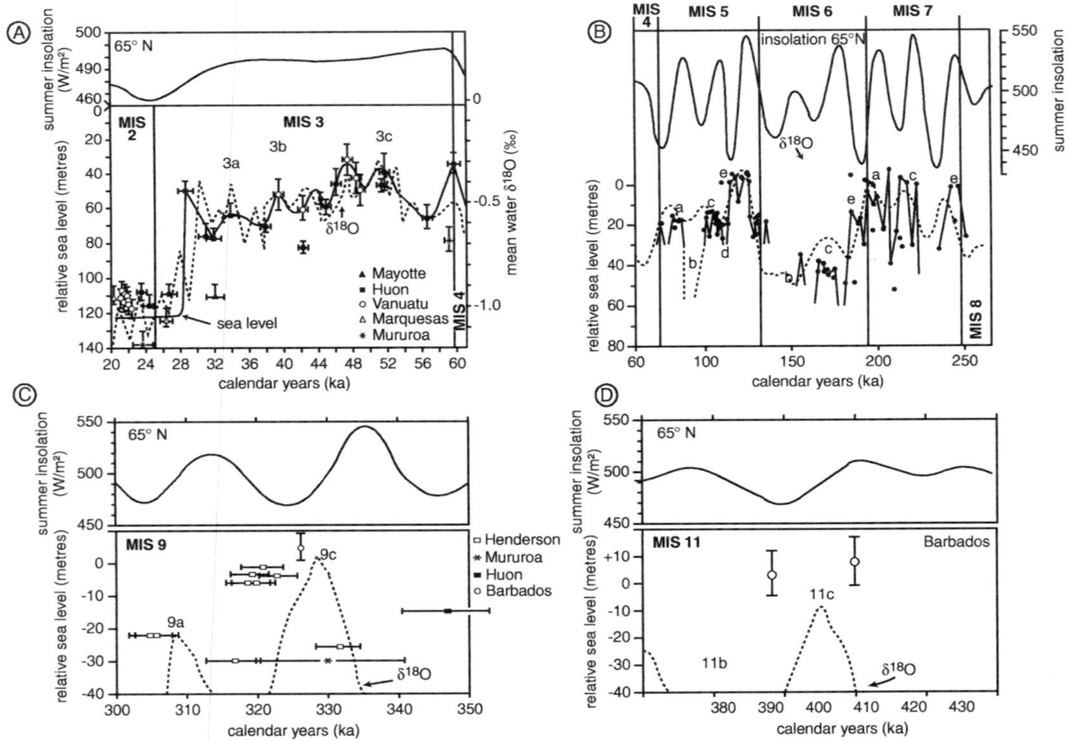

Figure 9.20 Relative sea-level reconstruction based on the elevation of reef terraces and associated deposits for the last 450 ka, compared to sea-level records derived from benthic foraminiferal oxygen isotopic ratios and the Northern Hemisphere summer insolation curve. Each dated coral sample is located on the graphs according to its age and elevation; vertical and horizontal error bars for each data point refer to elevation and age ranges respectively. (A) Stages MIS 2 and 3 (20–60 ka). The sea-level curve is from Cabioch & Ayliffe (2001), Camoin et al. (2001), Lambeck et al. (2002), Cutler et al. (2003) and Cabioch, Montaggioni, Frank, et al. (2008). (B) Stages MIS 4, 5, 6, 7 and 8 (60–260 ka). The sea-level curve showing the age and elevation of the highstands from Thompson and Goldstein (2005) and Henderson et al. (2006). (C) Stage MIS 9 (about 300–350 ka). Modified from Henderson et al. (2006) and Siddall et al. (2006). (D) Stage MIS 11 (about 350–410 ka). Modified from Siddall et al. (2006).

Florida. These regional variations in elevation are likely to reflect differential responses to glacio-hydroisostatic readjustments. In particular, the US western Atlantic continental margin, at or near to an intermediate-field setting, has been influenced by the large-scale regional effects of the North American ice sheets, driven by successive glacio-hydroisostatic submergence, emergence, forebulge and collapse during the late Pleistocene and probably earlier (Wehmiller et al., 2004).

In both Indo-Pacific and Caribbean reef sites, the existence of multiple subterraces (see Chapter 6, Section 6.4.4), representing rapid suborbital sea-level fluctuations during substages 5a and 5c demonstrate that Milankovitch orbital forcing was not the only factor controlling changes in global ice volume (Potter et al., 2004). Similar complex interplays between climate and land-ice evolution have also operated during the last interglacial and older glacial–interglacial cycles.

9.4.2.4. The last interglacial period

Siddall et al. (2006) and Hearty et al. (2007) provided outstanding compilations of data devoted to the last interglacial (MIS 5e) from a number of tropical and non-tropical locations, and thus a coherent history of the timing and magnitude of successive sea-level events.

Records from a variety of reef-related features indicate that the course of sea-level change was characterized by episodes of relative stability and transition (Figures 9.19 and 9.20B). The deglacial sea-level rise corresponding to the MIS 6/MIS 5e transition (Termination II) occurred before 130 ka, from a sea-level position approximately 120 m below the present level (Rohling et al., 1998). Gallup, Cheng, Taylor, and Edwards (2002) indicated that sea level was -18 ± 3 m below its present level at 135.8 ± 0.8 ka. The onset of the last interglacial period occurred at 128 ± 1 ka (Stirling, Esat, Lambeck, & McCulloch, 1998). From about 128 to 125 ka, sea level was relatively stable at elevations higher than 2.5 ± 1 m above the present position. It then fell slightly before rising to +3 to +4 m. The decrease in sea level at around 125 ka was related to the 'Intra-Eemian cooling event'. The interglacial period ended at 120–118 ka and was characterized by rapid sea-level fluctuations of +6 to +9 m. Average rates of sea-level rise were about 16 mm yr^{-1}. Such rapid rates require the disappearance of an ice sheet the size of Greenland in roughly four centuries (Rohling et al., 2008). Sea level fell from this late highstand position to about −60 m, the glacially induced elevation of MIS 5d.

These results indicate that orbital forcing alone cannot account for rapid changes in sea level, because summer insolation was at relatively low levels in northern high latitudes at the onset of Termination II, and later, at about 118–115 ka (Muhs et al., 2002). Multiple erosional and depositional features found at +6 to +9 m probably reflect successive abrupt sea-level oscillations, perhaps controlled by the regional forcings that caused the disintegration of the North Atlantic ice sheet. However, the increased water volume required for globally rising sea level during MIS 5e also implies an important collapse of the West Antarctic icecap. Similarly, the marked fall in sea level from 118 ka, probably required the rapid growth of both the North Atlantic and West Antarctic ice sheets.

9.4.2.5. Older glacial–interglacial cycles

Evidence of palaeo-sea levels older than the last interglacial is provided by raised reef terraces and reef sequences drilled in subsiding coastal areas, although both suffer limitations. Exposed and buried reef units tend to be severely altered, with alteration increasing with age. Radiometric dating of corals is usually problematic due to the poor preservation of aragonite. Other methods, such as electron spin resonance, are less accurate and yield dates with larger uncertainty (Siddall et al., 2006). A further limitation is that in most cases, drilling investigations typically only give access to interglacial highstand deposits (Alexander et al., 2001; Cabioch, Montaggioni, Thouveny, et al., 2008). Locally, however, extraction of lowstand reefs has been successful using inclined coring (Camoin et al., 2001).

Siddall et al. (2006) reviewed the literature devoted to interglacial sea levels in MIS 7–19. Three highstands have been identified during MIS 7, referring to as substages 7a, 7c and 7e, in order of increasing age (Figures 9.19 and 9.20B). Coral data from Barbados, revisited by Thompson and Goldstein (2005), suggest that MIS 7 ranges from approximately 190 to 245 ka. Estimates of sea level at that time are close to 6 m below the present level for Barbados (Schellmann & Radtke, 2004) and for Sumba Island, Indonesia (Pirazzoli et al., 1991). However, intraregional differences in elevation have been identified, probably arising from isostatic readjustment. There is a difference of about 20 m in relative sea level between the northern and southern Caribbean in MIS 5a (Potter & Lambeck, 2003). Similar uncertainty surrounds MIS 7c and 7e highstands. In both substages, sea level is likely to have been at elevations of −6 to −8 m, according to records from the Bahamas and western Australia.

Data from Henderson Island, South Pacific, display two sea-level peaks for MIS 9, substages 9a and 9c extending from 306±3 to 334±4 ka respectively (Figures 9.19 and 9.20C). However, full interglacial conditions may have been established as early as 343 ka, about 8 ka before the peak of summer insolation in the Northern Hemisphere (Henderson, Robinson, Cox, & Thomas, 2006). Sea level during MIS 9a was close to the present position, whereas during MIS 9c, it reached elevations somewhere between −3 and +8 m relative to the present level (Siddall et al., 2006). MIS 11 has been dated at approximately 395–415 ka and was maintained for 30–40 ka. Data from South Australia, Bermuda, Barbados and the Bahamas suggest that the highstand remained stable within ± 10 m of the present position (Figures 9.19 and 9.20D).

Little is known of sea-level positions during the interglacial periods older than MIS 11. Stages MIS 15 and 17 are dated at 600 ± 15 and 646 ± 37 to 718 ± 48 ka respectively (Thompson et al., 2003; Andersen et al., 2008). On Sumba Island, in Indonesia, the age of MIS 15 ranges from 584 ± 88 to 603 ± 90 ka. The uppermost raised reef terraces in this area have been inferred to have formed during highstands corresponding to MIS 17, 19, 21 and 23, by correlating their elevations with the astronomically calibrated benthic foraminiferan oxygen isotope record. Subsequently sea levels at MIS 13, 15 and 17 were estimated to have been lower than today, ranging from about 0 to 20 ± 10 m below the present level, whereas during MIS 23 and 27 sea-level positions may have been significantly lower, ranging between -50 m and present sea level. MIS 25 was regarded as within ± 5 m of the present position. The large uncertainties in height estimates arise from the extrapolation of estimated uplift rates (Pirazzoli et al., 1993). In addition, attempts to reconstruct sea-level elevations prior to 1.0 Ma have been made based on strontium isotope stratigraphy and the vertical distribution of diagenetic features in reef sequences extracted from a number of atolls (Quinn & Saller, 1997; Ohde et al., 2002).

As indicated by Siddall et al. (2006), the last nine interglacial highstands have differed in elevations and amplitudes of sea-level fluctuations, but also in their timing relative to the summer insolation maxima in the Northern Hemisphere. Factors other than solar orbital forcing may have governed the onset of deglacial sea-level rises, including southern summer insolation, increasing frequency and intensity of ENSO events, and changes in atmospheric CO_2 levels.

There is limited reef-based evidence for the heights reached by sea level during glacial and interstadial times before about 140 ka. On Barbados, sea-level indicators attributed to MIS 6e formed at depths ranging from $50 + 11$ to 47 ± 11 m below the present level during the interval from 176.1 ± 2.8 to 168.9 ± 1.4 ka (Scholz, Mangini, & Meischner, 2006). Camoin et al. (2001) identified a lowstand attributed to MIS 8 in sequences drilled on Mururoa, central Pacific, dated at about 270 ka and thus referred to substage MIS 8d. The palaeo-sea level was inferred to have been at 79–94 m below the present position, consistent with estimates from the Red Sea based on a hydraulic control model (Rohling et al., 1998).

9.5. Conclusions

Most of the investigative tools based on coral geochemistry have proven to be efficient, making coral skeletons almost ideal as archives of tropical climate variability. It is now well established that coral tracers have the potential to complement other natural archives of climate variability such as ice cores, tree rings, varved sediments and other marine organisms, providing monthly to subseasonal resolution. However, they suffer the disadvantage that data produced by single colonies can at best typically only encompass a maximum of about 400 years. However, the use of overlapping measurements may offer a more accurate picture of the variability of poorly documented climatic oscillations in the recent past, including variation in the Indian Ocean Dipole, in the NAO, the Antarctic Oscillation and ENSO. Unfortunately, coral-based proxy reconstructions suffer important limitations. Apart from the difficulty of interpreting the composite nature of some climate proxies (e.g. $\delta^{18}O$, Ba/Ca and $\delta^{13}C$), vital effects and diagenesis can alter the climate signatures. However, despite these limitations, the combined use of multiproxy records offers great promise for coral palaeoclimatology.

The Quaternary coral record only provides short-term insights into natural climate variability. Late Holocene to Pleistocene corals allow access to past climate histories through decades-to-century-long time windows at subseasonal to interannual resolution. To date, limited windows have been opened during particular periods that include the Holocene Climatic Optimum and the last glacial cycle. There are only sparse coral-based data from earlier periods and no record is continuous to modern times. Thus, it is important that these data are supplemented by the addition of high-resolution proxy records from other time intervals throughout the Quaternary. The possibility of periods devoid of significant changes in oscillations or experiencing different oscillation modes cannot be ruled out.

In the tropics, reconstructions of palaeo-sea level at local to global scales can be based on a variety of reef-related features. A number of difficulties and uncertainties arise when establishing palaeo-sea-level histories from such features, related to the depth ranges within which depositional and erosional features may have formed and the delayed response of reef growth to changes in sea level. An additional important limitation comes from dating techniques. Although U-series methods provide the most robust results, they decrease greatly in accuracy with increasing age of the coral material. Other dating methods are largely imprecise. These limitations may result in conflicting interpretations both within and between sites. However, more detailed knowledge of relationships between reef features and sea level may resolve many of these problems, particularly with data from reef terraces, either exposed or submerged, and from drilled reef sequences.

Mid–late Holocene sea-level history is demonstrated to have been driven by complex spatial and temporal patterns of interactions between climate and glacio-hydroisostasy. It is premature to claim what is the first-order forcing mechanism that has constrained mid-late Holocene sea-level variability. Defining this forcing requires further investigation of sea-level markers and long-term climatic coral records. The dephasing of the link between interglacial stages and insolation in the Northern Hemisphere, demonstrated by the coral record, calls into question the idea behind a strictly orbital control of high sea-level stands. The occurrence of suborbital climate variations strongly suggests that alternative processes may have operated over centennial-to-millennial time scales.

Assessing tropical climate variability during the Quaternary is critical to our understanding of present-day climate system functioning and improving climate predictions. Climate changes during glacial–interglacial cycles have been driven, at least in part, by boundary conditions and forcings different from those apparently operating today. Most reliable reconstructions of tropical palaeoclimate variability, based on both coral proxy data and simulations, require additional results from a variety of sites throughout the tropics, but especially from the Atlantic and the central and eastern Pacific. It is noteworthy that there are close similarities between the sea-level histories of the mid-late Holocene interval and the first half of the last interglacial period. As pointed out by Hearty et al. (2007), the last interglacial cycle was characterized by relative sea-level stability, abruptly followed by a dramatic deterioration of climate that raised sea level over an interval of only a few centuries. This may be used to gain insight into potential changes in the future.

CHAPTER TEN

Conclusions: Coral Reefs from the Past to the Future

This book is an attempt to provide a general outline of studies conducted in Quaternary coral reef systems worldwide. It is based on significant advances in the understanding of the Pleistocene to Holocene developmental history of reefs, as a result of recent investigations of marine sedimentology and ecology, paleobiology, paleobiogeography, coastal paleoceanography and paleoclimatology of the tropics. Such an approach leads to large-scale spatio-temporal insights into the conditions of reef development at times before the human occupation of reef sites.

10.1. The Historical Perspective

Quaternary coral reefs have been studied for over 150 years. Early investigations centred on their morphology but clearly included speculation as to their geological origins. For some years, although arguments concerning these matters continued, the emphasis has shifted a more thorough analysis of their biotas, ecology and regional characteristics and to their controlling environmental factors. This path led to more detailed work on the biology and physiology of individual organisms. Biology and physiology remain burgeoning fields and are of special importance where both individual organisms and the system are sensitive indices of climate change.

The geological significance of reef structures increased when it was realized that many major oil fields are in lithological associations with structures that resemble those of recent reefs. This has been one of the drivers for numerous shallow drilling programmes. However, these studies have so far delivered only generalized knowledge of how reefs form. Deep drilling, following in the footsteps of the Royal Society's attempts to prove Darwin's theory of subsidence, have shown that the majority of reef bodies are the cumulative result of deposition accompanying subsidence during the Pleistocene, Pliocene, and in some cases even the Miocene. A limiting factor in both 'shallow' and 'deep' drilling is the common lack of a robust chronology. However, it has clearly shown that growth has not been continuous and accretion has been punctuated by climatically driven changes in relative sea level.

10.2. The Role of Controlling Factors in Reef Growth and Distribution

No single phenomenon can satisfactorily account for changes in the patterns of reef distribution and development over time. A myriad of abiotic (environmental) and biotic factors appears to participate in governing the attributes of coral reefs and in maintaining the diversity of reef communities over time scales varying from decades to tens of thousands years.

10.2.1. Environmental Controls

The effects of environmental factors linked to changes in global climate and geography (land distribution patterns, oceanic circulation regimes) are likely to have caused changes in the diversity of reef biotas in the tropics throughout the Tertiary.

The influence of sea surface temperatures on reef growth, especially during glaciation and deglaciation events, actually remains unclear, even if field observations and laboratory experiments indicate that the optimal temperatures for efficient coral reproduction rates, recruitment and growth range from 25 to 27°C (Jokiel & Coles, 1990). By reference with the Last Glacial Maximum (LGM), the tropics appear not to have experienced temperatures significantly lower than today during Pleistocene glaciations. In contrast, nutrient supply seems to have been one of the major determinants, because upwelling activity was probably enhanced during phases of accelerated sea level rise, at least in some tropical areas. Hydrodynamic energy, particularly in areas subject to cyclones and tsunamis, has controlled the geometry and architecture of reefs. Reef architecture can be interpreted as the result of end-members of a hydrodynamically driven spectrum ranging from framework to detritus. Substrate availability, together with distance from centres of coral dispersal, is the most biogeographically limiting of all physical factors. The exact role of changes in atmospheric pCO_2 prior to the Industrial Revolution is not well known. Considering the carbon dioxide concentrations in the atmosphere over the last 400 ka (Petit et al., 1999), it is likely that reef coral calcification has not been significantly influenced. Other environmental forcing factors have acted only as modulators. While tectonics and antecedent topography contributed significantly to the determination of the distribution of coral reef systems throughout the Tertiary, in relation to the motion of lithospheric plates, these two factors have been of limited impact in post-Neogene reef development, as a result of the relatively stable configuration of the continents during the Quaternary. Similarly, high turbidity (and related low light levels), low salinity and elevated dust fluxes may only have been efficient inhibitors of reef growth at a few reef locations.

10.2.2. Biotic Controls

Biotic factors, including migration rates, dispersal and competitive abilities, may also determine species abundance and survival patterns and thus the biotic compositions of reefs. Quaternary reefs generally exhibit clear and repeated differences in zonation and species richness patterns. The taxonomic composition and diversity of coral assemblages has remained remarkably constant through time, despite experiencing multiple cycles of rapid turnover from the early Neogene to the Pliocene-Pleistocene. The most striking zonal feature is the generalized prominence of the genus *Acropora* in reef communities in time and space. In both the Caribbean and Indo-Pacific provinces, *Acropora*-rich communities are assumed to have been continuously dominant. The rise to dominance of branching Acropora has resulted in coral communities being structured as they are today by about the mid-Pleistocene.

10.2.3. Disturbances and Resilience of Reefs

Disturbance regimes are major determinants in the degree to which biotic factors can influence the structure and distribution of reef communities; their frequency and duration, together with intensity, control the resilience of local communities.

In the Holocene, coral populations appear to have been permanently luxuriant. Outside hurricane-swept areas, reef mass-destruction events were infrequent (about 1 disruption per 1–1.5 ka), placing these events in the frame of natural rates in healthy reef systems. Holocene instances demonstrate that the functional abilities of renewed coral communities were fully re-acquired following rapid coral re-settlement within a time range shorter than 100 years. Further studies of the disturbance suffered by Holocene reefs will undoubtedly help to place mortality events in a temporal frame to provide a better understanding of the nature of reef community structure (i.e. random or time-organized assemblages) and the long-term responses of reef systems to disruption.

10.3. The Fossil Record as a Proxy for the Future of Reefs

Assessing the nature and long-term environmental effects of mass-mortality episodes from the fossil coral record is a prerequisite for predicting the future of reef ecosystems. However, comparisons between changes in the community structure of modern reefs at scales of decades and those affecting their fossil counterparts have to be made cautiously. As emphasized by Perry et al. (2008), the nature and magnitude of any ecological phase-shift primarily determines the extent to which the shift is translated into an instantaneous impact on reef framework and sediment composition. In particular, a full understanding of reef growth history in sites experiencing ecological shifts today will require further analysis of detrital material (mainly sand and silt fractions) from cores. This may help to resolve possible non-anthropogenic ecological shift events that occurred prior to human settlement.

Taphonomic (postmortem) processes may also have severely affected reef substrates, and thus control their preservation. When first deposited, skeletal remains of reef dwelling organisms consist largely of aragonite and Mg-calcite. The sensitivity of these to diagenetic alteration depends upon local rainfall and the extent and duration of the exposure episode. The degree of preservation of the original reef community structure, and its potential to record short-term changes in the structure of fossil reef communities are controlled by the attributes of the organisms (the type of growth forms, original skeletal mineralogy and nature of live assemblages), wave exposure, rates and modes of burial, and the intensity of encrustation and boring. A variety of factors alter the degree of fidelity of death assemblages: including a greater degree of time-averaging (mixing different faunal generations); drastic changes in life communities over short-term scales, postmortem disturbance of ecological information; and the differential response of growth forms to taphonomic biases. As stressed by Aronson and Ellner (2007), determination of the meaning and importance of recent ecological shifts to the future of reefs is difficult as a consequence of taphonomic alteration. The absence of non-calcifying organisms that are not usually preserved in Holocene deposits, limits our ability to use the fossil record as a proxy. Nevertheless, the Quaternary reef record has to be regarded as providing a reliable baseline for predicting future trajectories of reef development. As emphasized in Chapter 3, the most appropriate use of Quaternary reefs as an environmental proxy is to assess the long-term changes in community structure patterns of modern reefs. The fossil record potentially represents an extensive, high-value, historical database to aid in understanding the ecology, taxonomy and evolution of modern reefs (Pandolfi, 1999). First, a large proportion of corals and associated reef taxa in the Quaternary record can be taxonomically identified with practically the same degree of accuracy as in modern reefs. Second, skeletal calcifying organisms appear in the form of discrete individuals from which relative abundance can be easily measured. Third, reef macrofaunas are generally deposited either in their growth position or slightly displaced from their life habitat, and finally trapped in the reef fabric. Thus, the composition of fossil assemblages is thought to express the species composition patterns of the original communities, although they have experienced time-averaging. Fourth, determining the ages of individuals, especially corals and molluscs, using high-resolution radiometric methods, provides an accurate time control on reef-building episodes.

10.4. Global Warming and the Future of Reefs

Recent coral mortality is viewed as unprecedented in the last millennia, reflecting coral disease, bleaching and the increasing effects of tropical storms. In the present context of global warming and increasing anthropogenic pressure, corals reefs worldwide and particularly those in the Western Atlantic continue to suffer accelerated and dramatic deterioration. As suggested by Aronson (2007), Caribbean reefs seem to be predisposed to experience a major crisis, because the coral taxa present are still in the process of recovering from high-turnover phases to which they had been subject during the Neogene. The rates of change of some disruptions operating at present have no counterparts in the Quaternary or even further back in the past. For instance, during the past 20 ka, from the LGM to the Industrial Revolution (mid-19th century), atmospheric CO_2 levels increased by approximately 40%, from about 200 to 280 ppmv. By comparison, within less than 200 years, since about 1850 AD, the increase in atmospheric pCO_2 has been of similar magnitude, changing from around 280 to 380 ppmv (Buddemeier, Gattuso, & Kleypas, 1998).

In such a world of increasing-CO_2, marine ecosystems dominated by calcifying organisms are subjected globally to increasing temperature, reduced aragonite saturation state and increased acidification (Kuffner, Andersson, Jokiel, Rodgers, & Mackenzie, 2007). More specifically, modern coral reefs are currently threatened, and are following similar trajectories of decline worldwide (Pandolfi et al., 2003). Reef areas are expected to become marginal with respect to aragonite saturation state and calcification potential. The rapid rise in pCO_2 in the atmosphere has resulted in a dramatic decline in calcification rates of skeletal organisms by about 30%. Some areas will probably drift towards the high-temperature borderline of growth. In addition, the effects of global warming in regions that today experience temperature conditions limiting reef growth are unlikely to promote significant latitudinal reef extension (Guinotte, Buddemeier, & Kleypas, 2003). Experimental data on coral tolerance to elevated temperatures and low-saturation states suggest that coral diversity and abundance on reefs may decline significantly from the 2050s if the atmospheric pCO_2 doubles or triples (Kleypas, Buddemeier, et al., 1999; Hoegh-Gulberg, 2005). The combination of high temperature and low-saturation state has probably not operated in the history of the reef phenomenon, during at least the last 5 million years (Guinotte et al., 2003). One of the best documented world-wide high-atmospheric CO_2 periods in the Phanerozoic was at the end of the Permian (251 million years ago) and was characterized by the most dramatic biological crisis (Kidder & Worsley, 2004; Kiehl & Shields, 2005). The disappearance of Paleozoic corals may have been triggered by a severe decrease in carbonate saturation state and a marked rise in temperature. Whilst this extinction event probably required 0.5–1 million years to be completed, the changes in reef community structure currently observed may result in the short term in reef demise (Sheppard, 2003). However, in spite of the deterioration rates of many modern reefs, calcifying organisms and communities may be able to adapt to changing environmental conditions. The reef coral fauna may change rather than disappear entirely, with some taxa offering greater tolerance to warming and coral bleaching than others (Hughes et al., 2003). Several lines of evidence indicate that the tropical belt has expanded over the last few decades, accompanied by a poleward migration of large-scale atmospheric circulation modes (Seidel, Fu, Randel, & Reichler, 2008). In response, corals have locally changed their geographic distribution, as observed today in Western Australia (Greenstein & Pandolfi, 2008) and in Florida (Precht & Aronson, 2004). However, according to Buddemeier, Kleypas, and Aronson (2004, p. 28), 'the pressing question is not, "can corals adapt?" but rather, "how fast and to what extent can they adapt?" Corals host a variety of zooxanthellate (*Symbiodinium*) clades in the form of different symbiotic coral–algal associations with different tolerances to environmental stress. Experiments show that in a given coral species or community, coral–algal combinations can change to the dominance of less temperature-sensitive symbionts (Rowan, 2004). Tests of the tolerance limits of communities to factor variability may generate data that provide a basis for the prediction of patterns of reef development, distribution and preservation over a range of spatial and temporal scales.

10.5. Prospective

From the foregoing, it appears that there are still critical needs for data on the nature and variability of reef responses to specific environmental parameters. Due to the relative scarcity of directly accessible Holocene and Pleistocene reef exposures, sustained efforts must be made to obtain high-quality records from inshore and offshore reef drilling programmes. Coring investigations on high-energy outer-reef margins and fore-reef slopes are required to aid in resolving some of the key issues to be faced in understanding reef anatomy: what have the respective roles of the different depositional processes and growth modes had in reef building? To what extent has upslope regression contributed to reefs maintaining pace with rising sea level? Can deepening-upward sequences be regarded systematically as transitional stages to reef drowning or can reefs develop successively as shallow-water, isolated backstepping units during rising sea level? These fundamental questions have been or will be addressed during IODP expeditions to coral reefs (particularly Expeditions 310-Tahiti Sea

Level and 325-Great Barrier Reef Environmental Changes). Additional objectives include the determination of the biological and geological responses of late Pleistocene reefs to former abrupt climate changes. These may provide a model to improve predictions of reef responses to future global warming (Camoin et al., 2007; Webster et al., 2008). However, to date, geographic coverage of the sites drilled is clearly not sufficient to provide meaningful reconstructions of reef growth history worldwide. Gaps in sample coverage are particularly accute in the Caribbean, the westernmost part of the Pacific, and in the western and eastern Indian Ocean. Access to submerged reef piles by coring will provide new insights into the distribution and development patterns, not only during the LGM and in deglacial times, but also during older glacial episodes. In addition to the examination of emergent reef deposits, land-based deep drilling operations, such as those recently conducted on the Nansha Islands, Southern China Sea (Multidisciplinary Oceanographic Expedition, 1992), the Great Barrier Reef (Alexander et al., 2001), Mururoa Atoll (Camoin et al., 2001), New Caledonia (Cabioch, Montaggioni, Thouveny, et al., 2008), known to be subsiding or relatively stable, obviously help to improve the database, reconstructing reef growth history back to the middle-to-early Pleistocene. However, completing all of these objectives will be facilitated by process-based computer simulations and modelling (for instance, see Webster, Wallace, et al., 2007).

The Quaternary coral reef record can only provide insights into natural climate variability through decades-to-century-long time windows at subseasonal to interannual resolution scales. To date, short-term windows have been opened during particular periods, including the Holocene Climatic Optimum and the last glacial cycle. There are only sparse coral-based data gained from periods prior to the last interglacial–glacial transition. Estimating tropical climate variability during the Quaternary is critical to a better understanding of present-day climate system functioning and to improve climate predictions, because during glacial–interglacial cycles the climate has been partly driven by boundary conditions and forcings different from those of the present day. It is now well established that coral geochemical tracers have the potential to complement other natural archives of climate variability (e.g. ice cores, tree rings, varved sediments, other marine organisms), in providing monthly to subseasonal resolution of change. Unfortunately, coral-based proxy reconstructions suffer important limitations linked to a variety of factors. The influence of vital effects and diagenetic alteration of the chemical coral signatures are still misunderstood. Despite these limitations, the combined use of multiproxy records offers great promise for coral paleoclimatology.

References

Abed, R. M. M., Golubic, S., Garcia-Pichet, F., Camoin, G. F., & Sprachta, S. (2003). Characterization of microbialite-forming cyanobacteria in a tropical lagoon: Tikehau Atoll, French Polynesia. *Journal of Phycology, 39*, 862–873.

Abram, N. J., Gagan, M. K., Liu, Z., Hantoro, W. S., McCulloch, M. T., & Suwargadi, B. W. (2007). Seasonal characteristics of the Indian Ocean dipole during the Holocene epoch. *Nature, 445*, 299–302.

Abram, N. J., Webster, J. M., Davies, P. J., & Dullo, W. C. (2001). Biological response of coral reefs to sea surface temperature variation: Evidence from the raised Holocene reefs of Kikai-jima (Ryukyu Islands, Japan). *Coral Reefs, 20*, 221–234.

Acevedo, R., Morelock, J., & Olivieri, R. A. (1989). Modification of coral reef zonation by terrigenous sediment stress. *PALAIOS, 4*, 92–100.

Adams, C. G., Lee, D. E., & Rosen, B. R. (1990). Conflicting isotopic and biotic evidence for tropical sea surface temperatures during the Tertiary. *Palaeogeography, Palaeoclimatology, Palaeoecology, 77*, 289–313.

Adams, J. E., & Rhodes, M. L. (1960). Dolomitization by seepage refluxion. *American Association of Petroleum Geologists Bulletin, 44*, 1912–1920.

Adey, W. H. (1975). The algal ridges and coral reefs of St Croix. *Atoll Research Bulletin, 187*, 1–67.

Adey, W. H. (1978). Coral reef morphogenesis: A multidimensional model. *Science, 202*, 831–837.

Adey, W. H. (1986). Coralline algae as indicators of sea-level. In: O. van de Plassche (Ed.), *Sea-level research: A manual for the collection and evaluation of data* (pp. 229–280). Norwich, U.K.: Geo Books.

Adey, W. H., & Burke, R. B. (1977). Holocene bioherms of Lesser Antilles: Geologic control of development. Studies in Geology (Vol. 4, pp. 67–81). Tulsa, OK: American Association of Petroleum Geologists.

Adey, W. H., & Macintyre, I. G. (1973). Crustose coralline algae: A re-evaluation in the geological sciences. *Geological Society of America Bulletin, 84*, 883–903.

Adey, W. H., & Vassar, M. J. (1975). Colonization, succession and growth rates of tropical crustose coralline algae (Rhodophyta, Cryptonemiales). *Phycologia, 14*, 55–69.

Adjas, A. (1988). *Sédimentologie comparée de quelques modèles lagonaires actuels des milieux récifaux coralliens du Pacifique* (Nouvelle-Calédonie, Polynésie). Unpublished Ph.D. thesis, University of Aix-Marseille I, 340 pp.

Adjas, A., Masse, J. P., & Montaggioni, L. F. (1990). Fine-grained carbonates in nearly closed reef environments: Mataiva and Takapoto atolls, Central Pacific Ocean. *Sedimentary Geology, 67*, 115–132.

Agegian, C.R. (1981). Growth of the branched coralline alga, *Porolithon gardineri* (Foslie) in the Hawaiian archipelago. *Proceedings of the Fourth International Coral Reef Symposium*, Manila, Vol. 2, pp. 419–423.

Ager, D. V. (1963). *Principles of Palaeoecology*. New York: McGraw-Hill, 371 pp.

Aguirre, J., Riding, R., & Braga, J. C. (2000). Diversity of coralline red algae: Origination and extinction patterns from the early Cretaceous to the Pleistocene. *Paleobiology, 26*, 651–667.

Aharon, P. (1991). Recorders of reef environment histories: Stable isotopes in corals, giant clams, and calcareous algae. *Coral Reefs, 10*, 71–90.

Aharon, P., Rasbury, M. & Murgulet, V. (Eds). (2006). Caves of Niue Island, South Pacific: Speleothems and water geochemistry. In: *Perspectives on karst geomorphology, hydrology, and geochemistry* (pp. 283–295), Geological Society of America (Special Publication 404).

Aharon, P., Socki, R. A., & Chan, L. (1987). Dolomitization of atolls by sea-water convection flow: Test of a hypothesis at Niue, South Pacific. *Journal of Geology*, *95*, 187–203.

Aharon, P., & Veeh, H. H. (1984). Isotope studies of insular phosphates explain atoll phosphatization. *Nature*, *308*, 614–617.

Aïssaoui, D. M., Buiges, D., & Purser, B. H. (1986). Model of reef diagenesis: Mururoa Atoll, French Polynesia. In: J. H. Schroeder, & B. H. Purser (Eds), *Reef diagenesis* (pp. 27–52). Berlin: Springer Verlag.

Aïssaoui, D. M., Buiges, D., & Purser, B. H. (1988). Magnesian calcite cements and their diagenesis: Dissolution and dolomitization, Mururoa Atoll. *Sedimentology*, *35*, 821–841.

Aïssaoui, D. M., & Purser, B. H. (1985). Reef diagenesis: Cementation at Mururoa Atoll. *Proceedings of the Fifth International Coral Reef Congress*, Tahiti, pp. 257–262.

Alderman, A. R., & Skinner, H. C. W. (1957). Dolomite sedimentation in the southeast of Australia. *American Journal of Science*, *255*, 561–567.

Alexander, I., Andres, M. S., Braithwaite, C. J. R., Braga, J. C., Davies, P. J., Elderfield, H., et al. (2001). New constraints on the origin of the Australian Great Barrier Reef: Results from an international project of deep coring. *Geology*, *29*, 483–486.

Alexander, I. A. (1996). Late Quaternary sedimentation off the Queensland continental margin (northeast Australia) in response to sea level fluctuations. Unpublished Ph.D. thesis, University of Edinburgh, 489 p.

Allan, J. R., & Matthews, R. K. (1982). Isotope signatures associated with early meteoric diagenesis. *Sedimentology*, *29*, 797–817.

Allemand, D., Ferrier-Pagès, C., Furla, P., Houlbrèque, F., Puverel, S., Reynaud, S., et al. (2004). Biomineralisation in reef-building corals: From molecular mechanisms to environmental control. *Comptes Rendus de l'Académie des Sciences, Paris, Paleovol*, *3*, 453–467.

Alley, R. B., & Ágústsdóttir, A. M. (2005). The 8 k event: Cause and consequences of a major Holocene abrupt climate change. *Quaternary Science Reviews*, *24*, 1123–1149.

Alley, R. B., Clark, P. U., Huybrechts, P., & Joughin, I. (2005). Ice-sheet and sea-level changes. *Science*, *310*, 456–460.

Allison, N., & Finch, A. A. (2004). High-resolution Sr/Ca records in modern *Porites lobata* corals: Effects of skeletal extension rate and architecture. *Geochemistry, Geophysics, Geosystems*, *5*, Q05001, doi: 10.1029/2004GC000696.

Allison, N., Finch, A. A., Tudhope, A. W., Newville, M., Sutton, S. R., & Ellam, R. M. (2005). Reconstruction of deglacial sea surface temperatures in the tropical Pacific from selective analysis of a fossil coral. *Geophysical Research Letters*, *32*, 1I7609, doi: 10.029/2005GL023183.

Allison, N., Finch, A. A., Webster, J. M., & Clague, D. A. (2007). Palaeoenvironmental records from fossil corals: The effects of submarine diagenesis on temperature and climate estimates. *Geochimica et Cosmochimica Acta*, *741*, 4693–4703.

Allmon, W. D. (2001). Nutrients, temperature, disturbance, and evolution: a model for the Late Cenozoic marine record of the western Atlantic. *Palaeogeography, Palaeoclimatology, Palaeoecology*, *166*, 9–26.

Allmon, W. D., Emslie, S. D., Jones, D. S., & Morgan, G. S. (1996). Late Neogene oceanographic change along Florida's west coast: Evidence and mechanisms. *Journal of Geology*, *104*, 143–162.

Almasi, M. N., Hoskin, C. M., Reed, J. K., & Milo, J. (1987). Effects of natural and artificial *Thalassia* on rates of sedimentation. *Journal of Sedimentary Petrology*, *57*, 901–906.

An, S.-I., Timmermann, A., Bejarano, L., Jin, F.-F., Justino, F., Liu, Z., et al. (2004). Modeling evidence for enhanced El Niño-southern oscillation amplitude during the Last Glacial Maximum. *Paleoceanography, 19*, PA4009, doi: 10.1029/2004PA001020.

Anderegg, D., Dodge, R.E., Swart, P.K., & Fisher, L. (1997). Barium chronologies from South Florida reef corals—Environmental implications. *Proceedings of the Eighth International Coral Reef Symposium*, Panama, Vol. 2, pp. 1725–1730.

Andersen, M. B., Stirling, C. H., Potter, E.-K., Halliday, A. N., Blake, S. G., McCulloch, M. T., et al. (2008). High-precision U-series measurements of more than 500,000 year old fossil corals. *Earth and Planetary Science Letters, 262*, 229–245.

Andrefouët, S., Miller-Karger, F. E., Robinson, J. A., Kranenburg, C. J., Torres-Pulliza, D., Spraggins, S. A., et al. (2006). Global assessment of modern coral reef extent and diversity for regional science and management application: A view from space. *Proceedings of the Tenth International Coral Reef Symposium*, Okinawa, Vol. 1, pp. 1732–1745.

Andrews, J. C., & Gentien, P. (1982). Upwelling as a source of nutrients for the Great Barrier Reef ecosystems: A solution to Darwin's question. *Marine Ecology Progress Series, 8*, 257–269.

Andrews, J. C., & Muller, H. (1983). Space-time variability of nutrients in a lagoonal patch reef. *Limnology and Oceanography, 28*, 215–227.

Andrie, C., Jean-Baptiste, J., Pierre, C., Dejardin, P., Fichez, R., Poupeau, J-J., et al. (1998). Pore water geochemistry and mixing processes within the Tahiti barrier reef. *Geochimica et Cosmochimica Acta, 62*, 2809–2822.

Angulo, R. J., Lessa, G. C., & de Souza, M. Ch. (2006). A critical review of mid- to late-Holocene sea-level fluctuations on the eastern Brazilian coastline. *Quaternary Science Reviews, 25*, 486–506.

Angwenyi, C. M., & Rudberg, L. (2005). Wave-driven circulation across the coral reef at Bamburi Lagoon, Kenya. *Estuarine, Coastal and Shelf Science, 63*, 447–454.

Anthony, K. R. N., & Connolly, S. R. (2004). Environmental limits to growth: Physiological niche boundaries of corals along turbidity light gradients. *Oecologia, 141*, 373–384.

Anthony, K. R. N., & Fabricius, K. E. (2000). Shifting roles of heterotrophy and autotrophy in coral energetics under varying turbidity. *Journal of Experimental Marine Biology and Ecology, 252*, 221–253.

Anthony, K. R. N., Hoogenboom, M. O., & Connolly, S. R. (2005). Adaptative variation in coral geometry and the optimization of internal colony light climates. *Functional Ecology, 19*, 17–26.

Aronson, R.B. (Ed.) (2007). *Geological approaches to coral reef ecology*. Ecological Studies (Vol. 192, 439 pp.). New York: Springer.

Aronson, R. B., & Precht, W. F. (1997). Stasis, biological disturbance, and community structure of a Holocene coral reef. *Paleobiology, 23*, 326–346.

Aronson, R. B., & Precht, W. F. (2001). Evolutionary paleoecology of Caribbean coral reefs. In: W. D. Allmon & D. J. Bottjer (Eds), *Evolutionary paleoecology: The ecological context of macroevolutionary change* (pp. 177–233). New York: Columbia University Press.

Aronson, R. B., & Ellner, S. P. (2007). Biotic turnover events on coral reefs: A probabilistic approach. In: R. B. Aronson (Ed.), *Geological approaches to coral reef ecology* (pp. 61–84). NewYork: Springer.

Aronson, R. B., Macintyre, I. G., Lewis, S. T., & Hilburn, N. L. (2005). Emergent zonation and geographic convergence of coral reefs. *Ecology, 86*, 2586–2600.

Aronson, R. B., Macintyre, I. G., Precht, W. F., Murdoch, T. J. T., & Wapnick, C. M. (2002). The expanding scale of species turnover events on coral reefs in Belize. *Ecological Monographs, 72*, 233–249.

Aronson, R. B., Macintyre, I. G., Wapnick, C. M., & O'Neill, M. W. (2004). Phase shifts, alternative states, and the unprecedented convergence of two reef systems. *Ecology, 85*, 1876–1891.

Aronson, R. B., Precht, W. F., & Macintyre, I. G. (1998). Extrinsic control of species replacement on a Holocene reef in Belize: The role of coral disease. *Coral Reefs, 17*, 223–230.

Asami, R., Yamada, T., Iryu, Y., Meyer, C. P., Quinn, T. M., & Paulay, G. (2004). Carbon and oxygen isotopic composition of a Guam coral and their relationships to environmental variables in the western Pacific. *Palaeogeography, Palaeoclimatology, Palaeoecology, 212*, 1–22.

Atkinson, M. J., Carlson, B., & Crow, G. L. (1995). Coral growth in high-nutrient, low-pH seawater: A case study of corals cultured at the Waikiki Aquarium, Honolulu, Hawaii. *Coral Reefs, 14*, 215–223.

Atkinson, M. J., Smith, S. V., & Stroup, E. D. (1981). Circulation in Enewetok Atoll Lagoon. *Limnology and Oceanography, 26*, 1074–1083.

Ayling, B. F., McCulloch, M. T., Gagan, M. K., Stirling, C. H., Andersen, M. B., & Blake, S. G. (2006). Sr/Ca and $\delta^{18}O$ seasonality in a *Porites* coral from the MIS 9 (339–303 ka) interglacial. *Earth and Planetary Science Letters, 248*, 462–475.

Babin, S. M., Carton, J. A., Dickey, T. D., & Wiggert, J. D. (2004). Satellite evidence of hurricane-induced phytoplankton blooms in an oceanic desert. *Journal of Geophysical Research, Oceans, 109*, CO3043.

Bach, S. D. (1979). Standing crop, growth and production of calcareous Siphonales (Chlorophyta) in a South Florida Lagoon. *Bulletin of Marine Science, 29*, 191–201.

Badiozamani, K., Mackenzie, F. T., & Thorstenson, D. C. (1977). Experimental carbonate cementation: Salinity, temperature and vadose-phreatic effects. *Journal of Sedimentary Petrology, 47*, 529–542.

Bagnato, S., Linsley, B. K., & Howe, S. S. (2005). Coral oxygen isotope records of interdecadal climate variations in the South Pacific Convergence Zone region. *Geochemistry, Geophysics, Geosystems, 6*, Q06001, doi: 10.1029/2004GC000879.

Bak, R. P. M. (1975). Ecological aspects of the distribution of reef corals in the Netherlands Antilles. *Bijdragen tot de Dierkunde, 45*, 181–190.

Bak, R. P. M. (1994). Sea urchin bioerosion on coral reefs: Place in the carbonate budget and relevant variables. *Coral Reefs, 13*, 99–103.

Bak, R. P. M., & Elgershuizen, J. (1976). Patterns of oil-sediment rejection in corals. *Marine Biology, 37*, 105–113.

Bak, R. P. M., & Nieuwland, G. (1995). Long-term change in coral communities along depth gradients over leeward reefs in the Netherlands Antilles. *Bulletin of Marine Science, 56*, 609–619.

Baker, J. C., Jell, J. S., Hacker, J. L. F., & Baublys, K. A. (1998). Origin of Recent Insular Phosphate Rock on a Coral Cay-Raine Island, Northern Great Barrier Reef, Australia. *Journal of Sedimentary Research, 68*, 1001–1008.

Baker, R. G. V., Haworth, R. J., & Flood, P. G. (2001). Inter-tidal fixed indicators of former Holocene sea levels in Australia: A summary of sites and a review of methods and models. *Quaternary International, 83–85*, 257–273.

Ball, M. M., Shinn, E. A., & Stockman, K. W. (1967). The geological effects of Hurricane Donna in south Florida. *Journal of Geology, 75*, 583–597.

Banerjee, P. K. (2000). Holocene and Late Pleistocene relative sea level fluctuations along the east coast of India. *Marine Geology, 167*, 243–260.

Barbosa, S. S., Byrne, M., & Kelaher, B. P. (2008). Bioerosion caused by foraging of the tropical chiton *Acanthopleura gemmata* at One Tree Reef, southern Great Barrier Reef. *Coral Reefs, 27*, 635–639.

Bard, E., Arnold, M., & Duplessy, J-C. (1991). Reconciling the sea level record of the last deglaciation with ^{18}O spectra from deep sea cores. In: J. J. Lowe (Ed.), *Radiocarbon*

dating: Recent applications and future potential (pp. 67–73). Cambridge: Quaternary Research Association.
Bard, E., Fairbanks, R. G., Arnold, M., & Hamelin, B. (1992). ^{230}Th/^{234}U and ^{14}C ages obtained by mass spectrometry on corals from Barbados (West Indies), Isabela (Galapagos) and Mururoa (French Polynesia). In: E. Bard & W. S. Broecker (Eds), *The last deglaciation: Absolute and radiocarbon chronologies* (pp. 103–110). Berlin: Springer-Verlag.
Bard, E., Hamelin, B., Arnold, M., Montaggioni, L. F., Cabioch, G., Faure, G., et al. (1996). Deglacial sea-level record from Tahiti corals and the timing of global meltwater discharge. *Nature, 385*, 241–244.
Bard, E., Hamelin, B., Fairbanks, R. G., & Zindler, A. (1990). Calibration of the ^{14}C timescale over the past 30,000 years using mass spectrometry U-Th ages from Barbados corals. *Nature, 345*, 405–410.
Bard, E., Hamelin, B., Pirazzoli, P. A., Arnold, M., Faure, G., Sumosusastro, P., et al. (1996). Pleistocene sea levels and tectonic uplift based on dating of corals from Sumba Island, Indonesia. *Geophysical Research Letters, 23*, 1473–1476.
Bar-Matthews, M., Wasserberg, G. J., & Chen, J. H. (1993). Diagenesis of fossil coral skeletons: Correlation between trace elements, textures and ^{234}U/^{238}U. *Geochimica et Cosmochimica Acta, 57*, 257–276.
Barnes, J., Bellamy, D. J., Jones, D. J., & Whitton, B. A. (1971). Morphology and ecology of the reef front of Aldabra. In: D. R. Stoddart, & M. Yonge (Eds), *Regional variation in Indian Ocean coral reefs. Symposia of the Zoological Society of London, 28*, 87–114.
Barnes, D. J. (Ed.) (1983). *Perspectives on coral reefs*. The Australian Institute of Marine Science, Contribution No. 200, Brian Clouston, Manuka, Australia, 277 pp.
Barnes, D. J., & Taylor, R. B. (2001). On the nature and causes of luminescent lines and bands in coral skeletons. *Coral Reefs, 19*, 221–230.
Barnes, D. J., Taylor, R. B., & Lough, J. M. (2003). Measurement of luminescence in coral skeletons. *Journal of Experimental Marine Biology and Ecology, 295*, 91–106.
Bar-Or, R., Erlick, C., & Gildor, H. (2008). The role of dust in glacial–interglacial cycles. *Quaternary Science Reviews, 27*, 201–208.
Bassett, S. E., Milne, G. A., Mitrovica, J. X., & Clark, P. U. (2005). Ice sheet and solid earth influences on far-field sea-level histories. *Science, 309*, 925–928.
Bathurst, R. G. C. (1958). Diagenetic fabrics of some British Dinantian limestones. *Liverpool and Manchester Geological Journal, 2*, 11–36.
Bathurst, R. G. C. (1964). The replacement of aragonite by calcite in the molluscan shell wall. In: J. Imbrie & N. D. Newell (Eds), *Approaches to Paleoecology* New York: Wiley.
Bathurst, R. G. C. (1966). Boring algae, micrite envelopes and lithification of molluscan biosparites. *Geological Journal, 5*, 15–32.
Bathurst, R. G. C. (1975). *Carbonate sediments and their diagenesis*. Developments in Sedimentology (Vol. 12, 658 pp.). Amsterdam: Elsevier.
Bathurst, R. G. C. (1983). Neomorphic spar versus cement in some Jurassic grainstones: Significance for evaluation of porosity evolution and compaction. *Journal of the Geological Society of London, 140*, 229–237.
Bayer, F. M., & Macintyre, I. G. (2001). The mineral component of the axis and holdfast of some gorgonian octocorals (Coelenterata: Anthozoa), with special reference to the family Gorgoniidae. *Proceedings of the Biological Society of Washington*, Vol. 114, pp. 309–345.
Beach, D. K. (1995). Controls and effects of subaerial exposure on cementation and development of secondary porosity in the subsurface of Great Bahama Bank. In: D. A. Budd, A. H. Saller, & P. M. Harris (Eds), *Unconformities and porosity in carbonate strata. American Association of Petroleum Geologists, Memoir, 63*, 1–33.
Beaman, R. J., Webster, J. M., & Wust, R. A. J. (2008). New evidence for drowned shelf edge reefs in the Great Barrier Reef, Australia. *Marine Geology, 247*, 17–34.

Beck, J. W., Edwards, R. L., Ito, E., Taylor, F. W., Récy, J., Rougerie, F., et al. (1992). Sea-surface temperature from coral skeletal strontium/calcium ratios. *Science, 257*, 644–647.

Beck, J. W., Récy, J., Taylor, F. W., Edwards, R. L., & Cabioch, G. (1997). Holocene sea surface temperature variations in the SW Pacific Ocean derived from coral Sr/Ca and oxygen isotope thermometry. *Nature, 385*, 705–707.

Beddows, P. A., Smart, P. L., Whitaker, F. F., & Smith, S. L. (2004). Heliothermic heating of density stratified Groundwater: An alternative interpretation to geothermal circulation. *Geological Society of America, Abstracts with Programs, 36*(5), 310, Denver Meeting.

Bellwood, D. R., Hughes, T. P., Connolly, S. R., & Tanner, J. (2005). Environmental and geometric constraints on Indo-Pacific coral reef biodiversity. *Ecology Letters, 8*, 643–651.

Bentor, Y. K. (Ed.). (1980). *Marine phosphorites: Geochemistry, occurrence genesis* (249 pp.), Society of Economic Paleontologists and Mineralogists (Special Publication 29).

Benway, H. M., & Mix, A. C. (2004). Oxygen isotopes, upper-ocean salinity, and precipitation sources in the eastern tropical Pacific. *Earth Planetary Science Letters, 224*, 493–507.

Benzie, J. A. H. (1999). Genetic structure of coral reef organisms: Ghosts of dispersal past. *American Zoologist, 39*, 131–145.

Berdin, R. D., Siringan, F. P., & Maeda, Y. (2004). Holocene sea-level highstand and its implications for the vertical stability of Panglao Island, southwest Bohol, Philippines. *Quaternary International, 115–116*, 27–37.

Berger, A. (1978). Long-term variations of calorific insolation resulting from the earth's orbital elements. *Quaternary Research, 9*, 139–167.

Berger, A. (1979). Long-term variations of daily insolation and Quaternary climatic changes. *Journal of Atmospheric Science, 35*, 2362–2367.

Berkeley, A., Perry, C. T., Smithers, S. G., Horton, B. P., & Taylor, K. G. (2007). A review of the ecological and taphonomic controls on foraminiferal assemblage development in intertidal environments. *Earth-Science Reviews, 83*, 205–230.

Berkelmans, R., & Willis, B. L. (1999). Seasonal and local spatial patterns in the upper thermal limits of corals on the inshore Central Great Barrier Reef. *Coral Reefs, 18*, 219–228.

Berner, R. A. (Ed.) (1971). *Principles of chemical sedimentology*. New York: McGraw Hill. 240 pp.

Bessat, F., & Buigues, D. (2001). Two centuries of variation in coral growth in a massive *Porites* colony from Moorea (French Polynesia): A response of ocean-atmosphere variability from south central Pacific. *Palaeogeography, Palaeoclimatology, Palaeoecology, 175*, 381–392.

Bicchi, E., Debenay, J. P., & Pagès, J. (2002). Relationship between benthic foraminiferal assemblages and environmental factors in atoll lagoons of the central Tuamotu Archipelago (French Polynesia). *Coral Reefs, 21*, 275–290.

Bice, D. (1988). Synthetic stratigraphy of carbonate platform and basin systems. *Geology, 16*, 703–706.

Bird, M. I., Fifield, L. K., Teh, T. S., Chang, C. H., Shirlaw, N., & Lambeck, K. (2007). An inflection in the rate of early mid-Holocene eustatic sea-level rise: A new sea-level curve from Singapore. *Estuarine and Coastal Shelf Science, 71*, 523–536.

Birkeland, Ch. (Ed.) (1997). *Life and death of coral reefs*. New York: Chapman & Hall. 536 pp.

Bishop, D., & Greenstein, B. J. (2001). The effects of Hurricane Floyd on the fidelity of coral life and death assemblages in San Salvador, Bahamas: Does a hurricane leave a signature in the fossil record? *Geological Society of America, Abstracts with Programs, 33*, A51.

Blanchon, P., & Eisenhauer, A. (2001). Multi-stage reef development on Barbados during the last interglaciation. *Quaternary Science Reviews, 20*, 1093–1112.

Blanchon, P., & Jones, B. (1997). Hurricane control on shelf edge-reef architecture around Grand Cayman. *Sedimentology, 44*, 479–506.

Blanchon, P., Jones, B., & Ford, D. C. (2002). Discovery of a submerged relic reef and shoreline off Grand Cayman: Further support for an early Holocene jump in sea level. *Sedimentary Geology, 147*, 253–270.

Blanchon, P., Jones, B., & Kalbfleisch, W. (1997). Anatomy of a fringing reef around Grand Cayman: Storm rubble, not coral framework. *Journal of Sedimentary Research, 67*, 1–16.

Blanchon, P., & Perry, C. T. (2004). Taphonomic differentiation of *Acropora palmata* facies in cores from Campeche Bank reefs, Gulf of Mexico. *Sedimentology, 52*, 53–76.

Blanchon, P., & Shaw, J. (1995a). Reef drowning during the last deglaciation: Evidence for catastrophic sea-level rise and ice-sheet collapse. *Geology, 23*, 4–8.

Blanchon, P., & Shaw, J. (1995b). Reef drowning during the last deglaciation: Evidence for catastrophic sea-level rise and ice-sheet collapse: Comment and Reply. *Geology, 23*(10), 958–959.

Bloom, A. L., Broecker, W. S., Chappell, J. M. A., Matthews, R. K., & Mesolella, K. J. (1974). Quaternary sea level fluctuations on a Tectonic coast: New $^{230}Th/^{234}U$ dates from the Huon Peninsula, New Guinea. *Quaternary Research, 4*, 185–205.

Boekschoten, G. J., & Best, M. B. (1988). Fossil and recent shallow-water corals from the Atlantic islands off Western Africa. *Zoologische Mededelingen, 62*, 99–112.

Böhm, F., Gussone, K., Einsenhauer, A., Dullo, W.-Ch., Reynaud, S., & Paytan, A. (2006). Calcium isotope fractionation in modern scleractinaian corals. *Geochimica et Cosmochimica Acta, 70*, 4452–4462.

Bond, G., Kromer, B., Beer, J., Muscheler, R., Evans, M. N., Showers, W., et al. (2001). Persistent solar influence on north Atlantic climate during the Holocene. *Science, 294*, 2130–2136.

Bond, G., Showers, W., Cheseby, M., Lotti, R., Almasi, P., deMenocal, P., et al. (1997). A pervasive millennial-scale cycle in North Atlantic Holocene and glacial climates. *Science, 278*, 1257–1266.

Bosellini, A., & Ginsburg, R. N. (1971). Form and internal structure of recent algal nodules (Rhodolites) from Bermuda. *Journal of Geology, 79*, 669–682.

Bosellini, F. R. (2006). Biotic changes and their control on Oligocene–Miocene reefs: A case study from the Apulia Platform margin (Southern Italy). *Palaeogeography, Palaeoclimatology, Palaeoecology, 241*, 393–409.

Bosellini, F. R., & Russo, A. (1988). The Oligocene *Actinacis* coral community of the southern Alps (Italy): Temperature vs. terrigenous control. *Proceedings of the Sixth International Coral Reef Symposium*, Australia, Vol. 3, pp. 385–391.

Bosence, D. W. J. (1983a). The occurrence and ecology of recent rhodoliths: A review. In: T. M. Peryt (Ed.), *Coated grains* (pp. 225–242). Berlin: Springer.

Bosence, D. W. J. (1983b). Description and classification of rhodoliths (rhodoids, rhodolites). In: T. M. Peryt (Ed.), *Coated grains* (pp. 214–224). Berlin: Springer.

Bosence, D. W. J. (1989). Biogenic carbonate production in Florida Bay. *Bulletin of Marine Science, 44*, 419–433.

Boss, S. K., & Liddell, W. D. (1987a). Patterns of sediment composition of Jamaican fringing reef facies. *Sedimentology, 34*, 77–87.

Boss, S. K., & Liddell, W. D. (1987b). Back-reef and fore-reef analogs in the Pleistocene of North Jamaica: Implications for facies recognition and sediment flux in fossil reefs. *PALAIOS, 2*, 219–228.

Bosscher, H., & Schlager, W. (1992). Computer simulation of reef growth. *Sedimentology, 3*, 503–512.

Bouchon, C. (1980). Quantitative study of the scleractinian coral communities of the Jordanian coast (Gulf of Aqaba, Red Sea): Preliminary results. *Tethys, 9*, 243–246.

Bouchon, C. (1985). Quantitative study of scleractinian coral communities of Tiahura reef (Moorea Island, French Polynesia). *Proceedings of the Fifth International Coral Reef Congress*, Tahiti, Vol. 6, pp. 279–284.

Bourrouilh-Le Jan, F. G., Carsin, J.-L., Niaussat, P.-M., & Thommeret, Y. (1985). Sédimentation phosphatée actuelle dans le lagon confiné de l'île de Clipperton (Océan Pacifique). Datations, sédimentologie et géochimie, Sciences Géologiques, Strasbourg, Vol. 77, pp. 109–124.

Bourrouilh-Le Jan, F., & Hottinger, L. C. (1988). Occurrence of rhodoliths in the tropical Pacific: A consequence of mid-Miocene paleo-oceanographic change. *Sedimentary Geology*, *60*, 355–367.

Bourrouilh-Le Jan, F. G., & Talandier, J. (1985). Sédimentation et fracturation de haute énergie en milieu récifalé-tsunamis, ouragans et cyclones et leur effets sur le sédimentologie et la géomorphologie d'un atoll: Motu et hoa à Rangiroa, Tuamotu, Pacifique SE. *Marine Geology*, *67*, 263–333.

Brachert, T. C. (1999). Non-skeletal carbonate production and stromatolite growth within a Pleistocene deep ocean (Last Glacial Maximum, Red Sea). *Facies*, *40*, 211–228.

Brachert, T. C., & Dullo, W. C. (1991). Laminar micrite crusts and associated foreslope processes, Red Sea. *Journal of Sedimentary Petrology*, *61*, 354–363.

Brachert, T. C., Reuter, M., Felis, T., Kroeger, K. F., Lohmann, G., Micheels, A., & Fassoulas, C. (2006). *Porites* corals from Crete (Greece) open a window into Late Miocene (10 Ma) seasonal and interannual climate variability. *Earth and Planetary Science Letters*, *245*, 81–94.

Bradley, R. S., Hughes, M. K., & Diaz, H. F. (2003). Climate in medieval time. *Science*, *302*, 404–405.

Bradshaw, C. (1997). Bioturbation of reefal sediments by crustaceans in Phuket, Thailand. *Proceedings of the Eighth International Coral Reef Symposium*, Panama, Vol. 2, pp. 1801–1806.

Braga, J. C., & Aguirre, J. (2004). Coralline algae indicate Pleistocene evolution from deep, open platform to outer barrier reef environments in the northern Great Barrier Reef margin. *Coral Reefs*, *23*, 547–558.

Braga, J. C., & Bassi, D. (2007). Neogene history of *Sporolithon* Heydrich (Corallinales, Rhodophyta) in the Mediterranean region. *Palaeogeography, Palaeoclimatology, Palaeoecology*, *243*, 189–203.

Braga, J. C., Bosence, D. W. J., & Steneck, R. S. (1993). New anatomical characters in fossil coralline algae and their taxonomic implications. *Paleontology*, *36*, 535–547.

Braithwaite, C. J. R. (1968). Diagenesis of phosphatic carbonate rocks on Remire, Amirantes, Indian Ocean. *Journal of Sedimentary Petrology*, *38*, 1194–1212.

Braithwaite, C. J. R. (1971). Seychelles reefs: Structure and development. In: D. R. Stoddart, & M. Yonge (Eds), *Regional variation in Indian Ocean coral reefs. Symposia of the Zoological Society of London* (Vol. 28, pp. 39–64).

Braithwaite, C. J. R. (1973). Settling behaviour related to sieve analysis of skeletal sands. *Sedimentology*, *20*, 251–262.

Braithwaite, C. J. R. (1975). Petrology of palaeosols and other terrestrial sediments on Aldabra, Western Indian Ocean. *Philosophical Transactions of the Royal Society of London, Series B*, *273*, 1–32.

Braithwaite, C. J. R. (1980). The petrology of oolitic phosphorites from Esprit (Aldabra), Western Indian Ocean. *Philosophical Transactions of Royal Society of London B*, *288*, 511–540.

Braithwaite, C. J. R. (1982). Patterns of accretion of reefs in the Sudanese Red Sea. *Marine Geology*, *46*, 297–325.

Braithwaite, C. J. R. (1983). Calcrete and other soils in quaternary limestones: Structures, processes and applications. *Journal of the Geological Society*, *140*, 351–363.

Braithwaite, C. J. R. (1984). Depositional history of the late Pleistocene limestones of the Kenya coast. *Journal of the Geological Society*, *141*, 685–699.

Braithwaite, C. J. R. (1991). Dolomites, a review of origins, geometry and textures. *Transactions of the Royal Society of Edinburgh: Earth Sciences*, *82*, 99–112.

Braithwaite, C. J. R. (1994). Quaternary oolites in the Indian Ocean. *Atoll Research Bulletin*, *420*, 1–9. Smithsonian Institution, Washington.

Braithwaite, C. J. R. (2005). *Carbonate Sediments and Rocks: A manual for Earth Scientists and Engineers*. Whittles Publishing, 164 pp.

Braithwaite, C. J. R., Dalmasso, H., Gilmour, M. A., Harkness, D. D., Henderson, G. M., Kay, R. L. F., et al. (2004). The Great Barrier Reef: The chronological record from a new borehole. *Journal of Sedimentary Research*, *74*, 298–310.

Braithwaite, C. J. R., & Montaggioni, L. F. (2009). The Great Barrier Reef: A 700,000 year Diagenetic history. *Sedimentology*, in press.

Braithwaite, C. J. R., Montaggioni, L. F., Camoin, G. F., Dalmasso, H., Dullo, W-C., & Mangini, A. (2000). The origins and development of Holocene reefs: A revisited model based on reef boreholes in the Seychelles, Indian Ocean. *International Journal for Earth Sciences*, *89*, 431–445.

Braithwaite, C. J. R., Taylor, J. D., & Kennedy, W. J. (1973). The evolution of an atoll, the depositional and erosional history of Aldabra. *Philosophical Transactions of the Royal Society, London, Series B*, *266*, 307–340.

Brander, R. W., Kench, P. S., & Hart, D. (2004). Spatial and temporal variations in wave characteristics across a reef platform, Warraber Island, Torres Strait, Australia. *Marine Geology*, *207*, 169–184.

Brandon, D. E. (1973). Waters of the Great Barrier Reef Province. In: O. A. Jones, & R. Endean (Eds), *Biology and Geology of Coral Reefs* (Vol.1, Geology 1, pp. 187–232) London: Academic Press.

Bricker, O. P. (Ed.). (1971). *Carbonate cements*. Studies in Geology. (Vol. 19, 376 pp.). Baltimore, MD: Johns Hopkins University Press.

Brocker, W. S., Clark, E., McCorkle, D., Peng, T. H., Hadjas, L., & Bonani, G. (1999). Evidence of a reduction in the carbonate ion content of the deep sea during the course of the Holocene. *Paleoceanography*, *14*, 744–752.

Broecker, W. S. (1963). Radioisotopes and large-scale oceanic mixing. In: M. N. Hill (Ed.), *The sea* (pp. 88–108). New York: Interscience.

Broecker, W. S. (1974). *Chemical oceanography*. New York: Harcourt-Brace Jovanovich, 214 pp.

Broecker, W. S. (1994). Massive iceberg discharges as triggers for global climate change. *Nature*, *372*, 421–424.

Broecker, W. S. (2000). Was a change in thermohaline circulation responsible for the Little Ice Age? *Proceedings of the National Academy of Sciences of the United States of America*, *97*, 1339–1342.

Broecker, W. S. (2002). Massive iceberg discharges as triggers for global climate change. *Nature*, *372*, 421–424.

Broecker, W. S. (2006). Was the Younger Dryas triggered by a flood? *Science*, *312*, 1146–1148.

Broecker, W. S., Thurber, D. L., Goddard, J., Ku, T. L., Matthews, R. K., & Mesolella, K. J. (1968). Milankovitch hypothesis supported by precise dating of coral reefs and deep-sea sediments. *Science*, *159*, 297–300.

Bromley, R. G. (1990). *Trace fossils: Biology and taphonomy*. London: Unwin Hyman, 280 pp.

Brooke, B. (2001). The distribution of carbonate eolianite. *Earth-Science Reviews*, *55*, 135–164.

Brooks, G. R., & Holmes, C. W. (1989). Recent carbonate slope sediments and sedimentary processes bordering a non-rimmed platform: Southwest Florida Continental Margin. In: P. D. Crevello, J. L. Wilson, J. L. Sarg & J. F. Read (Eds), *Controls on platform and basin development* (pp. 249–272), Society of Economic Paleontologists and Mineralogists (Special Publication 44).

Brown, B. E. (1997a). Coral bleaching: Causes and consequences. *Coral Reefs, 16* (Suppl.), S129–S138.

Brown, B. E. (1997b). Adaptations of reef corals to physical environmental stress. *Advances in Marine Biology, 31,* 221–299.

Buckley, H. E. (1951). *Crystal growth.* New York: Wiley, 571 pp.

Buczynski, C., & Chafetz, H. S. (1991). Habit of bacterially induced precipitates of calcium carbonate and influence of medium viscosity on mineralogy. *Journal of Sedimentary Petrology, 61,* 226–233.

Budd, A. F. (1989). Biogeography of Neogene Caribbean reef corals and its implications for the ancestry of Eastern Pacific reef corals. *Memoir of Association of Australasian Palaeontologists, 8,* 219–230.

Budd, A. F. (2000). Diversity and extinction in the Cenozoic history of Caribbean reefs. *Coral Reefs, 19,* 25–35.

Budd, A. F. (2006). The origin and divergence of modern Caribbean coral genera., *Geological Society of America, Abstracts with Programs, 38,* p. 534, Philadelphia Annual Meeting.

Budd, A. F., & Johnson, K. G. (1997). Coral reef community dynamics over 8 million years of evolutionary time: Stasis and turnover. *Proceedings of the Eighth International Coral Reef Symposium,* Panama, Vol. 1, pp. 423–428.

Budd, A. F., & Johnson, K. G. (1999). Origination preceding extinction during late Cenozoic turnover of Caribbean reefs. *Paleobiology, 25,* 188–200.

Budd, A. F., Johnson, K. G., & Stemann, T. A. (1996). Pleistocene turnover and extinctions in the Caribbean reef-coral fauna. In: J. B. C. Jackson, A. F. Budd & A. G. Coates (Eds), *Evolution and environment in tropical America* (pp. 168–204). Chicago: University of Chicago Press.

Budd, A. F., & Klaus, J. S. (2001). The origin and early evolution of the *Montastraea "annularis"* species complex (Anthozoa: Scleractinia). *Journal of Paleontology, 75,* 527–545.

Budd, A. F., Stemann, T. A., & Johnson, K. J. (1994). Stratigraphic distributions of genera and species of Neogene to Recent Caribbean reef corals. *Journal of Paleontology, 68,* 951–977.

Budd, A. F., Stemann, T. A., & Stewart, R. H. (1992). Eocene Caribbean reef corals: A unique fauna from the Gatuncillo Formation of Panama. *Journal of Paleontology, 66,* 570–594.

Budd, D. A. (1988). Petrographic products of freshwater diagenesis in Holocene ooid sands, Schooner Cays, Bahamas. *Carbonates and Evaporites, 3,* 143–163.

Budd, D. A. (1997). Cenozoic dolomites of carbonate islands: Their attributes and origin. *Earth-Science Reviews, 42,* 1–47.

Budd, D. A., & Land, L. S. (1990). Geochemical imprint of meteoric diagenesis in Holocene ooid sands, Schooner Cays, Bahamas; correlation of calcite cement geochemistry with extant groundwaters. *Journal of Sedimentary Research, 60,* 361–378.

Budd, D. A., & Vacher, H. L. (1991). Predicting the thickness of fresh-water lenses in carbonate paleo-islands. *Journal of Sedimentary Petrology, 61,* 43–53.

Buddemeier, R. W. (1997). Making light work of adaptation. *Nature, 388,* 229–230.

Buddemeier, R. W., Gattuso, J. P., & Kleypas, J. (1998). Rising CO_2 and marine calcification. *Land–Ocean Interactions in the Coastal Zone, Newsletter, 8,* 1–3.

Buddemeier, R. W., & Hopley, D. (1988). Turn-ons and turn-offs: Causes and mechanisms of the initiation and termination of coral reef growth. *Proceedings of the Sixth International Coral Reef Symposium,* Townsville, Australia, Vol. 1, pp. 253–261.

Buddemeier, R. W., Kleypas, J. A., & Aronson, R. B. (2004). Coral reefs and global climate change. Potential Contributions of Climate Change to Stresses on Coral Reef Ecosystems. Pew Center on Global Climate Change, Report, February, pp. 1–42.

Buddemeier, R. W., Maragos, J. E., & Knutson, D. W. (1974). Radiographic studies of reef coral exoskeletons: Rates and patterns of coral growth. *Journal of Experimental Marine Biology and Ecology*, *14*, 179–200.

Buddemeier, R. W., & Oberdorfer, J. A. (1986). Internal hydrology and geochemistry of coral reefs and atoll islands: Key to diagenetic variations. In: J. H. Schroeder & B. H. Purser (Eds), *Reef diagenesis* (pp. 91–111). New York: Springer-Verlag.

Buddemeier, R. W., & Oberdorfer, J. A. (1997). Hydrogeology of Enewetak Atoll. In: H. L. Vacher, & T. M. Quinn (Eds), *Geology and hydrogeology of carbonate islands*. Developments in Sedimentology (Vol. 54, pp. 667–692). Amsterdam: Elsevier.

Buigues, D. (1996). Mururoa and Fangataufa: Sea-level changes, karstification and the atoll morphology. *Réunion spécialisée de la Société Géologique de France, Paris*, *169*, 373–382.

Buigues, D. (1997). Geology and hydrogeology of Mururoa and Fangataufa, French Polynesia. In: H. L. Vacher, & T. M. Quinn (Eds), *Geology and hydrogeology of carbonate islands*. Developments in Sedimentology (Vol. 54, pp. 133–151). Amsterdam: Elsevier Science.

Burne, R. V., & Moore, L. S. (1987). Microbialites: Organosedimentary deposits of benthic microbial communities. *PALAIOS*, *2*, 241–254.

Burns, S. J., & Rossinsky, V., Jr. (1989). Late Pleistocene mixing zone dolomitization, southeastern Barbados, West Indies (Discussion). *Sedimentology*, *36*(6), 1135–1142.

Burrage, D. M., Steinberg, C. R., Skirving, W. J., & Kleypas, J. A. (1996). Mesoscale circulation features of the Great Barrier Reef region inferred from NOAA satellite imagery. *Remote Sensing of Environment*, *56*, 21–41.

Cabioch, G. (1988). *Récifs frangeants de Nouvelle-Calédonie (Pacifique Sud-Ouest)*. Aix-en-Provence: Publications de l'Université de Provence, 291 pp.

Cabioch, G. (2003). Postglacial reef development in the South-West Pacific: Case studies from New-Caledonia and Vanuatu. *Sedimentary Geology*, *159*, 43–59.

Cabioch, G., & Ayliffe, L. K. (2001). Raised coral terraces at Malakula, Vanuatu, Southwest Pacific, indicate high sea level during marine isotope stage 3. *Quaternary Research*, *56*, 357–365.

Cabioch, G., Banks-Cutler, K., Beck, D. J., Burr, G. S., Corrège, Th., Edwards, R. L., et al. (2003). Continuous reef growth during the last 23 cal kyr BP in a tectonically active zone (Vanuatu, southwest Pacific). *Quaternary Science Reviews*, *22*, 1771–1786.

Cabioch, G., Camoin, G. F., & Montaggioni, L. F. (1999). Post-glacial growth history of a French Polynesian reef tract, Tahiti, Central Pacific. *Sedimentology*, *46*, 985–1000.

Cabioch, G., Camoin, G. F., Webb, G. E., Le Cornec, F., Garcia Molina, M., Pierre, C., et al. (2006). Contribution of microbialites to the development of coral reefs during the last deglacial period: Case study from Vanuatu (South-West Pacific). *Sedimentary Geology*, *185*, 297–318.

Cabioch, G., Corrège, Th., Turpin, L., Castellaro, Ch., & Récy, J. (1999). Development patterns of fringing and barrier reefs in New Caledonia (south-west Pacific). *Oceanologica Acta*, *22*, 567–578.

Cabioch, G., Montaggioni, L. F., & Faure, G. (1995). Holocene initiation and development of New Caledonian fringing reefs, SW Pacific. *Coral Reefs*, *14*, 131–140.

Cabioch, G., Montaggioni, L. F., Faure, G., & Ribaud-Laurenti, A. (1999). Reef coralgal assemblages as recorders of paleobathymetry and sea level changes in the Indo-Pacific province. *Quaternary Science Reviews*, *18*, 1681–1695.

Cabioch, G., Montaggioni, L. F., Frank, N., Seard, C., Sallé, E., Payri, C., et al. (2008). Successive reef depositional events along the Marquesas foreslopes (French Polynesia) since 26 ka. *Marine Geology*, *254*, 18–34.

Cabioch, G., Montaggioni, L. F., Thouveny, N., Frank, N., Sato, T., Chazottes, V., et al. (2008). The chronology and structure of the western New Caledonian barrier reef tracts. *Palaeogeography, Palaeoclimatology, Palaeoecology, 268*, 91–105.

Cabioch, G., Taylor, F. W., Corrège, Th., Récy, J., Edwards, L. R., Burr, G. S., et al. (1999). Occurrence and significance of microbialites in the uplifted Tasmaloun reef (SW Espiritu Santo, SW Pacific). *Sedimentary Geology, 126*, 305–316.

Cabioch, G., Taylor, F. W., Récy, J., Edwards, L. P., Gray, S. C., Faure, G., et al. (1998). *Environmental and tectonic influence on growth and internal structure of a fringing reef at Tasmaloun (SW Espiritu Santo, New Hebrides island arc, SW Pacific)* (pp. 261–277), International Association of Sedimentologists (Special Publication 25).

Calvo, E., Marshall, J. F., Pelejero, C., McCulloch, M. T., Gagan, M. K., & Lough, J. M. (2007). Interdecadal climate variability in the Coral Sea since 1708 A.D. *Palaeogeography, Palaeoclimatology, Palaeoecology, 248*, 190–201.

Camoin, G. F., Cabioch, G., Eisenhauer, A., Braga, J. C., Hamelin, B., & Lericolais, G. (2006). Environmental significance of microbialites in reef environments during the Last Deglaciation. *Sedimentary Geology, 185*, 277–295.

Camoin, G. F., Colonna, M., Montaggioni, L. F., Casanova, J., Faure, G., & Thomassin, B. A. (1997). Holocene sea level changes and reef development in southwestern Indian Ocean. *Coral Reefs, 16*, 247–259.

Camoin, G. F., Ebren, P., Eisenhauer, A., Bard, E., & Faure, G. (2001). A 300,000 year coral reef record of sea level changes, Mururoa atoll (Tuamotu archipelago, French Polynesia). *Palaeogeography, Palaeoclimatology, Palaeoecology, 175*, 325–341.

Camoin, G. F., Gautret, P., Montaggioni, L. F., & Cabioch, G. (1999). Nature and environmental significance of microbialites in Quaternary reefs: The Tahiti paradox. *Sedimentary Geology, 126*, 271–304.

Camoin, G. F., Iryu, Y., McInroy, D. B., & the IODP Expedition 310 Scientists (2007). *IODP Proceedings of the Integrated Ocean Drilling Program*, Volume 310 Expedition Reports, Tahiti Sea Level: Washington, DC, Integrated Ocean Drilling Program Management International, Inc., doi:10.2204/iodp.proc.310.2007.

Camoin, G. F., & Montaggioni, L. F. (1994). High-energy coralgal stromatolite frameworks from Holocene reefs, Tahiti, French Polynesia. *Sedimentology, 41*, 655–676.

Camoin, G. F., Montaggioni, L. F., & Braithwaite, C. J. R. (2004). Late glacial to post glacial sea levels in the western Indian Ocean. *Marine Geology, 206*, 119–146.

Campbell, J. F. (1986). Subsidence rates of the southeastern Hawaiian islands determined from submerged terraces. *Geo-Marine Letters, 6*, 139–146.

Cane, M. A. (2005). The evolution of El Niño, past and future. *Earth and Planetary Science Letters, 230*, 227–240.

Cant, R. V., & Weech, P. S. (1986). A review of the factors affecting the development of Ghyben–Hertzberg lenses in the Bahamas. *Journal of Hydrology, 84*, 333–343.

Carannante, G., Esteban, M., Milliman, J. D., & Simone, L. (1988). Carbonate lithofacies as paleolatitude indicators: Problems and limitations. *Sedimentary Geology, 60*, 333–346.

Carew, J. L., Mylroie, J. E., 1994. Discussion of: Hearty, P. J., Kindler, P. (1993). New perspectives of Bahamian geology: San Salvador Island, Bahamas. *Journal of Coastal Research, 9*, 577–694; *Journal of Coastal Research, 10*, 1087–1094.

Carpenter, R. C. (1997). Invertebrate predators and grazers. In: Ch. Birkeland (Ed.), *Life and death of coral reefs* (pp. 198–248). New York: Chapman & Hall.

Carreiro-Silva, M., & McClanahan, T. R. (2001). Echinoid bioerosion and herbivory on Kenyan coral reefs: The role of protection from fishing. *Journal of Experimental Marine Biology and Ecology, 262*, 133–153.

Carriquiry, J. D., Risk, M. J., & Schwarcz, H. P. (1994). Stable isotope geochemistry of corals from Costa Rica as proxy indicator of El Niño/Southern Oscillation (ENSO). *Geochimica et Cosmochimica Acta, 58*, 335–351.

Carter, B. L., Simms, M. A., Moore, C. H., Roberts, H. H., & Lugo-Fernandez, A. (1989). Wave-forced hydrogeology and diagnetic responses in Tague reef, St. Croix, U.S. Virgin islands. *Proceedings of the Sixth International Coral Reef Symposium*, Townsville, Australia, Vol. 1, pp. 111–116.

Chadwick, O. A., Derry, L. A., Vitousek, P. M., Huebert, B. J., & Hedin, L. O. (1999). Changing sources of nutrients during four million years of ecosystem development. *Nature, 397*, 491–497.

Chafetz, H. S. (1986). Marine peloids: A product of bacterially induced precipitation of calcite. *Journal of Sedimentary Petrology, 56*, 812–817.

Chamberlain, J. A., & Graus, R. R. (1975). Water flow and hydromechanical adaptations of branched reef corals. *Bulletin of Marine Science, 25*, 112–115.

Chappell, J. (1974). Geology of coral terraces, Huon Peninsula, New Guinea: A study of Quaternary tectonic movements and sea-level changes. *Geological Society of America Bulletin, 85*, 553–570.

Chappell, J. (1983). Evidence for smoothly falling sea level relative to north Queensland, Australia, during the past 6,000 years. *Nature, 302*, 406–408.

Chappell, J. (1986). A revised sea-level record for the last 300,000 years from Papua New Guinea. *Search, 14*, 99–101.

Chappell, J. (2002). Sea level changes forced ice breakouts in the Last glacial cycle: New results from coral terraces. *Quaternary Science Reviews, 21*, 1229–1240.

Chappell, J., Omura, A., Esat, T., McCulloch, M., Pandolfi, J., Ota, Y., et al. (1996). Reconsiliation of late Quaternary sea levels derived from coral terraces at Huon Peninsula with deep sea oxygen isotope records. *Earth and Planetary Science Letters, 141*, 227–236.

Chappell, J., Ota, Y., & Berryman, K. (1993). Late Quaternary coseismic uplift history of Huon Peninsula. *Quaternary Science Reviews, 15*, 7–22.

Chappell, J., & Polach, H. (1991). Post-glacial sea level rise from a coral record at Huon Peninsula, Papua New Guinea. *Nature, 349*, 147–149.

Chappell, J., & Shackleton, N. J. (1986). Oxygen isotopes and sea level. *Nature, 324*, 137–140.

Chappell, J., & Veeh, H. H. (1978). Late Quaternary tectonic movements and sea-level changes at Timor and Atauro Island. *Geological Society of America Bulletin, 89*, 356–368.

Chardon, D., Austin, J. A., Jr., Cabioch, G., Pelletier, B., Saustrup, S., & Sage, F. (2008). Neogene history of the northeastern New Caledonia continental margin from multichannel reflection seismic profiles. *Comptes Rendus Geosciences, 340*, 68–73.

Charles, C. D., Hunter, D. E., & Fairbanks, R. G. (1997). Interaction between the ENSO and the Asian monsoon in a coral record of tropical climate. *Science, 277*, 925–928.

Chave, K. E., Deffeyes, K. S., Weyl, P. K., Garrels, R. M., & Thompson, M. E. (1962). Observations on the solubility of skeletal carbonates in aqueous solutions. *Science, 137*, 33–34.

Chave, K. E., Smith, S. V., & Roy, K. J. (1972). Carbonate production by coral reefs. *Marine Geology, 12*, 123–140.

Chazottes, V. (1996). Experimental study of bioerosion and sedimentogenesis in a reefal environment: Eutrophication effects (Réunion Island, Indian Ocean). *Comptes Rendus de l'Académie des Sciences, Paris, 323*, 787–794.

Chazottes, V., Le Campion-Alsumard, T., & Peyrot-Clausade, M. (1995). Bioerosion rates on coral reefs: Interactions between macroborers, microborers, and grazers. *Palaeogeography, Palaeoclimatology, Palaeoecology, 113*, 189–198.

Chazottes, V., Le Campion-Alsumard, T., Peyrot-Clausade, N., & Cuet, P. (2002). The effects of eutrophication-related alterations to coral reef communities on agents and rates of bio-erosion, Réunion Island, Indian Ocean. *Coral Reefs, 21*, 375–390.

Chazottes, V., Reijmer, J. J. G., & Cordier, E. (2008). Sediment characteristics in reef areas influenced by eutrophication-related alterations of benthic communities and bioerosion processes. *Marine Geology, 250*, 114–127.

Chen, C. A., Wallace, C. C., & Wolstenholme, J. (2002). Analysis of the mitochondrial 12S rRNA gene supports a two-clade hypothesis of the evolutionary history of scleractinian corals. *Molecular Phylogenetics and Evolution, 23*, 137–149.

Chen, Y., Brumby, S., Jacobson, G., Beckwith, A. J. L., & Polach, H. A. (1991). A novel application of the ESR method: Dating of insular phosphorites and reef limestone. *Quaternary Science Reviews, 11*, 209–217.

Chevalier, J. P. (1971). Les Scléractiniaires de la Mélanésie Française (Nouvelle Calédonie, Iles Chesterfield, Iles Loyauté, Nouvelles Hébrides). Expédition sur les Récifs Coralliens de Nouvelle Calédonie, Première Partie, Fondation Singer-Polignac, Paris, Vol. 5, pp. 5–307.

Chevalier, J.-P. (1972). Observations sur les chenaux incomplets appeles hoa dans les atolls des Tuamotus. *Proceedings of the Symposium on Corals and Coral Reefs 1969*, Mandapam Camp, pp. 477–488.

Chevalier, J. P. (1974). Aperçu sur les scléractiniaires des Iles Gambier. *Cahiers du Pacifique, 18*, 615–628.

Chevalier, J. P. (1975). Les Scléractiniaires de la Mélanésie Française (Nouvelle Calédonie, Iles Chesterfield, Iles Loyauté, Nouvelles Hébrides). Expédition sur les Récifs Coralliens de Nouvelle Calédonie, Deuxième Partie, Fondation Singer-Polignac, Paris, Vol. 57, pp. 5–407.

Chevalier, J.-P. (Ed.). (1977). *Second International Symposium on Corals and Fossil Coral Reefs, Paris. Mémoires du Bureau de Recherches Géologiques et Minières* (Vol. 89, 542 pp.).

Chevalier, J. P. (1979). La faune corallienne (Scléractiniaires et Hydrocoralliaires) de la Polynésie Française. *Cahiers du Pacifique, 2*, 129–151.

Chevalier, J. P. (1980). La faune corallienne de l'Ile de Tubuaï. *Cahiers de l'Indo-Pacifique, 2*, 55–68.

Chevalier, J. P., Denizot, M., Mougin, J. L., Plessis, Y., & Salvat, B. (1968a). Etude géomorphologique et bionomique de l'atoll de Mururoa (Tuamotu). *Cahiers du Pacifique, 12*, 1–144.

Chevalier, J.-P., Denizot, M., Mougin, J.-L., Plessis, Y., & Salvat, B. (1968b). Etude géomorphologique de l'atoll fermé de Taiaro. *Cahiers du Pacifique, 19*, 69–201.

Chevillon, C. (1996). Skeletal composition of modern lagoon sediments in New Caledonia: Coral, a minor constituent. *Coral Reefs, 15*, 199–207.

Chevillon, C., & Clavier, J. (1988). Sedimentological structure of the northern lagoon of New Caledonia. *Proceedings of the Sixth International Coral Reef Symposium*, Townsville, Australia, Vol. 3, pp. 425–430.

Chivas, A., Chappell, J., Polach, H., Pillans, B., & Flood, P. (1986). Radiocarbon evidence for the timing and rate of island development, beach-rock formation and phosphatization at Lady Elliot Island, Queensland, Australia. *Marine Geology, 69*, 273–287.

Choi, D. R., & Holmes, C. W. (1982). Foundations of Quaternary reefs in South-Central Belize lagoon, Central America. *American Association of Petroleum Geologists Bulletin, 66*, 2663–2681.

Choquette, P. W., & Pray, L. C. (1970). Geologic nomenclature and classification of porosity in sedimentary carbonates. *Bulletin of the American Association of Petroleum Geologists, 54*, 207–250.

Chung, G. S., & Swart, P. K. (1990). The concentration of uranium in freshwater vadose and phreatic cements in a Holocene ooid cay: A method of identifying ancient water tables. *Journal of Sedimentary Petrology, 60*, 735–746.

Clark, J. A., Farrell, W. E., & Peltier, W. R. (1978). Global changes in postglacial sea level: A numerical calculation. *Quaternary Research, 9*, 265–287.

Clark, P. U., Archer, D., Pollard, D., Blum, J. D., Rial, J. A., Brovkin, V., et al. (2006). The Middle Pleistocene transition: Characteristics, mechanisms, and implications for long-term changes in atmospheric pCO_2. *Quaternary Science Reviews, 25*, 3150–3184.

Clark, P. U., Marshall, S., Clarke, G., Licciardi, J., & Teller, J. (2001). Freshwater forcing of abrupt climate change during the last glaciation. *Science, 293*, 283–287.

Clark, P. U., Mitrovica, J. X., Milne, G. A., & Tamisiea, M. E. (2002). Sea level finger printing as a direct test for the source of global meltwater pulse 1a. *Science, 295*, 2438–2441.

Clark, P. U., & Mix, A. C. (2002). Ice sheets and sea level of the Last Glacial Maximum. *Quaternary Science Reviews, 21*, 1–7.

Clarke, G. K. C., Leverington, D. W., Teller, J. T., & Dyke, A. S. (2004). Paleohydraulics of the last outburst flood from glacial Lake Agassiz and the 8200 BP cold event. *Quaternary Science Reviews, 23*, 389–407.

Cobb, K. J., Charles, C. D., Cheng, H., Kastner, M., & Edwards, R. L. (2003). U/Th-dating living and young fossil corals from the central tropical Pacific. *Earth and Planetary Science Letters, 210*, 91–103.

Cobb, K. M., Charles, C. D., Cheng, H., & Edwards, R. L. (2003). El Niño/southern oscillation and tropical Pacific climate during the last millennium. *Nature, 424*, 271–276.

Cohen, A. L., & Hart, S. R. (2004). Deglacial sea surface temperatures of the western tropical Pacific: A new look at old corals. *Paleoceanography, 19*, PA4031, doi: 10.1029/2004PA001084.

Cohen, A. L., Smith, S. R., McCartney, M. S., & van Etten, J. (2004). How brain corals record climate: An integration of skeletal structure, growth, and chemistry of *Diploria labyrinthiformis* from Bermuda. *Marine Ecology Progress Series, 271*, 147–158.

Colby, N. D., & Boardman, M. R. (1989). Depositional evolution of a windward, high energy lagoon, Graham's Harbour, San Salvador, Bahamas. *Journal of Sedimentary Petrology, 59*, 819–834.

Cole, J. E. (1996). Coral records of climate change: understanding past variability in the tropical ocean-atmosphere. In: P. D. Jones, R. S. Bradley & J. Jouzel (Eds), *Climatic variations and forcing mechanisms over the last 2000 years* (pp. 331–354). Berlin: Springer.

Cole, J. E., Dunbar, R. B., McClanahan, T. R., & Mithiga, N. A. (2000). Tropical Pacific forcing of decadal SST variability in the western Indian Ocean over the past two centuries. *Science, 287*, 617–619.

Colgan, M. W. (1990). El Niño and the history of Eastern Pacific reef building. In: P. W. Glynn (Ed.), *Global Ecological Consequences of the 1982–83 El Niño-Southern Oscillation* (pp. 183–232). Amsterdam: Elsevier.

Colinvaux, P. A., Liu, K. B., de Oliveira, P., Bush, M. B., Miller, M. C., & Kannan, M. S. (1996). Temperature depression in the lowland tropics in glacial times. *Climatic Change, 32*, 19–33.

Collins, L. B., Zhao, J.-X., & Freeman, H. (2006). A high-precision record of mid-late Holocene sea level events from emergent coral pavements in the Houtman Abrolhos Islands, southwest Australia. *Quaternary International, 145–146*, 78–85.

Collins, L. B., Zhu, Z. R., & Wyrwoll, K.-H. (1998). Late Tertiary–Quaternary geological evolution of the Houtman Abrolhos carbonate platforms, northern Perth Basin. In: P. R. Purcell (Ed.), *The Sedimentary Basins of Western Australia 2. Proceedings of PESA Symposium*, Perth, pp. 647–663.

Collins, L. B., Zhu, Z. R., Wyrwoll, K.-H., & Eisenhauer, A. (2003). Late Quaternary structure and development of the northern Ningaloo Reef, Australia. *Sedimentary Geology, 159*, 81–94.

Collins, L. B., Zhu, Z. R., Wyrwoll, K.-H., Hatcher, B. G., Playford, P. E., Chen, J. H., et al. (1993). Late Quaternary evolution of coral reefs on a cool-water carbonate margin: The Abrolhos Carbonate platforms, southwest Australia. *Marine Geology, 110*, 203–212.

Colonna, M. (1994). *Chronologie des variations du niveau marin au cours du dernier cycle climatique (0–140 000 ans) dans la partie sud-occidentale de l'Océan Indien*. Unpublished Ph.D. thesis, University of Provence, 293 pp.

Colonna, M., Casanova, J., Dullo, W.-C., & Camoin, G. (1996). Sea-level changes and $\delta^{18}O$ record for the past 34,000 years from Mayotte Reef, Indian Ocean. *Quaternary Research*, *46*, 335–339.

Conaghan, P. J., Mountjoy, W. W., Edgecombe, D. R., Talent, J. A., & Owen, D. E. (1976). Nubrigyn algal reefs (Devonian), eastern Australia: Allochthonous blocks and megabreccias. *Geological Society of America Bulletin*, *87*, 515–530.

Coniglio, M., & Harrison, R. S. (1983). Holocene and Pleistocene Caliche from Big Pine Key, Florida. *Bulletin of Canadian Petroleum Geology*, *31*, 3–13.

Connell, J. H. (1978). Diversity in tropical rain forests and coral reefs. *Science*, *199*, 1302–1310.

Connell, J. H., Hughes, T. P., & Wallace, C. C. (1997). A 30-year study of coral abundance, recruitment, and disturbance at several scales in space and time. *Ecological Monographs*, *67*, 461–488.

Cooper, P. (1994). Ancient reef ecosystem expansion and collapse. *Coral Reefs*, *33*, 3–11.

Cooper, T. F., Uthicke, S., Humphrey, C., & Fabricius, K. E. (2007). Gradients in water column nutrients, sediment parameters, irradiance and coral reef development in the Whitsunday Region, central Great Barrier Reef. *Estuarine, Coastal and Shelf Science*, *74*, 458–470.

Coronado, C. C. J., Inglesias-Prieto, R., Sheinbaum, J., Lopez, M., & Ocampo-Torres, F. J. (2007). On the circulation in the Puerto Morelos fringing reef lagoon. *Coral Reefs*, *26*, 149–163.

Corrège, Th. (2006). Sea surface temperature and salinity reconstruction from coral geochemical tracers. *Palaeogeography, Palaeoclimatology, Palaeoecology*, *232*, 408–428.

Corrège, Th., Delcroix, Th., Récy, J., Beck, J. W., Cabioch, G., & Le Cornec, F. (2000). Evidence for stronger El Niño-southern oscillation (ENSO) events in a mid-Holocene massive coral. *Paleoceanography*, *15*, 465–470.

Corrège, Th., Gagan, M. K., Beck, J. W., Burr, G. S., Cabioch, G., & Le Cornec, F. (2004). Interdecadal variation in the extent of South Pacific tropical waters during the Younger Dryas event. *Nature*, *428*, 927–929.

Cortés, J. (1997). Biology and geology of eastern Pacific coral reefs. *Coral Reefs*, *16*(Suppl.), S39–S46.

Cortés, J. (Ed.) (2003). *Latin American coral reefs*. Amsterdam: Elsevier. 497pp.

Cortés, J., Macintyre, I. G., & Glynn, P. W. (1994). Holocene growth history of an Eastern Pacific fringing reef, Punta Islotes, Costa Rica. *Coral Reefs*, *13*, 65–73.

Cortés, J., & Risk, M. J. (1985). A reef under siltation stress: Cahuita, Costa Rica. *Bulletin of Marine Science*, *36*, 339–356.

Coudray, J., & Montaggioni, L. (1986). The diagenetic products of marine carbonates as sea-level indicators. In: O. van de Plassche (Ed.), *Sea-level research: A manual for the collection and evaluation of data* (pp. 311–360). Norwich, U.K.: Geo Books.

Coulbourn, W. T., & Resig, J. M. (1975). On the use of benthic foraminifera as sediment tracers in a Hawaiian Bay. *Pacific Science*, *29*, 99–115.

Coyne, M. K., Jones, B., & Ford, D. (2007). Highstands during Marine Isotope Stage 5: Evidence from the Ironshore formation of Grand Cayman, British West Indies. *Quaternary Science Reviews*, *26*, 536–559.

Crabbe, M. J. C., Wilson, M. E. J., & Smith, D. J. (2006). Quaternary corals from reefs in the Wakatobi Marine National Park, SE Sulawesi, Indonesia, show similar growth rates to modern corals from the same area. *Journal of Quaternary Science*, *21*, 803–809.

Crame, J. A. (1980). Succession and diversity in the Pleistocene coral reefs of the Kenya coast. *Palaeontology, 23*, 1–37.

Crame, J. A. (1981). Ecological stratification in the Pleistocene coral reefs of the Kenya coast. *Palaeontology, 24*, 609–646.

Crame, J. A. (1986). Late Pleistocene molluscan assemblages from the coral reefs of the Kenya coast. *Coral Reefs, 4*, 183–196.

Crame, J. A., & Rosen, B. R. (2002). Cenozoic palaeogeography and the rise of modern biodiversity patterns. In: J. A. Crame, & A. W. Owen (Eds), *Palaeobiogeography and biodiversity change: The Ordovician and Mesozoic–Cenozoic radiations* (pp. 153–168), Geological Society of London (Special Publication 194).

Crossland, C. J., & Barnes, D. J. (1974). The role of metabolic nitrogen in coral calcification. *Marine Biology, 28*, 325–332.

Crowley, T. J. (2000a). Causes of climate change over the past 1000 years. *Science, 289*, 270–277.

Crowley, T. J. (2000b). CLIMAP SSTs re-revisited. *Climate Dynamics, 16*, 241–255.

Crueger, T., Zinke, J., & Pfeiffer, M. (2009). Patterns of Pacific decadal variability recorded by Indian Ocean corals. *International Journal of Earth Sciences, 98*(1), 41–52.

Cuif, J.-P., & Sorauf, J. E. (2001). Biomineralization and diagenesis in the scleractinia: Part 1, biomineralization. *Bulletin of Tohoku University Museum, 1*, 144–151.

Cullis, C. G. (1904). The mineralogical changes observed in the cores of the Funafuti Borings: Coral Reef Report. In: T. G. Bonney (Ed.), *The atoll of Funafuti: Borings into a coral reef and the results: Being the report of the Coral Reef Committee of the Royal Society* (pp. 392–420). London, UK: The Royal Society, London.

Cumings, E. R., & Shrock, R. R. (1928). Niagaran coral reefs of Indiana and adjacent states and their stratigraphic relations. *Geological Society of America Bulletin, 39*, 579–620.

Cutler, K. B., Edwards, R. L., Taylor, F. W., Cheng, H., Adkins, J., Gallup, C. D., et al. (2003). Rapid sea-level fall and deep-sea temperature change since the last interglacial period. *Earth and Planetary Science Letters, 206*, 253–271.

Dahanayake, K., & Krumbein, W. E. (1985). Ultrastructure of a microbial mat generated phosphorite. *Mineralium Deposita, 20*, 260–265.

Dalmasso, H., Montaggioni, L. F., Bosence, D., & Floquet, M. (2001). Numerical modelling of carbonate platforms and reefs: Approaches and opportunities. *Energy Exploration & Exploitation, 19*, 315–345.

Daly, R. A. (1915). The glacial control theory of coral reefs. *Proceedings of the American Academy of Arts and Sciences, 51*, 155–251.

Daly, R. A. (1948). Coral Reefs: A review. *American Journal of Science, 246*, 193–207.

Dansgaard, W., White, J. W. C., & Johnsen, S. J. (1989). The abrupt termination of the Younger Dryas climate event. *Nature, 339*, 532–534.

D'Arrigo, R., Wilson, R., Palmer, J., Krusic, P., Curtis, A., Sakulich, J., et al. (2006). The reconstructed Indonesian warm pool sea surface temperatures from tree rings and corals: Linkages to Asian monsoon drought and El Niño–southern oscillation. *Paleoceanography, 21*, PA3005, doi: 10.1029/2005PA001256.

Darwin, C. (1842). *The structure and distribution of coral reefs*. London: Smith, Elder and Co., 214 pp.

Davies, D. J., Powell, E. N., & Stanton, R. J., Jr. (1989). Relative rates of shell dissolution and net sediment accumulation — A commentary: Can shell beds form by the gradual accumulation of biogenic debris on the sea floor. *Lethaia, 22*, 207–212.

Davies, P. J. (1974). Sub-surface solution unconformities at Heron Island, Great Barrier Reef. *Proceedings of the Second International Coral Reef Symposium*, Manila, Vol. 2: pp. 573–578.

Davies, P. J., Braga, J. C., Lund, M., & Webster, J. M. (2004). Holocene deep water algal buildups on the eastern Australian shelf. *PALAIOS, 19*, 598–609.

Davies, P. J., & Hopley, D. (1983). Growth facies and growth rates of Holocene reefs in the Great Barrier Reef. *Bureau of Mineral Resources Journal of Australian Geology and Geophysics, 8*, 237–251.

Davies, P. J., & Marshall, J. F. (1985). *Halimeda* bioherms — Low energy reefs, northern Great Barrier Reef. *Proceedings of the Fifth International Coral Reef Congress*, Tahiti, Vol. 5: pp. 1–7.

Davies, P. J., Marshall, J. F., & Hopley, D. (1985). Relationships between reef growth and sea level in the Great Barrier Reef. *Proceedings of the Fifth International Coral Reef Congress*, Tahiti, Vol. 3: pp. 95–103.

Davies, P. J., & Montaggioni, L. (1985). Reef growth and sea-level change: the environmental signature. *Proceedings of the Fifth International Coral Reef Congress*, Tahiti, Vol. 3: pp. 477–511.

De Deckker, P., Tapper, N. J., & Van der Kaars, S. (2002). The status of the Indo-Pacific Warm Pool and adjacent land at the Last Glacial Maximum. *Gobal and Planetary Change, 35*, 25–35.

Defarge, C., Trichet, J., Jaunet, A.-M., Robert, M., Tribble, J., & Sansone, F. J. (1996). Texture of microbial sediments revealed by cryo-scanning electron microscopy. *Journal of Sedimentary Research, 66*, 935–947.

Défarge, Ch., Trichet, J., Maurin, A., & Hucher, M. (1994). Kopara in Polynesian atolls: Early stages of formation of calcareous stromatolites. *Sedimentary Geology, 89*, 9–23.

Degauge-Michalski, F. M. (1990). Holocene sedimentary environments from two reef platforms, Chesterfield Islands (South-West Pacific): preliminary results. *Proceedings of the International Society for Reef Studies Meeting*, Nouméa, pp. 189–196.

Degauge-Michalski, F. M. (1993). *Croissance et évolution d'édifices récifaux du Pacifique occidental (Nouvelle-Calédonie, Chesterfield) à l'Holocène et au Pleistocène*. Unpublished Ph.D. thesis, University of Provence, 227 pp.

Delesalle, B. (1985). Mataiva Atoll, Tuamotu Archipelago. *Proceedings of the Fifth International Coral Reef Congress*, Tahiti, Vol. 1, pp. 269–322.

Delesalle, B., Galzin, R., & Salvat, B. (1985). French Polynesian coral reefs. *Proceeding of the Fifth International Coral Reef Congress*, Tahiti, Vol. 1, pp. 1–554.

Delworth, T. L. (1996). North Atlantic interannual variability in a coupled ocean-atmosphere model. *Journal of Climate, 9*, 2356–2375.

Deneufbourg, G. (1969). Les Forages de Mururoa. *Cahiers du Pacifique, 13*, 191–202.

Desenfant, F., Veron, A. J., Camoin, G. F., & Nyberg, J. (2006). Reconstruction of pollutant lead invasion into the tropical North Atlantic during the twentieth century. *Coral Reefs, 25*, 473–484.

DeVantier, L. M., & Done, T. J. (2007). Inferring past outbreaks of the crown-of-thorns seastar from scar patterns on coral heads. In: R. B. Aronson (Ed.), *Geological approaches to coral reef ecology* (pp. 85–125). New York: Springer.

de Villiers, S. (1999). Sea water strontium and Sr/Ca variability in the Atlantic and Pacific oceans. *Earth and Planetary Science Letters, 171*, 623–634.

de Villiers, S., Shen, G. T., & Nelson, B. K. (1994). The Sr/Ca-temperature relationship in coralline aragonite: Influence of variability in $(Sr/Ca)_{seawater}$ and skeletal growth parameters. *Geochimica et Cosmochimica Acta, 58*, 197–208.

Dickinson, W. R. (2000). Hydro-isostatic and tectonic influences on emergent Holocene paleoshorelines in the Mariana Islands, Western Pacific Ocean. *Journal of Coastal Research, 16*, 735–746.

Dickinson, W. R. (2001). Paleoshoreline record of relative Holocene sea levels on Pacific islands. *Earth-Science Reviews, 55*, 191–234.

Dickinson, W. R. (2003). Impact of mid-Holocene highstand in regional sea level on habitability of islands in Pacific Oceania. *Journal of Coastal Research, 19*, 489–502.

Dickinson, W. R. (2004). Impacts of eustasy and hydro-isostasy on the evolution and landforms of Pacific atolls. *Palaeogeography, Palaeoclimatology, Palaeoecology, 213*, 251–269.

Dickson, J. A. D. (1966). Carbonate identification and genesis as revealed by staining. *Journal of Sedimentary Petrology, 36*, 491–505.

DiMichele, W. A., Behrensmeyer, A. K., Olszewski, T. D., Labandira, C. C., Pandolfi, J. M., Wingh, S. L., et al. (2004). Long-term stasis in ecological assemblages: Evidence from the fossil record. *Annual Reviews of Ecology, Evolutionary and Systematics, 35*, 285–322.

Dodge, R. E., Aller, R. C., & Thomson, J. (1974). Coral growth related to resuspension of bottom sediments. *Nature, 247*, 574–577.

Dodge, R. E., Fairbanks, R. G., Benninger, L. K., & Maurasse, F. (1983). Pleistocene sea levels from raised coral reefs of Haiti. *Science, 219*, 1423–1425.

Dodge, R. E., & Gilbert, T. R. (1984). Chronology of lead pollution contained in banded coral skeletons. *Marine Biology, 82*, 9–13.

Dodge, R. E., & Vaisnys, J. R. (1977). Coral populations and growth patterns: Responses to sedimentation and turbidity associated with dredging. *Journal of Marine Research, 35*, 715–730.

Done, T. J. (1977). A comparison of units of cover in ecological classifications of coral communities. *Proceedings of the Third International Coral Reef Symposium*, Miami, Vol. 1, pp. 9–14.

Done, T. J. (1982). Patterns of distribution of coral communities across the Central Great Barrier Reef. *Coral Reefs, 1*, 95–108.

Done, T. J. (1983). Coral zonation: Its nature and significance. In: D. J. Barnes (Ed.), *Perspectives on coral reefs* (pp. 95–145). Manuka, Australia: The Australian Institute of Marine Science, Brian Clouston Publisher.

Done, T. J. (1992a). Phase shifts in coral reef communities and their ecological significance. *Hydrobiologia, 247*, 121–132.

Done, T. J. (1992b). Effects of tropical cyclone waves on ecological and geomorphological structures of the Great Barrier Reef. *Continental Shelf Research, 12*, 859–872.

Done, T. J., & Navin, K. F. (1990). Shallow-water benthic communities on coral reefs. In: T. J. Done & K. F. Navin (Eds), *Vanuatu marine resources: Report of a biological survey* (pp. 10–36). Townsville: Australian Institute of Marine Science.

Douillet, P. (1998). Tidal dynamics of the south-west lagoon of New Caledonia: Observations and 2D numerical modelling. *Oceanologica Acta, 21*, 69–79.

Douillet, P., Ouillon, S., & Cordier, E. (2001). A numerical model for fine suspended sediment transport in the southwest lagoon of New Caledonia. *Coral Reefs, 20*, 361–372.

Doyle, L. J., & Roberts, H. H. (Eds). (1988). *Carbonate-clastic transitions*. Developments in Sedimentology (Vol. 42, 304 pp.). Amsterdam: Elsevier.

Draschba, S., Pätzold, J., & Wefer, G. (2000). North Atlantic climate variability since AD 1350 recorded in $\delta^{18}O$ and skeletal density of Bermuda corals. *International Journal of Earth Sciences, 88*, 733–741.

Dravis, J. J. (1996). Rapidity of freshwater calcite cementation: Implications for carbonate diagenesis and sequence stratigraphy. *Sedimentary Geology, 107*, 1–10.

Drew, E. A., & Abel, K. M. (1985). Biology, sedimentology, and geography of the vast inter-reefal *Halimeda* meadows within the Great Barrier Reef Province. *Proceedings of the Fifth International Coral Reef Congress*, Tahiti, Vol. 5, pp. 15–20.

Druffel, E. R. M. (1981). Radiocarbon in annual coral rings from the eastern tropical Pacific Ocean. *Geophysical Research Letters, 8*, 59–62.

Druffel, E. R. M., & Griffin, S. (1993). Large variations of surface ocean radiocarbon: Evidence of circulation changes in the southwestern Pacific. *Journal of Geophysical Research, Oceans, 98*, 20249–20259.

Druffel, E. R. M., & Griffin, S. (1999). Variability of surface ocean radiocarbon and stable isotopes in the southwestern Pacific. *Journal of Geophysical Research, 104*, 23607D–23613D.

Dubinski, Z. (Ed.) (1990). *Coral Reefs. Ecosystems of the World* (Vol. 25, Amsterdam: Elsevier. 550 pp.

Dubois, N., Kindler, P., Spezzaferri, S., & Coric, S. (2008). The initiation of the southern central Great Barrier Reef: New multiproxy data from Pleistocene distal sediments from Marion Plateau (NE Australia). *Marine Geology, 250*, 223–233.

Dullo, W. C. (2005). Coral growth and reef growth: A brief review. *Facies, 51*, 33–48.

Dullo, W.-C., Camoin, G. F., Blomeier, D., Colonna, M., Eisenhauer, A., Faure, G., et al. (1998). Morphology and sediments of the fore-slopes of Mayotte, Comoro Islands: Direct observations from a submersible. In: G. F. Camoin, & P. J. Davies (Eds), *Reefs and carbonate platforms in the Pacific and Indian Oceans* (pp. 219–236), International Association of Sedimentologists (Special Publication 25).

Dullo, W.-Ch., & Montaggioni, L. F. (1998). Modern Red Sea coral reefs: A review of their morphologies and zonation. In: B. H. Purser & D. W. J. Bosence (Eds), *Sedimentation and tectonics in rift basins: Red Sea–Gulf of Aden* (pp. 583–594). London: Chapman & Hall.

Dullo, W.-Ch., Moussavian, E., & Brachert, Th. C. (1990). The foralgal crust facies of the deeper fore reefs in the Red Sea: A deep diving survey by submersible. *Geobios, 23*, 261–281.

Dumas, B., Hoang, C.-T., & Raffy, J. (2006). Record of MIS 5 sea-level highstands based on dated coral terraces at Haiti. *Quaternary International, 145–146*, 106–118.

Dunbar, G. B., & Dickens, G. R. (2003). Massive siliciclastic discharge to slopes of the Great Barrier Reef Platform during sea-level transgression: Constraints from sediment cores between 15°S and 16°S latitude and possible explanations. *Sedimentary Geology, 162*, 141–158.

Dunham, R. J. (1962). Classification of carbonate rocks according to depositional texture. In: W. E. Ham (Ed.), *Classification of carbonate rocks: Memoir of the American Association of Petroleum Geologists, 1*, 108–121.

Dunham, R. J. (1970). Stratigraphic reefs versus ecologic reefs. *Bulletin of the American Association of Petroleum Geologists, 54*, 1931–1932.

Dynesius, M., & Jansson, R. (2000). Evolutionary consequences of changes in species' geographical distributions driven by Milankovitch climate oscillations. *Proceedings of the National Academy of Sciences of the United States of America, 97*, 9115–9120.

Eakin, C. M. (1996). Where have all the carbonates gone? A model comparison of calcium carbonate budgets before and after the 1982-1983 El Niño at Uva Island in the eastern Pacific. *Coral Reefs, 15*, 109–119.

Easton, W. H., & Olson, E. A. (1976). Radiocarbon profile of Hanauma Reef, Oahu, Hawaii. *Geological Society of America Bulletin, 87*, 711–719.

Edinger, E. N., Kolasa, J., & Risk, M. J. (2000). Variation in within-site coral species diversity on coral reefs in three regions of Indonesia. *Diversity and Distributions, 6*, 113–127.

Edinger, E. N., Pandolfi, J. M., & Kelley, R. (2001). Diversity and taxonomic composition of Holocene reefs and modern reef coral life and death assemblages in Madang Lagoon, Papua New Guinea. *Paleobiology, 27*, 669–694.

Edinger, E. N., & Risk, M. J. (1994). Oligocene–Miocene extinction and geographic restriction of Caribbean corals: Roles of turbidity, temperature and nutrients. *PALAIOS, 9*, 576–598.

Edinger, E. N., & Risk, M. J. (1995). Preferential survivorship of brooding corals in a regional extinction. *Paleobiology, 21*, 200–219.

Edwards, F. J. (1987). Climate and oceanography. In: A. J. Edwards & S. M. Head (Eds), *Red Sea* (pp. 45–69). Oxford: Pergamon Press.

Edwards, L. R., Chen, J. H., Ku, T. L., & Wasserberg, G. J. (1987). Precise timing of the last interglacial period from mass-spectrometric determinations of thorium-230 in corals. *Science, 236*, 1547–1553.

Edwards, R. L., Beck, J. W., Burr, G. S., Donahue, D. J., Chappell, J. M. A., Bloom, A. L., et al. (1993). A large drop in atmospheric $^{14}C/^{12}C$ and reduced melting in the Younger Dryas, documented with ^{230}Th ages of corals. *Science*, *260*, 962–968.
El-Asmar, H. M. (1997). Quaternary isotope stratigraphy and paleoclimate of coral reef terraces, Gulf of Aqaba, south Sinai, Egypt. *Quaternary Science Reviews*, *16*, 911–924.
Ellis, J. P., & Milliman, J. D. (1985). Calcium carbonate suspended in Arabian Gulf and Red Sea waters: Biogenic, not "chemogenic". *Journal of Sedimentary Petrology*, *55*, 805–808.
Embry, A. F., & Klovan, J. E. (1972). Absolute water depth limits of Late Devonian paleoecological zones. *Geologische Rundschau*, *61*, 672–686.
Emery, D. (1996). Carbonate systems. In: D. Emery & K. J. Myers (Eds), *Sequence stratigraphy* (pp. 211–237). London: Blackwell Science Ltd.
Emery, K. O., Tracey, J. I., & Ladd, H. S. (1954). *Geology of Bikini and nearby atolls*. U.S. Geological Survey, Professional Papers, Vol. 260A, pp. 1–265.
Emiliani, C. (1955). Pleistocene temperatures. *Journal of Geology*, *63*, 538–575.
Engels, M. S., Fletcher, C. H., Field, M. E., Storlazzi, C. D., Grossman, E. E., Rooney, J. J. B., et al. (2004). Holocene reef accretion: South-west Molokai, Hawaii. *Journal of Sedimentary Research*, *74*, 255–269.
Enmar, R., Stein, M., Bar-Matthews, M., Sass, E., Katz, A., & Lazar, B. (2000). Diagenesis in live corals from the Gulf of Aqaba. I. The effect on paleo-oceanography tracers. *Geochimica et Cosmochimica Acta*, *64*, 3123–3132.
Esat, T. M., & Yokoyama, Y. (2006). Growth patterns of the last ice age coral terraces at Huon Peninsula. *Global and Planetary Change*, *54*, 216–224.
Esteban, M. (1996). An overview of Miocene reefs from Mediterranean areas: General trends and facies models. In: E. K. Franseen, M. Esteban, W. C. Ward & J.-M. Rouchy (Eds), *Models for carbonate stratigraphy from Miocene reef complexes of Mediterranean regions*. Concepts in Sedimentology and Paleontology (Vol. 5, pp. 3–53). Tulsa, OK: SEPM.
Esteban Cerdà, M. (1974). Caliche textures and "Microcodium". *Supplemento al Bolletin Societe Geologia Italiana*, *92*, 105–125.
Estrada Alvarez, L. M., Edinger, E. N., & Pandolfi, J. M. (2004). Taphonomy of modern corals from Madang Lagoon, Papua New Guinea. *Proceedings of the Canadian Paleontology Conference*, St. Andrews, New Brunswick, Vol. 2, pp. 23–28.
Evans, M. N., Kaplan, A., & Cane, M. A. (2000). Intercomparison of coral oxygen isotope data and historical sea surface temperature (SST): Potential for coral-based SST field reconstructions. *Paleoceanography*, *15*, 551–563.
Fabricius, K. E., & De'ath, G. (2001). Environmental factors associated with the spatial distribution of crustose coralline algae on the Great Barrier Reef. *Coral Reefs*, *19*, 303–309.
Fabricius, K. E., & Wolansky, E. (2000). Rapid smothering of coral reef organisms by muddy marine snow. *Estuarine, Coastal and Shelf Science*, *50*, 115–120.
Fagerström, J. A. (1987). *The evolution of reef communities*. New York: Wiley, 600 pp.
Fairbanks, R. G. (1989). A 17,000-year glacio-eustatic sea level record: Influence of glacial melting rates on the Younger Dryas event and deep-ocean circulation. *Nature*, *342*, 637–642.
Fairbanks, R. G., & Dodge, R. E. (1979). Annual periodicity of the $^{18}O/^{16}O$ and $^{13}C/^{12}C$ ratios in the coral *Montastrea annularis*. *Geochimica et Cosmochimica Acta*, *43*, 1009–1020.
Fairbanks, R. G., Mortlock, R. A., Chiu, T.-C., Cao, L., Kaplan, A., Guilderson, Th. P., et al. (2005). Radiocarbon calibration curve spanning 0 to 50,000 years BP based on paired $^{230}Th/^{234}U/^{238}U$ and ^{14}C dates on pristine corals. *Quaternary Science Reviews*, *24*, 1781–1796.
Falkland, A. C., & Woodroffe, C. D. (1997). Geology and hydrogeology of Tarawa and Christmas Island, Kiribati. In: H. L. Vacher & T. Quinn (Eds), *Geology and hydrogeology of carbonate islands*. Development in Sedimentology (Vol. 54. pp. 577–610). Amsterdam: Elsevier Science, B.V.

Fallon, S. J., McCulloch, M. T., & Alibert, C. (2003). Examining water temperature proxies in *Porites* corals from the Great Barrier Reef: A cross-shelf comparison. *Coral Reefs*, *22*, 389–404.

Fallon, S. J., McCulloch, M. T., van Woesik, R., & Sinclair, D. J. (1999). Corals at their latitudinal limits: Laser ablation trace element systematics in *Porites* from Shirigai Bay, Japan. *Earth and Planetary Science Letters*, *172*, 221–238.

Fallon, S. J., White, J. C., & McCulloch, M. T. (2002). *Porites* corals as recorders of mining and environmental impacts: Misima Island, Papua New Guinea. *Geochimica et Cosmochimica Acta*, *66*, 45–62.

Fanning, K. A., Byrne, R. H., Breland, J. A., II, Betzer, P. R., Moore, W. S., Elsinger, R. J., et al. (1981). Geothermal springs of the west Florida continental shelf: Evidence for dolomitization and radionucleide enrichment. *Earth and Planetary Science Letters*, *52*, 345–354.

Faure, G. (1982). *Recherches sur les peuplements de scléractiniaires des récifs coralliens de l'archipel des Mascareignes*. Unpublished D.Sc. thesis, University of Aix-Marseilles, vol. 1, 206 pp.; vol. 2, 230 pp.

Faure, G., & Laboute, P. (1984). Formations récifales: I. Définition des unités récifales et distribution des principaux peuplements de Scléractiniaires. L'atoll de Tikehau: Premiers Résultats. Office de Recherche Scientifique et Technique d'Outre-Mer. *Océanographie*, *22*, 108–136.

Fedorov, V. V., Rubinshteyn, I. G., Danilov, I. V., & Lanin, V. I. (1980). Bottom landscapes of Saya de Malha Bank in the Indian Ocean. *Oceanology*, *20*(4), 434–439.

Felis, Th., Lohmann, G., Kuhnert, H., Lorenz, S. J., Scholz, D., Patzöld, J., et al. (2004). Increased seasonality in the Middle East temperatures during the last interglacial period. *Nature*, *429*, 164–168.

Felis, Th., & Pätzold, J. (2004). Climate reconstructions from annually banded corals. In: M. Shiyomi, H. Kawahata, H. Koizumi, A. Tsuda & Y. Awaya (Eds), *Global environmental change in the ocean and on land* (pp. 205–227). Tokyo: Terrapub.

Felis, Th., Pätzold, J., & Loya, Y. (2003). Mean oxygen-isotope signatures in *Porites* spp. corals: Inter-colony variability and correction for extension-rate effects. *Coral Reefs*, *22*, 328–336.

Felis, Th., Pätzold, J., Loya, Y., Fine, M., Nawar, A. H., & Wefer, G. (2000). A coral oxygen record from the northern Red Sea documenting NAO, ENSO and North Pacific teleconnections on Middle East climate variability since the year 1750. *Paleoceanography*, *15*, 679–694.

Felis, Th., Pätzold, J., Loya, Y., & Wefer, G. (1998). Vertical water mass mixing and plankton blooms recorded in skeletal carbon isotopes of a Red Sea coral. *Journal of Geophysical Research*, *103*, 30731–30739.

Fernández-Díaz, L., Putnis, A., Prieto, M., & Putnis, C. V. (1996). The role of magnesium in the crystallization of calcite and aragonite in a porous medium. *Journal of Sedimentary Research*, *66*, 482–491.

Ferry, J., Kumar, P. B., Bronders, J., & Lewis, J. (1997). Hydrogeology of selected islands of Fiji. In: H. L. Vacher & T. Quinn (Eds), *Geology and Hydrogeology of Carbonate Islands*. Developments in Sedimentology (Vol. 54, pp. 763–782). Amsterdam: Elsevier.

Feuillet, N., Tapponnier, P., Manighetti, I., Villemant, B., & King, G. (2004). Differential uplift and tilt of Pleistocene reef platforms and Quaternary slip rate on the Morne-Piton normal fault (Guadeloupe, French West Indies). *Journal of Geophysical Research*, *109*, B02404, doi: 10.1029/2003JB002496.

Field, R. M., & Hess, H. H. (1933). A borehole in the Bahamas. American Geophysical Union, Transactions, 14th Annual meeting, pp. 234–245.

Finkh, A. E. (1904). Biology of the reef forming organisms of Funafuti Atoll, The Atoll of Funafuti. *Report of the Coral Reef Committee*, The Royal Society, London, pp. 125–150.

FitzPatrick, E. A. (1993). *Soil microscopy and micromorphology*. New York: Wiley, 394 pp.

Flamand, B., Cabioch, G., Payri, C., & Pelletier, B. (2008). Nature and biological composition of the New Caledonian barrier reef slopes. *Marine Geology, 2150*, 157–179.

Fleming, K., Johnston, P., Zwart, D., Yokoyama, Y., Lambeck, K., & Chappell, J. (1998). Refining the eustatic sea-level curve since the Last Glacial Maximum using far- and intermediate-field sites. *Earth and Planetary Science Letters, 163*, 327–342.

Fletcher, C., & Sherman, C. E. (1995). Submerged shorelines on Oahu, Hawaii: Archives of episodic transgression during the deglaciation? *Journal of Coastal Research, 17*, 141–152.

Flood, P. G. (1983). Coated grains from the Great Barrier Reef. In: T. M. Peryt (Ed.), *Coated grains* (pp. 561–565). Berlin: Springer.

Flood, P. G., Fagerstrom, J. A., & Rougerie, F. (1996). Interpretation of the origin of massive replacive dolomite within atolls and submerged carbonate platforms: Strontium isotopic signature ODP Hole 866A, Resolution Guyot, Mid-Pacific Mountains. *Sedimentary Geology, 101*, 9–13.

Flood, P. G., & Orme, G. R. (1988). Mixed siliciclastic/carbonate sediments of the northern Great Barrier Reef province, Australia. In: L. J. Doyle & H. H. Roberts (Eds), *Carbonate–clastic transitions* (pp. 175–206). Amsterdam: Elsevier.

Flood, P. G., & Scoffin, T. P. (1978). Reefal sediments of the Northern Great Barrier Reef. *Philosophical Transactions of the Royal Society of London, A, 291*, 55–71.

Flügel, E. (1988). *Halimeda*: Paleontological record and palaeoenvironmental significance. *Coral Reefs, 6*, 123–130.

Focke, J. W., & Gebelein, C. D. (1978). Marine lithification of reef rock and rhodolites at a fore-reef slope locality (−50 m) off Bermuda. *Geologie en Mijnbouw, 57*, 163–171.

Folk, R. L. (1959). Practical petrographic classification of limestones. *Bulletin of the American Association of Petroleum Geologists, 43*, 1–38.

Folk, R. L. (1965). Some aspects of recrystallization in ancient limestones. In: L. C. Pray, & R. C. Murray (Eds), *Dolomitization and Limestone Diagenesis* (pp. 14–48), Society of Economic Paleontologists and Mineralogists (Special Publication 13).

Folk, R. L., & Robles, R. (1964). Carbonate sands of Isla Perez, Alacran Reef Complex, Yucatan. *Journal of Geology, 72*, 255–292.

Foster, M. S. (2001). Rhodoliths: Between rocks and soft places. *Journal of Phylocology, 37*, 659–667.

Foster, R., Hagan, A., Perera, N., Gunwan, C. A., Silabab, I., Yaha, Y., et al. (2006). Tsunami and earthquake damage to coral reefs of Aceh, Indonesia, Reef Check Foundation, Pacific Palisades, California.

Fournier, F., Montaggioni, L. F., & Borgomano, J. (2004). Palaeoenvironments and high frequency cyclicity from Cenozoic South-East Asian shallow-water carbonates: A case study from the Oligo-Miocene buildups of Malampaya (Offshore Palawan, Phillippines). *Marine and Petroleum Geology, 21*, 1–21.

Frank, N., Turpin, L., Cabioch, G., Blamant, D., Tressens-Fedou, M., Colin, C., et al. (2006). Open system U-series ages of corals from a subsiding reef in New Caledonia: Implications for sea level changes and subsidence rate. *Earth and Planetary Science Letters, 249*, 274–289.

Frantz, B. R., Kashgarian, M., Coale, K. H., & Foster, M. S. (2000). Growth rate and potential climate record from a rhodolith using ^{14}C accelerator mass spectrometry. *Limnology and Oceanography, 45*, 1773–1777.

Freile, D., & Hillis, L. (1997). Carbonate productivity by *Halimeda incrassata* in a land proximal lagoon, Pico Feo, San Blas, Panama. *Proceedings of the Eighth International Coral Reef Symposium*, Townsville, Vol. 1, pp. 767–772.

Friedman, G. M. (1998). Rapidity of marine carbonate cementation — Implications for carbonate diagenesis and sequence stratigraphy: Perspective. *Sedimentary Geology, 119*, 1–4.

Friedman, G. M., Amiel, A. J., & Schneidermann, N. (1974). Submarine cementation in reefs: Examples from the Red Sea. *Journal of Sedimentary Petrology*, *44*, 816–825.

Frisia, S., Borsato, A., Fairchild, I. J., & McDermott, F. (2000). Calcite fabrics, growth mechanisms, and environments of formation in speleothems from the Italian Alps and Southwestern Ireland. *Journal of Sedimentary Research*, *70*, 1183–1196.

Frost, S. H. (1977a). Ecological controls of Caribbean and Mediterranean Oligocene reef coral communities. *Proceedings of the Third International Coral Reef Symposium*, Vol. 2, Miami, pp. 367–373.

Frost, S. H. (1977b). Oligocene reef coral biogeography Caribbean and western Tethys. *Deuxième Symposium sur les Coraux et les Récifs Coralliens Fossiles*, Paris, Mémoires du Bureau de Recherches Géologiques et Minières, 89, pp. 352–342.

Frost, S. H. (1977c). Miocene to Holocene evolution of Caribbean province reef building corals. *Proceedings of the Third International Coral Reef Symposium*, Vol. 2, Miami, pp. 353–359.

Frost, S. H. (1981). Oligocene reef coral biofacies of the Vicentin, northeast Italy. In: D. H. Toomey (Ed.), *European fossil reef models* (pp. 483–539), SEPM (Special Publication 30).

Frost, S. H., Weiss, M. P. & Saunders, J. B. (Eds) (1977). *Reefs and related carbonates — Ecology and sedimentology*. Studies in Geology (Vol. 4, 421 pp.). Tulsa, OK: The American Association of Petroleum Geologists.

Fujita, K., Shimoji, H., & Nagai, K. (2006). Paleoenvironmental interpretations of Quaternary reef deposits based on comparisons of 10 selected modern and fossil larger foraminifera from the Ryukyu Islands, Japan. *Island Arc*, *15*, 420–436.

Fukami, H., Omori, M., & Hatta, M. (2000). Phylogenetic relationships in the coral family Acroporidae, reassessed by inference from the mitochondrial genes. *Zoological Science*, *17*, 689–696.

Furnas, M. J., & Mitchell, A. W. (1996). Nutrient inputs into the central Great Barrier Reef (Australia) from subsurface intrusions of Coral Sea Waters: A two-dimensional displacement model. *Continental Shelf Research*, *16*, 1127–1148.

Furnas, M. J., & Smayda, T. J. (1987). Inputs of subthermocline waters and nitrate onto the Campeche Bank. *Continental Shelf Research*, 7, 161–175.

Gabrié, C., & Montaggioni, L. F. (1982a). Sediments from fringing reefs of Réunion Island, Indian Ocean. *Sedimentary Geology*, *31*, 281–301.

Gabrié, C., & Montaggioni, L. F. (1982b). Sedimentary facies from the modern coral reefs, Jordan Gulf of Aqaba, Red Sea. *Coral Reefs*, *1*, 115–124.

Gagan, M. K., Ayliffe, L. K., Beck, J. W., Cole, J. E., Druffel, E. R. M., Dunbar, R. B., et al. (2000). New views of tropical paleoclimates from corals. *Quaternary Science Reviews*, *19*, 45–64.

Gagan, M. K., Ayliffe, L. K., Hopley, D., Cali, J. A., Mortimer, G. E., Chappell, J., et al. (1998). Temperature and surface-ocean water balance of the mid-Holocene tropical western Pacific. *Science*, *279*, 1014–1018.

Gagan, M. K., Chivas, A. R., & Herczeg, A. L. (1990). Shelf-wide erosion, deposition and suspended sediment transport during cyclone Winifred, Central Great Barrier Reef, Australia. *Journal of Sedimentary Research*, *60*, 456–470.

Gagan, M. K., Hendy, E. J., Haberle, S. G., & Hantoro, W. (2004). Post-glacial evolution of the Indo-Pacific Warm Pool and El-Niño-Southern Oscillation. *Quaternary International*, *118–119*, 127–143.

Gagan, M. K., Sandstrom, M. W., & Chivas, A. R. (1987). Restricted terrestrial carbon input to the continental shelf during cyclone Winifred: Implications for terrestrial runoff to the Great Barrier Reef Province. *Coral Reefs*, *6*, 113–119.

Gallup, C. D., Cheng, H., Taylor, F. W., & Edwards, R. L. (2002). Direct determination of the timing of sea level change during Termination II. *Science*, *295*, 310–313.

Gardiner, F. S., Greenstein, B. J., & Pandolfi, J. M. (1995). Taphonomic analysis of Florida reef corals: The effect of habitat in preservation. *Geological Society of America, Abstracts with Programs, 27*, 46.

Gardiner, L. (2001). Stability of late Pleistocene reef mollusks from San Salvador, Bahamas. *PALAIOS, 16*, 372–386.

Garrels, R. M., & MacKenzie, F. T. (1971). *Evolution of sedimentary rocks*. New York: Norton & Co., 397 pp.

Garret, P., Smith, D. L., Wilson, O. A., & Patriquin, D. (1971). Physiography, ecology and sediments of two Bermuda patch reefs. *Journal of Geology, 79*, 647–668.

Gattuso, J. P., & Buddemeier, R. W. (2000). Ocean biogeochemistry: Calcification and CO_2. *Science, 407*, 313.

Gattuso, J. P., Frankignoulle, M., Bourge, I., Romaine, S., & Buddemeier, R. W. (1998). Effect of calcium carbonate saturation of seawater on coral calcification. *Global and Planetary Change, 18*, 37–46.

Gattuso, J. P., Frankignoulle, M., Smith, S. V., Ware, J. R., & Wollast, R. (1996). Coral reefs and carbon dioxide. *Science, 271*, 1298–1300.

Gautier, D. L. (Ed.) (1986). *Roles of Organic Matter in Sediment Diagenesis* (203 pp.), Society of Economic Paleontologists and Mineralogists (Special Publication 38).

Gautret, P. (2001). Biochemical features of intraskeletal organic matrices within sponge and coral aragonite: Implications for diagenetic pathways. *Bulletin Tohoku University Museum, 1*, 164–172.

Gautret, P., Camoin, G., Golubic, S., & Sprachta, S. (2004). Biochemical control of calcium carbonate precipitation in modern lagoonal microbialites, Tikehau Atoll, French Polynesia. *Journal of Sedimentary Research, 74*, 462–478.

Gautret, P., & Trichet, J. (2005). Automicrites in modern cyanobacterial stromatolitic deposits of Rangiroa, Tuamotu Archipelago, French Polynesia: Biochemical parameters underlaying their formation. *Sedimentary Geology, 178*, 55–73.

Geister, J. (1977). The influence of wave exposure on the ecological zonation of Caribbean coral reefs. *Proceedings of the Third International Coral Reef Symposium*, Miami, Vol. 1, pp. 23–29.

Geister, J. (1980). Calm-water reefs and rough-water reefs of the Caribbean Pleistocene. *Acta Palaeontologica Polonica, 25*, 541–556.

Getty, S. R., Asmerom, Y., Quinn, T. M., & Budd, A. F. (2001). Accelerated Pleistocene coral extinctions in the Caribbean Basin shown by uranium-lead (U-Pb) dating. *Geology, 29*, 639–642.

Gherardi, D. F. M., & Bosence, D. W. J. (2005). Late Holocene reef growth and relative sea-level changes in Atol Das Rocas, equatorial south Atlantic. *Coral Reefs, 24*, 264–273.

Gibbard, P., Cohen, K., & Ogg, J. (2008). Quaternary period. In: J. G. Ogg, J. Ogg & F. M. Gradstei (Eds), *A concise geologic time scale* (pp. 149–158). Cambridge: Cambridge University Press.

Gili, E., Skelton, P. W., Vicens, E., & Obrador, A. (1995). Corals to rudists — An environmentally induced assemblage succession. *Palaeogeography, Palaeoclimatology, Palaeoecology, 119*, 127–136.

Ginsburg, R. N. (1956). Environmental relationships of grain size and constituent particles in some south Florida carbonate sediments. *Bulletin of American Association of Petroleum Geologists, 10*, 2384–2427.

Ginsburg, R. N. (1964). *South Florida carbonate sediments*. Geological Society of America, Annual Meeting Field Trip Guidebook, Field Trip (Vol. 1, 72 pp.) (Reprinted as Ginsburg, 1972, Sedimenta II, University of Miami, Miami FL).

Ginsburg, R. N., & Choi, D. R. (1983). Quaternary reefs of Belize, Central America. In: J. T. Baker, R. M. Carter, P. W. Sammarco & K. P. Stark (Eds), *Proceedings of the Great Barrier Reef Conference*, pp. 27–42.

Ginsburg, R. N., & James, N. P. (1976). Submarine botryoidal aragonite in Holocene reef limestones. *Geology, 4*, 431–436.
Gischler, E. (1994). Sedimentation on three Caribbean atolls: Glovers Reef, Lighthouse Reef and Turneffe Islands, Belize. *Facies, 31*, 243–254.
Gischler, E. (2003). Holocene lagoonal development in the isolated carbonate platforms off Belize. *Sedimentary Geology, 159*, 113–132.
Gischler, E. (2007). Pleistocene facies of Belize barrier and atoll reefs. *Facies, 53*, 27–41.
Gischler, E. (2008). Accretion patterns in Holocene tropical coral reefs: Do massive coral reefs in deeper water with slowly growing corals accrete faster than shallower branched coral reefs with rapidly growing coral? *International Journal of Earth Sciences, 97*, 851–859.
Gischler, E., & Hudson, J. H. (1998). Holocene development of three isolated carbonate platforms, Belize, Central America. *Marine Geology, 144*, 333–347.
Gischler, E., & Hudson, J. H. (2004). Holocene development of the Belize Barrier Reef. *Sedimentary Geology, 164*, 223–236.
Gischler, E., Hudson, J. H., & Pisera, A. (2008). Late Quaternary reef growth and sea level in the Maldives (Indian Ocean). *Marine Geology, 250*, 104–113.
Gischler, E., & Lomando, A. J. (1997). Holocene cemented beach deposits in Belize. *Sedimentary Geology, 110*, 277–297.
Gischler, E., & Lomando, A. J. (1999). Recent sedimentary facies of isolated carbonate platforms, Belize-Yucatan system, Central America. *Journal of Sedimentary Research, 69*, 747–763.
Gischler, E., & Lomando, A. J. (2000). Isolated carbonate platforms of Belize, Central America: Sedimentary facies, Late Quaternary history and controlling factors. In: E. Insalaco, P. W. Skelton & T. J. Palmer (Eds), *Carbonate platform systems: Components and interactions* (pp. 135–146), Geological Society of London (Special Publication 178).
Gischler, E., & Oschmann, W. (2005). Historical climate variation in Belize (Central America) as recorded in scleractinian coral skeletons. *PALAIOS, 20*, 159–174.
Gischler, E., & Pisera, A. (1999). Shallow water rhodoliths from Belize reefs: Neues Jahrbuch für Geologie und Paläontologie. *Abhandlungen, 214*, 71–93.
Gischler, E., & Zingeler, D. (2002). The origin of carbonate mud in isolated carbonate platforms of Belize, Central America. *International Journal of Earth Sciences, 91*, 1054–1070.
Given, K. R., & Wilkinson, B. H. (1985). Kinetic control of morphology, composition, and mineralogy of abiotic sedimentary carbonates. *Journal of Sedimentary Petrology, 55*, 109–119.
Gladefelter, W. B. (1982). White band disease in *Acropora palmata*: Implications for the structure and function of shallow reefs. *Bulletin of Marine Science, 32*, 639–643.
Gladfelter, E. H. (1983). Skeletal development in *Acropora cervicornis* II. Diel patterns of calcium carbonate accretion. *Coral Reefs, 2*, 91–100.
Glenn-Sullivan, E. C., & Evans, I. (2001). The effects of time-averaging and taphonomy on the identification of reefal sub-environments using larger foraminifera: Apo Reef, Mindoro, Philippines. *PALAIOS, 16*, 399–408.
Glynn, P. W. (1973). Aspects of the ecology of coral reefs in the western Atlantic region. In: O. A. Jones & R. Endean (Eds), *Biology and geology of coral reefs* (Vol. II: Biology 1, pp. 271–325). New York: Academic Press.
Glynn, P. W. (1974). Rolling stones among the Scleractinia: Mobile coralliths in the Gulf of Panama. *Proceedings of the Second International Coral Reef Symposium*, Brisbane, Vol. 2, pp. 183–198.
Glynn, P. W. (1997). Eastern Pacific reef coral biogeography and faunal flux: Durham's dilemma revisited. *Proceedings of the Eighth International Coral Reef Symposium*, Panama, Vol. 1, pp. 371–378.
Glynn, P. W., & Ault, J. S. (2000). A biogeographic analysis and review of the far eastern Pacific coral reef region. *Coral Reefs, 19*, 1–23.

Glynn, P. W., & Macintyre, I. G. (1977). Growth rates and age of coral reefs on the Pacific coast of Panama. *Proceedings of the Third International Coral Reef Symposium*, Miami, Vol. 2, pp. 251–260.

Glynn, P. W., & Wellington, G. M. (1983). *Corals and coral reefs of the Galápagos Islands*. Berkeley, CA: University of California Press, 330 pp.

Goldberg, N. (2006). Age estimates and description of rhodoliths from Esperance Bay, Western Australia. *Journal of the Marine Biological Association of the United Kingdom*, *86*, 1291–1296.

Golubic, S. (1991). Modern stromatolites — A review. In: R. E. Riding (Ed.), *Calcareous algae and stromatolites* (pp. 541–561). Berlin: Springer-Verlag.

Golubic, S., Seong-Joo, L., & Browne, K. M. (2000). Cyanobacteria: Architects of sedimentary structures. In: R. E. Riding & S. M. Awramok (Eds), *Microbial sediments* (pp. 57–67). Berlin: Springer-Verlag.

Gonneea, M. E., & Paytan, A. (2006). Phase associations of barium in marine sediments. *Marine Chemistry*, *100*, 124–135.

González, L. A., Carpenter, S. J., & Lohmann, K. C. (1990). Inorganic calcite morphology: The roles of fluid chemistry and fluid flow. *Journal of Sedimentary Petrology*, *62*, 382–399.

González-Muñoz, M. T., Ben Chekroun, K., Ben Aboud, A., Arias, J. M., & Rodriguez-Gallego, M. (2000). Bacterially induced Mg-calcite formation: Role of Mg^{2+} in development of crystal morphology. *Journal of Sedimentary Research*, *70*, 559–564.

Goodkin, N. F., Hughen, K. A., Cohen, A. L., & Smith, S. R. (2005). Record of Little Ice Age sea surface temperatures at Bermuda using a growth-dependent calibration of coral Sr/Ca. *Paleoceanography*, *20*, PA4016, doi: 10.1029/2005PA001140.

Goodwin, I. D. (1998). Did changes in Antarctic ice volume influence late Holocene sea-level lowering? *Quaternary Science Reviews*, *17*, 319–332.

Goodwin, I. D., & Harvey, D. (2008). Subtropical sea-level history from coral microatolls in the Southern Cook Islands since 300 AD. *Marine Geology*, *253*, 14–25.

Goreau, T. F. (1959). The physiology of skeleton formation in corals. 1. A method for measuring the rate of calcium deposition by corals under different conditions. *Biological Bulletin*, *116*, 59–75.

Goreau, T. F., & Goreau, N. I. (1973). The ecology of Jamaican coral reefs. II. Geomorphology, zonation, and sedimentary phases. *Bulletin of Marine Science*, *23*, 399–464.

Goreau, T. F., & Hartman, W. D. (1963). Boring sponges as controlling factors in the formation and maintenance of coral reefs. In: R. F. Sognnaes (Ed.), *Mechanisms of Hard Tissue Destruction* (pp. 25–54), American Association for the Advancement of Science, Transactions (Publication 75), Washington, DC.

Goreau, T. F., & Land, L. S. (1974). Fore-reef morphology and depositional Processes, North Jamaica. In: L. F. Laporte (Ed.), *Reefs in Time and Space* (pp. 77–89), SEPM (Special Publication 18), Tulsa, OK.

Gorshkov, A. I., Baturin, G. N., Bogdanova, O. Yu., & Magazina, L. O. (2000). Sulphides in phosphorites of African Island: Morphological peculiarities and origin. *Lithology and Mineral Resources (Litologiya i Poleznye Iskopaemye)*, *35*(2), 183–188.

Grammer, G. M., Crescini, C., McNeil, D. F., & Taylor, L. H. (1999). Quantifying rates of syndepositional marine cementation in deeper Platform environments — new insights into a fundamental process. *Journal of Sedimentary Research*, *69*, 202–207.

Graus, R. R., Chamberlain, J. A., & Boker, A. M. (1977). Structural modification of corals in relation to waves and currents. In: S. H. Frost, M. P. Weiss & S. B. Saunders (Eds), *Reefs and related carbonates: Ecology and sedimentology*. Studies in Geology (Vol. 4, pp. 135–153). Tulsa, OK: American Association of Petroleum Geologists.

Graus, R. R., & Macintyre, I. G. (1989). The zonation pattern of Caribbean coral reefs as controlled by wave and light energy input, bathymetric setting and reef morphology. *Coral Reefs*, *8*, 9–18.

Graus, R. R., MacIntyre, I. G., & Herchenroder, B. E. (1984). Computer simulation of the reef zonation at Discovery Bay, Jamaica: Hurricane disruption and long-term physical oceanographic controls. *Coral Reefs, 3,* 59–68.

Graus, R. R., MacIntyre, I. G., & Herchenroder, B. E. (1985). Computer simulation of the Holocene facies history of a Caribbean fringing reef (Galeta Point, Panama). *Proceedings of the Fifth International Reef Congress*, Tahiti, Vol. 3, pp. 317–322.

Gray, S. C., Hein, J. R., Hausmann, R., & Radtke, U. (1992). Geochronology and subsurface stratigraphy of Pukapuka and Rakahanga atolls, Cook Islands: Late Quaternary reef growth and sea level history. *Palaeogeography, Palaeoclimatology, Palaeoecology, 91,* 377–394.

Greegor, R. B., Pingitore, N. E., Jr., & Lytle, F. W. (1997). Strontianite in coral skeletal aragonite. *Science, 275,* 1452–1454.

Greenstein, B. J. (1993). Is the fossil record of regular echinoids really so poor? A comparison of living and subfossil assemblages. *PALAIOS, 8,* 587–601.

Greenstein, B. J. (2007). Taphonomy: Detecting critical events in fossil reef-coral assemblages. In: R. B. Aronson (Ed.), *Geological approaches to coral reef ecology* (pp. 31–60). New York: Springer.

Greenstein, B. J., & Curran, H. A. (1997). How much ecological information is preserved in fossil reefs and how reliable is it? *Proceedings of the Eighth International Coral Reef Symposium*, Panama, Vol. 1, pp. 417–422.

Greenstein, B. J., Curran, H. A., & Pandolfi, J. M. (1998). Shifting ecological baselines and the demise of *Acropora cervicornis* in the western North Atlantic and Caribbean province: A Pleistocene perspective. *Coral Reefs, 17,* 249–261.

Greenstein, B. J., Harris, L. A., & Curran, H. A. (1998). Comparison of recent coral life and death assemblages to Pleistocene reef communities: Implications for rapid faunal replacement observed on modern reefs. *Carbonates and Evaporites, 13,* 23–31.

Greenstein, B. J., & Moffat, H. A. (1996). Comparative taphonomy of Holocene and Pleistocene corals, San Salvador, Bahamas. *PALAIOS, 11,* 57–63.

Greenstein, B. J., & Pandolfi, J. M. (1997). Preservation of community structure in modern reef coral life and death assemblages of the Florida Keys: Implications for the Quaternary fossil record of coral reefs. *Bulletin of Marine Science, 61,* 431–452.

Greenstein, B. J., & Pandolfi, J. M. (2003). Taphonomic alteration of reef corals: Effects of reef environment and coral growth form. II: The Florida Keys. *PALAIOS, 18,* 495–509.

Greenstein, B. J., & Pandolfi, J. M. (2008). Escaping the heat: range shifts of reef coral taxa in coastal Western Australia. *Global change Biology, 14,* 1–16.

Greenstein, B. J., Pandolfi, J. M., & Curran, H. A. (1998). The completeness of the Pleistocene fossil record: implications for stratigraphic adequacy. In: S. K. Donovan & C. R. C. Paul (Eds), *The adequacy of the fossil record* (pp. 75–109). New York: Wiley.

Greenstein, B. J., Pandolfi, J. M., & Moran, P. J. (1995). Taphonomy of crown-of-thorns starfish: Implications for recognizing ancient population outbreaks. *Coral Reefs, 14,* 91–97.

Greer, L., & Swart, P. W. (2006). Decadal cyclicity of regional mid-Holocene precipitation: Evidence from Dominican coral proxies. *Paleoceanography, 21,* PA2020, doi: 10.1029/2005PA001166.

Gregg, J. M., Howard, S. A., & Mazzullo, S. J. (1992). Early diagenetic recrystallization of Holocene (<3000 years old) peritidal dolomites, Ambergris Cay, Belize. *Sedimentology, 39,* 143–160.

Grigg, R. W. (1982). Darwin point: A threshold for atoll formation. *Coral Reefs, 1,* 29–34.

Grigg, R. W. (1983). Community structure, succession, and development of coral reefs in Hawaii. *Marine Ecology Progress Series, 11,* 1–14.

Grigg, R. W. (1988). Paleoceanography of coral reefs in the Hawaiian–Emperor chain. *Science, 240,* 1737–1743.

Grigg, R. W. (1997). Paleoceanography of coral reefs in the Hawaiian–Emperor chain–revisited. *Coral Reefs, 16,* S33–S38.

Grigg, R. W. (1998). Holocene coral reef accretion in Hawaii: A function of wave exposure and sea level history. *Coral Reefs*, *17*, 263–272.

Grigg, R. W., Grossman, E. E., Earle, S. A., Gittings, S. R., Lotts, D., & McDonough, J. (2002). Drowned reefs and antecedent karst topography, Au'au Channel, S.E. Hawaiian Islands. *Coral Reefs*, *21*, 73–82.

Grigg, R. W., & Hey, R. (1992). Paleoceanography of the tropical eastern Pacific ocean. *Science*, *255*, 172–178.

Grigor'ev, D. P. (1965). Ontogeny of Minerals (Ontogeniya Mineralov Izdatel'stro, L'vovskii Universitet, 1961) Israel Programmes for Scientific Translations, Jerusalem, 242 pp.

Gross, M. G., Milliman, J. D., Tracey, J. I., & Ladd, H. S. (1969). Marine geology of Kure and Midway Atolls, Hawaii: A preliminary report. *Pacific Science*, *23*, 17–25.

Grossman, E. E., Barnhardt, W. A., Hart, P., Richmond, B. M., & Field, M. E. (2006). Shelf stratigraphy and the influence of antecedent substrate on Holocene reef development, south Oahu, Hawaii. *Marine Geology*, *226*, 97–114.

Grossman, E. E., & Fletcher, C. H., III (2004). Holocene reef development where wave energy reduces accommodation space, Kailua Bay, Windward Oahu, Hawaii, U.S.A. *Journal of Sedimentary Research*, *74*, 49–63.

Grossman, E. E., Fletcher, C. H., III, & Richmond, B. M. (1998). The Holocene sea level highstand in the equatorial Pacific: Analysis of the insular paleosea-level database. *Coral Reefs*, *17*, 309–327.

Grottoli, A. G. (2000). Stable carbon isotopes ($\delta^{13}C$) in coral skeletons. *Oceanography*, *13*, 93–97.

Grottoli, A. G. (2001). Past climate from corals. In: J. Steele, S. Thorpe & K. Turekian (Eds), *Encyclopedia of ocean sciences* (pp. 2098–2107). London: Academic Press.

Grotolli, A. G. (2002). Effect of light and brine shrimp on skeletal $\delta^{13}C$ in the Hawaiian coral *Porites compressa*: A tank experiment. *Geochima et Cosmochimica Acta*, *66*, 1955–1967.

Grottoli, A. G., & Eakin, C. M. (2007). A review of modern coral $\delta^{18}O$ and $\delta^{14}C$ proxy records. *Earth Science Reviews*, *81*, 67–91.

Grottoli, A. G., Gille, S. T., Druffel, E. R. M., & Dunbar, R. D. B. (2003). Decadal timescale shift in a central equatorial Pacific coral radiocarbon record. *Radiocarbon*, *45*, 91–99.

Grottoli, A. G., & Wellington, G. M. (1999). Effect of light and zooplankton on skeletal $\delta^{13}C$ values in the eastern Pacific corals *Pavona clavus* and *Pavona gigantea*. *Coral Reefs*, *18*, 29–41.

Grumet, N. S., Abram, N. J., Beck, J. D., Dunbar, R. B., Gagan, M. K., Guilderson, T. P., et al. (2004). Coral radiocarbon records of Indian Ocean water mass mixing and wind-induced upwelling along the coast of Sumatra, Indonesia. *Journal of Geophysical Research*, *109*, doi: 10.1029/2003JC002087.

Guilcher, A. (1988). *Coral reef geomorphology. Coastal morphology and research*. Chichester: Wiley, 228 pp.

Guilderson, A., & Schrag, D. P. (1998). Radiocarbon variability in the western equatorial Pacific inferred from a high resolution coral record from Nauru Island. *Journal of Geophysical Research*, *103*, 24641–24650.

Guilderson, T. P., Fairbanks, R. G., & Rubenstone, J. L. (1994). Tropical temperature variations since 20,000 years ago: Modulating interhemispheric climate change. *Science*, *263*, 663–665.

Guilderson, T. P., Fairbanks, R. G., & Rubenstone, J. L. (2001). Tropical Atlantic coral oxygen isotopes: Glacial–interglacial sea surface temperatures and climate change. *Marine Geology*, *172*, 75–89.

Guilderson, T. P., Schrag, D. P., Goddard, E., Kashgarian, M., Wellington, G. M., & Linsley, B. K. (2000). Southwest subtropical Pacific surface water radiocarbon in a high-resolution coral. *Radiocarbon*, *42*, 249–256.

Guille, G., Goutière, G., Sornein, J. F., Buigues, D., Gachon, A., & Guy, C. (1996). The atolls of Mururoa and Fangataufa (French Polynesia). I. Geology–Petrology–Hydrology, from volcano to Atoll. Musée Océanographique, Monaco.

Guinotte, J. M., Buddemeier, R. W., & Kleypas, J. A. (2003). Future coral reef habitat marginality: Temporal and spatial effects of climate change in the Pacific basin. *Coral Reefs*, *22*, 551–558.

Gvirtzman, G. (1994). Fluctuations of sea level during the past 400,000 years: The record of Sinai, Egypt (northern Red Sea). *Coral Reefs*, *13*, 203–214.

Gvirtzman, G., & Friedman, G. M. (1977). Sequence of progressive diagenesis in coral reefs. In: S. H. Frost, M. P. Weiss & J. B. Saunders (Eds), *Reefs and related carbonates — Ecology and sedimentology*. Studies in Geology (Vol. 4, pp. 357–380). Tulsa, OK: American Association of Petroleum Geologists.

Halfar, J., Godinez-Orta, L., Riegl, B., Valdez-Holguin, J. E., & Borges, J. M. (2005). Living on the edge: High-latitude *Porites* carbonate production under temperate eutrophic conditions. *Coral Reefs*, *24*, 582–592.

Halfar, J., & Mutti, M. (2005). Global dominance of coralline red-algal facies: A response to Miocene oceanographic events. *Geology*, *33*, 481–484.

Halfar, J., Zack, T., Kronz, A., & Zachos, J. C. (2000). Growth and high-resolution paleoenvironmental signals of rhodoliths (coralline red algae): A new biogenic archive. *Journal of Geophysical Research*, *105*, 22 107-22, 116.

Halley, R. B., Vacher, H. L., & Shinn, E. A. (1997). Geology and hydrogeology of the Florida Keys. In: H. L. Vacher, & T. Quinn, (Eds), *Geology and hydrogeology of carbonate islands*. Development in Sedimentology (Vol. 54, 217–248). Amsterdam: Elsevier Science, B.V.

Hallock, P. (1981). Production of carbonate sediments by selected large benthic foraminifera on two Pacific coral reefs. *Journal of Sedimentary Petrology*, *51*, 467–474.

Hallock, P. (1988). The role of nutrient availability in bioerosion: Consequences to carbonate buildups. *Palaeogeography, Palaeoclimatology, Palaeoecology*, *63*, 275–291.

Hallock, P. (1997). Reefs and reef limestones in Earth history. In: C. Birkeland (Ed.), *Life and death of coral reefs* (pp. 13–42). London: Chapman & Hall.

Hallock, P., & Glenn, E. C. (1986). Larger foraminifera: A tool for paleoenvironmental analysis of Cenozoic carbonate depositional facies. *PALAIOS*, *1*, 389–398.

Hallock, P., Lidz, B. H., Cockey-Burkhard, E. M., & Donnelly, K. B. (2003). Foraminifera as bioindicators in coral reef assessment and monitoring: The FORAM index. *Environmental Monitoring and Assessment*, *81*, 221–238.

Hallock, P., & Schlager, W. (1986). Nutrient excess and the demise of coral reefs and carbonate platforms. *PALAIOS*, *1*, 389–398.

Hamelin, B., Ferrand, J. L., Alleman, L., Nicolas, E., & Véron, A. J. (1997). Isotopic evidence of pollutant lead transport for North America to the subtropical North Atlantic gyre. *Geochimica et Cosmochimica Acta*, *61*, 4423–4428.

Hamilton, H. G. H., & Brakel, W. H. (1984). Structure and coral fauna of East African reefs. *Bulletin of Marine Science*, *34*, 248–266.

Handford, C. R., & Loucks, R. G. (1993). Carbonate depositional sequences and systems tracts — Responses of carbonate platforms to relative sea-level changes. In: R. G. Loucks & J. S. Sarg (Eds), *Carbonate sequence stratigraphy. American Association of Petroleum Geologists, Memoir*, *57*, 3–41.

Hanebuth, T., Stattegger, K., & Grootes, P. M. (2000). Rapid flooding on the Sunda Shelf — A late glacial sea level record. *Science*, *288*, 1033–1035.

Hanshaw, B. B., & Back, W. (1980). Chemical mass wasting of the northern Yucatan Peninsula by groundwater dissolution. *Geology*, *8*, 222–224.

Hanshaw, B. B., Back, W., & Deike, R. G. (1971). A geochemical hypothesis for dolomitization by groundwater. *Economic Geology*, *66*, 710–724.

Hantoro, W. S., Pirazzoli, P. A., Jouannic, C., Faure, H., Hoang, C. T., Radtke, U., et al. (1994). Quaternary uplifted coral reef terraces on Alor Island, East Indonesia. *Coral Reefs, 13*, 215–223.

Hardie, L. A. (1987). Dolomitization: A critical review of some current views. *Journal of Sedimentary Petrology, 57*, 166–183.

Harmelin-Vivien, M. L. (1994). The effects of storms and cyclones on coral reefs: A review. *Journal of Coastal Research, Special Issue, Coastal Hazards, 12*, 211–231.

Harmelin-Vivien, M. L., & Laboute, P. (1986). Catastrophic impact of hurricanes on atoll outer slopes in the Tuamotu (French Polynesia). *Coral Reefs, 5*, 55–62.

Harney, J. N., & Fletcher, C. H. (2003). A budget of carbonate framework and sediment production, Kailua Bay, Oahu, Hawaii. *Journal of Sedimentary Research, 73*, 856–868.

Harney, J. N., Grossman, E. E., Richmond, B. M., & Fletcher, C. H. (2000). Age and composition of carbonate shoreface sediments, Kailua Bay, Oahu, Hawaii. *Coral Reefs, 19*, 141–154.

Harriott, V. J., & Banks, S. A. (2002). Latitudinal variation in coral communities in eastern Australia: A qualitative biophysical model of factors regulating coral reefs. *Coral Reefs, 21*, 83–94.

Harris, P. T., & Davies, P. J. (1989). Submerged reefs and terraces on the shelf edge of the Great Barrier Reef, Australia. Morphology, occurrence and implications for reef evolution. *Coral Reefs, 8*, 87–98.

Harris, P. T., Heap, A. D., Marshall, J. F., & McCulloch, M. (2008). A new coral reef province in the Gulf of Carpentaria, Australia: Colonisation, growth and submergence during the early Holocene. *Marine Geology, 251*, 85–97.

Harris, P. T., Heap, A. D., Wassenberg, Th., & Passlow, V. (2004). Submerged coral reefs in the Gulf of Carpentaria, Australia. *Marine Geology, 207*, 185–191.

Harrison, R. S. (1977). Caliche profiles: Indicators of near-surface subaerial diagenesis, Barbados, West Indies. *Bulletin of Canadian Petroleum Geology, 25*, 123–173.

Harrison, R. S., & Coniglio, M. (1985). Origin of the Pleistocene Key Largo Limestone, Florida Keys. *Canadian Society of Petroleum Geologists, Bulletin, 33*, 350–358.

Harris, W. H. (1971). *Groundwater-carbonate rock chemical interactions, Barbados*, W. I., Ph. D. dissertation, Brown University, Providence, R. I., 348 pp.

Harris, W. H., & Matthews, R. K. (1968). Subaerial diagenesis of carbonate sediments: Efficiency of the solution-precipitation process. *Science, 160*, 77–79.

Hart, D. E., & Kench, P. S. (2007). Carbonate production of an emergent reef platform, Warraber Island, Torres Strait, Australia. *Coral Reefs, 26*, 53–68.

Hart, S. R., & Cohen, A. L. (1996). An ion probe study of annual cycles of Sr/Ca and other trace elements in corals. *Geochimica et Cosmochimica Acta, 60*, 3075–3084.

Harvey, N. (1986). *The Great Barrier Reef: Shallow seismic investigations*. Monograph Series (Vol. 14, 103 pp), Department of Geography, James Cook University.

Hattin, D. E., & Warren, V. L. (1988). Stratigraphic analysis of a fossil *Neogoniolithon*-capped patch reef and associated facies, San Salvador, Bahamas. *Coral Reefs, 8*, 19–30.

Haug, G. H., & Tiedemann, R. (1998). Effect of the formation of the Isthmus of Panama on Atlantic thermohaline circulation. *Nature, 393*, 673–676.

Hays, J. D., Imbrie, J., & Shackleton, N. J. (1976). Variations in the earth's orbit: Pacemaker of the Ice Ages. *Science, 194*, 1121–1132.

Heap, A. D., Dickens, G. R., & Stewart, L. K. (2001). Late Holocene sediment in Nara Inlet, central Great Barrier Reef platform, Australia: Sediment accumulation on the middle shelf of a tropical mixed clastic/carbonate system. *Marine Geology, 176*, 39–54.

Heap, A. D., Dickens, G. R., Stewart, L. K., & Woolfe, K. J. (2002). Holocene storage of siliciclastic sediment around islands on the middle shelf of the Great Barrier Reef Platform, north-east Australia. *Sedimentology, 49*, 603–621.

Hearty, P. J. (1998). The geology of Eleuthera Island, Bahamas: A Rosetta Stone of Quaternary stratigraphy and sea-level history. *Quaternary Science Reviews*, *17*, 333–355.

Hearty, P. J., & Dai Pra, G. (1992). The age and stratigraphy of Quaternary coastal deposits along the Gulf of Taranto (south Italy). *Journal of Coastal Research*, *8*, 882–905.

Hearty, P. J., Hollin, J. T., Neumann, A. C., O'Leary, M. J., & McCulloch, M. T. (2007). Global sea-level fluctuations during the Last Interglaciation (MIS 5e). *Quaternary Science Reviews*, *26*, 2090–2112.

Hearty, P. J., & Kaufman, D. S. (2000). Whole-rock aminostratigraphy and Quaternary sea-level history of the Bahamas. *Quaternary Research*, *54*, 163–173.

Hearty, P. J., & Kindler, P. (1997). The stratigraphy and surficial geology of New Providence and surrounding Islands, Bahamas. *Journal of Coastal Research*, *13*, 798–812.

Heck, K. L., & McCoy, E. D. (1978). Long-distance dispersal and the reef-building corals of the eastern Pacific. *Marine Biology*, *48*, 349–356.

Hein, J. R., Gray, S. C., Richmond, B. M., & White, L. D. (1992). Dolomitization of Quaternary reef limestone, Aitutaki, Cook Islands. *Sedimentology*, *39*, 645–661.

Heinrich, H. (1988). Origin and consequences of cyclic ice rafting in the Northeast Atlantic Ocean during the past 130,000 years. *Quaternary Research*, *29*, 142–152.

Heiss, G. (1995). Carbonate production by scleractinian corals at Aqaba, Gulf of Aqaba, Red Sea. *Facies*, *33*, 19–34.

Heckel, P. H. (1974). Carbonate buildups in the geologic record: A review. In: L. F. Laporte (Ed.), *Reefs in Time and Space: Selected examples from the Recent and Ancient* (pp. 90–154). SEPM (Special Publication 18), Tulsa, OK.

Henderson, G. M., Robinson, L. F., Cox, K., & Thomas, A. L. (2006). Recognition of non-Milankovitch sea-level highstands at 185 and 343 thousand years ago from U-Th dating of Bahamas sediment. *Quaternary Science Reviews*, *25*, 3346–3358.

Hendy, E. J., Gagan, M. K., Alibert, C. A., McCulloch, M. T., Lough, J. M., & Isdale, P. J. (2002). Abrupt decrease in tropical Pacific sea surface salinity at the end of Little Ice Age. *Science*, *295*, 1511–1514.

Hendy, E. J., Gagan, M. K., & Lough, J. M. (2003). Chronological control of coral records using luminescent lines and evidence for non-stationary ENSO teleconnections in northeast Australia. *The Holocene*, *13*, 187–199.

Henny, R. W., Mercer, J. W., & Zbur, R. T. (1974). Near-surface geologic investigations at Eniwetok Atoll. *Proceedings of the Second International Coral Reef Symposium*, Manila, Vol. 2, pp. 615–626.

Herrera-Escalante, T., Lopez-Pérez, R. A., & Levte-Morales, G. E. (2005). Bioerosion caused by the sea urchin *Diadema mexicanum* (Echinodermata: Echinoidea) at Bahias de Huatulco, Western Mexico. *Revista de Biologia Tropical*, *53*(Suppl. 3), 263–273.

Hetzinger, S., Pfeiffer, M., Dullo, W.-Ch., Keelyside, N., Latif, M., & Zinke, J. (2008). Caribbean coral tracks Atlantic multidecadal oscillation and past hurricane activity. *Geology*, *36*, 11–14.

Hewins, M. R., & Perry, C. T. (2006). Bathymetric and environmentally influenced patterns of carbonate sediment accumulation in three contrasting reef settings, Danjugan Island, Philippines. *Journal of Coastal Research*, *22*, 812–824.

Highsmith, R. C. (1981). Coral bioerosion: Damage relative to skeletal density. *The American Naturalist*, *117*, 193–198.

Hildebrand, A. R., Pilkington, M., Connors, M., Ortíz-Alemanand, C., & Chavez, R. E. (1995). Size and structure of the Chicxulub crater revealed by horizontal gravity gradients and cenotes. *Nature*, *376*, 415–417.

Hillgärtner, H., Dupraz, C., & Hug, W. (2001). Microbially induced cementation of carbonate sands: Are micritic meniscus cements good indicators of vadose diagenesis? *Sedimentology*, *48*, 117–131.

Hillis, L. W. (1997). Coral reefs from a calcareous green alga perspective, and a first carbonate budget. *Proceedings of the Eighth International Coral Reef Symposium*, Townsville, Australia, Vol. 1, pp. 761–766.

Hillis, L. W. (2001). The calcareous reef alga *Halimeda* (Chlorophyta, Byropsidales): A Cretaceous genus that diversified in the Cenozoic. *Palaeogeography, Palaeoclimatology, Palaeoecology*, *166*, 89–100.

Hillis-Colinvaux, L. (1980). Ecology and taxonomy of *Halimeda*: Primary producers of coral reefs. *Advances in Marine Biology*, *17*, 1–327.

Hills, D. J., & Jones, B. (2000). Peyssonnelid rhodoliths from the late Pleistocene Ironshore Formation, Grand Cayman, British West Indies. *PALAIOS*, *15*, 212–224.

Hine, A. C., Hallock, P., Harris, M. W., Mullins, H. T., Belknap, D. F., & Jaap, W. C. (1988). *Halimeda* bioherms along an open seaway: Miskito Channel, Nicaraguan Rise, SW Caribbean Sea. *Coral Reefs*, *6*, 173–178.

Hodell, D. A., Mead, G. A., & Mueller, P. A. (1990). Variation in the strontium isotopic composition of seawater (8 Ma to present): Implications for chemical weathering rates and dissolved fluxes to the oceans. *Chemical Geology (Isotope Geosciences Section)*, *80*, 291–307.

Hoegh-Gulberg, O. (2005). Low coral cover in a high-CO_2 world. *Journal of Geophysical Research*, *11*, C09SO6, doi: 10.1029/2004JC002528.

Hoffmeister, J. E., & Ladd, H. S. (1944). The antecedent-platform theory. *Journal of Geology*, *52*, 388–402.

Hoffmeister, J. E., & Multer, H. G. (1968). Geology and origin of the Florida Keys. *Geological Society of America Bulletin*, *79*, 1487–1502.

Hohenegger, J., & Yordanova, E. (2001). Displacement of larger foraminifera at the western slope of Motobu Peninsula (Okinawa, Japan). *PALAIOS*, *16*, 53–72.

Hohenegger, J., Yordanova, E., Nakano, Y., & Tatzreiter, F. (1999). Habitats of larger foraminifera on the upper reef slope of Sesoko Island, Okinawa, Japan. *Marine Micropaleontology*, *36*, 109–168.

Hongo, C., & Kayanne, H. (2008). Holocene coral reef development under windward and leeward locations at Ishigaki Island, Ryuku Islands, Japan. *Sedimentary Geology*, *214*, 62–73.

Hopley, D. (1984). The Holocene "high-energy window" on the central Great Barrier Reef. In: B. G. Thom (Ed.), *Coastal Geomorphology in Australia* (pp. 135–150). Sydney: Academic Press.

Hopley, D. (1986a). Corals and reefs as indicators of paleo-sea levels, with special reference to the Great Barrier Reef. In: O. van de Plassche (Ed.), *Sea-level research: A manual for the collection and evaluation of data* (pp. 195–228). Norwich, UK: Geo Books.

Hopley, D. (1986b). Beach-rock as sea-level indicator. In: O. van de Plassche (Ed.), *Sea-level research: A manual for the collection and evaluation of data* (pp. 157–174). Norwich, UK: Geo Books.

Hopley, D. (1994). Continental shelf reef systems. In: R. W. G. Carter & C. D. Woodroffe (Eds), *Coastal evolution: Late Quaternary shoreline, morphodynamics* (pp. 303–340). Cambridge University Press.

Hopley, D., Graham, T. L., & Rasmussen, C. (1997). Submerged shelf-edge reefs, Great Barrier Reef, Australia. Recent advances in marine science and technology, *Proceedings of the PACON International Conference*, Honolulu, HI, pp. 305–315.

Hopley, D., & Partain, B. (1987). The structure and development of fringing reefs off the Great Barrier Reef Province. *Fringing Reef Workshop: Science, Industry and Management*, Magnetic Island, Australia, pp. 13–33.

Hopley, D., Slocombe, A. M., Muir, F., & Grant, C. (1983). Nearshore fringing reefs in north Queensland. *Coral Reefs*, *1*, 151–160.

Hopley, D., Smithers, S. G., & Parnell, K. E. (2007). *The geomorphology of the Great Barrier Reef: Development, diversity, and change*. Cambridge: Cambridge University Press, 532 pp.
Hopley, D. H. (1982). *The geomorphology of the Great Barrier Reef: Quaternary development of coral reefs*. New York: Wiley-Interscience, 453 pp.
Hottinger, L. (1983). Processes determining the distribution of larger foraminifera in space and time. *Utrecht Micropalaeontological Bulletin, 30*, 239–253.
Houck, J. E., Buddemeier, R. W., & Chave, K. E. (1975). Skeletal low-magnesium calcite in living scleractinian corals. *Science, 189*(4207), 997–999.
Hovland, M. (1990). Do carbonate reefs form due to fluid seepage? *Terra Nova, 2*(1), 8–18 http://www.sbg.ac.at/ipk/avstudio/pierofun/transcript/bioeros.pdf.
Hubbard, D. K. (1986). Sedimentation as a control of reef development: St. Croix. *U. S. V. I. Coral Reefs, 5*, 117–125.
Hubbard, D. K. (1988). Controls of modern and fossil reef development: Common ground for biological and geological research. *Proceedings of the Sixth International Coral Reef Symposium*, Townsville, Australia, Vol. 1, pp. 243–252.
Hubbard, D. K. (1992). Hurricane-induced sediment transport in open-shelf tropical systems — An example from St. Croix, U.S. Virgin Islands. *Journal of Sedimentary Petrology, 62*, 946–960.
Hubbard, D. K. (1997). Reefs as dynamic systems. In: Ch. Birkeland (Ed.), *Life and death of coral reefs* (pp. 43–67). New York: Chapman & Hall.
Hubbard, D. K., Burke, R. P., & Gill, I. P. (1985). Accretion in deep shelf-edge reefs, St. Croix, U.S.V.I. In: Crevello, P. D. & Harris, P. M. (Eds), *Deep-water carbonates: buildups, turbidites, debris-flows and chalks — A core workshop* (pp. 491–527), SEPM, Tulsa (Core Workshop 6).
Hubbard, D. K., Burke, R. B., & Gill, I. P. (1986). Styles of reef accretion along a steep, shelf-edge reef, St. Croix. U.S. Virgin Islands. *Journal of Sedimentary Petrology, 56*, 848–861.
Hubbard, D. K., Burke, R. B., & Gill, I. P. (1998). Where's the reef: The role of framework in the Holocene. *Carbonates and Evaporites, 13*, 3–9.
Hubbard, D. K., Gill, I. P., Burke, R. B., & Morelock, J. (1997). Holocene reef backstepping — Southwestern Puerto Rico Shelf. *Proceedings of the Eighth International Coral Reef Symposium*, Panama, Vol. 2, pp. 1779–1784.
Hubbard, D. K., Miller, A. I., & Scatturo, D. (1990). Production and cycling of calcium carbonate in a shelf-edge reef system (Ste Croix, U. S. Virgin Islands): Application to the nature of reef systems in the fossil record. *Journal of Sedimentary Petrology, 60*, 335–360.
Hubbard, D. K., Zankl, H., Van Heerden, I., & Gill, I. P. (2005). Holocene reef development along the northeastern St. Croix shelf, Buck Island, U.S. Virgin Islands. *Journal of Sedimentary Research, 75*, 97–113.
Hubbell, S. P. (1997). A unified theory of biogeography and relative species abundance and its application to tropical rain forests and coral reefs. *Proceedings of the Eighth International Coral Reef Symposium*, Panama, Vol. 1, pp. 33–42.
Hudson, J. D. (1977). Stable isotopes and limestone lithification. *Journal of the Geological Society of London, 133*, 637–660.
Hughen, K. A., Schrag, D. P., Jacobsen, S. B., & Hantoro, W. (1999). El Niño during the last interglacial period recorded by a fossil coral from Indonesia. *Geophysical Research Letters, 26*, 3129–3132.
Hughes, T. P. (Ed.) (1993). Preface. *Coral Reefs, Special Issue. Disturbance: effects on coral reef dynamics*, Berlin: Springer-Verlag, (Vol. 12(3–4), p. 115).
Hughes, T. P. (1994). Catastrophes, phase shifts and large-scale degradation of a Caribbean coral reef. *Science, 265*, 1547–1551.

Hughes, T. P., Baird, A. H., Dinsdale, E. A., Moltschaniwskyj, N., Pratchett, M. S., Tanner, J. E., et al. (1999). Patterns of recruitment and abundance of corals along the Great Barrier Reef. *Nature, 397*, 59–63.
Hughes, T. P., Baird, A. H., Belwood, D. R., Card, M., Connolly, S. R., Folke, C., et al. (2003). Climate change, human impacts, and the resilience of coral reefs. *Science, 301*, 929–933.
Humblet, M., & Iryu, Y. (2006). Variations of Pleistocene coral community structure in response to sea level change in the Ryukyu Islands. *Proceedings of the Twenty-First COE International Symposium, Climate Change: Past and Future, Poster Session Abstracts*, Sendai, Japan, pp. 38–39.
Humphrey, J. D. (1988). Late Pleistocene mixing zone dolomitization, southeastern Barbados, West Indies. *Sedimentology, 35*(2), 327–348.
Humphrey, J. D. (1989). Late Pleistocene mixing zone dolomitization, southeastern Barbados, West Indies (reply to discussion). *Sedimentology, 36*(6), 1138–1142.
Humphrey, J. D. (1997). Geology and hydrogeology of Barbados. In: H. L. Vacher, & T. M. Quinn (Eds), *Geology and hydrogeology of carbonate islands*. Developments in Sedimentology (Vol. 54, pp. 381–406). Amsterdam: Elsevier Science.
Humphrey, J. D., & Quinn., T. M. (1989). Coastal mixing zone dolomite, forward modeling, and massive dolomitization of platform margin carbonates. *Journal of Sedimentary Petrology, 59*, 438–454.
Humphrey, J. D., Ransom, K. L., & Matthews, R. K. (1986). Early meteoric diagenetic control of Upper Smackover production, Oaks Field, Louisiana. *American Association of Petroleum Geologists Bulletin, 70*, 70–85.
Hunt, C. D., Jr., & Peterson, F. L. (1980). Groundwater resources of Kwajalein Island, Marshall Islands. *Technical Report*, 126, pp. 1–91, University of Hawaii Water Resources Research Center.
Hunter, I. G., & Jones, B. (1996). Coral associations of the Pleistocene Ironshore formation, Grand Cayman. *Coral Reef, 15*, 249–267.
Hutchings, P. A. (1986). Biological destruction of coral reefs: A review. *Coral Reefs, 4*, 239–242.
Hutchinson, G. E. (1950). Survey of existing knowledge of biogeochemistry. 3. The biogeochemistry of vertebrate excretion. *Bulletin of American Museum of Natural History, 96*, 1–554.
Huybers, P. (2006). Early Pleistocene glacial cycles and the integrated summer insolation forcing. *Science, 313*, 508–511.
Ikeda, E., Iryu, Y., Sugihara, K., Ohba, H., & Yamada, T. (2006). Bathymetry, biota and sediments on the Hirota Reef, Tane-ga-shima: The northernmost coral reef in the Ruykyu Islands. *Island Arc, 15*, 407–419.
International Consortium, (2001). New constraints on the origins of the Australian Great Barrier Reef: Results from an International project of deep coring. *Geology, 29*, 483–487.
Iryu, Y., Nakamori, T., Matsuda, S., & Abe, O. (1995). Distribution of marine benthic organisms and its geological significance in the modern reef complex of the Ryukyu Islands. *Sedimentary Geology, 99*, 243–258.
Iryu, Y., Nakamori, T., & Yamada, T. (1998). Pleistocene reef complex deposits in the Central Ryukyus, south-western Japan (pp. 197–213). In: G. F. Camoin & P. J. Davies (Eds), *Reefs and Carbonate Platforms in the Pacific and Indian Oceans*. International Association of Sedimentologists (Special Publication 25).
Iryu, Y., Yamada, T., Matsuda, S., & Odawara, K. (2006). Pliocene to Quaternary carbonate sequence on Okinawa-Jima. *Proceedings of the Tenth International Coral Reef Symposium*, Okinawa, pp. 2022–2036.
Isdale, P. J. (1984). Fluorescent bands in massive corals record centuries of rainfall. *Nature, 310*, 578–579.

Isdale, P. J., Stewart, B. J., Tickle, K. S., & Lough, J. M. (1998). Palaeohydrological variation in a tropical river catchment: A reconstruction using fluorescent bands in corals of the Great Barrier Reef, Australia. *The Holocene, 8*, 1–8.

Isern, A. R., McKenzie, J. A., & Feary, D. A. (1996). The role of sea-surface temperature as a control on carbonate platform development in the western Coral Sea. *Palaeogeography, Palaeoclimatology, Palaeoecology, 124*, 247–272.

Jablonski, D. (1993). The tropics as a source of evolutionary novelty through geological time. *Nature, 364*, 142–144.

Jackson, J. B. C. (1992). Pleistocene perspectives on coral reef community structure. *American Zoologist, 32*, 719–731.

Jackson, J. B. C. (1994). Community unity. *Science, 264*, 1412–1413.

Jackson, J. B. C., Budd, A. F., & Pandolfi, J. M. (1996). The shifting balance of natural communities. In: D. Erwin, D. Jablonski & J. Lipps (Eds), *Evolutionary Paleobiology* (pp. 89–122). Chicago, IL: University of Chicago Press.

James, N. P. (1972). Holocene and Pleistocene calcareous crust (caliche) profiles; criteria for subaerial exposure. *Journal of Sedimentary Petrology, 25*, 489–522.

James, N. P. (1974). Diagenesis of scleractinian corals in the subaerial vadose environment. *Journal of Paleontology, 48*, 785–799.

James, N. P., & Choquette, P. E. (1984). Limestones — The meteoric diagenetic environment. Diagenesis 9. *Geoscience Canada, 11*, 161–194.

James, N. P., & Choquette, P. E. (1990). Limestone — The meteoric diagenetic environment. In: I. A. McIlreath, & Morrow, D. W. (Eds), *Diagenesis* (pp. 35–74). Geoscience Canada (Reprint Series 4).

James, N. P., & Ginsburg, R. N. (Eds) (1979a). *The seaward margin of Belize Barrier and Atoll Reefs* (191 pp.), International Association of Sedimentologists (Special Publication 3), Blackwells, Oxford.

James, N. P., & Ginsburg, R. N. (1979b). The geological setting of Belize reefs. In: N. P. James, & R. N. Ginsburg (Eds), *The seaward margin of Belize Barrier and Atoll reefs* (pp. 1–14), International Association of Sedimentologists (Special Publication 3), Blackwells, Oxford.

James, N. P., Ginsburg, R. N., Marszalek, D. S., & Choquette, P. (1976). Facies and fabric specificity of early subsea cements in shallow Belize (British Honduras) Reefs. *Journal of Sedimentary Petrology, 46*, 523–544.

James, N. P., Lindsay, B. L., Bone, Y., & Hallock, P. A. (1999). Subtropical carbonates in a temperate realm: Modern sediments on the southwest Australian shelf. *Journal of Sedimentary Research, 69*, 1297–1321.

James, N. P., Mountjoy, E., & Omura, A. (1971). An early Wisconsin Reef terrace at Barbados, West Indies, and its climatic implications. *Geological Society of America Bulletin, 82*, 2011–2018.

James, S. A. (2006). *Coral diversity and zonation on a Pleistocene reef, southeastern Jamaica.* Jamaica's Biodiversity Information Network, Institute of Jamaica, Natural History Division, 6 pp (http://www.jamaicachm.org.jm/Article/August2006.asp).

Jean-Baptiste, P., & Leclerc, A. M. (2000). Self-limiting geothermal convection in marine carbonate platforms. *Geophysical Research Letters, 27*, 743–746.

Jehl, C., & Rougerie, F. (1985). Phosphatogenèse en atolls polynésiens: la filiation mattes cyanobactériennes-phosphorites. *Oceanologica Acta, 18*, 79–93.

Jell, J. S., & Flood, P. J. (1978). *Guide to the geology of reefs of the Capricorn and Bunker Groups, Great Barrier Reef Province, with special reference to Heron Reef* (Vol. 8, pp. 1–85), Australian Sedimentologists Group, Department of Geology, University of Queensland, Brisbane.

Jenson, J. W., Keel, T. M., Mylroie, J. R., Mylroie, J. E., Stafford, K. W., Taborosi, D., & Wexel. (2006). *Karst of the Mariana Islands: The interaction of tectonics, glacio-eustasy,*

fresh-water/salt-water mixing in island carbonates (pp. 129–138). Geological Society of America (Special Paper 404).
Johns, H. D., & Moore, C. H. (1988). Reef to basin sediment transport using *Halimeda* as a sediment tracer, Grand Cayman Island, West Indies. *Coral Reefs, 6*, 187–194.
Johnson, D. P., Cuff, C., & Rhodes, E. (1984). Holocene reef sequences and geochemistry, Britomart Reef, Central Great Barrier Reef, Australia. *Sedimentology, 31*, 515–529.
Johnson, D. P., & Risk, M. J. (1987). Fringing reef growth on a terrigenous mud foundation, Fantome island, Central Great Barrier Reef, Australia. *Sedimentology, 34*, 275–288.
Johnson, J. H., Todd, R. M., Post, R., Cole, W. S., & Wells, J. W. (1954). *Bikini and nearby atolls: Part 4: Paleontology*. U.S. Geological Survey, Professional Paper, Vol. 206-Y, pp. 863–903.
Johnson, K. G. (2001). Middle Miocene recovery of Caribbean reef corals: New data from the Tamana Formation, Trinidad. *Journal of Paleontology, 75*, 513–526.
Johnson, K. G., & Pérez, M. E. (2006). Skeletal extension rates of Cenozoic Caribbean reef corals. *PALAIOS, 21*, 262–271.
Johnson, K. G., Todd, J. A., & Jackson, J. B. C. (2007). Coral reef development drives molluscan diversity increase at local and regional scales in the late Neogene and Quaternary of the southwestern Caribbean. *Paleobiology, 33*, 24–52.
Johnson, R. H. (1983). The saltwater–freshwater interface in the Tertiary limestone aquifer, southeastern Atlantic continental shelf of the USA. *Journal of Hydrology, 61*, 239–251.
Jokiel, P. L., & Coles, S. J. (1990). Response of Hawaiian and other Indo-Pacific reef corals to elevated temperatures. *Coral Reefs, 8*, 155–162.
Jones, A. T. (1995). Geochronology of drowned Hawaiian coral reefs. *Sedimentary Geology, 99*, 233–242.
Jones, B. (1991). Genesis of terrestrial oncoids, Cayman Islands, British West Indies. *Canadian Journal of Earth Sciences, 28*, 382–397.
Jones, B., & Hunter, I. G. (1991). Corals to rhodolites to microbialites — community replacement sequence indicative of regressive conditions. *PALAIOS, 6*, 54–66.
Jones, B., & Kahle, C. F. (1993). Morphology, relationship, and origin of fiber and dendrite calcite crystals. *Journal of Sedimentary Petrology, 63*, 1018–1031.
Jones, B., & Luth, R. W. (2002). Dolostones from Grande Cayman, British West Indies. *Journal of Sedimentary Research, 72*(4), 559–569.
Jones, B., Ng, K.-C., & Hunter, I. G. (1997). Geology and hydrology of the Cayman Islands. In: H. L. Vacher, & T. Quinn (Eds), *Geology and hydrogeology of carbonate islands*. Developments in Sedimentology (Vol. 54, pp. 453–473; 299–326). Amsterdam: Elsevier Science B.V
Jones, B., & Pemberton, S. G. (1988). *Lithophaga* borings and their influence on the diagenesis of corals on the Pleistocene Ironshore Formation of Grand Cayman Islands, British West Indies. *PALAIOS, 3*, 3–21.
Jones, O. A., & Endean, R. (Eds). (1973a). *Biology and geology of coral reefs. Vol. I/Geology 1*. New York: Academic Press. 400 pp.
Jones, O. A., & Endean, R. (Eds). (1973b). *Biology and geology of coral reefs. Vol. II/Biology 1*. New York: Academic Press. 480 pp.
Jones, O. A., & Endean, R. (Eds). (1976). *Biology and geology of coral reefs. Vol. III/Biology 2*. New York: Academic Press. 435 pp.
Jones, O. A., & Endean, R. (Eds). (1977). *Biology and geology of coral reefs. Vol. IV/Geology 2*. New York: Academic Press. 337 pp.
Jordan, C. F. (1973). Carbonate facies and sedimentation of patch reefs off Bermuda. *Bulletin of the American Association of Petroleum Geologists, 57*, 42–54.

Jouon, A., Lefebvre, J.-P., Douillet, P., Ouillon, S., & Schmied, L. (2009). Wind wave measurements and modelling in a fetch-limited semi-enclosed lagoon. *Coastal Engineering*, *58*, 599–608.

Juillet-Leclerc, A. (1999). El-Niño, present and past: Corals, paleoclimatic archives. *International Programme Geosphere–Biosphere, Newsletter*, CNRS, Paris, Vol. 9, pp. 22–28.

Juillet-Leclerc, A., Gattuso, J. P., Montaggioni, L., & Pichon, M. (1997). Seasonal variation of primary productivity and skeletal δ^{13}C and δ^{18}O in the zooxanthellate scleractinian coral *Acropora formosa*. *Marine Ecology Progress Series*, *157*, 109–117.

Juillet-Leclerc, A., Thiria, S., Naveau, P., Delcroix, T., Le Bec, N., Blamart, D., et al. (2006). SPCZ migration and ENSO events during the 20th century as revealed by climate proxies from a Fiji coral. *Geophysical Research Letters*, *33*, L17710, doi: 10.1029/2006GL025950.

Kan, H., Hori, N., & Ichikawa, K. (1997). Formation of a coral reef-front spur. *Coral Reefs*, *16*, 3–4.

Kan, H., Hori, N., Nakashima, Y., & Ichikawa, K. (1995). The evolution of narrow reef flats at high latitude in the Ryuku Islands. *Coral Reefs*, *14*, 123–130.

Karlson, R. H., & Hurd, L. E. (1993). Disturbance, coral reef communities, and changing ecological paradigms. *Coral Reefs*, *12*, 117–126.

Kato, M., Fukusawa, H., & Yasuda, Y. (2003). Varved lacustrine sediments of Lake Tougou-Ike, Western Japan, with reference to Holocene sea-level changes in Japan. *Quaternary International*, *105*, 33–37.

Kaufman, D. S., Ager, T. A., Anderson, N. J., Anderson, P. M., Andrews, J. T., Bartlein, P. J., et al. (2004). Holocene thermal maximum in the western Arctic (0–180°W). *Quaternary Science Reviews*, *23*, 529–560.

Kauffman, E. G., & Fagerström, J. A. (1993). The Phanerozoic evolution of reef diversity. In: R. E. Ricklefs & D. Schluter (Eds), *Species diversity in ecological communities* (pp. 315–329). Chicago, IL: University of Chicago Press.

Kayanne, H., Iijima, H., Nakamura, N., McClanahan, T. R., Behera, S., & Yamagata, T. (2006). Indian Ocean dipole index recorded in Kenyan coral annual density bands. *Geophysical Research Letters*, *33*, L19709, doi: 10.1029/2006GL027168.

Kayanne, H., Ishii, T., Matsumoto, E., & Yonekura, N. (1993). Late Holocene sea-level change on Rota and Guam, Mariana Islands, and its constraint on geophysical predictions. *Quaternary Research*, *40*, 189–200.

Kayanne, H., Yamano, H., & Randall, R. H. (2002). Holocene sea-level changes and barrier reef formation on an ocean island, Palau Islands, western Pacific. *Sedimentary Geology*, *150*, 47–60.

Kench, P. S. (1997). Currents of removal analysis of carbonate sediments. *Proceedings of the Eighth International Coral Reef Symposium*, Panama, Vol. 1, pp. 503–508.

Kench, P. S. (1998). Physical processes in a semi-enclosed Indian Ocean atoll. *Coral Reefs*, *17*, 155–168.

Kench, P. S., & Brander, R. W. (2006). Wave processes on coral reef flats: implications for reef geomorphology using Australian case studies. *Journal of Coastal Research*, *22*, 209–223.

Kench, P. S., Brander, R. W., Parnell, K. E., & McLean, R. F. (2006). Wave energy gradients across a Maldivian atoll: Implications for island geomorphology. *Geomorphology*, *81*, 1–17.

Kench, P. S., & McLean, R. F. (1997). A comparison of settling and sieve techniques for the analysis of bioclastic sediments. *Sedimentary Geology*, *109*, 111–119.

Kench, P. S., Nichol, S. L., Smithers, S. G., McLean, R. F., & Brander, R. W. (2008). Tsunami as agents of geomorphic change in mid-ocean reef islands. *Geomorphology*, *95*, 361–393.

Kennedy, D., & Woodroffe, C. D. (2004). Carbonate sediments of Elizabeth and Middleton Reefs close to the southern limits of reef growth in the southwest Pacific. *Australian Journal of Earth Sciences*, *51*, 847–857.

Kennedy, D. M. (2003). Surface lagoonal sediments on Lord Howe Island, Tasman Sea. *Journal of Coastal Research*, *19*, 57–63.

Kennedy, D. M., & Woodroffe, C. D. (2000). Holocene lagoonal sedimentation at the latitudinal limits of reef growth, Lord Howe Island, Tasman Sea. *Marine Geology*, *169*, 287–304.

Kennedy, D. M., & Woodroffe, C. D. (2002). Fringing reef growth and morphology: A review. *Earth Science Reviews*, *57*, 255–277.

Kidder, D. L., & Worsley, Th. R. (2004). Causes and consequences of extreme Permo-Triassic warming to globally equable climate and relation to Permo-triassic extinction and recovery. *Palaeogeography, Palaeoclimatology, Palaeoecology*, *203*, 207–237.

Kidwell, S. M., Best, M. M. R., & Kaufman, D. S. (2005). Taphonomic trade-offs in tropical marine death assemblages: Different time-averaging, shell loss, and probable bias in siliciclastic vs. carbonate facies. *Geology*, *33*, 729–732.

Kiehl, J. T., & Shields, Ch. A. (2005). Climate simulation of the latest Permian: Implications for mass extinction. *Geology*, *33*, 757–760.

Kiene, W. E. (1985). Biological destruction of experimental coral substrates at Lizard Island, Great Barrier Reef, Australia. *Proceedings of the Fifth International Coral Reef Congress*, Tahiti, Vol. 5, pp. 339–344.

Kiene, W. E. (1988). A model of bioerosion on the Great Barrier Reef. *Proceedings of the Sixth International Coral Reef Symposium*, Townsville, Australia, Vol. 3, pp. 449–454.

Kiene, W. E., & Hutchings, P. A. (1994). Bioerosion experiments at Lizard Island, Great Barrier Reef. *Coral Reefs*, *13*, 91–98.

Kiessling, W. (2002). Secular variations in the Phanerozoic reef ecosystem. In: W. Kiessling, E. Flügel & J. Golonka (Eds), *Phanerozoic reef patterns* (pp. 625–690), SEPM (Special Publication 72).

Kiessling, W., Aberhan, M., & Villier, L. (2008). Phanerozoic trends in skeletal mineralogy driven by mass extinctions. *Nature Geoscience*, *1*, 527–530.

Kiessling, W., & Baron-Szabo, R. C. (2004). Extinction and recovery patterns of scleractinian corals at the Cretaceous-Tertiary boundary. *Palaeogeography, Palaeoclimatology, Palaeoecology*, *214*, 195–223.

Kilbourne, K. H., Quinn, T. M., & Taylor, F. W. (2004). A fossil coral perspective on western tropical climate ~350 ka. *Paleoceanography*, *19*, PA1019, doi: 10.1029/2003PA000944.

Kilbourne, K. H., Quinn, T. M., Taylor, F. W., Delcroix, Th., & Gouriou, Y. (2004). El Niño-southern Oscillation-related salinity variations recorded in the skeletal geochemistry of a *Porites* coral from Espiritu Santo, Vanuatu. *Paleoceanography*, *19*, PA4002, doi: 10.1029/2004PA001033.

Kindler, P., Reyss, J.-L., Cazala, Ch., & Plagnes, V. (2007). Discovery of a composite reefal terrace of middle and late Pleistocene age in Great Inagua Island, Bahamas. Implications for regional tectonics and sea-level history. *Sedimentary Geology*, *194*, 141–147.

King, B. A., & Wolanski, E. J. (1992). Coastal dynamics along a rugged coastline. In: D. Prandle (Ed.), *Coastal and estuarine studies: Dynamics and exchanges in estuaries and the coastal zone* (Vol. 40, pp. 577–598). Washington, DC: American Geophysical Union.

Kinsey, D. W. (1983). Standards of performance in coral reef primary production and carbon turnover. In: D. J. Barnes (Ed.), *Perspectives on coral reefs* (pp. 209–220). Manuka, Australia: The Australian Institute of Marine Science, Brian Clouston Publisher.

Kinsey, D. W. (1985). Metabolism, calcification and carbon production. I. System level studies. *Proceedings of the Fifth International Coral Reef Congress*, Tahiti, Vol. 4, pp. 505–526.

Kinsey, D. W., & Davies, P. J. (1979). Effects of elevated nitrogen and phosphorous on coral reef growth. *Limnology and Oceanography*, *24*(5), 935–940.

Kinsey, D. W., & Hopley, D. (1991). The significance of coral reefs as global carbon sinks — Response to greenhouse. *Paleogeography, Palaeoclimatology, Palaeoecology, 89*, 363–377.

Kinsman, D. J. J. (1969). Interpretation of the strontium concentration in carbonate minerals and rocks. *Journal of Sedimentary Petrology, 39*, 486–508.

Kirtley, D., & Tanner, W. F. (1968). Sabellarid worms: Builders of a major reef type. *Journal of Sedimentary Petrology, 38*, 73–78.

Klaus, J. S., & Budd, A. F. (2003). Comparison of Caribbean coral reef communities before and after Plio-Pleistocene faunal turnover: Analyses of two Dominican Republic reef sequences. *PALAIOS, 18*, 3–21.

Kleemann, K. (2001). *Marine bioerosion*. University of Vienna, lecture notes, pp. 1–28.

Klein, R., Loya, Y., Gvirtzman, G., Isdale, P. J., & Susic, M. (1990). Seasonal rainfall in the Sinai Desert during the late Quaternary inferred from fluorescent bands in fossil corals. *Nature, 345*, 145–147.

Klein, R., Mokady, O., & Loya, Y. (1991). Bioerosion in ancient and contemporary corals of the genus *Porites*: Patterns and palaeoenvironmental implications. *Marine Ecology Progress Series*, 77, 245–251.

Klein, R., Tudhope, A. W., Chilcott, C. P., Pätzold, J., Abdulkarim, Z., Fine, M., et al. (1997). Evaluating southern Red Sea corals as a proxy record for the Asian monsoon. *Earth and Planetary Science Letters, 148*, 381–394.

Kleypas, J. A. (1996). Coral reef development under naturally turbid conditions: Fringing reefs near Broad Sound, Australia. *Coral Reefs, 15*, 153–167.

Kleypas, J. A. (1997). Modeled estimates of global reef habitat and carbonate production since the last glacial maximum. *Paleoceanography, 12*, 533–545.

Kleypas, J. A., Buddemeier, R. W., Archer, D., Gattuso, J. P., Langdon, Ch., & Opdyke, B. N. (1999). Geochemical consequences of increased atmospheric carbon dioxide on coral reefs. *Science, 284*, 118–120.

Kleypas, J. A., & Hopley, D. (1993). Reef development across a broad continental shelf, southern Great Barrier Reef, Australia. *Proceedings of the Seventh International Coral Reef Symposium*, Guam, Vol. 2, pp. 1129–1141.

Kleypas, J. A., McManus, J. W., & Menez, L. A. B. (1999). Environmental limits to coral reef development: Where we draw the line? *American Zoologist, 39*, 146–159.

Knowlton, N., & Jackson, J. B. C. (1994). New taxonomy and niche partitioning on coral reefs: Jack of all trades or master of some? *Trends in Ecology and Evolution, 9*, 7–9.

Knowlton, N., & Jackson, J. B. C. (2001). The ecology of coral reefs. In: M. D. Bertness, S. Gaines & M. E. Hay (Eds), *Marine community ecology* (pp. 395–422). Sunderland, MA: Sinauer. Chapter 15.

Knox, G. J. (1977). Caliche profile formation, Saldanha Bay (South Africa). *Sedimentology, 24*(5), 657–674.

Knutson, R. R., Tuleya, R. E., & Kurihasa, Y. (1998). Simulated increased of hurricane intensities in a CO_2 warmed climate. *Nature, 279*, 1018–1021.

Koelling, M., Webster, J. M., Camoin, G., Iryu, Y., Bard, E., & Seard, C. (2009). SEALEX — Internal reef chronology and virtual drill logs from a spreadsheet-based reef growth model. *Global and Planetary Change, 66*, 149–159.

Kohn, A. J., & Arua, I. (1999). An early Pleistocene molluscan assemblage from Fiji: Gastropod faunal composition, paleoecology and biogeography. *Palaeogeography, Palaeoclimatology, Palaeoecology, 146*, 99–145.

Kohout, F. A. (1965). A hypothesis concerning cyclic flow of salt water related to geothermal heating in the Floridan aquifer. *Transactions of the New York Academy of Sciences, Series 2, 28*, 249–271.

Kohout, F. A., Henry, H. R., & Banks, J. E. (1977). Hydrology related to geothermal conditions of the Floridan Plateau. In: K. L. Smith & G. M. Griffin (Eds), *The geothermal nature of the Floridan Plateau*. Bureau of Geology (Vol. 21, pp. 1–39). Florida Department of Natural Resources.

Konishi, K. (1975). Tidal channel rhodolites in a fringing reef complex of Ishigaki-jima, southwest Rykyus. *International Symposium on Fossil Algae, Abstracts*, Erlangen, p. 64.

Konishi, K. (1981). Alcyonarian spiculite: Limestone of soft corals. *Proceedings of the Fourth International Coral Reef Symposium*, Manila, Vol. 1, pp. 643–649.

Konishi, K., Omura, A., & Nakamichi, O. (1974). Radiometric coral ages and sea level records from the late Quaternary reef complexes of the Ryukyu Islands. *Proceedings of the Second International Coral Reef Symposium*, Brisbane, Australia, Vol. 2, pp. 595–613.

Kosnik, M. A., Hua, Q., Jacobsen, G. E., Kaufman, D. S., & Wust, R. A. (2007). Sediment mixing and stratigraphic disorder revealed by the age-structure of *Tellina* shells in Great Barrier Reef sediment. *Geology*, *35*, 811–814.

Kraines, S. B., Yanagi, T., Isobe, M., & Komiyama, H. (1998). Wind-wave driven circulation on the coral reef at Bora Bay, Miyako Island. *Coral Reefs*, *17*, 133–143.

Kuenen, P. H. (1947). *Two problems of marine geology, atolls and canyons* (Vol. DI 43, pp. 1–69). Amsterdam: K. Nederlands Akademie v Wet. Verh afd Nat.

Kuffner, I. B., Andersson, A. J., Jokiel, P. L., Rodgers, K. S., & Mackenzie, F. T. (2007). Decreased abundance of crustose coralline algae due to ocean acidification. *Nature Geoscience*, *1*, 114–117.

Kuhnert, H., Pätzold, J., Hatcher, B., Wyrwoll, K.-H., Eisenhauer, A., Collins, L. B., et al. (1999). A 200-year coral stable oxygen isotope record from a high latitude reef off Western Australia. *Coral Reefs*, *18*, 1–12.

Kundal, P., & Dharashivkar, A. P. (2005). Record of rhodoliths from Aramda Reef Member (Late Pleistocene to Holocene) of Chaya Formation, Dwarka-Okha area, Gujarat and their paleoenvironmental significance. *Current Science*, *88*, 1684–1689.

Laborel, J. (1986). Vermetid gastropods as sea-level indicators. In: O. van de Plassche (Ed.), *Sea-level research: A manual for the collection and evaluation of data* (pp. 281–310). Norwich, UK: Geo Books.

Lacroix, G. (2004). *Croissance récifale and contrôle environmental au cours du Quaternaire: l'exemple de la barrière récifale sud-occidentale de Nouvelle-Calédonie*. Unpublished Master thesis, Université de Provence, 74 pp.

Ladd, H. S. (1950). Recent Reefs. *American Association of Petroleum Geologists, Bulletin*, *34*, 203–214.

Ladd, H. S., Newman, W. A., & Sohl, N. F. (1974). Darwin Guyot, the Pacific's oldest atoll. *Proceedings of the Second International Coral Reef Symposium*, Brisbane, pp. 513–522.

Ladd, H. S., & Schlanger, S. O. (1960). *Drilling operations on Enewetok Atoll*. US Geological Survey Professional Paper, Vol. 260-Y, pp. 863–905).

Ladd, H. S., & Tracey, J. I., Jr. (1949). The problem of coral reefs. *Science Monthly*, *69*, 297–305.

Ladd, H. S., Tracey, J. I., & Gross, M. G. (1970). *Deep drilling on Midway Atoll*. US Geological Survey Professional Paper, Vol. 680-A, pp. 1–22.

LaJeunesse, T. C. (2005). "Species" radiations of symbiotic dinoflagellates in the Atlantic and Indo-Pacific since the Miocene–Pliocene transition. *Molecular Biology and Evolution*, *22*, 570–581.

Lambeck, K. (2002). Sea level change from Mid-Holocene to Recent time: An Australian example with global implications. In: J. X. Mitrovica, & B. L. A. Vermeersen (Eds), *Glacial Isostatic Adjustment and the Earth System: Sea Level, Crustal Deformation, Gravity and Rotation*. Monograph, Geodynamics Series (Vol. 29, pp. 33–50). Washington, DC: American Geophysical Union.

Lambeck, K. (2004). Sea-level change through the last glacial cycle: geophysical, glaciological and palaeogeographic consequences. *Comptes Rendus Geoscience*, *336*, 677–689.

Lambeck, K., & Chappell, J. (2001). Sea level change through the last glacial cycle. *Science, 292*, 679–686.
Lambeck, K., Yokoyama, Y., Johnston, P., & Purcell, A. (2000). Global ice volumes at the Last Glacial Maximum and early Lateglacial. *Earth and Planetary Science Letters, 181*, 513–527.
Lambeck, K., Yokoyama, Y., & Purcell, A. (2002). Into and out of the Last Glacial Maximum: Sea-level change during Oxygen Isotope Stages 3 and 2. *Quaternary Science Reviews, 21*, 343–360.
Lambrechts, J., Hanert, E., Deleersnijder, E., Bernard, P.-E., Legat, V., Remacle, J.-F., et al. (2008). A multi-scale model of the hydrodynamics of the whole Great Barrier Reef. *Estuarine, Coastal and Shelf Science, 79*, 143–151.
Land, L. S. (1973). Contemporaneous dolomitization of Middle Pleistocene reefs by meteoric water, North Jamaica. *Bulletin of Marine Science, 23*, 64–92.
Land, L. S. (1976). Early dissolution of sponge spicules from reef sediments, North Jamaica. *Journal of Sedimentary Petrology, 46*, 967–969.
Land, L. S. (1979). The fate of reef derived sediment on the north Jamaican island slope. *Marine Geology, 29*, 55–71.
Land, L. S., & Goreau, T. F. (1970). Submarine lithification of Jamaican reefs. *Journal of Sedimentary Petrology, 40*, 457–462.
Land, L. S., Mackenzie, F. T., & Gould, S. J. (1967). Pleistocene history of Bermuda. *Geological Society of America Bulletin, 78*, 993–1006.
Land, L. S., & Moore, C. H., Jr. (1977). Deep forereef and Upper Island Slope, North Jamaica. In: S. H. Frost, M. P. Weiss & J. B. Saunders (Eds), *Reefs and related carbonates: Ecology and sedimentology*. Studies in Geology (Vol. 4., pp. 53–65). Tulsa, OK: American Association of Petroleum Geologists.
Land, L. S., & Moore, C. H. (1980). Lithification, micritization and syndepositional diagenesis of biolithites on the Jamaican island slope. *Journal of Sedimentary Petrology, 50*, 357–370.
Langdon, C., Takahashi, T., Sweeney, C., Chipman, D., Goddard, J., Marubini, F., et al. (2000). Effect of calcium carbonate saturation state on the calcification rate of an experimental coral reef. *Global Biogeochemical Cycles, 14*(2), 639–654.
Langer, M. R., & Hottinger, L. (2000). Biogeography of selected "larger" foraminifera. *Micropaleontology, 46*, 105–126.
Langer, M. R., & Lipps, J. H. (2003). Foraminiferal distribution and diversity, Madang Reef and Lagoon, Papua New Guinea. *Coral Reefs, 22*, 143–154.
Langer, M. R., Silk, M. T., & Lipps, (1997). Global ocean carbonate and carbon dioxide production: The role of reef foraminifera. *Journal of Foraminiferal Research, 27*, 271–277.
Larcombe, P., & Carter, R. M. (2004). Cyclone pumping, sediment partitioning and the development of the Great Barrier Reef shelf system: A review. *Quaternary Science Reviews, 23*, 107–135.
Larcombe, P., Costen, A., & Woolfe, K. J. (2001). The hydrodynamics and sedimentary setting of nearshore coral reefs, central Great Barrier Reef shelf, Australia: Paluma Shoals, a case study. *Sedimentology, 48*, 811–835.
Lawrence, K. T., & Herbert, T. D. (2005). Late Quaternary sea-surface temperatures in the western Coral Sea: Implications for the growth of the Australian Great Barrier Reef. *Geology, 33*, 677–680.
Laxton, J. H., & Stablum, W. J. (1974). Sample design for quantitative estimation of sedentary organisms of coral reefs. *Biological Journal of the Linnean Society, 6*, 1–18.
Lea, D. W., Shen, G. T., & Boyle, E. A. (1989). Coralline barium records temporal variability in Equatorial Pacific upwelling. *Nature, 340*, 373–376.

Le Bec, N., Juillet-Leclerc, A., Corrège, Th., Blamart, D., & Delcroix, Th. (2000). A coral $\delta^{18}O$ record of ENSO driven sea surface salinity variability in Fiji (south-western tropical Pacific). *Geophysical Research Letters, 23*, 3897–3900.

Leclerc, A. M., Jean-Baptiste, P., & Texier, D. (1999). Density-induced water circulation in atoll coral reefs: A numerical study. *Limnology and Oceanography, 44*, 1268–1281.

Lees, A. (1964). The structure and origin of the Waulsortian (Lower Carboniferous) "reefs" of west central Eire. *Philosophical Transactions of the Royal Society of London, Series B, 247*, 483–531.

Lees, A. (1975). Possible influence of salinity and temperature on modern shelf carbonate sedimentation. *Marine Geology, 19*, 159–198.

Le Roy, P., Cabioch, G., Monod, B., Lagabrielle, Y., Pelletier, B., & Flamand, B. (2008). Late Quaternary history of the Nouméa lagoon (New Caledonia, South West Pacific) as depicted by seismic stratigraphy and multibeam bathymetry. A modern model of tropical rimmed shelf. *Palaeogeography, Palaeoclimatology, Palaeoecology, 270*, 29–45.

Lescinsky, H. L., Edinger, E., & Risk, M. J. (2002). Mollusc shell encrustation and bioerosion rates in a modern epeiric sea: Taphonomy experiments in the Java Sea, Indonesia. *PALAIOS, 17*, 171–191.

Lessios, H. A., Robertson, D. R., & Cubit, J. D. (1984). Spread of *Diadema* mass mortality through the Caribbean. *Science, 226*(4672), 335–337.

Lewis, J. B. (1977). Processes of organic production on coral reefs. *Biological Reviews, 52*, 305–347.

Lewis, J. B. (1984). The *Acropora* inheritance: A reinterpretation of the development of fringing reefs in Barbados. *Coral Reefs, 3*, 117–122.

Lewis, M. S. (1969). Sedimentary environments and unconsolidated carbonate sediments of the fringing coral reefs of Mahé, Seychelles. *Marine Geology, 7*, 89–94.

Lewis, M. S., & Taylor, J. D. (1966). Marine sediments and bottom communities of the Seychelles. *Philosophical Transactions of the Royal Society of London, Series A, 259*, 279–290.

Lewis, S. E., Wüst, R. A. J., Webster, J. M., & Shields, G. A. (2008). Mid-late Holocene sea-level variability in eastern Australia. *Terra Nova, 20*, 74–81.

Li, C., Jones, B., & Blanchon, P. (1997). Lagoon-shelf sediment exchange during storms: Evidence from foraminiferal tracers, Grand Cayman. *Journal of Sedimentary Research, 67*, 17–25.

Li, W.-X., Lundberg, J., Dickin, A., Ford, D. C., Schwarcz, H. P., McNutt, R., et al. (1989). High precision mass spectrometric uranium-series dating of cave deposits and implications for palaeoclimatic studies. *Nature, 339*, 534–536.

Liddell, W. D., Ohlhorst, S. L., & Boss, S. K. (1988). The significance of *Halimeda* as a space-occupier and sediment-producer, 1–750 m, North Jamaica. *Proceedings of the Sixth International Coral Reef Symposium*, Townsville, Australia, Vol. 3, pp. 127–132.

Liddell, W. D., Ohlhorst, S. L., & Coates, A. G. (1984). Modern and ancient carbonate environments of Jamaica. In: R. Ginsburg (Series Editor), *Sedimenta X* (101 pp.). The Comparative Sedimentology Laboratory, Rosenstiel School of Marine and Atmospheric Science, University of Miami, FL.

Lidz, B. H. (2004). Coral reef complexes at an atypical windward platform margin: Late Quaternary, southeast Florida. *Geological Society of America Bulletin, 116*, 974–988.

Lidz, B. H., & Hallock, P. (2000). Sedimentary petrology of a declining reef ecosystem, Florida reef tract (U.S.A.). *Journal of Coastal Research, 106*, 675–697.

Lidz, B. H., Reich, Ch. D., & Shinn, E. A. (2003). Regional Quaternary submarine geomorphology in the Florida Keys. *Geological Society of America Bulletin, 115*, 845–866.

Lidz, B. H., Shinn, E. A., Hine, A. C., & Locker, S. D. (1997). Contrasts within an outlier-reef system: Evidence for differential Quaternary evolution, South Florida windward margin, U.S.A. *Journal of Coastal Research, 13*, 711–731.

Lighty, R. G. (1977). Relict shelf-edge Holocene coral reef: Southeast coast of Florida. *Proceeding of the Third International Coral Reef Symposium*, Miami, Vol. 2, pp. 215–221.

Lighty, R. G. (1985). Preservation of internal porosity and diagenetic sealing of submerged Early Holocene Barrier Reef, Southeast Florida Shelf. In: N. Schneiderman & P. M. Harris (Eds), *Carbonate cements* (pp. 123–151), Society of Economic Paleontologists and Mineralogists (Special Publication 36).

Lighty, R. G., Macintyre, I. G., & Stuckenrath, R. (1978). Submerged early Holocene barrier reef south-east Florida Shelf. *Nature*, *276*, 59–60.

Lighty, R. G., Macintyre, I. G., & Stuckenrath, R. (1982). *Acropora palmata* reef framework: A reliable indicator of sea-level in the western Atlantic for the past 10,000 years. *Coral Reefs*, *1*, 125–130.

Linick, W., Damon, P. E., Donahue, D. J., & Jull, A. J. T. (1989). Accelerator Mass Spectrometry: The new revolution in radiocarbon dating. *Quaternary International*, *1*, 1–6.

Linsley, B. K., Messier, R. G., & Dunbar, R. B. (1999). Assessing between-colony oxygen isotope variability in the coral *Porites lobata* at Clipperton Atoll. *Coral Reefs*, *18*, 13–27.

Lisiecki, L. E., & Raymo, M. A. (2007). Plio-Pleistocene climate evolution: Trends and transitions in glacial cycle dynamics. *Quaternary Science Reviews*, *26*, 56–69.

Littler, M. M., & Doty, M. S. (1975). Ecological components structuring the seaward edges of tropical Pacific reefs: The distribution, communities and productivity of *Porolithon*. *Journal of Ecology*, *36*, 117–129.

Littler, M. M., Littler, D. M., & Hanisak, M. D. (1991). Deep-water rhodolith distribution, productivity and growth history at sites of formation and subsequent degradation. *Journal of Experimental Marine biology and Ecology*, *150*, 163–182.

Littler, M. M., Littler, D. S., & Lapointe, B. E. (1988). A comparison of nutrient-and light-limited photosynthesis in psammophytic versus epilithic forms of *Halimeda* (Caulerpales, Halimedaceae) from the Bahamas. *Coral Reefs*, *6*, 219–225.

Liu, Z., & Herbert, T. D. (2004). High-latitude influence on the eastern equatorial Pacific climate in the early Pleistocene epoch. *Nature*, *427*, 720–723.

Lloyd, R. M. (1971). Some observations on recent sediment alteration ("micritization") and the possible role of algae in submarine cementation. In: O. P. Bricker (Ed.), *Carbonate cements* (pp. 72–79). Baltimore, MD: The Johns Hopkins Press.

Lockwood, R., & Work, L. A. (2006). Quantifying taphonomic bias in molluscan death assemblages from the Upper Chesapeake Bay: Patterns of shell damage. *PALAIOS*, *21*, 442–450.

Longman, M. W. (1980). Carbonate diagenetic textures from nearsurface diagenetic environments. *Bulletin of the American Association of Petroleum Geologists*, *64*(4), 461–487.

Longman, M. W. (1981). A process approach to recognizing facies of reef complexes. In: D. F. Toomey (Ed.), *European fossil reef models* (pp. 9–40), SEPM (Special Publication 30), Tulsa, OK.

Lopez-Pérez, R. A. (2005). The Cenozoic hermatypic corals in the eastern Pacific: History of research. *Earth-Sciences Reviews*, *72*, 67–87.

Loubere, P. (2001). Nutrient and oceanographic changes in the Eastern Equatorial Pacific from the last full glacial to the present. *Global and Planetary Change*, *29*, 77–98.

Lough, J. M., & Barnes, D. J. (2000). Environmental controls on growth of the massive coral *Porites*. *Journal of Experimental Marine Biology and Ecology*, *245*, 225–243.

Lough, J. M., Barnes, D. J., & McAllister, F. A. (2002). Luminescent lines in corals from the Great Barrier Reef provide spatial and temporal records of reefs affected by land runoff. *Coral Reefs*, *21*, 333–343.

Loutre, M. F. (2003). Clues from MIS 11 to predict the future climate: A modeling point of view. *Earth and Planetary Science Letters*, *212*, 213–224.

Lowe, J. J., & Walker, M. (1997). *Reconstructing Quaternary environments*. London: Prentice Hall.

Loya, Y. (1972). Community structure and species diversity of hermatypic corals at Eilat (Red Sea). *Marine Biology, 13*, 100–123.

Loya, Y. (1978). Plotless and transect methods. In: D. R. Stoddart & R. E. Johannes (Eds), *Coral reefs: Research methods* (pp. 197–217). UNESCO Monographs on Oceanographic Methodology, Paris, Vol. 5.

Loya, Y., & Slobodkin, L. B. (1971). The coral reefs of Eilat (Gulf of Eilat, Red Sea). In: D. R. Stoddart & M. Yonge (Eds), *Regional variation in Indian Ocean coral reefs*. Symposia of the Zoological Society of London (Vol. 28, pp. 117–139).

Ludwig, K. R., Szabo, B. J., Moore, J. G., & Simmons, K. R. (1991). Crustal subsidence rate off Hawaii determined from $^{234}U/^{238}$ U ages of drowned coral reefs. *Geology, 19*, 171–174.

Lugo-Fernandez, A., Roberts, H. H., & Suhayda, J. N. (1998). Wave transformations across a Caribbean fringing-barrier Coral Reef. *Continental Shelf Research, 18*, 1099–1124.

Luick, J. L., Mason, L., Hardy, T., & Furnas, M. J. (2007). Circulation in the Great Barrier Reef Lagoon using numerical tracers and in situ data. *Continental Shelf Research, 27*, 757–778.

Lund, M., Davies, P. J., & Braga, J. C. (2000). Coralline algal nodules off Fraser Island, eastern Australia. *Facies, 42*, 25–34.

Lvovich, M. I. (1979). *World water resources and their future*. Washington, DC: American Geophysical Union.

Lyell, C. (1832). *Principles of geology: An attempt to explain the former changes in the earth's surface by reference to causes now in operation*. Vol. 2. London: John Murray, pp. 283–301.

Lyell, C. (1853). *Principles of geology or, the modern changes of the Earth and its inhabitants considered as illustrative of geology* (Book 3, Chapter 50, pp. 775–797). London: John Murray.

Ma, Z. B., Xiao, J. L., Zhao, X. T., Peng, Z. C., Xia, M., Zhang, G. P., et al. (2003). Precise U-series dating of coral reefs from the South China Sea and the high sea level during the Holocene. *Journal of Coastal Research, 19*, 296–303.

Macdonald, I. A., & Perry, C. T. (2003). Biological degradation of coral framework in a turbid lagoon environment, Discovery Bay, north Jamaica. *Coral Reefs, 22*, 523–535.

Macdonald, I. A., Perry, C. T., & Larcombe, P. (2005). Comments on "Rivers, runoff, and reefs" by McLaughlin et al. [Global Planetary Change 39 (2003) 191–199]. *Global and Planetary Change, 45*, 333–337.

Machel, H. (1988). Fluid flow direction during dolomite formation as deduced from trace-element trends. In: V. Shukla & P. A. Baker (Eds), *Sedimentology and geochemistry of dolostones* (pp. 115–125), Society of Economic Paleontologists and Mineralogists (Special Publication 43).

Machel, H. (2004). Concepts and models of dolomitization: A critical reappraisal. In: C. J. R. Braithwaite, G. Rizzi & G. Darke (Eds), *The geometry and petrogenesis of dolomite hydrocarbon reservoirs* (pp. 7–63), The Geological Society of London (Special Publication 235).

Macintyre, I. G. (1972). Submerged reefs of the eastern Caribbean. *Bulletin of the American Association of Petroleum Geologists, 5*, 19 53.

Macintyre, I. G. (1977). Distribution of submarine cements in a modern Caribbean fringing reef, Galeta Point, Panama. *Journal of Sedimentary Petrology, 47*, 503 516.

Macintyre, I. G. (1978). A hand-operated submersible drill for coring reef substrata. In: D. R. Stoddart & R. E. Johannes (Eds), *Coral reefs: Research methods* (pp. 75–80). UNESCO Monographs on Oceanographic Methodology, Paris, Vol. 5.

Macintyre, I. G. (1984a). Preburial and shallow-subsurface alteration of modern scleractinian corals. *Palaeontologica, 54*, 229–244.

Macintyre, I. G. (1984b). Extensive submarine lithification in a cave in the Belize Barrier Reef Platform. *Journal of Sedimentary Petrology, 54*, 221–235.

Macintyre, I. G. (1985). Submarine cements: The peloidal question. In: N. Schneidermann & P. M. Harris (Eds), *Carbonate cements* (pp. 109–121), SEPM (Special Publication 36).

Macintyre, I. G. (1988). Modern coral reefs of western Atlantic: New geological perspective. *American Association of Petroleum Geologists Bulletin, 72*, 1360–1369.

Macintyre, I. G. (1997). Re-evaluating the role of crustose coralline algae in the construction of coral reefs. *Proceedings of the Eighth International Coral Reef Symposium*, Panama, Vol. 1, pp. 725–730.

Macintyre, I. G. (2007). Demise, regeneration, and survival of some western Atlantic reefs during the Holocene transgression. In: R. B. Aronson (Ed.), *Geological approaches to coral reef ecology* (pp. 181–200). New York: Springer.

Macintyre, I. G., & Aronson, R. B. (2006). Lithified and unlithified Mg-calcite precipitates in tropical reef environments. *Journal of Sedimentary Research, 76*, 81–90.

Macintyre, I. G., Burke, R. B., & Stuckenrath, R. (1977). Thickest recorded Holocene reef section, Isla Perez, Alacran Reef, Mexico. *Geology, 5*, 749–754.

Macintyre, I. G., Burke, R. B., & Stuckenrath, R. (1981). Core holes in the outer fore reef off Carrie Fox Cay, Belize: A key to the Holocene history of the Belizian Barrier reef complex. *Proceeding of the Fourth Coral Reef Symposium*, Manila, Vol. 1, pp. 567–574.

Macintyre, I. G., & Glynn, P. W. (1976). Evolution of a modern Caribbean Fringing Reef, Galeta Point, Panama. *American Association of Petroleum Geologists Bulletin, 60*, 1054–1072.

Macintyre, I. G., Glynn, P. W., & Cortés, J. (1992). Holocene reef history in the Eastern Pacific: Mainland Costa Rica, Cano Island, Cocos Island, and Galapagos Islands. *Proceedings of the Seventh International Coral Reef Symposium*, Guam, Vol. 2, pp. 1174–1184.

Macintyre, I. G., Graus, R. R., Reinthal, P. N., Littler, M. M., & Littler, D. S. (1987). The barrier reef sediment apron: Tobacco Reef, Belize. *Coral Reefs, 6*, 1–12.

Macintyre, I. G., & Marshall, J. F. (1988). Submarine lithification in coral reefs: Some facts and misconceptions. *Proceedings of the Sixth International Coral Reef Symposium*, Townsville, Australia, Vol. 1, pp. 263–272.

Macintyre, I. G., Mountjoy, E. W., & D'Anglejan, B. F. (1968). An occurrence of submarine cementation of carbonate sediments off the west coast of Barbados, W.I. *Journal of Sedimentary Petrology, 38*, 660–664.

Macintyre, I. G., Precht, W. F., & Aronson, R. B. (2000). Origin of the Pelican Cays ponds, Belize. *Atoll Research Bulletin, 466*, 1–12.

Macintyre, I. G., Reid, R. P., & Steneck, R. S. (1996). Growth history of stromatolites in a Holocene fringing reef, Stocking Island, Bahamas. *Journal of Sedimentary Research, 66*, 231–242.

Macintyre, I. G., & Toscano, M. A. (2004). The Pleistocene limestone foundation below Twin Cays, Belize, Central America. *Atoll Research Bulletin, 511*, 1–16.

Macintyre, I. G., & Towe, K. M. (1976). Skeletal calcite in living scleractinian corals, microboring fillings, not primary skeletal deposits. *Science, 193*, 701–702.

Mackenzie, F. T., Kulm, L. D., Cooley, R. L., & Barnhart, J. T. (1965). *Homotrema rubrum* (Lamarck), a sediment transport indicator. *Journal of Sedimentary Research, 35*, 265–272.

Mackenzie, J. A., & Davies, P. J. (1993). Cenozoic evolution of carbonate platforms on the northeastern Australian margin. In: J. A. Mackenzie, P. J. Davies, A. Palmer-Julson, et al. (Eds), *Proceedings of the Ocean Drilling Program*, Scientific Results, Northeast Australian Margin, College Station, TX, Vol. 133, pp. 763–770.

Madin, J. S., Black, K. P., & Connolly, S. R. (2006). Scaling water motion on coral reefs: From regional to organismal scales. *Coral Reefs, 25*, 635–644.

Madin, J. S., & Connolly, S. R. (2006). Ecological consequences of major hydrodynamic disturbances on coral reefs. *Nature, 444*, 477–480.

Maeda, Y., Siringan, F., Omura, A., Berdin, R., Hosono, Y., Atsumi, S., et al. (2004). Higher-than-present Holocene mean sea levels in Ilocos, Palawan and Samar, Philippines. *Quaternary International, 115–116,* 15–26.

Maejima, Y., Matsuzaki, H., & Higashi, T. (2005). Application of cosmogenic ^{10}Be to dating soils on the raised coral reef terraces of Kikai Island, southwest Japan. *Geoderma, 126,* 389–399.

Maier, C., Felis, Th., Pätzold, J., & Bak, R. P. M. (2004). Effect of skeletal growth and lack of species effect in the skeletal oxygen isotope climate signal within the coral genus *Porites*. *Marine Geology, 207,* 193–208.

Maiklem, W. C. (1968). Some hydraulic properties of bioclastic carbonate grains. *Sedimentology, 10,* 101–109.

Maiklem, W. R. (1970). Carbonate sediments in the Capricorn Reef Complex, Great Barrier reef, Australia. *Journal of Sedimentary Petrology, 40,* 55–80.

Maliva, R. G. (1998). Skeletal aragonite neomorphism: Quantitative modelling of a two-water diagenetic system. *Sedimentary Geology, 121,* 179–190.

Maliva, R. G., Missimer, T. M., & Dickson, J. A. D. (2000). Skeletal aragonite neomorphism in Plio-Pleistocene sandy limestones and sandstones. *Sedimentary Geology, 136,* 147–154.

Mallela, J. (2007). Coral reef encruster communities and carbonate production in cryptic and exposed coral reef habitats along a gradient of terrestrial disturbance. *Coral Reefs, 26,* 775–785.

Mallela, J., & Perry, C. T. (2007). Calcium carbonate budgets for two coral reefs affected by different terrestrial runoff regimes, Rio Bueno, Jamaica. *Coral Reefs, 26,* 129–145.

Malone, M. J., & Baker, P. A. (1999). Temperature dependence of the strontium distribution coefficient in calcite: An experimental study from 40° to 200°C and application to natural diagenetic calcites. *Journal of Sedimentary Research, 69,* 216–223.

Malone, M. J., Baker, P. A., & Burns, S. J. (1996). Recrystallization of dolomite: An experimental study from 50–200°C. *Geochimica et Cosmochimica Acta, 60,* 2189–2207.

Manfrino, C. M., & Ginsburg, R. N. (2001). Pliocene to Pleistocene deposition history of the upper platform margin. In: R. N. Ginsburg (Ed.), *Subsurface geology of a prograding carbonate platform margin, Great Bahama Bank; Results of the Bahamas Drilling Project* (pp. 17–40), SEPM (Special Publication 70).

Mankiewicz, C. (1988). Occurrence and paleoecologic significance of *Halimeda* in late Miocene reefs, southeastern Spain. *Coral Reefs, 6,* 271–279.

Mann, P., Prentice, C. S., Burr, G., Pena, L., & Taylor, F.W. (1998). Tectonic geomorphology and paleoseismicity of the septentrional fault system, Dominican Republic. In: J. Dolan & P. Mann (Eds), *Active strike-slip and collisionial tectonics of the Northern Caribbean Plate boundary zone* (pp. 63–123), Geological Society of America (Special Paper 326).

Mann, P., Taylor, F. W., Lagoe, M. B., & Quarles, A. (1998). Late Quaternary uplift of the New Georgia Islands (Solomon island arc) in response to subduction of the Woodlark spreading center. *Tectonophysics, 295,* 259–306.

Mantua, N. J., & Hare, S. R. (2002). The Pacific Decadal Oscillation. *Journal of Oceanography, 58,* 35–44.

Maragos, M. (1974). Reef corals on Fanning Island. *Pacific Science, 28,* 247–255.

Maragos, M., & Jokiel, P. L. (1986). Reef corals of Johnson atoll, one of the world's most isolated reefs. *Coral Reefs, 4,* 141–150.

Marcantonio, F., Anderson, R. F., Higgins, S., Fleisher, M. Q., Stute, M., & Schlosser, P. (2001). Abrupt intensification of the SW Indian monsoon during the last deglaciation: Constraints from Th, Pa and He isotopes. *Earth and Planetary Science Letters, 184,* 505–514.

Marshall, J. F. (1983). Submarine cementation in a high-energy platform reef: One Tree Reef, Southern Great Barrier Reef. *Journal of Sedimentary Petrology, 53,* 1133–1149.

Marshall, J. F. (1988). Potential effects of oceanic deep waters on the initiation and demise of coral reefs. *Proceedings of the Sixth International Coral Reef Symposium*, Townsville, Australia, Vol. 3, pp. 509–512.

Marshall, J. F., & Davies, P. J. (1981). Submarine lithification on windward reef slopes: Capricorn-Bunker Group, Southern Great Barrier Reef. *Journal of Sedimentary Petrology*, *51*, 953–960.

Marshall, J. F., & Davies, P. J. (1982). Internal structure and Holocene evolution of One Tree Reef, southern Great Barrier Reef. *Coral Reefs*, *4*, 19–26.

Marshall, J. F., & Davies, P. J. (1988). *Halimeda* bioherms of the northern Great Barrier Reef. *Coral Reefs*, *6*, 139–148.

Marshall, J. F., & Jacobson, G. (1985). Holocene growth of a mid-Pacific atoll: Tarawa, Kiribati. *Coral Reefs*, *4*, 11–17.

Marshall, J. F., & McCulloch, M. T. (2001). Evidence of El Niño and the Indian Ocean Dipole from Sr/Ca derived SSTs form modern corals at Christmas Island, eastern Indian Ocean. *Geophysical Research Letters*, *28*, 3453–3456.

Marshall, J. F., & McCulloch, M. T. (2002). An assessment of the Sr/Ca ratio in shallow water hermatypic corals as a proxy for sea surface temperature. *Geochimica et Cosmochimica Acta*, *66*, 3263–3280.

Martin, G. D., Wilkinson, B. H., & Lohman, K. C. (1986). The role of skeletal porosity in aragonite neomorphism: *Strombus* and *Montastrea* from the Pleistocene Key Largo Limestone, Florida. *Journal of Sedimentary Petrology*, *56*, 194–203.

Martin, J. M., Braga, J. C., & Riding, R. (1997). Late Miocene *Halimeda* algal-microbial segment reef in the marginal Mediterranean Sorbas basin, Spain. *Sedimentology*, *44*, 441–456.

Martin, R. E. (1999). Taphonomy and temporal resolution of foraminiferal assemblages. In: B. K. S. Gupta (Ed.), *Modern foraminifera* (pp. 291–298). Dordrecht: Kluwer Academic Press.

Martin, R. E., & Liddell, W. D. (1988). Foraminiferal biofacies on a north coast fringing reef (1–75 m), Discovery Bay, Jamaica. *PALAIOS*, *3*, 298–314.

Martin, R. E., & Liddell, W. D. (1991). Taphonomy of foraminifera in modern carbonate environments: Implications for the formation of foraminiferal assemblages. In: S. K. Donovan (Ed.), *The processes of fossilization* (pp. 170–193). New York: Columbia University Press.

Martindale, W. (1992). Calcified epibionts as palaeoecological tools: Examples from the Recent and Pleistocene of Barbados. *Coral Reefs*, *11*, 167–177.

Marubini, F., & Davies, P. S. (1996). Nitrate increases zooxanthellae population density and reduces skeletogenesis in corals. *Marine Biology*, *127*, 319–328.

Masse, J. P. (1970). Contribution à l'étude des sédiments bioclastiques actuels du complexe récifal de l'Ile de Nossi-Bé (N. W. de Madagascar). *Recueils des Travaux de la Station Marine d'Endoume, hors série Supplément*, *10*, 229–251.

Masse, J. P., & Montaggioni, (2001). Growth history of shallow-water carbonates: Control of accommodation on ecological and depositional processes. *International Journal of Earth Sciences*, *90*, 452–469.

Masse, J. P., Thomassin, B., & Acquaviva, M. (1989). Bioclastic sedimentary environments of coral reefs and lagoons around Mayotte Island (Comoro Archipelago, Mozambique Channel, SW Indian Ocean). *Journal of Coastal Research*, *5*, 419–432.

Massel, S. R., & Done, T. J. (1993). Effects of cyclone waves on massive coral assemblages on the Great Barrier Reef: meteorology, hydrodynamics and demography. *Coral Reefs*, *12*, 153–166.

Masson, V., Vimeux, F., Jouzel, J., Morgan, V., Delmotte, M., Ciais, P., et al. (2000). Holocene climate variability in Antarctica based on 11 ice-core isotopic records. *Quaternary Research*, *54*, 348–358.

Matsuda, S. (1989). Succession and growth rates of encrusting crustose coralline algae (Rhodophyta, Cryptomeniales) in the upper fore-reef environment off Ishigaki Island, Ryukyu Islands. *Coral Reefs*, *7*, 185–195.

Matthews, B. J. H., Jones, A. C., Theodorou, N. K., & Tudhope, A. W. (1996). Excitation-emission-matrix fluorescence spectroscopy applied to humic acid bands in coral reefs. *Marine Chemistry*, *55*, 317–332.

Matthews, R. K. (1968). Carbonate diagenesis: equilibriation of sedimentary mineralogy to the subaerial environment: Coral Cap of Barbados, West Indies. *Journal of Sedimentary Petrology*, *38*, 1110–1119.

Matthews, R. K. (1974). A process approach to diagenesis of reefs and reef-associated sediments. In: L. F. Laporte (Ed.), *Reefs in time and space* (pp. 234–256), SEPM (Special Publication 18).

Matthews, R. K., & Frohlich, C. (1987). Forward modeling of bank-margin carbonate diagenesis. *Geology*, *15*, 673–676.

Maxwell, W. G. H. (1968). *Atlas of the Great Barrier Reef*. Amsterdam: Elsevier, 258 pp.

Maxwell, W. G. H., Day, R. W., & Fleming, P. J. G. (1961). Carbonate sedimentation on the Heron Island Reef, Great Barrier Reef. *Journal of Sedimentary Petrology*, *31*, 215–230.

Maxwell, W. G. H., & Swinchatt, J. P. (1970) Great Barrier Reef: Regional variation in a terrigenous-carbonate province. *Geological Society of America Bulletin*, *81*, 691–724.

Maxwell, W. G. W. (1973). Sediments of the Great Barrier Reef province. In: O. A. Jones & R. Endean (Eds), *Biology and geology of coral reefs, Vol. 1: Geology 1* (pp. 299–345). Academic Press.

Mazzullo, S. J. (2006). Late Pliocene to Holocene platform evolution in northern Belize, and comparison with coeval deposits in southern Belize and the Bahamas. *Sedimentology*, *53*, 1015–1047.

Mazzullo, S. J., Bischoff, W. D., & Teal, C. S. (1995). Holocene shallow-subtidal dolomitization by near-normal seawater, northern Belize. *Geology*, *23*, 341–344.

McClain, M. E., Swart, P. K., & Vacher, H. L. (1992). The hydrogeochemistry of early meteoric diagenesis in a Holocene deposit of biogenic carbonate. *Journal of Sedimentary Petrology*, *62*, 1008–1022.

McConnaughey, T. A. (1989). ^{13}C and ^{18}O isotopic disequilibrium in biological carbonates: I. Patterns. *Geochimica et Cosmochimica Acta*, *53*, 151–162.

McCook, L. J., Jompa, J., & Diaz-Pulido, G. (2001). Competition between corals and algae on coral reefs: A review of evidence and mechanisms. *Coral Reefs*, *19*, 400–417.

McCulloch, M. L., & Land, L. S. (1992). Dynamic hydrology and diagenesis of a submerged Pleistocene fringing reef, Discovery Bay, Jamaica. *Marine Geology*, *104*, 139–151.

McCulloch, M. T., & Esat, T. M. (2000). The coral record of last interglacial sea levels and sea surface temperatures. *Chemical Geology*, *169*, 107–129.

McCulloch, M. T., Fallon, S., Wyndham, T., Hendy, E., Lough, J., & Barnes, D. (2003). Coral record of increased sediment flow to the inner Great Barrier Reef since European settlement. *Nature*, *421*, 727–730.

McCulloch, M. T., Gagan, M. K., Mortimer, G. E., Chivas, A. R., & Isdale, P. J. (1994). A high resolution Sr/Ca and $\delta^{18}O$ coral record from the Great Barrier Reef, Australia, and the 1982–83 El Niño. *Geochimica et Cosmochimica Acta*, *58*, 2747–2754.

McCulloch, M. T., Mortimer, G., Esat, T., Xianhua, L., Pillans, B., & Chappell, J. (1996). High resolution windows into the early Holocene climate: Sr/Ca coral records from the Huon Peninsula. *Earth and Planetary Science Letters*, *138*, 169–178.

McCulloch, M. T., Tudhope, A. W., Esat, T. M., Mortimer, G. E., Chappell, J., Pillans, B., et al. (1999). Coral record of equatorial sea-surface temperatures during the penultimate deglaciation at Huon Peninsula. *Science*, *283*, 202–204.

McGregor, H. V., & Gagan, M. K. (2003). Diagenesis and geochemistry of *Porites* corals from Papua New Guinea: Implications for paleoclimate reconstruction. *Geochimica et Cosmochimica Acta, 64*, 3123–3132.

McGregor, H. V., & Gagan, M. K. (2004). Western Pacific coral $\delta^{18}O$ records of anomalous Holocene variability in the El Niño-Southern Oscillation. *Geophysical Research Letters, 31*, L11204, doi: 10.1029/2004GL019972.

McIlreath, I. A., & Morrow, D. W. (1990). *Diagenesis* (338 pp.). Geoscience Canada (Reprint Series 4).

McIlreath, I. A., & James, N. P. (1984). Carbonate slopes. In: R. G. Walker (Ed.), *Facies models* (pp. 245–257). Geoscience Canada (Reprint Series 1).

McKee, E. D., Chronic, J., & Leopald, E. B. (1959). Sedimentary belts in lagoon of Kapingamarangi Atoll. *American Association of Petroleum Geologists Bulletin, 43*, 501–562.

McKelvey, V. E. (1973). Abundance and distribution of phosphorus in the lithosphere. In: E. J. Griffith, A. Beeton, J. M. Spencer & D. T. Mitchell (Eds), *Environmental phosphorus handbook* (pp. 13–31). New York: Wiley.

McLean, R. F., Stoddart, D. R., Hopley, D., & Polach, H. (1978). Sea level change in the Holocene on the northern Great Barrier Reef. *Philosophical Transactions of the Royal Society of London, Series A, 291*, 167–186.

McLean, R. F., & Woodroffe, C. D. (1994). Coral atolls. In: R. W. G. Carter & C. D. Woodroffe (Eds), *Late Quaternary shoreline morphodynamics* (pp. 267–302). Cambridge: Cambridge University Press.

McManus, J. W. (1985). Marine speciation, tectonics and sea-level changes in Southeast Asia. *Proceedings of the Fifth International Coral Reef Symposium*, Tahiti, Vol. 4, pp. 133–138.

McMurtry, G. M., Tappin, D. R., Sedwick, P. N., Wilkinson, I., Fietzke, J., & Bellwood, B. (2007). Elevated deposits in Bermuda record a late Quaternary megatsunami. *Sedimentary Geology, 200*, 155–165.

McNeill, D., Eberli, G., Lidz, B. H., Swart, P. K., & Kenter, J. A. M. (2001). Chronostratigraphy of a prograded carbonate platform margin: A record of dynamic slope sedimentation, western Great Bahama Bank. In: R. N. Ginsburg (Ed.), *Subsurface geology of a prograding carbonate platform margin, Great Bahama Bank: Results of the Bahamas Drilling Project* (pp. 101–134), SEPM (Special Publication 70).

McNeill, D. F., Budd, A. F., & Borne, P. M. (1997). Earlier (Late Pliocene) first appearance of the Caribbean reef-building coral *Acropora palmata*: Stratigraphic and evolutionary implications. *Geology, 25*, 891–894.

Medina, M., Collins, A. G., Takaoka, T. L., Kuehl, J. V., & Boore, J. L. (2006). Naked corals: Skeleton loss in Scleractinia. *Proceedings of the National Academy of Sciences, 103*, 9096–9100.

Meibom, A., Mostefaoui, S., Cuif, J. P., Dauphin, Y., Houlbreque, F., Dunbar, R., et al. (2007). Biological forcing controls the chemistry of reef-building coral skeleton. *Geophysical Research Letters, 34*, L02601, doi: 10.129/2006GL028657.

Melim, L. A. (1996). Limitations on lowstand meteoric diagenesis in the Pliocene–Pleistocene of Florida and Great Bahama Bank: Implications for eustatic sea-level models. *Geology, 24*, 893–896.

Melim, L. A., Westphal, H., Swart, P. K., Eberli, G. P., & Munnecke, A. (2002). Questioning carbonate diagenetic paradigms: Evidence from the Neogene of the Bahamas. *Marine Geology, 185*, 27–53.

Mergner, H. (1971). Structure, ecology and zonation of Red Sea reefs (in comparison with south Indian and Jamaican reefs). In: D. R. Stoddart & M. Yonge (Eds), *Regional variation in Indian Ocean coral reefs*. Symposia of the Zoological Society of London (Vol. 28, pp. 141–161).

Mesolella, K. J. (1967). Zonation of uplifted Pleistocene coral reefs on Barbados, West Indies. *Science, 156*, 638–640.

Mesolella, K. J., Sealy, H. A., & Matthews, R. K. (1970). Facies geometries within Pleistocene reefs of Barbados, West Indies. *American Association of Petroleum Geologists Bulletin*, *54*, 1899–1917.

Meyer, C. P., Geller, J. B., & Paulay, G. (2005). Fine scale endemism on coral reefs: Archipelagic differentiation in turbinid gastropods. *Evolution*, *59*, 113–125.

Meyer, D. L., Bries, J. M., Greenstein, B. J., & Debrot, A. O. (2003). Preservation of *in situ* reef framework in regions of low hurricane frequency: Pleistocene of Curaçao and Bonaire, southern Caribbean. *Lethaia*, *36*, 273–285.

Meyer, D. L., & Meyer, K. B. (1986). Biostratinomy of recent crinoids (Echinodermata) at Lizard Island, Great Barrier Reef Australia. *PALAIOS*, *1*, 294–302.

Meyers, W. J. (1991). Calcite cement stratigraphy: An overview. In: C. E. Baker & O. C. Kopp (Eds), *Luminescence microscopy: Quantitative and qualitative aspects* (pp. 133–148), SEPM, Short Course.

Milankovitch, M. (1941). *Kanon der Erdbestrahlung und seine Andwendung auf das Eiszeiten-problem*. Belgrade: Royal Serbian Academy.

Miller, G. H., & Brigham-Grette, J. (1989). Amino acid geochronology: Resolution and precision in carbonate fossils. *Quaternary International*, *1*, 111–128.

Milliman, J. D. (1967). Carbonate sedimentation on Hogsty Reef, a Bahamian atoll. *Journal of Sedimentary Petrology*, *37*, 658–676.

Milliman, J. D. (1973). Caribbean coral reefs. In: O. A. Jones & R. Endean (Eds), *Biology and geology of coral reefs* (Vol. 1, Geology 1, pp. 1–50). New York: Academic Press.

Milliman, J. D. (1974). *Marine carbonates*. New York: Springer-Verlag, 375 pp.

Milliman, J. D. (1993). Production and accumulation of calcium carbonate in the ocean: Budget of a nonsteady state. *Global Biogeochemical Cycles*, 7, 927–957.

Milliman, J. D., & Droxler, A. W. (1996). Neritic and pelagic carbonate sedimentation in the marine environment: Ignorance is not a bliss. *Geologische Rundschau*, *85*, 496–504.

Milne, G. A. (2002). Recent advances in predicting glaciation-induced sea level changes and their impact on model applications. In: J. X. Mitrovica & B. L. A. Vermeersen (Eds), *Glacial isostatic adjustment and the earth system: Sea level, crustal deformation, gravity and rotation*, American Geophysical Union, Monograph, Geodynamics Series, Vol. 29, pp. 157–176, Washington, DC.

Milne, G. A., Long, A. J., & Bassett, S. E. (2005). Modelling Holocene relative sea-level observations from the Caribbean and South America. *Quaternary Science Reviews*, *24*, 1183–1202.

Milne, G. A., Mitrovica, J. X., & Schrag, D. (2002). Estimating past continental ice volume from sea-level data. *Quaternary Science Reviews*, *221*, 361–376.

Min, G. R., Edwards, R. L., Taylor, F. W., Récy, J., Gallup, C. D., & Beck, J. W. (1995). Annual cycles of U/Ca in coral skeletons and U/Ca thermometry. *Geochimica et Cosmochimica Acta*, *59*, 2025–2042.

Minnery, G. A. (1990). Crustose coralline algae from the Flower Garden Banks, Northwestern Gulf of Mexico: Controls on distribution and growth morphology. *Journal of Sedimentary Petrology*, *60*, 992–1007.

Minoura, K., & Nakamori, T. (1982). Depositional environment of algal balls in the Ryukyu Group, Rykyu Islands, southwestern Japan. *Journal of Geology*, *90*, 602–609.

Mitchell, J. T., Land, L. S., & Miser, D. E. (1987). Modern marine dolomite cement in a north Jamaican fringing reef. *Geology*, *15*, 557–560.

Mitrovica, J. X., & Milne, G. A. (2002). On the origin of late Holocene sea-level highstands within equatorial ocean basins. *Quaternary Science Reviews*, *21*, 2179–2190.

Mitrovica, J. X., & Peltier, W. R. (1991). On postglacial geoid subsidence over the equatorial oceans. *Journal of Geophysical Research*, *96*, 20053–20071.

Mitsuguchi, T., Matsumoto, E., Abe, O., Uchida, T., & Isdale, P. J. (1996). Mg/Ca thermometry in coral skeletons. *Science, 274*, 961–963.

Mitter, R. M., & Kriausakul, N. (1989). Calculation of amino acid racemization ages based on apparent parabolic kinetics. *Quaternary Science Reviews, 8*, 353–357.

Mokady, O., Lazar, B., & Loya, Y. (1996). Echinoid bioerosion as a major structuring force of Red Sea coral reefs. *Biological Bulletin, 190*, 367–372.

Monismith, S. G., Genin, A., Reidenbach, M. A., Yahel, G., & Koseff, J. R. (2006). Thermally driven exchanges between a coral reef and the adjoining ocean. *Journal of Physical Oceanography, 36*, 1332–1347.

Montaggioni, L. F. (1977). Structure interne d'un récif corallien holocène (Ile de La Réunion, Océan Indien). *Mémoires du Bureau de Recherches Géologiques et Minières, Paris, 89*, 456–466.

Montaggioni, L. F. (1978). *Recherches géologiques sur les complexes récifaux de l'archipel des Mascareignes (Océan Indien Occidental).* Unpublished D.Sc.thesis, University of Aix-Marseilles 2, 524 pp.

Montaggioni, L. F. (1979a). Environmental significance of rhodolites from the Mascarene reef province, Western Indian Ocean. *Bulletin du Centre de Recherche et d'Exploration-Production d'Elf-Aquitaine, 3*, 713–723.

Montaggioni, L. F. (1979b). Application du principe de Brunn à la détermination des variations du niveau marin au cours de l'Holocène: cas des îles Maurice et de la Réunion, Océan Indien. *Marine Geology, 31*, M29–M38.

Montaggioni, L.F. (1980). Alcyonarian spiculites in coral reefs. *Proceedings of the Twenty-Sixth International Geological Congress*, Paris, Vol. 2, 251 pp.

Montaggioni, L. F. (1981). Les associations de foraminifères dans les sédiments récifaux de l'archipel des Mascareignes (Océan Indien). *Annales de l'Institut Océanographique, Masson Editeur, Paris, 57*, 41–62.

Montaggioni, L. F. (1982). Pleistocene marine depositional environments from Mauritius Island, Indian Ocean. *Geobios, 15*, 161–179.

Montaggioni, L. F. (1985). Makatea Island, Tuamotu Archipelago. *Proceedings of the Fifth International Coral Reef Congress*, Tahiti, Vol. 1, pp. 103–158.

Montaggioni, L. F. (1988a). Miocene reef platforms from northwestern Tuamotu islands, Central Pacific. *American Association of Petroleum Geologists Bulletin, 72*, 224.

Montaggioni, L. F. (1988b). Holocene reef growth history in mid-plate high volcanic islands. *Proceedings of the Sixth International Coral Reef Symposium*, Townsville, Australia, Vol. 3, pp. 455–460.

Montaggioni, L. F. (1989). Le soulèvement polyphasé d'origine volcano-isostatique: clef de l'évolution post-oligocène des atolls du nord-ouest des Tuamotus (Pacifique Central). *Comptes Rendus de l'Académie des Sciences, Paris, Série II, 309*, 1591–1598.

Montaggioni, L. F. (2000). Postglacial reef growth. *Comptes Rendus de l'Académie des Sciences, Paris, Earth and Planetary Sciences, 331*, 319–330.

Montaggioni, L. F. (2005). History of Indo-Pacific reef systems since the last glaciation: Development patterns and controlling factors. *Earth-Science Reviews, 71*, 1–75.

Montaggioni, L. F., Behairy, A. K. A., El Sayed, M. K., & Yusuf, N. (1986). The modern reef complex, Jeddah area, Red Sea: A facies model for carbonate sedimentation on embryonic passive margins. *Coral Reefs, 5*, 127–150.

Montaggioni, L. F., Cabioch, G., Camoin, G., Bard, E., Ribaud-Laurenti, A., Faure, G., et al. (1997). Continuous record of reef growth over the past 14 k.y. on the mid-Pacific island of Tahiti. *Geology, 25*, 555–558.

Montaggioni, L. F., & Camoin, G. F. (1993). Stromatolites associated with coralgal communities in Holocene high-energy reefs. *Geology, 21*, 149–152.

Montaggioni, L. F., & Camoin, G. F. (1997). Geology of Makatea Island, Tuamotu Archipelago, French Polynesia. In: H. L. Vacher & T. M. Quinn (Eds), *Geology and hydrogeology of carbonate islands*. Developments in Sedimentology (Vol. 54, pp. 453–474). Amsterdam: Elsevier.

Montaggioni, L. F., Cuet, P., & Naïm, O. (1993). Effects of nutrient excess on a modern fringing reef (Réunion Island, Western Indian Ocean). Geological Implications. In: R. N. Ginsburg (Ed.), *Colloquium on the Global Aspects of Coral Reefs: Health, hazards, and history* (pp. 397–403). Miami, FL: University of Miami.

Montaggioni, L. F., & Faure, G. (1980). Les récifs coralliens de l'archipel des Mascareignes. Collection des Travaux de l'Université de la Réunion, 1–150.

Montaggioni, L. F., & Faure, G. (1997). Response of reef coral communities to sea level rise: A Holocene model from Mauritius (Western Indian Ocean). *Sedimentology*, *44*, 1053–1070.

Montaggioni, L. F., Gabrié, C., Naïm, O., Payri, C., Richard, G., & Salvat, B. (1987). The seaward margin of Makatea, an uplifted carbonate island (Tuamotus, Central Pacific). *Atoll Research Bulletin*, *299*, 1–18.

Montaggioni, L. F., Le Cornec, F., Corrège, Th., & Cabioch, G. (2006). Coral barium/calcium record of mid-Holocene upwelling activity in New Caledonia, South-West Pacific. *Palaeogeography, Palaeoclimatology, Palaeoecology*, *237*, 436–455.

Montaggioni, L. F., & Mahé, J. (1980). Facies characterization of reefal sediments from Mauritius Island (Western Indian Ocean) by correspondence factor analysis. *Oceanologica Acta*, *3*, 409–420.

Montaggioni, L. F., & Pirazzoli, P. A. (1984). The significance of exposed coral conglomerates from French Polynesia (Pacific Ocean) as indicators of recent relative sea-level changes. *Coral Reefs*, *3*, 29–42.

Montaggioni, L. F., Richard, G., Bourrouilh-Le Jan, F., Gabrié, C., Humbert, L., Monteforte, M., et al. (1985). Geology and marine biology of Makatea, an uplifted atoll, Tuamotu Archipelago, Central Pacific Ocean. *Journal of Coastal Research*, *1*, 165–171.

Montaggioni, L. F., & Venec-Peyré, M. Th. (1993). Shallow-water foraminiferal taphocoenoses at Site 821: Implications for the Pleistocene evolution of the Central Great Barrier Reef shelf, northeastern Australia. *Proceedings of the Ocean Drilling Program*, Scientific Results, Vol. 133, pp. 365–378.

Moore, C. H., Graham, E. A., & Land, L. S. (1976). Sediment transport and dispersal across the deep fore-reef and island slope (−55 m to −305 m), Discovery Bay, Jamaica. *Journal of Sedimentary Petrology*, *46*, 174–187.

Moore, J. G., Ingram, B. L., Ludwig, K. R., & Clague, D. A. (1996). Coral ages and island subsidence, Hilo drill hole. *Journal of Geophysical Research*, *101*, 11599–11605.

Moore, W. S. (1996). Large groundwater inputs to coastal waters revealed by 226Ra enrichments. *Nature*, *380*, 612–614.

Morimoto, M., Abe, O., Kayanne, H., Kurita, N., Matsumoto, E., & Yoshida, N. (2002). Salinity records for the 1997-98 El Niño from Western Pacific corals. *Geophysical Research Letters*, 29, 10.1029/2001GLO135021.

Morrow, D. W. (1982). Diagenesis 2: Dolomitization models and ancient dolostones. *Geoscience Canada*, *9*, 95–107.

Morse, J. W., & Mackenzie, F.T. (1990). *Geochemistry of sedimentary carbonates*. Developments in Sedimentology (Vol. 48, 707 pp.). Amsterdam: Elsevier.

Morse, J. W., Wang, Q., & Tsio, M. Y. (1997). Influences of temperature and Mg/Ca ratio on $CaCO_3$ precipitated from seawater. *Geology*, *25*, 85–87.

Morse, J. W., Zullig, J. J., Bernstein, L. D., Millero, F. J., Milne, P., Mucci, A., et al. (1985). Chemistry of calcium carbonate-rich shallow water sediments in the Bahamas. *American Journal of Science*, *285*, 147–185.

Morton, R. A., Gelfenbaum, G., Jaffe, B. E., & Gibbons, H. (2007). *Distinguishing Tsunami from Storm Deposits in the Geologic Record*. US Geological Survey, Monthly Newsletter: http://soundwaves.usgs.gov/2007/11/index.html

Morton, R. A., Richmond, B. M., Jaffe, B. E., & Gelfenbaum, G. (2006). Reconnaissance investigation of Caribbean extreme wave deposits: Preliminary observations, interpretations, and research directions, United States Geological Survey.

Moustafa, Y. A., Pätzold, J., Loya, Y., & Weber, G. (2000). Mid-Holocene stable isotope record of corals from the northern Red Sea. *International Journal of Earth Sciences, 88*, 742–751.

Moyer, R. P., Riegl, B., Banks, K., & Dodge, R. E. (2003). Spatial patterns and ecology of benthic communities on a high-latitude South Florida (Broward County, USA) reef system. *Coral Reefs, 22*, 447–464.

Mubarak, W. A.-M., & Kubryakov, A. I. (2000). Hydrological structure of waters of the Persian Gulf according to the data of observations in 1992. *Physical Oceanography, 11*, 459–471.

Muhs, D. R. (2006). Last Interglacial: Timing and environment (LITE). US Geological Survey, Denver, CO: http://esp.cr.usgs.gov/info/lite/

Muhs, D. R., Simmons, K. R., Kennedy, G. L., Ludwig, K. R., & Groves, L. T. (2006). A cool eastern Pacific Ocean at the close of the Last Interglacial complex. *Quaternary Science Reviews, 25*, 235–262.

Muhs, D. R., Simmons, K. R., & Steinke, B. (2002). Timing and warmth of the Last Interglacial period: New U-series evidence from Hawaii and Bermuda and a new fossil compilation for North America. *Quaternary Science Reviews, 21*, 1355–1383.

Muller, A., Gagan, M. K., & Lough, J. (2004). Effect of early marine diagenesis on coral reconstructions of surface-ocean $^{13}C/^{12}C$ and carbonate saturation state. *Global Biogeochemical Cycles, 18*, GB1033, doi: 10.1029/2003GB002112.

Muller, A., & Opdyke, B. N. (2000). Glacial-interglacial changes in nutrient utilization and paleoproductivity in the Indonesian throughflow sensitive Timor Through, easternmost Indian Ocean. *Paleoceanography, 15*, 85–94.

Muller-Parker, G., & D'Elia, C. F. (1997). Interactions between corals and their symbiotic algae. In: Ch. Birkeland (Ed.), *Life and death of coral reefs* (pp. 96–113). New York: Chapman & Hall.

Multer, H. G. (1988). Growth rate, ultrastructure and sediment contribution of *Halimeda incrassata* and *Halimeda monile*, Nonsuch and Falmouth Bays, Antigua, WI. *Coral Reefs, 6*, 179–186.

Multer, H. G., Gischler, E., Lundberg, J., Simmons, K., & Shinn, E. A. (2002). The Key Largo Limestone revisited: Pleistocene shelf-edge facies, Florida Keys, USA. *Facies, 46*, 229–272.

Multer, H. G., & Hoffmeister, J. E. (1968). Subaerial laminated crusts of the Florida Keys. *Geological Society of America Bulletin, 79*, 183–192.

Munk, W. H., & Sargent, M. C. (1954). Adjustment of Bikini Atoll to ocean waves. US Geological Survey, Professional Paper, Vol. 260-I, pp. 275–280.

Murphy, S. J., Hurlburt, H. E., & O'Brian, J. J. (1999). The connectivity of eddy variability in the Caribbean Sea, the Gulf of Mexico, and the Atlantic Ocean. *Journal of Geophysical Research, 104*, 1431–1453.

Murray, J. (1880). On the structure and origin of coral reefs and islands. *Proceedings of the Royal Society of Edinburgh, 10*, 505–518.

Mylroie, J. E., Jensen, J. W., Taborosi, D., Jocson, J. M. U., Vann, D. T., & Wexel, C. (2001). Karst features of Guam in terms of a general model of carbonate island karst. *Journal of Cave and Karst Studies, 63*, 9–22.

Nadaoka, K., Nihei, Y., Kumano, R., Yokobori, T., & Wakaki, K. (2001). A field observation on hydrodynamic and thermal environments of a fringing reef at Ishigaki Island under typhoon and normal atmospheric conditions. *Coral Reefs, 20*, 387–398.

Nadaoka, K., Nihei, Y., Wakaki, K., Kumano, , Kakuma, S., Moromizato, S., et al. (2001). Regional variation of water temperature around Okinawa coasts and its relationship to offshore thermal environments and coral bleaching. *Coral Reefs*, *20*, 373–384.

Naim, O. (1993). Seasonal responses of a fringing reef community to eutrophication (Réunion Island, Western Indian Ocean). *Marine Ecology Progress Series*, *99*, 137–151.

Nakada, M. (1986). Holocene sea levels in oceanic islands: Implications for the rheological structure of the Earth's mantle. *Tectonophysics*, *121*, 263–276.

Nakamori, T. (1986). Community structures of recent and Pleistocene hermatypic corals in the Ryukyu islands, Japan. Science Report of the Tohoku University, Second Series (Geology), Vol. 56, pp. 71–133.

Nakamori, T., Campbell, C. R., & Wallensky, E. (1995). Living hermatypic coral assemblages at Huon Peninsula, Papua New Guinea. *Journal of Geography (Japan)*, *104*, 743–757.

Nakamori, T., Iryu, Y., & Yamada, T. (1995). Development of coral reefs of the Ryukyu Islands (southwest Japan, East China Sea) during Pleistocene sea-level change. *Sedimentary Geology*, *99*, 215–231.

Nakamura, T., & Nakamori, T. (2007). A geochemical model for coral reef formation. *Coral Reefs*, *26*, 741–764.

NARA. (2005). *Rapid assessment of tsunami damage to coral reefs in Sri Lanka*. National Aquatic Resources Research and Development Agency, Mallakkuliya, Colombo, Sri Lanka.

Nassif, N., Pinna, N., Gehrke, N., Antonletti, M., Jager, C., & Colfen, H. (2005). Amorphous layer around aragonite platelets in nacre. *Proceedings of the National Academy of Sciences*, *102*, 12653–12655.

Nelsen, J. E., & Ginsburg, R. N. (1986). Calcium carbonate production by epibionts on *Thalassia* in Florida Bay. *Journal of Sedimentary Petrology*, *56*, 622–628.

Nelson, C. S., & Smith, A. M. (1996). Stable oxygen and carbon isotope compositional fields for skeletal and diagenetic components in New Zealand Cenozoic non-tropical carbonate sediments and limestones: A synthesis and review. *New Zealand Journal of Geology and Geophysics*, *39*, 93–107.

Nelson, R. C. (1996). Hydraulic roughness of coral reef platforms. *Applied Ocean Research*, *18*, 265–274.

Neumann, A. C., & Macintyre, I. G. (1985). Reef response to sea-level rise: Keep-up, catch-up or give-up. *Proceedings of the Fifth International Coral Reef Congress*, Tahiti, Vol. 2, pp. 105–110.

Neumeier, U. (1999). Experimental modelling of beachrock cementation under microbial influence. *Sedimentary Geology*, *126*, 35–46.

Neumeier, U., Bernier, P., Dalongeville, R., & Oberlin, C. (2000). Les variations holocènes du niveau marin mises en evidence par les caractères et la diagenèse des beach-rocks: Exemple de Damnono (Crète). *Géomorphologie*, *4*, 211–220.

Ninagawa, K., Kitahara, T., Toyoda, S., Hayashi, K., Nishido, H., Kinjo, M., et al. (2001). Thermoluminescence dating of the Ryukyu Limestone. *Quaternary Science Reviews*, *20*, 829–833.

Norris, R. M. (1953). Buried oyster reefs in some Texas bays. *Journal of Paleontology*, *27*, 569–576.

Notholt, A. J., & Jarvis, I. (Eds) (1990). *Phosphorite research and development* (326 pp.). The Geological Society of London (Special Publication 52).

Nott, J. (2003). The importance of pre-historic data and variability of hazard regimes in natural hazard risk assessment. *Natural Hazards*, *30*, 43–58.

Nott, J., & Hayne, M. (2001). High frequency of "super-cyclones" along the Great Barrier Reef over the past 5,000 years. *Nature*, *413*, 508–512.

Nunn, P. D. (1994). *Oceanic islands*. Oxford, UK: Blackwell, 413 pp.

Nunn, P. D., & Peltier, W. R. (2001). Far-field test of the ICE-4G (VM2) model of global isostatic response to deglaciation: Empirical and theoretical Holocene sea-level reconstructions for the Fiji Islands, Southwest Pacific. *Quaternary Research, 55*, 203–214.

Nyberg, J. (2002). Luminescence intensity in coral skeletons from Mona Island in the Caribbean Sea and its link to precipitation and wind speed. *Philosophical Transactions of the Royal Society of London, Series A, 360*, 749–766.

Oberdorfer, J. A., & Buddemeier, R. W. (1986). Coral Reef Hydrogeology: Field studies of water movement within a Barrier Reef. *Coral Reefs, 5*, 7–12.

Obrochta, S. P. (2004). *Australian Great Barrier Reef initiation timing constrained by seaward shallow-water sediment drift architecture (ODP Leg 194, Marion Plateau)*. Unpublished Master of Science thesis, University of South Florida, 56 pp.

Ogden, J. C. (1977). Carbonate sediment production by parrot fish on Caribbean reefs and sea grass beds. *Aquatic Botany, 2*, 103–116.

Ohde, S., & Elderfield, H. (1992). Strontium isotope stratigraphy of Kita-Daito-Jima Atoll, North Philippine Sea: Implications for Neogene sea-level and tectonic history. *Earth and Planetary Science Letters, 113*, 473–486.

Ohde, S., Greaves, M., Masuzawa, T., Buckley, H. A., Van Woesik, R., Wilson, P. A., et al. (2002). The chronology of Funafuti Atoll: Revisiting an old friend. *Philosophical Transactions of the Royal Society of London, Series A, 458*, 2289–2306.

Opdyke, B. N., & Walker, J. C. G. (1992). Return to the coral reef hypothesis: basin to shelf partitioning of $CaCO_3$ and its effect on atmospheric CO_2. *Geology, 20*, 733–736.

Opdyke, B. N., & Wilkinson, B. H. (1993). Carbonate mineral saturation state and cratonic limestone accumulation. *American Journal of Science, 293*, 217–234.

Orme, G. R. (1977a). The Coral Sea Plateau, a major reef province. In: O. A. Jones & R. Endean (Eds), *Biology and geology of coral reefs* (Vol. IV, Geology 2, pp. 267–306). New York: Academic Press.

Orme, G. R. (1977b). Aspects of sedimentation in the coral reef environment. In: O. A. Jones & R. Endean (Eds), *Biology and geology of coral reefs* (Vol. IV, Geology 2, pp. 129–182). New York: Academic Press.

Orme, G. R. (1985). The sedimentological importance of *Halimeda* in the development of back reef lithofacies, Northern Great Barrier Reef (Australia). *Proceedings of the Fifth International Coral Reef Congress*, Tahiti, Vol. 5, pp. 31–37.

Orme, G. R., & Flood, P. G. (1980). Sedimentation in the Great Barrier Reef Province, adjacent bays and estuaries. In: R. A. Henderson & P. J. Stephenson (Eds), *The geology and geophysics of northeastern Australia* (pp. 419–434). Townsville, Australia: Geological Society of Australia, Queensland Division.

Orme, G. R., & Salama, M. S. (1988). Form and seismic stratigraphy of *Halimeda* banks in part of the northern Great Barrier Reef Province. *Coral Reefs, 6*, 131–138.

Ostlund, H. G., & Stuiver, M. (1980). GEOSECS Pacific Radiocarbon. *Radiocarbon, 22*, 25–53.

Ota, Y., Chappell, J., Berryman, K., & Okamoto, Y. (1997). Late Quaternary paleolandslides on the coral terraces of Huon Peninsula, Papua New Guinea. *Geomorphology, 19*, 55–76.

Ota, Y., & Chappell, J. (1999). Holocene sea-level rise and coral reef growth on a tectonically rising coast, Huon Peninsula, Papua New Guinea. *Quaternary International, 55*, 51–59.

Ota, Y., Chappell, J., Kelley, R., Yonekura, N., Matsumoto, E., Nishimura, T., et al. (1993). Holocene coral reef terraces and coseismic uplift of Huon Peninsula, Papua New Guinea. *Quaternary Research, 40*, 177–188.

Ourbak, T., Corrège, Th., Malaizé, B., Le Cornec, F., Charlier, K., & Peypouquet, J.-P. (2006). A high resolution investigation of temperature, salinity, and upwelling activity proxies in corals. *Geochemistry, Geophysics, Geosystems*, 7, Q03013, doi: 10.1029/2005GC001064.

Pandolfi, J. M. (1992a). Successive isolation rather than evolutionary centres for the origination of Indo-Pacific reef corals. *Journal of Biogeography, 19*, 593–609.

Pandolfi, J. M. (1992b). A review of the tectonic history of New Guinea and its significance for marine biogeography. *Proceedings of the Seventh International Coral Reef Symposium*, Guam, Vol. 2, pp. 718–728.

Pandolfi, J. M. (1996). Limited membership in Pleistocene reef coral assemblages from the Huon Peninsula, Papua New Guinea: Constancy during global change. *Paleobiology, 22*, 152–176.

Pandolfi, J. M. (1999). Response of Pleistocene coral reefs to environmental change over long temporal scales. *American Zoologist, 39*, 113–130.

Pandolfi, J. M. (2001). Taxonomic and numerical scales of analysis in paleoecological data sets: Examples from the Pleistocene of the Caribbean. *Journal of Paleontology, 75*, 546–563.

Pandolfi, J. M. (2002). Coral reef ecology at multiple spatial and temporal scales. *Coral Reefs, 21*, 13–23.

Pandolfi, J. M. (2007). A new, extinct Pleistocene reef coral from the *Montastraea "annularis"* species complex. *Journal of Paleontology, 81*, 472–482.

Pandolfi, J. M., & Greenstein, B. J. (1997a). Taphonomic alteration of reef corals: Effects of reef environment and coral growth form I: The Great Barrier Reef. *PALAIOS, 12*, 27–42.

Pandolfi, J. M., & Greenstein, B. J. (1997b). Preservation of community structure in death assemblages of deep water Caribbean reef corals. *Limnology and Oceanography, 42*, 1505–1516.

Pandolfi, J. M., & Jackson, J. B. C. (1997). The maintenance of diversity on coral reefs: Examples from the fossil record. *Proceedings of the Eighth International Coral Reef Symposium*, Panama, Vol. 1, pp. 397–404.

Pandolfi, J. M., & Jackson, J. B. C. (2001). Community structure of Pleistocene coral reefs of Curaçao, Netherlands Antilles. *Ecological Monographs, 71*, 49–67.

Pandolfi, JM., & Jackson, J. B. C. (2007). Broad-scale patterns in Pleistocene coral reef communities from the Caribbean: Implications for ecology and management. In: R. B. Aronson (Ed.), *Geological approaches to coral reef ecology* (pp. 201–236). New York: Springer.

Pandolfi, J. M., Jackson, JB. C., & Geister, J. (2001). Geologically sudden natural extinction of two widespread Late Pleistocene Caribbean reef corals. In: J. B. C. Jackson, S. Lidgard & F. K. McKinney (Eds), *Evolutionary patterns: Growth, form, and tempo in the fossil record* (pp. 120–158). Chicago, IL: University of Chicago Press.

Pandolfi, J. M., Llewellyn, G., & Jackson, J. B. C. (1999). Pleistocene reef environments, constituent grains, and coral community structure: Curacao, Netherlands Antilles. *Coral Reefs, 18*, 107–122.

Pandolfi, J. M., & Michin, P. R. (1995). A comparison of taxonomic composition and diversity between reef coral life and dead assemblages in Madang Lagoon, Papua New Guinea. *Palaeogeography, Palaeoclimatology, Palaeoecology, 119*, 321–341.

Pandolfi, J. M., Tudhope, A. W., Burr, G., Chappell, J., Edinger, E., Frey, M., et al. (2006). Mass mortality following disturbance in Holocene coral reefs from Papua New Guinea. *Geology, 34*, 949–952.

Pandolfi, J. M., Bradbury, R. H., Sala, E., Hughes, T. P., Bjorndal, K. A., Cooke, R. G., et al. (2003). Global trajectories of the long-term decline of coral reef ecosystems. *Science, 301*, 955–958.

Parnell, K. E. (1986). Water movement within a fringing reef flat, Orpheus Island, North Queensland, Australia. *Coral Reefs, 5*, 1–6.

Parsons-Hubbard, K. (2005). Molluscan taphofacies in recent carbonate reef/lagoon systems and their application to sub-fossil samples from reef cores. *PALAIOS, 20*, 175–191.

Paulay, G. (1990). Effects of late Cenozoic sea-level fluctuations on the bivalve faunas of tropical oceanic islands. *Paleobiology, 16*, 415–434.

Paulay, G. (1991). Late Cenozoic sea level fluctuations and the diversity and species composition of insular shallow-water marine faunas. In: E. C. Dudley (Ed.), The unity of evolutionary biology. *Proceedings of the Fourth International Congress on Systematics and Evolutionary Biology*, Dioscorides Publisher, Portland, OR, pp. 184–193.

Paulay, G. (1994). Biodiversity on oceanic islands: Its origin and extinction. *American Zoologist, 34*, 133–144.

Paulay, G. (1996). Dynamic clams: Changes in the bivalve fauna of Pacific islands as a result of sea-level fluctuations. *American Malacologist Bulletin, 12*, 45–57.

Paulay, G. (1997). Diversity and distribution of reef organisms. In: C. E. Birkeland (Ed.), *Life and death of corals* (pp. 298–352). London: Chapman & Hall.

Paulay, G., & McEdward, L. R. (1990). A simulation model of island reef morphology: The effects of sea level fluctuations, growth, subsidence, and erosion. *Coral Reefs, 9*, 51–62.

Paulay, G., & Spencer, T. (1988). Geomorphology, palaeoenvironments and faunal turnover, Henderson island, S.E. Polynesia. *Proceedings of the Sixth International Coral Reef Symposium*, Townsville, Australia, Vol. 3, pp. 461–466.

Paull, C. K., Chanton, J. P., Neumann, A. C., Coston, J. A., & Martens, C. S. (1992). Indications of methane-derived carbonates and chemosynthetic organic carbon deposits: Examples from the Florida Escarpment. *PALAIOS*, 7, 361–375.

Payri, C., & Cabioch, G. (2004). The systematics and significance of coralline red algae in the rhodolith sequence of the Amédée 4 drill core (Southwest New Caledonia). *Palaeogeography, Palaeoclimatology, Palaeoecology, 204*, 187–208.

Payri, C. E. (1988). *Halimeda* contribution to organic and inorganic production in a Tahitian reef system. *Coral Reefs, 6*, 251–262.

Payri, C. E. (1997). *Hydrolithon reinboldii* rhodolith distribution, growth and carbon production of a French Polynesian reef. *Proceedings of the Eighth International Coral Reef Symposium*, Panama, Vol. 2, pp. 755–760.

Peltier, W. R. (2005). On the hemispheric origins of meltwater pulse 1a. *Quaternary Science Reviews, 24*, 1655–1671.

Peltier, W. R., & Fairbanks, R. G. (2006). Global glacial ice volume and Last Glacial Maximum duration from an extended Barbados sea level record. *Quaternary Science Reviews, 25*, 3322–3337.

Peltier, W. R., Farrell, W. E., & Clark, J. A. (1978). Glacial isostasy and relative sea level: A global finite element model. *Tectonophysics, 50*, 81–110.

Perrin, C. (1990). Genèse de la morphologie des atolls: le cas de Mururoa (Polynésie Française). *Comptes Rendus de l'Académie des Sciences, Paris*, Série II, Vol. 311, pp. 671–678.

Perrin, C. (2003). Compositional heterogeneity and microstructural diversity of coral skeletons: Implications for taxonomy and control on early diagenesis. *Coral Reefs, 22*, 109–120.

Perrin, C. (2004). Diagenèse précoce des biocristaux carbonatés: Transformations isominérales de l'aragonite corallienne. *Bulletin de la Société Géologique de France, 174*, 95–106.

Perrin, C., Bosence, D. W.J., & Rosen, B. R. (1995). Quantitative approaches to palaeozonation and palaeobathymetry of corals and coralline algae in Cenozoic reefs. In: D. W. J. Bosence & P. A. Allison (Eds), *Marine palaeo-environmental analysis from fossils* (pp. 181–229), Geological Society of London (Special Publication 83), Geological Society Publishing House, Bath, UK.

Perrin, C., & Cuif, J.-P. (2001). Ultrastructural controls on diagenetic patterns of scleractinian skeletons: Evidence at the scale of colony lifetime. *Bulletin Tohoku University Museum, 1*, 210–218.

Perrin, C. (1989). *Rôle des organismes dans l'édification et l'évolution de l'atoll de Mururoa.* Unpublished Ph.D. thesis, University of Paris-Sud, 301 pp.

Perrin, C. (2002). Tertiary: The emergence of modern reef ecosystems. In: W. Kiessling, E. Flügel & J. Golonka (Eds), *Phanerozoic reef patterns* (pp. 587–621), SEPM (Special Publication 72).

Perrin, C., Plaziat, J. C., & Rosen, B. R. (1998). The Miocene coral reefs and reef corals of the SW Gulf of Suez and NW Red Sea: Distribution, diversity and regional environmental controls. In: B. H. Purser & D. W. J. Bosence (Eds), *Sedimentation and tectonics of Rift basins: Red Sea — Gulf of Aden* (pp. 296–319). London: Chapman & Hall.

Perry, C. T. (1996). The rapid response to changes in community composition: Implications for time averaging and sediment accumulation. *Journal of Sedimentary Research, 66,* 459–467.

Perry, C. T. (1999). Reef framework preservation in four contrasting modern reef environments, Discovery Bay, Jamaica. *Journal of Coastal Research, 15,* 796–812.

Perry, C. T. (2000). Macroboring of Pleistocene coral communities, Falmouth Formation, Jamaica. *PALAIOS, 15,* 489–491.

Perry, C. T. (2001). Storm-induced coral rubble deposition: Pleistocene records of natural reef disturbance and community response. *Coral Reefs, 20,* 171–183.

Perry, C. T. (2005). Morphology and occurrence of rhodoliths in siliciclastic, intertidal environments from a high latitude reef setting, southern Mozambique. *Coral Reefs, 24,* 201–207.

Perry, C. T., & Bertling, M. (2000). Spatial and temporal patterns of macroboring within Mesozoic and Cenozoic coral reef systems. In: E. Insalaco, P. W. Skelton, & T. J. Palmer (Eds), *Carbonate platform systems: Components and interactions* (pp. 33–50), Geological Society of London (Special Publication 178).

Perry, C. T., & Hepburn, L. J. (2008). Syn-depositional alteration of coral reef framework through bioerosion, encrustation and cementation: Taphonomic signatures of reef accretion and reef depositional events. *Earth-Science Reviews, 86,* 106–144.

Perry, C. T., & Larcombe, P. (2003). Marginal and non-reef-building coral environments. *Coral Reefs, 22,* 427–432.

Perry, C. T., Spencer, T., & Kench, P. S. (2008). Carbonate budgets and reef production states: A geomorphic perspective on the ecological phase-shift concept. *Coral Reefs, 27,* 853–866.

Perry, C. T., & Taylor, K. G. (2006). Inhibition of dissolution within shallow water carbonate sediments: Impacts of terrigenous sediment input on syn-depositional carbonate diagenesis. *Sedimentology, 53,* 495–513.

Perry, C. T., Taylor, K. G., & Machent, P. G. (2006). Temporal shifts in reef lagoon sediment composition, Discovery Bay, Jamaica. *Estuarine Coastal and Shelf Science, 67,* 133–144.

Pestana, H. (1985). Carbonate sediment production by *Sargassum* epibionts. *Journal of Sedimentary Petrology, 55,* 184–186.

Petit, J. R., Jouzel, J., Raynaud, D., Barkov, N. I., Barnola, J. M., Basile, I., et al. (1999). Climate and atmospheric history of the past 420,000 years from the Vostok Ice Core, Antarctica. *Nature, 399,* 429–436.

Peyrot-Clausade, M., Le Campion-Alsumard, T., Harmelin-Vivien, M., Romano, J. C., Chazottes, V., Pari, N., et al. (1996). Bioerosion carbonate cycle: Approach of quantification of its processes in French Polynesia. *Mémoire de la Société Géologique de France, 169,* 119–128.

Peyrot-Clausade, M., & Chazottes, V. (2000). La bioérosion récifale et son rôle dans la sédimentogenèse à Mooréa (Polynésie Française) et à La Réunion. *Océanis, 26,* 275–310.

Pfeiffer, M., & Dullo, W.-Ch. (2006). Monsoon-induced cooling of the western equatorial Indian Ocean as recorded in coral oxygen isotope records from the Seychelles covering the period of 1840–1994 AD. *Quaternary Science Reviews*, *25*, 993–1009.

Pfeiffer, M., Timm, O., Dullo, W.-C., & Podlech, S. (2004). Oceanic forcing of interannual and multidecadal climate variability in the south-western Indian Ocean: Evidence from a 160 year coral isotopic record (La Réunion, 55°E, 21°S). *Paleoceanography*, *19*, PA4006, doi:10.1029/2003PA000964.

Philander, S. G. (1990). *El Niño, La Niña and the Southern Oscillation*. New York: Academic Press.

Phipps, C. V. G., Davies, P. J., & Hopley, D. (1985). The morphology of *Halimeda* banks behind the Great Barrier Reef east of Cooktown, Queensland. *Proceedings of the Fifth International Coral Reef Congress*, Tahiti, Vol. 5, pp. 27–30.

Phipps, C. V. G., & Roberts, H. H. (1988). Seismic characteristics and accretion history of *Halimeda* bioherms on Kalukalukuang Bank, eastern Java Sea (Indonesia). *Coral Reefs*, *6*, 149–160.

Pichler, T., & Dix, G. R. (1996). Hydrothermal venting within a coral reef ecosystem, Ambitle Island, Papua New Guinea. *Geology*, *24*, 435–438.

Pichler, T., & Veizer, J. (2004). The precipitation of aragonite from shallow-water hydrothermal fluids in a coral reef, Tutum Bay, Ambitle Island, Papua New Guinea. *Chemical Geology*, *207*, 31–45.

Pichon, M. (1974). Free living scleractinian coral communities in the coral reefs of Madagascar. *Proceedings of the Second International Coral Reef Symposium*, Brisbane, Australia, Vol. 2, pp. 173–181.

Pichon, M. (1978a). Recherches sur les peuplements à base de Scléractiniaires dans les récifs de la Baie de Tuléar (Sud-Ouest de Madagascar). *Atoll Research Bulletin*, *222*, 1–447.

Pichon, M. (1978b). Problems of measuring and mapping coral reef colonies. in: Coral reefs: Research methods. In: D. R. Stoddart & R. E. Johannes (Eds), *Coral reefs: Research methods*. UNESCO Monographs on Oceanographic Methodology, Paris, Vol. 5, pp. 219–230.

Pillai, C. S. G. (1971). Composition of the coral fauna of the southeastern coast of India and the Laccadives. In: D. R. Stoddart & M. Yonge (Eds), *Regional variation in Indian Ocean coral reefs*. Symposia of the Zoological Society of London (Vol. 28, pp. 301–327).

Pillai, C. S. G., & Scheer, G. (1976). Report of the stony corals from the Maldive archipelago. *Zoologica*, *126*, 1–83.

Piller, W. E. (1994). The Northern Bay of Safaga (Red Sea, Egypt): An actuopalaeontological approach. IV. Thin section analysis. *Beiträge zur Paläntologie von Österreich*, *18*, 1–73.

Piller, W. E., & Mansour, A. M. (1990). The Northern Bay of Safaga (Red Sea, Egypt): An actuopalaeontological approach. II. Sediment analyses and sedimentary facies. *Beiträge zur Paläntologie von Österreich*, *16*, 1–102.

Piller, W. E., & Rasser, M. (1996). Rhodolith formation induced by reef erosion in the Red Sea, Egypt. *Coral Reefs*, *15*, 191–198.

Piller, W. E., & Riegl, B. (2003). Vertical versus horizontal growth strategies of coral frameworks (Tulamben, Bali, Indonesia). *International Journal of Earth Sciences*, *92*, 511–519.

Pingitore, N. E., Jr. (1976). Vadose and phreatic diagenesis: Processes, products and their recognition in corals. *Journal of Sedimentary Petrology*, *46*, 985–1006.

Pingitore, N. E., Jr. (1982). The role of diffusion during carbonate diagenesis. *Journal of Sedimentary Petrology*, *52*, 27–39.

Pirazzoli, P. A. (1985). Leeward Islands, Maupiti, Tupai, Bora-Bora, Huahine, Society Archipelago. *Proceedings of the Fifth International Coral Reef Congress*, Tahiti, Vol. 1, pp. 17–72.

Pirazzoli, P. A. (1986). Marine notches. In: O. van de Plassche (Ed.), *Sea-level research: A manual for the collection and evaluation of data* (pp. 361–400). Norwich, UK: Geo Books.

Pirazzoli, P.A. (1991). *World atlas of Holocene sea-level changes*. Elsevier Oceanography Series (Vol. 58, 300 pp.). Amsterdam: Elsevier.

Pirazzoli, P. A. (1996). *Sea-level change: The last 20,000 years*. Chichester: Wiley, 211 pp.

Pirazzoli, P. A., & Montaggioni, L. F. (1986). Late Holocene sea-level changes in the Northwest Tuamotu Islands, French Polynesia. *Quaternary Research, 25*, 350–368.

Pirazzoli, P. A., & Montaggioni, L. F. (1988a). The 7,500-year Holocene sea level in French Polynesia: Geodynamic implications for mid-plate volcanic islands. *Proceedings of the Sixth International Coral Reef Symposium*, Townsville, Australia, Vol. 3, pp. 467–472.

Pirazzoli, P. A., & Montaggioni, L. F. (1988b). Late Holocene sea-level changes in French Polynesia. *Palaeogeography, Palaeoclimatology, Palaeoecology, 68*, 153–175.

Pirazzoli, P. A., Radtke, U., Hantoro, W. S., Jouannic, C., Hoang, C.-T., Causse, C., et al. (1991). Quaternary raised coral reef terraces on Sumba Island, Indonesia. *Science, 252*, 1834–1836.

Pirazzoli, P. A., Radkte, U., Hantoro, W. S., Jouannic, C., Hoang, C. T., Causse, C., et al. (1993). A one million-year-long sequence of marine terraces on Sumba Island, Indonesia. *Marine Geology, 109*, 221–236.

Pirazzoli, P. A., Reyss, J.-L., Fontugne, M., Haghipour, A., Hilgers, A., Kasper, H. U., et al. (2004). Quaternary coral-reef terraces from Kish and Qeshm Islands, Persian Gulf: New radiometric ages and tectonic implications. *Quaternary International, 120*, 15–27.

Pirazzoli, P. A., & Salvat, B. (1992). Ancient shorelines and Quaternary vertical movements on Rurutu and Tubuai (Austral Isles, French Polynesia). *Zeitschrift für Geomorphologie N.F., 36*, 431–451.

Pirazzoli, P. A., & Veeh, H. H. (1987). Age $^{230}Th/^{234}U$ d'une encoche émergée et vitesses de soulèvement quaternaire à Rurutu, îles Australes. *Comptes Rendus de l'Académie des Sciences, Paris*, Série II, 305(10), pp. 919–923.

Plaziat, J. C., Baltzer, F., Choukri, A., Conchon, O., Freytet, P., Orszag-Sperber, F., et al. (1998). Quaternary marine and continental sedimentation in the northern Red Sea and Gulf of Suez (Egyptian coast): Influences of rift tectonics, climatic changes and sea-level fluctuations. In: B. H. Purser & D. W. J. Bosence (Eds), *Sedimentation and tectonics in Rift Basins: Red Sea–Gulf of Aden* (pp. 537–573). London: Chapman & Hall.

Pochon, X., Montoya-Burgos, J. I., Stadelmann, B., & Pawlowski, J. (2006). Molecular phylogeny, evolutionary rates, and divergence timing of the symbiotic dinoflagellate genus *Symbiodinium*. *Molecular Phylogenetics and Evolution, 38*, 20–30.

Polach, H. A., Chappell, J. M. A., & Lovering, J. F. (1969). ANU radiocarbon date list III. *Radiocarbon, 11*, 245–252.

Pomar, L., & Hallock, P. (2007). Changes in coral-reef structure through the Miocene in the Mediterranean province: Adaptative versus environmental influence. *Geology, 35*, 899–902.

Pomar, L., & Hallock, H. (2008). Carbonate factories: A conundrum in sedimentary geology. *Earth Science Review, 87*, 134–169.

Potter, E.-K., Esat, T. M., Schellmann, G., Radtke, U., Lambeck, K., & McCulloch, M. T. (2004). Sub-orbital period sea-level oscillations during marine isotope substages 5a and 5c. *Earth and Planetary Science Letters, 220*, 191–204.

Potter, E.-K., & Lambeck, K. (2003). Reconciliation of sea-level observations in the Western North Atlantic during the last glacial cycle. *Earth and Planetary Science Letters, 217*, 171–181.

Potts, D. C. (1983). Evolutionary disequilibrium among Indo-Pacific corals. *Bulletin of Marine Science, 33*, 619–632.

Potts, D. C. (1984). Generation times and the Quaternary evolution of reef-building corals. *Paleobiology, 10*, 48–58.

Potts, D. C. (1985). Sea level fluctuations and speciation in scleractinia. *Proceedings of the Fifth International Coral Reef Symposium*, Tahiti, Vol. 4, pp. 127–132.

Potts, D. C., & Garthwaite, R. L. (1991). Evolution of reef building corals during periods of rapid global change. In: E. C. Dudley (Ed.), *The unity of evolutionary biology. Proceedings of the Fourth International Congress on Systematics and Evolutionary Biology*, Dioscorides Publisher, Portland, OR, pp. 170–178.

Potts, D. C., & Jacobs, J. R. (2002). Evolution of reef-building scleractinian corals in turbid environments: A paleo-ecological hypothesis. *Proceedings of the Ninth International Coral Reef Symposium*, Bali, Indonesia, Vol. 1, pp. 249–254.

Powers, L. A., Johnson, Th. C., Werne, J. P., Castaneda, I. S., Hopmans, E. C., Sinninghe Damsté, J. S., et al. (2005). Large temperature variability in the southern African tropics since the Last Glacial Maximum. *Geophysical Research Letters*, *32*, L08706, doi: 10.1029/2004GL022014.

Prakash Babu, C., Brumsack, H.-J., Schnetger, B., & Böttchert, M. E. (2002). Barium as a productivity proxy in continental margin sediments: A study from the eastern Arabian Sea. *Marine Geology*, *184*, 189–206.

Precht, W. F., & Aronson, R. B. (2004). Climate flickers and range shifts of reef corals. *Frontiers in Ecology and the Environment*, *6*, 307–313.

Precht, W. F., & Miller, S. L. (2007). Ecological shifts along the Florida Reef Tract: The past as the key to the future. In: R. B. Aronson (Ed.), *Geological approaches to coral reef ecology* (pp. 237–312). New York: Springer.

Presto, M. K., Ogston, A. S., Storlazzi, C. D., & Field, M. E. (2006). Temporal and spatial variability in the flow and dispersal of suspended-sediment on a fringing reef flat, Molokai, Hawaii. *Estuarine, Coastal and Shelf Science*, *67*, 67–81.

Pringault, O., de Wit, R., & Camoin, G. (2005). Irradiance regulation of photosynthesis and respiration in modern marine microbialites built by benthic cyanobacteria in a tropical lagoon (New Caledonia). *Microbial Ecology*, *49*, 604–616.

Purdy, E. G. (1974). Reef configurations: Cause and effect. In: L. F. Laporte (Ed.), *Reefs in time and space* (pp. 9–76), SEPM (Special Publication 18).

Purdy, E. G., & Bertram, G. T. (1993). *Carbonate concepts from the Maldives, Indian Ocean.* Studies in Geology (Vol. 34, pp. 1–56). Tulsa, OK: American Association of Petroleum Geologists.

Purdy, E. G., & Gischler, E. (2005). The transient nature of the empty bucket model of reef sedimentation. In: J. J. G. Reijmer & A. Immenhauser (Eds), *Sedimentology in the 21st century: a tribute to Wolfgang Schlager. Sedimentary Geology*, *175*, 35–47.

Purdy, E. G., Gischler, E., & Lomando, A. J. (2003). The Belize margin revisited. 2. Origin of Holocene antecedent topography. *International Journal of Earth Sciences*, *92*, 552–572.

Purdy, E. G., & Winterer, E. L. (2001). Origin of atoll lagoons. *Bulletin of the Geological Society of America*, *113*, 837–854.

Purdy, E. G., & Winterer, E. L. (2006). Contradicting barrier reef relationships for Darwin's evolution of reef types. *International Journal of Earth Sciences*, *95*, 143–167.

Purser, B. H. (Ed.) (1973). *The Persian Gulf.* Berlin: Springer. 471 pp.

Purser, B. H., & Siebold, E. (1973). The principal environmental factors influencing Holocene sedimentation and diagenesis in the Persian Gulf. In: B. H. Purser (Ed.), *The Persian Gulf: Holocene carbonate sedimentation and diagenesis in a shallow Epicontinental Sea* (pp. 1–9). Berlin: Springer.

Pusey III, W. C. (1975). Holocene carbonate sedimentation on Northern Belize shelf. In: K. F. Wantland & W. C. Pusey III (Eds), *Belize Shelf — Carbonate sediments, clastic sediments, and ecology.* Studies in Geology (Vol. 2, pp. 131–233). Tulsa, OK: American Association of Petroleum Geologists.

Quinn, T. M., Crowley, T. J., Taylor, F. W., Hénin, C., Joannot, P., & Join, Y. (1998). A multicentury stable isotope record from a New Caledonia coral: Interannual and decadal sea surface temperature variability in the southwest Pacific since 1657 AD. *Paleoceanography*, *13*, 412–426.

Quinn, T. M., & Matthews, R. K. (1990). Post-Miocene diagenetic and eustatic history of Enewetak Atoll: Model and data comparison. *Geology, 18*, 942–945.

Quinn, T. M., Lohmann, K. C., & Halliday, A. N. (1991). Sr isotopic variation in shallow-water carbonate sequences: Stratigraphic, chronostratigraphic, and eustatic implications of the record at Enewetok Atoll. *Paleoceanography, 6*, 371–385.

Quinn, T. M., & Saller, A. H. (1997). Geology of Anewetak Atoll, Republic of the Marshall Islands. In: H. L. Vacher & T. M. Quinn, (Eds), *Geology and hydrogeology of Carbonate Islands*. Developments in Sedimentology (Vol. 54, pp. 637–666.). Amsterdam: Elsevier.

Quinn, T. M., & Taylor, F. W. (2006). SST artifacts in coral proxy records produced by early marine diagenesis in a modern coral from Rabaul, Papua New Guinea. *Geophysical Research Letters, 33*, 1–4.

Quinn, T. M., Taylor, F. W., Halliday, A. N., Collot, J.-Y., & Greene, H. G. (1991). Sr isotopic dating of carbonate at Bougainville Guyot (Site 831), New-Hebrides Island arc. *Eos, 72*, 535.

Radtke, U., Grün, R., & Mangini, A. (1997). The Pleistocene coral reefs of Kikai-Jima (Ryukyu Islands, Japan) and their significance for palaeo sea-level reconstructions. *Ocean Research, 19*, 245–255.

Radtke, U., Grün, R., & Schwarcz, H. P. (1988). Electron spin resonance dating of Pleistocene coral reef tracts of Barbados (W.I.). *Quaternary Research, 29*, 197–215.

Radtke, U., Schellmann, G., Scheffers, A., Kelletat, D., Kromer, B., & Kasper, H. U. (2003). Electron spin resonance and radiocarbon dating of coral deposited by Holocene tsunami events on Curaçao, Bonaire and Aruba (Netherlands Antilles). *Quaternary Science Reviews, 22*, 1309–1315.

Rahman, M. A., & Oomori, T. (2008). Structure, crystallization and mineral composition of sclerites in the alcyonarian coral. *Journal of Crystal Growth, 310*, 3528–3534.

Ramsay, P. J., & Cooper, J. A. G. (2002). Late Quaternary sea-level change in South Africa. *Quaternary Research, 57*, 82–90.

Ramseyer, K., Miano, T. M., D'Orazio, V., Wildberger, A., Wagner, T., & Geister, J. (1997). Nature and origin of organic matter in carbonates from speleothems, marine cements and coral skeletons. *Organic Geochemistry, 26*, 361–378.

Rao, V. P., Kessarkar, P. M., Krumbein, W. E., Krajewski, K. P., & Schneiders, R. J. (2003). Microbial dolomite crusts from the carbonate platform off western India. *Sedimentology, 50*, 819–830.

Rao, V. P., Montaggioni, L. F., Vora, K. H., Almeida, F., Rao, K. M., & Rajagopalan, G. (2003). Significance of relic carbonate deposits along the central and southwestern margin of India for late Quaternary environmental and sea level changes. *Sedimentary Geology, 159*, 95–111.

Rasmussen, K. A., Macintyre, I. G., & Prufert, L. (1993). Modern stromatolite reefs fringing a brackish coastline, Chetumal Bay, Belize. *Geology, 21*, 199–202.

Raymo, M. E., & Huybers, P. (2008). Unlocking the mysteries of the ice ages. *Nature, 451*, 284–285.

Raymo, M. E., & Nisancioglu, K. (2003). The 41 kyr world: Milankovitch's other unsolved mystery. *Paleoceanography, 18*, 1011, doi: 10.1029/2002PA000791.

Read, J. F. (1974). Calcrete deposits and Quaternary sediments, Edel Province, Shark Bay, Western Australia. In: B. W. Logan, J. F. Read, G. M. Hagan, P. Hoffman, R. G. Brown, P. J. Woods & C. D. Gebelein (Eds), *Evolution and diagenesis of Quaternary carbonate sequences, Shark Bay, Western Australia. American Association of Petroleum Geologists, Memoir, 22*, 250–282.

Rees, S. A., Opdyke, B. N., Wilson, P. A., & Henstock, T. J. (2007). Significance of *Halimeda* bioherms to the global carbonate budget based on a geological sediment budget for the Northern Great Barrier Reef, Australia. *Coral Reefs, 26*, 177–188.

Reid, R. P., & Macintyre, I. G. (1988). Foraminiferal-algal nodules from the Eastern Caribbean: Growth history and implications on the value of nodules as paleoenvironmental indicators. *PALAIOS, 3,* 424–435.

Reid, R. P., & Macintyre, I. G. (1998). Carbonate recrystallization in shallow marine environments: A widespread diagenetic process forming micritized grains. *Journal of Sedimentary Research, 68,* 928–946.

Reid, R. P., & Macintyre, I. G. (2000). Microboring versus recrystallization: Further insight into the micritization process. *Journal of Sedimentary Research, 70,* 24–28.

Reid, R. P., Macintyre, I. G., Browne, K. M., Steneck, R. S., & Miller, Th. (1995). Modern marine stromatolites in the Exuma Cays, Bahamas: Uncommonly common. *Facies, 33,* 1–7.

Reid, R. P., Macintyre, I. G., & Post, J. E. (1992). Micritized skeletal grains in Northern Belize Lagoon: A major source of Mg-calcite mud. *Journal of Sedimentary Petrology, 62,* 145–156.

Reid, R. P., Visscher, P. T., Decho, A. W., Stolz, J. F., Bebout, B. M., Dupraz, C., et al. (2000). The role of microbes in accretion, lamination and lithification in modern marine stromatolites. *Nature, 406,* 989–992.

Reiss, Z., & Hottinger, L. (1984). *The Gulf of Aqaba: Ecological micropaleontology. Ecological studies 50.* Berlin: Springer, 354 pp.

Reitner, J. (1993). Modern cryptic microbialite/metazoan facies from Lizard Island (Great Barrier Reef, Australia). *Facies, 29,* 3–40.

Reitner, J., Gautret, P., Marin, F., & Neuweiler, F. (1995). Automicrites in a modern microbialite. Formation model via organic matrices (Lizard Island, Great Barrier Reef, Australia). *Bulletin de l'Institut Océanographique de Monaco, 14,* 237–263.

Renema, W., & Troelstra, S. R. (2001). Larger foraminifera distribution on a mesotrophic carbonate shelf in SW Sulawesi (Indonesia). *Palaeogeography, Palaeoclimatology, Palaeoecology, 175,* 125–146.

Repellin, P. (1977). Contribution à l'étude d'un récif corallien: le sondage "Colette" Atoll de Mururoa (Polynesie Francaise). *Cahiers du Pacifique, 20,* 1–209.

Resig, J. M. (2004). Age and preservation of *Amphistegina* (foraminifera) in Hawaiian beach sand: Implication for sand turnover rate and resource renewal. *Marine Micropaleontology, 50,* 225–236.

Reuer, M. K., Boyle, E. A., & Cole, J. C. (2003). A mid-twentieth century reduction in tropical upwelling inferred from coralline trace element proxies. *Earth and Planetary Science Letters, 210,* 437–452.

Revelle, R. (1954). Foreword to geology of Bikini and nearby atoll. US Geological Survey, Professional Paper, Vol. 260-A, pp. III–VII.

Reynaud, S., Ferrier-Pagès, C., Meibom, A., Mostefaoui, S., Mortlock, R., Fairbanks, R. G., et al. (2007). Light and temperature effects on Sr/Ca and Mg/Ca ratios in the scleractinian coral *Acropora* sp. *Geochimica et Cosmochimica Acta, 71,* 354–362.

Reynaud, S., Ferrier-Pagès, C., Sambrotto, R., Juillet-Leclerc, A., Jaubert, J., & Gattuso, J. P. (2002). Effect of feeding on the carbon and oxygen isotopic composition in the tissues and skeleton of the zooxanthellate coral *Stylophora pistillata. Marine Ecology Progress Series, 238,* 81–89.

Reynaud-Vaganay, S., Juillet-Leclerc, A., Jaubert, J., & Gattuso, J. P. (2001). Effect of light on skeletal $\delta^{13}C$ and $\delta^{18}O$, and interaction with photosynthesis, respiration and calcification in two zooxanthellate scleractinian corals. *Palaeogeography, Palaeoclimatology, Palaeoecology, 175,* 393–404.

Richards, H. C. & Hill, D. (1942). *Great Barrier Reef bores, 1926–1937 — Description, analyses and interpretations,* Reports of the Great Barrier Reef Committee, 5, 1–122.

Ricketts, B. D. (1980). Experimental investigation of carbonate precipitation in hydrated silica gels. *Journal of Sedimentary Petrology, 50,* 963–970.

Riding, R. (2002). Structure and composition of organic reefs and carbonate mud mounds: Concepts and categories. *Earth-Science Reviews, 58*, 163–231.

Riegl, B. (2001). Inhibition of reef framework by frequent disturbance: Examples from the Arabian Gulf, South Africa, and the Cayman Islands. *Palaeogeography, Palaeoclimatology, Palaeoecology, 175*, 79–101.

Riegl, B., & Piller, W. E. (1997). Distribution and environmental control of coral assemblages in Northern Safaga Bay (Red Sea, Egypt). *Facies, 36*, 141–162.

Riegl, B., Piller, & W. E. (2000). Reefs and coral carpets in the northern Red Sea as models for organism–environment feedback in coral communities and its reflection in growth fabrics. In: E. Insalaco, P. W. Skelton & T. J. Palmer (Eds), *Carbonate platform systems: Components and interactions* (pp. 71–88), Geological Society of London (Special Publication 178).

Riegl, B., Piller, W. E., & Rasser, M. (1996). Rolling stones: First record of an *Acropora anthocercis* (Brook) corallith from the northern Red Sea. *Coral Reefs, 15*, 149–150.

Riegl, B., & Velimirov, B. (1994). The structure of coral communities at Hurghada in the Northern Red Sea. *Marine Ecology, 15*, 213–233.

Riegl, B. M., Halfar, J., Purkis, S. J., & Godinez-Orta, L. (2007). Sedimentary facies of the eastern Pacific's northernmost reef-like setting (Cabo Pulmo, Mexico). *Marine Geology, 236*, 61–77.

Ries, J. B., Stanley, S. M., & Hardie, L. A. (2006). Scleractinian corals produce calcite, and grow more slowly, in artificial Cretaceous seawater. *Geology, 34*, 525–528.

Rimbu, N., Lohmann, G., Felis, T., & Pätzold, J. (2001). Arctic oscillation signature in Red Sea coral. *Geophysical Research Letters, 28*, 2959–2962.

Rimbu, N., Lohmann, G., Felis, Th., & Pätzold, J. (2003). Shift in ENSO teleconnections recorded by a northern Red Sea coral. *Journal of Climate, 16*, 1414–1422.

Roberts, H. H. (1976). *Carbonate sedimentation in a reef-enclosed lagoon, North Sound, Grand Cayman Island* (pp. 97–122), Geology Studies (Special Publication 4), Brigham Young University.

Roberts, H. H., Aharon, P., & Phipps, C. V. (1988). Morphology and sedimentology of *Halimeda* bioherms from the eastern Java Sea (Indonesia). *Coral Reefs, 6*, 161–172.

Roberts, H.H., & Macintyre, I.G. (Eds). (1988). *Halimeda.* Coral reefs, Special Issue, Vol. 6, pp. 121–279.

Roberts, H. H., Wilson, P. A., & Lugo-Fernandez, A. (1992). Biologic and geologic responses to physical processes: Examples from modern reef systems of the Caribbean-Atlantic Region. *Continental Shelf Research, 12*, 809–834.

Rodgers, K. A. (1989). Phosphatic limestones from Tuvalu (Ellice Islands). *Economic Geology, 84*, 2252–2266.

Rodgers, K. A., Easton, A. J., & Downes, C. J. (1982). The chemistry of carbonate rocks of Niue Island, South Pacific. *Journal of Geology, 90*, 645–662.

Rodgers, K. B., Aumont, O., Madec, G., Menkes, Ch., Blanke, B., Monfray, P., et al. (2004). Radiocarbon as a thermocline proxy for the eastern equatorial Pacific. *Geophysical Research Letters, 31*, L14314, doi: 10.1029/2004GL019764.

Roe, K. K., Burnett, W. C., & Lee, A. I. N. (1983). Uranium disequilibrium dating of phosphate deposits from the Lau Group, Fiji. *Nature, 302*, 603–606.

Roe, K. K., & Burnett, W. C. (1985). Uranium geochemistry and dating of Pacific island apatite. *Geochimica et Cosmochimica Acta, 49*, 1581–1592.

Roff, G. (2008). Corals on the move: Morphological and reproductive strategies of reef flat coralliths. *Coral Reefs, 27*, 343.

Rogers, C. S. (1990). Reponses of coral reefs and reef organisms to sedimentation. *Marine Ecology Progress Series, 62*, 185–202.

Rogers, C. S. (1993). Hurricanes and coral reefs: The intermediate disturbance hypothesis revisited. *Coral Reefs, 12*, 127–138.

Rögl, F. (1998). Palaeogeographic considerations for Mediterranean and Paratethys seaways (Oligocene to Miocene). *Naturhistorisches Museum in Wien, Annalen, 99*, 279–310.

Rohling, E. J., Fenton, M., Jorissen, F. J., Bertrand, P., Ganssen, G., & Caulet, J. P. (1998). Magnitudes of sea-level lowstands of the past 500,000 years. *Nature, 394*, 162–165.

Rohling, E. J., Grant, K., Hemleben, Ch., Siddall, M., Hoogakker, B. A. A., Bolshaw, M., et al. (2008). High rates of sea-level rise during the last interglacial period. *Nature Geoscience, 1*, 38–42.

Rollion-Bard, C., Chaussidon, M., & France-Lanord, C. (2003). pH control on oxygen isotopic composition of symbiotic corals. *Earth and Planetary Science Letters, 215*, 275–288.

Romano, S. L., & Cairns, S. D. (2000). Molecular phylogenetic hypothesis for the evolution of scleractinian corals. *Bulletin of Marine Science, 67*, 1043–1068.

Romano, S. L., & Palumbi, S. R. (1996). Evolution of scleractinian corals inferred from molecular systematics. *Science, 271*, 640–642.

Rosen, B.R. (1971). Principal features of reef coral ecology in shallow water environments, Mahé, Seychelles. In: D. R. Stoddart & M. Yonge (Eds), *Regional variation in Indian Ocean coral reefs*. Symposia of the Zoological Society of London (Vol. 28, pp. 163–184).

Rosen, B. R. (1975). The distribution of reef corals. *Report of the Underwater Association, 1*, 2–16.

Rosen, B. R. (1981). The tropical high diversity enigma: The coral's eye view. In: P. H. Greewood & P. L. Forey (Eds), *Chance, change and challenge; the evolving biosphere* (pp. 103–129). Cambridge: Cambridge University Press.

Rosen, B. R. (1984). Reef coral biogeography and climate through the late Cenozoic; just islands in the sun or a critical pattern of islands? In: P. Brenchley (Ed.), *Fossils and climate*. Chichester: Wiley, Geological Journal, Special Issue, Vol. 11, pp. 201–262.

Rosen, B. R. (1988). Progress, problems and patterns in the biogeography of reef corals and other tropical marine organisms. *Helgoländer Meeresuntersuchungen, 42*, 269–301.

Rosen, B. R. (1999). Palaeoclimatic implications of the energy hypothesis from Neogene corals of the Mediterranean. In: J. Agusti & L. Rook (Eds), *The evolution of Neogene terrestrial ecosystems in Europe* (pp. 309–327). Cambridge: Cambridge University Press.

Rosen, B. R. (2002). Biodiversity: Old and new relevance for palaeontology. *Geoscientist, 12*, 4–9.

Rosen, B. R., & Smith, A. B. (1988). Tectonics from fossils? Analysis of reef coral and sea urchin distributions from the late Cretaceous to Recent, using a new method. In: M. G. Audley-Charles & A. Hallam (Eds), *Gondwana and Tethys* (pp. 275–306). Oxford: Oxford University Press.

Rosenberg, E., Koren, O., Reshef, L., Efrony, R., & Zilber-Rosenberg, I. (2007). The role of microorganisms in coral health, disease and evolution. *Nature Reviews Microbiology, 5*, 355–362.

Rossinsky, V., Jr., Wanless, H. R., & Swart, P. K. (1992). Penetrative calcretes and their stratigraphic implications. *Geology, 20*, 331–334.

Rothfus, T.A., & Greenstein, B.J. (2001). Taphonomic evidence for late Pleistocene transitions in coral reef community composition, San Salvador, Bahamas. In: B. J. Greenstein & C. K. Carney (Eds), *Proceedings of the Tenth Symposium on the Geology of the Bahamas and other carbonate regions*, San Salvador, Gerace Research Center, pp. 152–162.

Rougerie, F., Fichez, R., & Dejardin, P. (1997). Geomorphology and hydrogeology of selected islands of French Polynesia: Tikehau (Atoll) and Tahiti (Barrier Reef). In: H. L. Vacher & T. M.Quinn (Eds), *Geology and hydrogeology of Carbonate Islands*. Developments in Sedimentology (Vol. 54, pp. 475–502). Amsterdam: Elsevier.

Rougerie, F., Jehl, C., & Trichet, J. (1997). Phosphorous pathways in atolls: Interstitial accumulation and carbonate-fluoro-apatite (CFA) precipitation. *Marine Geology, 139*, 201–217.

Rougerie, F., & Wauthy, B. (1986). Le concept d'endo-upwelling dans le fonctionnement des atolls-oasis. *Oceanologica Acta*, *9*, 133–148.
Rougerie, F., & Wauthy, B. (1993). The endo-upwelling concept: From geothermal convection to reef construction. *Coral Reefs*, *12*, 19–30.
Rowan, R. (2004). Thermal adaptation in reef coral symbionts. *Nature*, *430*, 742.
Roy, K., & Pandolfi, J. M. (2005). Responses of marine species and ecosystems to past climates. In: T. E. Lovejoy & L. Hannah (Eds), *Climate change and biodiversity* (pp. 160–175). New Haven, CT: Yale University Press.
Roy, K. J., & Smith, S. V. (1971). Sedimentation and coral reef development in turbid water: Fanning Lagoon. *Pacific Science*, *25*, 234–248.
Rutherford, S., & D'Hondt, S. (2000). Early onset and tropical forcing of 100,000-year Pleistocene glacial cycles. *Nature*, *408*, 7275.
Rützler, K., & Macintyre, I. G. (1978). Siliceous sponge spicules in coral reef sediments. *Marine Biology*, *49*, 147–159.
Rützler, K., & Macintyre, I.G. (1982). *The Atlantic Barrier Reef Ecosystem at Carrie Bow, Belize, I: Structure and Communities*. Smithsonian Contributions to Marine Sciences, Vol. 12, 539 pp. Smithsonian Institution Press, Washington.
Sagawa, N., Nakamori, T., & Iryu, Y. (2001). Pleistocene reef development in the southwest Ryukyu Islands, Japan. *Palaeogeography, Palaeoclimatology, Palaeoecology*, *175*, 303–323.
Saji, N. H., Goswami, B. N., Vinayachandran, P. N., & Yamagata, T. (1999). A dipole mode in the tropical Indian Ocean. *Nature*, *401*, 360–363.
Sakai, K., & Nishihira, M. (1981). Population study of the benthic foraminifer *Baculogypsina sphaerulata* on the Okinawan reef flat and preliminary estimation of its annual production. *Proceedings of the Fourth International Coral Reef Symposium*, Manila, Vol. 2, pp. 763–766.
Saller, A. H. (1984). Petrologic and geochemical constraints on the origin of subsurface dolomite, Enewetak Atoll: An example of dolomitization by normal sea water. *Geology*, *12*, 217–220.
Saller, A. H., & Moore, C. H., Jr. (1989). Meteoric diagenesis, marine diagenesis, and microporosity in Pleistocene and Oligocene limestones, Enewetak Atoll, Marshall islands. *Sedimentary Geology*, *63*, 253–272.
Saller, A. H., & Moore, C. H., Jr. (1991). Geochemistry of meteoric calcite cements in some Pleistocene limestones. *Sedimentology*, *38*, 601–621.
Sammarco, P. W. (1996). Comments on coral reef regeneration, bioerosion, biogeography, and chemical ecology: Future directions. *Journal of Experimental Marine Biology and Ecology*, *200*, 135–168.
Sandberg, P. A. (1983). An oscillating trend in Phanerozoic nonskeletal mineralogy. *Nature*, *305*, 19–22.
Sandberg, P. A., & Hudson, J. D. (1983). Aragonite relic preservation in Jurassic calcite-replaced bivalves. *Sedimentology*, *30*, 879–892.
Sandberg, P. A., Schneiderman, N., & Wunder, S. J. (1973). Aragonite ultrastructural relics in calcite-replaced Pleistocene skeletons. *Nature: Physical Sciences*, *245*, 133–134.
Sanders, D., & Baron-Szabo, R. C. (2005). Scleractinian assemblages under sediment input: Their characteristics and relation to the nutrient input concept. *Palaeogeography, Palaeoclimatology, Palaeoecology*, *216*, 139–181.
Sansone, F. J., Andrews, C. C., Buddemeier, R. W., & Tribble, T. W. (1988). Well point sampling of reef interstitial water. *Coral Reefs*, 7, 19–22.
Santini, F., & Winterbottom, R. (2002). Historical biogeography of Indo-western Pacific reef biota: Is the Indonesian region a centre of origin? *Journal of Biogeography*, *29*, 189–205.

Sanyal, A., Hemming, N. G., Hanson, G. N., & Broecker, W. S. (1995). Evidence for a higher pH in the glacial ocean from boron isotopes in foraminifera. *Nature, 373*, 234–236.

Sasaki, K., Omura, A., Miwa, T., Tsuji, Y., Matsuda, H., Nakamori, T., et al. (2006). Th-230/U-234 and C-14 dating of a lowstand coral reef beneath the insular shelf off Irabu Island, Ryukyus, southwestern Japan. *Island Arc, 15*, 455–467.

Sasaki, K., Omura, A., Murakami, K., Sagawa, N., & Nakamori, T. (2004). Interstadial coral reef terraces and relative sea-level changes during marine oxygen isotope stages 3-4, Kikai Island, central Ryukyus, Japan. *Quaternary International, 120*, 51–64.

Satapoomin, U., Phongsuwan, N., & Brown, B. E. (2006). A preliminary synopsis of the effects of the Indian Ocean tsunami on the coral reefs of western Thailand. *Phuket Marine Biological Center Research Bulletin, 67*, 77–80.

Sayles, R. W. (1931). Bermuda during the Ice Age. *Proceedings of the American Academy of Arts and Sciences, 66*, 381–486.

Scheer, G. (1971). Coral reefs and coral genera in the Red Sea and Indian Ocean. In: D. R. Stoddart & M. Yonge (Eds), *Regional variation in Indian Ocean coral reefs*. Symposia of the Zoological Society of London (Vol. 28, pp. 329–367).

Scheer, G. (1978). Application of phytosociologic methods. In: D. R. Stoddart & R. E. Johannes (Eds), *Coral reefs: Research methods*. UNESCO Monographs on Oceanographic Methodology, Paris, Vol. 5, pp. 175–196.

Scheffers, S., Scheffers, A., Radtke, U., Kelletat, D., Staben, K., & Bak, R. (2006). Tsunamis trigger long-lasting phase-shift in a coral reef ecosystem. *Zeitschrift für Geomorphologie, N.F., Supplement-Volume, 146*, 59–79.

Scheffers, S. R., Haviser, J., Browne, T., & Scheffers, A. (2009). Tsunamis, hurricanes, the demise of coral reefs and shifts in prehistoric human populations in the Caribbean. *Quaternary International, 195*, 69–87.

Scheffers, S. R., Scheffers, A., Kelletat, D., & Bryant, E. A. (2008). The Holocene paleo-tsunami history of West Australia. *Earth and Planetary Science Letters, 270*, 137–146.

Schellmann, G., & Radtke, U. (2004). A revised morpho- and chronostratigraphy of the late and middle Pleistocene coral reef terraces on Southern Barbados (West Indies). *Earth-Science Reviews, 64*, 157–187.

Schellmann, G., Radtke, U., Potter, E. K., Esat, T. M., & McCulloch, M. T. (2002). Comparison of ESR and TIMS U/Th dating of marine isotope stage (MIS) 5e, 5c, and 5a coral from Barbados — Implications for palaeo sea-level changes in the Caribbean. *Quaternary International, 120*, 41–50.

Schneider, R. R., Schulz, H. D., & Hensen, Ch. (1999). Marine carbonates: Their formation and destruction. In: H. D. Schulz & M. Zabel (Eds), *Marine geochemistry* (pp. 283–307). Berlin: Springer.

Schlanger, S. O. (1963). Subsurface geology of Enewetok Atoll. US Geological Survey Professional Paper, Vol. 260-B, pp. 991–1066.

Schlanger, S. O. (1965). Dolomite-evaporite relations on Pacific Islands. *Scientific reports of Tohoku University. Second Series (Geology), 62*, 15–29.

Schlager, W. (1981). The paradox of drowned reefs and carbonate platforms. *Bulletin of the Geological Society of America, 92*, 197–211.

Schlager, W. (1993). Accommodation and supply — A dual control on stratigraphic sequences. *Sedimentary Geology, 86*, 111–136.

Schlager, W., & James, N. P. (1978). Low-magnesian calcite limestones forming at the deep-sea floor, Tongue of the Ocean, Bahamas. *Sedimentology, 25*, 675–702.

Schneiderman, N., & Harris, P. M. (Eds) (1985). *Carbonate cements* (379 pp.), SEPM (Special Publication 36).

Scholz, D., & Mangini, A. (2007). How precise are U-series coral ages? *Geochimica et Cosmochimica Acta, 71*, 1935–1948.

Scholz, D., Mangini, A., & Felis, T. (2004). U-series dating of diagenetically altered fossil reef corals. *Earth and Planetary Science Letters*, *218*, 163–178.

Scholz, D., Mangini, A., & Meischner D. (2006). U-redistribution in fossil reef corals from Barbados, West Indies, and sea-level reconstruction for MIS 6.5. In: F. Sirocko, M. Clausen, M. F. Sanchez-Goni & Th. Litt (Eds), *The climate of past interglacials*. Developments in Quaternary Science (Vol. 7, pp. 19–139). Amsterdam: Elsevier.

Schönberg, C. H. L. (2002). Substrate effects on the bioeroding desmosponge *Cliona orientalis*. 1. Bioerosion rates. *Marine Ecology*, *23*, 313–326.

Schrimm, M., Heussner, S., & Buscail, R. (2002). Seasonal variations of downward particle fluxes in front of a reef pass (Moorea Island, French Polynesia). *Oceanologica Acta*, *25*, 61–70.

Schroeder, J. H. (1973). Submarine and vadose cements in Pleistocene Bermuda Reef Rock. *Sedimentary Geology*, *10*, 179–204.

Schroeder, J. H., & Purser, B. H. (Eds). (1986). *Reef diagenesis*. Berlin: Springer. 455 pp.

Schueth, J. D., & Frank, T. D. (2008). Reef foraminifera as bioindicators of coral reef health: Low Isles Reef, Northern Great Barrier Reef, Australia. *Journal of Foraminiferal Research*, *38*, 11–22.

Schuster, F. (2000). Oligocene and Miocene examples of *Acropora*-dominated palaeoenvironments: Mesohellenic Basin (NW Greece) and northern Gulf of Suez (Egypt). *Proceeding of the Ninth International Coral Reef Symposium*, Bali, Indonesia, Vol. 1, pp. 199–204.

Schwarcz, H. P., & Latham, A. G. (1989). Dirty calcites, 1. Uranium series dating of contaminated calcites using leachates alone. *Chemical Geology (Isotopes Geosciences Section)*, *80*, 35–43.

Scoffin, T. P. (1987). *Introduction to carbonate sediments and rocks*. Glasgow: Blackie, 274 pp.

Scoffin, T. P. (1992). Taphonomy of coral reefs: A review. *Coral Reefs*, *11*, 57–77.

Scoffin, T. P. (1993). The geological effects of hurricanes on coral reefs and the interpretation of storm deposits. *Coral Reefs*, *12*, 203–221.

Scoffin, T. P. (1997). Preservation and $CaCO_3$ budget of reef-flat corals, Phuket, South Thailand. *Proceedings of the Eighth International Coral Reef Symposium*, Panama, Vol. 2, pp. 1795–1800.

Scoffin, T. P., & Bradshaw, C. (2000). The taphonomic significance of endoliths in dead-versus live coral skeletons. *PALAIOS*, *15*, 248–254.

Scoffin, T. P., & Dixon, J. E. (1983). The distribution and structure of coral reefs: One hundred years since Darwin. *Biological Journal of Linnean Society*, *20*, 11–38.

Scoffin, T. P., & Le Tissier, M. D. A. (1998). Late Holocene sea level and reef-flat progradation, Phuket, South Thailand. *Coral Reefs*, *17*, 273–276.

Scoffin, T. P., & McLean, R. F. (1978). Exposed limestones of the Northern Province of the Great Barrier Reef. *Philosophical Transactions of the Royal Society of London, Series A*, *291*, 119–138.

Scoffin, T. P., Stearn, C. W., Boucher, D., Frydl, P., Hawkins, C. M., Hunter, I. G., et al. (1980). Calcium carbonate budget of a fringing reef on the west coast of Barbados, Part II — Erosion, sediments and internal structure. *Bulletin of Marine Science*, *30*, 475–508.

Scoffin, T. P., Stoddart, D. R., Tudhope, A. W., & Woodroffe, C. (1985). Rhodoliths and coralliths of Muri Lagoon, Rarotonga, Cook Islands. *Coral Reefs*, *4*, 71–80.

Scoffin, T. P., & Tudhope, A. W. (1985). Sedimentary environments of the central region of the Great Barrier Reef of Australia. *Coral Reefs*, *4*, 81–93.

Scoffin, T. P. & Tudhope, A. W. (1988). Shallowing-upwards sequences in reef lagoon sediments: Examples from the Holocene of the Great Barrier Reef and the Silurian of Much Wenlock, Shropshire, England. *Proceedings of the Sixth International Coral Reef Symposium*, Townsville, Australia, Vol. 3, pp. 479–484.

Scoffin, T. P., Tudhope, A. W., & Brown, B. E. (1989a). Corals as environmental indicators, with preliminary results from South Thailand. *Terra Nova*, *1*, 559–563.

Scoffin, T. P., Tudhope, A. W., & Brown, B. E. (1989b). Fluorescent and skeletal density banding in *Porites lutea* from Papua New Guinea and Indonesia. *Coral Reefs*, 7, 169–178.

Scoffin, T. P., Tudhope, A. W., Brown, B. E., Chansang, H., & Cheeney, R. F. (1992). Patterns and possible environmental controls of skeletogenesis of *Porites lutea*, South Thailand. *Coral Reefs*, *11*, 1–11.

Scott, G. A. J., & Rotondo, G. M. (1983a). A model for the development of types of atolls and volcanic islands on the Pacific Lithospheric Plate. *Atoll Research Bulletin*, *260*, 1–33.

Scott, G. A. J., & Rotondo, G. M. (1983b). A model to explain the differences between Pacific plate island-atoll types. *Coral Reefs*, *1*, 139–150.

Scott, R. W. (1995). Global environmental controls on Cretaceous reefal ecosystems. *Palaeogeography, Palaeoclimatology, Palaeoecology*, *119*, 187–199.

Schuster, F., & Wielandt, U. (1999). Oligocene and Early Miocene coral faunas from Iran: Palaeoecology and palaeobiogeography. *International Journal of Earth Sciences*, *88*, 571–581.

Segal, B., Evangelista, H., Kampel, M., Costa Goncalves, A., Simionatto Polito, P., & Alves dos Santos, E. (2008). Potential impacts of polar fronts on sedimentation processes at Abrolhos coral reef (South-West Atlantic Ocean/Brazil). *Continental Shelf Research*, *28*(4–5), 533–544.

Seidel, D. J., Fu, Q., Randel, W. J., & Reichler, Th. J. (2008). Widening of the tropical belt in a changing climate. *Nature Geoscience*, *1*, 21–24.

Seki, O., Ishiwatari, R., & Matsumoto, K. (2002). Millennial climate oscillations in the NE Pacific surface waters over the last 82 kyr: New evidence from alkenones. *Geophysical Research Letters*, *29*, 2144, doi: 10.1029/2002GL015200.

Servant, J. (2001). The 100 kyr cycle of deglaciation during the last 450 kyr: A new interpretation of oceanic and ice core data. *Global and Planetary Change*, *29*, 121–133.

Shackleton, N. J. (1967). Oxygen isotope analyses and Pleistocene temperatures re-assessed. *Nature*, *215*, 15–17.

Shackleton, N. J. (1977). The oxygen isotope record of the Late Pleistocene. *Philosophical Transactions of the Royal Society of London, Series B*, *280*, 169–182.

Shackleton, N. J. (1987). Oxygen isotopes, ice volume and sea level. *Quaternary Science Reviews*, *6*, 183–190.

Shearman, D. J., Khourik, J., & Taha, S. (1961). On the replacement of dolomite by calcite in some Mesozoic limestones from the French Jura. *Proceedings of the Geologists' Association*, *72*, 1–12.

Shen, C.-C., Lee, T., Chen, C.-Y., Wang, C.-H., Dai, C.-F., & Li, L.-A. (1996). The calibration of D [Sr/Ca] versus sea surface temperature relationship for *Porites* corals. *Geochimica et Cosmochimica Acta*, *60*, 3849–3858.

Shen, G. T., & Boyle, E. A. (1987). Lead in corals: Reconstruction of historical industrial fluxes to the surface ocean. *Earth and Planetary Science Letters*, *82*, 289–304.

Shen, G. T., & Boyle, E. A. (1988). Determination of lead, cadmium, and other trace metals in annually-banded corals. *Chemical Geology*, *67*, 47–62.

Shen, G. T., Boyle, E. A., & Lea, D. W. (1987). Cadmium in corals as a tracer of historical upwelling and industrial fallout. *Nature*, *328*, 794–795.

Shen, G. T., Campbell, T. M., Dunbar, R. B., Wellington, G. M., Colgan, M. W., & Glynn, P. W. (1991). Paleochemistry of manganese in corals from the Galapagos Islands. *Coral Reefs*, *10*, 91–100.

Shen, G. T., Cole, J. E., Lea, D. W., Linn, N. J., McConnaughey, T. A., & Fairbanks, R.G. (1992a). Surface ocean variability at Galapagos from 1936–1982: Calibration of geochemical tracers in corals. *Paleoceanography*, 7, 404–407.

Shen, G. T., & Dunbar, R. B. (1995). Environmental controls on uranium in reef corals. *Geochimica et Cosmochimica Acta*, *59*, 2009–2024.

Shen, G. T., Linn, L. J., Campbell, T. M., Cole, J. E., & Fairbanks, R. G. (1992b). A chemical indicator of trade wind reversal in corals from the Western Tropical Pacific. *Journal of Geophysical Research, 97*, 12689–12697.

Shen, G. T., & Sanford, C. L. (1990). Trace-element indicators of climate variability in reef-building corals. In: P. W. Glynn (Ed.), *Global ecological consequences of the 1982–83 El Niño-Southern Oscillation* (pp. 255–283). Amsterdam: Elsevier.

Sheppard, C. R. C. (2003). Predicted recurrences of mass coral mortality in the Indian Ocean. *Nature, 425*, 294–297.

Sheppard, C. R. C. & Sheppard, A. L. S. (1991). Corals and coral communities of Arabia. In: W. Büttiker, F. Krupp, V. Mahnert & W. Wittmer (Eds), *Fauna of Saudi Arabia* (Vols. 1, 2, pp. 1–170). Basel, Switzerland: Karger Libri AG.

Shinn, E. A. (1963). Spur and groove formation in the Florida Reef Tract. *Journal of Sedimentary Research, 33*, 291–303.

Shinn, E. A. (1969). Submarine lithification of Holocene carbonate sediments in the Persian Gulf. *Sedimentology, 12*(1–2), 109–144.

Shinn, E. A., Hudson, J. H., Halley, R. B. & Lidz, B. (1977). Topographic control and accumulation rate of some Holocene Coral reefs: South Florida and Dry Tortugas. *Proceedings of the Third International Coral Reef Symposium*, Miami, Vol. 2, pp. 1–7.

Shinn, E. A., Hudson, J. H., Robbin, D. M. & Lidz, B. (1981). Spurs and grooves revisited: Construction versus erosion Looe Key Reef, Florida. *Proceedings of the Fourth International Coral Reef Symposium*, Manila, Vol. 2, pp. 475–483.

Shinn, E. A., Smith, G. W., Prospero, J. M., Betzer, P., Hayes, M. L., Garrison, V., et al. (2000). African dust and the demise of Caribbean coral reefs. *Geophysical Research Letters, 27*, 3029–3032.

Sholkovitz, E., & Shen, G. T. (1995). The incorporation of rare earth elements in modern coral. *Geochimica et Cosmochimica Acta, 59*, 2749–2756.

Sibley, D. F., & Gregg, J. M. (1987). Classification of dolomite rock textures. *Journal of Sedimentary Petrology, 57*, 967–975.

Siddall, M., Chappell, J. & Potter, E. -K. (2006). Eustatic sea level during past interglacials. In: F. Sirocko, M. Clausen, M. F. Sanchez-Goni & T. Litt (Eds), *The climate of past interglacials: Developments in Quaternary Science* (Vol. 7, pp. 75–92). Amsterdam. Elsevier.

Siddall, M., Rohling, E. J., Almogi-Labin, A., Hemleben, Ch., Meischner, D., Schmelzer, I., & Smeed, D. A. (2003). Sea-level fluctuations during the last glacial cycle. *Nature, 423*, 853–858.

Silverman, J., Lazar, B., & Erez, J. (2007). Effect of aragonite saturation, temperature, and nutrients on the community calcification rate of a coral reef. *Journal of Geophysical Research, 112*, C05004, doi: 10.1029/2006JC003770.

Simms, M. (1984). Dolomitization by groundwater flow in carbonate systems: A historical perspective. *Gulf Coast Association of Geological Societies Transactions, 24*, 411–420.

Sinclair, D. J., Kinsley, L. P. J., & McCulloch, M. T. (1998). High resolution analysis of trace elements in corals by laser ablation ICP-MS. *Geochimica et Cosmochimica Acta, 62*, 1889–1901.

Sinclair, D. J., & McCulloch, M. T. (2004). Corals record low mobile barium concentrations in the Burdekin River during the 1974 flood: Evidence for limited supply to rivers? *Palaeogeography, Palaeoclimatology, Palaeoecology, 214*, 155–174

Sinclair, D. J., & Risk, M. J. (2006). A numerical model of trace-element coprecipitation in a physiochemical calcification system: Application to coral biomineralization and trace-element "vital effects". *Geochimica et Cosmochimica Acta, 70*, 3855–3868.

Slansky, M. (1986). *Geology of sedimentary phosphates. Translated by P. Cooper and A. Notholt. Studies in Geology*. London: North Oxford Academic, 210 pp.

Smart, C. C., Dawans, J. M., & Whitaker, F. F. (1988). Carbonate dissolution in a modern mixing zone, south Andros, Bahamas. *Nature, 335*, 811–813.

Smart, P. L., Beddows, P. A., Coke, J., Doerr, S., Smith, S. & Whitaker, F. F. (2006). Cave Development on the Caribbean Coast of the Yucatan Peninsula, Quintana Roo, Mexico. In: R. S. Harmon & C. M. Wicks (Eds), *Perspectives on Karst geomorphology, hydrology, and geochemistry* (pp. 105–128). Geological Society of America, Special Paper 404.

Smith, B. T., Frankel, E. & Jell, J. S. (1998). Lagoonal sedimentation and reef development on Heron Reef, southern Great Barrier Reef Province. In: G. F. Camoin & P. J. Davies (Eds), *Reefs and carbonate platforms in the Pacific and Indian Oceans* (pp. 281–294), International Association of Sedimentologists (Special Publication 25).

Smith, N. P. (2001). Weather and hydrographic conditions associated with coral bleaching: Lee Stocking Island, Bahamas. *Coral Reefs, 20*, 415–422.

Smith, S. V. (1983). Coral reef calcification. In: D. J. Barnes (Ed.), *Perspectives on coral reefs. The Australian Institute of Marine Science* (pp. 240–247). Manuka, Australia: Brian Clouston Publisher.

Smith, S. V., & Buddemeier, R. W. (1992). Global change and coral reef ecosystems. *Annual Review of Ecology and Systematics, 23*, 89–118.

Smith, S. V. & Kinsey, D. W. (1978). Calcification and organic carbon metabolism by carbon dioxide. In: D. R. Stoddart & R. E. Johannes (Eds). *Coral reefs: Research methods.* UNESCO, Paris, Monography on Oceanographic Methodology (Vol. 5, pp. 469–484).

Smith, T. J., Hudson, J. H., Robblee, M. B., Powell, G. W. N., & Isdale, P. J. (1989). Freshwater flow from the Everglades to Florida Bay: A historical reconstruction based on fluorescent banding in the coral *Solenastraea bournoni*. *Bulletin of Marine Science, 44*, 247–282.

Smithers, S. G., Hopley, D., & Parnell, K. E. (2006). Fringing and nearshore coral reefs on the Great Barrier Reef: Episodic Holocene development and future prospects. *Journal of Coastal Research, 22*, 175–188.

Smithers, S. G., & Larcombe, P. (2003). Late Holocene initiation and growth of a nearshore turbid-zone coral reef: Paluma Shoals, central Great Barrier Reef, Australia. *Coral Reefs, 22*, 499–505.

Smithers, S. G., & Woodroffe, C. D. (2001). Coral microatolls and 20th century sea level in the eastern Indian Ocean. *Earth and Planetary Science Letters, 191*, 173–184.

Smithers, S. G., Woodroffe, C. D., McLean, R. F. & Wallensky, E. P. (1992). Lagoonal sedimentation in the Cocos (Keeling) Islands, Indian Ocean. *Proceedings of the Seventh International Coral Reef Symposium*, Guam, Vol. 1, pp. 273–288.

Solomon, S., Qin, D. & Mannin, M. (Eds) (2007). IPCC 2007: Climate Change 2007. The Physical Science Basis. Contribution of Working Group I to the Fourth Assessment Report of the Intergovernmental Panel on Climate Change.

Songoni, C., Bard, E., & Rostek, F. (1998). Tropical sea-surface temperatures during the last glacial period: A view based on alkenones in Indian Ocean sediments. *Quaternary Science Reviews, 17*, 1185–1207.

Sorokin, Y. I. (1993). *Coral reef ecology. Ecological studies, 102.* Berlin: Springer-Verlag, 465 pp.

Southon, J., Kashgarian, M., Fontugne, M., Metivier, B., & Yim, W.-W. S. (2002). Marine reservoir corrections for the Indian Ocean and Southeast Asia. *Radiocarbon, 44*, 167–180.

Spencer, T. (1989). Sediments and sedimentary environments of Henderson Islands. *Atoll Research Bulletin, 324*, 1–7.

Sprachta, S., Camoin, G., Golubic, S., & Le Campion, Th. (2001). Microbialites in a modern lagoonal environment: Nature and distribution, Tikehau atoll (French Polynesia). *Palaeogeography, Palaeoclimatology, Palaeoecology, 175*, 103–124.

Stanford, J. D., Rohling, J. E., Hunter, S. E., Roberts, A. P., Rasmussen, S. O., Bard, E., et al. (2006). Timing of meltwater pulse 1a and climate responses to meltwater injections. *Paleoceanography, 21*, PA4103, doi: 10.1029/2006PA001340.

Stafford-Smith, M. G., & Ormond, R. F. G. (1992). Sediment-rejection mechanisms of 42 species of Australian scleractinian corals. *Marine and Freshwater Research, 43*, 683–705.

Stanley, G. D. (2001). *The History and Sedimentology of Ancient Reef Systems. Topics in Geobiology*. Vol. 17. New York: Kluwer Academic/Plenum Publishers, 458 pp.

Stanley, G. D., Jr. (2003). The evolution of modern corals and their early history. *Earth-Science Reviews*, *60*, 195–225.

Stanley, G. D., Jr. (2006). Photosymbiosis and the evolution of modern coral reefs. *Science*, *312*, 857–858.

Stanley, S. M., & Hardie, L. A. (1998). Secular oscillations in the carbonate mineralogy of reef-building and sediment-producing organisms driven by tectonically forced shifts in seawater chemistry. *Palaeogeography, Palaeoclimatology, Palaeoecology*, *144*, 3–19.

Scoffin, T. P. (1977). Carbonate budget of a fringing reef, Barbados. *Proceedings of the Third International Coral Reef Symposium*, Miami, Vol. 2, pp. 471–477.

Stearn, C. W., Scoffin, T. P., & Martindale, W. (1977). Calcium carbonate budget of a fringing reef on the west coast of Barbados: Part I — Zonation and productivity. *Bulletin of Marine Science*, *27*, 479–510.

Stearns, H. T. (1946). An integration of coral reef hypotheses. *American Journal of Science*, *244*, 245–262.

Steers, J. A., & Stoddart, D. R. (1977). The origin of fringing reefs, barrier reefs and atolls. In: O. A. Jones & R. Endean (Eds), *Biology and Geology of Coral Reefs* (Geology 2, pp. 212–227). New York: Academic Press.

Steinen, R. P., & Matthews, R. K. (1973). Phreatic versus vadose diagenesis. *Journal of Sedimentary Petrology*, *43*, 1012–1020.

Stemann, T. A., & Johnson, K. G. (1992). Coral assemblages, biofacies, and ecological zones in the mid-Holocene reef deposits of the Enriquillo Valley, Dominican Republic. *Lethaia*, *25*, 231–242.

Steneck, R. S. (1986). The ecology of coralline algal crusts: Convergent patterns and adaptative strategies. *Annual Review of Ecology and Systematics*, *17*, 273–303.

Steneck, R. S., Miller, T. E., Reid, R. P., & Macintyre, I. G. (1998). Ecological controls on stromatolite development in a modern reef environment: A test of the ecological refuge paradigm. *Carbonates and Evaporites*, *13*, 48–65.

Stienstra, P. (1985). Geology of a small rock-phosphate deposit, Ceru Colorado, Aruba, Netherlands Antilles. *Journal of Sedimentary Petrology*, *55*, 96–104.

Stirling, C. H., Esat, T. M., Lambeck, M. T., & McCulloch, M. T. (1998). Timing and duration of the Last Interglacial: Evidence for a restricted interval of widespread coral reef growth. *Earth and Planetary Science Letters*, *160*, 745–762.

Stirling, C. H., Esat, T. M., Lambeck, K., McCulloch, M. T., Blake, S. G., Lee, D.-C., & Halliday, A. N. (2001). Orbital forcing of the Marine Isotope Stage 9 Interglacial. *Science*, *291*, 290–293.

Stoddart, D. R. (1963). Effects of Hurricane Hattie on the British Honduras reefs and cays. October 30–31, 1961. *Atoll Research Bulletin*, *95*, 1–142.

Stoddart, D. R. (1964). Carbonate sediments of Half Moon Cay, British Honduras. *Atoll Research Bulletin*, *104*, 1–96.

Stoddart, D. R. (1969a). Ecology and morphology of recent coral reefs. *Biological Reviews*, *44*, 433–498.

Stoddart, D. R. (1969b). Reconnaissance geomorphology of Rangiroa Atoll, Tuamotu Archipelago, with list of vascular flora by M.-H. Sachet. *Atoll Research Bulletin*, *125*, 1–44.

Stoddart, D. R. (Ed.) (1971). *Coral reefs and islands and catastrophic storms. Applied coastal geomorphology*. (pp. 55–197). London: McMillan.

Stoddart, D. R. (1976). Continuity and crisis in the reef community. *Micronesia*, *12*, 1–9.

Stoddart, D. R. (Ed.) (2007). Tsunamis and coral reefs. *Atoll Research Bulletin*, *544*, 1–164.

Stoddart, D. R., & Fosberg, F. R. (1994). The Hoa of Hull Atoll and the problem of Hoa. *Atoll Research Bulletin*, *394*, 1–26.

Stoddart, D. R., & Scoffin, T. P. (1979). Microatolls: Review of form, origin and terminology. *Atoll Research Bulletin, 224*, 1–17.

Stoddart D. R. & Yonge, M. (Eds) (1971). *Regional variation in Indian Ocean Coral Reefs.* Symposia of the Zoological Society of London (Vol. 28, 584 pp.). London: Academic Press.

Stoddart, D. R., & Yonge, M. (Eds). (1978). *The Northern Great Barrier Reef.* (p. 172). London: Philosophical Transactions of the Royal Society of London.

Stoessell, R. K., Coke, J. G., & Easley, D. H. (2002). Localized thermal anomalies in haloclines of coastal Yucatan sinkholes. *Groundwater, 40*(4), 416–424.

Stolarski, J., Meibom, A., Przenioslo, R., & Mazur, M. (2007). A Cretaceous scleractinian coral with a calcitic skeleton. *Science, 318*, 92–94.

Stoll, H. M., & Schrag, D. P. (1998). Effects of Quaternary sea levels cycles on strontium in seawater. *Geochimica et Cosmochimica Acta, 62*, 1107–1118.

Storlazzi, C. D., McManus, M. A., Logan, J. B., & McLaughlin, B. E. (2006). Cross-shore velocity shear, eddies and heterogeneity in water column properties over fringing coral reefs: West Maui, Hawaii. *Continental Shelf Research, 26*, 401–421.

Storlazzi, C. D., Ogston, A. S., Bothner, M. H., Field, M. E., & Presto, M. K. (2004). Wave- and tidally-driven flow and sediment flux across a fringing coral reef: Southern Molokai, Hawaii. *Continental Shelf Research, 24*, 1397–1419.

Street, J. H., Knee, K. L., Grossman, E. E., & Paytan, A. (2008). Submarine groundwater discharge and nutrient addition to the coastal zone and coral reefs of leeward Hawai'i. *Marine Chemistry, 109*(3–4), 355–376.

Suchanek, T. (1985). Thalassinid shrimp burrows: Ecological significance of species-specific architecture. *Proceedings of the Fifth International Coral Reef Congress*, Tahiti, Vol. 5, pp. 205–210.

Sugihara, K., Masunaga, M., & Fujita, K. (2006). Latitudinal changes in larger benthic foraminiferal assemblages in shallow-water reef sediments along the Ryukyu Islands, Japan. *Island Arc, 15*, 437–454.

Sugihara, K., Nakamori, T., Iryu, Y., Saski, K., & Blanchon, P. (2003). Holocene sea-level change and tectonic uplift deduced from raised reef terraces, Kikai-jima, Ryukyu Islands, Japan. *Sedimentary Geology, 159*, 5–25.

Sugiyama, T. (1934). On the drilling on Kita-daito Island. Contributions to the Institute of Geology and Palaeontology, Tohoku Imperial University (Vol. 11, pp. 1–44).

Sugiyama, T. (1936). Second test drilling in Kita-daito-jima (in Japanese). Tohoku Imperial University, Faculty of Science, Institute for Geology and Paleontology, Contributions in Japanese Language (Vol. 25, pp. 1–34).

Sun, D., Gagan, M. K., Cheng, H., Scott-Gagan, H., Dykoski, C. A., Edwards, R. L., et al. (2005). Seasonal and interannual variability of the Mid-Holocene East Asian monsoon in coral $\delta^{18}O$ records from the South China Sea. *Earth and Planetary Science Letters, 237*, 69–84.

Sun, Y., Sun, M., Wei, G., Lee, T., Nie, B., & Yu, Z. (2004). Strontium contents of a *Porites* coral from Xisha Island, South China Sea: A proxy for a sea-surface temperature of the 20th century. *Paleoceanography, 19*, PA2004, doi: 10.1029/2003PA000959.

Sunagawa, I. (1982). Morphology of crystals in relation to growth conditions. In: C. R. Rodrigues & I. Sunagawa (Eds), *Crystal growth conditions in sedimentary environments* (Vol. 38, pp. 127–134). Madrid: Estudios Geológicos.

Suzuki, A., Gagan, M. K., De Deckker, P., Omura, A., Yukino, I., & Kawahata, H. (2001). Last Interglacial coral records of enhanced insolation seasonality and seawater ^{18}O enrichment in the Ryukyu Islands, northwest Pacific. *Geophysical Research Letters, 28*, 3685–3688.

Suzuki, A., Nakamura, T., Yamasaki, H., Minoshima, K., & Kawahata, H. (2008). Influence of water flow on skeletal isotopic composition in the coral *Pocillopora damicornis*. *Coral Reefs, 27*, 209–218.

Suzuki, A., & Kawahata, H. (2003). Carbon budget of coral reef systems: An overview of observations in fringing reefs, barrier reefs and atolls in the Indo-Pacific regions. *Tellus, 55B*, 428–444.

Suzuki, Y., Iryu, Y., Inagaki, S., Yamada, T., Aizawa, S., & Budd, D. (2006). Origin of atoll dolomites distinguished by geochemistry and crystal chemistry: Kita-daito-jima, northern Philippines Sea. *Sedimentary Geology, 183*, 181–202.

Swart, P. K. (1983). Carbon and oxygen isotope fractionation in scleractinian corals: A review. *Earth-Science Reviews, 19*, 51–80.

Swart, P. K., Dodge, R. E., & Hudson, H. J. (1996). A 240-year stable oxygen and carbon isotopic record in a coral from South Florida: Implications for the prediction of precipitation in Southern Florida. *PALAIOS, 11*, 362–375.

Swart, P. K., Elderfield, H., & Greaves, M. J. (2002). A high-resolution calibration of Sr/Ca thermometry using the Caribbean coral *Montastraea annularis*. *Geochemistry, Geophysics, Geosystems, 3*, 8402, doi: 10.1029/2002GC000306.

Swart, P. K., & Kramer, P. A. (1997). Geology of mud islands in Florida Bay. In: H. L. Vacher & T. Quinn (Eds), *Geology and hydrogeology of carbonate islands*. Developments in Sedimentology (pp. 249–274). New York: Elsevier.

Swart, P. K., White, K. S., Enfield, D., Dodge, R. E., & Milne, P. (1998). Stable oxygen isotopic composition of corals from the Gulf of Guinea as indicators of periods of extreme precipitation conditions in the sub-Sahara. *Journal of Geophysical Research, 103*, 27885–27891.

Swartz, J. H. (1958). *Geothermal measurements on Enewetok and Bikini Atolls*. US Geological Survey Professional Paper, Vol. 260-U, pp. 711–739.

Sy, J. C., Herrera, F. S., & McManus, J. W. (1981). Coral community structure of a fringing reef at Mactan Island, Cebu, Philippines. *Proceedings of the Fourth International Coral Reef Symposium*, Manila, Vol. 2, pp. 263–269.

Symonds, G., Black, K. P., & Young, I. R. (1995). Wave-driven flow over shallow reefs. *Journal of Geophysical Research, 100*, 2639–2648.

Takahashi, T., Kobe, M. & Kan, H. (1988). Relationship between reef growth and sea level on the northwestern coast of Kume Island, the Ryukyus: Data from drill holes on the Holocene coral reef. *Proceedings of the Sixth International Coral Reef Symposium*, Townsville, Australia, Vol. 3, pp. 491–496.

Takahashi, T., Kobe, M. & Nakamori, T. (1985). Coral reefs of the Ryukyu Islands: Reef morphology and reef zonation. *Proceedings of the Fifth International Coral Reef Congress*, Tahiti, Vol. 3, pp. 211–216.

Taniguchi, M., Burnett, W. C., Cable, J. E., & Turner, J. V. (2002). Investigation of submarine groundwater discharge. *Hydrological Processes, 16*, 2115–2129.

Tanner, J. E., Hughes, T. P., & Connell, J. H. (1994). Species coexistence, keystone species, and succession: A sensitivity analysis. *Ecology, 75*, 2204–2219.

Taviani, M. (1997). Post-Miocene coral reefs of the Red Sea: Glacio-eustatic controls. In: B. H. Purser & D. W. J. Bosence (Eds), *Sedimentation and tectonics in rift basins: Red Sea-Gulf of Aden* (pp. 580–588). London: Chapman & Hall.

Taviani, M. (1998). Stable tropics not so stable: Climatically driven extinctions of reef-associated molluscan assemblages (Red Sea and western Indian Ocean: Last interglaciation to present). In: G. F. Camoin & P. J. Davies (Eds), *Reefs and carbonate platforms in the Pacific and Indian Oceans* (pp. 69–76), International Association of Sedimentologists (Special Publication 25). Oxford: Blackwell Science.

Taylor, F. W., & Mann, P. (1991). Late Quaternary folding of coral reef terraces, Barbados. *Geology, 19*, 103–106.

Taylor, F. W., Mann, P., Bevis, M. G., Edwards, R. L., Cheng, H., Cutler, K. B., et al. (2005). Rapid forearc uplift and subsidence caused by impinging bathymetric features: Examples from the New Hebrides and Solomon arcs. *Tectonics*, *24*, TC6005, doi: 10.1029/2004TC001650.

Taylor, J. D. (1978). Faunal response to instability of reef habitats: Pleistocene molluscan assemblages of Aldabra. *Palaeontology*, *21*, 1–30.

Teal, C. S., Mazzullo, S. J., & Bischoff, W. D. (2000). Dolomitization of Holocene shallow-marine deposits mediated by sulfate reduction and methanogenesis in normal-salinity seawater, Northern Belize. *Journal of Sedimentary Research*, *70*, 649–663.

Tebbutt, G. E. (1975). Paleoecology and diagenesis of Pleistocene limestone on Ambergris Cay, Belize. In: H. C. Wantland & W. C. Pusey, III (Eds). *Belize shelf: Carbonate sediments, clastic sediments, and ecology*. Studies in Geology (Vol. 2, pp. 297–331). Tulsa, OK: American Association of Petroleum Geologists.

Teichert, C., & Fairbridge, R. W. (1948). Some coral reefs of the Sahul Shelf. *Geographical Reviews*, *38*, 222–249.

Terry, C. E. & Williams, J. J. (1969). The Idris 'A' bioherm and oilfield, Sirte Basin, Libya; its commercial development, regional Palaeocene geologic setting and stratigraphy. Exploration for petroleum in Europe and North Africa (pp. 31–48). London: Institute of Petroleum.

Thom, B. G. (1978). Shallow core drilling. In: D. R. Stoddart & R. E. Johannes (Eds), *Coral Reefs: Research Methods*. UNESCO Monographs on Oceanographic Methodology, Paris, Vol. 5, pp. 67–73.

Thomassin, B. (1978). Soft-bottom communities. In: D. R. Stoddart & R. E. Johannes (Eds), *Coral reefs: Research methods*. UNESCO Monographs on Oceanographic Methodology, Paris, Vol. 5, pp. 263–298.

Thompson, W. G., & Goldstein, S. L. (2005). Open-system coral ages reveal persistent suborbital sea-level cycles. *Science*, *308*, 401–404.

Thompson, W. G., Spiegelman, M. W., Goldstein, S. L., & Speed, R. C. (2003). An open-system model for U-series age determinations of fossil corals. *Earth and Planetary Science Letters*, *210*, 365–381.

Thorp, E. M. (1936). The sediments of the Pearl and Hermes Reefs. *Journal of Sedimentary Petrology*, *6*, 109–118.

Thorstenson, D. C., Mackenzie, F. T., & Ristvet, B. L. (1972). Experimental vadose and phreatic cementation of skeletal carbonate sand. *Journal of Sedimentary Petrology*, *42*, 162–167.

Tija, H. D. (1996). Sea-level changes in the tectonically stable Malay-Thai Peninsula. *Quaternary International*, *31*, 95–101.

Titlyanov, E. A., & Latypov, Y. Y. (1991). Light dependence in scleractinian distribution in the sublittoral zone of South China Sea Islands. *Coral Reefs*, *10*, 133–138.

Toggweiler, J. R., Dixon, K., & Broecker, W. S. (1991). The Peru upwelling and the ventilation of the South-Pacific thermocline. *Journal of Geophysical Research, Oceans*, *96*, 20467–20497.

Toomey, D. F. (Ed.) (1981). *European fossil reefs models* (546 pp), SEPM (Special Publication 30).

Toro-Farmer, G., Cantera, J. R., Londoño-Cruz, E., Orozco, C., & Neira, R. (2004). Distribution patterns and bioerosion of the sea urchin *Centrostephanus coronatus* (Diadematoida: Diadematidae) at the reef of Playa Blanca, Colombian Pacific. *Revista de Biologia Tropical*, *52*, 67–76.

Toscano, M. A., & Lundberg, J. (1998). Early Holocene sea-level record from submerged fossil reefs on the southeast Florida margin. *Geology*, *26*, 255–258.

Toscano, M. A., & Lundberg, J. (1999). Submerged late Pleistocene reefs on the tectonically-stable S.E. Florida margin: High-precision geochronology, stratigraphy, resolution of Substage 5a sea-level elevation, and orbital forcing. *Quaternary Science Reviews*, *18*, 753–767.

Toscano, M. A., & Macintyre, I. G. (2003). Corrected western Atlantic sea-level curve for the past 11,000 years based on calibrated ^{14}C dates from *Acropora palmata* framework and intertidal mangrove peat. *Coral Reefs, 22*, 257–270.

Tracey, J. I., & Ladd, H. S. (1974). Quaternary history of Eniwetok and Bikini Atolls, Marshall Islands. *Proceedings of the Second International Coral Reef Symposium*, Brisbane, Vol. 2, pp. 537–550.

Trenberth, K. E. (1984). Signal versus noise in the Southern Oscillation. *Monthly Weather Review, 112*, 326–332.

Tribble, G. W., & Randall, R. H. (1986). A description of the high-latitude shallow water coral community of Miyake-jima, Japan. *Coral Reefs, 4*, 151–159.

Trichet, J., & Fikri, A. (1997). Organic matter in the genesis of high-island atoll peloidal phosphorites: The lagoonal link. *Journal of Sedimentary Research, 67*, 891–897.

Trichet, J., Repellin, P., & Oustrière, P. (1984). Stratigraphy and subsidence of the Mururoa Atoll (French Polynesia). *Marine Geology, 56*, 241–257.

Trueman, N. A. (1965). The phosphate, volcanic and carbonate rocks of Christmas Island (Indian Ocean). *Journal of the Geological Society of Australia, 12*(2), 261–283.

Tsuji, Y. (1993). Tide influenced high energy environments and rhodolith-associated carbonate deposition on the outer shelf and slope off the Mikayo Islands, southern Ryukyu Island Arc, Japan. *Marine Geology, 113*, 255–271.

Tudhope, A. W. (1989). Shallowing-upwards sedimentation in a coral reef lagoon, Great Barrier Reef of Australia. *Journal of Sedimentary Petrology, 59*, 1036–1051.

Tudhope, A. W., Chilcott, C. P., McCulloch, T. M., Cook, E. R., Chappell, J., Ellam, R. M., et al. (2001). Variability in the El Niño-Southern Oscillation through a glacial-interglacial cycle. *Science, 291*, 1511–1517.

Tudhope, A. W., Lea, D. W., Shimmield, G. B., Chilcott, C. P., & Head, S. (1996). Monsoon climate and Arabian Sea coastal upwelling recorded in massive corals from Southern Oman. *PALAIOS, 11*, 347–361.

Tudhope, A. W., Lea, D. W., Shimmield, G. B., Chilcott, C. P., Scoffin, T. P., Fallick, A. E., et al. (1997). Climatic records from massive *Porites* corals in Papua New Guinea: A comparison of skeletal Ba/Ca, skeletal $\delta^{18}O$ and coastal rainfall. *Proceedings of the Eighth International Coral Reef Symposium*, Panama, Vol. 2, pp. 1719–1724.

Tudhope, A. W., & Scoffin, T. P. (1984). The effects of *Callianassa* bioturbation on the preservation of carbonate grains in Davies Reef Lagoon, Great Barrier Reef, Australia. *Journal of Sedimentary Petrology, 54*, 1091–1096.

Tudhope, A. W., & Scoffin, T. P. (1985). Sedimentary environments of the central region of the Great Barrier Reef of Australia. *Coral Reefs, 4*, 81–94.

Tudhope, A. W. & Scoffin, T. D. (1986). Nature and origin of carbonate mud on the central Queensland shelf. *Annual Meeting of the International Society for Reef Studies, Marburg (Germany), Abstracts*, pp. 50 51.

Tudhope, A. W. & Scoffin, T. P. (1988). The relative importance of benthic foraminiferans in the production of carbonate sediment of the central Queensland shelf. *Proceedings of the Sixth International Coral Reef Symposium*, Townsville, Australia, Vol. 2, pp. 583–588.

Tudhope, A. W., & Scoffin, T. P. (1994). Growth and structure of fringing reefs in a muddy environment, south Thailand. *Journal of Sedimentary Research, A, 64*, 752–764.

Tudhope, A. W., Scoffin, T. P., Stoddart, D. R. & Woodroffe, C. D. (1985). Sediments of Suwarrow atoll. *Proceedings of the Fifth International Reef Congress*, Tahiti, Vol. 6, pp. 611–616.

Turak, E., Veron, C., & Sanpanich, K. (2005). Post-Tsunami coral reef surveys of Thailand's north Andaman coast: Coral reef status after 26 December 2004 tsunami. New England Aquarium, Boston, U.S.A. *Journal of Coastal Research, 13*, 3711–3731.

Turcotte, D. L., & Bernthal, M. J. (1984). Synthetic coral-reef terraces and variations of Quaternary sea level. *Earth and Planetary Science Letters, 70*, 121–128.

Umezawa, Y., Miyajima, T., Kayanne, H., & Koike, I. (2002). Significance of groundwater nitrogen discharge into coral reefs at Ishgaki Island, southwest of Japan. *Coral Reefs, 21*, 346–356.

Urban, F. E., Cole, J. E., & Overpeck, J. T. (2000). Influence of mean climate change on climate variability from a 155-year tropical Pacific coral reef. *Nature, 407*, 989–993.

Uthicke, S., & Nobes, K. (2008). Benthic foraminifera as ecological indicators for water quality on the Great Barrier Reef. *Estuarine, Coastal and Shelf Science, 78*, 763–773.

Vacher, H. L., Bengtsson, T. O., & Plummer, L. N. (1990). Hydrology of meteoric diagenesis: Residence time of meteoric groundwater in island fresh-water lenses with application to aragonite-calcite stabilization in Bermuda. *Bulletin of the Geological Society of America, 102*, 223–232.

Vacher, H. L., Hearty, P. J. & Rowe, M. P. (Eds) (1995). Stratigraphy of Bermuda: Nomenclature, concepts, and status of multiple systems of classification. Terrestrial and Shallow Marine Geology of the Bahamas and Bermuda. Geological Society of America, Special Paper, Vol. 300, pp. 271–294.

Vacher, H. L., & Mylroie, J. E. (2002). Eogenetic Karst from the perspective of an equivalent porous medium. *Carbonates and Evaporites, 17*, 182–196.

Vacher, H. L., & Quinn, T. (Eds) (1997). *Geology and hydrogeology of carbonate islands: Developments in Sedimentology* (Vol. 54, 948 pp.). Amsterdam: Elsevier.

Vahrenkamp, V., Swart, P. K., & Ruiz, J. (1991). Episodic dolomitization of late Cenozoic carbonates in the Bahamas: Evidence from Strontium isotopes. *Journal of Sedimentary Petrology, 61*, 1002–1014.

Van Andel, T. H., & Veevers, J. J. (1967). *Morphology and sediments of the Timor Sea.* Bureau of Mines and Research, Geology and Geophysics, Department of National Development, Commonwealth of Australia, Bulletin, Vol. 63, pp. 1–173.

van der Plicht, J. (2002). Calibration of the ^{14}C timescale: Towards the complete dating range. *Netherlands Journal of Geosciences/Geologie en Mijnbouw, 81*, 85–96.

van Katwijk, M. M., Meier, N. F., van Loon, R., van Hove, E. M., Giesen, W. B. J. T., & van der Velde, G. (1993). Sabaki river sediment load and coral stress: Correlation between sediments and condition of the Malindi–Watamu reefs in Kenya (Indian Ocean). *Marine Biology, 117*, 675–683.

Van Oppen, M. J. H., McDonald, B. J., Willis, B., & Miller, D. J. (2001). The evolutionary history of the coral genus *Acropora* (Scleractinia, Cnidaria) based on a mitochondrial and a nuclear marker: Reticulation, incomplete lineage sorting, or morphological convergence. *Molecular Biology and Evolution, 18*, 1315–1329.

Van Tussenbroek, B. I., & van Dijk, J. K. (2007). Spatial and temporal variability in biomass and production of psammophytic *Halimeda incrassata* (Bryopsidales, Chlorophyta) in a Caribbean reef lagoon. *Journal of Phycology, 43*, 69–77.

van Woesik, R., Ayling, A. M., & Mapstone, B. (1991). Impact of cyclone "Ivor" on the GBR Australia. *Journal of Coastal Research*, 7, 551–558.

Van Woesik, R., & Done, T. J. (1997). Coral communities and reef growth in the southern Great Barrier Reef. *Coral Reefs, 16*, 103–115.

Vasconselos, C., & McKenzie, J. A. (1997). Microbial mediation of modern dolomite precipitation and diagenesis under anoxic conditions (Lagoa Vermelha, Rio de Janeiro, Brazil). *Journal of Sedimentary Research, 67*, 378–390.

Vecsei, A. (2001). Fore-reef carbonate production: Development of a regional census-based method and first estimates. *Palaeogeography, Palaeoclimatology, Palaeoecology, 175*, 185–200.

Vecsei, A. (2004). A new estimate of global reefal carbonate production including the fore-reefs. *Global and Planetary Change, 43*, 1–18.

Vecsei, A., & Berger, W. H. (2004). Increase of atmospheric CO_2 during deglaciation: Constraints on the coral reef hypothesis from patterns of deposition. *Global Biogeochemical Cycles, 18*, GB1035, doi: 10.1029/2003GB002147.

Veeh, H. H., & Burnett, W. C. (1978). Uranium-series dating of insular phosphorite from Ebon atoll, Micronesia. *Nature, 274*, 460–462.

Veeh, H. H., & Chappell, J. (1970). Astronomical theory of climatic change: Support from New Guinea. *Science, 167*, 862–865.

Vénec-Peyré, M. Th. (1991). Distribution of living benthic foraminifera on the back-reef and outer slopes of a high island (Moorea, French Polynesia). *Coral Reefs, 9*, 193–203.

Verrecchia, E. P., Freytet, P., Verrecchia, K. E., & Dumont, J.-L. (1995). Spherulites in calcrete laminar crusts: Biogenic $CaCO_3$ precipitation as a major contributor to crust formation. *Journal of Sedimentary Research, A, 65*, 690–700.

Veron, J. E. N. (1986). *Corals of Australia and the Indo-Pacific.* North Ryde, London: Angus and Robertson Publishers, 644 pp.

Veron, J. E. N. (1992a). Hermatypic corals of Japan. *Australian Institute of Marine Science Monographs Series, 10*, 1–433.

Veron, J. E. N. (1992b). Environmental control of Holocene changes to the world's most northern hermatypic coral outcrop. *Pacific Science, 46*, 405–425.

Veron, J. E. N. (1993). A biogeographic database of hermatypic corals: Species of the central Indo-Pacific genera of the world. *Australian Institute of Marine Science Monographs Series, 10*, 433.

Veron, J. E. N. (1994). Hermatypic corals of the Cocos (Keeling) Islands: A summary. *Atoll Research Bulletin, 409*, 1–21.

Veron, J. E. N. (1995). *Corals in space and time; the biogeography and evolution of the Scleractinia.* Sydney: University of South Wales Press, 321 pp.

Veron, J. E. N. (2000). *Corals of the world.* Australian Institute of Marine Science and CSIRO Publishing, Townsville, Australia, Vol. 1, pp. 1–463; Vol. 2, pp. 1–429; Vol. 3, pp. 1–490.

Veron, J. E. N., & Kelley, R. (1988). Species stability in reef corals of Papua New Guinea and the Indo-Pacific. *Memoir of the Association of Australasian Palaeontologists, 6*, 1–69.

Vieira, M. M., & De Ros, L. F. (2006). Cementation patterns and genetic implications of Holocene beachrocks from northeastern Brazil. *Sedimentary Geology, 192*, 207–230.

Villemant, B., & Feuillet, N. (2003). Dating open system by the ^{238}U–^{234}U–^{230}Th method: Application to Quaternary reef terraces. *Earth and Planetary Science Letters, 210*, 105–118.

Vogel, J. C. (1970). Gröningen radiocarbon dates, Part IX. *Radiocarbon, 12*, 444–471.

Vogel, K., Gektidis, M., Golubic, S., Kiene, W. E., & Radtke, G. (2000). Experimental studies on microbial bioerosion at Lee Stocking Island, Bahamas and One Tree Island, Great Barrier Reef, Australia: Implications for paleoecological reconstructions. *Lethaia, 33*, 190–204.

Von der Borch, C. (1965). The distribution and preliminary geochemistry of modern carbonate sediments of the Coorong area, South Australia. *Geochimica et Cosmochimica Acta, 29*, 781–799.

von Morlot, A. (1848). In: L. Cayeux, 1935. Etude Pétrographique des roches sédimentaires de France (les Roches Carbonatées). Première Edition, Masson, Paris, 463 p.

Waelbroeck, C., Labeyrie, L., Duplessy, J. C., McManus, J. F., Lambeck, K., Balbon, E., et al. (2002). Sea-level and deep-water temperature changes derived from benthic foraminifera isotopic records. *Quaternary Science Reviews, 21*, 295–305.

Walbran, P. D. (1994). The nature of the pre-Holocene surface, John Brewer Reef, with the implications for the interpretation of Holocene reef development. *Marine Geology, 122*, 63–79.

Walker, M. (2005). *Quaternary dating methods.* Chichester: Wiley, 286 pp.

Walker, S. E., Parsons-Hubbard, K., Powell, E., & Brett, C. E. (2002). Predation on experimentally deployed molluscan shells from shelf to slope depths in a tropical carbonate environment. *PALAIOS, 17*, 147–170.

Wallace, C. C. (1999). *Staghorn corals of the world: A revision of the coral genus* Acropora *(Scleractinia; Astrocoeniina; Acroporidae) worldwide, with emphasis on morphology, phylogeny and biogeography*. Melbourne: CSIRO Publishing, 421 pp.

Wallace, C. C., & Muir, P. R. (2005). Biodiversity of the Indian Ocean from the perspective of staghorn corals (*Acropora* spp.). *Indian Journal of Marine Science, 34*, 42–49.

Wallace, C. C., Pandolfi, J. M., Young, A., & Wolstenholme, J. (1991). Indo-Pacific coral biogeography: A case study from the *Acropora selago* group. *Australian Systematic Botany, 4*, 199–210.

Wallace, C. C., & Rosen, B. R. (2006). Diverse staghorn corals (*Acropora*) in high-latitude Eocene assemblages: Implications for the evolution of modern diversity patterns of reef corals. *Proceedings of the Royal Society, B273*, 975–982.

Wallace, R. J. & Schafersman, S. D. (1977). Patch reef ecology and sedimentology of Glovers Reef Atoll, Belize. In: S. F. Frost & J. B. Saunders (Eds), *Reefs and related carbonates*. Studies in Geology (Vol. 4, pp. 37–53). Tulsa, OK: American Association of Petroleum Geologists.

Walter, L. M., & Burton, E. A. (1990). Dissolution of recent platform sediments in marine pore fluids. *American Journal of Science, 290*, 601–643.

Wantland, K. F. (1977). Distribution of Holocene benthonic foraminifera on the Belize shelf. In: K. F. Wantland & W. C. Pusey Jr., (Eds). *Belize Shelf — Carbonate sediments, clastic sediments, and ecology*. Studies in Geology (Vol. 2, pp. 332–399). Tulsa, OK: American Association of Petroleum Geologists.

Wapnick, C. M., Precht, W. F., & Aronson, R. B. (2004). Millennial-scale dynamics of staghorn coral in Discovery Bay, Jamaica. *Ecology Letters*, 7, 354–361.

Ward, W. C., & Halley, R. B. (1985). Dolomitization in a mixing zone of near-seawater composition, Late Pleistocene, Northeastern Yucatan Peninsula. *Journal of Sedimentary Petrology, 55*, 407–420.

Wardlaw, B. R., & Henry, T. H. (1986). Physical stratigraphic framework, Pacific Enewetak Crater Exploration (PEACE) Program, Enewetak Atoll, Republic of the Marshall Islands. In: T. W. Henry & B. R. Wardlaw (Eds), *Part 3: Stratigraphic Analysis and other geologic and geophysical studies in vicinity of KOA and OAK craters*. US Geological Survey, Open File Report, 86–555: 2.1–2.36.

Warrlich, G. M. D., Waltham, D. A., & Bosence, D. W. J. (2002). Quantifying the sequence stratigraphy and drowning mechanisms of atolls using a new 3-D forward stratigraphic modelling program (Carbonate 3D). *Basin Research, 14*, 379–400.

Watanabe, T., Gagan, M. K., Corrège, Th., Scott-Gagan, H., & Hantoro, W. S. (2003). Oxygen isotopes systematics in *Diploastrea heliopora*: New coral archive of tropical climate. *Geochimica et Cosmochimica Acta, 67*, 1349–1358.

Webb, G. E. (1996). Was Phanerozoic reef history controlled by the distribution of non-enzymatically secreted reef carbonates (microbial carbonate and biologically induced cement). *Sedimentology, 43*, 947–971.

Webb, G. E., Baker, J. C., & Jell, J. S. (1998). Inferred syngenetic textural evolution in Holocene cryptic reefal microbialites, Heron Reef, Great Barrier Reef, Australia. *Geology, 26*, 355–358.

Webb, G. E., Jell, J. S., & Baker, J. C. (1999). Cryptic intertidal microbialites in beachrock, Heron Island, Great Barrier Reef: Implications for the origin of microcrystalline beachrock cement. *Sedimentary Geology, 126*, 317–334.

Weber, J. N., & Woodhead, P. M. J. (1972a). Carbonate lagoon and beach sediments of Tarawa Atoll, Gilbert Islands. *Atoll Research Bulletin, 157*, 1–29.

Weber, J. N., & Woodhead, P. M. J. (1972b). Temperature dependence of oxygen-18 concentration in reef coral carbonates. *Journal of Geophysical Research*, 77, 463–473.

Webster, J. M. (1999). *The response of coral reefs to sea level change: Evidence from the Ryukyu Islands and the Great Barrier Reef*. Unpublished Ph.D. thesis, University of Sydney, 328 p.

Webster, J. M., Beaman, R. J., Davies, P. J., Manning, P., Pizarro, O., Thomas, A. L., et al. (2008). From corals to canyons: The Great Barrier Reef margin. *EOS, Transactions, American Geophysical Union, 89*, 217–221.

Webster, J. M., Clague, D. A., & Braga, J. C. (2007). Support of the Giant Wave Hypothesis: Evidence from submerged terraces off Lanai, Hawaii. *International Journal of Earth Sciences, 96*, 517–524.

Webster, J. M., Clague, D. A., Braga, J. C., Spalding, H., Renema, W., Kelley, C., et al. (2006). Drowned coralline algal dominated deposits off Lanai, Hawaii: Carbonate accretion and vertical tectonics over the last 30 ka. *Marine Geology, 225*, 223–246.

Webster, J. M. & Davies, P. J. (2003). Coral variations in two deep drill cores from the Northern Great Barrier Reef: Significance for the Pleistocene development of the Great Barrier Reef. *Sedimentary Geology, Special Issue, 159*, 61–80.

Webster, J. M., Davies, P. J., & Konishi, K. (1998). Model of fringing reef development in response to progressive sea level fall over the last 7000 years (Kikai-Jima, Ryukyu islands, Japan). *Coral Reefs, 17*, 289–308.

Webster, J. M., Wallace, L., Silver, E., Potts, D., Braga, J. C., Renema, W., et al. (2004). Coralgal composition of drowned carbonate platforms in the Huon Gulf, Papua New Guinea; implications for lowstand reef development and drowning. *Marine Geology, 204*, 59–89.

Webster, J. M., Wallace, L. M., Clague, D. A., & Braga, J. C. (2007). Numerical modeling of the growth and drowning of Hawaiian coral reefs during the last two glacial cycles (0–250 kyr). *Geochemistry, Geophysics, Geosystems, 8*, Q03011, doi: 10.1029/2006GC001415.

Wehmiller, J. F., Simmons, K. R., Cheng, H., Edwards, R. L., Martin-McNaughton, J., York, L. L., et al. (2004). Uranium-series coral ages from the US Atlantic Coastal Plain-the "80 ka problem" revisited. *Quaternary International, 120*, 3–14.

Weinbauer, M. G., Brandstatter, F., & Velimirov, B. (2000). On the potential use of magnesium and strontium concentrations as ecological indicators in the calcite skeleton of the red coral (*Corallium rubrum*). *Marine Biology, 137*, 801–809.

Weiner, S., Levi-Kalisman, Y., Raz, S., Weiss, I. M., & Addadi, L. (2002). A basic strategy for biomineralization: Taking advantage of disorder. *Microscopy and Microanalysis, 8*(Suppl. 2), 164–165.

Wells, J.W. (1954). *Recent corals of the Marshall Islands, United States Geological Survey*, Professional Paper, Vol. 260-I, pp. 285–486.

Weydert, P. (1973). Morphologie et sédimentologie de la partie méridionale du grand récif de Tuléar (SW Madagascar): les ensembles sédimentaires de la pente interne. *Tethys, Suppl. 5*, 133–156.

Wheeler, C. W., & Aharon, P. (1991). Mid-oceanic carbonate platforms and oceanic dipsticks: Examples from the Pacific. *Coral Reefs, 10*, 101–114.

Wheeler, C. W. & Aharon, P. (1997). Geology and hydrology of Niue. In: H. L. Vacher & T. Quinn (Eds), *Geology and hydrogeology of Carbonate Islands*. Developments in Sedimentology (Vol. 54, pp. 453–473; 537–561). Amsterdam: Elsevier Science B.V.

Wheeler, C. W., Aharon, P., & Ferrell, R. E. (1999). Successions of Late Cenozoic platform dolomites distinguished by texture, geochemistry, and crystal chemistry: Niue, South Pacific. *Journal of Sedimentary Research, 69*, 239–255.

Whitaker, F., Smart, P., Hague, Y., Waltham, D., & Bosence, D. (1997). Coupled two-dimensional diagenetic and sedimentological modelling of carbonate platform evolution. *Geology, 25*, 175–178.

Whitaker, F. F., & Smart, P. L. (1990). Active circulation of saline ground waters in carbonate platforms: Evidence from the Great Bahama Bank. *Geology, 18*, 200–204.

Whitaker, F. F., & Smart, P. L. (1997). Hydrogeology and hydrology of the Bahamian archipelago. In: H. L. Vacher & T. M. Quinn (Eds), *Geology and hydrology of carbonate islands* (pp. 183–216). New York: Elsevier.

Whitaker, F. F., & Smart, P. L. (2007a). Geochemistry of meteoric diagenesis in carbonate islands of the northern Bahamas: 1. Evidence from field studies. *Hydrological Processes, 21*, 949–966.

Whitaker, F. F., & Smart, P. L. (2007b). Geochemistry of meteoric diagenesis in carbonate islands of the northern Bahamas: 2. Geochemical modelling and budgeting of diagenesis. *Hydrological Processes, 21*, 967–982.

Whitaker, F. F., Smart, P. L. & Jones, G. D. (2004). Dolomitization: From conceptual to numerical models. In: C. J. R. Braithwaite, G. Rizzi & G. Darke (Eds), *The geometry and petrogenesis of dolomite hydrocarbon reservoirs* (pp. 99–139), Geological Society (Special Publication 235), London.

Wigley, P. (1977). *Facies analysis of Holocene carbonate sediments and Tertiary(?) — Pleistocene limestones on and around Barbuda, West Indies*. Studies in Geology (Vol. 4, pp. 299–312). Tulsa, OK: American Association of Petroleum Geologists.

Wilkinson, C. (2004). *Status of coral reefs in the world*. Australian Institute of Marine Science, Townsville, Australia, Vol. 1, 301 pp.; Vol. 2, 538 pp.

Willis, B. L., van Oppen, M. J. H., Miller, D. J., Vollmer, S. T., & Ayre, D. J. (2006). The role of hybridization in the evolution of reef corals. *Annual Review of Ecology, Evolution and Systematics, 37*, 489–517.

Wilson, A. M., Sanford, W., Whitaker, F. F., & Smart, P. (2001). Spatial patterns of diagenesis during geothermal circulation in carbonate platforms. *American Journal of Science, 301*, 727–752.

Wilson, J. L. (1975). *Carbonate facies in geological history*. New-York: Springer-Verlag, 471 pp.

Wilson, M. E. J. (2002). Cenozoic carbonates in South-east Asia; implications for equatorial carbonate development. *Sedimentary Geology, 147*, 295–428.

Wilson, M. E. J., & Rosen, B. R. (1998). Implications of paucity of corals in the Paleogene of SE Asia: Plate tectonics or centre of origin? In: R. Hall & J. D. Holloway (Eds), *Biogeography and geological evolution of SE Asia* (pp. 165–195). Leiden: Backhuys Publishers.

Wilson, M. E. J., & Vecsei, A. (2005). The apparent paradox of abundant foramol facies in low latitudes: Their environmental significance and effect on platform development. *Earth-Science Reviews, 69*, 133–168.

Wilson, R., Tudhope, A., Brohan, Ph., Briffa, K., Osborn, Th., & Tett, S. (2006). Two-hundred-fifty years of reconstructed and modeled tropical temperatures. *Journal of Geophysical Research, 111*, C10007, doi: 10.1029/2005JC003188.

Winland, H. D. (1968). The role of high Mg calcite in the preservation of micrite envelopes and textural features of aragonitic sediments. *Journal of Sedimentary Petrology, 38*, 1320–1325.

Winter, A., Ishioroshi, H., Watanabe, T., & Oba, T. (2000). Caribbean sea surface temperatures: Two-to-three degrees cooler than present during the Little Ice Age. *Geophysical Research Letters, 20*, 3365–3368.

Winter, A., Paul, A., Nyberg, J., Oba, T., Lundberg, J., Schrag, D., et al. (2003). Orbital control of low-latitude seasonality. *Geophysical Research Letters, 30*, 1163, doi: 10.1029/2002GL016275.

Wolanski, E. (2001). *Oceanographic processes of coral reefs: Physical and biological links in the Great Barrier Reef* (356 pp.). Boca Raton, FL: Taylor and Francis CRC Press.

Wolanski, E., Richmond, R. H., Davis, G., Deleersnijder, E., & Leben, R. R. (2003). Eddies around Guam, an island in the Mariana Islands group. *Continental Shelf Research, 23*, 991–1003.

Wollast, R., & Reinhard-Derie, D. (1977). Equilibrium and mechanism of dissolution of Mg-calcites. In: N. R. Anderson & A. Malahoff (Eds), *The fate of fossil fuel CO_2 in the oceans* (pp. 479–493). New York: Plenum Press.

Wood, R. (1993). Nutrients, predation and the history of reef building. *PALAIOS, 8*, 526–543.

Wood, R. (1995). The changing biology of reef-building. *PALAIOS, 10*, 517–529.

Wood, R. (1999). *Reef evolution*. New York: Oxford University Press, 414 pp.

Woodroffe, C. D. (2005). Late Quaternary sea-level highstands in the central and eastern Indian Ocean: A review. *Global and Planetary Change, 49*, 121–138.

Woodroffe, C. D., Beech, M. R., & Gagan, M. K. (2003). Mid-late Holocene El Niño variability in the equatorial Pacific from coral microatolls. *Geophysical Research Letters, 30*, 1358, doi: 10.1029/2002GL015868.

Woodroffe, C. D., Kennedy, D. M., Hopley, D., Rasmussen, C. E., & Smithers, S. G. (2000). Holocene reef growth in Torres Strait. *Marine Geology, 170*, 331–346.

Woodroffe, C. D., Kennedy, D. M., Jones, B. G., & Phipps, C. V. G. (2004). Geomorphology and Late Quaternary development of Middleton and Elizabeth Reefs. *Coral Reefs, 23*, 249–262.

Woodroffe, C. D., & McLean, R. F. (1990). Microatolls and recent sea level change on atolls. *Nature, 334*, 531–534.

Woodroffe, C. D., Samosorn, B., Hua, Q., & Hart, D. E. (2007). Incremental accretion of a sandy island over the past 3000 years indicated by component-specific radiocarbon dating. *Geophysical Research Letters, 34*, L03602, doi: 10.1029/2006GL028875.

Woodroffe, C. D., Stoddart, D. R., Spencer, T., Scoffin, T. P., & Tudhope, A. W. (1990). Holocene emergence in the Cook Islands, South Pacific. *Coral Reefs, 9*, 31–39.

Woodroffe, S. A., & Horton, B. P. (2005). Holocene sea-level changes in the Indo-Pacific. *Journal of Asian Earth Sciences, 25*, 29–43.

Wooley, J. D. (1992). The incidence of hurricanes in the north coast of Jamaica since 1870: Are the classic reef descriptions atypical? *Hydrobiologia, 247*, 133–138.

Woolfe, K. J., & Larcombe, P. (1999). Terrigenous sedimentation and coastal reef growth: A conceptual framework. *Marine Geology, 155*, 331–345.

Wright, D. T., & Wacey, D. (2004). Sedimentary dolomite: A reality check. In: C. J. R. Braithwaite, G. Rizzi & G. Darke (Eds), *The geometry and petrogenesis of dolomite hydrocarbon reservoirs* (pp. 65–74), Geological Society (Special Publication 235), London.

Wright, V. P. (1994). Palaeosols in shallow marine carbonate sequences. *Earth-Science Reviews, 35*, 367–395.

Wu, J. (1994). Evaluation and models of Cenozoic sedimentation in the South China Sea. *Tectonophysics, 235*, 77–98.

Wust, G. (1964). *Stratification and circulation in the Antillean-Caribbean Basins*. Palisades, NY: Columbia University Press, 201 pp.

Wyndham, T., McCulloch, M., Fallon, S., & Alibert, C. (2004). High-resolution coral record of rare earth elements in coastal seawater: Biogeochemical cycling and a new environmental proxy. *Geochimica et Cosmochimica Acta, 68*, 2067–2080.

Yamamoto, K., Iryu, Y., Sato, T., Chiyonobu, S., Sagae, K., & Abe, E. (2006). Responses of coral reefs to increased amplitude of sea-level changes at the Mid Pleistocene Climatic Transition. *Palaeogeography, Palaeoclimatology, Palaeoecology, 241*, 160–175.

Yamano, H., Abe, O., Matsumoto, E., Kayanne, H., Yonekura, N., & Blanchon, P. (2003) Influence of wave energy on Holocene coral reef development: An example from Ishigaki Island, Ryuku Islands, Japan. *Sedimentary Geology, 159*, 27–41.

Yamano, H., Hori, K., Yamauchi, M., Yamagawa, O., & Ohmura, A. (2001). Highest-latitude coral reef at Iki Island, Japan. *Coral Reefs, 20*, 9–12.

Yamano, H., Kayanne, H., Matsuda, F., & Tsujii, Y. (2002). Lagoonal facies, ages and sedimentation in three atolls in the Pacific. *Marine Geology, 185*, 133–247.

Yamano, H., Kayanne, H., & Yonekura, N. (2001). Anatomy of a modern coral reef flat: A recorder of storms and uplift in the late Holocene. *Journal of Sedimentary Research, B, 71*, 295–304.

Yamano, H., Miyajima, T., & Koike, I. (2000). Importance of foraminifera for the formation and maintenance of a coral sand cay: Green Island, the Great Barrier Reef, Australia. *Coral Reefs, 19*, 51–58.

Yang, W., Mazzullo, S. J., & Teal, C. S. (2004). Sediments, tracts, and variations in sedimentation rates of Holocene platform carbonate sediments and associated deposits, northern Belize — implications for "representative" sedimentation rates. *Journal of Sedimentary Research, 74*, 498–512.

Yokoyama, Y., & Esat, T. M. (2004). Long-term variations of uranium isotopes and radiocarbon in the surface seawater recorded in corals. In: M. Shiyomi, H. Kawahata, H. Koizumi, A. Tsuda & Y. Awaya (Eds), *Global environmental change in the ocean and on land* (pp. 279–303). Tokyo: Terrapub.

Yokoyama, Y., Esat, T. Z., & Lambeck, K. (2001). Coupled climate and sea-level changes deduced from Huon Peninsula coral terraces of the last ice age. *Earth and Planetary Science Letters, 193*, 579–587.

Yokoyama, Y., Lambeck, K., De Dekkar, P., Johnston, P., & Fifield, L. K. (2000). Timing of the last glacial maximum from observed sea-level minima. *Nature, 406*, 713–716.

Yonekura, N., Ishii, T., Saito, Y., Maeda, Y., Matsushima, Y., Matsumoto, E., et al. (1988). Holocene fringing reefs and sea-level change in Mangaia Island, southern Cook Islands. *Palaeogeography, Paleoclimatology, Paleoecology, 68*, 177–188.

Yordanova, E. K., & Hohenegger, J. (2002). Taphonomy of larger foraminifera: Relationships between living individuals and empty tests on flat reef slopes (Sesoko, Japan). *Facies, 46*, 169–203.

Yu, K.-F., Zhao, J.-X., Liu, T.-S., Wei, G.-J., Wang, P.-X., & Collerson, K. D. (2004). High-frequency winter cooling and reef coral mortality during the Holocene climatic optimum. *Earth and Planetary Science Letters, 224*, 143–155.

Yu, K.-F., Zhao, J.-X., Shi, Q., Chen, T.-G., Wang, P.-X., Collerson, K. D., et al. (2006). U-series dating of dead *Porites* corals in the South China Sea: Evidence for episodic coral mortality over the past two centuries. *Quaternary Geochronology, 1*, 129–141.

Yu, K.-F., Zhao, J.-X., Shi, Q., & Meng, Q.-S. (2009). Reconstruction of storm/tsunami records over the last 4000 years using transported coral blocks and lagoon sediments in the Southern China Sea. *Quaternary International, 195*, 128–137.

Yu, K.-F., Zhao, J.-X., Wei, G.-J., Cheng, X.-R., Chen, T., Felis, Th., et al. (2005). $\delta^{18}O$, Sr/Ca and Mg/Ca records of *Porites lutea* corals from Leizhou Peninsula, northern South China Sea, and their applicability as paleoclimatic indicators. *Palaeogeography, Palaeoclimatology, Palaeoecology, 218*, 57–73.

Yu, K.-F., Zhao, J.-X., Wei, G.-J., Cheng, X.-R., & Wang, P.-X. (2005). Mid-Late Holocene monsoon climate retrieved from seasonal Sr/Ca and $\delta^{18}O$ records of *Porites lutea* corals at Leizhou Peninsula, northern coast of South China Sea. *Global and Planetary Change, 47*, 301–316.

Zachos, J. C., Obdyke, B. N., Quinn, T. M., Jones, C. E., & Halliday, A. N. (1999). Early Cenozoic glaciation, Antarctic weathering, and seawater $^{87}Sr/^{86}Sr$: Is there a link? *Chemical Geology, 161*, 165–180.

Zankl, H. (1993). The origin of high-Mg calcite microbialites in cryptic habitats of Caribbean coral reefs — Their dependence on light and turbulence. *Facies, 29*, 55–60.

Zankl, H. & Multer, H. G. (1977). Origin of some internal fabrics in Holocene reef rocks, St. Croix, U.S. Virgin Islands. *Proceedings of the Third International Coral Reef Symposium*, Miami, Vol. 2, pp. 127–133.

Zekster, I. S., Ivanov, V. A., & Meskheteli, A. V. (1973). The problem of direct groundwater discharge to the seas. *Journal of Hydrology, 20*, 1–36.

Zhao, J.-X., Neil, D. T., Feng, Y.-X., Yu, K.-F., & Pandolfi, J. (2009). High-precision U-series dating of very young cyclone-transported coral reef blocks from Heron and Wistari reefs, southern Great Barrier Reef, Australia. *Quaternary International, 195*, 122–127.

Zhu, Z. R., Marshall, J. & Chappell, J. (1988). Diagenetic sequences of reef-corals in the Later Quaternary raised coral reefs of the Huon Peninsula, New Guinea. *Proceedings of the Sixth International Coral Reef Symposium*, Townsville, Australia, Vol. 3, pp. 565–574.

Zinke, J., Dullo, W.-C., Heiss, G. A., & Eisenhauer, A. (2004). ENSO and Indian Ocean subtropical dipole variability is recorded in a coral record off southwest Madagascar for the period 1659–1995. *Earth and Planetary Science Letters, 228*, 177–194.

Zinke, J., Pfeiffer, M., Timm, O., Dullo, W.-C., & Davies, G. R. (2005). Atmosphere-ocean dynamics in the western Indian Ocean recorded in corals. *Philosophical Transactions of the Royal Society of London, Series A, 363*, 121–142.

Zinke, J., Reijmer, J. J. G., Taviani, M., Dullo, W.-Ch., & Thomassin, B. (2005). Facies and faunal assemblage changes in response to the Holocene transgression in the Lagoon of Mayotte (Comoro Archipelago, SW Indian Ocean). *Facies, 50*, 391–408.

Zinke, J., Reijmer, J. J. G., & Thomassin, B. A. (2001). Seismic architecture and sediment distribution within the Holocene barrier-reef complex of Mayotte (Comoro archipelago, SW Indian Ocean). *Palaeogeography, Palaeoclimatology, Palaeoecology, 175*, 343–368.

Zinke, J., Reijmer, J. J. G., & Thomassin, B. A. (2003). Systems tracts sedimentology in the lagoon of Mayotte associated with the Holocene transgression. *Sedimentary Geology, 160*, 57–79.

Zinke, J., Reijmer, J. J. G., Thomassin, B. A., & Dullo, W.-C. (2003). Postglacial flooding history of Mayotte Lagoon (Comoro Archipelago, southwest Indian Ocean). *Marine Geology, 194*, 181–196.

Zlatarski, V. N., & Estalella, N. M. (1982). *Les Scléractiniaires de Cuba*. Sofia: Editions de l'Académie Bulgare des Sciences, 472 pp.

Zubia, M., & Peyrot-Clausade, M. (2001). Internal bioerosion of *Acropora formosa* in Réunion (Indian Ocean): Microborer and macroborer activities. *Oceanologica Acta, 24*, 251–262.

Zundelevich, A., Lazar, B., & Ilan, M. (2007). Chemical versus mechanical bioerosion of coral reefs by boring sponges — Lessons from *Pione* cf. *vastifica*. *The Journal of Experimental Biology, 210*, 91–96.

Zuschin, M., Hohenegger, J., & Steininger, F. F. (2000). A comparison of living and dead molluscs on coral reef associated hard substrata in the northern Red Sea — Implications for the fossil record. *Palaeogeography, Palaeoclimatology, Palaeoecology, 159*, 167–190.

Zuschin, M., & Oliver, G. (2003). Fidelity of molluscan life and death assemblages on sublittoral hard substrata around Granitic Islands of the Seychelles. *Lethaia, 36*, 133–150.

Zuschin, M., & Stachowitsch, M. (2007). The distribution of molluscan assemblages and their postmortem fate on coral reefs in the Gulf of Aqaba (northern Red Sea). *Marine Biology, 151*, 2217–2230.

Subject Index

FSC
www.fsc.org
MIX
Papier | Fördert gute Waldnutzung
FSC® C083411

Kolibri 360 GmbH
Ferdinand-Jühlke-Straße 7
99095 Erfurt, Deutschland
produktsicherheit@kolibri360.de